C0-AUA-719

In this important volume, major events and personalities of twentieth century physics are portrayed through the recollections and historiographical works of one of the most prominent figures of European science. A former student of Enrico Fermi, and a leading personality of physical research and science policy in post-war Italy, Edoardo Amaldi devoted part of his career to documenting, both as witness and as historian, some significant moments of twentieth century science. The focus of the book is on the European scene, ranging from nuclear research in Rome in the 1930s to particle physics at CERN, and includes biographies of physicists such as Ettore Majorana, Bruno Touschek and Fritz Houtermans.

Edoardo Amaldi (Carpaneto, 1908 – Roma, 1989) was one of the leading figures in twentieth century Italian science. He was conferred his degree in physics at Rome University in 1929 and played an active role (as a member of the team of young physicists known as "the boys of via Panisperna") in the fundamental research on artificial induced radioactivity and the properties of neutrons, which won the group's leader Enrico Fermi the Nobel Prize for physics in 1938. Following Fermi's departure for the United States in 1938 and the disruption of the original group, Amaldi took upon himself the task of reorganising the research in physics in the difficult situation of post-war Italy. His own research went from nuclear physics to cosmic ray physics, elementary particles and, in later years, gravitational waves. Active research was for him always coupled to a direct involvement as a statesman of science and an organiser: he was the leading figure in the establishment of INFN (National Institute for Nuclear Physics) and has played a major role, as spokesman of the Italian scientific community, in the creation of CERN, the large European laboratory for high energy physics. He also actively supported the formation of a similar trans-national joint venture in space science, which gave birth to the European Space Agency. In these and several other scientific organisations, he was often entrusted with directive responsibilities. In his later years, he developed a keen interest in the history of his discipline. This gave rise to a rich production of historiographic material, of which a significant sample is collected in this volume.

EDOARDO AMALDI FOUNDATION SERIES

Vol. 1: Gravitational Wave Experiments
edited by E. Coccia, G. Pizzella & F. Ronga

Vol. 2: Gravitational Waves: Sources and Detectors
edited by I. Ciufolini & F. Fidecaro

Vol. 3: 20th Century Physics: Essays and Recollections
A Selection of Historical Writings by Edoardo Amaldi
edited by G. Battimelli & G. Paoloni

Forthcoming
Vol. 4: Gravitational Waves
edited by Liu' Catena

20th Century Physics: Essays and Recollections

A Selection of Historical Writings by Edoardo Amaldi

Editors

Giovanni Battimelli
Department of Physics
Rome University "La Sapienza"

Giovanni Paoloni
Scuola Speciale per Archivisti e Bibliotecari
Rome University "La Sapienza"

Published by

World Scientific Publishing Co. Pte. Ltd.
P O Box 128, Farrer Road, Singapore 912805
USA office: Suite 1B, 1060 Main Street, River Edge, NJ 07661
UK office: 57 Shelton Street, Covent Garden, London WC2H 9HE

Library of Congress Cataloging-in-Publication Data
Amaldi, Edoardo.
20th century physics : essays and recollections : a selection of historical writings / [by E. Amaldi.] ; editors, G. Battimelli, G. Paoloni.
p. cm. -- (Edoardo Amaldi Foundation series ; vol. 3)
ISBN 9810223692
1. Physics -- Italy -- History -- 20th century. 2. Amaldi, Edoardo.
I. Battimelli, Giovanni. II. Paoloni, Giovanni. III. Title.
IV. Series.
Q127.I8A53 1998
530'.0945'0904 -- dc20 96-1976
CIP

British Library Cataloguing-in-Publication Data
A catalogue record for this book is available from the British Library.

Printed in Singapore by Uto-Print

PREFACE

Edoardo Amaldi was a member of some of the most important scientific societies of our day. Upon his death in 1989, one of these societies, the prestigious Royal Society, invited Carlo Rubbia to write a biography on him which was published in 1991 in *Biographical Memoirs of Fellows of the Royal Society* (Volume 37). Here, Amaldi's scientific activities are described in great detail while his historical contributions, most of which are collected in this book, are dealt with only briefly. To treat the argument, Rubbia writes:

> Amaldi's contributions as a writer of history started relatively late in his career. He was a historian of his times. His studies of history began after War World II, under the spur of the events he had witnessed and of the reconstruction that he had guided, and then went on to cover events which, and scientists who, in his opinion, should not be forgotten...An entire article could be written on Amaldi's historical works, but unfortunately there is not sufficient room here. However, some of his major works must be mentioned. In 1968, he wrote on the history of CERN. One can only marvel at his ability to find time in a life already so full of activity to write works of great detail and commitment, such as *From the Discovery of the Neutron to the Discovery of Nuclear Fission*. A work that all those interested in the history of science should read is *The Years of Reconstruction* published in both Italian and English in 1979...Amaldi tells how important research was maintained in the most difficult conditions and how, little by little, Italian institutions were re-animated after the war. The style is simple and not unnecessarily erudite, yet one feels it has the rigour of a historian who bases his text strictly on his sources and who has no interest in sensationalism. Throughout *The Years of Reconstruction*, Amaldi is typically modest about his own contribution during this period.

The present volume will give the interested reader a complete overview of Edoardo Amaldi's historical works: the only important piece missing is the already quoted *From the Discovery of the Neutron to the Discovery of Nuclear Fission* (*Physics Reports*, Vol. 111, 1984), which is 300 pages long with 924 references and notes, and is quite technical. The reasons why this unique contribution is not reproduced here are therefore understandable. Still, it is worth stressing the unusual approach which Amaldi adopted in this review article. First, among the notes one finds short pieces describing historical developments, such as the "Anschluss" of Austria, the Nazification of Germany and the invasion of Ethiopia by Mussolini. Moreover, short biographic notes are provided on about twenty physicists who have contributed to the field but are not known worldwide. Finally, in several places Amaldi introduces some personal information which cannot be found elsewhere, the most interesting being the fact that he was the first to use the term "neutrino". I know it from himself and from Beppo Occhialini. The episode is recounted in reference 277 in the following words:

> The name "neutrino" (a funny and grammatically incorrect contraction of "little neutron" in Italian: neutronino) entered the international terminology through Fermi, who started to use it sometime between the conference in Paris in July 1932 and the Solvay Conference in October 1933, where Pauli used it. The word came out in a humorous conversation at the Istituto di via Panisperna. Fermi, Amaldi and a few others were present and Fermi was explaining Pauli's hypothesis about his "light neutron". For distinguishing this particle from the Chadwick neutron, Amaldi jokingly used this funny name, says Occhialini, who recalls having told this little story around Cambridge shortly afterwards.

Edoardo Amaldi recalled this episode to me a few times, stating that he could remember very well the corner in the laboratory where the people having this conversation were standing.

The 1984 *Physics Report* also contains a most interesting analysis of the scientific reasons and personal prejudices which postponed the discovery of nuclear fission by about five years. In particular, Amaldi discusses what was happening in Rome, where Fermi and his collaborators were uneasy about the "discovery" of transuranic elements but did not take seriously the fission process proposed in 1934 by Ida Noddack. He writes:

> I seem to remember some discussions among members of our group, including Fermi, in which the ideas of Ida. Noddack were hastily set aside because they involved a completely new type of reaction: fission. Enrico Fermi, and all of us at his school followed him, had always been very reluctant to invoke a new phenomena as soon as something new was observed: new phenomena have to be proven! As later developments have shown, a much more fruitful attitude would have been to try to test Noddack's suggestion and to eventually disprove it. But Fermi and all of us were, on that occasion, too conservative: an explanation of the "uranium case" in terms of what we had found for all lower values of Z was much simpler and therefore preferable. Two reasons or, maybe, two late excuses...She never tried, alone or with her husband, to carry out experiments on irradiated uranium, which they certainly could have done. Furthermore, in those years the Noddacks had fallen into some disrepute because of their claim to have discovered element Z = 43, which they had called "masurium".

As mentioned above, all the important writings are collected in this volume with the exception of *From the Discovery of the Neutron to the Discovery of Nuclear Fission*. We are grateful to Giovanni Battimelli and Giovanni Paoloni for the care and competence they have put in preparing this volume. Their introductory texts establish the general framework in which each contribution saw the light and the translations, where appropriate, render them accessible to all those who read English. It is a very positive initiative on the part of the Edoardo Amaldi Foundation (Piacenza, Italy) to publish this book in the "Latin" of today. Moreover, the publication takes place in the year when Europe is taking another crucial step towards unification, an ideal which Edoardo Amaldi has devoted much of his energy to and is beautifully summarized at the end of Rubbia's biography:

> Although Edoardo Amaldi did not like to be congratulated by colleagues on his appointments to important functions, since he considered himself first and foremost a man of science, there can be no doubt that he was one of Europe's greatest post-war scientific statesmen. It is above all as a man of vision, as a man guided by powerful and timely ideas, and as a man determined to build new scientific institutions that Edoardo Amaldi has left an indelible trace on history. Amaldi was inspired by two clear principles. First, he was convinced that science should not be pursued for military purposes...Secondly, Edoardo Amaldi was a dedicated European. He realized that no single European nation could hope to hold its own scientifically and technologically...These were not simply pious hopes or abstract ideas. Amaldi was determined to see them put into practice. This is clearly demonstrated by the key role he played in launching CERN and ESRO...Historian, teacher, man of ideals — all these facets of Edoardo Amaldi's life have been referred to, but it is also because of his human qualities that he will be remembered for. His exemplary coolness of mind, his total sincerity, his respect for the deserving, his unwavering intellectual and moral honesty, and his constant affection and commitment to his family were the hallmarks of this remarkable man.

This book offers, by reading Amaldi's own words, the occasion to delve into his way of viewing the people he liked and the events which he had participated in.

Ugo Amaldi

INTRODUCTION

It is not unusual for scientists to feel the need, at some point along their intellectual paths, to consign to posterity a memorial of their scientific activities by documenting their contributions to their fields of specialization or reflecting upon the advances in knowledge made during the historical period in which they lived and worked. This has given rise to that particular genre of scientific literature, halfway between memoirs and historiography in the strict sense of the word, which encompasses products of various depths and widely differing reliability, ranging from valedictory speeches at retirement to autobiographies.

As common as this practice of taking a retrospective look at one's personal and scientific course and setting it down on paper may be in the scientific world, we feel that in the case of the written works of Edoardo Amaldi, the exercise of memory was carried out with such rigor and intensity that the whole of his historiographic production merits special attention.

There are a number of reasons for this. First, Amaldi's unique position and the role he played on the Italian and European scientific scenes during this century. We are not simply referring to his direct contributions to the advancement of knowledge and to the acquisition of new scientific results: important as they have been (from the research on neutron-induced radioactivity and nuclear physics in the 1930s and 1940s to the cosmic ray investigations of the late 1940s and 1950s, and the more recent studies on gravitational waves), most often Amaldi's role is shadowed there by his appearing as just a name amongst several others, in the team effort so typical of modern experimental physics. But if one considers his contribution to the organization of research and the role he played in the field of scientific policy in the broad sense, then Edoardo Amaldi stands out as a protagonist in the scientific history of this century — and not only in Italy and physics. It is no exaggeration to say that the most important events connected to the dynamics of Italian and European scientific institutions from the postwar period up to the 1980s bore the mark of Amaldi's active intervention. Thus, interest in his memoirs is intrinsically linked, above all, to the importance of the events of which he was a protagonist, and the papers printed in this volume eloquently attest to that.

There is another reason why these works deserve particular attention. Contrary to what was the case in the United States in the period to which we are referring (approximately the half century from the 1930s to the 1980s), the literature available on developments in Italian science is extremely scarce and often unreliable, frequently responding more to the demands of commemoration than to the rigorous criteria of historical investigation. Only in recent years has an interest arisen in historiographic research into this particular aspect of contemporary Italian history and a greater concern shown for the conservation of the sources needed to carry it out in accordance with adequate criteria. As a result, Amaldi's works take on even more importance, as they are largely dedicated to, understandably, significant moments in Italian scientific life; some of these papers are still the only available source or in any case an indispensable reference for those undertaking a more thorough historiographic study of the events narrated in them.

This would be enough in itself to justify the existence of this collection of historical writings by Amaldi: they fill a gap — a gap relative to events of considerable importance. But we would like to make an even more important point: the degree of reliability of these writings, in strictly historiographic terms, is very unusual for this kind of publication. Amaldi rapidly got into the habit of collecting all available documentary evidence concerning the subject of his historiographic endeavors and it is difficult, especially in his later works, to find him making a committal statement or simply quoting a phrase or chronologically placing an event without having the necessary documentation to back it up. It is enough to take a quick look at the inventory of his personal files, now preserved at the Department of Physics of the University of Rome "La Sapienza", to realize how much material he used to collect and keep for his historical writings and that the amount gathered for each work increased progressively with the years, to the point where the documentation he had collected for the biography of Houtermans (on which he worked for over ten years) fills six archive boxes. Professional historians know that memoirs left by scientists are generally considered invaluable as direct testimonials, but at the same time unreliable unless backed up by independent documentary evidence. This is generally not the case for most of Amaldi's works (with some inevitable, minor exceptions), even though Amaldi was not a professional historian and did not consider himself one. With this characteristic rigor, Amaldi has rendered an important service to historians and has indirectly given his scientific colleagues an effective lesson in scientific method.

A glance at the bibliography of Amaldi's writings on the history of physics will quickly make the reader realize that his work in this area increased progressively through the years. As is to be expected, it reached its peak intensity in the last decade of his life, in concomitance with the gradual diminution in his more direct responsibilities in the fields of scientific work, the organization of research, and teaching. Without going into a presentation of the individual papers, which will be done in the introductions to the various sections into which this volume is divided, we would like to provide a hypothetical breakdown of his works into periods marked by particularly significant moments.

The papers in the first group, which date back to the mid-1950s and include the commemorations of Enrico Fermi and George Placzek, are clearly extemporaneous, being motivated by some special circumstance. It is, however, indicative of Amaldi's strong belief in the importance of the history of physics and of his awareness of the problem of conserving sources. As early as 1959, he wrote an article, albeit a short one, for *Physis*, a journal dedicated specifically to the history of physics, by presenting and underlining the value of Fermi's documents preserved in the Domus Galilaeana in Pisa. His first important contribution in the field of historiography came in 1966, however, with the volume dedicated to Ettore Majorana and put out by the Accademia dei Lincei. The project was conceived of and planned by Amaldi himself, who wrote a rich scientific biography for the occasion, and was supported not only by his own recollections, but also by an extensive collection of testimonials, records and writings by relatives and colleagues (including H. Bethe, W. Heisenberg, E. Feenberg and L. Rosenfeld). He also invited friends and colleagues (specifically A. Carrelli, L. Fermi, F. Rasetti, E. Segrè, E. Volterra and G. C. Wick) for a critical reading of the manuscript, thereby inaugurating a practice which he was to keep up for the rest of his life.

Amaldi was to write again about Majorana shortly afterwards, in 1968, on the thirtieth anniversary of the Sicilian physicist's disappearance. Meantime, his interests had extended from events in the 1930s to more recent occurrences; he undertook an initial reconstruction of the history of CERN and the European laboratory for high-energy physics which he had actively helped to establish in the early1950s. This broadening of his historiographic horizon found recognition in the lectures he gave at the summer school organized by the Società Italiana di Fisica in Varenna in 1972, an evident sign of the opening up towards the history of physics that characterized the Italian scientific community in those years, and in which Amaldi's direct influence was felt.

Towards the end of the 1970s, when Amaldi went into retirement (which certainly did not put an end to his involvement in active research), there was a significant increase in his historiographic production, and an even more significant increase in the terrain explored. Within the space of two years, he published four works (a study of the first research on radioactivity; a memorial to Enrico Persico; his fundamental work on the reconstruction years; and the biography of Bruno Touschek) which marked his definitive maturation as a historian. As was only natural, this enhanced ability as a historian soon found expression on the occasion of the fiftieth anniversary, in 1984, of the research on neutron-induced radioactivity carried out by Fermi's group to which Amaldi had belonged. The reconstruction, based only on personal recollections and firsthand reports typical of his early works, was now backed by the habitual act of comparison with sources and historical investigation based on rigorous criteria. One of the many papers from this period is the rich *From the Discovery of the Neutron to the Discovery of Nuclear Fission*, on which Amaldi had worked on since 1979 and constituted an entire volume of the 1984 *Physics Reports*.

In the last years of his life, this activity increased even more and his interest shifted from the strictly scientific to more general questions, in which the history of physics was integrated with history *tout court*. A significant example of how his usual attention to scientific events is accompanied by a strong civil sense is found in his speech delivered at the conference organized by the Accademia dei Lincei. In that speech, he touched upon the effects of racial laws on the Italian world of culture fifty years after their implementation.

Amaldi's death caught him in full swing and some mention must therefore be made of his uncompleted writings and his plans for future works. First , there is his excellent and impassioned biography of Fritz Houtermans, which was completed before his death and which we are pleased to include in this volume. As mentioned earlier, Amaldi had worked on the manuscript for almost a decade. However, Amaldi also had other more ambitious plans, traces of which were found in his files. In 1979, at a meeting of the Seminario di Storia della Scienza of the Faculty of Science at Rome University, Amaldi declared that he was planning to write the biography of Laura Fermi and, above all, a history of physics in Rome in the period between 1794 to 1968. To this end, he had already collected a large amount of documentation, especially on the history of the old Physics Institute in via Panisperna, from its foundation and first director, Pietro Blaserna, to the period which he personally witnessed when Orso Mario Corbino was its director. Some parts of the history of physics in Rome had more or less been thoroughly sketched out by Amaldi: a long manuscript describing the vicissitudes of Roman physicists during the Second World War (from 1938 to 1946) was found among his files, which contains much material that is supplementary to what is presented in his writings on the reconstruction years.

Amaldi did not consider himself a historian in the strictest sense of the word, and considered his writings on the history of physics to be a kind of retrospective reflection on his own research activity and the events of which he was a direct witness or protagonist. At the same time, though, he considered the history of physics an autonomous research field with its own institutional place in scientific faculties. His influence on the scientific community was decisive in helping to create a more open climate towards the discipline which led to the teaching of the history of physics in many of the physics institutes of major Italian universities in the 1960s and 1970s. This active involvement in the institutional vicissitudes of the discipline did not fail to reflect the quality of his writings and to leave a mark on his development, especially his attention to the problem of sources, their conservation and correct use in historiographic investigation.

It was during his lessons at the 57th course of the International School of Physics at Varenna, dedicated in 1972 to the *History of Twentieth Century Physics,* that Amaldi seems to have grasped the importance of archival sources for the first time. The subject he dealt with (*Personal Notes on Neutron Work in Rome in the 1930s and Post-War European Collaboration in High Energy Physics*) returned to some topics he had touched upon in preceding years — the life and works of Ettore Majorana and the origins of the CERN — but they were now set in a much broader and more articulated framework. Published in 1977, the text made ample reference not only to personal recollection and scientific literature, but also to correspondence and notebooks, used largely as illustrations along with pictures of instruments. This paper, which made use of a source not taken into consideration previously, marked a turning point in his way of reading and writing history, and was also the beginning of an unflagging concern for the problem of the conservation of contemporary historical-scientific heritage, of which archives constitute an important part.

Thus, Amaldi began to take interest in the Physics Museum at the University of Rome and promoted the collection and conservation of instruments and documentation related to Fermi's group. At the Physics Institute, Amaldi also encouraged the group working on the history of physics, which he had helped to set up, to recover the papers of Bruno Touschek and Enrico Persico, an undertaking in which he personally participated to some extent.

At the beginning of the 1980s, he promoted, along with Giovanni Battista Marini Bettolo, the establishment at the Accademia Nazionale delle Scienze (an important Italian scientific association founded in 1782 with important archives) of a Center for the History of Contemporary Science and had a copy of the *Archives for the History of Quantum Physics* deposited there. As the president of the Accademia Nazionale dei Lincei in the last years of his life, Amaldi endeavored (once again with Marini Bettolo) to preserve some archives of considerable value: those of Vito Volterra and Guglielmo Marconi which were donated to the Accademia by their heirs but left in abandon until he took interest in them.

Thus Amaldi showed a cultural depth and historical sensitivity which was relatively unprecedented among Italian scientists. These things were, however, in harmony with the effort to protect and enhance the Italian historical-scientific heritage which, after going through a period of crisis in the postwar period, started to pick up again in the second half of the 1970s. In fact, Amaldi's actions contributed significantly to this recovery. At the same time, he was well aware of the role this historical-scientific patrimony played as a founding element attesting to a tradition, and the contribution it could make in giving the scientific community possessing it a sense of identity: something which closely linked Amaldi the historian and archive keeper to Amaldi the research policymaker.

A complete list of the writings of Edoardo Amaldi on the history of science can be found at the end of this volume. Those appearing in this collection were selected on the basis of a number of general criteria. The first was an attempt to do justice to the variety of Amaldi's interests. As a result, we initially excluded only those works written for special occasions, which are clearly less interesting, and selected, in the event that the same subject had been dealt with more than once, the more complex and significant papers in order to avoid repetition. This was the case, for example, with his many works on Majorana and the history of CERN.

Given our intention to make this collection of writings accessible to a broader public not necessarily limited to people with specific scientific competences working in the field, we also decided to exclude papers of an extremely technical nature. Obviously, it is difficult to apply this criterion to a collection of pieces written by a physicist about the history of physics: Amaldi's is in any case a "physicist's history" and systematically involves technically complex passages. Although we tried to make a reasonable selection, this led to the exclusion of Amaldi's fundamental contribution to the *Physics Reports* of 1984, *From the Discovery of the Neutron to the Discovery of Nuclear Fission*, which is still the most complete and detailed reconstruction of the earliest developments in nuclear physics.

We were helped by a number of people in selecting the texts and preparing and editing them for publication. They cannot all be cited here and even if we tried, some would surely be omitted. Therefore, we will limit ourselves to mentioning with gratitude the support given to this undertaking and to our work by Professors Ugo Amaldi and Ferdinando Amman, and the constant stimulus in the form of discussion and common work offered by friends and colleagues in the Department of Physics at the University of Rome: Carlo Bernardini, Michelangelo De Maria, Lucia Orlando and Fabio Sebastiani. And special thanks to Bruno Pellizzoni, for the patience with which he made up for our technical shortcomings in the field of computer editing.

Giovanni Battimelli
Giovanni Paoloni

CONTENTS

20th Century Physics:
Essays and Recollections

Part I

From Nuclei to Particles: 50 Years of Physics in Italy

Section A

Nuclear Physics in Rome and the Italian Scientific Milieu Up to 1939

Amaldi's interest in physics research in Italy dates back to at least the beginning of the 1960s and probably coincides with the publication (1962) of the first volume of *Collected Papers by Enrico Fermi*, which reprinted Fermi's writings from his period in Italy, that is, from 1921 to 1938. The papers in this first part refer to physics in Italy before the departure of Enrico Fermi.

In the first work, *Neutron Work in Rome in 1934–36 and the Discovery of Uranium Fission*, Amaldi deals with the research on neutrons carried out by Fermi's group between 1934 and 1936. It was written for the first issue of the *Rivista di storia della scienza*, an editorial initiative undertaken by the Faculty of Science at the University of Rome "La Sapienza", in an attempt to enhance the contribution of historians with a scientific-disciplinary background to the development of the historiography of science which had started to blossom in Italy in the mid-1970s, and which led to an unprecedented opening up of studies on science in the second half of the 19th and the 20th centuries. A number of scholars who had been Amaldi's students took part in this initiative, and the subject and the place of publication are therefore significant. As for the paper itself, he returns to a subject he had dealt with in 1977 in a piece appearing in the following section of this volume, but examines it from a different point of view: going over the theoretical and experimental context which led Fermi to turn his group's efforts to research on neutrons, Amaldi throws light on the results obtained and the problems of interpretation which these posed, thus explaining why Fermi's group did not interpret these data in terms of uranium fission.

The second and third papers deal with the life and works of Ettore Majorana. The first of the two, *Ettore Majorana, Man and Scientist*, is the introduction to a collection of works by Majorana promoted by Amaldi and published by the Accademia Nazionale dei Lincei in 1966. This is Amaldi's first real historical work and belongs to an initial phase of research into the history of the Institute of Physics at the University of Rome, a study centering on the birth and activity of Fermi's group to which Amaldi drew various people from the institute. It already displays the particular characteristic of Amaldi's style of narration, alternating more specifically between biographical and institutional pieces with passages that go directly to the core of the scientific questions faced by Majorana. The last paragraph is devoted to the mystery of Majorana's disappearance in 1938, a mystery which was to draw considerable public attention. Not many years later, it was the subject of a book by a famous Italian writer, Leonardo Sciascia, but Sciascia's explanation of Majorana's disappearance — that he disappeared because he was concerned about the terrible implications of atomic research — were strongly criticized by Amaldi. This led to a journalistic controversy which brought much popularity to Fermi's young student but embittered Amaldi, who never again returned to the subject. Amaldi's article was immediately translated for publication as a historical introduction to the book *Strong and Weak Interactions: Present Problems* (1966), a collection of lectures given at the summer course at the International School of Physics in Erice on the forces acting inside the nuclear atom, a subject which Majorana had dedicated much of his attention to after 1932.

Later, Amaldi again focused his attention on Majorana's works. Included here is the last of such papers, published shortly before his death. This choice was made not only because it is the most recent, but also because it deals more with Majorana's initial and generally less well-known works. Amaldi concluded the paper by emphasizing the topicality of many of the subjects and problems broached by the great Sicilian physicist, and the need for a more thorough study of his work to discover the "profound links which have still not been entirely uncovered or completely understood".

The fourth paper, *Gian Carlo Wick During the Thirties*, is the text of a speech given in 1984 to the conference organized by the Scuola Normale di Pisa in honor of Wick, who spent the last years of his teaching career there. Amaldi's paper centers on the important contributions made by this brilliant theoretical scientist between 1932 and 1943 on subjects which brought him into close collaboration with the Rome group. Born and educated in Turin, where he received his degree, Wick specialized in Göttingen from 1930–1931 at the school of Max Born and in Leipzig at the school of Werner Heisenberg where, among other things, in 1934 Ettore Majorana spent a very important period in his brief scientific life. This is where Amaldi and Wick first met in 1931. They became closer, however, when Wick moved to Rome in 1933 to work as Fermi's assistant. In this capacity, the young man from Turin took part in the research on the slow neutron, working together with Fermi in 1934 and making some important theoretical contributions. In 1935–1936, when the group had started to disperse, he continued the experimental activity with Pontecorvo and Amaldi. Later transferred to Palermo and Padova, Wick finally returned to Rome in 1940 to take over the chair of theoretical physics left vacant by Fermi after his departure. Indeed, it was Fermi who suggested his successor's name. Thus Wick was at Amaldi's side in the difficult times experienced by the Physics Institute during the war and in the early postwar period. This made Wick's departure for the United States in 1946 even more painful for Amaldi who, committed as he was to trying to resuscitate physics in Italy, had turned down numerous offers of the same kind.

The fourth and fifth works deal with the consequences of the 1938 racial laws in Italy. Both texts date back to the same occasion: the conference on *Conseguenze culturali delle leggi razziali in Italia* organized by the Accademia Nazionale dei Lincei in 1988 on the fiftieth, anniversary of the implementation of the laws. Amaldi, who had just been elected president of the Accademia, had been strongly in favor of the conference and directly supervised its organization, promoting a series of documentary studies for the preparation of his speech and those of the other speakers. The racial laws introduced in the autumn of 1938 had been drafted by the government of Benito Mussolini within the framework of his policy to draw closer to Hitler's Germany. Preceded by a racist propaganda campaign of increasing intensity, they included a number of harassing measures which made life difficult for the Jews. In addition to those which prevented the Jews from running businesses and allowing the confiscation of their assets, there were others that had a direct impact on the world of culture: teachers of all levels in the schools, as well as university professors, lost their jobs; textbooks written by the Jews were withdrawn from circulation; and the Jews were no longer allowed to publish scientific papers under their own names. The problem for university professors was aggravated by the fact that there were people who took advantage of the positions vacated by them. The 1988 conference was dedicated to the consequences of those measures and it was obvious that Amaldi's focus of interest was in the field of physics. The studies he promoted led to the discovery in the archives of the Accademia dei Lincei of an exchange of letters between Einstein and the heads of the Accademia. In 1938, when the Italian academies were compelled by the law to expel their Jewish members, Einstein took a precise moral stance and in solidarity with his colleagues, resigned his membership. In 1939, the Accademia dei Lincei was closed down and incorporated into the Accademia d'Italia (set up by the regime in 1929) but only to be refounded — this time taking over the patrimony of the other academy — after the fall of fascism. At that point, Einstein informed the new president, Guido Castelnuovo, of his intention to once again take up his position as a foreign member.

Riv. Stor. Sci., *1* (1), 1984, 1-24 © Ed. Theoria

NEUTRON WORK IN ROME IN 1934-36 AND THE DISCOVERY OF URANIUM FISSION

EDOARDO AMALDI *

1. THE DISCOVERY OF ARTIFICIAL RADIOACTIVITY INDUCED BY NEUTRONS [1]

The paper by the Joliot-Curie announcing the discovery of artificial radioactivity induced by α-particles was presented by Jean Perrin to the 15 of January 1934 session of the Académie des Sciences in Paris [2]. At the end of the paper, after the presentation of their experimental results, these authors suggested the possibility of producing artificial radioactivity by irradiation with other particles. Experiments of this type were actually carried out successfully by Cockcroft, Gilbert and Walton with protons, and by Crane, Lauritsen and Harper and, independently, by Henderson, Livingstone and Lawrence with deuterons [3].

* E. Amaldi, Dipartimento di Fisica, Università di Roma "La Sapienza".

[1] The same subject, dealt with in different forms, can be found in: *Enrico Fermi Collected Papers*, 2 voll., Accademia Nazionale dei Lincei and The University of Chicago Press, Rome, vol. 1, 1962, *Italia 1921-1938*, (henceforth *FCP-I*), vol. 2, 1965, *1939-1954*; E. SEGRÈ, *Enrico Fermi, Physicist*, The University of Chicago Press, Chicago 1970; E. AMALDI, "Personal Notes on Neutron Work in Rome in the 30s", in *History of Twentieth Century Physics*, C. Weiner (ed.), Academic Press, New York and London 1977, 294-325.

[2] F. JOLIOT, I. CURIE, "Un nouveau type de radioactivité", *Compt. Rend. Acad. Sci.*, *198*, 1934, 254-256, séance du 15 janvier 1934; *Nature*, *133*, 1934, 201-202.

[3] J.D. COCKCROFT, C.W. GILBERT, E.T.S. WALTON, "Production of Induced Radioactivity by Protons", *Nature*, *13*, 1934, 328, dated February 24, 1934; H.R. CRANE, C.C. LAURITSEN, W.W. HARPER, "Artificial Production of Radioactive

After the first two papers by the Joliot-Curie [4] were read in Rome, Fermi, at the beginning of March 1934, suggested to Rasetti that they try to observe similar effects with neutrons using a polonium α-source that Rasetti had prepared shortly before [5].

About two weeks later several elements were irradiated and tested for activity by means of a thin-walled Geiger-Müller counter but the results were negative due to lack of intensity. Rasetti left for Morocco for a vacation while Fermi continued the experiments. The idea then occurred to Fermi that in order to observe a neutron induced activity it was not necessary to use a Poα + Be source. A much stronger Rnα + Be source could be employed, since its β- and γ-radiations (absent in Poα + Be sources) were no objection to the observation of a delayed effect. Radon sources were familiar to Fermi since they had been supplied previously by Professor G.C. Trabacchi [6] for use with a γ-ray spectrometer [7].

All one had to do was to prepare a similar source consisting of a glass bulb filled with beryllium powder and radon. When Fermi had his stronger neutron source (about 30 millicurie of Rn) he systematically bombarded the elements in the order of increasing atomic number, starting from hydrogen and following with lithium, beryllium, boron, carbon, nitrogen and oxygen, all with negative results. Finally, he was successful in obtaining a few counts on his Geiger-Müller counter when he bombarded fluorine and alluminium. These results

Substances", *Science*, *79*, 1934, 234-235, dated February 27, 1934; M.C. HENDERSON, M.S. LIVINGSTONE, E.O. LAWRENCE, "Artificial Radioactivity Produced by Deuton Bombardment", *Phys. Rev.*, *45*, 1934, 428-429, dated February 27, 1934.

[4] Ref. 2. and: F. JOLIOT, I. CURIE, "Séparation chimique des nouveaux radioéléments émetteurs d'électrons positifs", *Comp. Rend. Acad. Sci.*, *198*, 1934, 559-561, séance du 29 janvier 1934.

[5] F. RASETTI, "Su un forte preparato di Radio D ottenuto all'Istituto Fisico di Roma", *Ric. Scient.*, *5* (1), 1934, 3-5.

[6] G.C. Trabacchi (1884-1959) physicist, was director of the Laboratorio di Fisica of the Istituto Superiore di Sanità. He had a radon plant for medical use, which, however, was not always fully exploited. He generously provided the Rnα + Be sources used by Fermi and his collaborators from the beginning, for about two and half years, thus making possible all the work on "neutrons" carried out in those years at the Istituto di Fisica of the University of Rome.

[7] E. FERMI, F. RASETTI, "Uno spettrometro per raggi gamma", *Ric. Scient.*, *4* (2), 1933, 299-302; also in *FCP-I*, 549-552.

and their interpretation in terms of (n, α) reactions were announced in a letter to *La Ricerca Scientifica* on 25 March 1934 [8]. The title: "Radioattività indotta da bombardamento di neutroni - I" indicated his intention to start a systematic study of the phenomenon which would have brought to the publication of a series of similar papers.

Fermi wanted to proceed with the work as quick as possible. Therefore he asked Segrè and me to help with the experiments, as it appears also from the acknowledgement at the end of his second Letter to the Editor of *La Ricerca Scientifica* [9] where he reported preliminary results obtained in a number of other elements (Si, P, Cl, Fe, Cu, As, A, Te, J, Cr, Ba). A cable was sent to Rasetti asking him to come back from his vacation.

And now all at work. Fermi, helped a few days later by Rasetti, did good part of the measurements and calculations, Segrè secured the substances to be irradiated and the necessary equipment and later became involved in most of the chemical work. I took care of the construction of the Geiger-Müller counters and of what we now call electronics. This division of the activities, however, was not rigid at all and each of us participated in all phases of the work.

We immediately realized that we needed the help of a professional chemist. Fortunately we succeeded almost immediately in convincing Oscar D'Agostino to work with us. He had been a chemist in the laboratory of Trabacchi and at the time I am talking about, he held a fellowship in Paris at the laboratory of Madame Curie, where he was learning radiochemistry. He had come back to Rome for a few days during the Easter vacation, but we showed him our work and, on request of Fermi, he remained with us and never went back to Paris.

The results obtained during the first two weeks were summarized by Fermi in a Letter to *Nature* [10].

[8] E. FERMI, "Radioattività provocata da bombardamento di neutroni - I", *Ric. Scient.*, *5* (1), 1934, 283; also in *FCP-I*, 645-646 and 674-675 for its English translation.

[9] E. FERMI, "Radioattività provocata da bombardamento di neutroni - II", *Ric. Scient.*, *5* (1), 1934, 330-331; also in *FCP-I*, 647-648, and 676 for its English translation.

[10] E. FERMI, "Radioactivity Induced by Neutron Bombardment", *Nature*, *133*, 1934, 757, dated April 10, 1934; also in *FCP-I*, 702-703.

2. A SYSTEMATIC ATTACK TO THE ELEMENTS OF THE PERIODIC TABLE

During the successive months (April-July, 1934) our group published in rapid succession a series of experimental results [11].

Sixty elements (plus thorium and uranium which will be considered in Sect. 5) were irradiated with neutrons and in 35 (+ 2) of them at least one new radioactive product was discovered. The total number of new nuclides, with their characteristic halflives, amounted to 44 (plus at least 2 in Th and at least 4 in uranium). In 16 cases the chemical nature of the product of the reaction was identified by means of the "carrier technique".

The results were summarized in a comprehensive paper presented by Lord Rutherford to the Royal Society [12] as I will tell below (Sect. 3).

More detail were given in a few papers appeared in *Il Nuovo Cimento* some time later [13].

The neutron source consisted of a sealed glass tube about 6 mm in diameter and 15 mm in length, containing beryllium powder and radon in amounts up to 800 millicuries. According to ordinarily assumed yield of neutrons from beryllium, the number of neutrons emitted by this source

[11] E. FERMI, E. AMALDI, O. D'AGOSTINO, F. RASETTI, E. SEGRÈ, "Radioattività β provocata da bombardamento di neutroni - III", *Ric. Scient.*, *5* (1), 1934, 452-453, dated May 10, 1934; also in *FCP-I*, 649-650 and 677-678 for its English translation; E. FERMI, E. AMALDI, O. D'AGOSTINO, F. RASETTI, E. SEGRÈ, "Radioattività provocata da bombardamento di neutroni - IV", *Ric. Scient.*, *5* (1), 1934, 652-653, dated June 23, 1934; also in *FCP-I*, 651-652 and 679-680 for its English translation; E. FERMI, E. AMALDI, O. D'AGOSTINO, F. RASETTI, E. SEGRÈ, "Radioattività provocata da bombardamento di neutroni - V", *Ric. Scient.*, *5* (2), 1934, 21-22, dated July 12, 1934; also in *FCP-I*, 653-654 and 681-682 for its English translation.

[12] E. FERMI, E. AMALDI, O. D'AGOSTINO, F. RASETTI, E. SEGRÈ, "Artificial Radioactivity Produced by Neutron Bombardment", *Proc. Roy. Soc.*, *A146*, 1934, 483-500, communicated by Lord Rutherford, received July 25, 1934; also in *FCP-I*, 732-747.

[13] E. FERMI, "Radioattività provocata da bombardamento di neutroni", *Nuovo Cimento*, *11*, 1934, 429-441; also in *FCP-I*, 715-724; E. AMALDI, E. FERMI, F. RASETTI, E. SEGRÈ, "Nuovi radioelementi prodotti con bombardamento di neutroni", *Nuovo Cimento*, *11*, 1934, 442-451; also in *FCP-I*, 725-731; E. AMALDI, E. SEGRÈ, "Segno ed energia degli elettroni emessi da elementi attivati con neutroni", *Nuovo Cimento*, *11*, 1934, 452-460.

ought to be of the order of 1000 neutrons per second per millicurie [14]. These neutrons are distributed over a very wide range of energies from zero up to 7 or 8 million volts, besides a very small percentage having energies about twice as high as this limit.

The neutrons are mixed with a very intensive γ-radiation. This does not, however, produce any inconvenience, as the induced activity is tested after irradiation [...].

The emission of electrons from the activated substances was tested with Geiger-Müller counters about 5 cm in length and 1.4 cm in diameter. The walls of the counter were of thin aluminium foil, 0.1 to 0.2 mm in thickness. [...]

The substances to be investigated were generally put into the form of cylinders, which could be fitted round the counter in order to minimize the loss in intensity through geometrical factors. During irradiation the material was located as close as possible round the source. Substances which had to be treated chemically after irradiation were often irradiated as concentrated water solutions in a test tube. [...]

A preliminary investigation of the penetrating power of the β-rays of the new radio-elements has been carried out. For this purpose counters of the standard type were used, and the substance, instead of being put quite close to the counter, was shaped in the form of a cylinder of inner diameter somewhat larger than the diameter of the counter in order to allow cylindrical aluminium screens of different thicknesses to be interposed. In this way absorption curves of more or less exponential type were obtained. As the geometrical conditions of this absorption measure are different from the standard ones, and moreover, the number of impulses instead of the total ionization is computed, we checked the method by measuring the absorption coefficients of known radioactive substances; as expected, we found a difference (about 20%). The data are corrected for this factor.

In several cases the absorption by 2 mm of lead was not complete; this was assumed as a proof of the existence of a γ-radiation [15].

The results were sufficiently abundant to allow the beginning of a systematic classification of nuclear reactions produced by neutrons.

[14] This early estimate generally accepted in 1934 was shown to be too low by a factor between 50 and 100 about two years later: E. Amaldi, E. Fermi, "Sopra l'assorbimento e la diffusione dei neutroni lenti", *Ric. Scient.*, 7 (1), 1936, 454-503; also in *FCP-I*, 841-891; "On the Absorption and Slowing down of Neutrons", *Phys. Rev.*, *50*, 1936, 899-928; also in *FCP-I*, 892-942. [N.d.A.]

[15] E. Fermi, E. Amaldi, O. D'Agostino, F. Rasetti, E. Segrè, "Artificial Radioactivity Produced by Neutron Bombardment", *quot.*, 484-486, *FCP-I*, 733-734.

We had found that all elements, whatever their atomic weight, can be activated by neutrons. The nuclide product is sometimes an isotope of the target nucleus, on other occasions it has an atomic number lower by one or two units. From this point of view a marked difference was found in the behaviour of the light and heavy elements. For light elements the active product has in general Z smaller than that of the target nucleus, while for heavy elements (with the exception of uranium and thorium) the active product is always an isotope of the bombarded nucleus. The results obtained with the light elements can in general be explained as due to (n, α) and (n, p) reactions, in which the nuclear charge of the nucleus decreases by two or one unit, respectively:

$$ {}_Z X^A(n, \alpha)_{Z-2}Y^{A-3}; \qquad {}_{Z-2}Y^{A-3} \xrightarrow{(\beta)} {}_{Z-1}W^{A-3} + e^- + \bar{\nu}_e \tag{1} $$

$$ {}_Z X^A(n, p)_{Z-1}Y^A; \qquad {}_{Z-1}Y^A \xrightarrow{(\beta)} {}_Z W^A + e^- + \bar{\nu}_e \tag{2} $$

In these processes it is the outgoing particle which has to cross the electrostatic potential barrier; and the higher this is, the heavier is the residual nucleus. The energy of the neutrons emitted from Be bombarded with the α-particles from Po or Rn in equilibrium with its product (a few MeV, as we have seen above) is such that the penetrability of the electrostatic barrier turns out to be so small for heavy elements that the corresponding cross section is negligible.

The same argument applied to the incident particle explained why the artificial radioactivity produced by charged particles could be observed by the Joliot-Curie as well as by others only in the case of light elements [16].

Other points dealt with in the extensive paper of Fermi's group are: a) a very rough estimate of the cross section for the production of artificial radioactivity:

$$ \sigma = 2 \cdot 10^{-26}\, i \cdot A \quad (\mathrm{cm}^2) $$

where A is the mass number of the target nucleus and i the intensity of activation defined as the number of disintegrations per second

[16] See Ref. 2., 3., 4.

which take place in 1 gram of the element, placed at the distance of 1 cm from a neutron source consisting of 1 millicurie of radon (in equilibrium with its decay products) and beryllium powder.

b) The sign of the β-particles which was found to be negative in all tested cases [17] using the trochoide method of Tibeau [18]. This is what was expected for nuclides having an excess of neutrons with respect to the corresponding stable isotopes, as in general are those produced by (n, α), (n, p) and (n, γ) processes. For the same reasons, in the case of (α, n), (p, n), (γ, n) and $(n, 2n)$ reactions an emission of positrons, at least in general, was expected and actually observed.

c) A discussion of the nature of the nuclear reactions responsible of radionuclides isotopic with the target element. On this point I will come back in the next section.

The great efficiency of neutrons in producing nuclear reactions is due to a number of circumstances which more than compensate the disadvantage that (so long as a nuclear reactor is not available) they are only obtainable in rather limited numbers being themselves the products of another nuclear reaction. The factors in favour are all due to the electric neutrality:

a) They do not loss energy in ionizing matter but only in nuclear reactions. Therefore their efficiency does not change crossing the target as long as they do not undergo a nuclear collision.

b) As a consequence of a) the thickness of the target from which the secondary β-ray can reach the detector is limited by *their* penetration and not by the penetration of the primary particles, as happens for example in the case of α-particles. This fact alone gives a factor of the order of one hundred in favour of the neutrons.

c) Finally the cross section for any reaction is always a rather large fraction of the geometric cross section while in the case of charged particles the cross section is strongly reduced by the Gamow-factor representing the penetrability of the electrostatic potential barrier. This

[17] See E. Amaldi, E. Segrè, "Segno ed energia...", *quot.*

[18] J. Tibeau, "Déviation électrostatique et charge spécifique de l'électron positif", *Compt. Rend. Acad. Sci.*, *197*, 1933, 447-448; "Electrostatic Deflection of Positive Electrons", *Nature*, *32*, 1933, 480-481; "Focusing of Beams, Measurements of Charge to Mass Ratio, Study of Absorption and Conversion into Light", *Phys. Rev.*, *45*, 1934, 781-787.

last factor has obviously a greater importance if the atomic number Z of the target nucleus is large [19].

The importance of the work on artificial radioactivity produced by neutrons was generally recognized by all nuclear physicists. For example only one month after the beginning of this work, Lord Rutherford wrote to Fermi the letter shown in Fig. 1.

Two remarks may be in order at this point. The first is that our group was probably the first large physicists team working successfully for about two years in a very well organized way. The second that we were perhaps the first to introduce the use of preprints. In order to communicate rapidly our results to our colleagues we wrote almost weekly short letters in Italian to *La Ricerca Scientifica*, the journal of the Consiglio Nazionale delle Ricerche, and obtained what we would now call preprints of these letters that were mailed to a list of about forty of the most prominent and active nuclear physicists all over the world, and the letters appeared a couple of weeks later in the journal. This procedure was facilitated by the fact that my wife, Ginestra, was working at that time at *La Ricerca Scientifica.*

3. A DIRECT CONTACT WITH THE PHYSICISTS IN CAMBRIDGE.

The manuscript of the paper summarizing the work accomplished in Rome by summer 1934, was brought by Segrè and me to Lord Rutherford in Cambridge at the beginning of July 1934. When Fermi wrote to Rutherford if we could spend the summer at the Cavendish Laboratory, he answered the letter shown in Fig. 2 [20]. When, shortly

[19] E. AMALDI, "The Production and Slowing down of Neutrons", in *Handbuch der Physik*, S. Flügge (ed.), *38* (2), 1959, 1-659.

[20] The letter of Rutherford dated June 20th, 1934 reads as follows: Dear Professor Fermi, your letter has been forwarded to me in the country where I am taking a holiday. I saw your account in "Nature" of the effects on uranium. I congratulate you and your colleagues on a splendid piece of work. I have had two of my men Westcott and Bjerge repeating some of your experiments with an Em + Be tube and promised them when I returned in a week or two to try out for them the effect of the 2 million volts neutrons from the D + D reaction on a few elements. I myself have been naturally interested on the energy of the neutron required to start transmutations. We do not, however, propose at the moment to make systematic experiments in this direction as our installation is

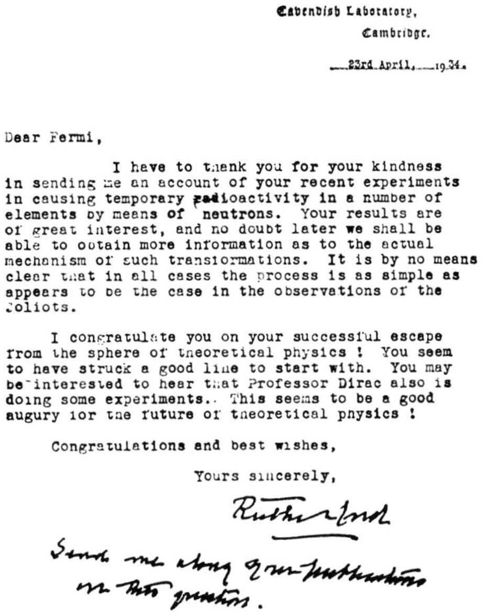

Cavendish Laboratory,
Cambridge.

23rd April, 1934.

Dear Fermi,

I have to thank you for your kindness in sending me an account of your recent experiments in causing temporary radioactivity in a number of elements by means of neutrons. Your results are of great interest, and no doubt later we shall be able to obtain more information as to the actual mechanism of such transformations. It is by no means clear that in all cases the process is as simple as appears to be the case in the observations of the Joliots.

I congratulate you on your successful escape from the sphere of theoretical physics ! You seem to have struck a good line to start with. You may be interested to hear that Professor Dirac also is doing some experiments. This seems to be a good augury for the future of theoretical physics !

Congratulations and best wishes,

Yours sincerely,

Rutherford

Send me along your publications on this question.

Fig. 1 - First letter of Rutherford to Fermi.

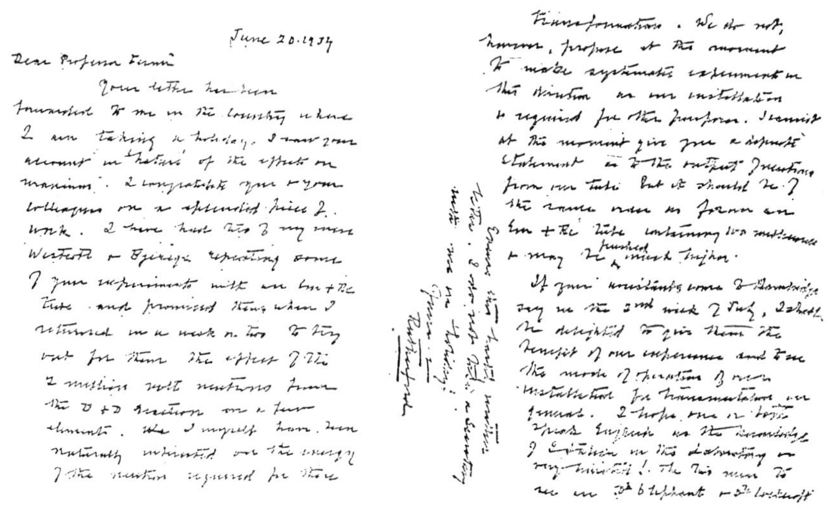

June 20.1934

Dear Professor Fermi

Your letter has been forwarded to me in the country where I am taking a holiday. I saw your account in 'Nature' of the effects on uranium. I congratulate you & your colleagues on a splendid piece of work. I have heard too of my men Westcott & Bjerge repeating some of your experiments with an Em + Be tube and promised them when I returned in a week or two to try out for them the effect of the 2 million volt neutrons from the D + D reaction on a few elements. We & myself have been naturally interested in the energy of the neutron required for these transformations. We do not, however, propose at the moment to make systematic experiment in that direction as our installation is required for other purposes. I cannot at the moment give you a definite statement as to the output of neutrons from our tube but it should be of the same order as from an Em + Be tube containing 100 millicuries & may be pushed much higher.

If your assistants come to Cambridge say in the 2nd week of July, I shall be delighted to give them the benefit of our experience and to see the mode of operation of our installation for transmutation in general. I hope one or both speak English as the knowledge of Italian in this laboratory is very limited! The two men to see are Dr Oliphant & Dr Cockcroft

Excuse this hasty writing [illegible]

Yours ever

Rutherford

Fig. 2 - Second letter of Rutherford to Fermi.

after our arrival we presented the manuscript to Lord Rutherford he immediately read it with attention, made several corrections to improve our English and turned it over the Royal Society. At our first encounter, at Segrè's question if it would be possible to obtain speedy publication, Rutherford answered: "What do you think I was the president of the Royal Society for?" Unfortunately, our understanding of Rutherford's English at the time was imperfect and we could not follow most of his remarks, many of which must have been humorous because he laughed, from time to time, and only then took the pipe out of his mouth [21].

At our arrival in Cambridge the Cavendish Laboratory appeared to us as the real capital for nuclear physics. Rutherford had moved from Manchester to Cambridge in 1920 and since then all the work of Rutherford and his collaborators and pupils had been carried out at the Cavendish.

The neutron was born there. At the time of our visit Rutherford was working with Oliphant on the $D + D$ reactions but with his strong personality dominated the whole laboratory. Chadwick was working with Maurice Goldhaber; they had discovered shortly before the photodisintegration of the deuteron. J. Cockroft, P.I. Dee, C.D. Ellis and N. Feather were there and F.W. Aston was going on improving the precision of his measurements of atomic masses. In his laboratory there was a young American, K.T. Bainbridge, who had recently made an important step forward in this fundamental technique. Going around in the Cavendish one could meet, from time to time, J.J. Thomson who had retired only a few years before.

On the trip to Great Britain I was accompanied by my wife. During the seven weeks we spent in Cambridge, Segrè and both of us had the pleasure to be invited once, perhaps two times, on Sunday

required for other purpose. I cannot at the moment give you definite statement as to the output of neutrons from our tube but it should be of the same order as from an Em + Be tube containing 100 millicurie and may be pushed much higher. If your assistants come to Cambridge say in the first week of July, I shall be delighted to give them the benefit of our experience and to see the mode of operation of our installations for transmutations in general. I hope one or both speak English as the knowledge of Italian in the laboratory is very modest. The two men to see are Dr. Oliphant and Dr. Cockroft. Excuse the hand written letter, I do not take a secretary with me in holiday! Yours sincerely, Rutherford.

[21] E. SEGRÈ, *Enrico Fermi, Physicist*, *quot.*

afternoon at Rutherford's house, Newnham Cottage in Queen's Road, for a cup of tea [22]. These very agreable receptions gave us the opportunity of establishing friendly relations with many of the physicists working at the Cavendish Laboratory and their wives. In London, on our trip to Cambridge, we visited P.M.S. Blackett at Birbeck College where we met also Otto Frisch [23] as well as many other interesting people.

On occasion of a one day trip from Cambridge to London, we met also Leo Szilard [24] with whom, however, on that occasion, we talked more politics than physics. The murder of the Austrian Chancellor Dollfuss by the Nazi in Vienna, the July 25, 1934, had opened in those days a period of acute world wide tension, which represented one of the steps towards the second World War.

Segrè and I visited systematically all research groups active at that time in the Cavendish, as well as Peter Kapitza who was in charge of the Mond Laboratory and had – not long before – constructed his famous rotating generator for producing extremely intensive magnetic fields [25].

It was in September 1934, *i.e.* less than one month after our visit to the Cavendish and Mond Laboratories that Kapitza left Cambridge for USSR to attend the Mendele'eff Congress. He had visited his country almost every summer since he had started his research activity in Cambridge. During these visits he gave lectures and advised on the construction of new institutes. It came, therefore, as a shock to his colleagues to learn, in October 1934, that Kapitza's return passport had been refused and that he had been ordered to begin the construction of a new laboratory in Russia. Such news made an enormous impression not only in Great Britain but everywhere in Europe and United States [26].

[22] M. Oliphant, *Rutherford Recollections of the Cambridge Days*, Elsevier Pbl. Co., Amsterdam 1972.

[23] O. Frisch, *What Little I Remember*, Cambridge University Press, Cambridge 1979.

[24] For the biography of Leo Szilard, see: E.P. Wigner, "Leo Szilard", in *Biographical Memoirs*, National Academy of Sciences of USA, *40*, 1969, 337-347.

[25] Pjotr Leonidovich Kapitza was awarded one half of the 1978 Nobel Prize for Physics for his discovery of superfluidity.

[26] The reasons underlying this action were given in the *News Cronicle* by the Soviet Embassy in London: "Pjotr Kapitza is a citizen of the USSR educated and

We had also the occasion of talking a few times with the well known technician of Lord Rutherford, Mr. G.R. Grove, who taught us some "know how" about the classical methods of particle detection. He even presented me two fluorescent screens similar to those used by Lord Rutherford in some of his famous experiments for observing the scintillations due to α-particles or protons.

I should say that we were rather surprised to see that in the whole Cavendish Laboratory the only people working on artificial radioactivity induced by neutrons were T. Bjerge and C.H. Westcott, the first from Danmark, the second one from Canada. But, after all, also from the second letter of Rutherford to Fermi (Fig. 2 and Ref. 20.) it appears that this research subject had a rather low priority in the program of experimental activities carried out at the Cavendish Laboratory.

After these visits to most of the laboratories I devoted some time to learn from Wynn Williams how to construct a linear amplifier for measuring the energy lost by a single α-particle or proton in the gas of a small ionization chamber [27]. Segrè and I, however, spent most of the time discussing our common problems with Bjerge and Westcott.

One of the questions that was not yet definitively settled in our paper to the *Proceedings of the Royal Society* was whether the reactions that produced an isotope of the target nuclide were (n, γ), *i.e.* radiative capture, or $(n, 2n)$:

$$_Z X^A(n, \gamma)_Z Y^{A+1}; \qquad _Z Y^{A+1} \xrightarrow{(\beta^-)} {}_{Z+1}W^{A+1} + e^- + \bar{\nu}_e \tag{3}$$

trained at the expenses of his country. He was sent to England to continue his studies... Now the time has arrived when the Soviet urgently needs all her scientists. So when Professor Kapitza came last summer, he was appointed as director of an important new research station which is now built in Moscow". "Professor P. Kapitza and the U.S.S.R.", *Nature, 135*, 1935, 755-756. Lord Rutherford in a letter to *The Times* of April 29th, 1935, expressed his concern about the whole story but shortly after he contributed to prepare an agreement concerning the sale to the Government of the USSR of the large generator for the production of strong magnetic fields, together with the associated apparatus and a duplicate of the hydrogen and helium liquifier to allow the continuation of Kapitza's work in Moscow. With the sum received new equipment was bought for the continuation of the work in Mond Laboratory. "Dr. Kapitza Apparatus and the U.S.S.R.", *Nature, 136*, 1935, 825.

[27] F.A.B. WARD, C.E. WYNN WILLIAMS, H.M. CAVE, "The Rate of Emission of Alfa Particles from Radium", *Proc. Roy. Soc., A125*, 1929, 713-730.

$$ {}_{Z}X^{A}(n,\,2n){}_{Z}Y^{A-1}; \qquad {}_{Z}Y^{A-1} \xrightarrow{(\beta^-)} {}_{Z-1}W^{A-1} + e^{+} + \bar{\nu}_e \qquad (4) $$

They could also have been processes of inelastic scattering of the incident neutron with formation of an isomer of the target nucleus. The last process, however, was not considered in summer 1934 and in any case it would have been excluded by the same experiments and considerations reported below.

The objections against the $(n, 2n)$ reactions (4) were based on the clearly endoenergetic character of this process. Therefore the (n, γ) reaction remained as the most natural interpretation. Furthermore two examples discussed in our extensive paper [28] were also in favour of (n, γ) processes in vanadine and manganese. Our comment, however, to this interpretations was the following:

This hypothesis, which would be in agreement with the observed fact of the emission of negative electron, [point b) of Sect. 2] gives rise, however, to serious theoretical difficulties when one tries to explain how a neutron can be captured by the nucleus in a stable or quasi-stable state. It is generally admitted that a neutron is attracted by a nucleus only when its distance from the centre of the nucleus is of the order of 10^{-12} cm. It follows that a neutron of a few million volts' energy can remain in the nucleus (*i.e.*, have a strong interaction with the constituent particles of the nucleus) only for a time of the order of 10^{-21} seconds; that is, of the classical time needed to cross the nucleus. The neutron is captured if, during this time, it is able to lose its excess of energy (*e.g.*, by emission of a γ-quantum). If one evaluates the probability of this emission process by the ordinary methods one finds a value much too small to account for the observed cross-sections. In order to maintain the capture hypothesis, one must then either admit that the probability of emission of a γ-quantum (or of an equivalent process as, for example, the formation of an electron-positron pair) should be much larger than is generally assumed; or that, for reasons that cannot be understood in the present theory, a nucleus could remain for at least 10^{-16} seconds in an energy state high enough to permit the emission of a neutron [29].

An important step for definitively proving that radiative capture (*i.e.*, a (n, γ) process) is just what happens, was made by Bjerge and

[28] E. Fermi, E. Amaldi, O. D'Agostino, F. Rasetti, E. Segrè, "Artificial Radioactivity Produced by Neutron Bombardment", *quot.*

[29] *Ibidem*, 497, *FCP-I*, 744.

Westcott, on Segrè's suggestion [30]. While we were in Cambridge they found that after neutron bombardment, sodium shows a weak activity, the period of which (about 10 h) is, within the error of measurement, the *same* as that of the long periods produced in magnesium and aluminium, according to the following (n, p) and (n, α) reactions [31]

$$_{12}\mathrm{Mg}(n, p)_{11}\mathrm{Na}; \qquad _{13}\mathrm{Al}^{27}(n, \alpha)_{11}\mathrm{Na}^{24}.$$

Aluminium is monoisotopic and therefore the final product of the (n, α) process necessarily should be $_{11}\mathrm{Na}^{24}$. Furthermore the nuclide $_{11}\mathrm{Na}^{22}$ was already known to emit positrons instead of negative electrons with a mean life of a few years (2.6 y).

Back to Rome Segrè and I, with the help of D'Agostino, completed the proof by showing that this long activity produced in Na was actually due to an isotope of the same elements [32].

A further argument tending to exclude the $(n, 2n)$ processes, is based on the conclusion that, as this is an endo-energetic process, one would expect a decrease of the intensity of its products by decreasing the energy of the incident neutrons. After a few attempts in this direction [33] which gave uncertain answers, Lise Meitner [34] succeeded in proving that the neutrons produced in Be by photoeffect of the γ-rays of Ra ($E \leq 0.670$ MeV) could produce artificial radioactivity in heavy elements as I, Ag and Au. This experimental result excludes that in these cases the involved processes are of the $(n, 2n)$ type. The general validity of such a conclusion was established soon afterwards by the results of experiments with slow neutrons. These were found to have cross sections

30 T. BJERGE, C.H. WESTCOTT, "Radioactivity Induced by Neutron Bombardment", *Nature, 134*, 1934, 286, dated August 14, 1934.

31 E. FERMI, E. AMALDI, O. D'AGOSTINO, F. RASETTI, E. SEGRÈ, "Artificial Radioactivity Produced by Neutron Bombardment", *quot.*

32 E. AMALDI, E. SEGRÈ, O. D'AGOSTINO, "Radioattività provocata da bombardamento di neutroni - VI", *Ric. Scient., 5* (2), 1934, 381-382, dated November 7, 1934.

33 T. BJERGE, C.H. WESTCOTT, "Radioactivity Induced by Bombardment with Neutrons", *Nature, 134*, 1934, 381-382, dated July 21, 1934.

34 L. MEITNER, "Über die Umwandlung der Elemente durch Neutronen", *Naturwiss., 22*, 1934, 759, dated October, 1934.

much larger than fast neutrons for (n, γ) processes but *not* for (n, p) and (n, α) reactions [35].

I have insisted on this particular problem because the proof that neutrons can undergo radiative capture with an appreciable cross section was an important piece of experimental evidence that the "one particle model" still very frequently used, was inadequate for describing simple nuclear processes as are those involving only incoming and outgoing neutral particles.

In September 1934 we found [36] a further case of "proved" radiative capture, which was based on the discovery of a new radioisotope of Al with a half life of 2.3 minutes. This by necessity had to be attributed to $_{13}Al^{28}$ because Al^{26} was known to emit positrons instead of electrons with a half life of about 7 seconds. But, when, a few days after our results were made known, I tried to repeat the measurements, I did not find anymore this new activity.

Unfortunately we had communicated our result to Fermi who, on his way back from South America (where he had spent the summer), was attending the International Conference on Physics in London [37]. He had even mentioned our experiment at one of the meetings. The fact that we were not able to confirm our result was hurriedly communicated to Fermi who was angry and embarassed at having presented at the Conference an erroneous result.

We were unhappy and confused because we were not able to understand the origin of our fault. This was the first hint of unexpected complications which were fully clarified in about one and a half month, with the discovery of slow neutrons [38].

[35] E. Fermi, E. Amaldi, B. Pontecorvo, F. Rasetti, E. Segrè, "Azione di sostanze idrogenate sulla radioattività provocata da neutroni - I", *Ric. Scient.*, 5 (2), 1934, 282-283, dated October 22, 1934; also in *FCP-I*, 757-758 and 761-762 for its English translation.

[36] E. Amaldi, E. Segrè, O. D'Agostino, "Radioattività provocata da bombardamento di neutroni - VI", *quot.*

[37] International Conference on Physics, London 1934 (a joint Conference organized by the International Union of Pure and Applied Physics and the Physical Society). Papers and Discussions in two volumes. Vol. I, *Nuclear Physics*, Cambridge University Press and The Physical Society, Cambridge 1935.

[38] E. Fermi, E. Amaldi, B. Pontecorvo, F. Rasetti, E. Segrè, "Azione di sostanze idrogenate sulla radioattività provocata da neutroni", *quot.*

4. THE DISCOVERY OF SLOW NEUTRONS

In September 1934, Fermi's group decided to try to construct a scale of activation (in arbitrary units) of the various elements irradiated under some kind of standard conditions. The work was assigned to me and Pontecorvo, one of our best students who had taken the degree (*laurea*) in July 1934 and after the summer vacations had joined the group.

We met, however, difficulties because apparently the activation depended on the materials surrounding the neutron source and the irradiated sample.

The study of the problem brought, at the end of October 1934, to the discovery of the action on the (n, γ) processes of surrounding hydrogeneous substances.

This effect was immediately interpreted by Fermi as due to the slowing down of neutrons through a succession of elastic collisions with the protons present in the surrounding medium, combined with an increase undergone by the (n, γ) cross section by decreasing the neutron energy [39].

Shortly later it was found that some substances like B, Cl, Co, Y, Rh, Ir, Ag, Cd etc. had very large radiative capture cross sections σ_c. The geometry was very poor, but the results of our measurements were sufficient for establishing that in some cases σ_c was 10^3 to 10^4 times greater than the geometric cross section of the nuclei.

Special attention was devoted for clarifying the nature of the processes involved in the capture of slow neutrons in a few special cases such as boron [40] and cadmium, characterized by large capture cross sections, followed by negligible activation [41].

These experiments were the first steps in the development of

39 *Ibidem*, 282.

40 E. AMALDI, O. D'AGOSTINO, E. FERMI, B. PONTECORVO, F. RASETTI, E. SEGRÈ, "Radioattività provocata da bombardamento di neutroni - VIII", *Ric. Scient.*, *6* (1), 1935, 123-125, dated January 15, 1935; also in *FCP-I*, 661-664 and 689-692 for its English translation; E. AMALDI, "Nuove radioattività provocate da neutroni. La disintegrazione del boro", *Nuovo Cimento*, *12*, 1935, 223-231; similar conclusions were reached by J. CHADWICK and M. GOLDHABER, "Disintegration by Slow Neutrons", *Nature*, *135*, 1935, 65; *Proc. Camb. Phil. Soc.*, *31*, 1935, 612-616.

41 E. AMALDI, O. D'AGOSTINO, E. FERMI, B. PONTECORVO, F. RASETTI, E. SEGRÈ, "Artificial Radioactivity Produced by Neutron Bombardment - II", *Proc. Roy.*

various slow neutrons detection devices based on the boron (n, α) disintegration and for the use of cadmium as a particularly convenient material for shielding and canalizing slow neutrons.

The explanation of these anomalous capture cross sections clearly required quantum mechanics as it was shown by Fermi, as well as by many other authors [42].

For particles of such a small velocity that their wave length λ is much larger than the radius R of the target obstacles, the upper limit of the cross section is not πR^2 but $\lambda^2/4\pi$ multiplied by a numerical factor smaller but, at least in some cases, close to 1.

The same type of reasoning brought Fermi as well as others to foresee a general law for the dependence of the capture cross section σ_c from the velocity of the neutron.

In a medium with n (slow) neutrons per cubic centimeter, the probability of capture per second of one of them by a nucleus is given by $\mathrm{P} = n \cdot \sigma_c \cdot v$. If one describes the process in a simple approach as that provided by the "one particle model", at the limit of $v \rightarrow 0$ (*i.e.* $\lambda \rightarrow \infty$) P should become a constant and therefore σ_c should follow the universal $1/v$-law

$$\sigma_c = \frac{k}{v}$$

Large values of the constant k were shown to be possible, thus providing a simple explanation for the anomalously large capture cross sections observed in some nuclides.

But how small was the velocity reached by the neutrons slowed down in water or paraffin at room temperature? An experiment aiming at establishing whether the neutrons were thermalized by comparing the activation of a piece of Rh surrounded by hydrocarbons at two

Soc., *A149*, 1935, 522-558, communicated by Lord Rutherford, received February 15, 1935; also in *FCP-I*, 765-794.

42 F. Perrin, W.M. Elsasser, "Théorie de la capture sélective des neutrons par certains noyaux", *Compt. Rend. Acad. Sci.*, *200*, 1935, 450-452; *J. Phys. Radium*, *6*, 1935, 194-202; F. Perrin, "Mécanisme de la capture des neutrons lents par les noyaux légers", *Compt. Rend. Acad. Sci.*, *200*, 1935, 1749-1751; G. Beck, L.H. Horsley, "Nonelastic Collision Cross Sections for Slow Neutrons", *Phys. Rev.*, *47*, 1935, 510; H.A. Bethe, "Theory of Disintegration of Nuclei by Neutrons", *Phys. Rev.*, *47*, 1935, 747-759.

different temperatures but with equal proton density, gave a negative result. Shortly afterwards Moon and Tillman, working at Imperial College in London [43], realized that such an effect can be observed only in the vicinity of the boundary between two media at different temperatures and succeeded in obtaining a clear evidence for the thermalization of neutrons.

Our results were again published in a series of letters to *La Ricerca Scientifica* and summarized in an extensive paper communicated by Lord Rutherford to the *Proceedings of the Royal Society* the February 15, 1935 [44].

Before and during summer 1935 mechanical experiments were carried out in Rome [45] and Copenhagen [46] which confirmed the very low velocities reached by slow neutrons and in some way paved the way to the construction by the Columbia University group [47] of the first mechanical selector. Its design followed very closely the principle employed in the instruments used for measuring the velocity of atomic and molecular beams, which in their turn were inspired by the Fizeau method for the laboratory determination of the velocity of light.

After summer 1935 the group in Rome was in great part dispersed and Fermi and I turned our attention to some results of Bjerge and Westcott [48] and Moon and Tillman [49] who had observed that the absorp-

[43] P.B. MOON, J.R. TILLMAN, "Evidence on the Velocity of Slow Neutrons", *Nature, 135*, 1935, 904, dated April 12, 1935.

[44] E. AMALDI, O. D'AGOSTINO, E. FERMI, B. PONTECORVO, F. RASETTI, E. SEGRÈ, "Artificial Radioactivity Produced by Neutron Bombardment - II", *quot.*

[45] E. AMALDI, O. D'AGOSTINO, E. FERMI, B. PONTECORVO, E. SEGRÈ, "Radioattività provocata da bombardamento di neutroni - X", *Ric. Scient.*, *6* (1), 1935, 581-584, dated June 14, 1935; also in *FCP-I*, 669-673 and 697-701 for its English translation; E. AMALDI, "Künstliche Radioaktivität durch Neutronen", *Phys. Zeit.*, *38*, 1937, 692-734.

[46] O.R. FRISCH, E.T. SORENSEN, "Velocity of Slow Neutrons", *Nature, 136*, 1935, 258, issue of August 17, 1935.

[47] J.R. DUNNING, G.B. PEGRAM, G.A. FINK, D.P. MITCHELL, E. SEGRÈ, "Velocity of Slow Neutrons by Mechanical Velocity Selector", *Phys. Rev.*, *48*, 1935, 704, dated October 7, 1935.

[48] T. BJERGE, C.H. WESTCOTT, "On the Slowing down of Neutrons in Various Substances Containing Hydrogen", *Proc. Roy. Soc.*, *A150*, 1935, 709-729, received the May 11, 1935.

[49] J.R. TILLMAN, B.P. MOON, "Selective Absorption of Slow Neutrons", *Nature, 36*, 1935, 66-67, dated June 27, 1935.

tion of slow neutrons by various elements was slightly dependent on the element used as detector. These results were in clear contradiction with the current theory of absorption of neutrons by nuclei, which, as I said before, predicted for all nuclei a capture cross section inversely proportional to the velocity of the neutron. This law was supposed to be valid for such a large energy interval as to certainly cover the energy range of slow neutrons.

Fermi and I repeated and extended the measurements carried out by the two British groups by measuring the absorption coefficient of 11 different elements in all possible combination with 7 detectors and confirmed the general rule that the absorption coefficient of a given element was greater when the same element was used as detector.

Since in the meantime the Columbia University group had shown that the neutrons absorbed by cadmium were, at least in great part of thermal energy, Fermi and I proceeded to repeat the same experiments on the slow neutrons filtered by a thick foil of cadmium. Thus, in November 1935, we established that under these conditions the self-absorption effect mentioned above was considerably enhanced in a number of elements such as Rh, Ag, and I. This result was an indication that several nuclides have absorption lines still in the slow neutron region but above thermal energy [50].

During the winter 1935-36 we carried out a systematic experimental study of the selective absorptions observed in Rh, Ag and I and of the neutrons endowed of the corresponding energies, arriving to the clarification of a number of problems.

We measured the neutron-proton elastic cross section at thermal as well as in the epithermal region and showed that they differ by a factor slightly greater than 3, and Fermi developed the theory of this effect produced by the chemical bond of the protons in the molecules of the moderator. We measured the diffusion length of thermal neutrons in paraffin and determined the radiative capture cross section of thermal neutrons by protons and Fermi showed that it was due to a magnetic dipole moment transition. We succeeded in estimating the energy of the resonances observed in Rh, Ag and I by two methods. The first one was based on the mean square distance travelled

[50] E. Amaldi, E. Fermi, "Sull'assorbimento dei neutroni lenti - I", *Ric. Scient.*, *6* (2), 1935, 344-347, issue of November 15-30, 1935; also in *FCP-I*, 811-815.

in the moderator from a fast neutron point source by the neutrons for reaching a specific resonance energy, the other on the assumption of the validity of the $1/v$-law in boron, formulated independently by Frisch and Placzek and Bethe, Livingston and Weeks [51].

We found that all the observed resonance lines were between 1 and 10 *eV*. Finally, in February 1936, we measured the width of the most important resonances of Rh, Ag and I and found they were all of the order of one tenth of electronvolt.

In the mean time two very important theoretical papers had appeared dealing with this ensemble of phenomena.

The first one was presented by Niels Bohr to the Danish Academy on January 26, 1936 and appeared in *Nature* on February 29 [52].

It contained the idea that by capture of the incident neutron a compound nucleus was formed in an excited state, the mean life of which was long enough for explaining the thin lines observed in the epithermal region.

The other paper, due to Breit and Wigner [53], was received by the *Physical Review* the February 15, 1936 and contained the derivation of the well known "one level Breit and Wigner formula", describing the resonance observed in capture (and scattering) of slow neutrons by nuclei with a single resonant level.

The experimental results obtained by Fermi and Amaldi again were published in a series of short letters to *La Ricerca Scientifica* and summarized in extensive papers appeared in *La Ricerca Scientifica* and the *Physical Review* [54]. The theoretical work carried out by Fermi was published in a long paper appeared only in *La Ricerca Scientifica* [55].

[51] G. Placzeck, O. Frisch, "Capture of Slow Neutrons", *Nature*, *137*, 1936, 357, issue of February 29, 1936; D.F. Weeks, M.S. Livingston, H.A. Bethe, "A Method for the Determination of the Selective Absorption of Slow Neutrons", *Phys. Rev.*, *49*, 1936, 471-473, received March 4, 1936.

[52] N. Bohr, "Neutron Capture and Nuclear Constitution", *Nature*, *137*, 1936, 344-348, issue of February 29, 1936.

[53] G. Breit, E. Wigner, "Capture of Slow Neutrons", *Phys. Rev.*, *49*, 1936, 519-531, received February 15, 1936, issue of April 1, 1936.

[54] E. Amaldi, E. Fermi, "Sopra l'assorbimento e la diffusione dei neutroni lenti", *Ric. Scient.*, 7 (1), 1936, 454-503, dated May 29, 1936, issue of 15-30 June 1936; also in *FCP-I*, 841-891; "On the Absorption and the Diffusion of Slow Neutrons", *Phys. Rev.*, *50*, 1936, 899-928, received September 5, 1936, issue of November 15, 1936; also in *FCP-I*, 892-942.

[55] E. Fermi, "Sul moto dei neutroni nelle sostanze idrogenate", *Ric. Scient.*, 7

5. The uranium puzzle and the discovery of fission

Before concluding this review I would like to add a few words about the work carried out by Fermi's group in 1934-35 on the problem of U under neutron bombardment.

In spring 1934, proceeding systematically according to increasing atomic number, we irradiated finally thorium and uranium. We observed a number of new activities which were not easily interpreted. Our first report was dated May 10, 1934. It contained the chemical proof that a 13 min half-life nuclide was not an isotope of U(92) or Th(90). In later publications we showed that it was not an isotope of any element between 82(Pb) and 92(U) with the exception of Po(84) and At(85).

Since we knew from all other heavy elements that under neutron irradiation they undergo (n, γ) reaction, we interpreted our results as due to a transuranic element $Z = 93$ produced by neutron capture by U^{238} which we suggested to be a β-emitter decaying in element 93, that we naively assumed to be an EkaRe:

$$U^{238}\,(n,\ \gamma)\,U^{239}$$

$$U^{239} \xrightarrow[13\ \mathrm{min}]{\beta^-} \mathrm{EkaRe}^{239}$$

We did not consider the isotope U^{235} because it was discovered only about one year later [56].

Shortly later we studied another body of $T_{1/2} = 100$ min (later 90 min) which also could be separated by the same natural elements as well as from the artificial 13 min body. We suggested that it could be an isotope of element 94, that we assumed to be an EkaOs.

This conclusion was criticized by Ida Noddack [57] who suggested that by bombardment of heavy elements with neutrons these nuclei could break in many larger pieces (large with respect to α-particles or protons), which are isotopes of known elements but not neighbouring

(2), 1936, 13-52, issue of 15-31 July, 1936; also in *FCP-I*, 943-979 and 980-1016 for its English translation.

56 A.J. Dempster, "Isotopic Constitution of Uranium", *Nature*, *136*, 1935, 180.

57 I. Noddack, "Über das Element 93", *Angewandte Chemie*, *47*, 1934, 653-655, dated September 10, 1934.

of those irradiated. She did not give any argument in favour of this interpretation and therefore her suggestion appeared as a speculation aiming more to point out a lack of rigor in the argument for the formation of element 93 than as a serious explanation of the observations. This remark seems to be supported by the fact that she did never try to do experiments on irradiated uranium as certainly she could have done.

At this point Hahn and Meitner entered the game and confirmed our conclusions for what concerns the transuranic elements [58].

We had worked on the problem from the end of April to the beginning of July 1934, *i.e.* less than three months and about two-three months more during the winter 1934-35. After the confirmation of our results from Hahn and Meitner we thought that they were better prepared than us for this kind of chemical work and abandoned the study of the "uranium puzzle" for concentrating our efforts on other themes such as the selective absorption of slow neutrons of a number of nuclides of medium atomic number.

Hahn, Meitner and Strassmann went on for years following the same lines in the experimental work and in the interpretation of their results.

In May 1937, they published a paper showing that three families of transuranic elements were necessary for interpreting the "uranium puzzle" [59]. The situation became even more serious during the successive eighteen months because of the discovery of 9 more radioactive bodies which, until November 8, 1938, were interpreted as new radionuclides of atomic number between 88(Ra) and 90(Th).

But in December of the same year Hahn and Strassmann found that some of the new radioactive nuclides they had previously attributed to radium isotopes could be separated from natural radium used as a carrier: in fractional distillation these nuclides did not follow radium but barium, the lower homologue of radium. They wrote: "[...] our radium isotopes have the chemical characteristic of barium. Speaking

[58] O. Hahn, L. Meitner, "Über die künstliche Umwandlung des Urans durch Neutronen", *Naturwiss.*, *23*, 1935, 37-38, dated 22 Dezember, 1934; *Ibidem*, 230-231, Eingegangen 2 Marz, 1935.

[59] O. Hahn, L. Meitner, F. Strassmann, "Über die Umwandlungsreihen des Urans, die durch Neutronenbestrahlung erzeugt werden", *Z. Phys.*, *106*, 1937, 249-270, Eingegangen 14 Mai, 1937.

as chemists, we even have to say that these new substances are barium, not radium"[60].

Fissione was discovered.

SOMMARIO

L'articolo ripercorre le fasi principali delle ricerche effettuate dal "gruppo Fermi" a Roma tra il 1934 e il 1935.

Dopo l'annuncio dato dai Joliot-Curie all'inizio del 1934 della scoperta della radioattività artificiale indotta da particelle α, ai primi di marzo Fermi suggerisce a Rasetti la possibilità di osservare effetti analoghi indotti da neutroni. Il 25 marzo una lettera alla *Ricerca Scientifica* annuncia i primi risultati del bombardamento sistematico degli elementi chimici in ordine crescente per numero atomico; nei mesi successivi (aprile-maggio) il gruppo pubblica in rapida successione una nutrita serie di nuovi risultati, al punto da consentire l'inizio di una classificazione sistematica delle reazioni nucleari prodotte da neutroni. Tra l'altro si rende evidente una differenza netta tra gli elementi leggeri e quelli pesanti: per i primi il prodotto attivo ha in genere uno Z piú piccolo di quello del nucleo originale, mentre per gli elementi pesanti, con l'eccezione dell'uranio e del torio, il prodotto attivo è sempre un isotopo del nucleo bombardato.

L'importanza di queste ricerche trova un generale riconoscimento all'estero, in particolare da parte di Lord Rutherford e del suo gruppo del Cavendish Laboratory. L'articolo descrive impressioni ed esperienze che Amaldi e Segrè raccolgono durante la loro visita a Cambridge nel luglio 1934.

A quel tempo non era ancora chiaro se le reazioni che producevano isotopi del nucleo bombardato erano del tipo (n, γ) o del tipo $(n, 2n)$. Diverse considerazioni e diversi risultati sperimentali fecero propendere per la prima ipotesi, anche se essa dimostrava che il modello "particella-singola" del nucleo era inadeguato.

La seconda fase delle ricerche del gruppo Fermi inizia nel settembre 1934 e porta quasi immediatamente alla scoperta della forte influenza sui processi (n, γ) delle sostanze idrogenate poste attorno alla sorgente. Fermi riconosce che l'effetto è dovuto al rallentamento dei neutroni in seguito alle collisioni elastiche con i protoni presenti nel mezzo e all'aumento della

[60] O. HAHN, F. STRASSMANN, "Über den Nachweis und das Verhalten der bei der Bestrahlung des Urans mittels Neutronen entstehenden Erdalkalimetalle", *Naturwiss.*, *27*, 1939, 11-15, Eingegangen 22 Dezember 1938, Heft 6 Januar 1939.

sezione d'urto (n, γ) al diminuire dell'energia dei neutroni. Nel 1935 la ricerca prosegue sia sul piano teorico sia sul piano sperimentale; in particolare viene analizzato il problema dell'assorbimento selettivo dei neutroni lenti con risultati che nell'inverno 1935-36 portano al chiarimento di molti aspetti dei processi di cattura radiativa. Subito dopo queste ricerche, nel febbraio 1936 appaiono i lavori teorici fondamentali di Bohr e di Breit e Wigner.

Nell'ultimo paragrafo dell'articolo vengono ricordati alcuni aspetti delle ricerche del gruppo Fermi, nello stesso periodo, sul comportamento "anomalo" dell'uranio bombardato da neutroni. Nella primavera del 1934 vennero irradiati torio e uranio con risultati di non facile interpretazione. L'ipotesi piú spontanea era che si aveva a che fare con la produzione di elementi transuranici con Z = 93 e Z = 94. Solo Ida Noddak criticò queste conclusioni, accennando in modo del tutto speculativo alla possibilità che si trattasse di una fissione dei nuclei pesanti. Hahn, Meitner e Strassmann invece accettarono le conclusioni dei fisici italiani e continuarono a porle a fondamento delle loro ricerche per alcuni anni ancora. Ma alla fine del 1938 Hahn e Strassmann pervennero alla scoperta della fissione.

ETTORE MAJORANA, MAN and SCIENTIST

Commemoration Speech

by

E. AMALDI

10

ETTORE MAJORANA, MAN AND SCIENTIST*)

E. Amaldi,

University of Rome,
Rome.

Ettore Majorana was born in Catania on 5 August, 1906, of a well-known professional family of that town. His father, an engineer, Fabio Massimo (b. in Catania 1875 - d. in Rome 1934), was the younger brother of Quirino Majorana (1871-1957), a well-known professor of experimental physics at the University of Bologna[1]. Fabio Massimo was for many years Director of the Telephone Service of Catania; moving to Rome in 1928, he was appointed Head of Division and a few years later Inspector-General of the Ministry of Communications. He married Miss Dorina Corso (b. in Catania 1876 - d. in Rome 1965), also of a Catanian family, and they had five children: Rosina, who later married Werner Schultze; Salvatore, a Doctor of Law interested in philosophical studies; Luciano, a civil engineer who specialized in aircraft construction but later devoted himself to the design and construction of instruments for optical astronomy, Ettore and the fifth and last, Maria, a musician and piano teacher.

Members of his family and their friends say that Ettore had already begun to show signs of a gift for arithmetic and numerical calculation when he was four years old: this revealed itself in his favourite game of multiplying in his head in a few seconds two three-figure numbers given to him by members of his family or their friends.

---

*) Translated, by kind permission of the Accademia Nazionale dei Lincei, from the volume: "La Vita e l'opera di Ettore Majorana", Accademia Nazionale dei Lincei, Rome 1966.

When one of them asked him to do a sum, little Ettore slipped under a table as if he wanted to isolate himself, and gave the answer from there a few seconds later.

By the time he was seven he had become such a well-known chess player that this was mentioned in the local newspaper. After having completed his first years of schooling at home, he went as a boarder to the Istituto Massimo in Rome[2], where he completed his elementary education and then went through secondary school in four years, jumping the fifth. When his family moved to Rome in 1921, he continued as a day boy at the Istituto Massimo during the first and second years of the "liceo", and for the third year went to the Liceo Statale Torquato Tasso, where in the summer term of 1923 he passed his "maturità classica" with high marks[3].

In the autumn of the same year Ettore entered the Biennio di Studi di Ingegneria*) at the University of Rome and began to follow the lectures and training courses regularly, passing the examinations with very high marks.

Among his fellow students was his brother Luciano, with whom he also spent a great deal of the time he devoted to leisure and to seeing their mutual friends: other fellow students were Emilio Segré, now Professor of Physics at the University of Berkeley in California, and Enrico Volterra, now Professor of Civil Engineering at the University of Houston, Texas.

After having completed the Biennio di Studi di Ingeneria, this group of students, all very brilliant, went to the "Scuola di Applicazione per gli Ingegneri**) in Rome.

Ettore went on to obtain high marks in all subjects except hydraulics, in which he failed.

---

*) Initial two-year science and engineering course.

**) School of engineering (final three year course).

At both the Biennio and the Scuola di Ingegneria he acted as consultant to all his companions for the solution of the most difficult problems, particularly in mathematics. Segré remembers, for instance, that while they were waiting for their turn in a descriptive geometry examination Majorana gave him an original and very subtle projective demonstration of the existence of Villarceau circles on a torus. When Segré went into the examination shortly afterwards, he told Professor Pittarelli that he had prepared a special subject: the professor, after a few questions, invited him to speak on his subject, which he did, in spite of the fact that, as far as he remembers, he had not even fully grasped it. The demonstration made a great impression on the professor and, according to Segré, was the main reason why he obtained 30 marks*).

While he was at the Scuola di Ingegneria, Majorana, together with some of his fellow students, grew very critical of the way in which some of the subjects were taught: he felt that too much time was spent on unnecessary detail and not enough on the general synthesis needed for serious and systematic scientific study. This deep-rooted conviction of his frequently gave rise to lively, and sometimes heated, discussions with some of the professors.

At the beginning of the second year of the Scuola di Ingegneria (the fourth university year) Emilio Segré decided to follow an earlier inclination of his and switch to physics. He had reached this decision during the summer of 1927 when he had made the acquaintance of Franco Rasetti, then a lecturer at the Physics Institute of the University of Florence. Through Rasetti, Segré had also made the acquaintance of Enrico Fermi, who was then 27 and had recently (November, 1926) been appointed extraordinary Professor of Theoretical Physics at the University of Rome[4]).

---

*) Full marks.

The creation of this new chair was due to the efforts of O.M. Corbino, Professor of Experimental Physics and Director of the Physics Institute of the University of Rome, who, realizing Enrico Fermi's exceptional qualities, had taken a series of steps to set up a modern school of physics in Rome.

I myself, who in June 1927 was at the end of my second year at the Biennio di Studi di Ingegneria, had decided to switch to physics as a result of an appeal made by Corbino during a lecture: he suggested that with the present ferment of ideas in the whole of Europe in the field of physics and with the appointment of Fermi as professor in Rome, exceptional prospects were opening up for promising young people who were willing to make a special effort in theoretical study and experimental work.

During the autumn and early winter of 1927 Emilio Segrè often talked about Ettore Majorana's exceptional qualities in the new circle of physicists which had in a few months grown up around Fermi, and at the same time he tried to persuade Ettore to follow his example, pointing out that the study of physics would be much more in line with his scientific aspirations and speculative gifts than that of engineering. Ettore Majorana took up physics at the beginning of 1928 after a talk with Fermi. A brief account of this talk will give a glimpse of Majorana's character.

He came to the Physics Institute in the Via Panisperna and was taken by Segrè to Fermi's office, where Rasetti was also present.

This was the first time I saw him. From a distance he looked slender with a timid, almost hesitant, bearing; close to, one noticed his very black hair, dark skin, slightly hollow cheeks and extremely lively and sparkling eyes. Altogether he looked like a Saracen.

Fermi was then working on the statistical model later known as the Thomas-Fermi model.

14

The discussion with Majorana soon turned to the research taking place at the Institute, and Fermi gave a broad outline of the model and showed Majorana reprints of his recent works on the subject, in particular the table showing the numerical values of the so-called Fermi universal potential.

Majorana listened with interest and, after having asked for some explanations, left without giving any indication of his thoughts or intentions. The next day, towards the end of the morning, he again came into Fermi's office and asked him without more ado to show him the table which he had seen for a few moments the day before. Holding this table in his hand, he took from his pocket a piece of paper on which he had worked out a similar table at home in the last twenty-four hours, transforming, as far as Segré remembers, the second-order Thomas-Fermi non-linear differential equation into a Riccati equation, which he had then integrated numerically. He compared the two tables and, having noted that they agreed, said that Fermi's table was correct: he then went out of the office and left the Institute. A few days later he switched over to physics and began to attend the Institute regularly.

Two events in his life during the 1926-1927 period will serve to illustrate certain other aspects of his character and youthful interests.

During these years Ettore and Luciano Majorana used to meet some of their friends and fellow students at the Caffé Faraglia (known as Faraglino) in the Via del Corso (then Corso Umberto I), opposite the offices of the "Giornale d'Italia". The young at that time were most interested in mountain climbing and winter sports, which were by way of being a novelty. The two Majorana brothers, who shared this enthusiasm, decided to try their luck in this field. They bought boots and other equipment and set out with two of their friends, medical students (Paolo Belelli and Aldo Coccia), without any training or preparation, for Monte Velino, which is not difficult as a winter climb but requires, nonetheless, a good deal of caution.

Before dawn on a winter's day in 1926 (or 1927) they left Magliano dei Marsi by moonlight and towards noon arrived near the summit, where the snow was so deep that they went in up to the waist. They then tried to continue on all fours but "not being Red Indians", as they afterwards said, they quickly realized that this was much too tiring and inefficient and turned back. It was only in the evening when they reached Massa d'Albe and stopped at an inn to rest that they learned from the locals that they had done something very foolish, in fact quite crazy, particularly in view of the enormous amount of snow covering the mountain at that time of year.

At the beginning of the summer of 1927, namely about six months before he switched from engineering to physics, while going for a drive in the family car (a Fiat 507 saloon) in the neighbourhood of Rome with his brother Luciano and three friends, Ettore, who had never learnt to drive, took the wheel: shortly afterwards the car left the road and overturned and it was a miracle that the five young people were virtually unhurt.

In order to avoid complications with the police who immediately came on the scene, since Ettore had no driving licence, the five young people said that the accident had occurred with Luciano and not Ettore at the wheel. The affair ended without too much trouble and the detached and natural way in which Ettore and Luciano talked about it gave them the reputation of being dare-devils with their fellow students. However, I recently learned from his brothers, that Ettore never wanted to drive again.

These episodes, and others which could be added, show his youthful interests apart from his studies, and the normality of his relations with other young people of both sexes.

2. Soon after taking up physics, Ettore Majorana impressed everyone with his lively mind, his insight and the range of his interests, which made him appear greatly superior to all his new companions. Being exceptionally penetrating and inexorable in his criticisms, he was nicknamed "the Great Inquisitor". In the same vein we called Fermi "the Pope", Rasetti "the Cardinal-Vicar" and so on[4,5].

His capacity for calculation was amazing. He not only did very complicated numerical calculations completely in his head but also in 20 or 30 seconds calculated definite integrals which were sufficiently complicated to require a considerable number of steps on the part of a clever mathematician: he also substituted algebraic or numerical limits and gave the final results directly.

One day Fermi and Majorana had a competition: this consisted in calculating an expression, if I remember rightly an integral, which Fermi was to calculate on the blackboard and Majorana in his head. While we all stood and watched in silence, Fermi wrote up one step after another at great speed until he had filled a normal sized blackboard. Majorana was facing in another direction with his eyes on the ground. When Fermi obtained the result and said: "I've got it!" Ettore replied "Me too!" and gave the numerical result.

In May and June 1928 while preparing for and sitting the university examinations, we had got into the habit of meeting before supper, between seven and eight in the evening, at the Casina delle Rose in the Villa Borghese. Besides Ettore Majorana, Giovanni Gentile junior, Emilio Segré and myself for the Physics Institute, there were Luciano Majorana, Giovanni Enriques, Giovanni Ferro-Luzzi and Gastone Piqué, all engineering students in the same year as Ettore. Sipping a drink or eating ice-cream, we talked over the preparation of the examinations or the last examination we had sat, or one of the physicists in the group talked about some atomic physics results which he had recently heard about, more often than not from

Fermi, or one of the engineering students discussed the properties of the electromagnetic field or one of its applications, or ran down the hydraulics professor, who was their bête noire. We also talked about literature. Ettore knew and appreciated the classics in general and preferred Shakespeare and Pirandello. We also talked of various cultural matters, a strong point with Ettore, and a bit about politics, but mostly about the Nobile expedition to the North Pole that had just taken place (March-May, 1928) and had given rise to the well-known sequence of events[6)].

Ettore's fellow engineering students often pulled his leg about his weakness in drawing: they described the scene when Ettore, during the projective geometry course, took a large sheet of paper and after some hours' work had traced on it his construction, which was quite in order from the point of view of principle, but extremely small and placed on the skew in a corner while the rest of the sheet remained untouched; they described the professor's expressions of reproach and impatience and the respectful but unperturbed and slightly absent air which Ettore assumed during this scene. Ettore listened to the accounts of this by his friends with a slightly amused air as though it had nothing to do with him, and at the end made some subtle and witty comment on the whole story and the account which had been given of it.

We resumed the habit of going to the Casina delle Rose, although not so regularly, in May and June of the following year, at the end of which we obtained our doctorate.

Ettore Majorana, Gabriello Giannini, who afterwards had a successful career as an electronic designer and manufacturer in the United States, and I, received our doctorate on the same day, 6 July, 1929. Ettore presented a thesis entitled "Sulla meccanica dei nuclei radioattivi" for which Fermi acted as sponsor, and he obtained 110/110 with distinction[7)]. Even some 40 years later it is striking to read his thesis for the clear way in which the problems relating to the structure of the nucleus and the theory of its alpha decay are set out and investigated.

During this period the Majorana family was upset by a trial, which created quite a stir at the time, in which an uncle of Ettore's was accused of burning a small child in order to inherit a legacy. The trial lasted several years: Ettore followed it closely and applied his keen and logical mind to gathering and arranging many of the facts which were to lead in 1932 to the full acquittal of the accused.

According to some of his friends[5)] this episode had a decisive influence on Ettore's attitude to life, but his brothers who all remember very clearly this period, categorically deny this.

After receiving his doctorate, Ettore continued to attend the Institute, where he more or less regularly spent two hours every morning, from 10.30 or 11 a.m. to 12.30 or 1 p.m., and a few hours in the afternoon from 5 to 7.30 p.m.

He spent his time in the library, where he mainly studied the works of Dirac, Heisenberg, Pauli, Weyl and Wigner.

The last two authors were perhaps the only ones for whom he expressed unqualified admiration. This was due, at least to a large extent, to his particularly lively, almost prophetic, interest in group theory and its application to physics. On many occasions during this period Majorana expressed the intention of writing a book on this subject; Segré even believes that he heard him say that he had already written one chapter of it. However, nothing was found among his papers which could be interpreted as a part or an outline of such a work, except the treatment of the properties of the rotation group, the Lorentz group and the two-dimensional unimodular group which are to be found in his papers (Notebooks Nos. 1, 2 and 17). Possible he did indeed write part of this book and destroy it later.

He was impressed most of all by the immensity of mathematics in general, as Segré recalls. In a conversation Majorana once pointed out to him that whereas physics could be summarized in a treatise such as the

19

"Handbuch der Physik", consisting at that time of about 35 volumes, a considerably larger work would be required in the case of mathematics.

However, he did not as a rule express his opinion spontaneously, but only when prompted by a conversation which interested him.

His rather reserved character was a sign of shyness, which made it difficult for him to establish relations with people whom he had known for a short time.

The judgments he passed on living scientists, even of the first rank, were nearly always exceedingly severe, so much so that one would be tempted to suspect him of quite exceptional conceit, were it not for the fact that he was highly critical of his own performance and voiced very harsh opinions about his own work; while his criticisms were toned down and even vanished completely in the case of his friends. Those who were close to him thus finally came to understand that this great severity was nothing more than a sign of his unsatisfied and tormented spirit. His apparent isolation from his fellows, not only in everyday life but also in his emotions, concealed a great sensitivity which led him to form very few friendships: when he did, however, his friendship had the depth which is characteristic of his native land.

He struck up a close friendship with me and an even closer one with Giovanni Gentile junior (1906-1942)[8)]. The latter had obtained his doctorate in physics at Pisa in November, 1927, and soon after had been appointed lecturer at the Physics Institute of the University of Rome, where he spent about six months of the academic year 1927-1928. It was during this time that he met and became friendly with Ettore, in collaboration with whom he later wrote his paper on the splitting of Roentgen terms (No. 1).

Fostered by an interest in similar subjects in their early research work, the friendship between Ettore Majorana and Giovanni Gentile junior was also based on their common Sicilian origin and consequently on closely comparable family backgrounds.

Gentile spent the next two academic years (1929-30 and 1930-31) in Germany, the first in Berlin with E. Schrödinger and the second in Leipzig with W. Heisenberg; returning to Italy in October, 1931, Gentile was appointed lecturer of Theoretical Physics at the University of Pisa and therefore moved to that city, but continued to make frequent visits to Rome so as to keep in close touch with all the physicists of the Institute in the Via Panisperna, particularly Ettore Majorana.

My friendship with Ettore Majorana began in the days when we used to go to the Casina delle Rose; since we lived fairly near each other we often used to go home together. After obtaining my doctorate I did my military service and then went to Germany for about ten months (January-October, 1931) to work in Leipzig under P. Debye on the diffraction of X-rays by liquids. I thus found myself again with Giovanni Gentile junior and met another young Italian theoretical physicist from Turin, Gian Carlo Wick.

On my return to Rome I began to work with George Placzek (1905-1955)[9], whom I had also known in Leipzig and who had moved to Rome for the academic year 1931-1932. We worked on the rotational spectrum of ammonia molecules observed in the Raman effect[10] to check the selection rules established for the symmetrical top rotator by G. Placzek and E. Teller.at that time[11]. Ettore Majorana had got into the habit of coming to the laboratory where I worked, at about 7.30 in the evening, and waiting until I had finished what I was doing so that we could go home together. When Gentile was in Rome he used to join us. Sometimes we were joined by a physics student, Miss Ginestra Giovene, who a few years later was to become my wife: in those days Ginestra attended the Institute, first as a student and then as a young graduate, and under the direction of Fermi and myself did, with a few other young people, numerical calculations of the eigenfunctions of the S states of the various elements using the modified Thomas-Fermi method[12].

During this period I became even more friendly with Ettore; I feel our friendship was based not only on our common interests but also on the great difference between our characters.

On 12 November, 1932, he obtained his university teaching diploma in theoretical physics*): he presented only five papers, but the board, composed of Enrico Fermi, Antonino Lo Surdo and Enrico Persico, was unanimous in recognizing that the candidate had "a complete mastery of theoretical physics"[13].

3. From the point of view of scientific production these years represent the first of the two phases of Ettore Majorana's regrettably short life as a researcher.

This first phase includes his papers Nos. 1 to 6, which all deal with problems of atomic and molecular physics: the second phase, represented by papers Nos. 7, 8 and 9, on the other hand, concerns nuclear physics problems or the properties of elementary particles.

The papers belonging to the first phase can be subdivided into three groups. The first consists of Nos. 1,3 and 5 and concerns atomic spectroscopy: the second includes Nos. 2 and 4, which deal with questions relating to the chemical bond. Finally, the third group consists of No. 6 only, on the problem of non-adiabatic spin flip in a beam of polarized atoms.

Paper No. 1, written in collaboration with Giovanni Gentile junior, concerns the splitting, induced by the electron spin, of the 3M Roentgen terms of gadolinium (Z = 64) and uranium (Z = 92) and of the P optical terms of caesium (Z = 55), and the calculation of the intensity of the lines of caesium. They based their calculations on perturbation theory and used the eigenfunctions of the electrons in the states considered, which were obtained by the Thomas-Fermi method.

---

*) He thus became "libero docente", equivalent to the German Privat dozent.

22

This paper fitted into the general framework of the activity at the Physics Institute in the Via Panisperna from 1928 to 1932, when, in addition to Fermi, most of us were applying the statistical method to various problems of atomic physics.

Paper No. 3 deals with the two lines which had recently been discovered in the helium spectrum and which could not be interpreted as a combination of known terms. P.G. Kruger, who had made this discovery, had proposed an interpretation whereby these two lines were due to transition from a normal state to two new "primed" states, namely states with two excited electrons which, being situated above the ionization energy of the helium atom, could produce transitions to states of the continuous spectrum, namely to spontaneous ionization processes. By means of an analysis based on the symmetry properties of the unperturbed eigenfunctions of the 16 states, which are obtained by combining two orbitals having a total quantum number of 2, and of perturbation calculations taken to second order terms, Ettore Majorana succeeded in confirming the interpretation originally proposed for one of the two lines, and in ruling out the possibility of the other being due to the helium atom.

In this paper it is very interesting to note the discussion of the symmetry properties of the various states with respect to electron-spin exchange (singlet and triplet states), spatial rotations (azimuthal or orbital angular momentum quantum number), axial rotations (magnetic quantum) and reflections with respect to the centre of force (parity).

Paper No. 5, the third and last of the papers on atomic spectroscopy, deals with the so-called incomplete P′ triplet of calcium. As was the case for paper No. 3, the primed terms are terms with two excited electrons: now, in the calcium spectrum at that time five groups of spectral lines had been observed, each consisting of six lines, due to the combination of normal terms and terms which could be arranged in series whose limit corresponds to calcium ionized once but with the external electron in the 3d excited state instead of the 1s state, as is the case of normal series. Some of these terms corresponding to configurations with two excited electrons are situated above the ionization potential and in spite of this are stable in that they do not give rise to appreciable spontaneous ionization. This stability, as Majorana pointed out, is due to the fact that in a non-relativistic approximation the terms concerned have symmetry characteristics which are such that transition to the continuous spectrum is strictly forbidden. However, the presence of the intrinsic magnetic moment of the electrons leads to slight instability, which can become appreciable only in exceptional circumstances. This is the case, as Majorana showed, for the J - 2 component of the anomalous triplets of zinc, cadmium and mercury, while for calcium experience shows that this component is absent or at least very weak.

Turning from the papers on atomic spectroscopy to those on the chemical bond, it should be noted that it was through his close study of these problems that Majorana mastered Heitler and London's[14)] quantum mechanical theory of the chemical bond, which was to be of great importance for his future research work. His thorough knowledge of the exchange mechanism of valence electrons, which forms the basis of the quantum mechanical theory of the homopolar chemical bond, was later to serve as the point of departure for the assumption that nuclear forces are exchange forces.

Paper No. 2 on the formation of the molecular ion of helium was prompted by spectroscopic observations, which had then recently been made and which had brought to light bands attributed to the molecular ion He_2^+. Majorana tackles the problem of the chemical reaction

$$He + He^+ \rightleftarrows He_2^+ \qquad (1)$$

showing that it can actually take place from an energy point of view. By means of a first approximation calculation, Majorana then found the equilibrium distance of the two nuclei in the He_2^+ ion, and obtained values for their oscillation frequency which agrees very well with those found experimentally.

However, much of the method used is apparently similar to that originally developed by Heitler and London to deal with the problem of the hydrogen molecule, the fact that only one of the two atoms which are bonded to form the He_2^+ ion is ionized, makes the problem substantially different from that of the hydrogen molecule and, in a more general way, from all the problems tackled until then by means of the homopolar valence theory.

In paper No. 4 Majorana makes a detailed study of the characteristics of a rather deep and anomalous even singlet term (called X term), which had recently been observed in the spectrum of the hydrogen molecule and which had been interpreted as a primed term.

He begins with the remark that the designation "primed term" is often -- and particularly in this case -- purely formal, in the sense that the designation of terms according to the states of the single electrons is useful for their enumeration and, above all, for the purpose of recognizing those symmetry characteristics of the state which are not broken by the interactions between electrons, but that such a designation does not generally allow valid conclusions to be drawn regarding the effective shape of the eigenfunctions. In the case of the hydrogen molecule the situation is completely different from that found in the case of central fields -- namely in the case of neutral atoms or atomic ions -- where it is generally possible to neglect the interaction between the electrons without altering the essential aspects of the phenomenon.

Majorana then poses the problem, limiting it to the even singlet states, which is the case for both the X state and the fundamental state of the H_2 molecule, dividing the configuration space into four regions, aa, ab, ba and bb, according to whether one or the other or both the two

25

electrons indicated by 1 and 2 are closer to the nucleus a or the nucleus b. Neglecting the interaction between the two electrons, the four possibilities mentioned above are equally represented in the states taken into consideration. However, once the interaction is introduced and taking into account that in the fundamental state, which had previously been successfully dealt with by Heitler and London, the major contribution to the eigenfunction comes from configurations belonging to the regions ab and ba, it is recognized qualitatively that there must exist a state which is orthogonal to the fundamental state, the main contribution to which comes from configurations belonging to the regions aa and bb.

While the configurations belonging to regions ab and ba may be considered to originate from the reaction between neutral atoms studied by Heitler and London

$$H + H \rightleftarrows H_2 , \tag{2}$$

whose unperturbed eigenfunction is of the type

$$\psi_{HL}(1, 2) = \varphi_a(1)\,\varphi_b(2) + \varphi_a(2)\,\varphi_b(1) ,$$

the configurations belonging to regions aa and bb are considered to originate from the reaction between one H^+ ion and one H^- ion:

$$H^+ + H^- \rightleftarrows H_2 , \tag{3}$$

whose eigenfunction, symmetrized with respect to the exchange of the two protons, is of the type

$$\psi_M = \Phi_a(1, 2) + \Phi_b(1, 2) .$$

Majorana calls the chemical bond existing in this case "pseudo-polar" in so far as it is a bond between two ions of opposite sign, but since the two components are equal the dipole electric moment changes sign with the exchange frequency, which is very high, and therefore cannot be observed.

The eigenfunctions ψ_{HL} and ψ_M are not orthogonal, but the eigenfunctions of the fundamental state and the anomalous state X should be represented in a first approximation by two orthogonal linear combinations of ψ_{HL} and ψ_M. Once this point has been cleared up we can immediately write the characteristic equation for determining the eigenvalues of the two states, the fundamental state and the excited state X. The coefficients appearing in this equation were in part already known from Heitler and London's paper; the others were calculated by Majorana, making use of semi-empirical expressions for the wave functions φ_i and Φ_i ($i = a$ or b). He thus found, in confirmation of the previous qualitative considerations, that the eigenfunctions of the fundamental state and the X state consist, mainly, the one of ψ_{HL} and the other of ψ_M; furthermore, he deduced that the equilibrium distance of the two protons in the fundamental state is about 2 Å, as against the 1.35 measured, and that the X state is separated from the fundamental state by 27,000 cm^{-1}, as against the 22,000 cm^{-1} measured. The last result, as pointed out by Majorana, is even better than would have been expected in view of the rather rough approximations made in the calculations.

The last paper in the period of work on molecular and atomic physics problems is No. 6. This deals with the problem of deriving the probability of spin-flip of the atoms of a polarized vapour beam moving in a rapidly varying magnetic field.

If the magnetic field slowly changes direction, the orientation of the atom follows adiabatically the direction of the field; but what happens if the variation in the direction of the field is not adiabatic? In particular, what happens if the atom passes near to a point where the magnetic field is zero?

This problem was discussed in the early days of quantum mechanics by Darwin and by Landé[15]; Stern, who a few years earlier with Gerlach had done the famous experiments on spatial quantization, undertook to make a quantitative check of the theoretical predictions of the transition probability in the non-adiabatic case. The problem can be described as follows: one has a quantized system (such as, for instance, an atom or a molecule) with a total angular momentum J and a magnetic quantum number m with respect to the direction of a magnetic field which is constant in time; the problem is to find out the final state or states of the system if the field begins to vary rapidly with time in value and direction, according to a certain vector function $\vec{H}(t)$. At Stern's suggestion, Güttinger[16] had made calculations for the case of a magnetic field rotating uniformly with constant intensity, since this situation corresponded to an experimental device which had been prepared and tried out, without much success, by Phipps and Stern[17]. This device had then been handed over to O. Frisch and E. Segré, when the latter had arrived in Hamburg, for them to carry on with the experiment. Segré had realized that they could not fully carry out Stern's plan, owing to some technical difficulties and had then found a different way of carrying out the non-adiabatic transition by making the beam of vapour pass near to a point where the field was zero.

However, it was necessary to establish a suitable quantitative theory for the new device; for this he had turned to Majorana, who had written his paper No. 6, which was then used by Frisch and Segré to interpret their measurements[18].

In his paper Majorana shows that the total effect of a variable magnetic field $\vec{H}(t)$ on a particle with an angular momentum $J = \frac{1}{2}$ and a given component m along the z axis, can be described as a sudden rotation by an angle α of the total angular momentum.

This angle is obtained by solving the equation of motion (whether classical or quantum mechanical) and its most important property is that it depends only on the gyromagnetic ratio g and on $\vec{H}(t)$, but is independent of the initial value of the magnetic quantum m.

It follows that after the rotation of the angle α, the system is no longer in a state of well-defined spatial quantization, with respect to the original direction of the field, but should be described by means of a wave packet consisting of the superposition of $2J+1$ states, each characterized by a different value of the magnetic quantum number m' and a corresponding probability amplitude. The square of the modulus $W(m, m', \alpha)$ of this latter represents the transition probability for each m', for which Majorana gives a definite expression. The problem was, in fact, tackled and solved by Majorana with extraordinary elegance and conciseness for the special case when $J = \frac{1}{2}$; but, as he points out, the method can easily be extended to the general case of any J. This was done in 1937 by Rabi and re-discussed and presented in a more evident way from a physics point of view in 1945 by Bloch and Rabi[19)], who thus contributed in making the results obtained by Majorana 13 years earlier more widely known.

This paper of Majorana's has thus remained a classic on non-adiabatic spin-flip processes, and as such is often quoted at the present time; his results, generalized to arbitrary angular momenta as mentioned above, have since provided the principle, serving as the basis for the experimental method for flipping neutron spin with an r.f. field, a method which is used both in the analysis of polarized neutron beams[20)] and in all polarized neutron spectrometers used for studying magnetic structures[21)].

The above gives some idea of the high quality of these papers, but their class is even more evident after their careful study: they show a thorough knowledge of experimental data down to the smallest details and an ease which was quite unusual, particularly at that time, in using the symmetry properties of the states to simplify problems or choose the most suitable approximation for solving each problem quantitatively, this latter capacity being, no doubt, at least partly due to his exceptional gifts for calculation.

4. Majorana's interest in nuclear physics, which had already been evident in his thesis, was greatly strengthend by the appearance of the classical papers which were to lead to the discovery of the neutron at the beginning of 1932. His renewed interest was actually in tune with the new general orientation of the whole Institute in the Via Panisperna, where for some years there had been talk of the advisability of gradually abandoning atomic physics, the field in which everyone had worked for some years, and concentrating the main research effort on nuclear physics.

These ideas were beginning to take practical shape towards the end of 1931 and so, on my return from Leipzig in October, I was pleased to accept the task of systematically presenting, in a series of seminars, the content of the classic treatise by Rutherford, Chadwick and Ellis[22) which I had begun to study a few months earlier. Fermi, Majorana, Rasetti, Segré and a few others attended the seminars. My lecture was often interrupted by observations of the most varied kind from members of the group, which gave rise to long discussions and gave Fermi the opportunity of developing extempore the theory of some of the phenomena mentioned. Majorana listened in silence and only exceptionally took part in the discussion, making observations which were almost always extremely penetrating.

The attendance at the seminars quickly fell off: in November Segré went to work in Hamburg at the Institute directed by O. Stern, to learn the technique of molecular beams, and Rasetti went to the Kaiser Wilhelm Institut für Physik in Dahlem, Berlin, directed by Lise Meitner, to work on the penetrating radiation emitted by beryllium bombarded with α particles, which had been discovered in 1930 by W. Bothe and H. Becker. At the same time Fermi and I in Rome had begun to construct a cloud chamber in order to become familiar with the most important techniques then in use for studying radioactivity and transmutations.

Towards the end of January, 1932, we began to receive the issues of the "Comptes rendus" containing the famous notes by F. Joliot and I. Curie on the penetrating radiation discovered by Bothe and Becker.

In the first of these notes it was shown that the penetrating radiation emitted by Be under bombardment with polonium α particles could transfer kinetic energies of about 5 million electronvolts to the proton present in small layers of various hydrogenated materials (such as water or cellophane). In order to interpret these observations the Joliot-Curies at first[23)] put forward the hypothesis that the phenomenon was similar to the Compton effect, namely that the incident photon undergoes an elastic collision with a proton; they had calculated, by applying the laws of energy and momentum conservation, that the incident photons should have had an energy of about 50 million electronvolts in order to be able to transfer such high energy to a proton. However, they had very soon realized that when Klein and Nishina's formula was applied to the protons, the cross-section was too small by many orders of magnitude, and had suggested that the effect observed was due to a new type of interaction between γ rays and protons, different from that responsible for the Compton effect[24)].

When Ettore read these notes he said, shaking his head: "They haven't understood a thing. They are probably recoil protons produced by a heavy neutral particle". A few days later we got, in Rome, the issue of "Nature" containing the letter to the editor from Chadwick[25)] dated 17 February, 1932, entitled "Possible existence of a neutron", in which he demonstrated the existence of the neutron on the basis of a classical series of experiments, in which recoil nuclei of some light elements (such as nitrogen, for instance,) were observed in addition to recoil protons.

In order to understand how Ettore could guess this discovery, which was suggested but certainly not demonstrated by the Joliot-Curie results, it should be remembered that he was familiar, through a paper published a few years earlier by Giovanni Gentile junior, with the nuclear model proposed in 1927 by Lord Rutherford[26] in an attempt to make the value of the radius of the uranium nucleus, deduced from the deviations observed in the elastic scattering of α particles with respect to the predictions based on Coulomb repulsion alone (3.2×10^{-12} cm), agree with the value deduced from the energy of the α particles emitted by radioactive nuclei ($\geq 6\times 10^{-12}$ cm). According to Rutherford, the nucleus consisted of a central part with a positive charge Ze, around which turned neutral satellites kept on their circular orbits by attractive forces originating from the electric polarization to which the satellites were subjected under the effect of the central electric field.

Rutherford had actually suggested in 1920 that if the nucleus consisted of protons and electrons, inside it there might be neutral particles consisting of a proton and an electron closely bound together; this idea had remained so much alive in Cambridge that some of Rutherford's students and colleagues, and particularly Chadwick, had tried several times between 1920 and 1932 to show experimentally that there existed neutral particles with a mass of the same order as, or greater than, that of the proton[27].

In 1928 Gentile had shown[28] the inconsistency of the model suggested by Rutherford the year before; however, the idea that there might exist in nature neutral particles of sub-atomic dimensions had, as it were, also remained in the air, in Rome.

Soon after Chadwick's discovery, various authors understood that the neutron must be one of the components of the nucleus[29] and began to propose various models which included α particles, protons, electrons and neutrons.

The first to publish the idea that the nucleus consists solely of protons and neutrons was probably Iwanenko[30]. Neither I nor his other friends questioned remember whether Ettore Majorana came to this conclusion independently. What is certain is that before Easter of that year he tried to work out a theory on light nuclei, assuming that they consisted solely of protons and neutrons (or neutral protons as he then said) and that the former interacted with the latter through exchange forces. He also reached the conclusion that these exchange forces must act only on the space co-ordinates (and not on the spin) if one wanted the α particle, and not the deuteron, to be the system saturated with respect to binding energy.

He talked about this outline of a theory to his friends at the Institute, and Fermi, who had at once realized its interest, advised him to publish his results as soon as possible, even though they were partial. However, Ettore would not hear of this, because he considered his work to be incomplete. Thereupon, Fermi, who had been invited to participate in the physics conference which was to take place in July of that year in Paris in the wider framework of the Fifth International Conference on Electricity, and who had chosen as his subject the properties of the atomic nucleus, asked Majorana for permission to mention his ideas on nuclear forces. Majorana forbade Fermi to mention them but added that if he really must he should say they were the ideas of a well-known professor of electrical engineering who, among others, was to be present at the Paris conference and whom Majorana considered to be a living example of how not to carry out scientific research[31)].

Thus, on 7 July, Fermi presented his report in Paris on "The Present State of the Physics of the Atomic Nucleus"[32)] without mentioning the type of force which was subsequently called "Majorana force" and which had actually been thought of, although in a crude form, some months earlier.

The issue of the "Zeitschrift für Physik" dated 19 July, 1932 contained Heisenberg's first paper[33)] on "Heisenberg exchange forces", namely forces involving the exchange of both the space and spin co-ordinates.

This paper made a great impression in the scientific world; it was the first attempt to put forward a theory of the nucleus which, although incomplete and imperfect, succeeded in overcoming some of the theoretical difficulties which had so far seemed insurmountable. Everyone at the Physics Institute of the University of Rome was extremely interested and full of admiration for Heisenberg's results, but at the same time disappointed that Majorana had neither published nor even allowed Fermi to mention his ideas at an international conference. Heisenberg's paper tackled the problem from a wider and fuller point of view but Ettore Majorana had completely understood, or so at least it appeared to us, the consequences of the action of the exchange forces in so far as the binding energy of light nuclei was concerned.

Fermi again tried to persuade Majorana to publish something, but all his efforts and those of his friends and colleagues were in vain. Ettore replied that Heisenberg had now said all there was to be said and that, in fact, he had probably even said too much. Finally, however, Fermi succeeded in persuading him to go abroad, first to Leipzig and then to Copenhagen, and obtained a grant from the National Research Council for his journey, which began at the end of January, 1933 and lasted six or seven months.

His aversion to publishing or making known his results in any way, which is evident from this episode, was part of his general attitude. Sometimes in the course of conversation with a colleague he would say, almost casually, that he had the previous evening made calculations or worked out the theory of a phenomenon which was not clear, and which had come to his or one of his friends' notice in the last few days. During the subsequent discussion, in which he was always very laconic, at a certain point Ettore would draw from his pocket the cigarette packet (he was a heavy smoker) on which he had written in small but neat writing the main equations of his theory, or a table of numerical results. He copied on the blackboard the part of the results which was necessary for elucidating the problem and then, when the discussion was over and the last cigarette smoked, screwed up the packet and threw it into the waste paper basket.

5. Meanwhile, in winter 1932-1933, Eugene Feenberg arrived in Rome from Harvard University, where he had been studying under Professor E.C. Kemble. He had a travelling scholarship for graduate students from that university, with which he spent about three months in Rome and one or two in Leipzig. His stay in Europe was interrupted by an unexpected invitation to return to the United States by the authorities of Harvard University, who were worried about the political situation which was developing in Germany: within a few months Hitler had succeeded in suppressing civil rights and democratic liberty and finally seizing power[34)].

During his stay in Europe Feenberg was writing his Ph.D. thesis on electron scattering by neutral atoms[35)], the paper in which, among other things, he established the optical theorem without, however, realizing its interest and scope[36)].

Feenberg and Majorana were immediately drawn to each other but did not succeed in establishing close working relations, since neither of them could speak the other's language. Feenberg had bought a small English/Italian dictionary with which he did his best, but the result of his efforts, made with honesty and perseverance, was extremely modest; they therefore sat in the same part of the library of the Institute in the Via Panisperna, studying at the same table and communicating with each other only at long intervals by means of some formula or other written on a piece of paper, while reading some page of recent publications.

Before leaving for Leipzig, Majorana published another paper, No. 7, on the Relativistic Theory of Particles with Arbitrary Intrinsic Angular Momentum. This is his first paper dealing with elementary particles and not groups of particles like atoms and nuclei.

35

$$+ \Sigma_r Z_r \frac{\partial q_r}{\partial z} - L\left(q, \dot{q}, \frac{\partial q}{\partial x}, \frac{\partial q}{\partial y}, \frac{\partial q}{\partial z}\right)$$

$$\frac{\partial R}{\partial q_i} = -\frac{\partial L}{\partial q_i} = -\dot{p}_i - \frac{\partial X_i}{\partial x} - \frac{\partial Y_i}{\partial y} - \frac{\partial Z_i}{\partial z}$$

$$\frac{\partial R}{\partial p_i} = \dot{q}_i \; ; \; \frac{\partial R}{\partial X_i} = \frac{\partial q_i}{\partial x} \; ; \; \frac{\partial R}{\partial Y_i} = \frac{\partial q_i}{\partial y} \; ; \; \frac{\partial R}{\partial Z_i} = \frac{\partial q_i}{\partial z}$$

$$\delta \int \Big[\Sigma_r p_r \dot{q}_r + \Sigma_r X_r \frac{\partial q_r}{\partial x} + \Sigma_r Y_r \frac{\partial q_r}{\partial y} + + \Sigma_r Z_r \frac{\partial q_r}{\partial z} - R(q, p, X, Y, Z) \Big] dx\,dy\,dz\,dt$$

$$(\delta q_i, \delta p_i, \delta X_i, \delta Y_i, \delta Z_i)$$

Dirac real.

$$\alpha_x = \rho_1 \sigma_x , \; \alpha_y = \rho_3 , \; \alpha_z = \rho_1 \sigma_z , \; ~~\beta = \rho_2~~$$

$$\beta = -\rho_1 \sigma_y$$

$$\left[\frac{W}{c} + (\alpha, p) + \beta mc\right] U = 0 \qquad \text{senza } e. \qquad U = \bar{U}$$

$$\psi = U + iV$$

$$\left[\frac{W}{c} + (\alpha, p) + \beta mc\right] U + i\,\frac{e}{c}\left[\varphi + (\alpha, A)\right] V = 0$$

$$\left[\frac{W}{c} + (\alpha, p) + \beta mc\right] V - i\,\frac{e}{c}\left[\varphi + (\alpha, A)\right] U = 0$$

$$\beta' = -i\beta \; ; \; \mu = \frac{2\pi mc}{h} \; ; \; \varepsilon = \frac{2\pi e}{hc} \qquad \left(= \frac{1}{137\,e}\right)$$

$$\left[\frac{1}{c}\frac{\partial}{\partial t} - (\alpha, \text{grad}) + \beta'\mu\right] U + \varepsilon\left[\varphi + (\alpha, A)\right] V = 0$$

$$\left[\frac{1}{c}\frac{\partial}{\partial t} - (\alpha, \text{grad}) + \beta'\mu\right] V - \varepsilon\left[\varphi + (\alpha, A)\right] U = 0$$

Page of Notebook No. 13 in which Ettore Majorana began to develop the the theory later to be known as the Majorana neutrino theory.

In January Majorana left for Leipzig, where he remained until the beginning of the summer: from Leipzig he went to Copenhagen, where he spent about three months, and from there back to Leipzig, where he stayed for about a month before finally returning to Italy at the beginning of the autumn in 1933.

Both in Leipzig and Copenhagen he remained very much alone, partly because of his natural aversion to making new acquaintances, but also because, although he understood German well, he spoke it rather badly.

In those days Leipzig was one of the major centres of modern physics; W. Heisenberg had gathered round him a group of exceptional young physicists including F. Bloch, F. Hund, R. Peierls and, among the visitors, E. Feenberg, D.R. Inglis and E.G. Uhlenbeck.

Feenberg remembers attending one of Heisenberg's seminars on nuclear forces, in which Heisenberg also mentioned the contribution made by Majorana to this subject: he said that the author was present and invited him to say something about his ideas, but Ettore refused. When he left the seminar, Uhlenbeck told Feenberg how much he admired Majorana's penetrating ideas which had been mentioned by Heisenberg.

During this period Majorana became friendly with Heisenberg, for whom he always had a great admiration and a feeling of friendship. It was Heisenberg who persuaded him without difficulty by the sheer weight of his authority to publish his paper on nuclear theory, which appeared in the same year, both in the "Zeitschrift für Physik" (No. 8a) and in "Ricerca Scientifica" (No. 8b).

Heisenberg recognized Majorana's exceptional capacity for research and also the difficulty which he seemed to have in establishing relations with people whom he had recently met, and with the outside world in general.

In Copenhagen, which if not the greatest was certainly one of the most important physics centres at that time, Ettore met Niels Bohr, C. Møller, L. Rosenfeld and many others. At that time Placzek was also in Copenhagen and Majorana was friendly with him, as he had already known him for several years.

Rosenfeld remembers that he never saw Majorana without Placzek and says that he heard his voice only once; they had all three gone to a café near the Institute [a café which in the slang of the Copenhagen Institute was called "Unter den Quanten"[37)]]. Placzek and Rosenfeld were talking to each other and Majorana was listening. Rosenfeld said that he would like to have certain notes made by Placzek, who replied that he was perfectly willing to lend them to him, but that they were very long and that he had only one copy. Rosenfeld then said in German: "Don't you worry about that. My wife can copy them". At this point, Majorana said unexpectedly in Italian, without moving a muscle of his face: "She is the ideal wife".

In July the whole Majorana family went by car to see Ettore in Leipzig.

During the time that he spent abroad, Majorana was very struck by the efficiency of German organization and by the high level of the German economy and he conceived a great admiration for Germany, which he expressed on serveral occasions, in particular in a letter to Emilio Segré: this letter was unfortunately lost, together with other documents sent from Italy to the United States, when the Andrea Doria sank on 25 July, 1956.

When he returned to Rome in the autumn of 1933, Ettore was not in good health, because of gastritis which he had developed in Germany. It is not clear what caused this, but the family doctors attributed it to nervous exhaustion.

He began to attend the Institute in the Via Panisperna only at intervals, and after some months no longer came at all: he tended more and more to spend his days at home immersed in study for a quite extraordinary number of hours.

At that time he was more interested in political economy, politics the fleets of various countries and their respective power, and the constructional characteristics of the ships than in physics. At the same time his interest in philosophy, which had always been great, increased and prompted him to reflect deeply on the works of various philosophers, particularly Schopenhauer.

It was probably at this time that he wrote the paper on the value of statistical laws in physics and the social sciences which was found among his papers by his brother Luciano, and was published after his disappearance (No. 10) by Giovanni Gentile junior.

In addition to these old and new interests he found a new one in medicine, a subject which he perhaps tackled in order to understand the symptoms and significance of his illness.

A considerable number of attempts by Giovanni Gentile junior, Emilio Segré and myself to bring him back to living a normal life met with no success. I remember that in 1936 he rarely left the house, not even to go to the barber's, and his hair was therefore abnormally long; during this period some of his friends who had been to see him sent him a barber, in spite of his protests. However, none of us succeeded in finding out whether he was still doing theoretical physics research; I believe he was, but I have no proof.

6. Meanwhile, various other young physicists were coming to maturity in the field of theoretical physics. Gian Carlo Wick, who had obtained his doctorate at the University of Turin with Somigliana, had come to Rome after spending some time in Göttingen and Leipzig when I was there;

Giulio Racah, who had obtained his doctorate in Florence with Enrico Persico, divided his time between Florence, Rome and Zürich, where he worked under Pauli's guidance; Giovanni Gentile junior, who has already been mentioned; Leo Pincherle, who had studied in Bologna and had then come to Rome, and Gleb Wataghin, who had emigrated from Russia to Italy and had studied in Turin, where he had been teaching and working for years.

The time had now come for a new competition for a chair in theoretical physics. The first and only competition for chairs in this subject had been held in 1926, and had resulted in Enrico Fermi being appointed to the chair in Rome, and Enrico Persico to the one in Florence. The competition was advertised at the beginning of 1937 at the request of the University of Palermo, followed upon a request from Segré, who in the meantime had become Professor of Experimental Physics there.

The problem naturally was to make Ettore enter the competition, since he did not seem to want to do so, and in any event had not published any physics papers for some years. Fermi and various friends tried to persuade him, and finally Majorana was convinced that he should take part in the examination, and he sent his paper on the "Symmetrical theory of the electron and the positron" (No. 9) for publication in the "Nuovo Cimento". The Board of Examiners for the competition for the Chair in theoretical physics at the University of Palermo was appointed by the Ministry of National Education, as laid down by the Fascist laws in force at that time, and was made up as follows: Antonio Carrelli, Enrico Fermi, Orazio Lazzarino, Enrico Persico and Giovanni Polvani. The other candidates besides Ettore Majorana were the five physicists mentioned in the first paragraph. The board first met in October, 1937, but it was soon invited by the Ministry of National Education to suspend its work for the purpose of granting a request by Senator Giovanni Gentile[38)] to appoint (according to the terms of Article 8 of the R.D.L., 20 June, 1935, n. 1071) the candidate Ettore Majorana as ordinary Professor of Theoretical Physics in the Royal University of Naples. The above article referred to special merit; it had been introduced a few years earlier for the purpose of allowing Guglielmo Marconi to be appointed Professor of Electromagnetic Wave Theory at the University of Rome without competing for the Chair.

Soon after the appointment of Ettore Majorana to Naples, the board resumed its work, and with the number of candidates reduced to five, proceeded to examine their qualifications, unanimously reaching agreement on the following short list[38]: 1) Gian Carlo Wick, 2) Giulio Racah and 3) Giovanni Gentile junior.

7. Ettore Majorana's major contribution to science consists of papers Nos. 8 and 9.

Paper No. 8, of whose origin I have already given some account, should be studied in connection with the well-known notes by Heisenberg on the same subject. In the first, dated 7 June, 1932, Heisenberg[33] starts from the model which had already been taken into consideration by other writers, according to which the nucleus is a system of protons and neutrons only, and the neutrons, like the proton, a particle with spin ½ which follows the Fermi statistics: he tries then to define the nature of the forces acting between these particles. On the basis of the analogy with the case of the homopolar forces of chemistry, Heisenberg assumes that between protons and neutrons there is an exchange force similar to that which gives rise to the molecular hydrogen ion (H_2^+), while for neutron pairs he introduces attractive forces of normal type (Wigner forces, as they were to be called later) and for the proton pairs Coulomb repulsion only.

In order to write the Hamiltonian of the system consisting of Z protons and N neutrons ($Z + N = A$), Heisenberg introduces for the first time, as a mere formal artifice, isotopic spin and the corresponding projection operators related to n-p, n-n and p-p pairs. It is not easy to know whether at that time he had, at least partly, guessed the importance which this formalism would assume later.

Its systematic application from 1936 onwards to an increasing number of new categories of phenomena which were gradually discovered and studied experimentally in sufficient detail, led to isotopic spin being accepted as a new physical quantity, whose conservation is a generally valid principle in the sense that it is respected in all processes involving strong interactions only[39)].

From the study of the Hamiltonian thus constructed, Heisenberg by qualitative reasoning deduces some interesting properties of the neutron-proton system, particularly of the deuteron, and concludes by denying the existence of the corresponding bound state of the system consisting of two neutrons, as a consequence of Pauli's principle. He then observes that the α particle must correspond to a complete shell but does not attempt to explain this fact, suggested by the study of the mass defects, by assuming that the neutron-proton forces are exchange forces. The rest of the paper is devoted to discussing the properties of intermediate and heavy nuclei, their binding energy and their beta and alpha instability. This leads him to describe, semi-qualitatively, the behaviour of the binding energy of the nuclei as a function of Z and N, a subject to which he returns in the second paper[40)], dated 30 July, 1932, where he discusses in a form destined to be more or less final, the difference of energy between nuclei with Z and N both even, with Z or N odd and with Z and N both odd.

It is very interesting now to see how he succeeded in arriving at conclusions which were substantially correct, although they were semi-qualitative, in spite of the fact that at that time the beta processes with emission of positrons had not yet been observed, since they were discovered by the Joliot-Curies only at the beginning of 1934[41)]. In the same paper Heisenberg discusses the scattering of γ rays by nuclei and the properties of the neutron, which, if it is regarded as a system composed of one electron and one proton so strongly bound as to have linear dimensions much smaller than the nucleus, gives rise to a considerable number of conceptual difficulties. The first was that the neutron should then have obeyed Bose and not Fermi statistics, since it consisted of an even number of particles which were governed by Fermi statistics, or fermions as they are now called. In addition to this difficulty, which had already been mentioned in his first paper and which, if I remember rightly, had also been mentioned by Majorana in his conversations with friends in the early spring of 1932, Heisenberg pointed out another deriving from the principle of uncertainty; the energy of the electron bound with the proton to form the neutron should be of the order of 137 mc^2 if the dimensions of the neutron are of the order of e^2/mc^2.

Both the problem of the scattering of gamma rays by nuclei and that of the nature of the neutron are taken up and investigated by Heisenberg in his third paper[42] dated 22 December, 1932. He puts forward two alternative hypotheses on the nature of the neutron: the first is that the neutron is an elementary particle without structure, as assumed by Perrin[43,29]. However, in order to explain the beta decay of the nuclei it would then be necessary for the latter to contain electrons in addition to neutrons and protons. There thus arose, although on a smaller scale, the same difficulty deriving from the principle of uncertainty, which had cast doubt on the second hypothesis on the structure of the neutron put forward by Heisenberg, according to which the neutron consisted of an electron and a proton and as such was likely to interact through exchange forces with the proton. The situation appeared to Heisenberg to be so serious as to prompt him to advance the hypothesis at the end of his third paper that quantum mechanics might be inadequate to describe the phenomena which take place on a nuclear scale.

Today, we know that this is not true and that this difficulty, and many others which are not mentioned here, was to disappear in 1935 with the hypothesis put forward by Yukawa[44] and confirmed a few years later by experiment[45], that the particles exchanged between any two nucleons, namely protons and neutrons, are not electrons but mesons, namely particles with zero spin (which are therefore governed by Bose statistics) whose mass is 280 times greater than that of the electron.

A large part of Heisenberg's third paper is devoted to extending the Thomas-Fermi method to the case of the nucleus, assuming for this purpose a Hamiltonian which differs from that given in the first paper only in so far as, in addition to the exchange forces acting between protons and neutrons, there are also ordinary forces.

Heisenberg deduces the expression of the energy of the system in its lowest state, imposing as a condition that the probable value of the Hamiltonian be stationary with respect to arbitrary variations of the density ρ of the particles. At first, the Hamiltonian was calculated by taking for the kinetic energy of the protons and neutrons, as in the Thomas-Fermi method, the expression valid for a totally degenerate Fermi gas not subjected to external fields, and a potential energy of the form described above.

Under these conditions, however, it is not possible to avoid the collapse of the nucleus as a result of the fact that the nuclear forces tend to reduce to zero, or at least to an extremely small value, the distance between the two particles of each neutron-proton pair. In order to overcome this drawback, Heisenberg had to introduce a cut in the potential energy at the shortest distances, a fore-runner of what, at a much later date, was to be called the repulsive core[46)].

After having introduced a minimum distance between neighbouring protons and neutrons, a maximum density ρ_0 of these particles was obtained, whose inverse was none other than the co-volume which appears in the Van der Waals state equation used for describing, although qualitatively, the behaviour of a fluid also in the liquid state.

Majorana's paper No. 8a (and 8b), dated 3 March, 1933 (and 11 May, 1933), Leipzig, appeared at this point. After re-examining the hypothesis based on the Heisenberg model, Majorana reconsidered the difficulties relating to the structure of the neutron and reached the conclusion that in the current state of knowledge, the only thing to do was to try to establish the law of interaction between proton and neutron on the basis of criteria of simplicity only, but in such a way as to reproduce as correctly as possible the most general and characteristic properties of the nuclei. After having discussed some of these and having observed that the nucleus behaved like a piece of extensive and inpenetrable "nuclear matter", whose different parts interacted only upon immediate contact, and having particularly stressed the fact that the experimental data clearly showed a law of proportionality of both the binding energy and the volume with respect to the total number of nucleons present, Majorana introduced an exchange potential acting on the space co-ordinates alone, whose sign was such as to be attractive in states with even angular momentum. He justified his choice by observing that by doing this, two important results were obtained; the first was that both neutrons of the α particle exercise an attraction on each proton. The second was that in the approximation in which Coulomb repulsion is neglected, the eigenfunction of the α particle is totally symmetrical in the co-ordinates of the centres of mass of the four component particles, as it is reasonable to expect since it represents a complete shell.

Turning from the α particle to heavier nuclei, the additional nucleons are forced, because of Pauli's principle, to enter into more excited states and since the exchange energy is high only in the case of particles in the same (or almost the same) orbital state, it is concluded that saturation both of the binding energy and the density is practically reached in the case of the α particles.

There are two essential differences with respect to Heisenberg's scheme; the first is that the exchange forces concern only the space co-ordinates and not the spin co-ordinates: the second is the sign of Majorana's exchange potential with respect to that used by Heisenberg. As a consequence of these modifications, it becomes possible to explain the saturation phenomena without introducing a repulsive core as Heisenberg had done. Incidentally, I should like to note that in comparing his Hamiltonian to that of Heisenberg, Majorana states that he wishes to avoid using the "inconvenient" formalism of isotopic spin (ρ spin according to the symbols used at that time by Heisenberg).

After these qualitative considerations, Majorana writes the Hamiltonian as the sum of the kinetic energy of all the particles, of the Coulomb repulsion energy between all the proton pairs and of the energy due to the exchange forces between all the proton-neutron pairs: he calculates the corresponding eigenvalue, taking for the eigenfunction of the system the product of two totally antisymmetric eigenfunctions (for the simultaneous exchange of both the space and the spin co-ordinates), the one relating to the protons and the other to the neutrons. He thus reaches the conclusion that if Coulomb interaction is neglected, the energy of the system is proportional to the number of particles of which it is composed; in other words he shows quantitatively how, with only Majorana exchange forces, it is possible to obtain the saturation both of the density and the binding energy of the nuclei[47].

The situation has changed considerably today. Above all, we know that for energies between a few tens and several hundreds of MeV, the angular distribution in the centre-of-mass system of neutrons, scattered in neutron-proton collision, is approximately symmetric with respect to the equatorial plane. This experimental result means that the interaction between a proton and a neutron consists of two more or less equal parts, one of which has the nature of Wigner forces and the other of Majorana exchange forces. Under these conditions, however, the Majorana forces are no longer sufficient to ensure saturation, namely to prevent the collapse of the nucleus. It is therefore necessary to introduce a repulsive core[48] which, moreover, seems to be also necessary to explain the angular distribution, in the centre-of-mass system, of protons scattered in proton-proton collision at energies between a few tens and a few hundreds of MeV.

This does not alter the fact that the Majorana exchange forces have finally remained an essential term in all the empirical or semi-empirical representation of the nuclear forces, which moreover find a natural qualitative explanation in all the meson field theories.

Paper No. 7 on the relativistic theory of particles with arbitrary intrinsic angular momentum was written about two years earlier than the paper on nuclear forces; however, I have chosen to reverse the order in which I have discussed these two papers because the basic idea of paper No. 8 is connected with the papers of the earlier period, in particular papers Nos. 2 and 4, whereas No. 7 is the first expression of a different and very important aspect of Majorana's cultural make-up and scientific interests, which some time later fully revealed themselves in paper No. 9.

Paper No. 7, although not very well known, and in some ways outside the main line of the historical development of elementary particle physics, appears to be very interesting for the following reasons which emerged from discussions that I had the pleasure of having with Professor M. Fierz[49)] and Professor D.M. Fradkin, and from the paper that the latter has written on this subject[50)]:

a) Majorana's paper is the first attempt to construct a relativistically invariant theory of arbitrary half integer or integer spin particles.

b) It is mathematically correct and contains the first recognition, development and application of the (actually the simplest, as correctly stated by Majorana) infinite dimensional unitary representations of the Lorentz group.

c) The theory appears to be outside the main line of successive developments for two main reasons: the first is that from the beginning Majorana set himself the problem of constructing a relativistically invariant linear theory for which the eigenvalues of the mass were all positive. This point of view was justified at the time when the paper was written (summer 1932), since the news of the discovery of the positron by C.D. Anderson had not yet arrived in Rome[51]. The second is that Majorana unlike most subsequent authors does not require the dispersion relation

$$\left(\frac{W}{c}\right)^2 = p^2 + (m_0 c)^2 \tag{4}$$

to be satisfied as an operator equation for the components of the eigenfunctions of the linear differential equation of Dirac form

$$\left[\frac{W}{c} + \vec{\alpha} \cdot \vec{p} - \beta m_0 c\right]\psi = 0 \ , \tag{5}$$

which is the starting point of his theory.

d) In a reference frame where the momentum is not zero, Majorana's solution of Eq. (5) has an infinite number of components labelled with two indices ψ_{jm}; m includes all values between ± j, while j takes all non-negative values either of the integer (j = 0, 1, 2 ...) or of the half integer (j = 1/2, 3/2, 5/2, ...) series.

e) The interpretation of the result given under item (d) appears, today, very difficult from a physics point of view; each set of 2j + 1 components ψ_{jm} (j fixed, m = j, j - 1, ..., - j) seems to correspond to a particle of different spin, with the result that the most natural interpretation of ψ is that of a wave function representing simultaneously either all possible bosons or all possible fermions. By changing the reference frame, the various components are mixed together. However, in the rest frame [see point (g)], where most authors define the spin properties of a particle, Majorana's wave function corresponds to, and mathematically transforms like, a particle with a single spin whose value depends on which eigenfunction is selected.

f) Majorana's theory provides a mass spectrum formula according to which the mass is determined by the value of the spin j

$$\frac{m_0}{j + \frac{1}{2}} \quad , \tag{6}$$

where m_0 is an assigned constant.

g) Majorana shows that in the non-relativistic limit ($v/c \to 0$), the particular set of components ψ_{sm} corresponding to the rest-mass eigenvalue characterized by $j = s$ (s fixed), is the only finite one since the components ψ_{s+1m}, ψ_{s-1m} are of order v/c, the components ψ_{s+2m}, ψ_{s-2m} of order $(v/c)^2$, etc.

The main line of Majorana's procedure can be summarized as follows. As said before, he starts from the single linear differential equation (5), it being understood that the wave function ψ can have "a priori" any number of components, possibly even an infinite number. Since in the original Dirac theory of the electron the negative masses arise from the fact that the operator β has eigenvalues ± 1, Majorana set himself the problem of the possible existence of a relativistic invariant linear theory for which the eigenvalues of β are all positive.

Then he notices that one of the conditions for the relativistic invariance of Eq. (5) is that the Lagrangian density appearing in the variational principle from which Eq. (5) is derived

$$\delta \int \psi^* \left[\frac{W}{c} + \vec{\alpha} \cdot \vec{p} - \beta m_0 c \right] \psi \, dv \, dt = 0 \ ,$$

must also be invariant. But the density contains the term

$$\psi^* \beta \psi \tag{7}$$

which in its turn should be invariant. If one now requires β to have only positive eigenvalues, then the term (7) must be positive definite and it becomes possible to make a (non-unitary) transformation $\psi \to \varphi$ such that

$$\psi^* \beta \psi \to \varphi^* \varphi \ .$$

Once the linear differential equation for φ is obtained,

$$\left[\gamma_0 \frac{W}{c} + \vec{\gamma} \cdot \vec{p} - m_0 c\right]\varphi = 0 \ , \qquad (8)$$

Majorana returns to the Lorentz invariance of $\varphi^* \varphi$ and notices that this property can certainly be achieved if one represents the Lorentz transformation in terms of unitary operators. With this in mind he constructs the generators of the infinitesimal Lorentz transformations so that they operate on the co-ordinates ct, x, y, z and deduces the commutation relations which must be satisfied by the generators in any representation. At this point he gives an infinite dimensional hermitian representation of the Lorentz transformation generators [item (b)].

Then he goes back to the Lagrangian density expressed in terms of the wave function φ, and again insists on the invariance of the Lagrangian density under Lorentz transformation. This requirement implies certain commutation relations between the γ matrices with the generators which are sufficient to determine the infinite dimensional representation of the γ matrices.

Majorana thus solves the problem of obtaining an explicit form of wave equation which is relativistically invariant and involves only positive energy eigenvalues. The eigenvalues of the infinite dimensional matrix β found by Majorana are:

$$\frac{1}{j + \frac{1}{2}} \ ,$$

where j ranges over all values in either the integer $(j = 0, 1, 2, \ldots)$ or the half integer $(j = \frac{1}{2}, \frac{3}{2}, \frac{5}{2}, \ldots)$ series; this implies that the eigenvalues of the mass are given by Eq. (6).

The paper is completed by mentioning the existence of imaginary mass eigenvalues in the theory, deriving the expression for the eigenfunction in a plane wave state, and discussing the incorporation in the theory of both an electromagnetic field (by means of the usual replacement: $p_\mu \rightarrow p_\mu - e/c\, A_\mu$) and its extension to include the case of particles with anomalous magnetic moment (by means of the addition of a Pauli-type term).

As pointed out by Fradkin, Majorana was not interested in the idea of a mass spectrum, probably because too few elementary particles were known in 1932. Furthermore, he did not try to suggest any interpretation of the wave function ψ, such as that mentioned under item (e).

However, Majorana emphasized item (g) by observing that in the non-relativistic limit the infinite dimensional wave function has only $2s + 1$ non-vanishing components, each of which satisfy the Schrödinger equation, in agreement with the Pauli electron-spin theory.

The situation is in some way similar to that found in the original Dirac theory of the electron where, in the non-relativistic limit, the "small" components of the positive energy eigenfunction vanish. Majorana admitted that in the case of non-zero velocity, transitions between these $2s + 1$ states and all other states begins to become possible for sufficiently strong interactions, and consequently he restricted the applicability of the theory to interactions sufficiently weak to cause no transitions. The argument is again similar to the one used in the Dirac electron theory at that time to forbid transitions to the (unphysical) negative energy states.

Majorana's paper attracted little attention. The same problem was tackled by Dirac[52)], Klein[53)], Petiau[54)] and Proca[55)] in 1936, but none of these authors quotes Majorana. All subsequent papers[53-56)] refer to the article by Dirac and ignore Majorana's work completely, or almost completely.

The main features of the present relativistic theory of particles of arbitrary spin have been set out, developed, and applied to a few particular cases in three fundamental papers which appeared in 1939.

The first is by Fierz[57], who gave a complete account of the problem in the absence of external fields, succeeding in establishing, among other results, the relationship between spin and statistics. The second paper is that by Fierz and Pauli[58] which extends the results of Fierz's paper to the case of particles moving in an electromagnetic field, and treats in detail the cases of spin 3/2 and 2, with special regard to particles of zero mass. The only quotation of Majorana's work is made incidentally in the last of the three papers mentioned above, the fundamental paper on the meson field by Kemmer[59].

The two basic differences between Majorana's theory and the theory developed by the other authors have already been mentioned under item (c). While the requirement that the eigenvalues of the mass shall all be positive has its justification in the frame of the facts experimentally established at the time the paper was written, Majorana was also justified, from the standpoint of the mathematical requirements of special relativity theory, in assuming only the linear wave equation (5), without simultaneously requiring the satisfaction of the dispersion relation (4).

Dirac, and also most subsequent authors[60], obtained a linear wave equation of the Dirac type [Eq. (5)], starting from a second-order equation, and by factorizing the dispersion relation which was required to be satisfied for each component of the wave function. As pointed out in 1945 by Bhabha[61], who however does not quote Majorana, if one starts from a linear equation and does not require the dispersion relationship to be satisfied, then it is not necessary to subject the wave function to subsidiary conditions.

It may be recalled that in the present form of the theory, these conditions serve to separate the particles of different spin; for example, in the case of a vector field they ensure that particles of spin $s = 0$ are not mixed with particles of spin $s = 1$.

A remark should be added on the mathematical aspects of Majorana's paper. The same matrix elements as those given by Majorana for the infinite dimensional Lorentz group representations were given by Weyl[62] in his book -- first published in 1928 -- in connection with the selection and intensity rules of dipole transition for an atom in Schrödinger quantum mechanics. Majorana was surely acquainted with Weyl's work; however, recalling his way of working, I think he derived these results himself and recognized later that the same matrices played a part in the dipole transition problem.

The problem of the infinite dimensional representations of the Lorentz group was discussed by Wigner in 1939[63], Dirac in 1945[64] and Wigner again in 1948[65]. Dirac apparently was not aware of Majorana's work, while Wigner quotes it more or less incidentally. Wigner's point of view is different from that of Majorana; instead of assuming a definite wave equation and a postulated set of eigenfunctions, Wigner assumes only the existence of a relativistically invariant linear manifold, that is, a set of states which map into a superposition of themselves under the influence of a Lorentz transformation. This invariant theoretic approach is more general and more rigorous than Majorana's but gives less information (especially kinematic relations), and it seems more complex. Corson's discussion in his book[66] is based on Wigner's treatment.

Further detailed investigations of the infinite representation have been made by various authors[67] who, apparently were not aware of Majorana's paper. The same is true for those who, in recent years, have unwittingly reconstructed, discussed and generalized Majorana's theory[68].

Apart from its mathematical and historical interest, this paper shows Majorana's sensitivity to fundamental problems, which he tried to tackle independently, considerably in advance of the majority of contemporary physicists.

The problem of particles with arbitrary spin was regarded in 1932 as a mere mathematical curiosity: today at a distance of thirty years, it can be appreciated how important this is, not only because its study led to the clarification of fundamental problems such as, for instance, that of the relation between spin and statistics in general, but also because it is now known that there exist in nature complex particles with arbitrary spin, such as the nuclei and the so-called resonances[69]; in fact, there are so many of the latter that one cannot help spontaneously wondering whether all the known particles so far considered as elementary do not really have a complex structure[70].

In paper No. 9 on the symmetrical theory of the electron and the positron, Majorana begins by observing that the Dirac relativistic theory[71] which had led to the prediction of the existence of the positron, shortly afterwards confirmed by experiment[51], was founded on the Dirac equation, which is completely symmetrical with respect to the charge sign of the fermion concerned: but this symmetry was partly lost in the subsequent development of the theory, which described the vacuum as a situation in which all the negative energy states were occupied and all the positive energy states unoccupied. The excitation of a fermion from one of the negative energy states to a positive energy state left a hole endowed with positive energy, which could be interpreted as the antifermion. Thus, the excitation of a fermion from a negative to a positive energy state is equivalent to the creation of a fermion-antifermion pair. This asymmetric postulation also implied the need to cancel, without any valid justification of principle, certain infinite constants due to the negative energy states, such as charge density.

The new contribution made by Majorana in 1937 consisted in having discovered a representation of the Dirac matrices γ_k ($k = 1, 2, 3, 4$) which has the following properties[72]:

54

i) Contrary to what happens in the original Dirac representation (where γ_2 and γ_4 are real, and γ_1 and γ_3 imaginary) in Majorana's representation the four γ_k matrices have the same properties of reality of the components of the vector $x \equiv \vec{r}$, ict.

ii) In the representation defined in (i), the Dirac equation relating to a free fermion has real coefficients and therefore its solutions are divided into a real part and an imaginary part, each of which satisfies the said equation separately. However, each of these real φ solutions, on account of its reality, has two very important properties. The first is that is cannot give rise to an electric current-charge 4 vector,

$$j_k(x) = \bar{\varphi}(x)\gamma_k\varphi(x) \qquad (k = 1, 2, 3, 4)$$

(where in second quantization theory $\bar{\varphi} = \varphi^+\gamma_4$), since one always has $j_k(x) = 0$. In fact for $\varphi^* = \varphi$, one has

$$j_k(x) = \varphi(x)\gamma_4\gamma_k\varphi(x) = (\gamma_4\gamma_k)_{\rho\sigma}\varphi_\rho(x)\varphi_\sigma(x) .$$

However, it follows from the rules of anticommutation of a fermion field

$$[\varphi_\rho(x), \varphi_\sigma(y)]_+ = 0 ,$$

that $\varphi_\rho(x)\,\varphi_\sigma(x)$ is antisymmetrical in the exchange of the ρ and σ indices, while $(\gamma_4\gamma_k)_{\rho\sigma}$ is symmetrical[73]. The 4 vector $j_k(x)$ is thus the product of a symmetric and an antisymmetric tensor and is, therefore, equal to zero.

It follows that the real solutions of the Dirac equation must correspond to fermions with neither electric charge nor magnetic moment.

The second consequence of the reality of the fermion field φ is that the corresponding field operator must be Hermitian, so that its degrees of freedom are reduced by half and the distinction between fermion and antifermion disappears.

Majorana suggested in his paper that the neutron or the neutrino, or both, were particles of this type, namely neutral particles which identify themselves with the corresponding antiparticles.

iii) From the study of the Dirac equation relating to a fermion located in an electromagnetic field in Majorana's representation, it follows that in order to represent a charged particle it is sufficient to take a combination of two real solutions φ_1, φ_2, as for instance:

$$\psi = \frac{1}{\sqrt{2}} (\varphi_1 + i\varphi_2)$$

$$\bar{\psi} = \frac{1}{\sqrt{2}} (\varphi_1 - i\varphi_2) .$$

The fermion field ψ thus defined obviously gives rise to an electric current-charge 4 vector j_k not identically zero, owing to the interference terms between the two real fields φ_1 and φ_2. Moreover, it has the property now recognized for a scalar field, namely that the charge conjugated field operator (namely the operator which describes a particle of opposite charge to that of the particle concerned) is obtained by applying the operation of Hermitian conjugation to the operator ψ.

In reality at the present time no neutral particle of the type described in (ii) is known: in particular it is known from an experiment carried out in 1939 that the neutron has a magnetic moment different from zero[74)], and from others carried out in 1956 that both the antineutron[75)] and the antineutrino[76,77)] are particles distinct from the neutron and the neutrino, respectively. Nevertheless, Majorana's neutrino, characterized by the equality

$$\nu_M = \bar{\nu}_M , \qquad (9)$$

has played an important part in the physics of weak interactions also in the last ten years.

In the beta-decay theory, originally put forward by Fermi in 1934[78], the neutral particles emitted in neutron-proton and proton-neutron transitions (and therefore associated with the emission of the e^- and e^+, respectively) were different from each other:

$$n \rightarrow p + e^- + \bar{\nu}_e \tag{10a}$$

$$p \rightarrow n + e^+ + \nu_e \; . \tag{10b}$$

Which of these should be called neutrino and which antineutrino was simply a question of nomenclature until an observable physical property was found by which they could be distinguished. Equations (10) are written in the way currently accepted today owing to the principle of conservation of the lepton number, which we shall discuss later[79].

In 1937 with the introduction of Majorana's neutrino Eq. (9), Fermi's theory should have been modified in such a way that the same particle would have been involved in the two elementary processes of beta decay

$$\begin{aligned} n &\rightarrow p + e^- + \nu_M \\ p &\rightarrow n + e^+ + \nu_M \; . \end{aligned} \tag{11}$$

In paper No. 9, Majorana is fully aware of this and observes that such a change could be made immediately.

The theory thus modified leads to various consequences which can be checked experimentally. A first consequence concerns the double beta-decay process[80] which is responsible, for instance, for the transition from ${}^{48}_{20}Ca$ to ${}^{48}_{22}Ti$. In Majorana's neutrino theory this reaction takes place according to the scheme

$${}^{48}_{20}Ca \rightarrow {}^{48}_{22}Ti + 2e^- \; , \tag{12}$$

while in Fermi's original theory, or rather in the similar theory based on (10), the process should also involve two antineutrinos

$$^{48}_{20}\mathrm{Ca} \rightarrow {}^{48}_{22}\mathrm{Ti} + 2e^- + 2\bar{\nu}_e \;. \tag{13}$$

The origin of the differences between (12) and (13) is clear if the transition $^{48}_{20}\mathrm{Ca} \rightarrow {}^{48}_{22}\mathrm{Ti}$ is considered to be a second-order process: whereas in Majorana's theory the neutrino emitted with the first electron can be reabsorbed when the second is emitted, since it follows from (11) that the emission and the absorption are equivalent to each other; this process is not possible in Fermi's theory.

However, the mean life calculated for process (12) is several orders of magnitude smaller than that calculated for process (13). Experiments carried out some years later[81)] showed that the mean life of the $^{48}_{20}\mathrm{Ca} \rightarrow {}^{48}_{22}\mathrm{Ti}$ process is so long as to be in open contradiction with the assumption (9).

The other consequence concerns the processes induced by neutrinos; in the hypothesis based on equality (9), the neutrinos emitted during processes which take place in a nuclear reactor are all identical to each other and capable of producing both the following processes:

$$\nu_M + n \rightarrow p + e^- \tag{14a}$$

$$\nu_M + p \rightarrow n + e^+ \tag{14b}$$

whereas, according to (10), the only possible processes are

$$\nu_e + n \rightarrow p + e^- \tag{15a}$$

$$\bar{\nu}_e + p \rightarrow n + e^+ \;. \tag{15b}$$

Now experience has shown that in the vicinity of a reactor where there are intense antineutrino fluxes, processes are observed in which positive electrons[76)], but not negative electrons[77)] are emitted, in open contradiction with (14) and in agreement with (15), since the very great majority of processes which take place in a reactor involve the emission of e^- [see Eq. (10a)].

Very recently it has been discovered[82,83] that the neutrinos emitted in pion decay

$$\pi^+ \to \mu^+ + \nu_\mu \qquad (16a)$$

$$\pi^- \to \mu^- + \bar{\nu}_\mu , \qquad (16b)$$

and presumably one of the two neutrinos emitted in muon decay

$$\mu^- \to e^- + \bar{\nu}_e + \nu_\mu \qquad (17a)$$

$$\mu^+ \to e^+ + \nu_e + \bar{\nu}_\mu , \qquad (17b)$$

are different from those emitted in the beta decay, so far discussed. This assertion originates from the experimental observation that the neutrinos emitted in pion decay, according to the two processes (16), interacting with the nucleons are not capable of giving rise to the processes[82,83] (15) but only to the processes

$$\nu_\mu + n \to p + \mu^- \qquad (18a)$$

$$\bar{\nu}_\mu + p \to n + \mu^+ . \qquad (18b)$$

However, even the neutrinos of this new type ν_μ are different from the corresponding antineutrinos $\bar{\nu}_\mu$, which is demonstrated by the experimental fact that the ν_μ, defined as the neutral particles with spin ½ and extremely small mass (presumably zero) which are emitted in the decay of π^+ in association with a μ^+, according to (16a), give rise only to processes of the type (18a) and not to processes of the type (18b)[84]: on the contrary, the neutral particles emitted in (16b) give rise to processes of type (18b) but not (18a). In other words, the processes observed with the neutrinos emitted in pion decay also take place as if the principle of lepton conservation were valid. It may be violated at most with a frequency of an order of a few per cent[85].

In spite of this, Majorana's paper No. 9 has been and still is of considerable importance in so far as it has contributed to throwing light on the properties of the neutrino and, in a wider sense, of the fermion.

In the first place, Majorana's work explained the relation between Hermitian conjugation of the fermion field and conjugation of the electric charge of the corresponding particle, as pointed out under item (iii) on page 55 : in all the representations different from Majorana's representation, the operation of charge conjugation involves manipulation of the spin indices in addition to the operation of Hermitian conjugation.

To this an observation by B. Touschek should be added, namely that in Majorana's paper No. 9 it is clearly implied that the second quantization is necessary for the physical interpretation of the Dirac equation, a fact which in 1937 many people did not know.

It should finally be pointed out that the interest in Majorana's theory revived soon after the discovery by Lee and Yang[86] of the non-conservation of parity in weak interactions and its experimental confirmation[87], and even more in relation to the development of the two-component theory of the neutrino[88].

This theory is based on the experimental fact that the helicity of the neutrino is always equal to -1 :

$$\mathcal{H}(\nu_e) = \frac{\vec{\sigma}\cdot\vec{p}}{|\vec{\sigma}|\;|\vec{p}|} = -1\,,$$

namely the neutrino ν_e, defined as the particle emitted in association with the positron according to Eq. (10b), always has its spin antiparallel to its momentum $\vec{p}$[89].

Indirect arguments show that the antineutrino appearing in Eq. (10a) always has a helicity

$$\mathcal{H}(\bar{\nu}_e) = +1\,,$$

namely, its spin is always parallel to its momentum. Similar conclusions are also valid for the neutrino ν_μ [90].

The simplest way of interpreting this situation is to admit that of the four possible states for a particle governed by the Dirac equation (the two orientations of the spin with respect to the particle momentum and the two states corresponding to a particle and an anti-particle) only two play a part in the pehnomena observed. This remark was the starting point of the two component theory [88]; at the same time a close study was made of its relations with Majorana's theory [72,91-93].

It is now known that there exist a priori various ways of describing in a relativistic manner a neutrino (namely a particle of spin ½, zero charge and zero mass) with two states only. One description is provided by the two-component theory, based on the discovery that a neutral particle with spin ½ and mass zero can be described ny a spinor with two components [94]. Another is the theory proposed by Majorana; whereas before the discovery of the non-conservation of parity, Majorana's theory based on identity (9) led to the non-conservation of the lepton number and therefore had to be rejected, the fact that the neutrino and the antineutrino <u>always</u> have opposite helicity, makes it possible to identify these two objects with the two states of ν_M having helicity -1 and $+1$, respectively.

This possibility of reducing the lepton number of the neutrino to its helicity subsists in Majorana's theory only in the case of a strictly zero mass. However, contrary to the two-component theory, Majorana's theory does not require the neutrinos to have a mass exactly equal to zero. A small neutrino mass ($m_\nu \ll m_e$) would, however, involve a small deviation from the conservation of the lepton number which cannot at present be excluded on the basis of the available experimental data: and, after all, at the present time we are only on the threshold of neutrino physics.

Having analysed in detail the various Majorana papers, there now naturally arises the difficult problem of trying to evaluate his general stature as a scientist, in so far as this is possible. There is no doubt that he had an extraordinary gift for mathematics, an exceptionally keen analytical mind, and a most acute critical sense. Perhaps it was this highly developed critical sense, together with a certain lack of balance on the human side, that interfered with his capacity for creative synthesis and prevented him from reaching a level of scientific productivity comparable with that reached at the same age by the major contemporary physicists. This does not alter the fact that the choice of some of the problems which he tackled and the methods which he used and, more generally speaking, the choice of the mathematical means of dealing with them, showed a natural tendency to be ahead of his time which, in some cases, was almost prophetic.

8. Having been appointed Professor of Theoretical Physics at Naples in November, 1937, Ettore Majorana moved to that city at the beginning of January the following year, and lodged at first at the Albergo Patria and then at the Albergo Bologna which overlooked the Via de Pretis, a very busy street.

In Naples he struck up a friendship with Antonio Carrelli, Professor of Experimental Physics and Director of the Physics Institute of that University.

In a letter to his mother dated 11 January, 1938, he told her that he was to begin his lecturers two days later, on Thursday 13 at 9 a.m., and that he had succeeded in making the opening of the course completely informal; he added "This is another reason why I don't advise you to come." Later in the letter he wrote "I found a letter from the Rector which had been waiting for me for two months, in which he informed me of my appointment on account of my well-known and exceptional talent, and since I could not get in touch with him I wrote him a letter in an equally stilted style." For anyone who knew him personally, this phrase was full of light and graceful irony directed at the law containing that phrase, at the Rector, but above all at himself. The rest of the letter is in the same vein, when he talks about the Physics Institute in Naples, some colleagues whom he mentions in a friendly way, and their assistants.

At the same time, this letter shows clearly his warm affection for and strong attachment to his mother and brothers.

It should also be said, incidentally, that in spite of his plain neglect of religious observance and his extreme reserve on subjects of this kind, everything points to the fact that his early education had left him with an essentially religious turn of mind.

In Naples, as in Rome, he led an extremely sheltered life: in the morning when he had a lecture to give he went to the Institute, and in the late afternoon he went for long walks in the busiest parts of the city.

He gave his lectures with the same care and skill that he had always shown in all his duties in the past. The manuscript of his lectures on quantum mechanics shows that he taught this subject in a very similar way to that used nowadays.

After his lectures he went to Carrelli's study and had long discussions with him on various current physics problems. At that time he never mentioned the research work that he was doing: Carrelli remembers that he had, nevertheless, the impression that he was trying to do something very exacting which he did not want to talk about. In any event, he did not do this work at the Institute but in his hotel room.

Carelli remembers particularly vividly a long conversation on Majorana's neutrino theory, which gave him the impression that he attached much more importance to these results than to those on the nuclear forces.

These conversations also revealed that he did not believe at all that the quantum mechanical theories had attained their definitive formulation; he pointed out several flaws in quantum mechanics which, according to him, had actually received very little quantitative experimental confirmation, except for the calculation of the helium energy levels.

In Naples, as in Rome in earlier years, Majorana was tormented by his illness, which finally inevitably had an effect on his temper and even on his character. This perhaps explains the excessive disappointment which he felt, according to Carrelli, when, after a few months' teaching, he realized that very few of the students were capable of following and appreciating his lectures, which were always on a very high level.

On 26 March, 1938, Carrelli was most astonished to receive from Ettore Majorana a telegram from Palermo, in which he told him not to worry about what he had said in the letter he had sent. Carrelli awaited the arrival of the letter, posted in Palermo a few hours before the telegram was sent. In it Ettore Majorana wrote very coldly and with great determination that he found life in general, and his own in particular, absolutely useless and that he had therefore decided to commit suicide. The letter was unfortunately lost, but one sentence which remained fixed in Carrelli's mind, ran more or less as follows: "I am not a young girl from one of Ibsens plays, you understand, the problem is much greater than that." The letter ended with warm greetings to Carrelli and thanks for the friendship which he had shown him in the last few months.

Carelli, upset by this letter, at once rang up Fermi, who got in touch with Ettore's brother Luciano in Rome: the latter immediately went to Naples, where he began feverishly to search for information about Ettore. The enquiries made in Palermo and Naples revealed that Ettore had left Naples for Palermo on the steamer of the Società Tirrenia in the night of 23 March, and had arrived in Palermo where he had spent a couple of days and from where, on the 25th, he had sent both the letter and the telegram to Carrelli. On the evening of the same day, he had boarded the steamer for Naples. Professor Michele Strazzeri of the University of Palermo saw him that night on board and in fact just as dawn was breaking, and the steamer was entering the Bay of Naples, he saw him sleeping in his cabin. A sailor said that he had seen him a stern after passing Capri and not long before the ship berthed in Naples. According to the Naples Office of the Società Tirrenia, Majorana's ticket from Palermo to Naples was found among those handed in upon disembarkation at Naples, but this information was never definitely confirmed.

Enquiries continued for another three months: they were conducted mostly by Luciano Majorana and Francesco Maria Dominedo, son of Emilia, younger sister of Quirino, and Fabio Majorana, then Professor of Commercial Law at the University of Sienna and later Professor of Maritime Law at the University of Rome and Senator of the Italian Republic.

The enquiry was continued for a further three months by both the police and the carabinieri, and Mussolini took a personal interest in it; Ettore's mother had written to him enclosing a letter from Fermi, in which, among other things, he said[95]: "I have no hesitation in saying, and it is no way an exaggeration, that of all the Italian and foreign scholars whom I have had the opportunity of knowing, Majorana is the one whose depth most impressed me. Capable of developing boldly hypotheses and, at the same time of criticising his own work and that of others, highly skilled in calculation and a mathematician of great depth, who never lost sight of the true nature of the physics problems behind the veil of figures and mathematical techniques, Ettore Majorana was highly endowed with that rare combination of gifts which go to make a typical theoretician of the first rank."

The family offered a reward of 30,000 lire, a considerable sum at the time, to anybody who could give news of Ettore and for months published in the leading newspapers an appeal to Ettore to return home:

the Vatican tried to find out whether he had entered a monastery. However, all these attempts were fruitless. No trace was ever found: it was only discovered that a few days before Ettore Majorana left for Palermo, a young man who appeared very upset and whose somatic and mental characteristics seemed to his family to correspond to those of Ettore, went to the Chiesa del Gesù Nuovo in Naples near the Albergo Bologna. Moreover, Father de Francesco, ex-Father Provincial of the Jesuits, who received the young man, thought that he recognized him from the photograph of Ettore shown to him by the family.

The young man asked Father de Francesco if he could "try the religious life", an expression which according to the brothers should be understood as "go into retreat". They did not in fact think that he meant by this phrase that he had a religious vocation, but simply that he wished to spend some time in meditation. On receiving the reply that they would willingly offer him hospitality, but only for a short time -- since, if it was to be a final decision it would be necessary, according to the rules of the Order, to enter the Novitiate -- the young man replied: "Thank you. Excuse me", and left.

The most likely assumption for his friends was that he had thrown himself into the sea: but all the experts on the waters of the Bay of Naples maintain that the sea would sooner or later have washed up his body.

It was only some 30 years later that someone who had never known him, or who had known him only very superficially, imagined that he had been kidnapped or had left the country in connection with a supposed case of atomic ispionage. However, for anyone who was familiar with the atmosphere in nuclear physics at the time and who knew Ettore Majorana, this assumption is not only devoid of all foundation, but it is completely absurd from both an historical and a human point of view. A few years after his disappearance, when discussing the subject with mutual friends,

Fermi observed that with his intelligence, once he had decided to disappear or to make his body disappear, Majorana would certainly have succeeded.

Nothing further was discovered: all his friends and relatives felt a sense of deep sorrow at the loss of this man who was gentle, reserved and averse to outward show, so deeply affectionate although profoundly unhappy; a sense of frustration for all that his mind had not produced but would certainly have produced if it had not been for his absurd disappearance; and above all a sense of deep admiration and astonishment at this man and thinker who passed among us so rapidly, like a character from Pirandello, laden with problems which he bore alone; a man who succeeded admirably in finding the answer to some of nature's problems, but who sought in vain for the meaning of life, his own life, even though it was infinitely richer in promise than that of the great majority of other men.

BIBLIOGRAPHY

1) E. Perucca, Commemorazione del Socio Quirino Majorana, "Rend. Accad. Lincei", 25, 2° semestre, p. 354. For a list of publications and some biographical data see also the "Annuario Generale dell'Accademia Nazionale dei XL", Rome, 31 (1953).

2) "Istituto Parificato Massimiliano Massimo" directed by Jesuit fathers.

3) The "Diploma di Maturità Classica" certificate (equivalent to general certificate of education) released on 11 May 1964 by the Liceo Ginnasio Statale Torquato Tasso shows the following marks (full mark: 10) : Italian: written 7, oral 8; Latin: written 7, oral 8; Greek: written 7, oral 7; History and Geography: 8; Philosophy: 7; Mathematics: 9; Physics: 9; Natural History: 7; Gymnastics: 8.

4) Further details are given in the biographical note on Enrico Fermi by Emilio Segré appearing in Vol. 1 of "Enrico Fermi, Note e Memorie (Collected papers)" Accademia Nazionale dei Lincei and the University of Chicago Press, Rome (1962).

5) L. Fermi. "Atoms in the Family", the University of Chicago Press (1954); see also the Italian translation "Atomi in famiglia", Mondadori (1954).

6) In the second expedition to the North Pole with an airship, organized and led by U. Nobile (in the first, which had taken place in 1926 under the leadership of R. Amundsen, Nobile had piloted the airship), while returning from the second flight over the Pole, the airship "Italia" crashed on to the ice, leaving some castaways. As a result, 8 members of the expedition died, in addition to 6 people who tried to go to the aid of the stranded men; among these was R. Amundsen.

7) The certificate of his Doctorate in Physics released on 15 May, 1964, by the University of Rome, shows the following marks: Algebra: 30; Analytical and Projective Geometry: 30, distinction; Applied Chemistry: 27; Dynamics: 30, distinction; Engineering Drawing: 18, (minimum for a pass); Probability Calculus: 27; Descriptive Geometry: 30; Physics: 30; Advanced Physics: 30, distinction; Geophysics: 30, distinction; Physics Laboratory: 30; Mathematical Physics; 30, distinction.

8) G. Polvani, Giovanni Gentile junior, "Reale Istituto Lombardo di scienze e lettere", 75, fasc. 11 (1941-42); C. Salvetti, Giovanni Gentile junior, "Rendiconti del Seminario Matematico e Fisico di Milano, 16 (1942).

9) L. Van Hove, George Placzek (1905-1955), Institute for Advanced Study, Princeton, New Jersey (1955); E. Amaldi, George Placzek, Ric. Scientifica, 26, 2037 (1956).

10) E. Amaldi and G. Placzek, Naturwiss, 20, 521 (1932); Z. Physik, 81, 259 (1933).

11) G. Placzek and E. Teller, Z. Physik, 81, 209 (1933).

12) E. Fermi and E. Amaldi, Mem. Acc. Italia, 6, 119 (1934).

13) The report of the Board of Examiners for the university teaching diploma in theoretical physics is published in the: "Bollettino del Ministero dell'Educazione Nazionale", Anno 60°, Vol. II, N. 27, 6 July 1933, Part II, Atti di Amministrazione, p. 2341.

14) W. Heitler and F. London, Z. Physik, 44, 455 (1927); F. London, Z. Physik, 46, 455 (1927); 50, 24 (1928); W. Heitler, Z. Physik, 46, 47 (1927); 47, 835 (1925); Z. Physik, 31, 185 (1930).

15) C.G. Darwin, Proc. Roy. Soc. (London), A 117, 258 (1928); A. Landé, Naturwiss., 17, 634 (1929).

16) P. Güttinger, Z. Physik, 73, 169 (1931).

17) T.E. Phipps and O. Stern, Z. Physik, 73, 185 (1931).

18) R. Frisch and E. Segré, Nuovo Cimento, 10, 78 (1933); Z. Physik, 80, 610 (1933).

19) I.I. Rabi, Phys. Rev., 51, 652 (1937); F. Bloch and I.I. Rabi, Rev. Mod. Phys., 17, 237 (1945).

20) See for instance: C.R. Stanford, T.E. Stephenson, L.N. Cochran and S. Bernstein, Phys.Rev., 94, 374 (1954).

21) See for instance: R. Nathans, C.G. Shull, G. Shirane and A.A. Andresen, J.Chem.Phys.Solids, 10, 953 (1959).

22) Sir E. Rutherford, J. Chadwick and C.D. Ellis, Radiations from Radioactive Substances, Cambridge University Press (1930).

23) I. Curie, C.R.Acad.Sci, Paris, 193, 1412 (1932); F. Joliot, C.R.Acad.Sci., Paris, 193, 1415 (1932).

24) I. Curie and F. Joliot, C.R.Acad.Sci., Paris, 194, 708 (1932).

25) J. Chadwick, Nature (London), 129, 322 (1932); see also by the same author, Proc.Roy.Soc. (London), A 136, 692 (1932).

26) Sir E. Rutherford, Phil.Mag., 4, 580 (1927).

27) J. Chadwick, Some personal Notes on the Search for the Neutron, p. 159, Vol. I, Actes du X Congrés International d'Histoire des Sciences, Ithaca, N.Y. (1962), Herman, Parigi (1964).

69

28) G. Gentile, Rend.Acc.Lincei, 7, 346 (1928).

29) F. Perrin, C.R.Accad.Sci.Paris, 195, 236 (1932); see also: W. Heisenberg, Reports et Discussions du 7ème Congrès de l'Institut International de Physique Solvay, p. 289 (1934).

30) D. Iwanenko, Nature (London), 129, 798 (1932).

31) This same anectode is mentioned in the short historical introduction by E. Segré, which appears on page 488, Vol. I of "Enrico Fermi", Note e Memorie", quoted in Ref. 4.

32) Comptes Rendus de la Première Section du Congrès International d'Electricité, Gauthier-Villars, Paris (1932) Vol. II, p. 789. The same article was published by Fermi in Italian in "Ricerca Scientifica", 3 (2), 101 (1932) and appears on page 489 of Volume I of "Enrico Fermi, Note e Memorie" quoted in Ref. 4.

33) W. Heisenberg, Z.Physik, 77, 1 (1932).

34) The speed with which events moved in Germany in 1933 is really surprising. On 30 January, after a long government crisis, Hindenburg constitutionally appointed Hitler Chancellor of the Reich; on 27 February the Reichstag fire was engineered by the Nazis and attributed by them to the communists. On 28 February, taking advantage of the impression produced by the Reichstag fire, Hitler made Hindenburg sign a document suppressing the articles of the Constitution guaranteeing individual and civil liberty. On 5 March new elections were held, and on 21 March the first meeting of the new Reichstag of the Third Reich was held; the Nazis did not yet have a majority, but they had enough power to succeed during the next few days in putting Hitler finally in power.

35) E. Feeberg, Phys.Rev.40, 40 (1932); 42, 17 (1933).

36) The optical theorem was established by N. Bohr, R. Peierls and G. Placzek, Nature (London), 144, 202 (1939), who among other things revealed its full significance.

37) The flippant name "Unter den Quanten" originates by analogy from "Unter den Linden" (Under the lime trees), the name of the famous Berlin street full of hotels, cafés, etc., particularly dear to the inhabitants of Berlin.

38) The report of the Board of Examiners for appointment of an extraordinary professor to the Chair of Theoretical Physics of the Royal University of Palermo, which gives an outline of what I have mentioned, is published in the "Bollettino del Ministero dell'Educazione Nazionale" Anno 65°, Vol. I, N. 6, 10 February 1938, Part II, Atti di Amministrazione, p. 280.

39) The concept of isotopic spin in fact acquired the status of a veritable physical quantity with the discovery that strong interactions are independent of electric charge. This law can be formulated by saying that in all processes involving only strong interactions, the total isotopic spin of the final state is always equal to the total isotopic spin of the initial state, and that the Hamiltonian of the system is invariant with respect to rotations in isotopic spin space.

It was in 1936 that, on the basis of a comparison of the results of n-p and p-p collision experiments, B. Cassen and E.U. Condon [Phys.Rev. 50, 846 (1936)] and G. Breit and E. Feenberg [Phys.Rev. 50, 850 (1936] began to talk about the charge independence of the nuclear forces. The first two writers based their paper on the idea that the particles making up the nucleus are characterized by five variables: the three space coordinates, the third component of the spin and a charge variable, which they took to be identical with the third component of isotopic spin, following the example of Heisenberg and of Fermi, who had made use of it in his beta decay theory [Nuovo Cimento 11, I (1934) and Z.Physik 88, 161 (1934)].

The connection between isotopic spin and charge independence was further investigated later by E. Wigner in Phys.Rev. 51, 106 and 970 (1937), where he discussed the consequences of charge indipendence on nuclear levels, and in a paper, written in collaboration with E. Feenberg [Rep.Progr.Phys. 8, 274 (1941)], on beta transitions. Meanwhile N. Kemmer [Proc.Camb.Phys.Soc. 34, 354 (1938)] had extended the concept of isotopic spin to meson theory and had studied the implications of charge independence independently of perturbation theory.

Although at that time some physicists were already willing to consider isotopic spin as the physical quantity necessary for the correct representation of the charge indepndence laws of strong interactions, this conclusion received universal recognition after the first experiments by Hildebrand [Phys.Rev. 89, 1090 (1953)] on the angular distribution of the π^0 emitted in n-p collision and its comparison with the results of similar experiments on the production of π^+ in p-p collision.

For further details see, for istance, : The Proceedings of the III Annual Rochester Conference on high-energy nuclear physics, Dec. 18-20, 1952, Interscience Publishers Inc., New York (1953); the article by M. Gell-Mann and K.W. Watson, Annual Rev.Nucl.Sci. 4, 219 (1954); and the book by H.A. Bethe and F. de Hoffmann, Mesons and fields, Vol. II Cap. III B,Row, Peterson and Co, Evanston, Ill. (1956).

40) W. Heisenberg, Z.Physik 78, 156 (1932).

41) I. Curie and F. Joliot, C.R.Acad.Sci.Paris, 198, 559 (1934); J.Phys.Radium, 5, 153 (1934).

42) W. Heisenberg, Z.Physik, 80, 587 (1933).

43) F. Perrin, C.R.Acad.Sci.Paris, 194, 1343 (1932).

44) H. Yukawa, Proc.Phys.Math.Soc.Japan, 17, 48 (1935); see also: H. Yukawa and C. Kikuchi, Birth of the Meson Theory, Am.J.Phys., 18, 154 (1950).

45) See for example: S.C. Williams, Introduction to elementary particles, Academic Press, New York (1962).

46) R. Jastrow, Phys.Rev. 81, 165 (1951).

47) The most complete discussion of saturation phenomena, which was possible two years after the papers by Heisenberg and Majorana, is to be found in paragraph 7 of the article by H.A. Bethe and R.F. Bacher, Rev.Mod.Phys. 8, 82 (1936).

48) For a discussion of this point see Chapter III of J.M. Blatt and V. Weisskopf, Theoretical Nuclear Physics, John Wiley and Sons, New York (1952).

49) I should like to thank Prof. M. Fierz of the E.T.H. Zürich, for a very enlightening discussion on paper N. 7.

50) I should also like to thank Prof. D.M. Fradkin of the Wayne State University, Detroit, Michigan, for various discussions on the significance of paper N. 7 and for having placed at my disposal the manuscript of his paper, Comments on a paper by Majorana concerning elementary particles, Am.J.Phys. 34, 314 (1966) on the basis of which I have discussed this subject.

51) C.D. Anderson, Science 76, 238 (1932); Phys.Rev. 43, 491 (1933). The above-mentioned issue of Science is dated 9 September and contains Anderson's article (dated 1 September) in which he mentions positive electrons for the first time. In Anderson's paper which appeared in August of the same year [Phys.Rev. 41, 405 (1932)], which is often quoted as that announcing the discovery of the positron, positive particles are mentioned but they are said to be protons.

It should also be added that the Journal "Science" was not received in the library of the Physics Institute of the University of Rome and from a study of the 1932 manuscripts in the archives of the "Nuovo Cimento", it seems that Majorana's article reached the editor of the review in Bologna at the beginning of the summer of that year.

52) P.A.M. Dirac, Proc.Roy.Soc. A 155, 477 (1936).

53) O. Klein, Ark.Mat.Astr.Fis. A 25, 15 (1936).

54) G. Petiau, Memories; Ac.Soc.Roy.Belgique, 16, N. 2, 1-116 (1936).

55) A. Procat, C.R.Acad.Sci.Paris, 202, 1490 (1936); J.Phys.Radium, (VII) 7, 347 (1936); 9, 61 (1938).

56) R.I. Duffin, Phys.Rev. 54, 1114 (1938).

57) M. Fierz, Helv.Phys.Acta, 12, 3 (1939).

58) M. Fierz and W. Pauli, Proc.Roy.Soc. A 173, 211 (1939).

59) N. Kemmer, Proc.Roy.Soc. A 173, 91 (1939).

60) For a brief summary of the papers of this first period see Chapter 6 of the book by G. Wentzel, Quantum theorie der Wellenfelder, Franz Deutike, Vienna (1943); Quantum Theory of Fields, Interscience Publ., New York (1949).

61) H.J. Bhabha, Rev.Mod.Phys. 17, 200 (1945).

62) H. Weyl, Gruppentheorie und Quantummechanik, Verlag von S. Hirzel, Leipzig (1918), p. 160 of the first edition.

63) E.P. Wigner, Ann.Math. (Princeton), 40, 149 (1939).

64) P.A.M. Dirac, Proc.Roy.Soc. A 183, 284 (1945).

65) E.P. Wigner, Z.Physik, 124, 665 (1948).

66) E.M. Corson, Introduction to tensors, spinors and relativistic wave-Equations, Blackie and Sons Ltd, Glashow (1953), Cap. V, pp. 140-199.

67) I.M. Gelfand and M.A. Naimark, Journ.of Phys. (USSR), X, 93 (1946); Harish-Chandea, Proc.Roy.Soc. A 189, 372 (1947); V. Bargmann, Ann.Math. 48, 568 (1947).

68) I.M. Gelfand and A.M. Yaglow, Zh.experim. i Teor.Fiz. 18, 707 (1948); English trans.: TT-345 National Research Council of Canada (1953); V.L. Ginzburg, Acta Phys.Polon. 15, 163 (1956); I.M. Gelfand, R.A. Minlos and Z.Ya Shapiro, Representations of the Rotations and Lorentz Group and their applications, Pergamon Press, Oxford 1963, translated by G. Cummins and T. Boddington from the Russian (Fizmatgiz,Moscow, 1958); M. Naimark, Linear Representations of the Lorentz Group, Pergamon Press, Oxford (1964), translated by A. Swinfen and O.J. Marstrand from the Russian (Fizmatgiz, Moscow, 1958).

69) M. Gell-Mann and A.H. Rosenfeld, Annual Rev.Nucl.Sci. 7, 407 (1957); R.H. Dalitz, Annual Rev.Nucl.Sci. 13, 339 (1963); M. Gell-Mann and Y. Nee'man, The Eightfold Way, Benjamin, New York (1964).

70) Today a few attempts are being made to interpret each of the many sub-nuclear particles with strong interactions as a bound state of the others.

73

71) P.A.M. Dirac, Proc.Camb.Phil.Soc. 30, 150 (1934).

72) B. Touschek, Rendiconti della Scuola Internazionale di Fisica"Enrico Fermi" XI Corso, page 40, 1956. I should like to take the opportunity of thanking Professors B. Touschek and N. Cabibbo for various discussions concerning this paper of Majorana's.

73) If the index T is taken to mean "transposed" one has:

$$(\gamma_4 \gamma_\kappa)_{\sigma\rho} = (\gamma_4 \gamma_\kappa)^T_{\rho\sigma} = (\gamma_4 \gamma_\kappa)_{\rho\sigma} \quad .$$

74) L. Alvarez and F. Bloch, Phys.Rev. 67, 111 (1940).

75) B. Cork, G.R. Lambertson, O.P. Piccioni and W.A. Wenzel, Phys.Rev. 104, 1193 (1956).

76) C.L. Cowan, F. Reines, F.B. Harrison, H.W. Kruse and A.D. Mc Guire, Science, 124, 103 (1956).

77) R. Davis, Bull.Am.Phys.Soc. 1, 219 (1956); C.O. Muelhause and S.O. Oleska, Phys.Rev. 105, 1332 (1957).

78) E. Fermi, Nuovo Cimento, 11, 1 (1934); Z.Physik, 88, 161 (1934).

79) E. Konopinski and H.M. Mahmoud, Phys.Rev.92, 1045 (1953); T.D. Lee and C.N. Yang, Phys.Rev. 105, 1671 (1956). If every lepton (e^-, ν_e, μ^-, ν_μ) is given the lepton number ℓ = + 1 and every anti-lepton (e^+, $\bar{\nu}_e$, μ^+, $\bar{\nu}_\mu$) the lepton number ℓ = - 1 and all the other particles ℓ = 0, the principle of lepton conservation maintains that "the only processes which can take place are those in which the algebraic sum of the lepton numbers of the particles taking part in them is conserved". One can immediately see that processes (10) and (13) respect this principle while processes (11), (12) and (14) violate it.

80) F. Furry, Phys.Rev. 56, 1184 (1939); see also the review by: G.F. Dell'Antonio and E. Fiorini, Supplem.Nuovo Cimento, 17, 132 (1960).

81) E. Der Mateosan and M. Goldhaber, Bull.Am.Phys.Soc. 9, 717 (1964); Brookhaven Nat. Lab. Rep. 9791 (1965).

82) G. Danby, J.M. Gaillard, K. Goulianos, L.M. Lederman, N. Mistry, M. Schwartz and J. Steinberger, Phys.Rev.Letters, 9, 36 (1962); 10, 260 (1963).

83) G. Bernardini, G. von Dardel, P. Egli, H. Faissner, F. Ferrero, C. Franzinetti, S. Fukui, J.M. Gaillard, H.J. Gerber, B. Hahm, R.R. Hillier, V. Kaftanov, F. Krienen, M. Reinharz and R.A. Salmeron, Proceedings of the Sienna International Conference on elementary particles, October 1963; Vol. I, p. 571, Società Italiana di Fisica, Bologna (1963). J.K. Bienlein, A. Böhm, G. von Dardel,

H. Faissner, F. Ferrero, J.M. Gaillard, H.J. Gerber, B. Halm, V. Kaftanov, F. Krienen, M. Reinharz, R.A. Salmeron, P.G. Seiler, A. Staude, J. Stein and H.J. Steiner, Phys.Letters, 13, 80 (1964).

84) H.H. Bingham, H. Burmeister, D.C. Cundy, P.G. Innocenti, A. Lecourtois, R. Møllerud, G. Myatt, M. Paty, D.H. Perkins, C.A. Ramm, S. Schultze, H. Sletten, K. Soop, R.G.P. Voss and H. Yoshiki, Proceedings of the Sienna International Conference on Elementary particles, October 1963; Vol. I, p. 555, Società Italiana di Fisica, Bologna (1963). M.M. Block, H. Burmeister, D.C. Cundy, B. Eiben, C. Franzinetti, J. Keren, R. Møllerud, G. Myatt, M. Nikolic, A. Orkin-Lecortois, M. Paty, D.H. Perkins, C.A. Ramm, K. Schultze, H. Sletten, K. Soop, R. Stump, W. Venus and H. Yoshiki, Phys.Letters, 12, 281 (1964).

85) In the neutrino beam used by all the above-mentioned authors, besides the neutrinos emitted in process (16 a) there is actually a fraction of the order of a few percent of antineutrinos produced by process (16) and about 0.4% of ν_e neutrinos produced by the process $K^+ \to e^+ + \pi^0 + \nu_e$.

86) T.D. Lee and C.N. Yang, Elementary particles and weak interaction. BNL 443; U.S.Department of Commerce Office of Technical Sciences, Washington D.C. October 1957.

87) C.S. Wu, E. Amber, R.W. Hayward, D.D. Hoppes and R.P. Hudson, Phys.Rev. 105, 1413 (1957); R.L. Garwin, L.M. Lederman and M. Wenirich, Phys. Rev. 105, 1415 (1957); J.I. Friedman and V.L. Telegdi, Phys.Rev. 105, 1681 (1957).

88) T.D. Lee and C.N. Yang, Phys.Rev. 105, 1671 (1957); L. Landau, Nucl.Phys. 3, 127 (1957); A. Salam, Nuovo Cimento, 5, 299 (1957).

89) M. Goldhaber, L. Grodzins and A.W. Sunyar, Phys.Rev. 109, 1015 (1958).

90) M. Bardon, P. Franzini and J. Lee, Phys.Rev.Letters, 7, 23 (1961).

91) J. Serpe, Physica, 18, 295 (1952).

92) L.A. Radicati and B. Touschek, Nuovo Cimento, 5, 1693 (1957).

93) K. Case, Phys.Rev. 107, 307 (1957).

94) H. Weyl, Z.Physik, 56, 330 (1929).

95) This letter from Fermi, of which there remains a hand-written copy made by Luciano Majorana before it was sent to Mussolini, is written in a very different style from Fermi's usual one. This may possibly be partly due to the dramatic circumstances in which it was written and partly to the fact that he was writing to Mussolini.

Fermi's opinion of Ettore Majorana, was, however, expressed on various other occasions and particularly in a conversation which he had with Giuseppe Cocconi during the days following Ettore's

75

disappearance and about which Cocconi himself at my request sent me a letter dated 18 July, 1965. Soon after taking his doctorate at the end of 1937 in Milan, G. Cocconi had come to Rome and in January 1938 had begun to work with E. Fermi and G. Bernardini on the products of meson decay in cosmic radiation.

He thus had the opportunity of seeing Majorana once when he had come to the Institute to see Fermi, and then he had been present, in the dramatic days after Ettore's disappearance, during some of Fermi's various telephone calls and conversations.

It was at this time that Fermi said something to Cocconi which ran more or less as follows: "Because, you see, there are various kinds of scientists in the world; the second and third-rate ones who do their best but do not get very far. There are also first-rate men who make very important discoveries which are of capitol importance for the development of science. Then there are the geniuses like Galileo and Newton. Well, Ettore Majorana was one of these. Majorana had greater gifts than anyone else in the world; unfortunately he lacked one quality which other men generally have: plain common sense".

SCIENZA IN PRIMO PIANO

ETTORE MAJORANA, A CINQUANT'ANNI DALLA SUA SCOMPARSA

Edoardo Amaldi
Dipartimento di Fisica
Università «La Sapienza» di Roma

1. – Sono passati cinquant'anni dalla scomparsa di Ettore Majorana e uno sguardo retrospettivo alla sua opera scientifica è piú che naturale e doveroso(1).

Per chi non lo abbia presente, ricorderò che essa consiste in 9 lavori, il primo dei quali era stato inviato, per la stampa, da O. M. Corbino, allora Direttore dell'Istituto di Fisica di Via Panisperna, alla R. Accademia dei Lincei, a cui era pervenuto il 24 luglio 1928, e che l'ultimo è apparso nel fascicolo di aprile 1937 de *Il Nuovo Cimento*, cioè un anno prima della sua scomparsa.

Vorrei anche ricordare che tale produzione può essere divisa in due fasi; la prima comprende i primi 6 lavori [1-6] che si riferiscono tutti a problemi di fisica atomica e molecolare, mentre la seconda ne comprende 3 soli, dei quali uno [8] concerne le forze che si esercitano fra i protoni e i neutroni in un nucleo, mentre gli altri due ([7] e [9]) riguardano la trattazione quantistica e relativistica delle particelle subnucleari.

La fama di Ettore Majorana, come è ben noto, è dovuta prevalentemente a questi tre ultimi lavori; ma anche i sei lavori della sua prima fase produttiva sono molto interessanti, sia per le tematiche affrontate che per i metodi da lui impiegati.

2. – Cominciamo, dunque, dai lavori appartenenti alla prima fase produttiva. Essi possono venir ulteriormente divisi in 3 gruppi.

Il primo gruppo è costituito da 3 lavori ([1], [3] e [5]) che riguardano problemi di spettroscopia atomica; mentre il secondo gruppo comprende 2 lavori ([2] e [4]) che trattano aspetti ignorati o quasi, a quell'epoca, del legame chimico. Il terzo gruppo infine comprende un sol lavoro [6], che verte sul problema del ribaltamento dello spin (spin-flip) non adiabatico in un fascio di atomi polarizzati.

Senza entrare in un esame dettagliato dei 3 lavori di spettroscopia atomica, vorrei ricordare che il primo [1], fatto in collaborazione con Giovanni Gentile jr, presenta i risultati del calcolo dello sdoppiamento dovuto allo spin dell'elettrone di alcuni termini Röntgen (i termini $3M$ del gadolinio ($Z=64$) e dell'uranio ($Z=92$)) e ottici (i termini P del cesio ($Z=55$)) e l'intensità delle righe del cesio. Esso è svolto introducendo nelle espressioni dedotte a mezzo della teoria delle perturbazioni le funzioni d'onda dei singoli stati elettronici calcolate con il metodo statistico di Thomas-Fermi(2). Questo lavoro rientra nel quadro dell'attività dell'Istituto di quell'epoca; infatti negli anni dal 1928 al 1932, oltre a Fermi, anche la maggior parte di noi applicava il metodo statistico di Thomas-Fermi a diversi problemi di fisica atomica(3).

Gli altri due lavori di spettroscopia atomica sono assai piú originali e interessanti sia dal punto di vista della scelta dell'argomento — allora molto avanzato — che da quello metodologico. Essi riguardano entrambi stati 13
atomici *doppiamente eccitati* (o accentati come si diceva allora), cioè stati con due elettroni eccitati in livelli tali da rendere energeticamente possibili processi di autoionizzazione spontanea, ossia transizioni a stati finali consistenti in uno ione positivo monovalente e un elettrone libero (effetto Auger).

Ciò che oggi maggiormente colpisce di questi due lavori è la modernità dell'impostazione data da Majorana, che basa tutta la sua discussione sulle proprietà di simmetria dei diversi stati che si ottengono combinando fra loro due orbitali, rispetto allo scambio degli spin degli elettroni (stati di singoletto e di tripletto), alle rotazioni spaziali (numero quantico orbitale), rispetto alle rotazioni assiali (quanto magnetico) e alle riflessioni nel centro di forza (parità).

Il primo di questi due lavori [3] riguarda due righe nell'estremo ultravioletto ($\lambda=320.4$

e 357.5 Å) che erano state scoperte recentemente nello spettro dell'elio e che non erano interpretabili come combinazioni di termini conosciuti. Queste righe sono importanti anche perché si osservano nella corona solare e di conseguenza hanno un notevole interesse astrofisico.

Lo scopritore (P. G. Kruger) aveva proposto d'interpretare queste righe come dovute a transizioni dai termini normali $1s2p^3P_{0,1,2}$ e $1s2s^1S_0$ ai termini doppiamente eccitati $2p2p^3P_{0,1,2}$ e $2s2s^1S_0$. Questi due termini finali sono assai piú alti del limite normale dell'elio ed è quindi energeticamente possibile la ionizzazione spontanea (o effetto Auger). Grazie a calcoli perturbativi spinti al 2° ordine e alle proprietà di simmetria dei singoli stati sopra elencate, Majorana giunge a confermare l'interpretazione proposta da Kruger per la prima di queste due righe, ma ad escluderla per la seconda in base a considerazioni energetiche.

Mi faceva notare Ugo Fano, in una discussione che ho avuto il piacere di fare recentemente con lui (19-9-87) su questi lavori spettroscopici, che nella letteratura internazionale questo lavoro di Majorana viene spesso ignorato, mentre si cita un lavoro di Wu di tre anni dopo nel quale anche si giunge ad interpretare la prima delle due righe qui discusse come dovuta alla transizione $1s2p^3P_0 - 2p2p^3P$ servendosi di una forma modificata del metodo variazionale in cui due «costanti di schermaggio, una per ciascuno dei due elettroni», vengono introdotte come parametri aggiusta-
14 bili ([4]).

Il terzo lavoro di spettroscopia atomica [5] riguarda invece i tripletti doppiamente eccitati incompleti, cioè tripletti doppiamente eccitati che si trovano al di sopra del limite delle serie normali ma sono stabili se si trascurano le correzioni relativistiche. Essi però acquistano una leggera instabilità determinata dal momento magnetico intrinseco degli elettroni, la quale, in circostanze particolari, può tuttavia assumere un'importanza considerevole tanto da far scomparire una componente (o piú) di un tripletto doppiamente eccitato, il quale, di conseguenza, si presenta come «incompleto». I calcoli di Majorana, che si riferiscono agli atomi di Zn, Cd e Hg, combinati nuovamente con l'uso delle proprietà di simmetria degli stati, sono in pieno accordo con i corrispondenti risultati sperimentali.

Passando ora ai due lavori sul legame chimico va notato fin d'ora che fu proprio attraverso lo studio approfondito di questi problemi che Ettore Majorana s'impadroní della teoria del legame chimico, circostanza questa che doveva risultare di grande importanza per la sua futura attività di ricerca. La sua conoscenza approfondita del meccanismo di scambio degli elettroni di valenza, che è alla base della teoria quantistica del legame chimico omeopolare, enunciata per la prima volta da Heitler e London nel 1927 ([5]), costituirà piú tardi il punto di partenza per l'ipotesi che le forze nucleari siano forze di scambio.

Il primo di questi due lavori [2] risale al 1931 e riguarda la formazione dello ione molecolare di elio. In esso Majorana parte da osservazioni spettroscopiche, allora recenti, che avevano messo in evidenza alcune bande attribuite allo ione molecolare, e affronta il problema della reazione chimica che porta alla sua formazione, giungendo a dimostrare che essa può effettivamente aver luogo dal punto di vista energetico. Con un calcolo di prima approssimazione, Majorana trova quindi la distanza di equilibrio dei due nuclei nello ione He_2^+, che risulta in buon accordo con il valore misurato sperimentalmente, valuta l'energia del sistema (allora ancora non misurata) e la frequenza di oscillazione della molecola ionizzata, il cui valore si accorda molto bene con il valore sperimentale.

Alla fine del lavoro Majorana ringrazia Fermi per essergli stato largo di consigli e aiuti e Gentile per l'interesse con cui aveva seguito il suo lavoro.

L'altro lavoro sul legame chimico [4] studia in dettaglio le caratteristiche di un termine anomalo pari di singoletto $(2p\sigma)^2\,{}^1\Sigma_g$ (detto termine X) piuttosto profondo che era stato recentemente osservato nello spettro della molecola d'idrogeno e che era stato interpretato come dovuto a uno stato in cui due elettroni sono eccitati. Questo termine, come fa vedere Majorana, in realtà non può essere designato come un termine con due elettroni eccitati descrivibile in termini dei numeri quantici dei singoli elettroni.

Questa descrizione, come egli scrive, «se giova alla loro numerazione e al riconoscimento di quei caratteri di simmetria che non sono turbati dall'interazione [fra elettroni], non permette da sola di trarre conclusioni attendibili sulla forma effettiva delle autofunzioni; le cose stanno qui ben diversamente che nel caso di campi centrali, dove è in generale possibile astrarre dall'interdipendenza dei movimenti degli elettroni (polarizzazione), senza perdere di vista l'essenziale».

Il termine della molecola d'idrogeno in questione può essere pensato, solo in una descrizione assai schematica, come la sovrapposizione di due stati: uno proveniente dalla reazione fra due atomi neutri d'idrogeno (H+H) già trattata da Heitler e London, e l'altro proveniente da un nuovo processo di formazione della stessa molecola in cui due ioni idrogeno, uno positivo e l'altro negativo ($H^+ + H^-$), si legano insieme.

Il legame chimico che interviene in questo secondo caso viene chiamato da Majorana «pseudopolare» in quanto è vero che si tratta di un legame fra due ioni di segno opposto, ma, a causa dell'eguaglianza dei componenti, il momento elettrico cambia segno con la frequenza di scambio che è assai elevata, e pertanto non è osservabile.

Partendo da uno stato iniziale sovrapposizione di quelli relativi a questi due processi, e prendendo in considerazione solo stati pari (dato che quelli dispari non danno origine a stati stabili), Majorana trova due soli stati finali legati: lo stato fondamentale della molecola di H_2 e lo stato anomalo $(2p\sigma)^2\ {}^1\Sigma_g$, il primo dei quali deriva prevalentemente dalla unione H+H, mentre il secondo deriva prevalentemente dalla reazione pseudopolare $H^+ + H^-$.

L'ultimo lavoro del periodo di attività dedicato a problemi di fisica atomica e molecolare riguarda il calcolo della probabilità di ribaltamento dello spin (spin-flip) degli atomi di un raggio di vapore polarizzato quando questo si muove in un campo magnetico rapidamente variabile [6].

Se il campo magnetico cambia lentamente di direzione, l'orientamento dell'atomo segue adiabaticamente la direzione del campo. Ma che cosa succede se la variazione di direzione del campo non è adiabatica? In particolare cosa succede se l'atomo passa in prossimità di un punto ove il campo magnetico è zero?

Il problema era stato discusso fin dai primi tempi della meccanica quantistica da Darwin e da Landé([6]); e Stern, che aveva fatto qualche anno prima con Gerlach le famose esperienze sulla quantizzazione spaziale, si era proposto di verificare le probabilità di transizione nel caso non adiabatico.

Nel suo lavoro Majorana dimostra che l'effetto globale di un campo magnetico variabile $H(t)$ su di un corpuscolo di momento angolare $J=1/2$ e data componente m lungo l'asse delle z può venire descritto come una brusca rotazione di un angolo α del momento angolare totale.

Questo angolo è ottenuto da una soluzione dell'equazione del moto (sia classica che quantistica) e la sua proprietà piú importante è che esso dipende solo dal rapporto giromagnetico e dal campo magnetico, ma è indipendente dal valore iniziale del quanto magnetico m.

Il problema fu in realtà trattato e risolto da Majorana, con straordinaria eleganza e concisione, solo per il caso particolare $J=1/2$; ma, come egli stesso fa notare, il metodo si presta facilmente ad un'estensione al caso generale di J qualsiasi. Tale estensione fu fatta nel 1937 da Rabi e ridiscussa e posta in forma fisicamente piú evidente nel 1945 da Bloch e Rabi([7]), i quali contribuirono cosí a diffondere i risultati che Majorana aveva trovato tredici anni prima.

Questo lavoro di Majorana è rimasto un classico della trattazione dei processi di ribaltamento non adiabatico e come tale viene correntemente citato; i suoi risultati, estesi come si è detto sopra, hanno costituito successivamente il principio su cui è basata la realizzazione sperimentale del metodo usato per ribaltare lo spin dei neutroni con un campo a radiofrequenza, metodo impiegato sia nell'analisi di fasci di neutroni polarizzati, sia in tutti gli spettrometri a neutroni polarizzati usati nello studio delle strutture magnetiche.

Da quanto ho detto, ma ancor piú da un esame approfondito di questi lavori, si resta colpiti dalla loro alta classe: essi rivelano una 15
profonda conoscenza dei dati sperimentali anche nei piú minuti dettagli, una disinvoltura, non comune a quell'epoca, nello sfruttare le proprietà di simmetria degli stati per semplificare i problemi o per la scelta della piú opportuna approssimazione per risolvere quantitativamente i singoli problemi, qualità quest'ultima che senza alcun dubbio derivava, almeno in parte, dalle sue eccezionali doti di calcolatore.

3. – Come ho già ricordato fin dall'inizio, il maggior contributo scientifico di Ettore Majorana è tuttavia rappresentato dalla seconda fase della sua produzione, la quale, come anche ho già detto, comprende tre lavori: il lavoro sulle forze nucleari oggi dette alla Majorana, il lavoro sulle particelle di momento intrinseco arbitrario e il lavoro sulla teoria simmetrica dell'elettrone e del positone.

Per inquadrare storicamente il primo di questi tre lavori, ricorderò che la scoperta del neutrone era stata annunciata da James Chadwick con una Lettera a *Nature* ricevuta il 13 febbraio 1932 e che già prima di Pasqua dello stesso anno Majorana aveva cercato di sviluppare una teoria dei nuclei leggeri assumendo che essi fossero costituiti solo di protoni e neutroni[8] che interagiscono fra loro con forze di scambio alla Majorana. Aveva anche tentato di fare una teoria dei nuclei piú leggeri, teoria descrivibile come un primitivo modello a strati (shell model); ma, avendo incontrato difficoltà quando, procedendo in ordine di numero atomico crescente, era arrivato a $Z=6$, aveva abbandonato questa linea di lavoro e si era rifiutato di pubblicare i suoi risultati anche sulle forze di scambio, nonostante le pressanti e ripetute insistenze di Fermi e di tutti gli altri membri dell'Istituto[1].

Nel fascicolo di luglio dello stesso anno 1932 apparve il primo dei tre lavori di Heisenberg[9], il quale era giunto per conto proprio a ideare forze di scambio che operavano però su tutte le coordinate del protone e del neutrone interagenti; cioè sulle coordinate di spin oltre che su quelle spaziali.

Solo verso la fine del 1932 Fermi riuscí a convincere Majorana a recarsi all'estero con una borsa di studio del CNR che Fermi stesso gli aveva fatto assegnare. Nel mese di gennaio 1933 Majorana partí per Lipsia ove rimase fino all'inizio dell'estate. Da Lipsia andò a Copenhagen ove trascorse circa tre mesi
16 e di lí di nuovo a Lipsia, ove si fermò per circa un altro mese, prima di ritornare a Roma all'inizio dell'agosto 1933.

Nel periodo passato a Lipsia si legò di amicizia con Heisenberg e fu Heisenberg, con la sua autorità, a convincere Ettore a pubblicare il suo lavoro sulle forze nucleari, in quanto, grazie alla loro diversa natura, davano alcuni risultati nettamente superiori a quelli già pubblicati da Heisenberg stesso.

Il lavoro di Majorana apparve, come spedito da Lipsia, in tedesco [8*a*] con la data del 3 marzo 1933, e in italiano [8*b*] con quella dell'11 maggio 1933. Le prime tre pagine del testo tedesco, quasi senza formule e molto discorsive, sono dedicate alle varie scelte fatte per giungere alla sua espressione per le forze nucleari. Esse sono piú ampie e interessanti delle prime due pagine del testo italiano, il quale differisce di pochissimo da quello tedesco in tutte le altre parti. Dopo un riesame delle ipotesi del modello di Heisenberg, Majorana riprende in considerazione le difficoltà relative alla struttura del neutrone sollevate da Heisenberg: il neutrone è una particella composta da un protone + un elettrone strettamente legati, o una particella neutra a se stante? E giunge alla conclusione che per il momento la sola cosa da fare è di cercare di stabilire la legge d'interazione fra protone e neutrone sulla base di criteri di semplicità scelti in modo tale da permettere di riprodurre il piú correttamente possibile le proprietà piú generali e caratteristiche dei nuclei. Dopo aver discusso alcune di queste e aver osservato che i nuclei si comportano come pezzi di materia nucleare estesa e impenetrabile, le cui diverse parti interagiscono essenzialmente solo a contatto, e dopo aver sottolineato il fatto che i dati sperimentali mostrano chiaramente una legge di proporzionalità sia dell'energia di legame che del volume con il numero dei protoni e neutroni presenti, Majorana introduce un'energia potenziale di scambio che opera *solo* sulle 3 coordinate spaziali, il cui segno deve essere tale da risultare attrattivo negli stati di momento angolare pari. Egli giustifica questa scelta osservando che, cosí facendo, si ottengono due importanti risultati: il primo è che entrambi i neutroni presenti nella particella alfa interagiscono con ciascuno dei due protoni. Il secondo è che, nell'approssimazione in cui si può trascurare la repulsione coulombiana, l'autofunzione della particella alfa è totalmente simmetrica rispetto allo scambio delle coordinate dei centri di massa dei quattro corpuscoli componenti, come è naturale attendersi, se questo, come tutto sembrava indicare, corrisponde ad un guscio o anello completo.

Passando dalla particella alfa a nuclei piú pesanti, a causa del principio di Pauli, i protoni e i neutroni che vengono aggiunti sono costretti ad andare in stati piú eccitati, e poiché l'energia di scambio è elevata solo nel caso di corpuscoli che stanno nello stesso (o quasi nello stesso) stato orbitale, ne segue che la saturazione sia dell'energia di legame che della densità è praticamente raggiunta nel caso della particella alfa.

Le differenze essenziali rispetto allo schema di Heisenberg sono che le forze di scambio di Majorana riguardano solo le coordinate spaziali e che la funzione $J(r)$ usata da Majorana ha il segno opposto a quella usata da Heisenberg.

In queste condizioni le proprietà di simmetria delle autofunzioni sono tali che è possibile spiegare i fenomeni di saturazione senza introdurre un taglio (cut-off) dell'energia potenziale di scambio, come era stato costretto a fare Heisenberg per evitare il collasso del nucleo.

Dopo queste considerazioni Majorana scrive la hamiltoniana di un nucleo come somma delle energie cinetiche di tutti i protoni e neutroni componenti, dell'energia di repulsione coulombiana fra tutte le coppie di protoni e dell'energia potenziale di scambio fra tutte le coppie di protoni-neutroni. Calcola quindi l'autovalore del sistema composto da Z protoni ed N neutroni nel caso limite di Z ed N molto grandi (ossia nel caso della materia nucleare) nel suo stato fondamentale. Per far questo egli usa come autofunzione globale il prodotto di due autofunzioni antisimmetriche (nello scambio simultaneo delle coordinate spaziali e di quelle di spin), delle quali una si riferisce ai protoni e l'altra ai neutroni. Egli giunge cosí alla conclusione che, se si trascura la repulsione coulombiana, l'energia del sistema è proporzionale al numero delle particelle componenti; egli dunque dimostra che con forze di *scambio alla Majorana* è possibile ottenere la saturazione sia dell'energia di legame che della densità di un nucleo senza introdurre alle brevi distanze un taglio dell'energia di scambio.

Questo elegante risultato di Majorana fu tuttavia di non lunga durata. Quando a partire dalla seconda metà degli anni 40, con lo sviluppo degli acceleratori, si cominciò ad estendere le misure di sezione d'urto elastico differenziale protone-neutrone ad energie fra 50 e 100 MeV, si trovò che la distribuzione angolare, nel centro di massa, è simmetrica rispetto al piano equatoriale con massimi praticamente eguali a 0° e 180°.

Una simile distribuzione suggerisce che il potenziale che descrive l'interazione debba essere una mistura al 50% di forze alla Wigner (cioè forze ordinarie) e di forze di scambio alla Majorana ([10]). Ma un'interazione nucleare che contenga soltanto il 50% di forze alla Majorana non è piú sufficiente per dar luogo alla saturazione. Questa può essere ottenuta solo introducendo un nocciolo repulsivo alla brevi distanze, ossia facendo uso di un tipo di forze ben noto dal caso delle forze interatomiche nelle molecole e nei solidi, ma che molti fisici teorici degli anni trenta non gradivano per ragioni estetiche. Per esempio, Majorana nella parte iniziale del suo testo tedesco accenna a questa ovvia possibilità per ottenere la saturazione delle forze nucleari, ma aggiunge: «Eine solche Lösung des Problems ist aber von ästhetischen Standpunkt aus unbefriedigend, ... (una simile soluzione del problema è, però, insoddisfacente dal punto di vista estetico, ...)».

Questi suoi gusti, o forse pregiudizi, estetici appaiono anche in un altro punto di questo lavoro, ma questa volta si tratta di un fatto puramente formale. Quando Majorana, dopo aver scritto la sua espressione dell'energia di scambio, passa al confronto con quella di Heisenberg, premette di farlo «eliminando l'incomoda coordinata di ϱ-spin, ciò che è possibile se si riguardano, anche formalmente, i protoni e i neutroni come particelle differenti». Come tutti sanno, il ϱ-spin di Heisenberg era semplicemente quello che assai piú tardi fu chiamato lo spin isotopico del nucleone, la cui importanza cominciò a manifestarsi con la dimostrazione nel 1936 dell'indipendenza dalla carica delle forze nucleari ([11]) e il riconoscimento della sua importanza nella discussione delle proprietà dei nuclei fatta da Wigner nel 1937 ([12]). Heisenberg aveva introdotto questa variabile solo come un artificio formale e da un punto di vista rigorosamente metodologico Majorana, a quell'epoca, aveva ragione di voler evitare l'uso di questo nuovo algoritmo, ma la storia successiva ha mostrato che, come talvolta accade, la scelta di Heisenberg, forse per caso, o forse per intuizione geniale, era quella giusta.

La storia dello sviluppo successivo delle nostre idee sulle forze nucleari è troppo complessa e lunga per poter essere ricordata an- 17
che per sommi capi. È chiaro tuttavia che nel 1935 con l'idea di Yukawa ([13]) che le forze nucleari che si esercitano fra due nucleoni abbiano origine dallo scambio di un corpuscolo di massa intermedia detto mesone, il quale entra nella teoria come quanto del campo mesonico, si realizza quella che si potrebbe chiamare l'unificazione delle forze ordinarie, o di Wigner, con quelle di scambio, sia di Heisenberg che di Majorana, in quanto il presentarsi dell'uno o dell'altro caso è determinato solo dall'esistenza di vari stati di carica del messaggero scambiato. In questo quadro concettuale anche le forze di Wigner sono, in un certo senso, forze di scambio mediate da un messaggero di carica elettrica nulla.

Il risultato piú caratteristico di questa teoria, e in un certo senso quello piú generale e definitivo, è che il «range» o raggio d'azione delle forze è legato alla massa m del bosone scambiato dalla ben nota relazione di Yukawa

$$d=\frac{\hbar}{mc}. \tag{1}$$

L'applicazione di queste idee generali e universalmente accettate è semplice, e fu fatta fin dall'origine da Yukawa e da altri, solo nel caso di due nucleoni in moto relativo con velocità piccola rispetto a c, e che si trovano a grande distanza l'uno dall'altro. In queste condizioni infatti affinché la teoria dia risultati in soddisfacente accordo con i dati sperimentali è sufficiente prendere in considerazione lo scambio di un sol mesone π o pione, in quanto questo, fra tutti i mesoni osservati in natura, è quello dotato di energia di quiete piú piccola

$$m_\pi c^2=139.57 \text{ MeV},$$

e pertanto, per la relazione di Yukawa (1), il corrispondente contributo delle forze nucleari è quello di range piú lungo:

$$d_\pi=1.41\cdot 10^{-15}\text{m}.$$

Ma quando si va a distanze piú brevi le cose si complicano e un'interpretazione completa e definitiva dei risultati sperimentali non è ancora attuata.

18 La figura che vi mostro (fig. 1) è presa da un articolo di rassegna del 1985 di S. O. Bäckman, G. E. Brown e J. A. Niskanen intitolato «L'interazione fra due nucleoni e il problema dei molti corpi nel caso del nucleo» ([14]). Essa indica, in maniera schematica, quali sono i processi che determinano il potenziale dovuto al campo mesonico alle diverse distanze fra i due nucleoni interagenti. Alle grandi distanze il potenziale è determinato dallo scambio di un sol pione (bosone di massa piú piccola, pseudoscalare), mentre alle distanze intermedie si è costretti ad interpretare l'interazione osservata come dovuta allo scambio di mesoni scalari dotati di varie masse abbastanza elevate, consistenti in realtà in sistemi virtuali di due pioni accoppiati in stati con S (spin) $=0$ e T (spin isotopico) $=0$ che coinvolgono una o piú risonanze Δ.

Nell'approccio teorico a cui si riferisce questo articolo di rassegna, alle piccole distanze i messaggeri scambiati sono bosoni vettoriali, ma ormai da anni vi sono solidi argomenti per ritenere che già a distanze dell'ordine $(1\div 1.5)\times 10^{-15}$m $(=(1\div 1.5)$ fm) la struttura a quark dei nucleoni cominci a manifestarsi e che fra le particelle scambiate si debba cominciare a considerare i gluoni. Questi, come è noto, sono i messaggeri delle forze che si esercitano fra i quark, forze che ai grandi momenti trasferiti (distanze brevi) mostrano una riduzione della costante di accoppiamento (la cosiddetta libertà asintotica) che, in qualche modo, simula un effetto di taglio o cut-off classico, molto graduale ([15]).

Aggiungendo a questi modelli a quark il solo contributo dovuto allo scambio di un

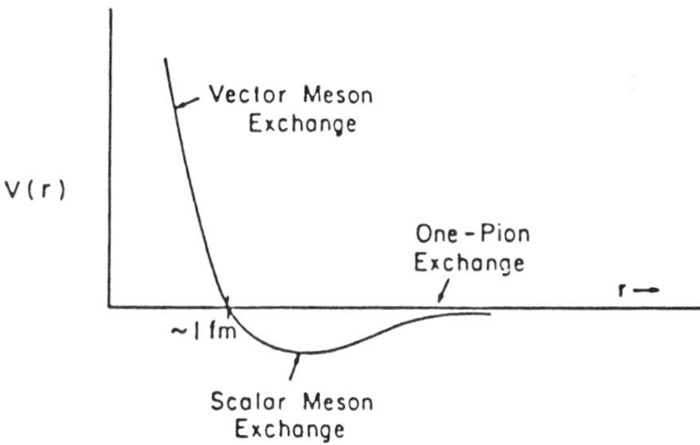

Fig. 1. – Schematizzazione del potenziale nucleone-nucleone ([14]).

pione, si ottengono dei potenziali locali, funzioni dell'energia, del tutto equivalenti ai potenziali fenomenologici come quelli di Reid ([16]) ma soprattutto quello di Parigi ([17]) per i canali S di singoletto e per i canali S e D di tripletto accoppiati, mentre per i canali P danno ancora solo risultati qualitativi.

Come è noto, tutti questi potenziali fenomenologici contengono termini sia centrali che tensoriali di scambio alla Heisenberg, alla Majorana e anche alla Bartlett (che operano solo sulle coordinate di spin) ([18]), anche se di solito vengono oggi presentati in forme tali che queste distinzioni non sono evidenti a vista.

D'altra parte si deve anche osservare che la fisica nucleare degli anni trenta non si estendeva oltre i 10 MeV di energia cinetica relativa e pertanto riguardava solo le grandi distanze, cioè solo la zona dove vale con ottima approssimazione il cosiddetto «one-pion exchange model».

Per concludere questa discussione del lavoro di Majorana sulle forze nucleari, vorrei anche ricordare che la sua traduzione in inglese è pubblicata alla fine del libro sulle forze nucleari del fisico inglese Brink([19]), insieme alla ben nota discussione, presieduta da Lord Rutherford alla Royal Society, subito dopo la scoperta del neutrone, ad un articolo di Niels Bohr, della stessa epoca, sui problemi della struttura nucleare, al I e III lavoro di Heisenberg sulle sue forze di scambio e loro applicazione, ad un lavoro di Wigner sul difetto di massa dell'elio e a pochi altri celebri lavori tutti relativi alla struttura del nucleo.

Passiamo ora al lavoro sulla teoria relativistica di particelle con momento intrinseco arbitrario [7], il quale precede di circa due anni il lavoro sulle forze nucleari che ho già discusso. Ho preferito invertire l'ordine in cui esamino questi due lavori perché il lavoro sulle forze nucleari, concettualmente, si ricollega ai lavori del periodo precedente sul legame chimico, mentre il lavoro sulle particelle di spin arbitrario costituisce la prima espressione di un diverso aspetto molto importante della struttura culturale e degli interessi scientifici di Majorana che, qualche tempo dopo, si manifesteranno pienamente nell'ultimo suo lavoro sulla teoria simmetrica degli elettroni e dei positoni.

Il lavoro sulla teoria relativistica delle particelle di spin arbitrario quando apparve, nel 1932, rimase praticamente sconosciuto per due ragioni. Esso fu pubblicato solo in italiano e inoltre trattava un problema prematuro sia rispetto agli interessi dominanti fra i fisici dei primi anni trenta sia rispetto alle conoscenze di risultati sperimentali relativi alle particelle elementari. Il lavoro, infatti, fu concepito, scritto e inviato da Majorana a *Il Nuovo Cimento*, e forse anche pubblicato da questo periodico([20]), prima che a Roma giungesse notizia della scoperta del positone annunciata da C. D. Anderson in un articolo pubblicato su *Science* e datato 1 settembre 1932([21]). Pertanto il lavoro di Majorana era destinato a rimanere, ed è effettivamente rimasto, fuori dalla linea principale di sviluppo della teoria delle particelle elementari. Ciò nonostante, come cercherò di mostrare, esso è molto interessante.

L'origine e lo scopo del lavoro di Majorana vengono chiariti nell'introduzione che si estende su poco piú di due facciate.

L'autore comincia con l'osservare che la teoria di Dirac, che descrive correttamente l'elettrone, fa uso di una funzione d'onda a quattro componenti delle quali, quando si passa a velocità sufficientemente piccole, due assumono valori trascurabili, mentre le altre due ubbidiscono, in prima approssimazione, all'equazione di Schrödinger. In modo analogo una particella di momento angolare intrinseco s arbitrario è descritta nell'approssimazione non relativistica della meccanica quantistica da un complesso di $2s+1$ funzioni d'onda che soddisfano separatamente l'equazione di Schrödinger.

Dopo alcune considerazioni generali sul passaggio dal caso relativistico a quello di particelle che si muovono con velocità $v \ll c$, Majorana osserva che «L'equazione d'onda, in assenza di campo, di una particella materiale deve avere secondo Dirac la forma

$$\left[\frac{W}{c} + (\boldsymbol{\alpha} \cdot \boldsymbol{p}) - \beta\, mc\right] \psi = 0» \qquad (2)$$

e che equazioni di questo tipo «presentano una difficoltà di principio» consistente nel fatto che per ogni valore del momento p «l'energia W può avere due valori che differiscono per il segno

$$W = \pm\sqrt{m^2c^4 + c^2p^2}»; \qquad (3)$$

e aggiunge subito l'osservazione conclusiva dell'introduzione: «L'indeterminazione del segno può essere in realtà superata, usando equazioni del tipo fondamentale (2), solo se la funzione d'onda ha infinite componenti che *non* si lasciano spezzare in tensori o spinori finiti».

19

Le ragioni per cui questo lavoro di Majorana è ancora oggi molto interessante sono emerse in modo particolare dalle discussioni che io ho avuto il piacere d'intrattenere circa 20 anni fa con il professor M. Fierz ed il professor D. M. Fradkin([1]) e dal lavoro che il professor Fradkin ha pubblicato su questo argomento([22]).

1) Il lavoro di Majorana costituisce il primo tentativo di costruire una teoria relativisticamente invariante di particelle con spin arbitrario, sia intero che semintero.

2) Esso rappresenta una teoria matematicamente corretta e contiene il primo riconoscimento, sviluppo e impiego delle rappresentazioni unitarie a infinite dimensioni del gruppo di Lorentz. Si tratta, anzi, della piú semplice di tali rappresentazioni, come asserisce correttamente Majorana.

3) La teoria è in qualche modo fuori dalla linea principale degli sviluppi successivi, essenzialmente per due ragioni; la prima è che Majorana si era posto fin da principio il problema di costruire una teoria lineare, relativisticamente invariante, nella quale tutti gli autovalori della massa fossero positivi. Questo punto di vista era giustificato nel momento in cui fu scritto il lavoro (estate 1932) in quanto la notizia della scoperta del positone da parte di C. D. Anderson non era ancora giunta a Roma. La seconda ragione è che Majorana, a differenza della maggior parte degli autori successivi, non richiede che la relazione di dispersione che lega l'energia all'impulso del corpuscolo

$$(W/c)^2 = p^2 + (mc)^2$$

sia soddisfatta, come equazione operatoriale, dalle componenti delle autofunzioni dell'equazione differenziale lineare della forma di Dirac che costituisce il punto di partenza della sua teoria.

4) Nel sistema di riferimento in cui la quantità di moto non è nulla la soluzione dell'equazione trovata da Majorana ha infinite componenti, individuate da due indici ψ_{jm}. L'indice m assume tutti i valori (interi o seminteri) compresi fra $\pm\ j$, e j tutti i valori non negativi appartenenti o alla serie dei numeri interi ($j = 0$, 1, 2, ...) o a quella dei numeri seminteri ($j = 1/2$, 3/2, 5/2, ...).

5) Il risultato precedente ci appare oggi di
20 difficile interpretazione fisica in quanto ciascun gruppo di $2j+1$ componenti ψ_{jm} (j fisso e $m = j, j-1, ..., -j$) sembra corrispondere ad una particella di diverso spin. Pertanto l'interpretazione piú naturale di ψ è quella di una funzione d'onda che rappresenta simultaneamente o tutti i possibili bosoni, o tutti i possibili fermioni. Se si cambia il sistema di riferimento le varie componenti si mischiano fra di loro. Tuttavia, nel sistema di riferimento del centro di massa, ove la maggior parte degli autori definiscono le proprietà dello spin di un corpuscolo, la funzione d'onda di Majorana corrisponde a (e si trasforma matematicamente come) quella di una particella di spin ben definito, il cui valore dipende dall'autofunzione prescelta.

6) La teoria di Majorana fornisce uno spettro di massa, ossia una formula che esprime la massa dei vari corpuscoli in funzione del loro spin j, facendo intervenire una sola costante assegnata m_0:

$$m_j = m_0/(j+1/2).$$

7) Majorana dimostra che al limite non relativistico ($v/c \to 0$), le $2s+1$ componenti ψ_{jm} che corrispondono all'autovalore della massa di quiete caratterizzato da $j = s$ (s fissato) sono le sole ad avere valori finiti, in quanto le componenti ψ_{s+1m} e ψ_{s-1m} sono di ordine v/c, le componenti ψ_{s+2m} e ψ_{s-2m} di ordine $(v/c)^2$, e cosí via.

Come ha osservato Fradkin, non sembra che Majorana fosse interessato all'idea di uno spettro di massa, e questo probabilmente si spiega con il fatto che nel 1932 si conosceva un numero troppo piccolo di particelle elementari. Egli inoltre non cercò di suggerire un'interpretazione della funzione d'onda ψ come quella a cui si è accennato al punto 5).

Majorana sottolineò invece il punto 7) osservando che, al limite non relativistico, l'autofunzione a infinite componenti ψ possiede solo $2s+1$ componenti non nulle, ciascuna delle quali separatamente soddisfa l'equazione di Schrödinger, in accordo con la teoria di Pauli dello spin dell'elettrone.

Questo lavoro di Majorana è rimasto praticamente sconosciuto. Lo stesso problema fu affrontato da Dirac, Klein, Petiau e Proca nel 1936 e Duffin nel 1938, ma nessuno di questi autori([23]) cita Majorana. I lavori successivi si ricollegano tutti a quello di Dirac e ignorano completamente, o quasi, il contributo di Majorana.

Le linee principali della presente teoria relativistica delle particelle di spin arbitrario sono state tracciate, sviluppate e applicate ad alcuni casi importanti in tre lavori fondamentali apparsi nel 1939. Il primo di questi è il lavoro di Fierz([24]) che ha trattato in modo completo la teoria relativistica di un corpuscolo di spin qualsivoglia in assenza di campi esterni, giungendo a stabilire, fra l'altro, la relazione che intercorre fra spin e statistica. Il secondo lavoro è quello di Fierz e Pauli([25]) che estende i risultati del lavoro precedente al caso di corpuscoli che si muovono in presenza di un campo elettromagnetico, trattando in modo completo i casi di spin 3/2 e 2, con particolare riguardo a corpuscoli di massa nulla. L'unica citazione del lavoro di Majorana è fatta incidentalmente nell'ultimo di questi tre lavori, ossia nel fondamentale articolo di Kemmer([26]) sul campo mesonico.

Un'ultima osservazione sugli aspetti matematici di questo lavoro. Gli stessi elementi di matrice dati da Majorana per le rappresentazioni a infinite dimensioni del gruppo di Lorentz sono presentate da Weyl, nel suo libro apparso per la prima volta nel 1928([27]), in connessione con le regole di selezione e il calcolo dell'intensità delle transizioni atomiche di dipolo trattate nell'ambito della meccanica quantistica di Schrödinger. Majorana conosceva certamente il libro di Weyl; ricordando il suo modo di lavorare([1]) io ritengo che egli sia pervenuto indipendentemente a questi risultati e che poi abbia riconosciuto che le stesse matrici intervenivano nel problema delle transizioni di dipolo.

Il problema delle rappresentazioni a infinite dimensioni del gruppo di Lorentz è stato discusso da Wigner nel 1938([28]), da Dirac nel 1945([29]) e di nuovo da Wigner nel 1948([30]). Dirac, a quanto pare, non era a conoscenza del lavoro di Majorana, mentre Wigner, nella sua classica analisi del gruppo di Poincaré, lo cita quasi incidentalmente. La discussione fatta da Corson nel suo libro([31]) è basata sulla trattazione di Wigner.

I numerosi e approfonditi studi delle rappresentazioni a infinite dimensioni del gruppo di Lorentz fatte successivamente da vari autori non fanno riferimento a Majorana. Esso non viene citato nemmeno da quegli autori([32]) che in anni recenti hanno ricostruito, discusso e generalizzato la teoria di Majorana, senza rendersene conto([33]). Fa eccezione il lavoro di Barut e Kleinert([34]) che fa specifico riferimento ai risultati di Majorana.

A parte il suo interesse matematico e storico, questo lavoro dimostra la sensibilità di Majorana a problemi fondamentali che egli cercava di affrontare in maniera autonoma con notevole anticipo sulla maggior parte dei fisici contemporanei. Il fatto poi che oggi si sappia che i nucleoni e i mesoni sono sistemi di quark, in diversi stati quantici, descritti in maniera assai soddisfacente dalla quanto-cromo-dinamica (QCD) e che le forze nucleari abbiano in realtà origine dalle forze che si esercitano fra i quark grazie allo scambio di gluoni, non può ridurre la nostra ammirazione per l'immaginazione di Ettore Majorana e per la sua abilità nel maneggiare l'equazione di Dirac([35]).

Veniamo ora all'ultimo lavoro di Ettore Majorana, quello intitolato «Teoria simmetrica dell'elettrone e del positrone» [9], lavoro che rimane come il suo piú importante contributo scientifico. Esso apparve solo in italiano, nel fascicolo di aprile 1937 de *Il Nuovo Cimento*. In tempi relativamente recenti esso è stato tradotto in inglese da Luciano Maiani e pubblicato in questa lingua sul periodico giapponese Soryushiron Kenkyu (1981) [9].

Questo lavoro comincia con l'osservazione che la teoria relativistica di Dirac, che ha portato alla previsione dell'esistenza del positone, poco dopo confermata dall'esperienza, s'impernia sull'equazione di Dirac che è completamente simmetrica rispetto al segno della carica del fermione considerato; ma che tale simmetria va in parte perduta nello sviluppo successivo della teoria che descrive il vuoto come una situazione in cui tutti gli stati di energia negativa sono occupati e tutti quelli di energia positiva liberi. L'eccitazione di un fermione da uno degli stati di energia negativa a uno di energia positiva lascia una lacuna dotata di energia positiva, che può venir interpretata come l'antifermione. In tal modo il processo di eccitazione di un fermione da uno stato di energia negativa a uno di energia positiva equivale alla creazione di una coppia fermione-antifermione. Questa impostazione asimmetrica porta come conseguenza anche la necessità di cancellare, *senza alcuna sana giustificazione di principio*, alcune costanti in finite dovute agli stati di energia negativa, come, per esempio, la densità di carica elettrica.

21

Come si vede, le preoccupazioni di Majorana, anche se manifestate in modo piú esplicito e circonstanziato, erano all'epoca in cui fu scritto il lavoro [9] ancora quelle che figurano nella parte introduttiva del suo lavoro sulla teoria delle particelle di spin arbitrario del 1932. Ed era a questi aspetti, e ad altri analoghi, delle equazioni della fisica quantistica e relativistica che si riferiva la frase che Ettore diceva ai colleghi, ma talvolta anche a parenti od amici che non appartenevano alla ristretta cerchia dei fisici: «La fisica è su una cattiva strada» o qualcosa di equivalente.

Tali inconvenienti sono evitati nella teoria proposta da Majorana in questo lavoro, in cui egli propone una nuova rappresentazione delle matrici di Dirac γ_μ (μ=1, 2, 3, 4), la quale ha le seguenti proprietà:

1) A differenza di ciò che accade nella rappresentazione originaria di Dirac, nella

rappresentazione di Majorana le quattro matrici γ_μ hanno le stesse proprietà di realità del quadrivettore $x_\mu \equiv \mathbf{r}$, ict; o, se si adottano coordinate spazio-temporali tutte reali, associate ad una metrica pseudoeuclidea, sono tutte quattro reali.

2) In questa rappresentazione l'equazione di Dirac relativa a un fermione libero è a coefficienti reali e pertanto le sue soluzioni si spezzano in una parte reale e una immaginaria, ciascuna delle quali soddisfa separatamente detta equazione. Ma ciascuna di queste soluzioni reali, proprio come conseguenza della sua realità, gode di due proprietà molto importanti: la prima è che essa dà origine a un quadrivettore corrente-carica elettrica sempre eguale a zero. Ne segue che le soluzioni reali dell'equazioni di Dirac debbono corrispondere a fermioni privi sia di carica elettrica che di momento magnetico. La seconda conseguenza della realità del campo fermionico ψ è che il corrispondente operatore di campo deve essere hermitiano, cosicché i suoi gradi di libertà sono ridotti alla metà e scompare la distinzione fra fermione e antifermione.

Majorana nel suo lavoro suggeriva che il neutrone o il neutrino, o entrambe queste particelle, fossero proprio corpuscoli di questo tipo ossia corpuscoli neutri che s'identificano con i corrispondenti anticorpuscoli.

3) Dall'esame dell'equazione di Dirac relativa ad un fermione posto in un campo elet-
22 tromagnetico, scritta nella rappresentazione di Majorana, segue che per rappresentare un corpuscolo carico basta prendere una combinazione ψ di due soluzioni reali. Il campo fermionico cosí definito dà ovviamente luogo ad un quadrivettore corrente-carica elettrica non identicamente nullo grazie ai termini d'interferenza fra i due campi reali: esso inoltre gode della proprietà nota per un campo scalare, che l'operatore di campo coniugato rispetto alla carica (ossia l'operatore che descrive un corpuscolo di carica opposta a quella del corpuscolo considerato) si ottiene applicando all'operatore ψ l'operatore di coniugazione hermitiana.

In realtà non si conosce nessun corpuscolo del tipo descritto al punto 2): in particolare si sa, da esperienze eseguite nel 1939([36]), che il neutrone ha un momento magnetico diverso da zero e, da altre eseguite nel 1956([37]), che l'antineutrone è un corpuscolo ben distinto dal neutrone.

Il problema di sapere se neutrini di Majorana, cioè corpuscoli caratterizzati dall'eguaglianza

$$\nu_{\mathrm{M}} = \bar{\nu}_{\mathrm{M}}, \tag{4}$$

esistano in natura, oppure non esistano affatto, non ha ancora ricevuto una risposta definitiva.

La questione è di grande importanza in quanto vi sono fenomeni che si possono osservare con i neutrini di Majorana (ν_{M}) diversi da quelli che si possono osservare con i neutrini di Dirac (ν_{D}). La ragione di tale differenza è chiara: mentre ai neutrini di Dirac si può attribuire un numero leptonico, per esempio, pari a +1 per il neutrino e a −1 per l'antineutrino, ciò non è possibile per i neutrini di Majorana a causa dell'eguaglianza (4). Ne segue che se esistono dei neutrini di Majorana cade la legge di conservazione del numero leptonico e, in assenza di questa legge restrittiva, il numero di tutti i possibili processi in cui intervengono leptoni aumenta notevolmente.

Majorana era cosciente che, per neutrini che soddisfano l'eguaglianza (4), la teoria di Fermi del decadimento beta([38]) doveva essere modificata. Nel suo lavoro egli fa un'osservazione a questo riguardo in connessione con il lavoro di Gian Carlo Wick([39]) che aveva esteso la teoria di Fermi al caso del decadimento con emissione di positoni osservato dai Joliot-Curie al principio del 1934. Majorana scrive infatti che, se i neutrini sono del tipo da lui stesso suggerito, «la teoria [del decadimento beta] può essere, ovviamente modificata in modo che l'emissione β, sia negativa che positiva, venga sempre accompagnata dall'emissione del neutrino» [9].

Tre mesi dopo la pubblicazione di questo bellissimo lavoro di Majorana, Giulio Racah pubblicò un lavoro([40]) in cui mostra che il postulare la simmetria fra particelle e antiparticelle dà luogo a modifiche della teoria del decadimento beta di Fermi e che l'aggiunta del postulato dell'identità fra neutrino e antineutrino porta direttamente e necessariamente alla teoria di Majorana. Egli fa anche osservare per primo il diverso comportamento del neutrino ν_{M} dal neutrino ν_{D}, non solo nel decadimento beta, ma anche nelle reazioni provocate da neutrini.

Gli argomenti sviluppati da Racah in questa seconda parte del suo lavoro, assai importanti nel 1937, richiesero tuttavia una revisione quando, circa venti anni dopo, si scoprí che i neutrini posseggono un'elicità ben definita.

La determinazione sperimentale della natura del neutrino (cioè se si comporta come un neutrino di Dirac o come un neutrino di Majorana) è ancora possibile ma richiede un approccio piú sottile.

Il metodo principale per rispondere a questo quesito nel caso del ν_e consiste nello studiare sperimentalmente il decadimento beta doppio, ossia la transizione di un nucleo radioattivo di numero atomico Z, con emissione di 2 elettroni, ad un nucleo di numero atomico Z' che differisce da Z per 2 unità.

Prima di accennare in maniera sintetica ai risultati di questi esperimenti vorrei ricordare che l'interesse per la teoria di Majorana si ravvivò notevolmente dopo la scoperta della non conservazione della parità([41]) ed il successo della teoria a due componenti per l'interpretazione di vaste classi di fenomeni([42]). Vari autori, in particolare McLennan e Case, mostrarono l'equivalenza di queste due teorie per neutrini di massa rigorosamente nulla([43]).

A priori esistono vari modi per descrivere in maniera relativistica un neutrino (cioè una particella di spin 1/2, carica e massa zero) dotato di due soli stati. Un modo è quello fornito dalla teoria a due componenti, basato sulla scoperta che il neutrino e l'antineutrino hanno sempre elicità ben definite ed opposte e pertanto possono essere descritti da uno spinore a 2 componenti (spinore di Weyl). Un'altra teoria è quella di Majorana che, come abbiamo già detto, viola la conservazione del numero leptonico.

Questo secondo approccio era in sostanza una teoria da rigettare prima della scoperta della non conservazione della parità. Ma una volta dimostrato che il neutrino ν_e osservato nel decadimento beta ha sempre elicità -1 e l'antineutrino $\bar{\nu}_e$ sempre l'elicità $+1$, diventa possibile identificare questi due oggetti con un ν_M rispettivamente con elicità -1 e $+1$. Questa proprietà, combinata con l'interazione $(V-A)$ che — contenendo solo correnti leptoniche sinistrorse — conserva il numero leptonico, è equivalente alla conservazione del numero leptonico.

Tuttavia la possibilità di ridurre il numero leptonico del neutrino alla sua elicità sussiste nella teoria di Majorana solo nel caso di massa del neutrino rigorosamente zero. Una piccola massa del neutrino ($m_\nu \ll m_e$) è però compatibile con la teoria di Majorana, ma non con la teoria a due componenti. Un neutrino con $m_\nu \neq 0$ può essere solo o un neutrino di Majorana, o un neutrino di Dirac a quattro componenti. In entrambi i casi il valore della massa si presenterebbe come il parametro che rompe l'elicità del neutrino, la quale non può piú essere rigorosamente eguale a -1 per il neutrino e $+1$ per l'antineutrino, salvo che al limite $v \to c$.

Nel primo caso (ν_M) ne seguirebbe automaticamente una (piccola) deviazione dalla conservazione del numero leptonico, mentre nel secondo (ν_D), dato che l'elicità è indipendente dal numero leptonico, questa legge si salva anche in queste condizioni. Una deviazione dall'elicità perfetta dovuta a una piccola massa può essere messa in evidenza solo con esperienze piuttosto raffinate fatte oggi soprattutto sui ν_e da molti gruppi nel mondo, in particolare dal gruppo dell'Università di Milano guidato da Ettore Fiorini.

Oltre a questo tipo di rottura dell'elicità perfetta del neutrino, indicata talvolta come *rottura implicita*, potrebbe esistere anche una *rottura esplicita*, avente origine dal fatto che, in aggiunta all'interazione dovuta a correnti leptoniche sinistrorse (interazione $(V-A)$ pura), vi sia una frazione, magari molto piccola, d'interazione dovuta a correnti destrorse.

Questi problemi possono venir affrontati dal punto di vista sperimentale in maniera approfondita solo nel caso del ν_e facendo uno studio accurato del decadimento β doppio. In particolare misurando lo spettro somma 23
E_1+E_2 delle energie dei due elettroni emessi. Dalla relazione generale presentata da Ettore Fiorini al 8th International Workshop on Grand Unification tenuta a Siracuse (N.Y.) nell'aprile 1987([44]) emerge che gli esperimenti eseguiti da vari gruppi, compreso quello dell'Università di Milano, sulla transizione $(0^+ \to 0^+)$ fra gli stati fondamentali del germanio e selenio 76

$$^{76}_{32}\mathrm{Ge} \to ^{76}_{34}\mathrm{Se} + 2e^-, \qquad (5)$$

sono quelli che fino ad oggi hanno permesso di stabilire i limiti piú bassi per i parametri relativi alla violazione del numero leptonico.

L'osservazione dello spettro E_1+E_2 permette di distinguere tre casi che differiscono in quanto nel decadimento (5) i due elettroni sono emessi simultaneamente a i) 2 $\bar{\nu}_e$ di Dirac o Weil, cosicché lo spettro è continuo (decadimento a 4 corpi + il nucleo residuo) con un massimo a circa 1/3 di $\varepsilon_m = (E_1 + E_2)_{\max}$, il valore di ε_m è dato dalla differenza di energia fra gli stati nucleari della transizione (5); ii) nessun altro corpuscolo, cosicché lo spettro si riduce ad una riga di energia ε_m (decadimento a 2 corpi + il nucleo di rinculo: teoria di Majorana); iii) un bosone di massa e carica nulle (detto majorone([45]); decadimento a 3 corpi + il nucleo residuo) cosicché lo spettro è continuo con un massimo a circa 4/5 di ε_m.

Fino ad ora nessun gruppo sperimentale ha osservato processi del tipo previsto dalla teoria di Majorana. Essi invece concordano sui seguenti limiti superiori per valori dei parametri che rompono la conservazione del numero leptonico:

a) per la massa

$$\langle m_{\nu_e} \rangle < (1 \div 2) \text{ eV}, \tag{6a}$$

in cui, si noti, figura il valor medio ($\langle\ \rangle$) della massa del neutrino per tener conto del fatto che, fra i molti approcci teorici, ve ne è uno almeno in cui si postula che il ν_e possa esistere in due stati corrispondenti ai due diversi autovalori di *CP*;

b) per il valor medio della frazione η dell'interazione dovuta a correnti leptoniche destrorse ($V+A$), rispetto a quelle sinistrorse, certamente dominanti,

$$\langle \eta_{\nu_e} \rangle < 10^{-6} \div 10^{-8} \tag{6b}$$

24 per i processi $0^+ \rightarrow 0^+$ previsti in base alla teoria di Majorana; l'ampio intervallo di valori che figura a destra della (6b) ha origine esclusivamente dall'impiego di diversi modelli teorici per il calcolo dell'elemento di matrice della transizione nucleare $0^+ \rightarrow 0^+$ nel processo $^{76}\text{Ge} \rightarrow {}^{76}\text{Se}$.

Si noti che i limiti superiori (6) per i valori di $\langle m_{\nu_e} \rangle$ ed $\langle \eta_{\nu_e} \rangle$ sono inferiori l'uno per circa un ordine di grandezza e l'altro per circa quattro ordini di grandezza rispetto ai corrispondenti valori di quattro anni fa. Esperienze ancora piú accurate sono previste in un avvenire non lontano.

Si può quindi concludere che, nel caso del ν_e, non vi è alcuna evidenza che esso possa essere un neutrino di Majorana. Per il ν_μ non esistono metodi sperimentali che permettano di stabilire limiti cosí bassi per le eventuali deviazioni dalla conservazione del numero leptonico. La situazione è ancora assai piú lontana da un chiarimento definitivo per quanto riguarda il ν_τ. Sia l'uno che l'altro di questi neutrini potrebbero avere una massa molto maggiore del limite superiore (6a) stabilito oggi per il ν_e.

Come è noto, il problema della massa del neutrino è di grandissimo interesse anche per l'interpretazione di un certo numero di fatti astrofisici e per le oscillazioni fra i diversi tipi di neutrini oggi noti (ν_e, ν_μ, ν_τ) suggerite già nel 1958 da B. Pontecorvo([46]), ma che fino ad ora nessuno è riuscito ad osservare.

A conclusione di questo rapido ricordo dell'opera scientifica di Ettore Majorana vorrei accennare ad una proprietà molto interessante degli spinori di Majorana γ_μ messa in luce recentemente da Budinich ed esplorata piuttosto in dettaglio da Budinich stesso e vari collaboratori. Si tratta di vari lavori, non ancora pubblicati a stampa, ma circolati come note interne della SISSA.

Il matematico francese Élie Cartan (1869-1951), a cui si deve la scoperta degli spinori (1913), li concepí e definí come vettori equivalenti a piani isotropi in spazi complessi ordinari, cioè piani i cui vettori sono di lunghezza nulla e a due a due ortogonali fra loro.

In realtà, vettori di modulo nullo erano già stati considerati da lungo tempo (da Weierstrass, Bianchi ed Eisenhart), all'origine di superfici minime nello spazio ordinario e nello spazio-tempo.

Budinich ha mostrato([47,48]) che seguendo la concezione di Cartan degli spinori è facile generare superfici minime in spazi ordinari e nello spazio-tempo partendo da spinori complessi. Sostituendo, nella stessa costruzione nello spazio-tempo, agli spinori complessi di Dirac gli spinori reali di Majorana, al posto di superfici minime Budinich ottiene le corde relativistiche classiche([47]) che ubbidiscono alla equazione del moto

$$\frac{\partial^2 x_\mu}{\partial \sigma^2} - \frac{\partial^2 x_\mu}{\partial \tau^2} = 0, \tag{7}$$

dove le x_μ sono le coordinate dello spazio-tempo, $d\sigma$ è l'arco infinitesimo di corda e $d\tau$ l'intervallo di tempo proprio.

A proposito di questo risultato Budinich fa le seguenti considerazioni:

All'ovvio carattere geometrico elementare

degli spinori di Majorana, corrispondono in natura gli oggetti fisici piú elementari che si conoscano, e cioè

1) i neutrini privi di massa e di carica elettrica, rappresentabili a mezzo di spinori a due componenti di Weyl[49],
2) le corde relativistiche da essi generate.

Secondo Budinich, questo risultato può, in qualche modo, essere considerato come un ulteriore argomento a favore del carattere fondamentale che viene oggi attribuito alle corde relativistiche. Sempre secondo Budinich si può anche mostrare che l'azione delle equazioni di Eulero-Lagrange (7) per le corde è spinorialmente equivalente alla lagrangiana (quadrilineare nei fermioni chirali) di Fermi delle interazioni deboli[50].

Questo interessante problema, tutt'ora in corso di studio da parte del gruppo di Trieste, sta a mostrare come nell'opera scientifica di Ettore Majorana ci possano essere germi di connessioni profonde che ancora oggi non sembrano essere state scoperte o completamente comprese.

Bibliografia

(1) Per un'esposizione piú ampia, anche se in parte sorpassata per quanto riguarda il commento all'opera scientifica di E. Majorana, vedi E. Amaldi: *La vita e l'opera di Ettore Majorana*, Accademia Nazionale dei Lincei (1966) e la traduzione in inglese: *Ettore Majorana, man and scientist* in *Strong and Weak Interactions, Present Problems, International School of Physics Ettore Majorana, Erice, June 19-July 4, 1966*, edited by A. Zichichi (Academic Press, New York and London, 1966), p. 10. Presentazioni parziali e abbreviate fatte dallo stesso autore sono: *Ricordo di Ettore Majorana, G. Fis.*, **9**, 300 (1968); *L'opera scientifica di Ettore Majorana, Physis*, **10**, 173 (1968).

(2) L. H. Thomas: *Calculation of atomic fields, Proc. Cambridge Philos. Soc.*, **33**, 542 (1927); E. Fermi: *Un metodo statistico per la determinazione di alcune proprietà dell'atomo, Rend. Lincei*, **6**, 602 (1927).

(3) E. Fermi: *Über die Anwendung der Statistichen Methode auf die Probleme des Atombau*, in *Quantentheorie und Chemie, Leipziger Vörtrage*, edited by H. Falkenhagen (Leipzig, 1928), p. 95.

(4) Ta-You Wu: *Energy states of doubly excited helium, Phys. Rev.*, **46**, 239 (1934). Il Dr. Wu era un ben noto allievo e collaboratore di S. Goudsmidt.

(5) W. Heitler and F. London: *Wechselwirkung neutraler Atom und homöpolare Bindung nach der Quantenmechanik, Z. Phys.*, **44**, 455 (1927).

(6) C. G. Darwin: *Free motion in the wave mechanics, Proc. R. Soc. London, Ser. A*, **117**, 258 (1928); A. Landè: *Polarization von Materiewellen, Naturwissenschaften*, **17**, 634 (1929).

(7) I. I. Rabi: *Space quantization in a gyrating magnetic field, Phys. Rev.*, **51**, 652 (1937); F. Bloch and I. I. Rabi: *Atoms in variable magnetic fields, Rev. Mod. Phys.*, **17**, 237 (1945).

(8) Che i neutroni, al pari dei protoni, fossero tra i costituenti dei nuclei era stato detto dallo stesso Chadwick e da molti altri. Il primo a dire che queste due particelle fossero i *soli* costituenti del nucleo fu probabilmente D. Iwanenko: *The neutron hypothesis, Nature (London)*, **129**, 798 (1932), datato 21 aprile 1932. Non sono sicuro che Majorana sia giunto indipendentemente a questa stessa conclusione, ma lo ritengo probabile.

(9) W. Heisenberg: *Über den Bau der Atomkern I, Z. Phys.*, **77**, 1 (1932), datato 7 giugno 1932; *Über den Bau der Atomkern II, Z. Phys.*, **78**, 156 (1932), datato 30 luglio 1932; *Über den Bau der Atomkern III, Z. Phys.*, **80**, 587 (1933), datato 22 dicembre 1933.

(10) Questa osservazione sta alla base, fra l'altro, del cosiddetto potenziale di Serber (*Nuclear reactions at high energy, Phys. Rev.*, **72**, 1114 (1947), che è uno dei piú semplici ed è stato spesso usato per calcoli orientativi.

(11) G. Breit, E. U. Condon and R. D. Present: *Theory of scattering of protons by protons, Phys. Rev.*, **50**, 825 (1936); B. Cassen and E. U. Condon: *On nuclear forces, Phys. Rev.*, **50**, 846 (1936); G. Breit and E. Feenberg: *The possibility of the same form of specific interaction for all nuclear particles, Phys. Rev.*, **50**, 850 (1936).

(12) E. Wigner: *On the consequences of the symmetry of the nuclear Hamiltonian on the spectroscopy of nuclei, Phys. Rev.*, **51**, 106 (1937).

(13) H. Yukawa: *On the interaction of elementary particles I, Phys. Math. Soc. Jpn.*, **17**, 48 (1935).

(14) S. O. Bäckman, G. E. Brown and J. A. Niskanen: *The nucleon-nucleon interaction and the nuclear many-body problem, Phys. Rep.*, **124**, 1 (1985). La stessa figura si trova anche a pag. 7 del libro G. E. Brown and A. D. Jackson: *The Nucleon-Nucleon Interaction* (North-Holland Publishing Company, Amsterdam, 1976).

(15) Vedi, per esempio, R. Mignani and D. Prosperi: *Constituent Quark Models for Hadrons and Nuclei*, lectures given at the First International Course on Condensed Matter, International Center of Physics, Bogotà (Columbia) July 7-18, 1986 (in corso di stampa).

(16) R. V. Reid: *Logical phenomenological nucleon-nucleon potential, Ann. Phys. (N.Y.)*, **50**, 411 (1968).

(17) Fra le piú recenti pubblicazioni relative al cosiddetto potenziale di Parigi, ove è data anche la bibliografia di articoli precedenti sullo stesso argomento, vedi M. La Combe, B. Loiseau, J. M. Richard and R. Vinh Mau: *Parametrization of the Paris N-N potential, Phys. Rev. C*, **21**, 861 (1980).

(18) J. H. Bartlett, jr: *Exchange forces and the structure of the nucleus, Phys. Rev.*, **49**, 102 (1936).

(19) D. M. Brink: *Nuclear Forces* (Pergamon Press, Oxford, 1965).

(20) Purtroppo nel volume 9 de *Il Nuovo Cimento* non ci sono informazioni circa la data in cui la Redazione ricevette il manoscritto del lavoro [7] né sul mese a cui si riferisce il fascicolo in cui il lavoro apparve.

(21) C. D. Anderson: *The apparent existence of easily deflectable positives, Science*, **76**, 238 (1932), datato 1 settembre 1932; oltre a questo breve articolo, Anderson spedí successivamente un articolo piú esteso alla *Physical Review* che lo ricevette il 28 febbraio 1933 ma apparve nel fascicolo di marzo dello stesso anno (*Positive electron, Phys. Rev.*, **43**, 491 (1933).

(22) D. M. Fradkin: *Comments on a paper by Majorana concerning elementary particles, Am. J. Phys.*, **34**, 314 (1966).

(23) P. A. M. Dirac: *Relativistic wave equation, Proc. R. Soc. London, Ser. A*, **155**, 447 (1936); O. Klein: *Eine Veralgemeinerung der Diracschen relativistischen Wellengleichung, Ark. Mat. Astr. Fis. A*, **25**, n. 15, 1 (1936); G. Petieau: *Contribution à la théorie des équations d'ondes corpuscolaires, Mem. Acad. Soc. R. Belg.*, **16**, n. 2, 1 (1936); A. Proca: *Sur les équations fondamentales des particules élémentaires, C.R. Acad. Sci. Paris*, **202**, 1490 (1936); *Sur la théorie ondulatoire des électrons positifs et negatifs, J. Phys. Radium*, **7**, 347 (1936); *Théorie non relativiste des particules a spin entier, J. Phys. Radium*, **9**, 61 (1936); R. I. Duffin: *On the characteristic matrixes of covariants systems, Phys. Rev.*, **54**, 1114 (1938).

(24) M. Fierz: *Über die relativistiche Theorie kräftfreier Teilchen mit bebiebigen Spin, Helv. Phys. Acta*, **12**, 3 (1939).

(25) M. Fierz and W. Pauli: *On relativistic wave equations for particles of arbitrary spin in an electromagnetic field, Proc. R. Soc. London, Ser. A*, **173**, 211 (1939).

(26) N. Kemmer: *The particle aspect of meson theory, Proc. R. Soc. London, Ser. A*, **173**, 91 (1939).

(27) H. Weyl: *Gruppentheorie und Quantummechanik* (Verlag von S. Hirzel, Leipzig, 1928). p. 160 della 1ª edizione.

(28) E. P. Wigner: *On unitary representations of the inhomogeneous Lorentz group, Ann. Math., Princeton 2nd series*, **40**, 149 (1939).

(29) P. A. M. Dirac: *Unitary representations of the Lorentz group, Proc. R. Soc. London, Ser. A*, **183**, 284 (1945).

(30) E. P. Wigner: *Relativistische Wellengleichungen, Z. Phys.*, **124**, 665 (1947).

(31) E. M. Corson: *Introduction to Tensors, Spinors and Relativistic Wave-Equations* (Blackie and Sons Ltd., Glasgow, 1953), cap. V, p. 140.

25

(32) I. M. GEL'FAND and M. A. NAIMARK: *Unitary representations of the Lorentz group, J. Phys. (URSS)*, **10**, 93 (1946); HARISH-CHANDRA: *Infinite irreducible representations of the Lorentz group, Proc. R. Soc. London, Ser. A*, **189**, 372 (1947); V. BARGMANN: *Irreducible unitary representations of the Lorentz group, Ann. Math., Princeton 2nd series*, **48**, 568 (1947).

(33) I. M. GEL'FAND and A. M. YAGLOM: *General relativistic invariant equations and infinite dimensional representations of the Lorentz group, Ž. Eksp. Teor. Fiz.*, **18**, 703 (1948); traduzione inglese: TT-345 National Research Council of Canada (1953); V. L. GINZBURG: *On relativistic wave equations with a mass spectrum, Acta Phys. Pol.*, **15**, 163 (1956); I. M. GEL'FAND, R. A. MINLOS and Z. YA. SHAPIRO: *Representations of the Rotations and Lorentz Group* (Pergamon Press, New York, N.Y. 1963); G. FELDMAN and P. T. MATTHEUS: *Multimass field, spin and statistics, Phys. Rev.*, **154**, 1241 (1967).

(34) A. O. BARUT and H. KLEINERT: *Current operators and Majorana equations for the hydrogen atom from dynamic groups, Phys. Rev.*, **157**, 1180 (1967).

(35) Nota aggiunta subito dopo il LXXIII Congresso della SIF. In una conversazione che abbiamo avuto alla fine della mia presentazione di questa relazione al Congresso della SIF a Castel dell'Ovo in Napoli, Federico Capasso ha richiamato alla mia memoria l'esistenza di lavori riguardanti le connessioni fra questo lavoro di Majorana e la fenomenologia nota come poli e traiettorie di Regge. E. Recami, a cui mi rivolsi subito per avere informazioni piú precise in proposito, mi ha fatto conoscere il dr. Enrico Giannetto, dell'Università di Catania, come la persona, a sua conoscenza, piú informata su questo tema. Qualche giorno dopo il dr. Giannetto, di passaggio a Roma, mi ha brevemente illustrato una serie di articoli su questo argomento. Si vedano, per esempio, gli articoli di Y. NAMBU, C. FRONSDAL e altri, nel volume *Elementary Particle Theory: Relativistic Groups and Analiticity*, edited by N. SVARTHOLM, *Proc. Engl. Nobel Symposium, May 19-25, 1968* (Aspernågarden, Almquist and Wixell, Stoccolma, 1968) e gli articoli di A. O. BARUT, I. T. TODOROV e altri nei *Proceedings of the 1967 International Conference on Particles and Fields*, edited by C. R. HAGEN, G. GURALNIK and V. A. MATHUR (Interscience, New York, N.Y., 1968).

(36) L. W. ALVAREZ and F. BLOCH: *A quantitative determination of the neutron moment in absolute nuclear magnetons, Phys. Rev.*, **57**, 111 (1960).

(37) B. CORK, G. R. LAMBERTSON, O. P. PICCIONI and W. A. WENZEL: *Antineutrons produced from antiprotons in charge exchange collision, Phys. Rev.*, **104**, 1193 (1956).

(38) E. FERMI: *Tentativo di una teoria della emissione dei raggi beta, Ric. Sc.*, **4**(2), 491 (1933); *Versuch einer Theorie der β-Strahlen I, Z. Phys.*, **88**, 161 (1934).

(39) G. C. WICK: *Gli elementi radioattivi di F. Joliot e I. Curie, Rend. Lincei*, **19**, 319 (1934).

(40) G. RACAH: *Sulla simmetria tra particelle e antiparticelle, Nuovo Cimento*, **14**, 322 (1937).

26 (41) T. D. LEE and C. N. YANG: *Elementary Particles and Weak Interactions*, BNL 443 U.S., Department of Commerce Office of Technical Sciences, Washington, D.C., 1957.

(42) L. LANDAU: *On the conservation laws of weak interactions, Nucl. Phys.*, **3**, 127 (1957); A. SALAM: *On parity conservation and neutrino mass, Nuovo Cimento*, **5**, 299 (1957); T. D. LEE and C. N. YANG: *Parity non-conservation and a two component theory of the neutrino, Phys. Rev.*, **105**, 1671 (1957).

(43) J. SERPE: *Sur la théorie abrégée des particules de spin 1/2, Physica*, **18**, 295 (1952); L. A. RADICATI and B. TOUSCHEK: *On the equivalence theorem for the massless neutrino, Nuovo Cimento*, **5**, 1623 (1957); J. MCLENNAN: *Parity non conservation and the theory of the neutrino, Phys. Rev.*, **106**, 821 (1957); K. CASE: *Reformulation of the Majorana theory, Phys. Rev.*, **107**, 307 (1957).

(44) E. FIORINI: *The Present and Future of Double Beta Decay*, discorso generale presentato alla *Eight Workshop on Grand Unification, Syracuse, 16-18 April, 1987*.

(45) G. GELMINI and M. RONCADELLI: *Left-handed neutrino mass scale and spontaneously broken lepton number, Phys. Lett. B*, **99**, 411 (1981); H. M. GEORGI, S. L. GLASHOW and S. NUSINOV: *Unconventional model of neutrino masses, Nucl. Phys. B*, **193**, 297 (1981).

(46) B. PONTECORVO: *Inverse beta processes and non-conservation of lepton charge, Sov. Phys. JEPT*, **34**, 172 (1958).

(47) P. BUDINICH: *On the Possible Role of Cartan Spinors in Physics*, SISSA 60/86/E.P.

(48) P. BUDINICH, L. DOBROVOSKI and P. FURLAN: *Minimal Surfaces and Strings from Spinors: a Realization of the Cartan Programme*, SISSA 59/86/E.P.

(49) R. E. MARSHAK, RIAZUDIN and C. P. RESAN: *Theory of Weak Interactions* (Interscience, New York, N.Y., 1969), p. 71.

(50) P. BUDINICH and M. RIGOLI: in stampa.

Note scientifiche di Ettore Majorana

[1] *Sullo sdoppiamento dei termini Röntgen e ottici a causa dell'elettrone rotante e sulla intensità delle righe del cesio*, in collaborazione con G. GENTILE JUNIOR, *Rend. Acc. Lincei*, **8**, 229 (1928).

[2] *Sulla formazione dello ione molecolare di He, Nuovo Cimento*, **8**, 22 (1931).

[3] *I presunti termini anomali dell'elio, Nuovo Cimento*, **8**, 78 (1931).

[4] *Reazione pseudopolare fra atomi di idrogeno, Rend. Acc. Lincei*, **13**, 58 (1931).

[5] *Teoria dei tripletti P' incompleti, Nuovo Cimento*, **8**, 107 (1931).

[6] *Atomi orientati in campo magnetico variabile, Nuovo Cimento*, **9**, 43 (1932).

[7] *Teoria relativistica di particelle con momento intrinseco arbitrario, Nuovo Cimento*, **9**, 335 (1932).

[8*a*] *Über die Kerntheorie, Z. Phys.*, **82**, 137 (1933).

[8*b*] *Sulla teoria dei nuclei, Ric. Sci.*, **4**(1), 559 (1933).

[9] *Teoria simmetrica dell'elettrone e del positrone, Nuovo Cimento*, **5**, 171 (1937). Per la traduzione in inglese, fatta da L. MAIANI, vedi *Soryushiron Kenkiu* (giugno 1981), p. 149.

ETTORE MAIORANA, FIFTY YEARS AFTER HIS DISAPPEARANCE

Edoardo Amaldi
Department of Physics
University of Rome "La Sapienza"

1. - Fifty years have passed since the disappearance of Ettore Majorana and an overview of his scientific work is now in order.(1) I would like to remind those who are not acquainted with his work, that he published nine papers in all. The first, sent to the Accademia dei Lincei by O. M. Corbino, director of the Physics Institute in Via Panisperna at the time, was received on July 24, 1928. The last appeared in *Il Nuovo Cimento* in April 1937, that is, a year before he disappeared.

Majorana's works can be divided into two periods: the first includes the first six papers [1-6] which refer to problems of atomic and molecular physics; the second includes only 3 papers of which one [8] deals with the forces between protons and neutrons in a nucleus [8], and the other two ([7] and [9]) are quantistic and relativistic treatises on subnuclear particles.

As is well known, Ettore Majorana's reputation is based mainly on these last three works, but the first six are also of great interest in terms of the subjects broached and the approaches adopted.

2. - Let's start with the works belonging to the first period. They can be further divided into three groups: the first is composed of three papers ([1], [3] and [5]) dealing with problems of atomic spectroscopy; the second consists of two papers ([2] and [4]) on aspects of chemical bonds that were almost unknown at that time. The third group is made up of only one paper [6], focussed on the problem of non-adiabatic spin-flip in a polarized atom beam.

Without going into a detailed examination of the three papers on atomic spectroscopy, I would like to recall that the first paper[1], written together with Giovanni Gentile, jr., presents the results of calculations of the splitting caused by electron spin of some Röntgen terms (the *3M* terms of gadolinium (Z=64) and uranium (Z=92)) and optic terms (the *P* terms of caesium (Z=55)) and the intensity of the lines of caesium. This is done by introducing the wave functions of the individual electronic states calculated using the Thomas-Fermi statistical method(2) into the expressions deduced using the perturbation theory. This work was within the framework of the Institute's activity at that time. In fact, most of us, along with Fermi, were applying the Thomas-Fermi statistical method to various problems of atomic physics.(3)

The other two papers in atomic spectroscopy are far more original and interesting in terms of both choice of subject--very advanced for the time--and methodology. Both papers deal with *doubly excited* (or "accented" as they said at the time) atomic states, that is, states with two excited electrons in levels that energetically enable processes of spontaneous self-ionization or rather, transitions to final states consisting of one monovalent positive ion and one free electron (Auger effect).

The most striking feature of these two papers is the modernness of Majorana's approach. He bases his discussion entirely on the symmetry properties of the different states obtained by combining two orbitals, with respect to electron spin exchange (singlet or triplet states), spatial rotation (quantum orbital number), axial rotation (magnetic quantum) and reflections in the center of force (parity).

The first of these papers [3] deals with two extreme ultraviolet lines (λ=320.4 and 357.5 Å) that had just been discovered in the helium spectrum and could not be interpreted as combinations of known terms. The same lines can be observed in the solar corona and as such are also of great astrophysical interest.

The man who discovered them (P. G. Kruger) had suggested interpreting the lines as the result of the transitions of normal terms $1s2p^3P_{0,1,2}$ and $1s2s^1S_0$ to the doubly excited terms $2p2p^3P_{0,1,2}$ and $2s2s^1S_0$. The last two terms are much higher than helium's normal limit, making spontaneous ionization (or the Auger effect) possible. Using perturbation calculations taken to the 2nd order and the symmetry properties of the individual states mentioned above, Majorana confirmed Kruger's interpretation for the first of these lines but ruled it out for the second on the grounds of energy considerations.

In a recent (September 19, 1987) discussion on these works on spectroscopy, Ugo Fano pointed out that this paper by Majorana is often ignored in international literature, while a paper by Wu which appeared three years later is frequently cited. In that paper, Wu uses a modified form of the variational method, in which two "screening constants, one for each of the two electrons" are introduced as adjustable parameters, to arrive at the conclusion that the first of the two lines is due to the $1s2p^3P_0$-$2p2p^3P$ transition.(4)

The third paper on atomic spectroscopy [5] deals with incomplete doubly excited triplets, that is, doubly excited triplets that are above the limit of normal series but are stable if relativistic corrections are ignored. The triplets nevertheless take on a slight instability produced by the electrons' intrinsic magnetic momentum and this instability can, in particular circumstances, become so intense that it causes the disappearance of one (or more) of the components of a doubly excited triplet, which then appears to be "incomplete". Majorana's calculations on Zn, Cd and Hg atoms, done again applying properties of symmetry of states, are in full agreement with corresponding experimental results.

Turning to the two papers on chemical bonds, it is interesting to note that it was through this work that Ettore Majorana acquired the command of the chemical bonding theory that was to be of such great service to him in his later research activity. His profound knowledge of the mechanism of valence electron exchange, which provided the basis for the quantum theory of homopolar chemical bonding, first set down by Heitler and London in 1927,(5) later constituted the starting point for his hypothesis that nuclear forces are exchange forces.

The first of these two papers [2] dates back to 1931 and concerns the formation of the molecular helium ion. Majorana starts out from spectroscopic observations which had only shortly before revealed bands attributed to the molecular ion. He then turns to the problem of the chemical reaction leading to its formation, and demonstrates that it is feasible from an energetic point of view. Using a first approximation calculation, Majorana finds the equilibrium distance between the two nuclei of the He^{+}_{2} ion, which turns out to be in good agreement with the value measured experimentally and then assesses the energy of the system (still not measured at that time) and the oscillation frequency of the ionized molecule, which is close to the value found experimentally.

At the end of the paper, Majorana thanks Fermi for his generous advice and help and Gentile for the interest he took in the work.

The other paper on chemical bonding [4] is a detailed study of the characteristics of an even anomalous singlet term (known as term X) which had been discovered shortly before in the spectrum of the hydrogen molecule and was interpreted as owing to a state in which two electrons are excited. But as Majorana points out, the term cannot be considered as one with two excited electrons described in terms of the quantum numbers of the individual electrons.

He writes that while this description "is handy for the numbering and identification of those symmetrical characteristics that are not disturbed by interaction [among electrons], it does not, in itself, make it possible to draw reliable conclusions about the real shape of the eigenfunctions; in this case, things are quite different from the case of central fields, where it is generally possible to abstract from the interdependence of electron movements (polarization) without losing sight of the essential".

The term of the hydrogen molecule in question may be described very schematically as the superposition of two states: one deriving from the reaction between two neutral hydrogen atoms (H + H) already dealt with by Heitler and London, and the other deriving from a new process of formation of the same molecule in which two hydrogen ions, one positive and the other negative (H^{+} + H^{-}), bond together.

Majorana calls the chemical bond that forms in this case "pseudopolar", in that it is a bond between two opposite ions, but because of the components' equality, the electric moment changes sign with the frequency of exchange, which is high, and is therefore not observable.

Starting out from an initial state of superposition of the states relative to these two processes, and considering only even states (as odd ones do not produce stable states), Majorana finds only two final states that are bonded: the fundamental state of the H_2 molecule and the anomalous state. The former derives mainly from the union of H + H, while the latter derives primarily from the pseudopolar reaction H^{+} + H^{-}.

The last paper from the period dedicated to problems of atomic and molecular physics concerns calculation of the probability of spin-flip of atoms in a ray of polarized vapour when this is moved in a rapidly variable magnetic field [6].

If the magnetic field changes direction slowly, the orientation of the atom follows the direction of the field adiabatically. But what happens if the change in field direction is not adiabatic? In particular, what happens if the atom passes close to a point in which the magnetic field is zero?

The problem had been discussed from the early times of quantum mechanics by both Darwin and Landé;(6) and Stern, who had performed the famous spatial quantization experiments with Gerlach, had set out to verify the probability of transition in the event of non-adiabatic change.

In this paper, Majorana shows that the global effect of a variable magnetic field *H(t)* on a particle with angular momentum J=1/2 and a given component m along the z axis may be described as an abrupt rotation by an angle α of the total angular momentum.

This angle is obtained by solving the equations of motion (both classic and quantistic) and its most important property is that it depends only on the gyromagnetic relations and the magnetic field and is independent of the initial value of the magnetic quantum m.

Actually, Majorana only worked out his extremely elegant and concise solution for the special case J= 1/2, but as he points out, the method can easily be extended to any value of J. Such an extension was carried out in 1937 by Rabi and reelaborated and rewritten in a physically more obvious form by Bloch and Rabi in 1945,(7) who thereby contributed to the diffusion of the results that Majorana had obtained thirteen years earlier.

This paper by Majorana has become a classic in the treatment of non-adiabatic spin-flip and is correctly cited as such. His results, once extended as mentioned above, later formed the basis for the experimental method used to flip neutron spin with a radio-frequency field, the method employed both for analysis of polarized neutron beams and in all polarized neutron spectrometers for the study of magnetic structures.

Although a thorough study would bring it home even more clearly, my brief description suffices to point out the excellent quality of these papers. They reveal a thorough knowledge of even the most minute details of experimental data and a facility, which was uncommon at the time, in using the symmetry properties of states to simplify problems or choose the appropriate approximation for quantitative solution of problems. The latter was without a doubt a result of his exceptional mathematical talents.

3. - As I pointed out at the beginning, Ettore Majorana's greatest scientific contribution came during his second period of work, in which he produced three papers: one on nuclear forces now known as Majorana forces, one on particles with arbitrary intrinsic moment, and one on the electron-positron symmetry theory.

The historical background to the first of these three works is the announcement of the discovery of the neutron by James Chadwick in a letter to *Nature* received February 13, 1932. Before Easter that same year, Majorana had already tried to develop a theory of light nuclei that assumed that nuclei are made up only of protons and neutrons([8]) which interact with each other through Majorana exchanges forces. He had also attempted to work out a theory of light nuclei, which may be described as a primitive shell model, but when he encountered difficulties at $Z = 6$ while proceeding in increasing atomic number, he gave up this line of work and refused to publish his findings--even those on exchange forces--despite the urgings of Fermi and other members of the Institute.([1])

In July 1932, Heisenberg published the first of three works,([9]) in which he also hypothesized exchange forces which acted, however, on all coordinates of the interacting proton and neutron, that is, on those of spin as well as those of space.

Only towards the end of 1932 was Fermi able to convince Majorana to go abroad on a CNR grant which Fermi had arranged for him. In January 1933, Majorana left for Leipzig, where he remained until early summer. He then spent approximately three months in Copenhagen before returning to Leipzig, where he stayed another month before coming back to Rome at the beginning of August 1933.

During his stay in Leipzig, Majorana became friends with Heisenberg and it was this authoritative figure who finally convinced Majorana to publish his work on nuclear forces. Heisenberg maintained that the different nature of the forces hypothesized by Majorana produced some results that were clearly superior to those that he had published.

Majorana's work appeared in German [8a] (as sent from Leipzig) on March 3, 1933 and in Italian [8b] on May 11, 1933. The first three pages of the German text, extremely discursive and almost void of formulas, are dedicated to the various choices made in arriving at his expression for nuclear forces. They are more detailed and interesting than the introductory section of the Italian version, but the rest of the two texts differ very little. After reexamining the hypotheses underlying Heisenberg's model, Majorana turns to the queries relative to the structure of the neutron raised by Heisenberg: is the neutron a particle composed of a closely bonded proton and electron or is it a self-standing neutral particle? His conclusion is that all that can be done for the time being is to try to establish the law of interaction between the proton and the neutron on the basis of criteria of simplicity chosen to make it possible to reproduce the general and specific properties of nuclei as correctly as possible. After discussing some of these and observing that nuclei behave like bits of extended and impenetrable nuclear matter whose various parts essentially only interact on contact, and after underlining the fact that experimental data clearly point to a law of proportionality of both bonding energy and volume to the number of protons and neutrons, Majorana introduces a potential exchange energy that acts only on the 3 spatial coordinates. The sign of this energy must be such as to be attractive in states of even spin. He justifies his choice by pointing out that it achieves two important results: the first is that both neutrons present in the alpha particle interact with the two protons; the second is that in the approximation in which Coulomb repulsion can be neglected, the eigenfunction of the alpha particle is totally symmetrical with respect to the inversion of coordinates of the centers of mass of the four component particles, as is to be expected if it corresponds to a complete shell or ring, as everything seems to indicate.

Moving on from the alpha particle to heavier nuclei, Pauli's principle calls for the protons and neutrons added to enter more excited states. But since exchange energy is high only when particles are in the same (or nearly the same) orbital state, it follows that saturation of both bonding energy and density is practically achieved only in the case of the alpha particle.

The essential differences with respect to Heisenberg's model are that Majorana's exchange forces act only on the spatial coordinates and that the $J(r)$ function used by Majorana is of the opposite sign from the one used by Heisenberg.

Under these conditions, the symmetry properties of the eigenfunction are such that the phenomena of saturation can be explained without introducing a cut-off of potential exchange energy, as Heisenberg was forced to do to avoid collapse of the nucleus.

After these considerations, Majorana writes the Hamiltonian of a nucleus as the sum of the kinetic energies of all component protons and neutrons, of the energy of Coulomb repulsion between all proton pairs and of the potential exchange energy between all proton-neutron pairs. He then calculates the eigenvalue of the system composed of Z protons and N neutrons in the limiting case of very large Z and N (that is, in the case of nuclear matter) in its fundamental state. In order to do so, he uses as a global eigenfunction the product of two anti-symmetrical (during the simultaneous inversion of spatial and spin coordinates) eigenfunctions, one of which refers to the protons and the other to the neutrons. He thus comes to the conclusion that, neglecting Coulomb repulsion, the energy of the system is proportional to the number of particles composing it. He then demonstrates that Majorana exchange forces make it possible to obtain the saturation of both the bonding energy and the density of a nucleus without introducing a cut-off of exchange energy for short distances.

Majorana's elegant result did not stand up long. When the construction of the first accelerators in the second half of the forties allowed for the extension of measurement of proton-neutron differential elastic cross section to energies between 50 and 100 MeV, it was found that the angular distribution at the center of mass is symmetrical with respect to the equatorial plane with maximums practically equal to 0° and 180°.

This kind of distribution suggests that the potential describing the interaction must be a 50-50 mixture of Wigner's forces (that is, ordinary forces) and Majorana exchange forces.([10]) But a nuclear interaction with only 50% Majorana forces can no longer achieve saturation. This can only be obtained by introducing a repulsive core at close distance, that is, by using a kind of force now well known from the case of interatomic forces in molecules and solids but to which theoretical physicists in the thirties objected for aesthetic reasons. For example, in the first part of the German text, Majorana mentions this obvious possibility for obtaining saturation of nuclear forces. He then adds, however, "Eine solche Lösung des Problems ist aber vom ästhetischen Standpunkt aus unbefriedigend, . . . " (such a solution of the problem is, however, unsatisfactory from an aesthetic point of view).

These aesthetic preferences or perhaps they should be called biases also appear in another point of this paper, but this time in a purely formal issue. After having set down his expression for exchange energy, Majorana compares it with that of Heisenberg, stating beforehand that he will do so "eliminating the awkward ρ-spin coordinate, which may be done if the protons and the neutrons are regarded, even formally, as different particles". As is well known, Heisenberg's ρ-spin was simply what later became known as isotopic nucleon spin, the importance of which started to become manifest with the demonstration of the independence of charge in nuclear forces in 1936([11]) and the recognition of its importance in the discussion of the properties of nuclei carried out by Wigner in 1937.([12]) But Heisenberg had introduced the variable as a formal device and from Majorana's rigorously methodological point of view, he was right in wanting to avoid use of the new algorithm at that time. As sometimes happens, however, Heisenberg's choice, whether by chance or by genial intuition, proved to be right.

The story of the subsequent development of ideas on nuclear forces is too long and complex to be told or even summarized here. Yet it is clear that Yukawa's idea,([13]) advanced in 1935, that the nuclear forces exerted between two nucleons originate in the exchange of a particle of intermediate mass called a meson, which fits into the theory as a quantum of the mesonic field, brought about what might be called the unification of ordinary forces, or Wigner's forces, with the exchange forces of both Majorana and Heisenberg. In fact, whether it is the one case or the other to appear is determined by the different states of charge of the messenger exchanged. In this conceptual framework, Wigner's forces are in some ways exchange forces mediated by a messenger with zero electrical charge.

The most characteristic and in a certain sense general and definitive result of this theory is that the range of forces is linked to the mass m of the exchanged boson by the well known Yukawa relation

$$d = \frac{\hbar}{mc}. \qquad (1)$$

The application of these general and universally accepted ideas is simple and was immediately carried out by Yukawa and others, but only in the case of two nucleons in relative motion with low velocity with respect to c and at a great distance from one another. In order for the theory to provide results in these conditions that are in good accordance with experimental data, it is enough to consider the exchange of a single π meson or pion in that it is the meson with the lowest rest energy of all mesons observed in nature:

$$m_\pi c^2 = 139.57 \text{ MeV}$$

Thus for the Yukawa relation ([1]), the corresponding contribution of nuclear forces is the one with the longer range:

$$d_\pi = 1.41 \cdot 10^{-15} \text{ m}$$

But for shorter distances, things become more complicated and a complete and definitive interpretation of experimental results has still not been provided.

I have taken Figure 1 from a 1985 review article by S. O. Bäckman, G. E. Brown and J. A. Niskanen entitled "The nucleon-nucleon interaction and the nuclear many-body problem".([14]) It schematically illustrates the processes that determine the potential due to the mesonic field at different distances between the two interacting nucleons. At great distances, the potential is determined by the exchange of one single pion (pseudoscalar boson with small mass), while at intermediate distances, the interaction observed must be interpreted as the result of the exchange of scalar mesons endowed with relatively large mass, in reality consisting of virtual systems of two pions paired in states with S (spin) = 0 and T (isotopic spin) = 0 and involving one or more Δ resonances.

In the theoretical approach to which this review article refers, the messengers exchanged at short distances are vector bosons, but sound arguments have been put forward for years now that suggest that the quark structure of nucleons already begins to manifest itself at distances in the order of (1 -1.5) x 10^{-15}m (= 1-1.5 fm) and that gluons must be considered among the particles exchanged. As is known, gluons are the messengers of the forces exerted between quarks. When transferred moment is large (short distances), these forces show a reduction in the coupling constant (the so-called asymptotic freedom), which in some way simulates a classic, very gradual cut-off effect.(15)

If the contribution due to exchange of a pion is added to these quark models, the resulting local potentials as functions of energy are equivalent to phenomenological potentials such as the Reid potential(16) and above all the Paris potential(17) for the *S* channels of singlets and the *S* and *D* channels of coupled triplets; only qualitative results are available for the *P* channels.

As is known, all these phenomenological potentials contain both central and tensorial exchange terms proposed by Heisenberg, Majorana and Bartlett (which act only on spin coordinates)(18) even though they are usually presented in such a way today that these distinctions are not immediately evident.

Then again, it must be recalled that nuclear physics in the thirties did not go beyond 10 MeV of relative kinetic energy and therefore only dealt with large distances, that is, the area in which the so-called "one-pion exchange model" holds with excellent approximation.

To conclude this discussion I would like to point out that the English translation of Majorana's paper on nuclear forces can be found appended to the book on nuclear forces by English physicist Brink,(19) together with the well known discussion of the Royal Society chaired by Lord Rutherford following the discovery of the neutron, an article from the same period by Niels Bohr on the problems of nuclear structure, the first and third works by Heisenberg on exchange forces and their application, a paper by Wigner on the mass defect of helium, and several other famous works on the structure of the nucleus.

Let's go on to the paper on the relativistic theory of particles with arbitrary spin,[7] which preceded the paper on nuclear forces just discussed by about two years. I have inverted the order in which I have examined them because the paper on nuclear forces links back conceptually to the work carried out in the preceding period on chemical bonds, while the paper on particles with arbitrary spin is the first expression of a different and very important aspect of Majorana's cultural and scientific interests, which was to manifest itself fully in the paper on the theory of electron and positron symmetry.

The paper on the relativistic theory of arbitrary spin particles remained practically unknown after publication in 1932 for two reasons: it was published only in Italian and it dealt with a problem that was premature with respect to both the dominant interests of physicists in the early thirties and the knowledge of experimental data on elementary particles at that time. In fact, the paper was conceived, written and sent by Majorana to *Il Nuovo Cimento* and perhaps even published in that journal,(20) before news of the discovery of the positron by C. D. Anderson reached Rome via the *Science* article dated September 1, 1932.(21) As a result, it was destined to remain, as it did, outside of the mainstream of development of elementary particle theory. I will try to demonstrate here that it is nevertheless of great interest.

The background and aims of the paper are explained in the first two pages. The author begins by observing that the theory of Dirac, which correctly describes the electron, uses a wave function with four components, of which two, at sufficiently low velocity, become negligible, while the other two obey the Schrödinger equation in first approximation. In the same way, a particle with arbitrary spin s can be described in a quantum mechanics non-relativistic approximation by a complex of $2s + 1$ wave functions which individually satisfy the Schrödinger equation.

After a few general considerations on the transition from the relativistic case to the case of particles that move with velocity $v<<c$, Majorana observes that "In the absence of a field, the wave equation of a particle of matter must, according to Dirac, take the form

$$\left[\frac{W}{c}+(\boldsymbol{\alpha}\cdot\boldsymbol{p})-\beta\, mc\right]\psi=0\text{»} \qquad (2)$$

and that equations of this kind "present a basic difficulty" which is that for each value of the moment p , "the energy W can have two values which differ in sign

$$W=\pm\sqrt{m^2c^4+c^2p^2}\text{»}; \qquad (3)$$

He immediately adds the final introductory observation: "The indeterminateness of the sign may be overcome by using the fundamental kind of equation (2), only if the wave function has infinite components that *cannot* be broken down into finite tensors and spinors."

The reasons for the continuing interest of this work became particularly evident to me from a discussion in which I had the pleasure of participating about twenty years ago with Prof. M Fierz and Prof. D. M. Fradkin([1]) and from the paper that Prof. Fradkin has published on the subject.([22])

1) The paper by Majorana is the first attempt to develop a relativistically invariant theory for particles with arbitrary spin, both integer and semi-integer.

2) It represents a correct mathematical theory and contains the first recognition, development and use of unitary infinite dimensional representations of the Lorentz group. It is, in fact, the simplest of those representations, as Majorana correctly claims.

3) The theory somehow remained outside of the mainstream of subsequent developments for two reasons: the first is that Majorana had set himself the goal of working out a linear, relativistically invariant theory in which all mass eigenvalues are positive. This point of view was justified by the fact that news of the discovery of the positron by D. C. Anderson had not yet reached Rome when the paper was written (summer 1932). The second reason is that Majorana, unlike most subsequent authors, does not call for the dispersion relation linking the energy to the particle's momentum

$$(W/c)^2 = p^2 + (mc)^2$$

to be satisfied as an operational equation by the components of the eigenfunctions of the Dirac linear differential equation from which his theory started out.

4) In the system of reference in which momentum is not equal to zero, the solution to the equation found by Majorana has infinite components identified by the two indices ψ_{jm}. The index m takes on all values (integers and semi-integers) between $+-j$, while j takes on all non-negative values belonging either to the series of integers (j= 0, 1, 2,...) or to the series of semi-integers (j = 1/2, 3/2, 5/2, ...).

5) This result is difficult to interpret today in that each group of $2j + 1$ components ψ_{jm} (j set and $m = j, j -1, ..., -j$) seems to correspond to a particle with different spin. Therefore, the most natural interpretation for ψ is as a wave function which simultaneously represents either all bosons or all possible fermions. If the system of reference is changed, the various components mix. Nevertheless, in the system of reference of the center of mass, in which most authors define the properties of particle spin, Majorana's wave function corresponds to (and transforms mathematically like) the wave function of a particle with defined spin, the value of which depends on the chosen eigenfunction.

6) Majorana's theory provides a mass spectrum, that is, a formula which expresses the mass of the various particles as a function of their spin j, bringing in only one given constant m_o:

$$m_j = m_o/(j + 1/2)$$

7) Majorana demonstrates that at the non-relativistic limit ($v/c \rightarrow 0$), the $2s + 1$ ψ_{jm} components that correspond to the eigenvalue of the rest mass characterized by $j = s$ (for a set value of s) are the only ones that have finite values in that the ψ_{s+1m} and ψ_{s-1m} components are in the order of v/c, the components ψ_{s+2m} and ψ_{s-2m} in the order of $(v/c)^2$, and so on.

As Fradkin has observed, Majorana did not seem interested in the idea of a mass spectrum and this can probably be explained by the fact that too few elementary particles were known in 1932. Furthermore, he did not attempt to suggest an interpretation of the wave function similar to the one described in point 5).

Instead, Majorana emphasized point 7), observing that at the non-relativistic limit, the eigenfunction with infinite components has only $2s + 1$ non-zero components, each of which separately satisfies Schrödinger's equation in accordance with Pauli's theory of electron spin.

This paper by Majorana remains practically unknown. The same problem was examined by Dirac, Klein, Petiau and Proca in 1936 and by Duffin in 1938,[(23)] but none of these authors cite Majorana. Subsequent works refer back to Dirac and completely or almost completely overlook Majorana's contribution.

The main lines of this relativistic theory of arbitrary spin particles were laid down, developed and applied to some important cases in three fundamental papers that appeared in 1939. The first, by Fierz,[(24)] is a thorough examination of the relativistic theory of a particle with any spin in the absence of external fields, establishing a relation between spin and statistics. The second paper is by Fierz and Pauli[(25)] and extends the results of the previous paper to the case of particles in an electromagnetic field, exhaustively exploring the cases of 3/2 and 2 spin, in particular for particles with zero mass. The only coincidental reference to Majorana's work is in the third paper, the fundamental article by Kemmer on the mesonic field.(26)

A last observation is on the mathematical aspects of this paper: Majorana uses the same matrix elements for the infinite dimensional representation of the Lorentz group that Weyl uses in his book which came out in 1928(27) in connection with the rules of selection and calculation of the intensity of the dipolar atomic transitions dealt with in Schrödinger's quantum mechanics. Majorana was definitely familiar with Weyl's book, but thinking back on the way he worked,(1) I believe that he achieved these results independently and then recognized that the same matrices were involved in the problem of dipolar transitions.

The problem of infinite dimensional representation of the Lorentz group was discussed by Wigner in 1938,(28) by Dirac in 1945,(29) and again by Wigner in 1948.(30) Dirac, it seems, was not aware of Majorana's work, while Wigner, in his classic analysis of the Poincaré group, cites him almost incidentally. Corson's discussion in his book(31) is based on Wigner's work.

The numerous and thorough studies of infinite dimensional representations of the Lorentz group carried out later by various authors make no mention of Majorana. He is not even cited by those authors,(32) who have more recently, without realizing it, reconstructed, discussed and generalized Majorana's theory.(33) The only exception is the paper by Barut ad Kleinert,(34) which makes specific reference to his results.

In addition to its mathematical and historic interest, this paper shows Majorana's sensitivity to fundamental problems, which he approached in an autonomous manner and much earlier than most of his contemporaries. The fact that we now know that nucleons and mesons are systems of quarks in different quantum states that are described very satisfactorily by quantum chromodynamic (QCD) and that nuclear forces originate from the forces exerted between quarks, thanks to the exchange of gluons, does not detract from our admiration for Ettore Majorana's imagination and ability in manipulating Dirac's equation.(35)

Let us now examine the last paper by Ettore Majorana, the one entitled "Symmetrical theory of the electron and positron"[9], which remains his most important scientific contribution. The paper appeared in Italian only in the April 1937 issue of *Il Nuovo Cimento*. Later, it was translated into English by Luciano Maiani and published in that language in the Japanese periodical *Soryushiron Kenkyu* (1981)[9].

This paper starts with an observation regarding Dirac's relativistic theory, which led to the prediction of the existence of the positron, later confirmed by experiment. Dirac's theory is based on the Dirac equation which is completely symmetrical with respect to the sign of the charge of the fermion considered. That symmetry, however, is partially lost in the subsequent development of the theory which describes a vacuum as a situation in which all negative energy states are occupied and all positive energy states are free. Excitation of a fermion from one of the negative energy states to one of the positive states leaves a gap endowed with positive energy, which may be interpreted as an antifermion. Thus, the process of excitation of a fermion from a state of negative energy to one of positive energy is equivalent to the creation of a fermion-antifermion pair. This asymmetrical approach calls for the elimination *without any reasonable justification in principle* of some infinite constants resulting from the negative energy state, such as the density of electric charge.

It is clear that the concerns Majorana addressed in this paper [9], although now manifested more explicitly and in more detail, are the same as those set down in the introductory part of his 1932 paper on the theory of particles with arbitrary spin. And it was to these and other analogous aspects of the equations of quantum and relativistic physics that the phrase which Ettore repeated to his colleagues and sometimes even to his friends and relatives outside of the close sphere of physics referred: "Physics is on the wrong course" or something to that effect.

Those difficulties were avoided in the theory Majorana proposed in this paper, which described a new representation of the Dirac matrices γ_μ (μ = 1, 2, 3, 4) with the following properties:

1) Unlike the original Dirac representation, in Majorana's representation the four matrices γ_μ have the same properties of realness as the four-vector $x_\mu \equiv \mathbf{r}, ict$; or, if real spatial-temporal coordinates, associated to a pseudo-Euclidian metric system, are adopted, then all four are real.

2) In this representation, the Dirac equation relative to a free fermion has real coefficients and therefore its solutions break down into a real and an imaginary part, each of which satisfies said equation separately. But since they are real, each of these solutions has two very important properties: the first is that it gives rise to an electric current-charge four-vector that is always equal to zero. It follows that the real solutions to the Dirac equations must correspond to fermions lacking both electrical charge and magnetic moment. The second consequence of the real nature of the fermionic field is that the corresponding field operator must be Hermitian so as to reduce by half its degrees of freedom and to eliminate the distinction between fermion and antifermion.

In his paper, Majorana suggested that the neutron or the neutrino, or both, were particles of this kind, that is, neutral particles that identify with corresponding antiparticles.

3) Examination of Dirac's equation relative to a fermion in an electromagnetic field written in the Majorana representation shows that it is enough to take a combination of two real solutions to represent a charged particle. The fermionic field thus defined obviously gives rise to an electric current-charge four-vector not perfectly equal to zero thanks to the interference terms between the two real fields: furthermore, it has the properties known for a scalar field, which the field operator conjugated with respect to the charge (that is the operation describing a particle with a charge opposite to the charge of the particle considered) obtains applying the Hermitian conjugation operator to the operator ψ.

Actually, we know of no particle of the kind described in point 2): in particular, experiments carried out in 1939([36]) showed that the neutron magnetic moment is not equal to zero, while others carried out in 1956([37]) revealed that the antineutron is a separate particle from the neutron.

The problem of whether or not Majorana neutrinos, that is particles characterized by the equation

$$\nu_M = \bar{\nu}_M, \tag{4}$$

exist in nature has still not been solved definitively.

The question is of great importance in that the phenomena that can be observed with Majorana neutrinos are different from those that can be observed with the Dirac neutrinos. The reason for this difference is obvious: while Dirac neutrinos may be attributed a leptonic number, for example, equal to +1 for the neutrino and -1 for the antineutrino, this cannot be done for Majorana neutrinos because of equality (4). It follows that if Majorana neutrinos existed, lepton conservation would be violated and, in the absence of this restrictive law, the number of possible processes in which leptons are involved would increase considerably.

Majorana was aware of the fact that Fermi's theory of beta decay(38) would have to be changed if neutrinos were to satisfy equation (4). He makes a comment to this effect in the paper in connection with the work carried out by Gian Carlo Wick(39) extending Fermi's theory to the case of decay with positron emission observed by Joliot-Curie in early 1934. Majorana wrote that, if the neutrinos were of the kind he suggested, "the theory [of beta decay] can be changed so that both negative and positive β emission is always accompanied by the emission of a neutrino" [9].

Three months after publication of this beautiful paper by Majorana, a paper by Giulio Racah came out (40), in which he demonstrates that postulating the symmetry of particles and antiparticles means changing Fermi's theory of beta decay and that if the identity between the neutrino and the antineutrino is postulated as well, this necessarily leads directly to Majorana's theory. He was also the first to observe the difference in the behaviour of neutrino M and that of neutrino D, not only in beta decay but also in reactions caused by neutrinos.

Yet, the arguments developed in the second part of Racah's paper, which was very important in 1937, had to be revised approximately twenty years later when it was discovered that neutrinos have a well-defined helicity.

Discovery of the nature of the neutrino (that is, whether the neutrino behaves like a Dirac neutrino or a Majorana neutrino) may still be achieved experimentally, but requires a more subtle approach.

The principal way of answering this question in the case of ν_e is by experimental study of double beta decay, that is, the transition of a radioactive nucleus of atomic number Z with the emission of 2 electrons to a nucleus with atomic number Z' which differs from Z by two units.

Before briefly mentioning the results of these experiments, I would like to recall that interest in Majorana's theory increased considerably after the discovery of the non-conservation of parity(41) and the subsequent success of the two component theory in interpreting a broad variety of phenomena.(42) Various authors, in particular McLennan and Case, have shown that these two theories are equivalent for massless neutrinos.(43)

There are a number of ways of describing a neutrino (a massless and chargeless particle with 1/2 spin) with only two states relativistically. One way is provided by the two component theory, based on the discovery that the neutrino and the antineutrino always have well defined and opposite helicity and can therefore be described as one two-component spinor (Weyl's spinor). The other is based on Majorana's theory which, as mentioned, violates lepton conservation.

Basically, the second approach was rejected prior to the discovery of the non-conservation of parity. But once it was proven that the ν_e neutrino observed during beta decay always has a helicity of -1 and that the $\bar{\nu}_e$ antineutrino always has a helicity of +1, it became possible to identify these two objects with a ν_M, with a helicity of -1 and +1 respectively. This property, combined with the $(V - A)$ interaction which, as it contains only lefthand leptonic currents, conserves the leptonic number, is equivalent to lepton conservation.

Nevertheless, Majorana's theory only considers the possibility of reducing the leptonic number of the neutrino to its helicity in the case of massless neutrinos. Small mass ($m_\nu << m_e$) for the neutrino is compatible with Majorana's theory but not with the two-component theory. A neutrino with m_ν = O can only be either a Majorana neutrino or a four-component Dirac neutrino. In both cases, the mass value would be a parameter breaking the neutrino's helicity, which can no longer be equal to -1 for the neutrino and +1 for the antineutrino, except at the limit $v \to c$.

The first case (ν_M) involves a (slight) deviation from lepton conservation, while in the second case (ν_D), as helicity is independent of the lepton number under these conditions, the law is respected. The deviation from perfect helicity caused by a tiny mass may be detected only with the most sophisticated experiments now being carried out above all on ν_e by many groups throughout the world, in particular, the group headed by Ettore Fiorini at the University of Milan.

Besides this break in the perfect helicity of the neutrino, sometimes called *implicit break*, there may also, however, be an *explicit break* caused by the fact that, in addition to the interaction due to lefthand lepton currents (pure $(V - A)$ interaction), a very tiny fraction of interaction may also be due to righthand currents.

These problems can only be thoroughly investigated experimentally for the case of ν_e. This requires study of double beta decay, in particular, measurement of the spectrum sum $E_1 + E_2$ of the energies of the two electrons emitted. In his general report presented to the 8th International Workshop on Grand Unification held in Syracuse, New York in April 1987,[(44)] Ettore Fiorini revealed that the experiments carried out by various groups, including the one at the University of Milan, on transition (0+ 0+) between the fundamental states of germanium and selenium 76

$$^{76}_{32}\mathrm{Ge} \to ^{76}_{34}\mathrm{Se} + 2e^-, \qquad (5)$$

have made it possible to establish the lower limits of the parameters relative to violation of lepton conservation.

By observing the $E_1 + E_2$ spectrum, three cases may be distinguished which differ in that during decay (5), the two electrons are emitted simultaneously with i) 2 Dirac or Weyl $\bar{\nu}_e$, so that the spectrum is continuous (4 body decay + residual nucleus) with a maximum at approximately 1/3 of $\varepsilon_m = (E_1 + E_2)_{max}$, the value of ε_m given by the energy difference between the nuclear transition states (5), ii) no other particle, so that the spectrum is reduced to one ε_m energy line (two body decay + recoil nucleus: Majorana's theory); iii) one massless and chargeless boson (known as a majorone; ([45]) three body decay + residual nucleus) so that the spectrum is continuous with a maximum at approximately 4/5 of ε_m.

To date, no experimental group has observed processes of the kind predicted by Majorana's theory. Instead they agree on the following upper limits for the values of the parameters that break the lepton number:

a) for mass

$$\langle m_{\nu_e} \rangle < (1 \div 2)\ \text{eV}, \tag{6a}$$

which includes the average value (< >) of the neutrino mass to take account of the fact that one of the theoretical approaches postulates that it can exist in two states corresponding to two different eigenvalues of *CP*;

b) for the average value of the fraction η of interaction due to righthanded lepton currents ($V + A$), as compared to the certainly more dominant lefthanded currents,

$$\langle \eta_{\nu_e} \rangle < 10^{-6} \div 10^{-8} \tag{6b}$$

for the processes $0^+ \to 0^+$ called for by Majorana's theory; the broad interval of values on the righthand side of (6b) originates exclusively from the use of different theoretical models in calculating the matrix element of the nuclear transition $0^+ \to 0^+$ in the $^{76}Ge \to {}^{76}Se$ process.

It is important to note that the upper limits (6) of $<m_{\nu_e}>$ and $<\eta_{\nu_e}>$ are inferior, one by one order of magnitude, the other by approximately four orders of magnitude, to the corresponding values of four years ago. Even more accurate experiments are predicted in the not too distant future.

It may therefore be concluded that, in the case of ν_e, there is no evidence that it may be a Majorana neutrino. For ν_μ, on the other hand, no experimental methods exist by which such low limits for the possible violation of lepton conservation may be established. Even farther from a definitive clarification is the situation concerning ν_τ. Both of these neutrinos could have a mass that is much greater than the upper limit (6a) now established for ν_e.

As is known, the problem of neutrino mass is of extreme interest for interpretation of several astrophysical phenomena and for the oscillation among the various kinds of neutrinos known today, suggested by B. Pontecorvo in 1958,([46]) but still unobserved.

In winding up this brief review of Ettore Majorana's scientific works, I would like to mention a very interesting property of the Majorana spinors recently discovered by Budinich and explored in detail by him and his group in a number of papers that have not yet been published, but which have circulated within SISSA.

The French mathematician Elie Cartan (1869-1951), who discovered spinors in 1913, defined them as vectors equivalent to isotropic planes in complex ordinary spaces, that is, planes whose vectors have no length and are, in pairs, orthogonal to each other.

Actually vectors with null modulus had already been considered for a long time (by

Weierstrass, Bianchi and Eisenhart) as the origin of minimum surfaces in ordinary space and in space-time.

Budinich has demonstrated(47,48) that by following Cartan's conception of spinors it is easy to generate minimum surfaces in ordinary spaces and in space-time starting from complex spinors. By replacing Dirac's complex spinors with Majorana's real spinors in the same space-time frame, Budinich obtains, instead of minimum surfaces, classic relativistic strings(47) that obey the following equation of motion

$$\frac{\partial^2 x_\mu}{\partial \sigma^2} - \frac{\partial^2 x_\mu}{\partial \tau^2} = 0, \qquad (7)$$

where x_μ are the space-time coordinates, $d\sigma$ is the infinitesimal arc of the string and $d\tau$ the proper time interval.

Budinich expresses the following considerations on these results: to the obviously elementary geometric nature of Majorana's spinors correspond, in nature, the most elementary physical objects known, that is

1) massless and chargeless neutrinos that can be represented by Weyl's two-component spinors;(49)

2) the relativistic strings generated by them.

According to Budinich, this result may in some ways be considered as another argument in favour of the fundamental nature which is now attributed to relativistic strings. He also feels that it can be shown that the action of the Eulero-Lagrange equations (7) for strings is equivalent in terms of spinors to Fermi's Lagrangian equation (quadrilinear for the chiral fermions) of weak interactions.(50)

This interesting problem, still under study by the group in Trieste shows that the scientific work of Ettore Majorana contains profound links which have still not been entirely uncovered or completely understood.

(1) For a more thorough review, albeit with a somewhat outdated commentary, of the works of E. Majorana , see E. Amaldi: *La vita e l'opera di Ettore Majorana*, Accademia Nazionale dei Lincei (1966) and the translation in English: *Ettore Majorana, man and scientist* in *Strong and Weak Interactions, Present Problems, International School of Physics Ettore Majorana, Erice, June 19-July 4, 1966*, edited by A. ZICHICHI (Academic Press, New York and London, 1966), p. 10. The following are partial and abridged presentations by the same author: *Ricordo di Ettore Majorana, G. Fis*, **9**, 300 (1968); *L'opera scientifica di Ettore Majorana, Physics*, **10**, 173 (1968).

(2) L. H. THOMAS: *Calculation of atomic fields, Proc. Cambridge Philos. Soc.*, **33**, 542 (1927); E. FERMI: *Un metodo statistico per la determinazione di alcune proprietà dell'atomo, Rend. Lincei*, **6**, 602 (1927).

(3) E. FERMI: *Über die Anwendung der Statischtischen Methode auf die Probleme des Atombau*, in *Quantentheorie und Chemie, Leipziger Vorträge*, edited by H. Falkenhagen (Leipzig, 1928), p. 95.

(4) TA-YOU WU: *Energy states of doubly excited helium, Phys. Rev.*, **46**, 239 (1934). Dr. Wu was a well known student and collaborator of S. Goudsmidt.

(5) W. HEITLER and F. LONDON: *Wechselwirkung neutraler Atom und homopoläre Bindung nach der Quantenmechanik, Z. Phys.*, **44**, 455 (1927).

(6) C.G.DARWIN: *Free motion in the wave mechanics, Proc. R. Soc. London, Ser. A*, **117**, 258 (1928); A. LANDE': *Polarization von Materiewellen, Naturwissenschaften*, **17**, 634 (1929).

(7) I.I. RABI: *Space quantization in a gyrating magnetic field, Phys. Rev*, **51**, 652 (1937); F. BLOCH and I.I. RABI: *Atoms in variable magnetic fields, Rev. Mod. Phys*, **17**, 237 (1945).

(8) Chadwick and many others had stated that the nucleus is made up of neutrons and an equal number of protons. The first to state, however, that these two particles are the *only* components of the nucleus was probably D. IWANENKO: *The neutron hypothesis, Nature (London)*, **129**, 798 (1932), dated April 21, 1932. I am not sure whether Majorana came to the same conclusion independently, but I feel that it is likely.
(9) W. HEISENBERG: *Über den Bau der Atomkern I, Z. Phys.*, **77**, 1 (1932), dated June 7, 1932; *Über den Bau der Atomkern II, Z. Phys.*, **78**, 156 (1932), dated July 30 1932; *Über den Bau der Atomkern III, Z. Phys.*, **80**, 587 (1933), dated December 22, 1933.
(10) This observation underlies, among other things, the so-called Serber potential (*Nuclear reactions at high energy, Phys. Rev.*, **72**, 1114 (1947), which is one of the simplest and has often been used for approximate calculations.
(11) G. BREIT, E. U. CONDON and R. D. PRESENT: *Theory of scattering of protons by protons, Phys. Rev.*, **50**, 825 (1936); B. CASSEN and E. U. CONDON: *On nuclear forces, Phys. Rev.*, **50**, 846 (1936); G. BREIT and E. FEENBERG: *The possibility of the same form of specific interaction for all nuclear particles, Phys. Rev.*, **50**, 850 (1936).
(12) E. WIGNER: *On the consequences of the symmetry of the nuclear Hamiltonian on the spectroscopy of nuclei, Phys. Rev.*, **51**, 106 (1937).
(13) H. YUKAWA: *On the interaction of elementary particles I, Phys., Math. Soc. Jpn.*, **17**, 48 (1935).
(14) S. O. BÄCKMAN, G. E. BROWN and J. A. NISKANEN: *The nucleon-nucleon interaction and the nuclear many-body problem, Phys. Rep.*, **124**, 1 (1985). The same figure is also found on page 7 of the book by G. E. BROWN and A.D. JACKSON: *The Nucleon-Nucleon Interaction* (North Holland Publishing Company, Amsterdam, 1976).
(15) See, for example, R. MIGNANI and D. PROSPERI: *Constituent Quark Models for Hadrons and Nuclei*, lectures given at the First International Course on Condensed Matter, International Center of Physics, Bogotà (Colombia), July 7-18, 1986 (forthcoming).
(16) R. V. REID: *Logical phenomenological nucleon-nucleon potential, Ann. Phys. (N. Y.)*, **50**, 411, (1968).
(17) Among the most recent publications on the so-called Paris potential, is M. LA COMBE, B. LOISEAU, J. M. RICHARD and R. VINH MAU: *Parametrization of the Paris N-N potential, Phys. Rev. C.*, **21**, 861 (1980). It also provides a bibliography of preceding articles on the same subject.
(18) J.H. BARTLETT, jr.: *Exchange forces and the structure of the nucleus, Phys. Rev.*, **49**, 102 (1936).
(19) D. M. BRINK: *Nuclear Forces* (Pergamon Press, Oxford, 1965).
(20) Unfortunately, volume 9 of *Il Nuovo Cimento* contains no information regarding the date on which the manuscript[7] was received, or the month to which the issue in which it was published refers.
(21) C.D. ANDERSEN: *The apparent existence of easily deflectable positives, Science*, **76**, 238 (1932), dated September 1, 1932. Andersen also sent a more exhaustive article to the *Physical Review* (received on February 28, 1933) which appeared in the March 1933 issue (*Positive electrons, Phys. Rev.*, **43**, 491 (1933).
(22) D. M. FRADKIN: *Comments on a paper by Majorana concerning elementary particles, Am. J. Phys.*, 34, 314 (1966).
(23) P. A. M. DIRAC: *Relativistic wave equation. Proc R. Soc. London, Ser. A*, **155**, 447 (1936); O. KLEIN: *Eine Veralgemeinerung der Diracschen relativistischen Wellengleichung, Ark. Mat. Astr. Fis. A.*, **25**, n. 15, 1 (1936); G. PETIAU: *Contribution à la théorie des équations d'ondes corpuscolaires, Mem. Acad. Soc. R. Belg.*, **16**, n. 2, 1 (1936): A. PROCA: *Sur les équations fondamentales des particules élémentaires, C. R. Acad. Sci. Paris*, **202**, 1490 (1936); *Sur la théorie ondulatoire des électrons positifs et negatifs, J. Phys. Radium*, **7**, 347 (1936); *Théorie non relativiste des particules a spin entier, J. Phys. Radium*, **9**, 61 (1936); R. I. DUFFIN: *On the characteristic matrixes of covariant systems, Phys. Rev.*, **54**, 1114 (1938).
(24) M. FIERZ: *Über die relativistische Theorie kräftfreier Teilchen mit beliebigen Spin, Helv. Phys. Acta*, **12**, 3 (1939).
(25) M. FIERZ and W. PAULI: *On relativistic wave equations for particles of arbitrary spin in an electromagnetic field, Proc. R. Soc. London, Ser. A*, **173**, 211 (1939).
(26) N. KEMMER: *The particle aspect of meson theory , Proc. R. Soc. London, Ser. A*, **173**, 91 (1939).
(27) H. WEYL: *Gruppentheorie und Quantummechanik* (Verlag von S. Hirzel, Leipzig, 1928), p. 160 of the 1st edition.
(28) E. P. WIGNER: *On unitary representation of the inhomogeneous Lorentz group, Ann. Math., Princeton 2nd Series*, **40**, 149 (1939).
(29) P. A. M. DIRAC: *Unitary representations of the Lorentz group, Proc. R. Soc. London, Ser. A*, **183**, 284 (1945).
(30) E.P. WIGNER: *Relativistische Wellengleichungen, Z. Phys.*, **124**, 665 (1947).
(31) E. M. CORSON: *Introduction to Tensors, Spinors and Relativistic Wave-Equations* (Blackie and Sons Ltd., Glasgow, 1953), chap. V, p. 140.
(32) I. M. GEL'FAND and M. A. NAIMARK: *Unitary representations of the Lorentz group, J. Phys. (USSR)*, **10**, 93 (1946); HARISH-CHANDRA: *Infinite irreducible representations of the Lorentz group, Proc. R. Soc. London, Ser. A*, **189**, 372 (1947); V. BARGMANN: *Irreducible unitary representations of the Lorentz group, Ann. Math., Princeton 2nd series*, **48**, 568 (1947).
(33) I. M. GEL'FAND and A. M. YAGLOM: *General relativistic invariant equations and infinite dimensional representations of the Lorentz group, Z. Eksp. Teor. Fiz*, **18**, 703 (1948); English translation: TT-345 National

Research Council of Canada (1953); V. L. GINZBURG: *On relativistic wave equations with a mass spectrum, Acta Phys. Pol*, **15**, 163 (1956); I. M. GEL'FAND, R. A. MINLOS and Z. YA. SHAPIRO: *Representations of the Rotations and Lorentz Group* (Pergamon Press, New York, N.Y. 1963); G. FELDMAN and P. T. MATTHEUS: *Multimass field, spin and statistics, Phys. Rev.*, **154**, 1241 (1967).
(34) A. O. BARUT and H. KLEINERT: *Current operators and Majorana equations for the hydrogen atom from dynamic groups, Phys. Rev.*, **157**, 1180 (1967).
(35) Note added after the LXXIII Congress of the Italian Physics Society. During a conversation following my presentation of this paper, Federico Capasso drew my attention to the existence of papers regarding the connection between Majorana's works and Regge's phenomenology of poles and trajectories. E. Recami, to whom I turned for more precise information, introduced me to Dr. Enrico Giannetto of the University of Catania as the person most knowledgeable on this matter. A few days later, while passing through Rome, Dr. Giannetto provided me with the names of a number of articles on the subject. See, for example, the articles by Y. NAMBU, C. FRONSDAL and others, in the volume *Elementary Particle Theory: Relativistic Groups and Analiticity,* edited by N. SVARTHOLM, *Proc. Engl. Nobel Symposiun, May 19-25, 1968* (Aspenagarten, Almquist and Wixell, Stockholm, 1968) and the articles by A. O. BARUT, I. T. TODOROV and others in *Proceedings of the 1967 International Conference on Particles and Fields*, edited by C. R. HAGEN, G. GURALNIK and V. A. MATHUR (Interscience, New York, N. Y., 1968).
(36) L. W. ALVAREZ and F. BLOCH: *A quantitative determination of the neutron moment in absolute nuclear magnetons, Phys. Rev.*, **57**, 111 (1960).
(37) B. CORK, G. R. LAMBERTSON, O. P. PICCIONI and W. A. WENZEL: *Antineutrons produced from antiprotons in charge exchange collision, Phys. Rev.*, **104**, 1193 (1956).
(38) E. FERMI: *Tentativo di una teoria della emissione dei raggi beta, Ric. Sc.*, **4**(2), 491 (1933); *Versuch einer Theorie der -Strahlen I, Z. Phys.*, **88**, 161 (1934).
(39) G. C. WICK: *Gli elementi radioattivi di F. Joliot e I. Curie, Rend. Lincei*, **19**, 319 (1934).
(40) G. RACAH: *Sulla simmetria tra particelle e antiparticelle, Nuovo Cimento*, **14**, 322 (1937).
(41) T. D. LEE and C. N. YANG: *Elementary Particles and Weak Interactions*, BNL 443 US Department of Commerce Office of Technical Sciences, Washington, DC, 1957.
(42) L. LANDAU: *On the conservation laws of weak interactions, Nucl. Phys.*, **3**, 127 (1957); A. SALAM: *On parity conservation and neutrino mass, Nuovo Cimento*, **5**, 299 (1957); T. D. LEE and C. N. YANG: *Parity non-conservation and a two component theory of the neutrino, Phys. Rev.*, 105, 1671 (1957).
(43) J. SERPE: *Sur la théorie abrégée des particules de spin 1/2, Physica*, **18**, 295 (1952); L. A. RADICATI and B. TOUSCHEK: *On the equivalence theorem for the massless neutrino, Nuovo Cimento*, **5**, 1623 (1957); J. MCLENNAN: *Parity non conservation and the theory of the neutrino, Phys. Rev.*, **106**, 821 (1957); K. CASE: *Reformulation of the Majorana theory, Phys. Rev.*, **107**, 307 (1957).
(44) E. FIORINI: *The Present and Future of Double Beta Decay,* general report presented to the *Eighth Workshop on Grand Unification, Syracuse, April 16-18, 1987.*
(45) G.GELMINI and M. RONCADELLI: *Left-handed neutrino mass scale and spontaneously broken lepton number, Phys. Lett. B*, **99**, 411 (1981); H. M. GEORGI, S. L. GLASHOW and S. NUSINOV: *Unconventional model of neutrino masses, Nucl. Phys. B*, **193**, 297 (1981).
(46) B. PONTECORVO: *Inverse beta processes and non-conservation of lepton charge, Sov. Phys, JEPT*, **34**, 172 (1958).
(47) P. BUDINICH: *On the Posible Role of Cartan Spinors in Physics*, SISSA 60/86/E.P.
(48) P. BUDINICH, L. DOBROVOSKI and P. FURLAN: *Minimal Surfaces and Strings from Spinors: a Realization of the Cartan Programme,* SISSA 59/86/E.P.
(49) R. E. MARSHAK, RIAZUDIN and C.P. RESAN: *Theory of Weak Interactions* (Interscience, New York, N.y., 1969), p. 71.
(50) P. BUDINICH and M. RIGOLI: at press.

EDOARDO AMALDI

Gian Carlo Wick during the Thirties

Mister Chairman, dear Wick, Ladies and Gentlemen!

Of all people present here today to honour you, Gian Carlo Wick, I am the person who has known you for the longest time. We met for the first time at the beginning of 1931 in Leipzig.

I had passed my « laurea » examination at the University of Rome in 1929. I had done the military service, and at the beginning of January 1931, I went to Leipzig, with a grant from my native town. The Institute of Physics of the University of Leipzig was a great center of research and had been suggested to me by my young professors Fermi and Rasetti. The Institute was directed by Peter Debye who led the experimental researches along two lines: the investigation of the diffraction of X-rays by molecules in the gaseous and in the liquid state, and the measurement of the dielectric constant of polar molecules as a function of the frequency of the electric field. Werner Heisenberg had the chair of theoretical physics, Felix Bloch was his assistant, Friedrich Hund was professor of spectroscopy, and for example C. F. v. Weizsäcker was a very gifted student. From time to time H. Bethe, G. Placzek, E. Teller, V. Weisskopf and many other brilliant theoreticians paid visits to the Heisenberg group which lasted sometimes weeks or even months.

Talking with some people of the Institute I learned pretty soon that there were two other young Italians working in the theoretical group directed by Heisenberg. They were Gian Carlo Wick and Giovanni Gentile junior (b. 1906) who, unfortunately, died prematurely in 1942.

I still remember very vividly my first encounter with Gian Carlo. I was working in a room on the underground floor, where the X-ray plants were placed, and Gian Carlo came in looking for me since he had found a message I had left for him shortly before.

We immediately started to exchange views on the city of Leipzig, on the « Leipziger Institut » and its extraordinary staff and about our personal problems and specific interests.

Thus I learned that Gian Carlo had studied under professor Somigliana at the University of Turin, where he had passed his laurea examination in 1930, and that towards the end of the same year, with a « Fano grant of the University of Turin », he had spent a semester in Göttingen in the Institute directed by Max Born where he had established friendly relations with W. Heitler and L. Nordheim. He had decided to use the rest of his grant to spend another semester in Leipzig. We became immediately great friends and our relations became even closer when, in autumn 1932, from Turin he went to Rome as assistant to Fermi.

Among the many papers that Gian Carlo published before leaving Italy for the United States in March 1946, I will recall only a few that seem to me to be sufficient to illustrate his scientific production. In 1933, for example, he published, in Italian and German, as we used to do in those times, a paper on the magnetic moment of a rotating hydrogen molecule /1/. The magnetic moment of this molecule had been recently measured by Frisch and Stern [1] who used molecular beams at such a low temperature that all the parahydrogen molecules had rotational quantum number $J = 0$ and therefore their magnetic moment is zero because the sum of the antiparallel spin of the two protons is null and there is no contribution originating from the rotation of the molecule. The magnetic moment of the orthohydrogen molecules, on the contrary, is the sum of two terms, both of the order of a nuclear magneton. The first term is the sum of the intrinsic magnetic moments of the two protons (which are parallel to each other), while the other one originates from the rotation of the molecule. The derivation from the experimental results of the magnetic moment of the proton required the subtraction of the magnetic moment due to the rotation. This was computed by G. C. Wick who showed that the pertinent current does not correspond to a rigid rotation and derived, by appropriate approximate methods, an estimate of an upper and a lower limit for its value. These results were used by Stern and collaborators in the derivation of the magnetic moment of the proton, which for the first time, was found to amount to 2.5 nuclear magnetons with an error of about 10%.

The need for such an important correction was avoided in the molecular beam resonant method developed shortly later by Rabi et al. [2]. The external magnetic field used by these authors was so strong that each constituent magnetic moment was decupled from the others and precessed independently around the direction of the field.

In another paper, which also appeared in 1933, Wick proposes a method /2/ for recognizing whether the nuclear forces acting between a neutron and a proton are ordinary forces or exchange forces of the kind proposed shortly before by Heisenberg and Majorana [3]. If the energy of the incident neutrons is so high that the corresponding de Broglie wave length is comparable to the range of the nuclear forces, it becomes hard to deflect the incident particles and small deflections are more likely than the large ones. But exchange forces interchange the nature of the particles and the incident neutron, which would tend to continue in the forward direction, becomes a proton and the outgoing neutron tends to go backward in the center of mass reference frame.

The artificial radioactivity induced by alpha particles was discovered by the French authors. In agreement with Pauli's suggestion it assumed as dioactive bodies which emit positrons instead of electrons [4]. Fermi's theory of beta decay was conceived of and developed at the end of 1933 [5] and appeared—purely accidentally—at about the same time as the paper by the French authors. In agreement with Pauli's suggestion it assumed as basic process the transformation of a neutron into a proton with the emission of an electron and a neutrino. Wick immediately pointed out /3/ that Fermi's theory contained naturally also the possibility of the inverse process: « transformation of a proton into a neutron and destruction of an electron and a neutrino. For such a process to take place, however,—wrote Wick—it is essential that in the vicinity of the nucleus there be a certain density of neutrinos. Just this density is provided by the neutrinos of negative energy; the destruction of one of them is equivalent to the formation of a particle (neutrino's hole) perfectly analogous to the neutrino. If the electron which is absorbed by the proton is an electron of negative energy, one has the emission of a positron. It is natural to identify this phenomenon with that observed by Curie and Joliot. If, on the contrary, the destroyed electron ... is one of the K, L, M, ... electrons belonging to the external structure of the atom ... one has the emission of X-rays or of Auger electrons i.e. a phenomenon, which in our case can be observed only with considerable difficulties » /3/.

After these general considerations Wick develops in all detail Fermi's theory for positrons emitters. He does not, on the contrary, enter into the formal details of the new phenomenon he predicted, the capture of orbital electrons which was envisaged, shortly after, also by Bethe and Peierls [6]. The detailed theory of this type of beta instability was developed in 1935 by Yukawa and Sakata [7], who quote Fermi's and Wick's papers. It was observed for the first time in 1938 by L. Alvarez [7] who studied various elements, with particular attention to the case of $^{67}Ga \rightarrow {}^{67}Zn$.

After the recognition that the behaviour of electrons is described, if not exactly, at least with great accuracy, by means of the Dirac equation, it was natural to assume that all other particles, different from electrons but with spin 1/2, could be described by the same equation. But if one applies the Dirac equation to the nucleon, one finds that the proton must have a magnetic moment equal to a nuclear magneton μ_N and the neutron a magnetic moment equal to zero. Therefore when the experiments by Stern et al. [1] and Rabi et al. [2] showed that the magnetic moment of the proton is about 2.8 μ_N it was great surprise.

In order to explain this serious difficulty Wick proposed, at the beginning of 1935 /4/, to describe the proton existing in nature, referred to as the « physical proton », as a mixture of states. For part of the time it is really a proton (a bare proton) with magnetic moment equal to μ_N, but for a fraction τ of the time it is virtually dissociated according to the process responsible for positron-decay i.e. into a « bare neutron », a positron and a neutrino

$$ {}_1p^1 \rightleftarrows {}_0n^1 + e^+ + \nu \tag{1} $$

so that the observed value of μ_p is given by

$$ \mu_p = \frac{\mu_N + \tau\mu_{e^+}}{1+\tau} \tag{2} $$

where μ_{e^+} is the magnetic moment of the positron (i.e. a Bohr magneton).

Wick knew of course that Tamm as well as other authors had shown, not long before [9], that the exchange interaction deduced from Fermi's Hamiltonian is many orders of magnitude smaller than that derived from the binding energy of nuclei, if the value of the coupling constant is determined from beta decay. He knew also of Heisenberg's remark that the exchange interaction deduced by Tamm corresponds to a « Heisenberg exchange », while many good reasons had been presented by Majorana in favour of a « Majorana exchange ».

What Wick does is to start from the expression given by Tamm for the exchange interaction, point out that it should become important for momenta of the exchanged particles of the order of $137\, m_e c$, and try to estimate the value of τ under these conditions by using the relation (2). He finds that τ should not be much smaller than 1, but rather in the range 1/20 to 1/2.

In 1938 the Fermi process (1) was replaced by Frölich, Heitler and Kemmer [10] with the Yukawa process

$$ {}^1_1p \rightleftarrows {}^1_0n + Y^+ \qquad (Y^+ = \text{Yukawa boson}) $$

but Wick's idea of explaining the anomalous magnetic moment of the physical proton (and neutron) as due to the fact that the « observed nucleons » are mixtures of virtual states, remained valid and is still accepted today.

In 1936, at the time of the study of the nuclear resonances of nuclei for slow neutrons, Wick worked experimentally for a few months in collaboration with Bruno Pontecorvo /5/ on the back scattering of the neutrons belonging to a few different intervals of energy (or groups in the terminology of the time) by a number of materials such as Al, Mg, Fe, Ni, Cu,...Pb. They observed a 100 % increase or more in the case of Fe and Pb, and that in both cases the back scattering was very large not only for the thermal neutrons (group C), but also for the neutrons of energy equal to the first resonance of Rh ($E_R \simeq 1$ eV, $\Gamma \simeq 0.1$ eV).

These interesting results were correctly interpreted by the authors as due to the fact that the energy lost by a resonance neutron in a large angle scattering against a heavy nucleus is small compared to the width Γ of the resonance. Such a behaviour is completely different from what is observed with hydrogenous scatterers, which give a much greather energy loss for collision.

Gian Carlo published also a series of theoretical papers on the slowing down of neutrons and their diffusion in hydrogenous materials /6/. The second of these papers (1936), is a short letter which appear in the *Physical Review*, presenting the probability distribution that neutrons whose energy has been reduced from the initial value E_0 to E by moving in a hydrogenous medium have undergone just n collisions against protons. The distribution is only slightly asymmetric and very close to a Gaussian distribution. Its maximum is very close to the value

$$n = \ln \frac{E_0}{E}$$

which had been published by Fermi but without explaining its derivation.

The most important of these papers is the one which appeared in 1943 in *Zeitschrift für Physik*. It contains a very interesting approximate metod for deriving, with the desired accuracy, the solutions of stationary plane problems treated with the Boltzmann equation describing the diffusion through a moderator of thermal neutrons assumed to move at a constant velocity taken equal to the average thermal velocity. The method applied, for example, to the computation of the angular distribution of the thermal neutrons emerging from the plane surface of a moderator, gives, in Wick's third approximation, a result almost identical with that of an exact but much more laborious computation.

The last of these papers, in collaboration with M. Verde, was completed in 1943, but appeared in Physical Review only in 1947 because of war conditions. It contains the presentation of the stationary solution of the transport equation describing the diffusion and slowing down of neutrons in an infinite homogeneous medium surrounding a point source of fast neutrons, for a few simple laws of dependence on energy of the neutron mean free path.

In 1937 Wick developed also the basic principles of the theory of the scattering of slow neutrons by crystals /7/. The importance of this phenomenon had been pointed out, in spring 1926, by Elsasser [11], and the first experimental observations were made shortly later by Preiswerk and von Halban in Paris [12] and by Mitchell and Powers at Columbia University [13].

For the interaction of the incident neutron with the nuclei of the atoms bound in the lattice Wick uses the pseudo-potential approach developed by Fermi [14] for dealing with the scattering of neutrons by protons bound in molecules. While Fermi had used protons oscillating with a single frequency, Wick attributes to the crystal a discrete spectrum of frequencies and notices, from the beginning of his paper, that the improvement he obtains is in some way comparable to that introduced by Debye with respect to the first theory of Einstein of the specific heat of crystals. Wick showed that crystals are transparent to ultraslow neutrons. Both authors, i.e. Fermi and Wick, considered also processes in which the neutron exchanges energy with the scatterer (molecule or crystal), sometimes exciting and sometimes deexciting one of its eigenfrequencies, a subject that, years later, became of primary importance as a tool for investigating the dynamic properties of crystalline lattices. In his last publication on this subject Wick takes into account also the influence of the isotopic composition of the elements constituting the crystal and the spin of the corresponding nuclei.

In a letter to the Editor of *Nature* which appeared at the end of 1938 /8/ Wick illustrates the meaning of the expression

$$\varrho = \frac{\hbar}{mc} \tag{3}$$

given by Yukawa for the range ϱ of the nuclear forces in terms of the mass m of the meson exchanged between the two interacting nucleons. The approach followed is in some way inspired by the arguments used, many years before, by Bohr in connection with Gamow's theory of alpha-decay.

On the occasion of a trip to Copenhagen, Wick mentioned his considerations to Bohr, who liked them and encouraged Wick to publish them. By making use of Heisenberg's uncertainty principle, Wick shows that the expression (3) can be interpreted as the upper limit for the distance to which the virtual transitions involved in the emission and absorption of mesons by the two interacting nucleons can make themselves felt without contradiction with the energy principle.

In connection with the problem of the so-called isomer's series of transuranic elements, Wefelmeyer [15] had proposed that, because of the Coulomb repulsion, the very heavy nuclei could possibly exist in forms strongly different from the spherical one. This qualitative idea was taken up by Weizsäcker who, in 1939 [16], evaluated the energy of a charged liquid drop having the shape of a rotational ellipsoide. With the discovery of fission, at the beginning of 1939, the problem of the isomer's series of the transuranic elements disappeared, but the problem of possible equilibrium shapes of a heavy nucleus different from a sphere became interesting in connection with the study of the collective motions of a drop which can bring it to fission. In a paper of 1939 Wick showed /9/ that configurations of equilibrium of a charged liquid drop different from a sphere do not exist.

In another paper of the year 1939, Wick derives the cross sections for the excitation of electric and magnetic dipole and electric quadrupole transitions induced by relativistic electrons /10/.

The possibility of processes of this type had been envisaged by Weisskopf in 1938 [17], and in 1939 Pontecorvo and Lazard [18] in Paris had succeeded in exciting by means of harh X-rays a metastable state of ^{115}In. Wick's paper, however, is the first quantitative approach to the state of theory of this kind of processes, except for a paper of 1935 by Bethe and Peierls concerning the photo-and electro-disintegration of the deuteron by electric dipole transition [19].

In 1941 Wick studied the propagation of a de Broglie wave in a material medium by methods similar to those classical in optics; in particular he examined the interference between the incident wave and the secondary waves emitted by the medium molecules, applied his general formalism to the case of slow neutrons, and deduced, among others, the relation between the elastic cross section per unit solid angle in the forward direction and the ratio of the elastic cross section and the de Broglie wavelength of the neutrons /11/.

A problem of interest at the end of the 30's was that of the influence of the density of the medium on the energy lost in collisions against its atoms by a charged particle moving through it. In investigating the interaction of charged particles with atoms it is permissible to treat them as isolated only when te medium is a gas. When the particle travels in a condensed material, we can still consider the atoms as isolated in the case of close collisions but we cannot do so when the impact parameter is larger than tha atomic separation. For such distant collisions one has to take into account the screening of the electric field of the passing particle by the atoms of the medium. The screening reduces the interaction and decreases, therefore, the energy loss. The theory of this phenomenon was investigated by Fermi in 1939 [20], who caracterized the medium by means of its static dielectric constant. Immediately Wick /12/ made a more refined analysis by considering the resonance frequencies characteristic of the medium and in successive papers applied its general formulas to te cases of carbon and lead, the two materials usually employed by experimentalist working in cosmic rays. « Wick's curves » were reproduced in various books on cosmic rays and were used by people working on the penetrating component of cosmic rays. Similar results were obtained, at about the same time, also by Halpern and Hall in the United States [21].

In those years the interest in the density effect mentioned above had been stimulated by the experiments that a few groups had started to carry out in 1939 on the hard component of cosmic rays. The corresponding particles, discovered by Neddermayer and Anderson and by Street and Stevenson in 1936 [22], were called mesotrons, and showed a larger absorption in air than, for example, in carbon taken in equal amount per unit area. The density effect was found to represent an important correction but did not explain the difference observed.

In 1938 Kulenkampff [23] proposed as an explanation of this anomaly that mesotrons are unstable particles as suggested by Yukawa's theory and that their meanlife is comparable with their time of flight in the atmosphere. While in condensed absorbers decay would play a negligible role, mesotrons travelling in the atmosphere would disappear by spontaneous decay before reaching the end of their range. In subsequent years Bruno Rossi and collaborators in United States [24], Gilberto Bernardini and collaborators in Italy [25], as well as other authors [26] made a systematic investigation of this effect and of a few other related phenomena.

In those years Gian Carlo Wick joined Bernardini's group and contributed considerably to many of their papers regarding the amomalous absorption of the hard component of cosmic rays and the determination of the ratio τ/mc^2 of the mesotron /13/, the investigation of the equilibrium conditions of the electronic and mesotronic component at various altitudes /14/, and the experimental determination of the positive excess of the mesotrons present in the cosmic radiation /15/. Wick's contribution to these important works did not consist only of theoretical advice and detailed computations such as those of the trajectories of the mesotrons inside the magnetized iron of the magnetic lenses, but also of analysis of the data and, occasionaly, the taking of data, especially at mountain altitudes, where Gian Carlo Wick from time to time liked to spend even weeks.

In autumn 1937 Wick was nominated professor of theoretical physics at the University of Palermo where Emilio Segrè had had the chair of experimental physics since autumm 1935. About one year later, he passes to the same chair at the University of Padua, at the invitation of Bruno Rossi, who had been there since 1933, in the chair of experimental physics.

The racial laws promulgated by the fascist governemt on July 14, 1938 determined, however, the departure from Italy during the autumn of the same year of Enrico Fermi. Bruno Rossi, Emilio Segrè, Giulio Racah, Ugo Fano and a few other still young but very promising physicists.

When, in October 1940, Fermi did not return from the United States to Italy and the Minister of National Education considered him « resigned » we could arrange for Wick to move to the chair of theoretical physics of the University of Rome, as had been suggested to the Faculty by Fermi himself.

The majority of the courses given by Gian Carlo in those years were devoted to quantum mechanics, which he always presented in a very transparent and at the same time rather deep form. He devoted, however, the academic year 1940-41 to a detailed discussion of all the experiments and theories that during the XIX-th century had dealt with the study of the velocity of light and the search for evidence of its compositition with the velocity of the observer with respect to the ether. The lecture room was always crowded not only with students but also junior and senior staff of the Institute who enjoyed listening to a detailed and deep discussion of the arguments that paved the way the advent of restricted relativity. In response to the unanimous desire of audience, the notes of these lectures of Wick were published in a very simple and inexpensive form /16/.

In those years Gian Carlo and I planned to write a book on neutron physics. About two chapters were written and printed in a very simple form /17/, but our common work was interrupted by the departure of Gian Carlo for the United States in 1946. I used part of them for the article that I wrote about twelve years later for the *Handbuch der Physik*.

The qualities that always impressed me most in Gian Carlo as a physicist were his clarity of ideas and the depth of his understanding of any phenomenon and of its theoretical representation. He carries this clarity of ideas also outside of the domain of physics or of the natural sciences. When we were young I had the impression that he had a similar clear vision of the political problems we were facing in Italy in those years. His antifascism, inherited, so to say, from his mother Barbara Allason, had extremely deep human roots as well as a very solid rational structure and was supported by a wide culture which rose well above any local ambition or nationalistic pride.

I have learned a lot from Gian Carlo when we were young as well as at later times. I regretted it deeply when he left Italy and Europe; I enjoyed enormously when a few years ago I heard for the first time that he was coming back, that he was coming to Pisa, to the Scuola Normale Superiore. I was very glad to think that a few generations of well selected students of physics would have the opportunity to know Wick, to learn some physics from him, and to appreciate his personal style.

Many thanks, Gian Carlo!

G. C. WICK'S PAPERS QUOTED HERE

/1/ *Sul momento magnetico di una molecola di idrogeno*, Nuovo Cimento, **9**, 118 (1933);
Über das magnetische moment eines rotierenden Wasserstoffmoleküls, Zeit. f. Phys., **85**, 25 (1933).

/2/ *Sull'interazione dei neutroni coi protoni*, Ric. Scient., **4** (1), 585 (1933);
Über die Wechselwirkung zwischen Neutronen und Protonen, Zeit. f. Phys., **84**. 799 (1933).

/3/ *Sugli elementi radioattivi di F. Joliot e I. Curie*, Rend. Accad. Lincei, **19**, 319 (1934).

/4/ *Teoria dei raggi beta e momento magnetico del protone*, Rend. Accad. Lincei, **21**, 170 (1935).

/5/ *Sulla diffusione dei neutroni* (in collaboration with B. PONTECORVO), I, Ric. Scient., **7** (1), 134 (1936); II, ibidem, **7** (1), 220 (1936).

/6/ *Sulla diffusione dei neutroni lenti*, Rend. Accad. Lincei, **23**, 774 (1936);
On the slowing down of neutrons, Phys. Rev., **49**, 192 (1936);
Über eben Diffusionsprobleme, Zeit. f. Phys., **121**, 702 (1943);
Application of Fokker-Planck equation of the energy spectrum of thermal neutrons, Phys. Rev., **70**, 103 (1946);
Some stationary distributions of neutrons in an infinite medium (in collaboration with M. VERDE), Phys. Rev., **71**, 852 (1947).

/7/ *Sulla diffusione dei neutroni nei cristalli*, Ric. Scient., **8**, 400 (1937);
Über die Streung der Neutronen an Atomgittern I, Phys. Zeit., **38**, 403 (1937); II, ibidem, 689.

/8/ *Range of the nuclear forces in Yukawa's theory*, Nature (London), **142**, 993 (1938).

/9/ *Sulla stabilità del modello nucleare a goccia allungata*, Nuovo Cimento, **16**, 229 (1939).

/10/ *Eccitazione dei nuclei mediante elettroni veloci*, Ric. Scient., **11**, 49 (1940).

/11/ *Sulla propagazione di un'onda di de Broglie in un mezzo materiale*, Mem. Classe Fis. Mat. Nat. Acad. Italia, **13**, 1203 (1942).

/12/ *Osservazioni sul frenamento delle particelle veloci*, Ric. Scient., **11**, 273 (1939); **12**, 858 (1941); Nuovo Cimento, **1**, 302 (1943).

/13/ *Anomalous absorption of hard component of cosmic rays in air* (in collaboration with M. Ageno, G. Bernardini, N. B. Cacciapuoti, B. Ferretti), Phys. Rev., **57**, 945 (1940).

/14/ *Condizioni di equilibrio delle componenti elettronica e mesonica della radiazione cosmica alle varie altezze* (in collaboration with G. Bernardini, B. N. Cacciapuoti, B. Ferretti, O. Piccioni), Mem. Classe Scien. Fis. Mat. Nat. Accad. Italia, **11** (1), 471 (1941);
Genetic relation between electronic and mesotronic components of cosmic rays near and above sea level (in collaboration with G. Bernardini, B. N. Cacciapuoti, B. Ferretti, O. Piccioni), Phys. Rev., **58**, 1017 (1940).

/15/ *Sull'eccesso positivo della radiazione cosmica* (in collaboration with G. Bernardini, M. Conversi, E. Pancini), Ric. Scient., **12**, 127 (1941);
Positive excess in mesotron spectrum (in collaboration with G. Bernardini M. Conversi, E. Pancini), Phys. Rev., **60**, 535 (1941);
Researches on the magnetic deflection of the hard component of cosmic rays (in collaboration with G. Bernardini, M. Conversi, E. Pancini, E. Scrocco), Phys. Rev., **68**, 109 (1945).

/16/ *Introduzione alla teoria della Relatività. Lezioni di Fisica Teorica*, Anno Accademico 1944-45, Veschi (Roma), 1945.
In consultation with G. C. Wick and M. Conversi, who followed this course as a student, I arrived at the conclusion that it should have been given in 1940-41 and that the academic year (1944-45) printed on the frontispiece is that of its publication.

/17/ *Appunti di Fisica Nucleare I*, Anno Accademico 1944-45 (in collaboration with E. Amaldi), pp. 3-422 (Tipo-litografia Romolo Piola, Roma);
Appunti di Fisica Nucleare II, Anno Accademico 1945-46 (in collaboration with E. Amaldi), pp. 3-435 (Tipo-litografia Romolo Piola, Roma).

REFERENCES

[1] R. Frisch and O. Stern: *Über die magnetische Ablenkung Wasserstoffmolekülen und das magnetische Moment des Proton I*, Zeit. f. Phys., **85**, 4-16 (1933);
I. Estermann and O. Stern: ibidem, 17-24.

[2] I. I. Rabi, J. M. B. Kellogg and J. R. Zacharias: *The magnetic moment of the proton*, Phys. Rev., **46**, 157 (1934).

[3] W. Heisenberg: *Über den Bau der Atomkerne I*, Zeit. f. Phys., **77**, 1 (1932); II, ibidem, **78**, 156 (1932); III, ibidem, **80**, 587 (1933);
E. Majorana: *Über die Kerntheorie*, Zeit. f. Phys., **82**, 137 (1933); *Sulla teoria dei nuclei*, Ric. Scient., **4** (1), 559 (1933).

[4] F. Joliot and I. Curie: *Un nouveau type de radioactivité*, C. R. Acad. Sci. Paris, **198**, 254 (1934); *Separation chimique des nouveau radioelements émitteurs d'éléctrons positivs*, C. R. Acad. Sci. Paris, **198**, 559 (1934).

[5] E. Fermi: *Tentativo di una teoria dei Raggi Beta*, Ric. Scient., **4** (2), 491 (1933); Nuovo Cimento, **11**, 1 (1934); *Versuch einer Theorie der β-Strahlen I*, Zeit. f. Phys., **88**, 161 (1934).

[6] H. Bethe and R. Peierls: *The Neutrino*, Nature (London), **133**, 689 (1934).

[7] H. Yukawa and S. Sakata: *On the Theory of β-Disintegration and the Allied Phenomena*, Proc. Phys. Mat. Soc. Japan, **17**, 467 (1935); ibidem, **18**, 128 (1936).

[8] L. W. Alvarez: *The Capture of Orbital Electrons*, Phys. Rev., **54**, 486 (1938).

[9] Ig. Tamm: *Exchange Forces between Neutrons and Protons and Fermi's Theory*, Nature (London), **133**, 981 (1934);
D. Iwanenko: *Interaction of Neutrons and Protons*, Nature (London), **133** 981 (1934);
A. Nordsick: *Neutron Collisions and the Beta Ray Theory of Fermi*, Phys. Rev., **46**, 234 (1934);
W. Heisenberg: *Bemerkung zur Theorie des Atomkerns*, in « Peter Zeeman, 1865 - 25 Mai 1935 (Martin Nijkoff, 's-Gravenhage, 1935), p. 108;
C. F. v. Weizsäcker: *Über die Spinabhangigkeit der Kernkräfte*, Zeit. f. Phys., **102**, 572 (1936);
M. Fierz: *Zur Fermischen Theorie des β-Zerfalls*, Zeit. f. Phys., **104**, 553 (1936).

[10] H. Fröhlich, W. Heitler and N. Kemmer: *On the nuclear forces and the magnetic moment of the proton*, Proc. Roy. Soc. A, **166**, 154 (1938).

[11] W. M. Elsasser: *Sur la diffraction des neutrons lents par les substances cristallines*, C. R. Acad. Sci. Paris, **202**, 1929 (1936).

[12] P. Preiswerk and H. v. Halban: *Preuve experimental de la diffraction des neutrons*, C. R. Acad. Sci. Paris, **203**, 73 (1936);
P. Preiswerk: *Ein Neutronenbeugungexperiment*, Helv. Phys. Acta, **10**, 400 (1937).

[13] B. P. Mitchell and P. N. Powers: *Bragg Reflection of Slow Neutrons*, Phys. Rev., **50**, 486 (1936).

[14] E. Fermi: *Sul moto dei neutroni nelle sostanze idrogenate*, Ric. Scient., **7** (2), 13 (1936).

[15] W. Wefelmeyer: *Ein Model der Transurane*, Naturwiss., **27**, 110 (1939).

[16] C. F. v. Weizsäcker: *Zum Wefelmeyerschen Model der Transurane*, Naturwiss., **27**, 133, 277 (1939).

[17] V. F. Weisskopf: *Excitation of Nuclei by Bombardment with charged Particles*, Phys. Rev., **53**, 1018 (1938).

[18] B. Pontecorvo and A. Lazard: *Isomerie nucléaire produite par le rayons* X *du spectre continu*, C. R. Acad. Sci. Paris, **208**, 99 (1939).

[19] H. A. Bethe and R. Peierls: *Quantum Theory of the Diplon*, Proc. Roy. Soc. A, **148**, 146 (1935).

[20] E. Fermi: *The Absorption of Mesotrons in Air and in Condensed Materials*, Phys. Rev., **56**, 412 (1939); *The Ionization Loss of Energy in Gases and in Condensed Materials*, Phys. Rev., **57**, 485 (1940).

[21] O. Halpern and H. Hall: *Energy Losses of Fast Mesotrons and Electrons in Condensed Materials*, Phys. Rev., **57**, 459 (1940); *The Ionization Loss of Energy of Fast Charged Particles in Gases and Condensed Bodies*, Phys. Rev., **73**, 477 (1948).

[22] S. H. Neddermeyer and C.D. Anderson: *Note on theNature of Cosmic Ray Particles*, Phys. Rev., **51**, 884 (1937);
J. C. Street and E. C. Stevenson: *Penetrating corpuscular component of the Cosmic Radiation*, Phys. Rev., **51**, 1005 (1937); ibidem, **52**, 1003 (1937).

[23] H. Kulemkampff: *Bernerkung über die durchdringenden Komponente der Ultrastrahlung*, Verh. d. Deutsch. Phys. Ges., **19** (3), 92 (1938).

[24] B. Rossi, N. Hilbery and J. B. Hoag: *The Variation of the Hard Component of Cosmic Rays with Height and the Disintegration of the Mesotron*, Phys. Rev. **57**, 461 (1940);
B. Rossi and D. B. Hall: *Variation of the Rate of Decay of Mesotrons with Momentum*, Phys. Rev., **59**, 223 (1941);
B. Rossi, K. Greisen, J. C. Stearns, D. K. Froman and P. Koontz: *Further Measurements of the Mesotron Lifetime*, Phys. Rev., **61**, 675 (1942).

[25] For these researches see the concluding paper: G. Bernardini: *Über anomale Absorption in Luft und die Lebensdauer des Mesons*, Zeit. f. Phys., **120**, 413 (1942).

[26] W. M. Nielson, C. M. Ryerson, L. W. Nordheim and K. Z. Morgan: *A Measurement of Mesotron Lifetime*, Phys. Rev., **57**, 158 (1940); ibidem, **59**, 547 (1941);
M. A. Pomerantz: *The Instability of the Meson*, Phys. Rev., **57**, 3 (1940);
M. V. Neher and H. G. Stever: *The Mean Lifetime of the Mesotron from Electroscope Data*, Phys. Rev., **58**, 766 (1940).

EDOARDO AMALDI

IL CASO DELLA FISICA

1. INTRODUZIONE

Mi sembra giusto, direi quasi doveroso, iniziare il mio contributo a questo convegno con qualche considerazione più ampia, fatta da vari studiosi, in particolare da Emilio Segrè, purtroppo scomparso da meno di venti giorni. Oltre ad una grande figura della fisica attiva in rapido sviluppo, Egli è stato un notevole e profondo cultore della storia della fisica degli ultimi 200 anni.

Sia Emilio Segrè (1) che altri prima di lui, hanno notato che la fisica in Italia, dopo lo splendido periodo di Alessandro Volta, a cavallo dell'anno 1800, è entrata in una fase di quasi stagnazione. Naturalmente vi sono state figure come quelle di Macedonio Melloni (1798-1852) e di Ottaviano Mossotti (1791-1863), ma il più grande, Amedeo Avogadro (1776-1856) e, più tardi, Stanislao Cannizzaro (1826-1910), vanno considerati più come appartenenti alla storia universale della chimica che a quella della fisica anche se, a quell'epoca, non era facile fare tale distinzione.

Ma soprattutto dopo l'unificazione d'Italia, a differenza di ciò che accadde per le matematiche, i fisici italiani contribuirono relativamente poco al progresso della loro scienza. Non è certo il caso che mi dilunghi in questa sede in una discussione delle possibili cause di questa arretratezza o, forse meglio, di questo ritardo di sviluppo.

Mi limiterò a ricordare che storici e sociologi ne hanno studiato le cause. Alcuni lo hanno attribuito allo sforzo richiesto dal «Risorgimento», che avrebbe in qualche modo logorato tutte le energie intellettuali del Paese, altri lo collegano allo sviluppo industriale dell'Italia, rispetto a quello dell'Europa del Nord, altri lo attribuiscono al sistema scolastico eccessivamente letterario e retorico, ed io confesso di propendere per questa interpretazione, altri ancora cercano spiegazioni connesse al Marxismo.

Ma lasciamo ad altri queste considerazioni e veniamo ai fatti. Inizierò con il ricordare che alla fine dello scorso secolo il maggior fisico d'Italia fu Augusto Righi (1850-1921), il cui maggior contributo fu l'estensione verso le alte frequenze dell'opera di Heinrich Hertz (1857-1894); vi fu, contemporaneamente o poco dopo, nel campo della tecnologia, l'opera di primo piano

(1) E. SEGRÈ, *The Fermi School in Rome*, «Eur. J. Phys.», 9 (1988), 83-87.

di Guglielmo Marconi (1874-1937) che ricevette il dovuto riconoscimento con il Premio Nobel per la Fisica nel 1909.

Più tardi vi sono stati fisici come Orso Mario Corbino (1876-1937), Antonio Garbasso (1871-1933) e Luigi Puccianti (1875-1952) che hanno dato contributi importanti, ancora oggi citati nella letteratura scientifica.

Ma nessuno nel campo della fisica ha dato in quegli anni un contributo confrontabile con quello dei matematici dell'Università di Padova, Gregorio Ricci Curbastro (1853-1925) e Tullio Levi-Civita (1873-1941) che, attorno al 1901, hanno sviluppato il Calcolo Assoluto, o Analisi Tensoriale, la quale, a partire dal 1913, costituì lo strumento matematico utilizzato da Albert Einstein per costruire e portare a compimento la sua teoria della *Relatività Generale*.

Anche l'insegnamento universitario della fisica era piuttosto modesto, come si può facilmente riconoscere dal fatto che fin verso la metà degli anni 20 la teoria della *Relatività Ristretta* veniva insegnata in qualche corso di matematica, ma non in quelli di fisica e che la *Meccanica Quantistica* veniva praticamente ignorata in tutte le università italiane.

2. La nascita dei gruppi di Roma e Firenze e la loro quasi distruzione

La situazione cominciò a cambiare in maniera sostanziale dopo la prima guerra mondiale con l'apparire di Enrico Fermi (1901-1954) che, diventato professore di fisica teorica a Roma nel 1926, aveva creato rapidamente una scuola di fisica moderna, all'Istituto di Fisica di Via Panisperna diretto da O.M. Corbino (2). Un secondo gruppo era nato poco dopo all'Istituto di Fisica di Arcetri, diretto da A. Garbasso, e di cui, inizialmente, facevano parte Bruno Rossi, Gilberto Bernardini e Giuseppe Occhialini.

La nascita di questi due gruppi non era casuale ma almeno in parte la conseguenza dell'azione di Corbino, che, nel 1926, era riuscito a far bandire dal Ministero dell'Educazione Nazionale il primo concorso a cattedra per la Fisica Teorica.

I vincitori del concorso erano stati Enrico Fermi primo, Enrico Persico secondo ed Aldo Pontremoli terzo, che erano stati chiamati a ricoprire le cattedre rispettivamente delle Università di Roma, Firenze e Milano.

(2) *a*) *Enrico Fermi: Note e Memorie*, 2 voll. Accademia Nazionale dei Lincei e The University of Chicago Press, Rome; Vol. 1, 1962, Italia 1921-1938, Vol. 2, 1965, USA, 1939-1954. *b*) E. Segrè, *Enrico Fermi, fisico*, Zanichelli, Bologna (1971); *Enrico Fermi, physicist*, The University of Chicago Press (1970). *c*) E. Amaldi, *Personal Notes on Neutron Work in Rome in the 30s*, in *History of Twentieth Century Physics*, C. Weiner (ed.), Academic Press, New York e Londra, 1977, 294-325. *d*) Laura Fermi, *Atomi in famiglia*, Zanichelli, Bologna (1955); *Atoms in the Family*, The University of Chicago Press (1954). *e*) B. Pontecorvo, *Fermi e la fisica moderna*, Editori Riuniti, Roma (1972). Traduzione dal russo della introduzione scritta da B. Pontecorvo a *Le opere scientifiche di Enrico Fermi*, Mosca, Nauka, 1971.

Pontremoli (1896-1928) si era laureato in fisica nel 1920 a Roma, dopo aver interrotto gli studi universitari per partecipare alla prima guerra mondiale come volontario. Dopo essere stato assistente di O.M. Corbino all'Istituto di Via Panisperna dal 1920 al 1924, era stato chiamato all'Università di Milano, ove aveva fondato l'Istituto di Fisica di quella Università, che da circa sessant'anni porta il suo nome. Purtroppo meno di due anni dopo la sua nomina a professore di ruolo a Milano era scomparso nella seconda spedizione al Polo Nord con dirigibile, organizzata e diretta da Umberto Nobile (3).

Persico (1900-1969) era stato amico di Fermi fin dal Liceo e, negli ultimi anni, era stato aiuto di Corbino (4). A Firenze egli cominciò subito a svolgere il suo corso di Fisica Teorica e le sue lezioni, raccolte da Bruno Rossi, giovane assistente di Garbasso, e Giulio Racah, ancora studente, e stampate a spese dell'Istituto di Corbino, diventò il testo su cui tutti i giovani dell'epoca studiarono la nascente meccanica quantistica. Franco Rasetti, che era stato aiuto di Garbasso a Firenze, ben presto passò all'Università di Roma, dove diventò aiuto di Corbino e più tardi (1928) professore di spettroscopia.

In breve tempo il gruppo di Firenze si allargò con l'aggiunta di Giulio Racah (1909-1965), che divideva il suo tempo fra Firenze, Roma e Zurigo, ove lavorava con W. Pauli, Daria Bocciarelli (n. 1910), Lorenzo Emo Capodilista (1909-1973), Sergio De Benedetti (n. 1912) e Manlio Madò (n. 1912).

L'atmosfera caratteristica dell'Istituto di Fisica di Arcetri in quegli anni emerge in maniera particolarmente efficace dagli Atti della tavola rotonda su «Arcetri dagli anni '20 agli anni '30» che si è svolta nei locali del Dipartimento di Fisica dell'Università di Firenze (in Arcetri) il 4 dicembre 1987 (5).

Fra l'autunno 1926 e il dicembre 1938, a Roma, Enrico Fermi, con l'aiuto di Franco Rasetti per la parte sperimentale, riuscì a creare un numero assai notevole di allievi, sia sperimentali che teorici.

Gli sperimentali, in ordine di arrivo all'Istituto di Fisica di Via Panisperna, sono stati: Emilio Segrè (1905-1989), Edoardo Amaldi (n. 1908), Bruno Pontecorvo (n. 1913) proveniente da Pisa, Eugenio Fubini (n. 1914) proveniente da Torino, Mario Ageno (n. 1915) proveniente da Genova, Giuseppe Cocconi (n. 1913) da Milano, Oreste Piccioni (n. 1916) da Pisa e Marcello Conversi (1917-1988). I giovani teorici: Ettore Majorana (1906-1938), Giovanni Gentile

(3) Al ritorno dal suo secondo volo sul Polo Nord di questa seconda spedizione (1928), il dirigibile Italia si era appesantito per la formazione di uno strato di ghiaccio, fino a precipitare sulla banchisa distruggendosi e lasciando sul pack polare alcuni naufraghi, fra cui Pontremoli. Nelle vicende che seguirono, trovarono la morte 8 membri della spedizione, oltre a 6 persone che avevano tentato di portare loro soccorso.

(4) E. Amaldi e F. Rasetti, *Ricordo di Enrico Persico*, in *Celebrazioni Lincee*, *115* (1979), e «Il Giornale di Fisica», *20* (1979), 235-269.

(5) Tavola rotonda su: *Arcetri dagli anni '20 agli anni '30* (partecipanti: E. Amaldi, G. Bernardini, D. Bocciarelli, M. Mandò, G. Occhialini, B. Rossi), Arcetri, 4 dicembre 1987, a cura di Alberto Bonetti e Manlio Mandò.

jr. (1906-1942), Renato Einaudi (1909-1976) e Gian Carlo Wick (n. 1909) gli ultimi due provenienti da Torino, Giulio Racah (1909-1965) proveniente da Firenze, Leo Pincherle (1910-1976) da Bologna, U. Fano (n. 1912) ancora da Torino e Piero Caldirola (1914-1984) da Pavia.

Un caso anomalo fu quello di Salvatore E. Luria (n. 1912), un giovane medico torinese che voleva imparare la fisica o per lo meno i suoi metodi. Amico e coetaneo di Fano si consultò con Fermi e Rasetti che lo incoraggiarono nelle sue intenzioni (6). Fu messo nel mio studio e lui si mise ad approfondire le sue conoscenze soprattutto sotto l'influenza di Rasetti. Fra Luria e me fu fatto un patto di mutua assistenza. Io l'avrei aiutato a dipanare qualche passaggio di matematica a lui poco famigliare, e lui mi avrebbe prestato una, o due mani, quando nel montaggio di un esperimento le mie fossero risultate insufficienti.

I due gruppi di fisici delle Università di Roma e Firenze acquistarono rapidamente fama internazionale lavorando in campi di ricerca diversi ma vicini, con lavori e scoperte che sono rimaste nella storia della fisica universale. Il gruppo di Firenze si dedicò prevalentemente allo studio della radiazione cosmica. Una descrizione molto affascinante dei risultati ottenuti in quegli anni è presentata nei primi capitoli del libro autobiografico di Bruno Rossi «Momenti nella vita di uno scienziato» recentemente pubblicato dall'editore Zanichelli (7).

Qui mi limiterò a ricordare: il metodo di coincidenze di Rossi (1930), usato per vari decenni dai fisici di tutto il mondo, la scoperta della produzione di particelle secondarie da parte dei raggi cosmici (1932), la curva di transizione, o curva di Rossi, dovuta alla produzione di questi secondari (1933), l'analisi della radiazione cosmica in una componente molle e una componente dura e il fatto che quest'ultima attraversa anche un metro di piombo. Infine lo studio dell'effetto Est-Ovest provocato dal campo magnetico terrestre previsto dal Rossi e da lui studiato insieme a Sergio De Benedetti in una spedizione in zona equatoriale (Asmara, 1934). Quest'ultima attività fu svolta dopo il trasferimento di Rossi all'Università di Padova, come chiarirò fra poco.

Il gruppo di Roma si occupò invece di vari problemi di fisica atomica e molecolare e, a partire dal 1934, di problemi di fisica nucleare.

Fra i molti contributi alla fisica atomica e molecolare qui mi limiterò a ricordare il modello statistico di atomo di Thomas-Fermi, ideato da Fermi (e poco prima da Thomas) e applicato a vari problemi specifici da lui stesso, da Rasetti, Majorana, Gentile, Segrè e Pincherle, e perfezionato sotto vari aspetti da Fermi e Amaldi; lo studio sperimentale dell'effetto Raman dei livelli rotazionali delle molecole biatomiche simmetriche e dei cristalli (Rasetti) e la teoria dell'effetto Raman nei cristalli di Fermi, l'effetto Raman dei livelli

(6) SALVATORE EDOARDO LURIA, *A Slot Machine, A Broken Test Tube - An Autobiography*, Harper and Row Publishers, New York 1984.

(7) B. ROSSI, *Momenti della vita di uno scienziato*, Zanichelli, Le Ellissi, 1987.

rotazionali dell'ossido di carbonio e dell'ammoniaca (Amaldi e G. Placzek), lo studio sperimentale delle righe proibite degli alcalini (Segrè), la teoria delle strutture iperfini (Fermi e Segrè), lo studio sperimentale dell'effetto Stack-Lo Surdo degli stati di Rydberg e la scoperta dello spostamento subito da questi stati per la presenza di un gas estraneo (Amaldi e Segrè) e la teoria di questo effetto fatta da Fermi, vari eleganti lavori di spettroscopia e sul legame chimico di Ettore Majorana. Fondamentali per lo sviluppo e diffusione di questo insegnamento sono stati i libri di Fermi e Persico (8).

Nel campo della fisica nucleare debbo ricordare la teoria delle forze di scambio di Majorana, la teoria della disintegrazione beta di Fermi, la estensione di questa teoria fatta da G.C. Wick, vari altri eleganti contributi di questo stesso autore, alcuni lavori di elettrodinamica di Racah, e la teoria simmetrica dell'elettrone e del positrone di Majorana.

Sul piano sperimentale è stata particolarmente importante la scoperta (Fermi, 1934) della radioattività provocata da neutroni, seguita da un lavoro sperimentale sistematico (svolto da Fermi, Amaldi, D'Agostino – chimico –, Pontecorvo, Rasetti, Segrè), che ha portato a individuare quasi cinquanta nuovi corpi radioattivi artificiali, alcuni dei quali prodotti nel bombardamento dell'uranio, la scoperta della possibilità di rallentare i neutroni, e della straordinaria efficacia dei neutroni lenti così ottenuti nell'indurre processi di cattura radiativa.

A partire dal 1936 il libro di Rasetti sulla fisica nucleare (9) fu per molti anni il testo fondamentale a livello mondiale, soprattutto per le parti più moderne di questo capitolo della fisica, accanto ai trattati classici di Lord Rutherford e collaboratori, Maria Sklodowska Curie e St. Meyer e E. von Schweidler.

Le attività di tutti i componenti dei due gruppi di Roma e Firenze erano tuttavia solo agli inizi, quando il «Manifesto in difesa della Razza» pubblicato il 14 luglio 1938 segnava l'apertura di un periodo di molti mesi durante i quali il governo Fascista promulgava, a intervalli di poche settimane, una serie di Decreti-legge che venivano definiti di protezione della razza ariana.

Entrambi i gruppi di Roma e Firenze venivano gravemente decimati, quasi distrutti (10). I professori di Fisica che lasciarono la cattedra in seguito alle leggi razziali ed emigrarono all'estero furono cinque: Bruno Rossi, Emilio Segrè e Giulio Racah in quanto, in seguito a tali leggi avevano perso il posto e i diritti civili in Italia. Enrico Fermi in quanto sua moglie, Laura Capon, ap-

(8) E. FERMI, *Fisica atomica*, Zanichelli, Bologna (1928); ID., *Molecole e Cristalli*, Zanichelli, Bologna (1934); E. PERSICO, *Lezioni di Meccanica Ondulatoria redatte da B. Rossi e G. Racah*, CEDAM, Padova (1935); ID., *Fondamenti di Meccanica Atomica*, Zanichelli, Bologna (1939); *Fundamentals of Quantum Mechanics*, tradotto in inglese da G.M. Temmer, Prentice Hall, Inc., New York (1950).

(9) F. RASETTI, *Il nucleo Atomico*, Zanichelli, Bologna (1936); *Elements of Nuclear Physics*, Prentice Hall, Inc., New York (1936).

(10) E. AMALDI, *Gli anni della ricostruzione*, «Il Giornale di Fisica», *20* (1979), 186-225.

parteneva a una famiglia ebrea di Venezia e pertanto era colpita da tali leggi, ed anzi il loro stesso matrimonio era diventato illegale; infine Franco Rasetti perché, dopo le leggi razziali, non voleva più vivere in un paese così incivile.

I timori di questi colleghi non erano infondati. A titolo di esempio ricorderò che la madre di Emilio Segrè, Amelia Treves, figlia di un noto architetto fiorentino, e l'Ammiraglio della Marina Militare Italiana Augusto Capon (n. 1872), padre di Laura Fermi, furono arrestati di sorpresa dalle SS nelle loro abitazioni in Roma il 16 ottobre 1943 e deportati. Essi scomparvero, forse per morte causata dagli stenti e maltrattamenti durante il viaggio, o uccisi all'arrivo in Germania o poco dopo, secondo ben note metodologie naziste.

Una sorte analoga toccò anche ad Egle Segrè (n. 1899), prima delle tre figlie di Gino Segrè, zio di Emilio. Si era sposata con Edgardo Levy, commerciante, e dal loro matrimonio erano nati due figli: Mariuccia ed Enzo. Durante la guerra erano sfollati da Torino a Tradate dove, l'8 novembre 1943, le SS avevano arrestato Egle e i figli, rispettivamente di 21 e 16 anni, e li avevano deportati ad Auschwitz dove Enzo era stato separato dalla madre e dalla sorella. Dopo la guerra solo Enzo era ritornato: di Egle e Mariuccia non si seppe mai più nulla.

Altri fisici ebrei italiani che non erano ancora in cattedra ma che emigrarono alla stessa epoca, e nel seguito raggiunsero posizioni universitarie, o sociali eminenti sono: Leo Pincherle, Ugo Fano, Sergio De Benedetti e Bruno Pontecorvo nel campo della fisica, Eugenio Fubini nel campo della fisica applicata, e Mario Salvadori e Roberto Fano nel campo dell'ingegneria.

Questa mia lista è incompleta anche per i fisici, ma soprattutto per i fisici applicati e ingegneri. A ciò va aggiunto che altri ancora, colpiti dalle leggi razziali, non lasciarono l'Italia perché in età o troppo giovanile o troppo avanzata o per motivi di famiglia o altro, ne subirono tutte le conseguenze per così dire in sede. Essi sono molti e meriterebbero altrettanto di essere ricordati. Esclusivamente per ragioni di tempo sono costretto a limitarmi solo ad alcuni di quelli che sono emigrati.

Debbo però prima chiarire come nel corso degli anni 30, prima delle leggi razziali, il quadro dei professori universitari di fisica avesse subito vari mutamenti per il normale corso della vita universitaria e per le vicende della vita dei singoli individui.

Persico, avendo avuto, nell'autunno del 1930, l'offerta della cattedra di Fisica Teorica dell'Università di Torino, l'aveva accettata e l'Università di Firenze aveva sopperito affidando per incarico la Fisica Teorica prima a Bruno Rossi e, a partire dal 1932-33 a Giulio Racah. Il 23 marzo 1933 Antonio Garbasso era scomparso dopo lunga malattia e gli era succeduto Laureto Tieri (1879-1952), il quale peraltro non era riuscito a stabilire gli stessi rapporti di reciproca comprensione e affabilità che esistevano fra il suo predecessore e i fisici della nuova generazione.

Nell'autunno 1932 un primo concorso a cattedra di Fisica Sperimentale era stato vinto da Bruno Rossi che era stato chiamato all'Università di Padova.

Un secondo concorso della stessa materia dell'autunno 1935 era stato vinto da Emilio Segrè che era stato chiamato all'Università di Palermo, e un terzo, dell'autunno 1936 aveva portato in cattedra Amaldi e Bernardini, che erano stati chiamati rispettivamente a Cagliari e Camerino. Ma il 23 gennaio 1937 moriva improvvisamente Orso Mario Corbino (11), e la Facoltà di Scienze dell'Università di Roma chiamava E. Amaldi a succedergli. G. Bernardini veniva chiamato alla cattedra di Fisica Superiore dell'Università di Bologna nell'autunno 1938.

Emilio Segrè, subito dopo aver preso il suo posto a Palermo, riuscì a convincere la Facoltà di Scienze di quella Università a chiedere un concorso a cattedra per la Fisica Teorica, il secondo dopo quello del 1926 (!) e, con questo il solo di questa materia fra le due guerre mondiali. Come conseguenza di tale concorso, nell'autunno 1937 Ettore Majorana fu nominato professore ordinario di Fisica Teorica all'Università di Napoli (12) per meriti speciali. L'articolo di legge utilizzato (Art. 8 del R.D.L. 20 giugno 1935, n. 1071) era stato fatto qualche anno prima allo scopo di rendere possibile la nomina di Guglielmo Marconi alla cattedra di Onde Elettromagnetiche dell'Università di Roma, senza concorso. La terna per il concorso dell'Università di Palermo era risultata subito dopo: Gian Carlo Wick primo, Giulio Racah secondo e Giovanni Gentile jr terzo, che erano stati chiamati alle cattedre di Fisica Teorica di Palermo, Pisa e Milano. Ettore Majorana era poi scomparso nella primavera 1938.

3. La distruzione quasi totale del gruppo di Roma

Veniamo alle conseguenze delle leggi razziali in Italia.

Fermi lasciò l'Italia con la famiglia il 6 dicembre 1938 per recarsi a Stoccolma (13) e giunse a New York ai primi di gennaio 1939. Ufficialmente egli aveva accettato di insegnare alla Columbia University come «visiting professor» per un anno accademico, ma già qualche settimana prima aveva informato Rasetti e me che la sua partenza sarebbe stata definitiva. Al Pupin Laboratory, ove aveva sede ed ancor oggi ha sede, in parte almeno, il Department of Physics di quella Università, Fermi si mise a lavorare con Herbert L. Anderson per chiarire vari aspetti del fenomeno della fissione dell'uranio scoperto a Berlino-Dahlem da Otto Hahn e F. Strassmann alla fine del 1938,

(11) E. Amaldi, E. Segreto, *Corbino O.M.*, in *Dizionario Biografico degli Italiani, 28*, 760-766, Istituto dell'Enciclopedia Italiana. E. Amaldi, *Corbino O.M.*, in *Scienziati e tecnologi contemporanei, 1*, 264-266, Mondadori, Milano 1974; v. anche nota 2*b*.

(12) E. Amaldi, *La vita e l'opera di Ettore Majorana*, Accademia Nazionale dei Lincei (1966); *Ettore Majorana, man and scientist*, in *Strong and Weak Interactions – Present Problems*, Academic Press, New York (1966); v. anche nota 2*d*.

(13) Amaldi, *Gli anni della ricostruzione*... cit.

cioè poche settimane prima. Il risultato del loro lavoro fu ben presto tenuto segreto in vista delle possibili applicazioni pacifiche e forse militari. Il gruppo allargato con l'aggiunta di altri ricercatori, fu trasferito alla Università di Chicago nella primavera del 1942 ove, il 2 dicembre dello stesso anno, riuscì a mettere in funzione per la prima volta un reattore nucleare di potenza piccola, ma non nulla, che dimostrava la possibilità di produrre energia per scopi civili utilizzando il fenomeno della fissione nucleare.

Da Chicago Fermi si trasferì ai Laboratori di Los Alamos, diretti da Robert Oppenheimer, ove ebbe funzioni di consulente per la costruzione della prima bomba atomica. Finita la guerra, nel 1946, Fermi tornò all'Università di Chicago ove rimase fino alla morte (1954). In questa ultima fase della sua vita Fermi dapprima lavorò sulle proprietà dei neutroni lenti e il loro modo di interagire con la materia, servendosi di fasci di queste particelle prodotti da reattori nucleari di ricerca; poi, dopo la scoperta dei mesoni pigreco, studiò l'urto di queste nuove particelle contro nucleoni scoprendo così la prima risonanza fra particelle elementari. Altri lavori dello stesso periodo comprendono una originale teoria dell'origine dei raggi cosmici e una teoria statistica della produzione di nuove particelle in processi d'urto di energia elevata.

Come traspare anche da questo breve cenno biografico era più che naturale e praticamente inevitabile che, dopo la scoperta della fissione dell'uranio, Fermi si mettesse a lavorare sulle possibili applicazioni pacifiche dell'energia nucleare. Ciò, probabilmente, sarebbe accaduto anche se non fosse scoppiato il secondo conflitto mondiale.

Il passaggio dalle applicazioni civili a quelle militari fu invece certamente determinato dal fatto che tale conflitto era già in atto in Europa e che, secondo quanto si sapeva sia in Europa che oltremare, gli scienziati tedeschi si erano messi a lavorare segretamente in questa direzione. Lo stesso argomento veniva ovviamente usato, specularmente, anche dagli scienziati del Terzo Reich, che ignorassero per cecità politica, deformazione nazionalistica o scelta di parte, qualsiasi valutazione giuridico-morale del Governo del loro paese.

Enrico Fermi fu certo fra i primi scienziati del mondo occidentale a trovarsi di fronte al gravissimo problema morale di decidere se partecipare o meno alle ricerche sulle applicazioni militari dell'energia nucleare. La decisione andava presa da lui come da altri di cui parlerò nel seguito in un momento in cui il nemico avanzava su tutti i fronti e l'esito finale del conflitto era molto incerto. Erano tutte persone pacifiche, le informazioni erano scarse, ma per molti di loro sufficienti per ritenere un dovere verso il mondo libero di partecipare alla sua difesa: il 2 settembre 1939 le truppe di Hitler e di Stalin avevano iniziato l'invasione della Polonia dagli opposti confini. Circa un anno dopo Hitler aveva occupato buona parte della Francia e messo la Gran Bretagna in stato d'assedio. Non molti mesi dopo le truppe tedesche avevano invaso la Danimarca, la Norvegia e i Balcani e avanzavano profondamente in Russia e in Africa. Sembrava proprio che Hitler potesse giungere a dominare il mondo.

In tutte le fasi della sua vita Fermi ha prodotto un numero incredibile di allievi. Questa sua attività di maestro durante il periodo americano è stata illustrata da uno dei suoi collaboratori statunitensi, Albert Wattenberg, il quale ha pubblicato un articolo su questo argomento sull'European Journal of Physics del 1988 (14).

Fra le varie decine di allievi sia teorici che sperimentali di Fermi durante il periodo americano vi sono 6 Premi Nobel: tre teorici e tre sperimentali. I teorici sono: i due cinesi, naturalizzati americani, T.D. Lee e C.N. Yang (1957), lo statunitense Murray Gell-Mann (1969) e gli sperimentali E. Segrè e O. Chamberlain (1959) e J. Steinberger (1988). Chamberlain per altro era stato in precedenza allievo di Emilio Segrè e dopo la scoperta dell'antiprotone, fatta nel 1955, seguitò ad essere un suo stretto collaboratore.

Rasetti, scapolo, lasciò l'Italia, insieme alla madre Adele Galeotti, da Napoli il 2 luglio 1939 con la motonave Vulcania, da cui sbarcò a New York dieci giorni dopo. Dopo la partenza di Fermi, Rasetti ed io avevamo fatto alcuni lavori insieme, in particolare avevamo studiato l'emissione istantanea di raggi gamma ed elettroni da parte del gadolinio in seguito alla cattura di un neutrone lento. Ma in primavera Franco aveva ricevuto un'offerta inaspettata dall'Università Cattolica in Canada. Più precisamente l'Università Laval di Quebec aveva recentemente istituito la Facoltà di Scienze e Ingegneria e aveva offerto a Rasetti la direzione del Dipartimento di Fisica. Tale «*posizione* – come Rasetti scrisse anni dopo – *presentava vari aspetti attraenti, ed anche in considerazione della situazione internazionale che andava rapidamente deteriorandosi, io decisi di accettare* [*l'offerta*]» (15). Ufficialmente egli aveva accettato un posto di «visiting professor» per la durata di un anno. Da New York Franco Rasetti, insieme a sua madre, proseguì immediatamente per Quebec.

Se mi è lecita una breve parentesi personale, anch'io ero partito insieme ai Rasetti con la stessa motonave, ma avevo lasciato in Italia una moglie, due figli e un terzo in arrivo. Ufficialmente andavo negli Stati Uniti per studiare la costruzione di un ciclotrone da installare all'Esposizione Universale di Roma del '42, studio che infatti feci (16), ma speravo di trovare un posto in quel Paese e di portarci la famiglia, se non subito per lo meno entro un anno

(14) A. WATTENBERG, *The Fermi School in the United States*, «Eur. J. Phys.» 9 (1988), 88-93.

(15) Ai primi di aprile del 1977, Franco Rasetti, in procinto di lasciare Roma per stabilirsi a Waremme in Belgio, città di origine della moglie, mi ha consegnato 18 pagine dattiloscritte in inglese intitolate: «Biographical Notes and Scientific Work of Franco Rasetti», che dovrò rendere note nella loro interezza dopo la sua morte. Ritengo tuttavia di poter utilizzare qualche sua frase in quei casi in cui non voglio correre il rischio di alterare il suo pensiero.

(16) E. AMALDI, *Studio dei ciclotroni negli Istituti degli Stati Uniti*, Reale Accademia d'Italia, Viaggi di studio promossi dalla Fondazione Volta. Vol. VI, 1941.

o due. Ma il posto non lo trovai (17), la guerra in Europa scoppiò ai primi di settembre, le autorità italiane rifiutarono il visto per i miei famigliari, ed io rientrai in Italia, di nuovo con la motonave Vulcania, che mi sbarcò a Napoli il 14 ottobre 1939.

A Quebec non solo Rasetti fondò il Dipartimento di Fisica di quella Università ma in tre mesi organizzò un laboratorio dove si poteva fare insegnamento e ricerca sui neutroni lenti ed entro due anni concluse la prima misura, sia pur grossolana, della vita media del muone (18), a cui seguirono, non molto dopo, quella di Auger et al. a Parigi (19), Rossi e Nereson a Cornell University (20) e Conversi e Piccioni a Roma (21).

Nel gennaio 1943 H. von Halban e G. Placzek offersero a Rasetti un posto nel gruppo di scienziati inglesi e francesi che in Canada avevano cominciato a sviluppare l'energia nucleare per scopi militari. Dopo matura riflessione Rasetti declinò l'offerta e come scrisse più tardi «*ci sono poche decisioni prese nella mia vita di cui abbia meno ragioni di rimpiangere. Io ero convinto che nulla di buono potesse mai seguire dalla costruzione di nuovi e più mostruosi mezzi di distruzione* ...» (22).

Nell'autunno 1947 Rasetti accettò l'offerta di una cattedra di Fisica alla John Hopkins University, in Baltimora, ove rimase fino al 1967. Nel 1949 sposò la signora di origine belga Marie Madeleine Hennin, che aveva due figli di 16 e 12 anni, nati da un precedente matrimonio.

Le attività di ricerca in fisica a John Hopkins ebbero un inizio lento e non raggiunsero mai una alta intensità per tutta una serie di ragioni: in primo luogo le ricerche in fisica nucleare, di bassa ed alta energia, stavano cambiando dalla tradizionale scala dei laboratori universitari a quella di imprese semi-industriali, trasformazione che a una persona individualista come Rasetti, che aveva sempre cercato di far tutto da solo o quasi, non andava a genio; in secondo luogo per avere un contratto di ricerca da qualsiasi ente

(17) Le persone che si interessarono seriamente del mio caso furono: I.I. Rabi alla Columbia University, E.U. Condon, Direttore dei Laboratori di Ricerca della Westinghouse, Pittsburg, Penn., e M.A.Tuve del Department of Terrestrial Magnetism della Carnegie Institution, Washington, DC. Quando, alla domanda se avevo perso il posto in Italia, rispondevo di no, tutti mi esprimevano simpatia ma mi facevano presente che c'erano molti altri che erano in condizioni ben peggiori delle mie e che era urgente sistemare. Ed io, ovviamente, riconoscevo che avevano pienamente ragione.

(18) F. Rasetti, *Disintegration of slow mesons*, «Phys. Rev.», *59* (1941) 613; *ibidem*, 60 (1942), 198-204.

(19) P. Auger, R. Maze, R. Chaminade, *Une mesure directe de la vie moyenne du méson au repos*, «C.R. Acad. Sci. Paris», *214* (1942), 266-269; R. Chaminade, R. Freon, R. Maze, *Nouvelle mesure directe de la vie moyenne du méson au repos*, «C.R.Acad.Sci.Paris», *218* (1944), 402-404.

(20) B. Rossi, N.C. Nereson, *Experimental Determination of the Disintegration Curve of Mesotrons*, «Phys. Rev.», *62* (1942), 417-422; *Further Measurements on the Disintegration Curve of Mesotrons*, «Phys. Rev.», *64* (1943), 199-201.

(21) M. Conversi, O. Piccioni, *Sulla disintegrazione dei mesoni lenti*, «Nuovo Cimento», *2* (1944), 71-87: *On the Disintegration of the Mesons*, «Phys. Rev.», *70* (1946), 874-881.

(22) V. nota 15.

governativo in USA, il direttore del contratto doveva avere una «clearance» cioè una dichiarazione ufficiale che egli non era un rischio per la sicurezza dello stato e questa condizione valeva anche se si trattava di fare un lavoro completamente *non classificato*. E Rasetti considerava una offesa intollerabile contro la dignità e la libertà scientifica, il sottomettersi ad una investigazione delle proprie opinioni politiche, associazioni passate e presenti e così via. In terzo luogo alla John Hopkins University i fondi per la ricerca si potevano ottenere quasi esclusivamente tramite contratti con enti governativi.

Egli quindi si ridusse ad adempiere scrupolosamente ai suoi doveri didattici in fisica, e concentrò la sua attività di ricercatore in altri campi quali la *geologia* e la *paleontologia*.

Rasetti aveva avuto sempre un grande interesse per le scienze naturali. Si era, per esempio, occupato in modo molto serio di entomologia estendendo e completando una collezione di insetti, soprattutto coleotteri, iniziata dal padre. In particolare scoprì tre nuovi coleotteri cavernicoli, uno, nel 1917, in una grotta delle Alpi Apuane e due, nel 1938, il primo in una grotta vicino a Ravello, l'altro in una grotta in Friuli.

Nel 1939, prima di partire dall'Italia, Rasetti aveva donato questa collezione di trentamila esemplari, al Museo Civico di Zoologia di Roma, situato dietro al Giardino Zoologico.

Il suo interesse per la geologia e la paleontologia iniziò nel 1939, a Quebec, si estese a tutto il periodo trascorso a Baltimora (1947-67) e ai primi anni del suo successivo soggiorno italiano. Le sue ricerche furono concentrate sui trilobiti del periodo Cambriano; cominciò con l'esame degli strati cambriani del Quebec e in 8 anni, con assistenti ed amici, esaminò diecine di tonnellate di roccia raccogliendo e preparando migliaia di esemplari. Dal 1941 in poi estese queste ricerche alle Montagne Rocciose della British Columbia e ad alcune regioni degli Stati Uniti e dal 1967 agli strati cambriani dell'Iglesiente (Sardegna). Sui trilobiti del Cambriano Rasetti aveva già pubblicato oltre 40 lavori nel 1952, quando la National Academy of Sciences degli Stati Uniti gli assegnò il Premio Internazionale Charles D. Walcott per il miglior lavoro pubblicato sul Cambriano nel mondo in un quadriennio.

Complessivamente Rasetti ha raccolto e preparato personalmente più di 20.000 fossili del Cambriano che ha distribuito in parti pressoché eguali (2500) fra i seguenti musei:

- United States National Museum (Washington, DC),
- Geological Survey of Canada,
- British Museum (Natural History),
- Musée de Paléontologie de l'Université Laval (Quebec).

Un quantitativo pressoché eguale di fossili del Cambriano della Sardegna (~2000) è stato da lui depositato nel Museo di Paleontologia del Servizio Geologico d'Italia, Roma, in attesa che venga creato un Museo Nazionale di Scienze Naturali.

Altre due imprese naturalistiche riguardanti la botanica sono state portate avanti da Rasetti per anni lavorando indefessamente – a sue spese – durante i periodi primaverili ed estivi verso la fine del periodo americano e soprattutto nel periodo italiano successivo con prosecuzione per la seconda anche nel periodo trascorso in Belgio.

La prima è una collezione di diapositive a colori della *flora alpina*, al di sopra del limite della foresta, che contiene più del 97% delle piante facenti parte del programma e da cui è stato estratto il volume: *Franco Rasetti: «I fiori delle Alpi»*, pubblicato dall'Accademia Nazionale dei Lincei (1980), con il contributo della Banca Commerciale Italiana, che contiene, oltre a 572 riproduzioni a colori, un testo scientificamente corretto, moderno e ricco di informazioni ma adatto per un vasto pubblico.

La seconda impresa botanica è una collezione di diapositive a colori delle *Orchidee italiane* (che rappresentano l'80% delle orchidee europee). Si teme che non si riesca a fare in tempo a pubblicare un volume analogo al precedente, ma su questo argomento. Tuttavia Rasetti ha già fatto (prima di lasciare l'Italia per il Belgio) le seguenti donazioni (23):

– una serie completa (~4000 diapositive a colori) della *Flora Alpina* all'Istituto Botanico dell'Università di Firenze;

– un'altra serie completa della stessa al Centro Linceo;

– una collezione di ~2000 diapositive a colori delle orchidacee italiane al Centro Linceo;

– una serie di 200-300 diapositive a colori delle orchidacee dell'Est e Sud degli Stati Uniti e del Canada al Centro Linceo;

– ha dato, durante gli anni di lavoro botanico, all'Erbario Nazionale di Firenze varie centinaia di piante che non esistevano nell'erbario o erano rappresentate da esemplari insoddisfacenti;

– ha depositato presso di me (momentaneamente) ~300 diapositive stereoscopiche in bianco e nero di montagne e scalate alpine.

4. L'azzeramento poco dopo la nascita dei gruppi di Padova, Palermo e Pisa

Passiamo ora ai fisici che all'epoca delle leggi razziali in Italia erano andati in cattedra da poco tempo, procedendo in ordine di anzianità.

Bruno Rossi, era stato nominato professore di fisica sperimentale all'Università di Padova nell'autunno 1932, ove in breve tempo era riuscito a costruire un nuovo istituto ottimamente attrezzato per la ricerca. Egli, dopo le leggi

(23) Questi dati mi sono stati dettati da Franco Rasetti su mia richiesta quando Ginestra ed io, alle ore 11 di sabato 2 aprile 1977 andammo a trovare Franco e Madelaine nel loro appartamento a Via Salaria 300, avendoci Franco comunicato l'intenzione di trasferirsi in Belgio alla fine del mese.

razziali era riuscito a partire, insieme alla moglie Nora Lombroso, per Copenhagen, già il 12 ottobre 1938. Essi avevano potuto affrontare una partenza così improvvisa grazie al residuo di una precedente borsa di studio per l'estero della Reale Accademia d'Italia, che Bruno aveva utilizzato per trascorrere alcuni mesi nel laboratorio del Duca Maurice De Broglie a Parigi.

Emilio Segrè, che era stato nominato professore di fisica sperimentale a Palermo nell'autunno 1935, all'inizio dell'estate 1938 era andato a Berkeley, California, per lavorare sugli isotopi a vita media breve dell'elemento 43, il tecnezio, che lui stesso, insieme al collega Carlo Perrier (1886-1948), aveva scoperto nel 1937 a Palermo (24).

Visto l'andamento che aveva preso la politica in Italia, e più in generale in Europa, egli aveva deciso di restare negli Stati Uniti, aveva scritto alla moglie Elfride Spiro di chiudere casa senza fretta e di raggiungerlo in California insieme al loro primo figlio Claudio, di non ancora due anni (25).

Elfride era una giovane donna tedesca non nuova a partenze improvvise anche più drammatiche. Non erano ancora passati 4 anni da quando era scappata da Breslavia per sottrarsi alle persecuzioni di Hitler. La differenza era che prima della partenza da Palermo molti colleghi del marito si recarono ad esprimerle la loro solidarietà ed amicizia e le varie mogli la aiutarono efficacemente nelle sgradevoli e laboriose operazioni di imballaggio e spedizione della mobilia, anche dopo la sua partenza da Palermo.

Le carriere di Bruno Rossi ed Emilio Segrè negli Stati Uniti sono state entrambe estremamente brillanti e diverse. Entrambe sono cominciate con non piccole difficoltà dato che non è stato facile neppure per loro avere un posto di «full professor».

Rossi da Copenhagen dove si era recato lasciando l'Italia andò, con l'aiuto di Niels Bohr, a Manchester (GB), a lavorare nel laboratorio di P.M.S. Blackett (1897-1974). Dopo circa un anno i Rossi poterono trasferirsi all'Università di Chicago e più tardi alla Cornell University (Itacha, N.Y.). Durante la guerra e precisamente dal 1943 al 1946 Bruno lavorò ai Laboratori di Los Alamos come capo di uno dei gruppi sperimentali. Finita la guerra, nel 1946, ebbe un posto di professore al Massachussets Institute of Technology ove rimase fino ai limiti di età, e dove ora è professore emerito.

Emilio Segrè invece rimase sempre a Berkeley salvo il periodo 1943-46 che passò a Los Alamos come capo di un altro gruppo sperimentale. Egli ebbe un posto definitivo di «full professor» all'Università di Berkeley solo nel 1946 al suo ritorno da Los Alamos (26).

(24) C. Perrier, E. Segrè, *Radioactive Isotope of Element 43*, «Nature (London)», *140*, (1937), 193-194; *Some Chemical Properties of Element 43. I*, «Jour.Chem.Phys.», *5*, (1937), 712-716; *II*, *ibidem*, 7 (1939), 155-156.

(25) E. Segrè, *Fifty years up and down a strenuous and scenic trail*, «Ann.Rev.Nucl.Sci», *31* (1981), 1-18.

(26) *Ibid.*.

L'attività scientifica di Bruno Rossi nel periodo passato a Chicago fu dedicata al confronto fra l'assorbimento subito dalla componente penetrante dalla radiazione cosmica nell'atmosfera con quello subito in uno strato di grafite di egual massa per unità di superficie e alla determinazione del rapporto della vita media τ dei mesotroni, costituenti la componente penetrante, alla loro energia di quite mc^2.

Alla Cornell University egli eseguì insieme a Nereson una misura diretta di considerevole precisione della vita media τ dei mesotroni (27). Tale misura seguiva di poco le osservazioni pionieristiche di Rasetti (28) e precedeva di alcuni mesi quelle fatte in Italia da M. Conversi e O. Piccioni (29). Sempre alla Cornell University, in parte con Greisen, faceva uno studio sperimentale degli sciami elettronici e un confronto fra i risultati di calcoli piuttosto raffinati e l'esperienza, che è rimasto un classico nella letteratura scientifica di questo argomento.

Dopo la guerra, al M.I.T., egli riprendeva lo studio dei raggi cosmici dedicandosi soprattutto all'individuazione delle particelle secondarie che questi producono in diversi materiali servendosi della tecnica della camera a nebbia.

In questo periodo il gruppo di Rossi diventa un centro di ricerca di importanza internazionale ove vanno a lavorare giovani e meno giovani, non solo da varie università italiane, ma molti dalla Francia, soprattutto dall'École Politechnique di Parigi e dal Giappone. Numerosi scienziati illustri di questi paesi ricordano con piacere e ammirazione il periodo passato nel gruppo di Rossi al M.I.T.

Alla fine degli anni 50 lascia i raggi cosmici e si dedica alla ricerca spaziale, campo in cui riesce a creare una nuova scuola. Prima studia il plasma interplanetario e in un secondo tempo scopre (1963) l'emissione di raggi X da parte della stella Scorpio X-1. Fra i collaboratori in questa ricerca c'era Riccardo Giacconi, e fra gli ospiti era tornato il giapponese Minoru Oda, che già aveva lavorato con Rossi sui raggi cosmici e che più tardi è stato per anni il capo di tutta la ricerca scientifica spaziale giapponese.

Il Premio Wolf per la Fisica 1987 è stato dato a B. Rossi e R. Giacconi per aver scoperto che Scorpio X-1 è una sorgente di raggi X ed a H. Friedman per aver mostrato con una esperienza di occultamento dietro la Luna, che la sorgente scoperta dal gruppo di M.I.T. è puntiforme.

L'attività scientifica di Emilio Segrè durante il periodo americano, comprende la dimostrazione sperimentale, data in collaborazione con Glenn Seaborg, allora giovanissimo, dell'esistenza di uno stato isomerico di 6 ore dell'isotopo 99 del tecnezio (elemento 43) prodotto dal decadimento del molibdeno 99 (di 67 h di periodo) che nel seguito ha acquistato una straordinaria importanza nella diagnostica e terapia del cervello e del cuore; ha

(27) V. nota 20.
(28) Vedi nota 18.
(29) Vedi nota 21.

poi sviluppato (in collaborazione con Halford e Seaborg nel 1939) la estensione del metodo di Szilard-Chalmers alla separazione di isomeri nucleari, e, nei primi mesi del 1941 (insieme a Kennedy, Seaborg e Wahl) è riuscito a produrre piccole quantità del nuclide più tardi chiamato $_{94}Pu^{239}$ e a misurare la corrispondente sezione d'urto per fissione con neutroni lenti. Tali risultati permisero di stabilire non solo la possibilità di produrre questo isotopo del plutonio ma anche che esso può essere usato come sostituto dell'uranio 235 a tutti gli effetti applicativi.

Durante la guerra a Los Alamos il gruppo di Segrè fu incaricato di studiare la fissione spontanea dell'uranio e del plutonio.

Tornato a Berkeley egli riuscì a dimostrare la influenza del legame chimico sulla costante di decadimento nucleare per cattura degli elettroni K del Be^7 dovuta al cambiamento della densità elettronica nella regione spaziale occupata dal nucleo.

Successivamente con un numeroso gruppo di collaboratori ha fatto, negli anni 1949-51, uno studio sperimentale su larga scala dell'urto neutrone-protone e protone-protone e, quando il Bevatrone di Berkeley ha cominciato a funzionare, ha costruito insieme a O. Chamberlain, C.E. Wiegand e T.J. Ypsilantis uno spettrometro di massa dotato di particolari prestazioni che gli permise di stabilire l'esistenza fra le particelle prodotte dall'urto di protoni di 6.3 GeV (accelerati con il Bevatrone) contro nuclei in quiete, di un numero piccolo ma chiaramente osservabile di *protoni negativi* (ottobre 1955). Che si trattasse di antiprotoni nel senso di Dirac, ossia di corpuscoli capaci di annichilarsi con altrettanti protoni, fu dimostrato in un esperimento eseguito con la tecnica delle emulsioni nucleari dallo stesso gruppo allargato con l'aggiunta di G. Goldhaber et al. a Berkeley e di E. Amaldi et al. a Roma. Lo studio dei processi di annichilamento tenne occupati per vari anni numerosi fisici di Berkeley e Roma. Per il risultato ottenuto nell'ottobre 1955 E. Segrè e O. Chamberlain ricevettero il Premio Nobel per la Fisica nel 1959.

Sia Bruno Rossi che Emilio Segrè sono autori di vari libri di fisica di notevole livello (30). Inoltre Emilio Segrè ha pubblicato, oltre ad un volume dedicato alla vita e all'opera di Enrico Fermi (32), due bellissimi volumi di storia della fisica (31).

(30) B. ROSSI, H.H. STAUB, *Ionization Chambers and Counters*, McGraw-Hill Book Company Inc., New York (1949); ID., *High Energy Particles*, Prentice Hall, Inc., New York (1952); ID., *Cosmic Rays*, McGraw-Hill Book Company Inc., New York (1964); ID.*Optics*, Addison and Wesley Publishing Company, Inc., Reading, Mass. (1957); traduzione in italiano: *Ottica*, Tamburini editore, Milano (1971). E. SEGRÈ (editor), *Experimental Nuclear Physics*, in tre volumi, John Wiley & Sons, Inc., New York (1953, 1953, 1959); ID., *Nuclei and Particles*, Benjamin, Inc., New York (1964), tradotto in italiano, *Nuclei e Particelle*, Zanichelli, Bologna (1964).

(31) E. SEGRÈ, *Personaggi e scoperte della fisica contemporanea*, Biblioteca Est, Mondadori (1976); traduzione dall'inglese: *From X-rays to Quarks*, Friedman, S. Francisco (1978); ID. *Personaggi e scoperte della fisica classica*, Biblioteca Est, Mondadori (1983); tradotto dall'inglese: *From falling bodies to radiowaves*, Friedman, S. Francisco (1983).

Giulio Racah, aveva dato vari importanti contributi scientifici fin dal periodo italiano iniziale; fra questi ricorderò un lavoro sulle strutture iperfini degli spettri atomici, fatto sotto l'influenza di Persico e Fermi e vari lavori di elettrodinamica quantistica, con particolare riguardo al calcolo delle sezioni d'urto per i processi di bremsstrahlung e di creazione di coppie, fatti all'inizio degli anni trenta, sotto l'influenza di Pauli e Fermi.

Successivamente aveva chiarito il significato della teoria di Majorana dei neutrini e la possibilità di stabilire con appropriate esperienze se i neutrini osservati in natura sono neutrini alla Dirac o alla Majorana. Aveva poi cominciato ad occuparsi degli spettri atomici complessi. Essendo stato ternato nel concorso a cattedre di Fisica Teorica dell'autunno 1937, fu, poco dopo, nominato professore all'Università di Pisa, una sede particolarmente importante anche per gli allievi interni della Scuola Normale Superiore.

Meno di 12 mesi dopo la nomina perdeva il suo posto di lavoro in seguito alle leggi razziali; ma solo nel 1939, quando la seconda guerra mondiale stava per scoppiare, emigrava in Israele ove ben presto cominciava a insegnare e lavorare all'Università Ebraica di Gerusalemme (33), di cui, parecchi anni più tardi diventò Rettore. Dopo la sua morte il suo nome è stato dato all'Istituto di Fisica dell'Università di Gerusalemme, ove ogni anno, viene tenuta una «Racah Lecture» da uno studioso invitato a parlare di un argomento scientifico di attualità.

Il suo maggior contributo scientifico è contenuto in quattro lavori (tre del 1942-43 ed uno del 1949) riguardanti gli stati antisimmetrici di sistemi composti di n corpuscoli di spin 1/2 tutti dotati dello stesso momento angolare orbitale *l*. Nell'ultimo di questi lavori Racah si serve della teoria dei gruppi. Queste applicazioni di tale teoria alla spettroscopia sono state descritte dall'autore in una serie di lezioni che egli fece all'Institute for Advanced Studies di Princeton nel 1950. Poiché erano molto richieste esse furono ristampate una prima volta dall'Institute di Princeton, una seconda volta dalla Società Italiana di Fisica, una terza dal CERN e, infine, nel 1965, sono state pubblicate nei *Springer Tracts in Modern Physics*.

Come ha scritto il suo allievo e ben noto fisico teorico Talmi: «*Racah was the man who established theoretical physics in Israel*», in quanto «*con studenti e collaboratori, egli ha trattato tutti gli spettri* [*atomici*] *in vista di ottenere una comprensione teorica di tutti gli spettri atomici misurati. Egli ha dato anche importanti contributi al campo della spettroscopia nucleare, sviluppando prevalentemente i metodi generali*» (34).

(32) Vedi nota 2*b*.

(33) IGAL TALMI, *In memoriam Giulio Yoel Racah*, «Proc. Israel Acad. Sci. and Human», Section of Sciences, No. 2, read at Memorial Meeting, 26 October 1965, Jerusalem, 1966.

(34) *Ibid.*

5. L'EMIGRAZIONE DEI FISICI PIÙ GIOVANI

In questa parte del mio contributo parlerò di alcuni casi di giovani fisici che pur essendo molto promettenti, al tempo delle leggi razziali, non erano ancora andati in cattedra, sostanzialmente per ragioni di età.

Il primo di cui voglio parlare è *Leo Pincherle* (1910-1976), nipote del matematico Salvatore Pincherle di Bologna, il quale, dopo essersi laureato in Fisica a Bologna venne a lavorare all'Istituto di Via Panisperna che frequentò dal 1932 al 1935. Durante i tre anni accademici successivi (1935-38) Pincherle tenne per incarico il corso di Fisica Teorica all'Università di Padova.

In seguito alle leggi razziali italiane, nel 1938 Leo si trasferì insieme alla moglie, Nora Cameo e al figlio Guido (n. 1937), a Zurigo, dove nacque il secondo figlio, Aldo Italo, che morì a 8 mesi come conseguenza dei disagi derivanti dalla drammatica partenza dall'Italia. Da Zurigo i Pincherle si trasferirono in Inghilterra ove Leo cominciò a lavorare nel 1939 al King's College di Londra con il Professore H.T. Flint.

Con l'entrata dell'Italia in guerra (1 giugno 1940) Leo Pincherle viene internato. Ma dopo cinque mesi, su richiesta dell'Università, può riprendere il suo lavoro a King's College.

Il figlio Ugo nasce a Malvern nel 1950. Durante la guerra Leo Pincherle insegnò, come lecturer (professore incaricato) a quello stesso College in Bristol e quindi a Londra (35). In quel difficile periodo egli trovò anche il tempo di andare a fare lezione, come visiting lecturer, ai Politecnici di Chelsea e Regent Street.

Nel 1948 Leo Pincherle fu nominato Principal Scientific Officer al Telecommunication Research Establishment (TRE), Great Malvern, che ben presto diventò il più importante laboratorio britannico specializzato in semiconduttori. Fu in questo laboratorio che Pincherle diede un considerevole contributo consistente nel calcolo delle funzioni d'onda degli elettroni nei solidi, argomento che diventò la sua specialità come campo sia di ricerca che di insegnamento. In particolare sono importanti il lavoro del 1953 sulla struttura a bande del solfuro di piombo (Bell et al) in cui il metodo a celle di Wigner-Seitz viene applicato per la prima volta ad un semiconduttore biatomico, e che è il primo lavoro classico in questo campo, ed il lavoro teorico fondamentale sulle applicazioni della teoria dei gruppi apparso nella forma di «*TRE Memoranda*».

Nel 1955 Pincherle tornò all'Università di Londra al Bedford College, prima come lecturer e, dal 1969 in poi, come Professore di Fisica Matematica. Il suo corso su «*The Band Structure of Semiconductors*» tenuto nel 1963 alla

(35) ERNST SONDHEIMER, *Leo Pincherle*, «Nature (London)», *266* (1977), 202. A.K. JONSCHER, *Leo Pincherle (1910-1976)*, «Solid-State Electr.», *657* (1977). Vedi anche: «The Time» del 4 novembre 1976 e il «Resto del Carlino» del 26 novembre 1976.

Scuola Internazionale Enrico Fermi di Varenna, e un secondo corso su analogo argomento tenuto a Perugia nel 1965, sono stati presentazioni classiche della teoria a bande che lo hanno portato alla pubblicazione nel 1971 del suo autorevole volume «*Electronic Energy Bands in Solids*» (MacDonald and Co. Londra).

Per molti anni Leo Pincherle è stato l'editor (o come si dice in italiano, il direttore) della edizione italiana del giornale culturale inglese *Endeavour*.

Per concludere mi par giusto seguire Ernest Sondheimer nella parte finale del suo cenno biografico. «*The picture of Leo Pincherle the scientist cannot be separated from that of Leo Pincherle the man. He was devoted his family for whom his too-early death is a tragic loss. His sense of humour, dry and delicious, showed in his fund of stories about well-known contemporaries – always amusing and never malicious. His quite demeanour concealed plenty of fire and imagination, as you soon discovered when you partnered him at bridge. He was the most knowledgeable of men on fundamental aspects of European culture and history and wore his learning lightly; his love of music and the arts was intense, his taste discriminating. He was, in the best sense, both very Italian and very English, combining in his person that intertionalism in science and that European spirit which it is so necessary to keep alive today*» (36). Per sua volontà è seppellito a Bologna.

Ugo Fano (n. 1912), figlio del matematico Gino Fano di Torino, e primo cugino di Giulio Racah, dopo l'esame di laurea in Matematica a Torino, venne nell'autunno 1934 a Roma a lavorare sotto la guida di Fermi, salvo un interludio di un anno (1936-37) passato a Lipsia con W. Heisenberg. Dopo un seminario di Pascual Jordan sull'analisi fisica dei fenomeni genetici provocati da radiazioni, Fermi suggerì a Fano di specializzarsi in questo argomento. Jordan lo presentò ai maggiori cultori di questo campo (in particolare H.J. Müller). Ciò spiega come mai Fano, dopo aver pubblicato vari lavori di spettroscopia atomica, di fisica nucleare, sulla termodinamica dei nuclei e sulla fissione dell'uranio, si sia poi dedicato a vari problemi di genetica con particolare riguardo alla teoria delle mutazioni indotte in drosofila da irraggiamento con radiazioni.

Fu attraverso a questo canale che nel 1939-40 egli ottenne un primo appoggio negli Stati Uniti nel Washington Biophysical Institute.

Nel 1940 Ugo Fano sposò Lilla Lattes, studente in ingegneria al Politecnico di Milano, figlia di Leone Lattes, noto Professore di Medicina Legale dell'Università di Pavia. Ugo e Lilla hanno avuto due figlie, Mary Giacomoni di Chicago e Virginia Ghattas di Wallesley, Mass., ciascuna delle quali ha avuto due figli.

A parte il periodo 1944-45 durante il quale lavorò in un laboratorio di ricerca balistica dell'Esercito degli Stati Uniti, dal 1940 al 1946 seguitò a fare

(36) *Ibid.*

ricerca presso il Dipartimento di Genetica della Carnegie Institution di Washington, Cold Spring Harbor e il Dipartimento di Fisica della Columbia University. Dal 1946 al 1966 ha lavorato come capo della teoria delle radiazioni del National Bureau of Standards (NBS) e poi come Senior Research Fellow sempre al NBS. Nel 1966 è diventato professore di fisica all'Università di Chicago, ove, successivamente, è stato Chairman del Dipartimento di Fisica dal 1972 al 74 e professore emerito dal 1982 in poi. È membro della National Academy of Sciences degli Stati Uniti dal 1976 e Dottore Honoris Causa della Queen's University di Belfast (1978) e dell'Università Pierre e Marie Curie di Parigi (1979).

Al National Bureau of Standards si è occupato in modo particolare degli aspetti teorici della dosimetria dei raggi X ed ha sviluppato un vasto programma sulla penetrazione e diffusione delle radiazioni ionizzanti e sul meccanismo del loro modo di agire. Più tardi è stato assegnato all'ufficio del Direttore del National Bureau of Standards con la possibilità di fare ricerca indipendente ma al tempo stesso con il compito di sovraintendere a tutte le attività teoriche dell'Ente. Nel 1960, con l'installazione al NBS di apparecchiature per la produzione e utilizzo di luce di sincrotrone e di tecniche di spettroscopia elettronica si è aperta a Fano la possibilità di individuare e interpretare tutta una serie di nuovi fenomeni che ha determinato uno straordinario periodo di nuova attività.

La teoria degli atomi e delle molecole aveva portato a soluzioni approssimate della equazione di Schrödinger che erano adatte per il ristretto campo di fenomeni che si conoscevano negli anni 30. Ma la descrizione dei nuovi fenomeni che si osservano con le tecniche moderne ha richiesto un nuovo approccio che comprende la identificazione dei loro parametri critici e la loro valutazione, *prima*, semi-empiricamente, e, *poi*, derivandoli da primi principi. In particolare, a lui è dovuta la comprensione di stati localizzati «*risonanti*» negli atomi a energie maggiori di quelle di prima ionizzazione (*stati di Fano*).

Questa impresa ha portato a sviluppare nuovi concetti teorici e nuove tecniche a cui Ugo Fano ha dato molti importanti contributi nelle sue più di 230 pubblicazioni scientifiche.

Egli oggi è considerato un maestro in questo campo come risulta dai numerosi riconoscimenti ricevuti e dalle molte università americane ed europee ove viene invitato a far lezioni e conferenze.

Sergio De Benedetti (n. a Firenze, 1912), ha cominciato la sua carriera a Padova con Bruno Rossi, di cui, come ho già detto, fu collaboratore in occasione della spedizione del 1934 all'Asmara per studiare l'effetto Est-Ovest della radiazione cosmica.

In seguito alle leggi razziali si trasferì in Francia ove fu accolto come ricercatore al Laboratorio Curie dell'Università di Parigi, ed ove rimase fino al 1940. Fu in questo laboratorio che incontrò Bruno Pontecorvo che vi si era trasferito attorno a Pasqua del 1936. Trasferitosi negli Stati Uniti nel 1940 ebbe la posizione di Research Associate presso la Bartol Research Foundation dal 1940 al 1943, di Associate Professor al Kenyon College, Gambier, Ohio, nel 1943-44, di Senior Physicist alla Monsanto Chem. Co., Dayton, Ohio, nel 1944-45, di Principal Physicist ai Oak Ridge National Laboratories dal 1946 al 48, di Associate Professor alla Washington University, St. Louis, nel 1948-49, e infine Professor of Physics al Carnegie Mellon University nel 1949 ove è rimasto per il resto della sua carriera, salvo l'anno accademico 1956-57 passato all'Università di Torino come Fulbright Fellow.

I suoi lavori riguardano gli isotopi radioattivi a vita media breve, i positroni, e la loro annichilazione, la violazione della parità, gli atomi mesici e lo studio dell'effetto Mössbauer. Ha pubblicato un bel libro, intitolato «*Nuclear Interactions*» (John Wiley, 1964) ed ha prodotto un notevole numero di allievi fra i quali almeno otto professori universitari e parecchi altri con elevate posizioni in industrie private come i Bell Telephone Laboratories, l'IBM e la Westinghouse.

Bruno Pontecorvo è un caso diverso dagli altri in quanto lasciò l'Italia già nella primavera 1936, cioè due anni prima del Manifesto della razza. Quando verso la fine di marzo 1937 ci trovammo con alcuni amici comuni in Austria (37) ed io gli espressi le mie preoccupazioni per i molteplici segni di un crescente antisemitismo in Italia, Bruno ci tenne a sottolineare che egli stava all'estero da tempo per antifascismo, e che considerava l'antisemitismo solo come un aspetto secondario di una situazione politica ben più grave.

Queste considerazioni di cui riconosco la fondatezza, non mi sembra possano giustificare una sua esclusione dalla panoramica che cerco di dare di un certo ambiente in una certa situazione.

Dopo aver fatto il biennio di ingegneria all'Università di Pisa espresse il desiderio di passare a fisica e il fratello maggiore Guido, biologo e genetista (38), gli consigliò di trasferirsi a Roma dove si iscrisse al terzo anno di fisica e si laureò, molto brillantemente a vent'anni, nel luglio 1934.

A partire dai primi di settembre dello stesso anno entrò a far parte del gruppo sperimentale di Fermi che studiava la radioattività artificiale provocata da neutroni e partecipò così a varie scoperte a cominciare da quella del rallentamento dei neutroni. Durante l'autunno 1935 e l'inverno 1935-36, fece insieme a Gian Carlo Wick, una serie di esperienze sulla diffusione elastica ed anelastica dei neutroni da parte di vari nuclei medi e pesanti. Avendo ricevuto un Premio del Ministero della Educazione Nazionale per il suo contri-

(37) All'incontro sciistico che ebbe luogo a Vent, in Austria, parteciparono: Bruno Pontecorvo e Paul Ehrenfest jr., provenienti da Parigi, e S. Fubini, G. Racah, F. Rasetti, G.C. Wick ed E. Amaldi provenienti dall'Italia.

(38) BRUNO PONTECORVO, *Una nota autobiografica*, 82-87 in «Scienza e Tecnica», Annuario della EST 88/89, Mondadori.

buto alle ricerche sui neutroni, egli se ne servì per trasferirsi a Parigi, attorno a Pasqua 1936, ove fu accolto all'Istituto Curie, a lavorare con Frédéric Joliot Curie.

Successivamente passò al College de France ove si occupò di isomeria nucleare, quasi da solo; fece la previsione che dovessero esistere casi di nuclei isomeri stabili e riuscì a dimostrarne sperimentalmente un primo esempio nel caso del cadmio eccitato per urto di neutroni veloci. Fece anche la previsione che i fotoni emessi nella transizione fra stati isomerici fossero fortemente convertiti e quindi cercò e trovò alcuni casi interessanti di isomeria in cui ciò si verifica, fra cui quello del Rh. Analoghi risultati furono ottenuti alla stessa epoca da Seaborg e Segrè a Berkeley. Insieme ad A. Lazard, Pontecorvo riuscì infine a produrre (1939) isomeri beta-stabili ($^{115}In^*$) mediante irraggiamento di nuclei stabili con raggi X di alta energia (~3 MeV).

Per la scoperta di questo fenomeno, a cui Joliot diede il nome di «fosforescenza nucleare», Bruno Pontecorvo ricevette il Premio Curie-Carnegie.

A Parigi si sposò con una ragazza svedese da cui ebbe tre figli, il primo dei quali nato in Francia, gli altri due più tardi in Canada.

Nel 1940, dopo la disfatta della Francia, fuggì da Parigi in bicicletta con Marianne e Gillo piccolissimo. Da Lisbona si imbarcò per gli Stati Uniti ove fu assunto da una società privata americana con sede in Oklaoma.

Nel 1941 lavorando per questa società ideò un nuovo metodo di carotaggio, il *carotaggio neutronico* basato sulle proprietà dei neutroni lenti. Non molto dopo accettò l'offerta di andare a lavorare allo sviluppo delle applicazioni dell'energia nucleare con il gruppo anglo-franco-canadese, di cui facevano parte molti dei fisici da lui conosciuti a Parigi (P. Auger, B. Goldsmith, H.v. Halban, L. Kowarski, G. Placzeck). In Canada, prima a Montreal e poi a Chalk River, partecipò alla progettazione e costruzione del reattore nucleare NRX a uranio naturale e acqua pesante in qualità di dirigente degli aspetti fisici del progetto.

Dopo la guerra, Pontecorvo dal Canada andò in Gran Bretagna, ove lavorò, per un periodo relativamente breve, al centro atomico di Harwell. Nel settembre 1950 Pontecorvo andò in URSS con tutta la famiglia e cominciò a lavorare al centro di ricerche di Dubna come capo della divisione di fisica sperimentale del Laboratorio per i Problemi Nucleari.

Fu in Canada che Pontecorvo cominciò a lavorare in fisica delle particelle subnucleari in particolare sulle interazioni deboli o interazioni di Fermi.

Subito dopo le esperienze di Conversi, Pancini e Piccioni, Pontecorvo fu il primo (1947) a formulare la legge dell'universalità delle interazioni deboli, enunciata poco dopo, indipendentemente, anche da altri (O. Klein, G. Puppi). Guidato da questa legge riuscì a prevedere diverse proprietà del muone in seguito dimostrate corrette da esperimenti di vari autori tra cui Pontecorvo stesso.

Alcuni degli esperimenti di Pontecorvo risalgono al periodo di Chalk River (Canada, 1948-49), mentre altri sono stati fatti assai più tardi (1958 e 1961) a Dubna. Nel 1959 suggerì l'uso dei neutrini prodotti nel decadimento di pioni di alta energia per estendere le nostre conoscenze nel campo delle interazioni deboli. Questo approccio, suggerito successivamente anche da altri autori, è diventato quello correntemente impiegato nei grandi laboratori come il Brookhaven National Laboratory, il Fermilab, il CERN, ed altri.

La serie di suggerimenti estremamente acuti e importanti che Pontecorvo ha continuato a fare nel campo delle interazioni deboli è molto lunga e ricca e in molti casi ha portato a risultati di grande importanza, ma la loro esposizione coinvolgerebbe una serie di concetti fisici non facilmente presentabili in forma comprensibile in questa sede.

Pontecorvo sia in Canada che in URSS è stato un maestro di qualità non comuni e al tempo stesso un ambasciatore della cultura e del costume italiano di grande rilievo.

Fra l'altro ha diffuso fra i giovani dei vari ambienti scientifici con cui è venuto a contatto in URSS, interessi sportivi e amore per la natura di grande limpidezza e valore morale.

6. Qualche esempio di emigrazione di giovani ingegneri e di un giovane fisico applicato

In questa parte del mio contributo presenterò insieme due esempi di giovani ingegneri che hanno dedicato la maggior parte della loro attività all'insegnamento universitario (Mario Salvadori e Roberto Fano) e il caso di un giovane fisico che è stato portato più dai casi della vita connessi con la sua emigrazione che dalle sue scelte personali, a diventare un fisico applicato e, tramite questo canale, un alto consulente del Governo degli Stati Uniti (Eugenio Fubini). Se le circostanze fossero state diverse avrebbe potuto diventare, altrettanto bene, un brillante fisico sperimentale o forse teorico.

Mario Salvadori (n. a Roma nel 1907) da Riccardo Salvadori, cattolico, ed Ermelinda Alatri, si laureò in ingegneria civile nel 1930 e in matematica pura nel 1933 all'Università di Roma. Fummo compagni di corso al primo biennio universitario e la nostra amicizia fu rinsaldata da una stagione alpinistica (estate 1927) dedicata ad arrampicarci sulle Dolomiti di Cortina e dintorni insieme a suo fratello minore, Giorgio Salvadori.

Nel 1933-34 Mario Salvadori fu l'ultimo allievo in fotoelasticità di E.G. Coker, professore al University College di Londra, ove si era recato con una borsa di studio della Lega delle Nazioni, prima negata e poi assegnata dal governo fascista.

Libero docente in scienza delle costruzioni nel 1937 e assistente ordinario nel 1938, aveva iniziato una carriera universitaria con la pubblicazione di 15 memorie a carattere tecnico fra il 1933 e il 1938 e l'insegnamento di un corso libero sulla teoria delle piastre e volte, quando le prime avvisaglie della politica antisemita lo convinsero a lasciare l'Italia per gli Stati Uniti nel gennaio del 1938, utilizzando una borsa di studio vinta in un concorso indetto dal Ministero delle Comunicazioni, che poté vincere perché «non discriminato» essendo figlio di padre cattolico.

Dopo un anno di impiego come ingegnere in una fabbrica di giocattoli, fu chiamato alla Columbia University di New York nel gennaio del 1940 come insegnante di matematica applicata nella scuola di ingegneria. Durante i difficili anni della seconda guerra mondiale, come membro del gruppo Statunitense di Giustizia e Libertà, partecipò alla preparazione dei materiali di propaganda antifascista spedito clandestinamente in Italia e illustrò le attività partigiane al pubblico americano sia alla radio che in numerosi convegni.

Dal 1942 al 1945 collaborò (a sua insaputa) al Manhattan Project, dedicato alla realizzazione della prima bomba atomica, ma si rifiutò, dopo la fine della guerra, di partecipare a ricerche di carattere militare.

La sua carriera universitaria gli permise di pubblicare 13 libri di matematica applicata e di scienza delle costruzioni, due dei quali, «*Metodi Numerici in Ingegneria*» e «*Le Strutture architettoniche*», sono stati tradotti in 14 lingue e adottati come libri di testo e di studio in numerose università europee e non europee, comprese varie in Russia, Cina e Giappone.

Mario Salvadori fu chiamato nel 1954 con il titolo di professore di architettura alla Princeton University, dove insegnò fino al 1959, quando la Columbia University richiese la sua opera anche nella facoltà di architettura.

Ventisei degli 8.000 studenti di Mario Salvadori sono oggi professori nelle più rinomate università americane ed estere, e la Columbia University lo ha onorato con la nomina a «*Great Teacher*», l'invito a tenere una delle prestigiose conferenze chiamate «*University Lectures*», ed il conferimento di una laurea honoris causa nel 1980. Ha posto termine all'insegnamento alla Columbia University nel 1988 con la nomina a James Renwick Professor Emeritus of Civil Engineering e Professor Emeritus of Architecture.

Nel 1960 Mario Salvadori fu ternato in un concorso per una cattedra di Scienza delle Costruzioni in Italia e chiamato dall'Università di Palermo, ma rinunciò alla cattedra.

Durante gli stessi anni e fin dal 1959, Mario Salvadori è entrato a far parte come socio dello studio di progettazione e ricerche ingegneristiche Weidlinger Associates di New York, Boston, Washington D.C. e San Francisco, di cui è ora Consigliere Delegato. Durante gli ultimi 30 anni Mario Salvadori ha partecipato alla progettazione strutturale in tutte le parti del mondo di importanti edifici, collaborando con architetti di fama mondiale come Gropius, Breuer, Saarinen e molti altri. A seguito della sua attività professionale, Mario Salvadori è stato premiato da associazioni ingegneristiche americane e nominato membro honoris causa della American Society of Civil Engineers, della American Architects, e membro della National Academy of Engineering.

Da 14 anni Mario Salvadori si dedica quasi esclusivamente all'insegnamento della matematica e delle scienze nelle scuole elementari e medie dei quartieri più poveri di New York, sotto l'egida del Salvadori Educational Center on the Built Environment, per il quale ha scritto un manuale per insegnanti, un libro per bambini e ha preparato 10 pellicole su nastro magnetico proiettate regolarmente dalle stazioni televisive degli Stati Uniti. Le autorità scolastiche della città di New York hanno stabilito 8 «*Scuole Salvadori*», dove sono stati adottati i suoi metodi educativi, e lo ha onorato con riconoscimenti vari.

Eugenio Fubini (n. a Torino nel 1913), figlio del matematico di Torino Guido Fubini, ha fatto il primo biennio di fisica all'Università di Torino ed il secondo biennio a Roma ove si è laureato nel 1933 con una tesi di cui Fermi fu relatore. Fra il 1935 e il 1938 Fubini fu impiegato dell'Istituto Nazionale di Elettrotecnica di Torino. Trasferitosi negli Stati Uniti lavorò dal 1939 al 1942 come ingegnere presso la Columbia Broadcasting System di New York.

Dal 1942 al 1945 Fubini è stato impegnato come research associate del Harvard University Radio Research Laboratory nella progettazione, sviluppo e funzionamento di apparati radio, e radar per contro-misure, riconoscimento e inseguimento.

Egli fu consulente scientifico e osservatore tecnico dell'Esercito e della Marina degli Stati Uniti nel Teatro Europeo di Operazioni durante il 1943 e il 1944 e partecipò alla realizzazione dei sistemi elettronici di riconoscimento e disturbo per l'invasione dell'Italia e della Francia meridionale. Durante il 1944 e il 1945 fu in Inghilterra con l'Ottava Forza Aerea USA, a capo della ricognizione e delle contro-misure elettroniche. Nel 1945 prestò servizio anche come consulente speciale delle contro-misure elettroniche presso l'Ufficio delle Comunicazioni Aeree del Dipartimento della Guerra.

In quegli anni Eugenio Fubini ebbe una vita molto movimentata che lo portò in Africa del Nord, Corsica, Italia centrale e Gran Bretagna. Al suo ritorno sposò Jane Elisabeth (in casa Betty) Mathmer, figlia del «dean» dell'Università del Massachussets a Amherst, che aveva conosciuto già nel 1941 e che lo aveva aspettato. Betty gli diede sei figli: cinque femmine e un maschio. Eugenio e Betty possedevano 5 acri (circa 4000 m^2) di terreno. Ad ogni figlio giunto in prossimità di dodici anni essi regalavano un cavallo da cavalcare, ma che il giovane aveva l'impegno di alimentare e governare giornalmente nella sempre crescente scuderia di famiglia. I ragazzi cavalcavano sul terreno dei genitori ove oggi vanno a cavalcare i nipoti.

Eugenio Fubini entrò a far parte dell'Airborne Instruments Laboratory (AIL), Melville, New York, nel 1945, come ingegnere e lavorò allo sviluppo dei componenti a microonde, dei rivelatori magnetici degli apparati elettronici di prova, degli apparecchi antidisturbo, delle antenne, dei rivelatori di direzione e dei sistemi di ricognizione. Ricoprì vari incarichi in questa società, che diventò in seguito la Airborne Laboratory Division della Cutler-Hammer Corporation. Nel 1960 fu nominato Vicepresidente della Research and Systems Engineering Division dell'AIL Division.

Nel marzo del 1961 Fubini entrò a far parte dell'Ufficio del Defense Research and Engineering for Research and Information Systems del Segretario della Difesa, allorché il Presidente Kennedy lo nominò Assistant Secretary alla Difesa. In quest'ultimo incarico le sue responsabilità comprendevano la direzione dell'intera rete delle ricerche militari e dei problemi di sviluppo, la supervisione della National Security Agency e la revisione della programmazione di tutte le apparecchiature di comando, controllo e comunicazione del Dipartimento della Difesa.

In seguito della nomina da parte del Presidente Kennedy, Eugenio Fubini prestò servizio come Assistant Secretary alla Difesa e Vice Direttore del Defense Research and Engineering fino al giugno 1965, quando si dimise per diventare uno dei Vice Presidenti dell'International Business Machines Corporation (IBM). Egli fu promosso Vice Presidente e Group Executive nel febbraio 1966 con la responsabilità della Research Division, del Science Research Associates, Inc., del Production Systems Department e dell'Instructional Systems Development Department dell'IBM.

Nell'aprile 1969 ha dato le dimissioni dall'IBM ed è diventato un consulente privato sia di industrie che del governo.

Egli è l'autore di circa 40 pubblicazioni tecniche, e ha preso 11 brevetti. Ha ricevuto titoli «honoris causa» del Brooklyn Polytechnic, dal Rensselaer Polytechnic Institute, e dal Pratt Institute, numerose decorazioni e medaglie per la Difesa, ha fatto corsi di lezioni alla Harvard University (1956) ed ha fatto parte di numerose commissioni scientifiche e tecniche. È membro della National Academy of Engineering degli USA.

Roberto M. Fano (n. a Torino nel 1917) fratello minore di Ugo Fano, ha seguito gli studi di ingegneria al Politecnico di Torino negli anni 1935-39 e li ha completati al Massachusetts Institute of Technology (MIT), Cambridge, Mass: SB nel 1941, Sc. D. nel 1947. La sua carriera universitaria si è svolta prevalentemente al MIT dove è passato da Teaching Assistant (1941-43) a Instructor, Research Associate, Associate Professor (1955-56) e Professor (1956-1984) e dal 1984 in poi Professor Emeritus.

Ha iniziato come Electric Engineer ed in seguito è diventato un Computer Scientist che si è dedicato alla progettazione, costruzione ed impiego di grandi progetti come il Project MAC di cui è stato Founding Director (1963-68).

Nel 1974-75 è stato Visiting Scientist del Laboratorio di Ricerca di Zurigo dell'IBM, Department of Electrical Engineering, ed ha svolto attività di consulenza per varie fondazioni (Ford Foundation 1961-63, Agnelli Foundation 1972-74 ecc.) e compagnie industriali (Radio Corporation 1961-63).

È membro della National Academy of Engineering dal 1973 e della National Academy of Sciences dal 1978.

È autore, insieme a R.B. Adler e L.J. Chu, di due bei volumi: «*Electromagnetic Fields, Energy and Forces*» e «*Electromagnetic Transmission and Radiation*» pubblicati da Wieley & Sons nel 1966 e autore da solo, del volume

«*Transmissions and Informations*» pubblicato nel 1961 dal M.I.T. Press e Wiley & Sons, che è stato tradotto in russo (1963), giapponese (1965) e tedesco (1966).

7. Qualche riflessione conclusiva

Per concludere vorrei osservare come, prima le leggi razziali, poi la seconda guerra mondiale con tutte le sue conseguenze, immediate e lontane, abbiano portato alla distruzione completa o quasi di questi gruppi di fisici italiani e a una loro dispersione nel mondo, certo non prevedibile ai tempi di via Panisperna, neppure da persone lungimiranti come O.M. Corbino. Molti di loro sono stati posti di fronte a decisioni di una gravità senza precedenti ed hanno preso le loro meditate decisioni. Ciascuno dei fisici emigrati, anche senza volerlo, è poi diventato un ambasciatore della cultura e dello stile italiano ad alto livello. Ma ciò che appare ancor più evidente, riconsiderato a distanza di tempo, è l'enorme potenzialità intellettuale, la vasta competenza scientifica e l'impagabile capacità di insegnamento universitario e post-universitario che le leggi razziali hanno fatto fuggire dall'Italia e riversare altrove. È vero che fortunatamente molte delle persone di cui ho parlato dopo la guerra sono tornate, per periodi più o meno lunghi, ad insegnare nel nostro Paese. Fermi tornò in Italia, per la prima volta, per partecipare ad un Convegno Internazionale tenuto a Como nel settembre 1948 (39) e una seconda, nell'estate 1954 per tener un celebre corso di lezioni alla Scuola Internazionale di Varenna (40), che fu l'ultimo suo insegnamento prima di morire circa quattro mesi dopo.

Bruno Rossi ed Emilio Segrè, dopo essere andati a riposo dalle loro istituzioni statunitensi all'età di circa 65 anni, sono tornati ad insegnare nelle nostre Università.

Bruno Rossi è stato titolare del Corso di Complementi di Fisica Generale dell'Università di Palermo negli anni accademici 1974-75 e 1975-76 ed Emilio Segrè titolare del Corso di Fisica Nucleare, indirizzo generale, dell'Università di Roma negli anni accademici 1973-74 e 1974-75, corso che ha tenuto come professore fuori ruolo anche nell'anno accademico 1977-78. A parte questi corsi sia Bruno Rossi che Emilio Segrè sono tornati in Italia quasi ogni anno e hanno dato conferenze o lezioni particolari in molte università italiane. E anche Fano è tornato spesso ad insegnare a Roma e altrove, Pincherle a Varenna e Perugia, De Benedetti a Torino e Pontecorvo, a partire dal 1978, è tornato in Italia quasi ogni anno per due o tre mesi. Noi siamo tutti loro molto grati.

(39) Enrico Fermi, *Conferenze di Fisica Atomica*, Accademia Nazionale dei Lincei, Fondazione Donegani (1950).

(40) Enrico Fermi, *Lectures on Pions and Nucleons*, B.T. Feld editor, «Nuovo Cimento», *2*, Suppl. (1955), 17-95.

Tuttavia è impossibile dimenticare né si deve dimenticare, quel che è accaduto alla fine degli anni 30, per i danni materiali, fisici e morali causati alle persone, per lo scompiglio provocato nelle famiglie di tutti coloro che sono stati direttamente o indirettamente colpiti dalle leggi razziali, ma anche per i danni assai meno dolorosi, meno appariscenti sul piano personale, ma certo non meno gravi dal punto di vista sociale, che tali leggi hanno inflitto soprattutto alle nuove generazioni di quelli che le leggi razziali stesse fingevano di difendere ma che in realtà hanno danneggiato gravemente, privandoli di tanti maestri dotati di conoscenze così approfondite e personalmente rielaborate, di capacità di ricerca e insegnamento così fuori del comune.

Edoardo Amaldi

THE CASE OF PHYSICS

1. Introduction

It seems fitting, I would almost say proper, to start my talk to this conference with a few general considerations made by various experts and in particular Emilio Segrè, who unfortunately passed away not even twenty days ago. In addition to his important role in the rapidly developing field of active physics, he was also a profound scholar of the history of physics in the last 200 years.

Both Emilio Segrè[1] and others before him noted that after the "golden age" of Alessandro Volta at the turn of the 18th century, physics in Italy entered a period of stagnation. Of course there were figures like Macedonio Melloni (1789-1852) and Ottaviano Mossotti (1791-1863), but the most famous of all, Amedeo Avogadro (1776-1856), and later Stanislao Cannizzaro (1826-1910) must be considered as belonging to the world of chemistry rather than that of physics, even if it was difficult to make the distinction at the time.

It was mainly after the unification of Italy, however, that Italian physicists, unlike Italian mathematicians, contributed little to the progress of their science. This is certainly not the place to go into a discussion of the possible causes for this backwardness or perhaps I should say delay in development.

I will merely point out that historians and sociologists have studied the causes. Some feel that it was the effort of the "Risorgimento" (the unification of Italy) that exhausted the intellectual energies of the country. Some believe it to have been related to Italy's industrial development as compared to industrial development in northern Europe. Others blame it on an overly literary and rhetorical school system--and I confess that I tend towards this interpretation--while others yet have explanations connected to Marxism.

Let us leave these considerations to others and turn to the facts. At the end of the last century, the most prominent Italian physicist was Augusto Righi (1850-1921), whose greatest contribution was the extension of the work of Heinrich Hertz (1857-1894) towards high frequencies. Contemporaneously, or only a short time later, there was Guglielmo Marconi's (1874-1937) outstanding work in the field of technology, for which he duly received the 1909 Nobel Prize for Physics.

Later there were physicists like Orso Mario Corbino (1876-1937), Antonio Garbasso (1871-1933) and Luigi Puccianti (1875-1952), who made important contributions still cited in the scientific literature today.

---

[1] E. SEGRE', *The Fermi School in Rome*, "Eur. J. Phys." 9 (1988), 83-87.

But no one in the field of physics in those years produced works comparable to those of the mathematicians of the University of Padua, Gregorio Ricci Curbastro (1853-1925) and Tullio Levi-Civita (1873-1941). Around 1901, Curbastro and Levi-Civita developed absolute calculus or tensor analysis, which was the mathematical instrument that Albert Einstein used to develop and complete his theory of general relativity.

The teaching of physics at university level was not particularly good either, as attested to by the fact that until the mid-twenties the special theory of relativity was taught only in a few mathematics courses and no physics courses and that quantum mechanics was practically ignored in all Italian universities.

2. The Rise of the Roman and Florentine Groups and their Near Destruction

The situation started to change considerably after the First World War with the appearance of Enrico Fermi (1901-1954). After taking over the chair of theoretical physics at the University of Rome in 1926, he rapidly turned the Physics Institute in Via Panisperna directed by O. M. Corbino into a modern school of physics.[2] Another group including Bruno Rossi, Gilberto Bernardini and Giuseppe Occhialini sprang up soon thereafter at the Physics Institute in Arcetri directed by A. Garbasso.

The formation of these two groups was not coincidental and was at least partly the result of Corbino's pressure on the Italian Ministry of Education to hold the first competitive examination for the chair of theoretical physics in 1926.

The winners of the examination were Enrico Fermi, first, Enrico Persico, second, and Aldo Pontremoli, third, who received positions at the Universities of Rome, Florence and Milan, respectively.

Pontremoli (1896-1928) received his degree in physics from the University of Rome in 1920, having interrupted his studies to serve as a volunteer in the First World War. After working as an assistant to O. M. Corbino at the Institute in Via Panisperna from 1920 to 1924, he moved to the University of Milan, where he founded the Institute of Physics which has now borne his name for almost sixty years. Unfortunately, less than two years after he took over the chair in Milan, he was killed during the second dirigible expedition to the North Pole organized and directed by Umberto Nobile.[3]

---

[2] *a) Enrico Fermi: Note e Memorie*, 2 volumes, Accademia Nazionale dei Lincei and The University of Chicago Press, Rome; Vol. 1, 1962, Italia 1921-1938, Vol. 2, 1965, USA, 1939-1954, *b)* E. SEGRE', *Enrico Fermi, fisico*, Zanichelli, Bologna (1971); *Enrico Fermi, physicist*, The University of Chicago Press (1970), *c)* E. AMALDI, *Personal Notes on Neutron Work in Rome in the 30s*, in *History of Twentieth Century Physics*, C. Weiner (ed.), Academic Press, New York and London, 1977, 294-325, *d)* LAURA FERMI, *Atomi in famiglia*, Zanichelli, Bologna (1955); *Atoms in the Family*, The University of Chicago Press (1954), *e)* B. PONTECORVO, *Fermi e la fisica moderna*, Editori Riuniti, Rome (1972). Translation from the Russian of the introduction written by B. Pontecorvo for *Le opere scientifiche di Enrico Fermi*, Moscow, Nauka, 1971.

[3] During the return from the second expedition to the North Pole, the dirigible Italia became weighed down by ice and crashed on the pack. Eight members of the expedition including Pontremoli died, as well as six members of the rescue team sent to help them.

Persico (1900-1969) was a friend of Fermi's from their school days and was also an assistant to Corbino before moving to Florence.[4] Upon his arrival in Tuscany, he immediately began to teach theoretical physics and the text of his lectures, which were compiled by Bruno Rossi, Garbasso's young assistant, and Guilio Racah, still a student, and printed at the expense of Corbino's Institute, became *the* textbook in the nascent field of quantum mechanics for all Italian physics students. Franco Rasetti, who had been associate professor (*aiuto*) under Garbasso in Florence, moved to Rome where he held the same position under Corbino before becoming professor of spectroscopy in 1928.

The group in Florence quickly grew with the addition of Giulio Racah (1909-1965), who divided his time between Florence, Rome and Zurich, where he worked with W. Pauli; Daria Bocciarelli (born 1910); Lorenzo Emo Capodilista (1909-1973); Sergio De Benedetti (born 1912) and Manlio Mandò (born 1912).

The climate typical of the Physics Institute in Arcetri in those years clearly emerges from the Proceedings of the round table "Arcetri dagli anni '20 agli anni '30" (Arcetri from the twenties to the thirties) held at the Physics Department of the University of Florence (in Arcetri) on December 4, 1987.[5]

Between autumn 1926 and winter 1938, Enrico Fermi--with the help of Franco Rasetti for the experimental part--managed to shape a remarkable number of students in both experimental and theoretical physics.

The experimental physicists (in order of their arrival at the Institute in Via Panisperna) were: Emilio Segrè (1905-1989), Edoardo Amaldi (born 1908), Bruno Pontecorvo (born 1913) from Pisa, Eugenio Fubini (born 1914) from Turin, Mario Ageno (born 1915) from Genova, Giuseppe Cocconi (born 1913) from Milan, Oreste Piccioni (born 1916) from Pisa and Marcello Conversi (1917-1988). The young theoretical physicists included Ettore Majorana (1906-1938), Giovanni Gentile jr. (1906-1942), Renato Einaudi (1909-1976) and Gian Carlo Wick (born 1919), the last two of whom came from Turin, Giulio Racah (1909-1965) from Florence, Leo Pincherle (1910-1976) from Bologna, Ugo Fano (born 1912) also from Turin and Piero Caldirola (1914-1984) from Pavia.

An unusual case was Salvatore E. Luria (born 1912), a young physician from Turin who was interested in learning physics or at least its methods. As Ugo Fano's friend, he consulted Fermi and Rasetti who encouraged him to go ahead.[6] He was given a desk in my office and there he started to study, mainly under the guidance of Rasetti. Luria and I made a pact for mutual assistance: I was to help him with mathematics, which was unfamiliar to him, and he would lend me a hand when mine were insufficient for setting up an experiment.

---

[4] E. AMALDI e F. RASETTI, *Ricordo di Enrico Persico*, in *Celebrazioni Lincee, 115* (1979), and "Il Giornale di Fisica", *20* (1979), 235-269.

[5] Round table discussion on *Arcetri dagli anni '20 agli anni '30* (participants: E. Amaldi, G. Bernardini, D. Bocciarelli, M. Mandò, G. Occhialini, B. Rossi), Arcetri, December 4, 1987, edited by Alberto Bonetti and Manlio Mandò.

[6] SALVATORE EDOARDO LURIA, *A Slot Machine, A Broken Test Tube - An Autobiography*, Harper and Row Publishers, New York, 1984.

The two groups of physicists at the Universities of Rome and Florence soon gained international renown in related but different fields of research with papers and discoveries that have left their mark on the history of physics. The Florentine group dedicated its attention mainly to the study of cosmic rays. A very alluring description of the results obtained in those years is given in the first chapters of Bruno Rossi's autobiographical book "Momenti nella vita di uno scienziato", recently published by Zanichelli.[7]

Let me simply recall Rossi's method of coincidences (1930), used by physicists throughout the world for decades; the discovery of the production of secondary particles by cosmic rays (1932); the transition or Rossi curve, due to the production of these particles (1933); and analysis of the soft and hard components of cosmic rays and the fact that the latter will go through a meter of lead. Finally, study of the East-West effect caused by the Earth's magnetic field, predicted by Rossi and studied by him and Sergio De Benedetti in an expedition to the equatorial zone (Asmara, 1934). This activity was carried out after Rossi's transfer to the University of Padua, which I will refer to shortly.

The group in Rome, on the other hand, was concerned with various problems of atomic and molecular physics and, as of 1934, nuclear physics.

Of the many contributions made to atomic and molecular physics, I will mention only the Thomas-Fermi statistical atomic model, worked out by Fermi (and shortly before him by Thomas) and applied to various specific problems by Fermi, Rasetti, Majorana, Gentile, Segrè and Pincherle, and perfected in many ways by Fermi and Amaldi; experimental study of the Raman effect of the rotational levels of symmetrical biatomic molecules and crystals (Rasetti); Fermi's theory of the Raman effect in crystals; the Raman effect of the rotational levels of the oxides of carbon and ammonia (Amaldi and G. Placzek); the experimental study of the forbidden lines of alkalines (Segrè), the theory of hyperfine structures (Fermi and Segrè); the experimental study of the Stark-Lo Surdo effect of Rydberg states and the discovery of the shift in these states caused by the presence of an extraneous gas (Amaldi and Segrè) and Fermi's theory of this effect; and various elegant papers on spectroscopy and chemical bonds by Ettore Majorana. Fundamental in the development and spread of these teachings were the books by Fermi and Persico.[8]

In the field of nuclear physics, I must recall Majorana's theory of exchange forces, Fermi's theory of beta disintegration, the extension of this theory by G. C. Wick, various other elegant works by the latter, some papers on electrodynamics by Racah, and the theory of electron and positron symmetry by Majorana.

Particularly important from an experimental point of view were Fermi's discovery (1934) of the radioactivity caused by neutrons, followed by systematic experimental work (carried out by Fermi, Amaldi, D'Agostino--a chemist--Pontecorvo, Rasetti and Segrè) which led to the identification

---

[7] B. ROSSI, *Momenti della vita di uno scienziato*, Zanichelli, Le Ellissi, 1987.

[8] E. FERMI, *Fisica atomica*, Zanichelli, Bologna (1928); ID., *Molecole e Cristalli*, Zanichelli, Bologna (1934); E. PERSICO, *Lezioni di Meccanica Ondulatoria redatte da B. Rossi e G. Racah*, CEDAM, Padua (1935); ID., *Fondamenti di Meccanica Atomica*, Zanichelli, Bologna (1939); *Fundamentals of Quantum Mechanics*, translated into English by G. M. Temmer, Prentice Hall, Inc., New York (1950).

of almost fifty new artificial radioactive bodies, some of which are produced by bombarding uranium, the discovery of neutron slowdown and the extraordinary ability of the slow neutrons thus obtained to bring about processes of radiative capture.

From 1936 onwards, Rasetti's book on nuclear physics[9] came into use as the fundamental text around the world, especially for the more modern aspects of physics, alongside the classic texts by Lord Rutherford et al., Maria Sklodowska Curie and St. Meyer, and E. von Schweidler.

In any case, the activity of all members of the two groups in Rome and Florence was only just getting under way when the "Manifesto in difesa della Razza" (Manifesto in defence of the race) was made public on July 14, 1938, marking the beginning of a period of many months in which the Fascist government promulgated at more or less weekly intervals a series of decree laws defined as protecting the Aryan race.

The groups in both Rome and Florence were decimated, almost destroyed.[10] Five professors of physics emigrated: Bruno Rossi, Emilio Segrè and Giulio Racah who lost their jobs and civil rights in Italy as a result of those laws; Enrico Fermi, whose wife, Laura Capon, came from a Venetian Jewish family and was therefore affected by the laws--their very marriage had become illegal; and Franco Rasetti who no longer wanted to live in such an uncivil country after the introduction of the racial laws.

The fears of these colleagues were not unfounded. Let me just mention that Emilio Segrè's mother, Amelia Treves, daughter of a well-known Florentine architect, and Augusto Capon, Laura Fermi's father, born in 1872 and Admiral of the Italian Navy, were taken by surprise by the SS in their homes in Rome on October 16, 1943, arrested and deported. They never returned, although it is not known whether they died of the hardships and abuse suffered during the trip or whether they were killed upon arrival in Germany or later by the well-known Nazi methods.

The same fate was shared by Egle Segrè (born 1899), the first of the three daughters of Gino Segrè, Emilio's uncle. She was married to Edgardo Levy, a businessman, and had two children: Mariuccia and Enzo. During the war they evacuated Turin and moved to Tradate, where on November 8, 1943, the SS arrested Egle and her children, 21 and 16 years old at the time, and deported them to Auschwitz. Enzo, who was immediately separated from his mother and sister, was the only one to return after the war: nothing was ever again heard of either Egle or Mariuccia.

Other Jewish Italian scientists who did not yet have a chair, but emigrated at that time, and later achieved emininent positions in universities or society abroad were Leo Pincherle, Ugo Fano, Sergio De Benedetti and Bruno Pontecorvo in physics, Eugenio Fubini in applied physics and Mario Salvadori and Roberto Fano in engineering.

This list is incomplete for physicists, but above all for applied physicists and engineers. In addition, there were other scientists who were affected by the racial laws but could not leave Italy because they were too young or too old or had family ties or whatnot, and therefore suffered all the

[9] F. RASETTI, *Il nucleo atomico*, Zanichelli, Bologna (1936); *Elements of Nuclear Physics*, Prentice Hall, Inc., New York (1936).

[10] E. AMALDI, *Gli anni della ricostruzione*, "Il Giornale di Fisica", *20* (1979), 186-225.

consequences in loco, so to speak. They were many in number and deserve to be remembered, but the constraints of time force me to speak only of some of those who emigrated.

First of all, though, I must explain that some reshuffling in teaching posts in Italian universities had taken place during the thirties prior to the introduction of the racial laws as a result of the normal course of university affairs and personal decisions.

In autumn 1930, Persico was offered and accepted the chair of theoretical physics at the University of Turin. He was replaced at the University of Florence first by Bruno Rossi and later, as of 1932-33, by Giulio Racah. On March 23, 1933, Antonio Garbasso died after a long illness and was succeeded by Laureto Tieri (1879-1952) who did not, however, manage to establish the same kind of relationships of mutual understanding and friendship that had existed between his predecessor and the new generation of physicists.

In fall 1932, the first competitive exam for the chair of experimental physics was won by Bruno Rossi who was called to the University of Padua. A second competitive examination in the same field was held in autumn 1935 and won by Emilio Segrè who went to the University of Palermo. The third in the autumn of 1936 brought Amaldi and Bernardini to chairs in Cagliari and Camerino, respectively. But on January 23, 1937, Orso Mario Corbino died unexpectedly,[11] and the Faculty of Science of the University of Rome offered the job to E. Amaldi. G. Bernardini took over the chair of superior physics at the University of Bologna in autumn 1938.

Almost immediately after his arrival in Palermo, Emilio Segrè managed to convince the Faculty of Science of that university to hold a competitive examination for a chair of theoretical physics. That exam, the second one in theoretical physics since 1926 (!), turned out to be the only one in that field held between the two world wars. The shortlist for the competitive exam at the University of Palermo was Gian Carlo Wick, first, Giulio Racah, second, and Giovanni Gentile jr., third, who took over the chairs of theoretical physics in Palermo, Pisa and Milan, respectively. At the same time, Ettore Majorana was named full professor of theoretical physics at the University of Naples on special merit.[12] The article of the law used for this purpose (Art. 8 of the RDL June 20, 1935, no. 1071) had been introduced some years earlier to make it possible to bring Guglielmo Marconi to the chair of electromagnetic waves at the University of Rome without an examination. Ettore Majorana then disappeared in spring of 1938.

3. The almost total destruction of the Rome group

Let us turn to the consequences of the racial laws in Italy.

---

[11] E. AMALDI, E. SEGRETO, *Corbino O. M.*, in *Dizionario Biografico degli Italiani, 28*, 760-766, Istituto dell'Enciclopedia Italiana. E. AMALDI, *Corbino O. M.*, in *Scienziati e tecnologi contemporanei, I*, 264-266, Mondadori, Milan 1974, see also endnote 2*b*.

[12] E. AMALDI, *La vita e l'opera di Ettore Majorana*, Accademia Nazionale dei Lincei (1966); *Ettore Majorana, man and scientist*, in *Strong and Weak Interactions - Present Problems*, Academic Press, New York (1966); see also endnote 2*d*.

Enrico Fermi left Italy with his family on December 6, 1938, heading for Stockholm,[13] and reached New York in early January 1939. Officially, he had accepted the position of "visiting professor" at Columbia University for one academic year, but a few weeks before his departure he had informed Rasetti and me that he was leaving for good. At the Pupin Laboratory, where the Department of Physics of that university was located then and is still in part located today, Fermi started work with Herbert L. Anderson investigating several aspects of the phenomenon of uranium fission discovered in Berlin-Dahlem by Otto Hahn and F. Strassmann at the end of 1938, that is, only a few weeks earlier. The results of their work were kept secret in view of the peaceful and possibly military applications. The group was enlarged and transferred to the University of Chicago in spring 1942, and there, on December 2nd of the same year, a nuclear reactor of small--but not neglegible--power was put into operation for the first time, demonstrating the possibility of using nuclear fission for the production of energy for civilian uses.

From Chicago, Fermi moved to the Los Alamos Laboratories directed by Robert Oppenheimer, where he worked as a consultant for the construction of the first atomic bomb. After the war, in 1946, Fermi returned to the University of Chicago and remained there until his death (1954). In this last part of his life, he worked first on the properties of slow neutrons and their interactions with matter, using beams of these particles produced by research nuclear reactors and then, after the discovery of the pi meson, on the collision of these new particles against nucleons, thus discovering the first resonance between elementary particles. Other work from the same period includes an original theory on the origin of cosmic rays and a statistical theory on the production of new particles during high energy collisions.

As is obvious from this brief biographical summary, it was only natural and almost inevitable that Fermi should start to work on peaceful applications of nuclear energy after the discovery of uranium fission. And this would probably have happened had the Second World War not broken out.

Instead the shift from civilian to military applications was no doubt determined by the fact that the conflict was already under way in Europe and that, as far as was known in Europe and overseas, German scientists were working secretly in that direction. The same argument was obviously used in the same way by the scientists of the Third Reich who, out of political blindness, contorted nationalism or simply a choice of sides, had no legal or moral reservations about the government of their country.

Enrico Fermi was certainly one of the first scientists in the Western world to be faced with the very profound moral dilemma of having to decide whether or not to participate in research on the military applications of nuclear energy. He and others that I will speak of later had to make the decision at a time when the enemy was advancing on all fronts and the final outcome of the conflict was not certain. They were all peace-loving people. Information was scarce, but it was sufficient to convince many of them that it was their duty towards the free world to participate in its defence: on

---

13 AMALDI, *Gli anni della ricostruzione . . .* cit.

September 2, 1939, Hitler's and Stalin's troops started to invade Poland from opposite sides. Approximately a year later, Hitler occupied a large part of France and laid siege to Great Britain. Not many months later, German troops marched into Denmark, Norway and the Balkans and made considerable advances in Russia and Africa. It looked as though Hitler might succeed in conquering the world.

At all times in his life, Fermi produced an incredible number of students. His teaching activity in the United States has been described by one of his American co-workers, Albert Wattenberg, who published an article on the subject in the European Journal of Physics in 1988.[14]

Of Fermi's many students in both experimental and theoretical physics, six went on to receive the Nobel Prize: three theoretical physicists, the two naturalized Chinese Americans T. D. Lee and C. N. Yang (1957) and the American Murray Gell-Mann (1969), and three experimental physicists, E. Segrè and O. Chamberlain (1959) and J. Steinberger (1988). Chamberlain was also a student of Emilio Segrè and after the discovery of the anti-proton in 1955 became one of his closest co-workers.

Franco Rasetti, the bachelor, sailed from Naples aboard the Vulcania with his mother Adele Galeotti on July 2, 1939, reaching New York ten days later. After Fermi's departure, Rasetti and I had done some work together, in particular, we had studied the instantaneous emission of gamma rays and electrons by gadolinium after capture of a slow neutron. But in spring, Franco received an unexpected offer from a Catholic university in Canada, namely Laval University, Quebec, which had recently set up a Faculty of Science and Engineering and wanted Rasetti to head the Department of Physics. "The position," as Rasetti recalled many years later, "had a number of attractive features, and also considering the international situation which was rapidly deteriorating, I decided to accept [the offer]."[15] Officially, he had accepted the position of "visiting professor" for one year. From New York, Franco Rasetti and his mother continued directly on to Quebec.

If I may make a personal aside here, I would like to mention that I, too, was on the Vulcania when it sailed out of Naples. But I had left behind a wife, two children and a third that was on its way. Officially I was on my way to the United States to study the construction of a cyclotron to be installed at the 1942 Universal Exposition in Rome. And this I did,[16] but I hoped at the same time to be able to find a job in that country so that I could move there with my family within a year or two. I was unable to find a job, though,[17] and the war in Europe broke out in early September and the Italian authorities refused to give my family a visa, so I returned to Italy, again aboard the Vulcania,

---

[14] A. WATTENBERG, *The Fermi School in the United States*, "Eur. J. Phys.", *9* (1988), 88-93.

[15] In early April 1977, Franco Rasetti, who was about to leave Rome to go and live in Waremme, Belgium, the home town of his wife, gave me 18 typewritten pages in English entitled: "Biographical Notes and Scientific Works of Franco Rasetti", which I am to publish only after his death. I nevertheless feel that I can draw from this source in those cases in which I do not want to risk altering his thoughts.

[16] E. AMALDI, *Studio dei ciclotroni negli Istituti degli Stati Uniti*, Reale Accademia d'Italia, Viaggi di studio promossi dall Fondazione Volta, Vol. VI, 1941.

[17] The people who seriously tried to help me were: I. I. Rabi of Columbia University, E. U. Condon, head of the Westinghouse Research Laboratories, Pittsburg, Penn.; and M. A. Tuve of the Department of Terrestrial Magnetism of the Carnegie Institution, Washington, DC. Upon my answer no to the question of whether or not I had lost my job in Italy, they explained that there were many others who were much worse off than me and to whom they had to give priority. I obviously had to admit that they were perfectly right.

arriving at Naples on October 14, 1939.

In Quebec, Rasetti not only founded the Physics Department of that university, but in three months set up a laboratory in which he could teach and do research on slow neutrons. In fact, within two years he had made the first rough measurements of the average life of a muon,[18] which were soon followed by those of Auger et al. in Paris,[19] Rossi and Nereson at Cornell University,[20] and Conversi and Piccioni in Rome.[21]

In January 1943, H. von Halban and G. Placzek asked Rasetti to join the group of English- and French-speaking scientists that had started to develop nuclear energy for military purposes in Canada. After much deliberation, Rasetti turned down the offer. As he wrote later, "I have taken very few decisions in my life that I have less reason to regret. I was convinced that nothing good could come of the construction of new and more monstrous means of destructions . . . "[22]

In autumn 1947, Rasetti accepted the job of professor of physics at Johns Hopkins University in Baltimore, where he remained until 1967. In 1949, he married Marie Madeleine Hennin, a woman of Belgian origin, who had two children, aged 16 and 12, from a previous marriage.

His research activity at Johns Hopkins started slowly and never became very intense for a number of reasons. First, research in low and high energy nuclear physics was changing from the traditional scale of university laboratories to semi-industrial ventures and this transformation was not appreciated by an individualist like Rasetti, who had always tried to do everything or almost everything on his own. Second, in order to receive a research contract from any American government agency, the director of the project was required to have "clearance", that is, an official declaration that he was not a security risk, even when the project was in no way classified. Rasetti considered the investigation of his political opinions, past and present associations, etc. an intolerable insult to scientific dignity and freedom. Third, research funds at Johns Hopkins could be obtained almost exclusively by contract with government agencies.

Therefore, he finally decided to do no more than scrupulously carry out his teaching duties in physics and concentrate his research activity in other fields such as geology and paleontology.

Rasetti had always had an intense interest in the natural sciences. For example, he had done some serious work in entomology, extending and completing his father's insect--mainly coleopter--collection. He had also discovered three new cavernicolous coleopters, one in 1917, in a cave in the Apuan Alps, and two in 1938, the first in a cave near Ravello and the other in a cave in Friuli.

In 1939, before leaving Italy, Rasetti donated his collection of 30,000 exemplars to the Rome

---

[18] F. RASETTI, Disintegration of slow mesons, "Phys. Rev.", *59* (1941) 613; *ibidem, 60* (1942) 198-204.

[19] P. AUGER, R. MAZE, R. CHAMINADE, *Une mesure directe de la vie moyenne du méson au repos*, "C. R. Acad. Sci. Paris", *214* (1942), 266-269; R. CHAMINADE, R. FREON, R. MAZE, *Nouvelle mesure directe de la vie moyenne du méson au repos*, "C. R. Acad. Sci. Paris", *218* (1944), 402-404.

[20] B. ROSSI, N.C. NERESON, *Experimental Determination of the Disintegration Curve of Mesotrons*, "Phys. Rev.", *62* (1942), 417-422; *Further Measurements on the Disintegration Curve of Mesotrons*, "Phys. Rev.", *64* (1943), 199-201.

[21] M. CONVERSI, O. PICCIONI, *Sulla disintegrazione dei mesoni lenti*, "Nuovo Cimento", *2* (1944), 71-87; *On the Disintegration of the Mesons*, "Phys. Rev.", *70* (1946), 874-881.

[22] See endnote 15.

Museo Civico di Zoologia located in the Rome Zoo.

His interest in geology and paleontology started in 1939 in Quebec and continued throughout his stay in Baltimore (1947-67) to the first years after his return to Italy. His research, which was concentrated on trilobites of the Cambrian period, began with examination of the Cambrian layers in Quebec. In eight years he and his assistants and friends screened tons of rock, gathering and preparing thousands of samples. From 1941, he extended his investigations to the Rocky Mountains in British Columbia and some regions of the United States and in 1967 to the Cambrian layers of the Iglesiente (Sardinia). He had already published more than 40 papers on Cambrian trilobites by 1952, when the United States National Academy of Sciences gave him the Charles D. Walcott International Award for the best paper on the Cambrian Period published anywhere in the world in the last four years.

All in all, Rasetti personally gathered and prepared more than 20,000 fossils from the Cambrian which he distributed in almost equal parts (2500 each) among the following museums:

- United States National Museum (Washington, DC),
- Geological Survey of Canada,
- British Museum (Natural History),
- Musée de Paléontologie de l'Université Laval (Quebec).

He left an almost equal amount of Cambrian fossils from Sardinia (2000) to the Museo di Paleontologia of the Servizio Geologico d'Italia in Rome for exhibition when a National Museum of Natural Sciences is opened.

For years, Rasetti also worked untiringly during spring and summer--and at his own expense--on two botanical studies. Both were begun during the last years of his stay in the United States and continued later in Italy. He even continued work on the second after moving to Belgium.

The first is a collection of colour slides of Alpine flowers found above the tree line. This collection contains 97% of the plants included in the project from which the book *Franco Rasetti, I fiori delle Alpi* was taken. Published by the Accademia Nazionale dei Lincei (1980) with funding from the Banca Commerciale Italiana, the volume contains over 572 colour prints, accompanied by a scientifically correct, modern and informative text suited to a larger audience.

The second botanical undertaking is a collection of colour slides of Italian orchids (which account for 80% of European orchids). It is feared that there may not be enough time to publish a book analogous to the previous one on this subject, but before leaving for Belgium, Rasetti had already made the following donations:[23]

- a complete series (4000 colour slides) of *Flora Alpina* to the Botanic Institute of the University of Florence;
- another complete series to the Centro Lincei;

[23] This information was supplied by Franco Rasetti on my request when Ginestra and I went to visit him and Madeleine in their apartment in Via Salaria 300 at 11 o'clock on April 2, 1977, after Franco had told us of his intention to move to Belgium at the end of the month.

- a collection of 2000 colour slides of Italian orchids to the Centro Lincei;
- a series of 200-300 colour slides of orchids of the eastern and southern United States and Canada to the Centro Lincei;
- samples of hundreds of plants to the Erbario Nazionale in Florence, plants which the herbarium lacked or of which the samples were inadequate.

He also momentarily left 300 black and white stereoscopic slides of mountains and climbs with me.

4. The destruction shortly after their formation of the groups in Padua, Palermo and Pisa

Let us now look, in order of age, at the physicists who, at the time of the entry into force of the racial laws, had just received professorships.

Bruno Rossi was named professor of experimental physics at the University of Padua in autumn 1932, where he soon managed to have a new institute built and equipped for research. After the promulgation of the racial laws, he left for Copenhagen with his wife Nora Lombroso on October 12, 1938. They were able to leave so quickly by taking advantage of a last part of a scholarship for study abroad from the Reale Accademia d'Italia, the main part of which Bruno had used to spend time working in the laboratory of Duke Maurice De Broglie in Paris.

Emilio Segrè, who had taken over the chair of experimental physics at the University of Palermo in the autumn of 1935, left for Berkeley, California in early summer 1938 to work on the short mean life isotopes of element 43, technetium, which he and his colleague Carlo Perrier (1886-1948) had discovered in Palermo in 1937.[24]

Given the political trend in Italy and more generally in Europe, Segrè decided to remain in the United States and wrote to his wife Elfride Spiro to close down their apartment and to come over to California with their first son Claudio, not yet two years old.[25]

Elfride was a young German woman who had some experience in sudden and even more dramatic departures. No more than four years earlier, she had fled Breslau to escape Hitler's persecutions. The difference this time was that before leaving Palermo, many of her husband's colleagues came to express their solidarity and friendship and their wives helped her in the unpleasant and laborious job of packing and sending their furniture, even after she had left Palermo.

The careers in the United States of Bruno Rossi and Emilio Segrè were both extremely brilliant but different. Both suffered initial difficulties because it was not easy even for them to find positions as "full professors" right away.

With the help of Niels Bohr in Copenhagen, Rossi's first stop after leaving Italy, he finally found a job in the laboratories of P.M.S. Blackett (1897-1974) in Manchester, Great Britain. After

---

[24] C. PERRIER, E. SEGRE', *Radioactive Isotope of Element 43*, "Nature" (London), *140* (1937), 193-194; *Some Chemical Properties of Element 43, I*, "Jour. Chem. Phys.", *5* (1937), 712-716, *II*, *ibidem*, *7* (1939), 155-156.

[25] E. SEGRE', *Fifty years up and down a strenuous and scenic trail*, "Ann. Rev. Nucl. Sci.", *31* (1981), 1-18.

approximately a year, the Rossis were able to move to the University of Chicago and later to Cornell University (Ithaca, NY). During the war, that is from 1943 to 1946, Bruno worked at the Los Alamos Laboratories as head of one of the experimental groups. After the war in 1946, he became professor at the Massachussets Institute of Technology where he remained until retirement and still holds the position of professor emeritus.

Emilio Segrè, on the other hand, stayed in Berkeley for good with the exception of the time spent from 1943 to 1946 at Los Alamos as head of another experimental group. He only received his full professorship from Berkeley, however, in 1946 after his return from Los Alamos.[26]

While in Chicago, Bruno Rossi's scientific activity was concentrated on comparison of the absorption of the hard component of cosmics rays by the atmosphere with absorption by a layer of graphite of equal mass per unit surface area and on determination of the relationship between the mean life of the mesotrons making up the hard component and their rest energy mc^2.

Working with Nereson at Cornell University, he achieved a direct measurement of considerable precision of the mean life of the mesotron,[27] which came only shortly after the pioneering observations of Rasetti[28] and preceded by a few months those carried out in Italy by M. Conversi and O. Piccioni.[29] Still at Cornell University, in part with Greisen, he did an experimental study of electron showers and compared the results obtained from rather sophisticated calculations with experimental results, a paper which has become a classic of scientific literature on this subject.

After the war, Rossi returned to the study of cosmic rays at MIT, especially identification of the secondary particles they produce in different materials using a cloud chamber.

In this period, Rossi's group became a center of international importance drawing young and experienced researchers not only from Italian universities, but also from Japanese and French institutes, especially the Ecole Politechnique in Paris. Many famous scientists from these countries recall with pleasure and admiration the time spent with Rossi's group at MIT.

At the end of the fifties, Rossi turned his attention from cosmic rays to space research, a field in which he formed a new school. He first studied interplanetary plasma and later (1963) discovered the emission of x-rays by the star Scorpio X-1. His co-workers in this study included Riccardo Giacconi and guest researcher Japanese Minoru Oda, who had worked with Rossi on cosmic rays and was later head of the Japanese space research programme for many years.

In 1987, the Wolf Prize for Physics went to B. Rossi and R. Giacconi for their discovery of Scorpio X-1's emission of x-rays and to H. Friedman for having demonstrated in a blacking out experiment behind the moon that the source discovered by the group at MIT was punctiform.

The scientific activity of Emilio Segrè during his stay in the United States included the experimental demonstration that decay of molybdenum 99 (with a 67 h period) produces a 6-hour isomeric state in the 99 isotope of technetium (element 43). Carried out together with Glenn

---

[26] *Ibid.*
[27] See endnote 20.
[28] See endnote 18.
[29] See endnote 21.

Seaborg, then still very young, this experiment later acquired extraordinary importance in the diagnosis and treatment of brain and heart disorders. In 1939, together with Halford and Seaborg, he extended the Szilard-Chambers method to the separation of nuclear isomers and in early 1941, with Kennedy, Seaborg and Wahl, managed to produce tiny quantities of the nuclide later called $_{94}Pu^{239}$ and measure the corresponding cross section for fission with slow neutrons. These results made it possible to establish not only that this isotope of plutonium could be produced but also that it could be used as a substitute for uranium 235 in all applications.

During the war, Segrè's group at Los Alamos studied spontaneous fission of uranium and plutonium.

On his return to Berkeley, Segrè demonstrated the effect of chemical bonding on the nuclear decay constant for capture of the K electrons of Be^7 due to the change in electron density in the spatial region occupied by the nucleus.

Between 1949 and 1951, Segrè carried out a large-scale experimental study of neutron-proton and proton-proton collisions with a large group of co-workers. When Berkeley's Bevatron came into operation, he built a mass spectrometer together with O. Chamberlain, C. E. Wiegand and T. J. Ypsilantis, whose specific characteristics allowed him to establish subsequently (October 1955) that a small but clearly observable number of *negative protons* were produced by the collision of 6.3 GeV protons (accelerated by the Bevatron) against nuclei at rest. That these were antiprotons in the sense intended by Dirac, that is, bodies capable of being annihilated by as many protons, was demonstrated experimentally using nuclear emulsions by the same group plus G. Goldhaber et al. in Berkeley and by E. Amaldi et al. in Rome. Study of the process of annihilation kept many physicists in Berkeley and Rome busy for years. In 1959, E. Segrè and O. Chamberlain received the Nobel Prize for Physics for the result obtained in October 1955.

Both Bruno Rossi and Emilio Segrè wrote several excellent books on physics.[30] Emilio Segrè also wrote a book on the life and works of Enrico Fermi[31] and two beautiful volumes on the history of physics.[32]

Giulio Racah, had already made various important scientific contributions when he was in Italy. Among these I would like to recall a study of the hyperfine structures of atomic spectra, carried out under the guidance of Persico and Fermi and various papers on quantum electrodynamics, especially calculation of cross sections for Bremsstrahlung processes and the creation of pairs, written in the early thirties following Pauli and Fermi.

Racah later clarified the meaning of Majorana's theory on neutrinos and wrote that

---

[30] B. ROSSI, H. H. STAUB, *Ionization Chambers and Counters,* McGraw-Hill Book Company Inc., New York (1949); ID., *High Energy Particles,* Prentic Hall, Inc., New York (1952); ID., *Cosmic Rays,* McGraw-Hill Book Company Inc., New York (1964); ID., *Optics,* Addison and Wesley Publishing Company, Inc., Reading, Mass. (1957); translation in Italian, *Ottica,* Tamburini editore, Milan (1971). E. SEGRE' (ed.), *Experimental Nuclear Physics,* in three volumes, John Wiley & Sons, Inc., New York (953, 1953, 1959); ID., *Nuclei and Particles,* Benjamin, Inc., New York (1964), translated into Italian, *Nuclei e Particelle,* Zanichelli, Bologna (964).

[31] See endnote 2*b*.

[32] E. SEGRE', *Personaggi e scoperte della fisica contemporanea,* Biblioteca Est, Mondadori (1967); translation from the English, *From X-rays to Quarks,* Friedman, S. Francisco (978); ID., *Personaggi e scoperte della fisica classica,* Biblioteca Est, Mondadori 81983); translated from the English, *From falling bodies to radiowaves,* Friedman, S. Francisco (1983).

appropriate experiments could establish whether the neutrinos observed in nature are Dirac neutrinos or Majorana neutrinos. He had then started to investigate complex atomic spectra. After being short listed in the competitive examination for the chair of theoretical physics in autumn 1937, he was soon named professor at the University of Pisa, a particularly prestigious position, considering that students also came from the Scuola Normale Superiore.

Less than one year later, he lost his job as a result of the racial laws, but only in 1939, that is, on the eve of the Second World War, did he immigrate to Israel. There he soon started to teach and work at the Hebrew University in Jerusalem,[33] of which he later became the rector. After his death, the Physics Institute of the university was named in his honour and each year an international scholar is invited to give the "Racah Lecture" on some topical scientific subject.

Racah's greatest scientific contribution is contained in four papers (three dating back to 1942-43 and one from 1949) on the antisymmetrical states of systems composed of n bodies of 1/2 spin, all with the same orbital angular momentum *l*. In the last of these papers, Racah uses the group theory. In 1950, he described this application of the theory to spectroscopy in a series of lectures given at the Institute of Advanced Studies at Princeton. Request for these lectures was so great that they were first reprinted by the Institute in Princeton and later by the Società Italiana di Fisica and CERN. Finally, they were published in the *Springer Tracts in Modern Physics* in 1965.

As his student Talmi, a well known theoretical physicist, wrote, "Racah was the man who established theoretical physics in Israel" in that "he studied all the atomic spectra with his students and co-workers in an attempt to obtain a theoretical understanding of all measured atomic spectra. He also made important contributions in the field of nuclear spectroscopy, mainly by developing general methods."[34]

5. The Emigration of Younger Physicists

I would now like to speak about some young physicists who, despite promising careers at the time of promulgation of the racial laws, were not yet professors, mainly because of age.

The first is Leo Pincherle (1910-1976), grandson of the Bolognese mathematician Salvatore Pincherle. After receiving his degree in physics in Bologna, Pincherle worked from 1932 to 1935 at the Physics Institute in Via Panisperna. In the next three years, he taught a course in theoretical physics at the University of Padua.

Following the entry into force of the racial laws, Pincherle moved to Zurich in 1938 with his wife, Nora Cameo, and their first son, Guido (born in 1937). Their second son, Aldo Italo, was born in Switzerland but died after only 8 months as a result of the hardships incurred by their dramatic departure from Rome. From Zurich, the Pincherle's moved on to London, where Leo started work in

33 IGAL TALMI, *In memoriam Giulio Yoel Racah*, "Proc. Israel. Acad. Sci. and Human.", Section of Sciences, No. 2, read at Memorial Meeting, 26 October 1965, Jerusalem, 1966.
34 *Ibid.*

1939 with Prof. H. T. Flint at King's College in London.

Upon Italy's entry into war (June 1, 1940), Leo Pincherle was interned, but was released after five months upon request from the university and was able to return to his work.

His son Ugo was born in Malvern in 1950. During the war, Pincherle worked as a lecturer at the same college in Bristol and later in London.[35] In that difficult period, he also found the time to work as a visiting lecturer at the Polytechnical Schools in Chelsea and Regent Street.

In 1948, Leo Pincherle was named Principal Scientific Officer of the Telecommunications Research Establishment (TRE), in Great Malvern, which soon became the most important British laboratory specialized in semiconductors. It was in this laboratory that Pincherle made an important contribution to the calculation of the wave functions of electrons in solids, a subject which became his specialty in both research and teaching. Of particular importance are a paper published in 1953 on the band structure of lead sulphite (Bell et al.), in which the Wigner-Seitz cell method is applied to a biatomic semiconductor for the first time and which constitutes the first classic work in this field, and the fundamental theoretical work on the application of the group theory which appeared as a *TRE Memorandum*.

In 1955, Pincherle returned to the University of London, Bedford College, first as a lecturer and then, from 1969 onwards, as professor of mathematical physics. His course on the "Band Structure of Semiconductors" given in 1963 at the Enrico Fermi International School in Varenna and a second course on the same subject given in Perugia in 1965 were classic presentations of the band theory which led him to publish his authoritative volume "Electronic Energy Bands in Solids" (MacDonald and Co. London) in 1971.

For many years, Leo Pincherle was the editor of the English cultural journal *Endeavour*.

To conclude, it seems fitting to cite the final words of Ernst Sondheimer's biographical article on Leo Pincherle: "The picture of Leo Pincherle the scientist cannot be separated from that of Leo Pincherle the man. He was devoted to his family for whom his too-early death is a tragic loss. His sense of humour, dry and delicious, showed in his fund of stories about well-known contemporaries--always amusing and never malicious. His quiet demeanour concealed plenty of fire and imagination, as you soon discovered when you partnered him at bridge. He was the most knowledgeable of men on fundamental aspects of European culture and history and wore his learning lightly; his love of music and the arts was intense, his taste discriminating. He was, in the best sense, both very Italian and very English, combining in his person that internationalism in science and that European spirit which it is so necessary to keep alive today."[36] As he wished, he was buried in Bologna.

Ugo Fano (born in 1912), son of Turinese mathematician Gino Fano and first cousin of Giulio Racah came to Rome in autumn 1934 after having studied mathematics in Turin. In Rome, Fano worked under Fermi, except one year (1936-37) which he spent in Leipzig with W.

[35] ERNST SONDHEIMER, *Leo Pincherle*, "Nature (London)", *266* (1977), 202. A. K. JONSCHER, *Leo Pincherle (1910-1976)*, "Solid-State Elettr.", *657* (1977). See also "The Times" of November 4, 1976 and the "Resto del Carlino" of November 26, 1976.

[36] *Ibid.*

Heisenberg. Following a seminar by Pascual Jordan on the physical analysis of genetic phenomena caused by radiation, Fermi suggested that he specialize in this field. Jordan introduced him to the most outstanding scholars in the field, in particular, H. J. Müller. This explains why Fano, after publishing various works on atomic spectroscopy, nuclear physics, the thermodynamics of nuclei and uranium fission, later turned his attention to various problems of genetics with special concern for the theory of mutations induced in drosophila by radiation.

It was through this channel that Fano first obtained support in the United States from the Washington Biophysical Institute in 1939-40.

In 1940, Fano married Lilla Lattes, engineering student at the Politecnico in Milan and daughter of Leone Lattes, well known professor of forensic medicine at the University of Pavia. Ugo and Lilla had two children, Mary Giacomoni, who lives in Chicago and Virginia Ghattas, who lives in Wallesley, Mass., both of whom in turn had two children.

Aside from the years 1944-45, in which he worked in a US Army laboratory for ballistic research, from 1940 to 1946, Fano carried out research at the Department of Genetics of the Carnegie Institution in Washington, Cold Spring Harbor, and the Physics Department of Columbia University. From 1946 to 1966, he worked as head of the radiation theory section and later as Senior Research Fellow of the National Bureau of Standards (NBS). In 1966, Fano became professor of physics at the University of Chicago, where he was Chairman of the Department of Physics from 1972 to 1974 and professor emeritus from 1982 onwards. He was made a member of the US National Academy of Sciences in 1976 and received honorary degrees from Queen's University, Belfast, in 1978 and from the Université Pierre et Marie Curie, Paris, in 1979.

At the National Bureau of Standards, he was particularly concerned with the theoretical aspects of X-ray dosimetry and developed a wide-ranging programme on the penetration and diffusion of ionizing rays and the underlying mechanisms. Later, he was assigned to the office of the director of the National Bureau of Standards, allowing him to carry out independent research while supervising all of the agency's theoretical activities. In 1960, with the installation of equipment for the production and use of synchrotron light and electronic spectroscopy techniques, he was able to identify and interpret a whole new series of phenomena, resulting in an extraordinarily intense period of new activity.

The theory of atoms and molecules had led to approximate solutions to the Schrödinger equation that were suitable for the limited range of phenomena known in the thirties. But description of the new phenomena observed using modern techniques called for a new approach that included identification of their critical parameters and their evaluation, *first* semi-empirical and *then* derived from fundamental principles. In particular, Fano helped explain localized "resonant" states of atoms at energies higher than those required for initial ionization (Fano states).

This effort led to the development of new theoretical concepts and new techniques, to which Ugo Fano made important contributions with his more than 230 scientific publications.

Today, Ugo Fano is considered a master in this field, as is attested to by the many awards received and the many invitations to speak and lecture received from US and European universities.

Sergio De Benedetti (born in Florence in 1912) started his career in Padua with Bruno Rossi, with whom he participated in the expedition to Asmara in 1934 to study the East-West effect of cosmic rays.

Following the introduction of the racial laws, De Benedetti moved to France and worked as a researcher at the Curie Laboratory of the University of Paris until 1940. It was at this lab that he met Bruno Pontecorvo, who had also moved to France around Easter 1936. After moving to the United States in 1940, he worked as research associate for the Bartol Research Foundation from 1940 to 1943, associate professor at Kenyon College, Gambier, Ohio, in 1943-44, senior physicist at the Monsanto Chem. Co., Dayton, Ohio, in 1944-45, principal physicist at the Oak Ridge National Laboratories from 1946 to 1948, associate professor at Washington University, St. Louis, in 1948-49 and, finally, professor of physics at the Carnegie Mellon University in 1949, where he remained for the rest of his career with the exception of academic year 1956-57, spent at the University of Turin as a Fulbright Fellow.

His work centered around short mean life radioactive isotopes, positrons and their annihilation, the violation of parity, mesic atoms and the Mössbauer effect. He wrote a book entitled *Nuclear Interactions* (John Wiley, 1964) and produced a large number of students, of which at least eight went on to become university professors and many others reached high-ranking positions in private industries such as Bell Telephone Laboratories, IBM and Westinghouse.

Bruno Pontecorvo is slightly different from all the rest in that he left Italy in spring of 1936, that is, two years before the racial manifesto. When we met with some mutual friends in Austria in March 1937[37] and I expressed my concern for the many signs of increasing anti-Semitism in Italy, Bruno emphasized that he was living abroad because he was against Fascism and that he considered anti-Semitism only a secondary aspect of a much more serious political situation. Although well-founded, I do not think that these considerations justify leaving Pontecorvo out of the general picture that I am trying to paint of a certain environment at a certain time.

After completing his two years in engineering at the University of Pisa, Pontecorvo decided that he wanted to switch to physics and his older brother Guido, biologist and geneticist,[38] advised him to transfer to Rome, where he registered for third year physics and managed to finish his degree brilliantly in July 1934 at the age of twenty.

In September of the same year, he started to work in Fermi's experimental group, which was studying artificial radioactivity caused by neutrons, and therefore participated in a number of discoveries, starting with neutron slowdown. In the fall of 1935 and the winter of 1935-36, he carried out a series of experiments with Gian Carlo Wick on the elastic and inelastic scattering of neutrons by various medium and heavy nuclei. He subsequently used a monetary award received from the Italian Ministry of Education for his contribution to the research on neutrons to move to Paris around Easter 1936, taking a job offered him at the Curie Institute to work with Frédéric Joliot

---

[37] Participating in the ski holiday in Vent, Austria, were Bruno Pontecorvo and Paul Ehrenfest jr. who had come from Paris, and S. Fubini, G. Racah, F. Rasetti, G. C. Wick and E. Amaldi who had come from Italy.

[38] BRUNO PONTECORVO, *Una nota biografica*, 82-87 in "Scienza e Tecnica", Annuario della EST 88/89, Mondadori.

Curie.

He later moved to the College de France, where he turned almost alone to the study of nuclear isomerism. He predicted that there should be cases of nuclear isomerism and managed to demonstrate it experimentally in the case of cadmium excited by collisions with fast neutrons. He also predicted that the photons emitted during transition between isomeric states would be strongly converted and then sought and found some interesting cases of isomerism in which this happened, including that of the Rh. Similar results were obtained at almost the same time by Seaborg and Segrè in Berkeley. Together with A. Lazard, Pontecorvo finally managed to produce (1939) beta-stable isomers ($^{115}In^{*}$) by irradiating stable nuclei with high energy X-rays (-3 MeV). The discovery of this phenomenon, which Joliot called "nuclear phosphorescence", earned Bruno Pontecorvo the Curie-Carnegie Award.

While in Paris, Pontecorvo married a Swedish girl with whom he had three children: the first was born in France; the other two in Canada.

In 1940, after the downfall of France, Pontecorvo fled Paris on bicycle with Marianne and baby Gillo. From Lisbon, he sailed for the United States, where he found a job in a private American company headquartered in Oklahoma.

While working for this company, he developed a new method of well logging, neutron well logging, based on the properties of slow neutrons. A short time later, he accepted an offer to work on the development of applications of nuclear energy with the Anglo-French-Canadian group in which many of the physicists he had known in Paris (P. Auger, B. Goldsmith, H. v. Halban, L. Kowarski, G. Placzek) were participating. In Canada, he worked first in Montreal and later at Chalk River in the project involving design and construction of the NRX natural uranium and heavy water nuclear reactor for which he was in charge of the physical aspects.

After the war, Pontecorvo left Canada and went to England, where he worked for a relatively short time at the Harwell atomic center. In September 1950, Pontecorvo moved with his entire family to the Soviet Union and started to work at the Dubna research center as head of the experimental physics division of the nuclear laboratory.

It was in Canada that Pontecorvo started work on subnuclear particle physics, in particular, on weak interactions and Fermi interactions. Immediately after the experiments by Conversi, Pancini and Piccioni, Pontecorvo was the first (1947) to formulate the law of the universality of weak interactions, stated independently shortly afterwards by others (O. Klein, G. Puppi). Guided by this law, he managed to predict various properties of muons later proven correct by experiments carried out by himself and others.

Some of Pontecorvo's experiments date back to the Chalk River period (Canada, 1948-49), while others were done much later (1958 and 1961) at Dubna. In 1959, he suggested using neutrons produced by the decay of high energy pions to extend knowledge in the field of weak interactions. This approach, later suggested by others, is commonly used in major laboratories such as Brookhaven National Laboratory, the Fermilab, CERN, etc. today.

Pontecorvo made many other extremely acute and important suggestions in the field of weak interactions, many of which led to significant results, but in order to describe them I would have to go into some physical concepts that are not easy to present here in a comprehensible form.

In both Canada and the Soviet Union, Pontecorvo was an exceptional teacher and an outstanding ambassador of Italian culture and customs. Among other things, he is known to have spread a serene interest in sports and love of nature of great moral value among the young people in the various scientific circles with which he came into contact in the USSR.

6. A few examples of emigration of young engineers and applied physicists

In this part of my talk I will refer to two young engineers who dedicated most of their time to university teaching (Mario Salvadori and Roberto Fano) and one young physicist who, more as a result of the circumstances of life than of personal decisions, became an applied physicist and ended up as a consultant to the United States Administration (Eugenio Fubini). If circumstances had been different, he could just as well have become a brilliant experimental or theoretical physicist.

Mario Salvadori was born in Rome in 1907 of a Jewish mother, Ermelinda Alatri, and a Catholic father, Riccardo Salvadori. He completed his degree in civil engineering in 1930 and another in pure mathematics in 1933, both at the University of Rome. We met during our first two years at university and became good friends after a mountaineering holiday in the summer of 1927 spent climbing in the Dolomites near Cortina with his younger brother, Giorgio.

In 1933-34, Mario Salvadori was the last student in photoelasticity of E. G. Coker, professor at the University College of London, where he had gone to study with a scholarship from the League of Nations, initially rejected and later assigned by the Fascist government.

Libero docente in structures in 1937 and assistant lecturer (*assistente ordinario*) in 1938, Salvadori's university career was well on its way with the publication of 15 technical papers between 1933 and 1938 and a course of lectures on the theory of plates and vaults, when the first signs of anti-Semitic policy convinced him to leave Italy. He set out for the United States in January 1938 on a scholarship from the Ministry of Communications, which he was able to win because "he had not been discriminated against" since his father was Catholic.

After spending a year working as an engineer in a toy factory, he was called to Columbia University in 1940 to teach applied mathematics at the school of engineering. During the difficult years of the Second World War, as a member of the US branch of Giustizia and Libertà, he helped to prepare anti-Fascist propaganda material which was sent to Italy clandestinely, and explained the activity of the resistance movement to the American public in a number of radio programmes and conferences.

From 1942 to 1945, he worked (unwittingly) on the Manhattan Project, dedicated to the construction of the first atomic bomb, but refused to take part in military research after the end of the war.

His university career allowed him to publish 13 books on applied mathematics and structures, of which two, *Metodi Numerici in Ingegneria* and *Le strutture architettoniche*, have been translated into 14 languages and adopted as textbooks by numerous European and non-European (Russian, Chinese and Japanese) universities.

In 1954, Mario Salvadori was offered the position of professor of architecture at Princeton University, where he taught until 1959, when Columbia University asked him to join its faculty of architecture.

Twenty-six of Mario Salvadori's 8000 students are now professors at some of the best known American and foreign universities and he was honoured by Columbia University with the distinction of "Great Teacher", an invitation to hold one of the prestigious conferences called "University Lectures", and an honorary degree in 1980. He stopped teaching at Columbia University in 1988 with the title of James Renwick Professor Emeritus of Civil Engineering and Professor Emeritus of Architecture.

In 1960, Mario Salvadori was shortlisted in the Italian competitive examination for the chair of structures and was offered a professorship at the University of Palermo, which he turned down.

In 1959, he became a partner in the Weidlinger Associates engineering design and research company in New York, Boston, Washington DC and San Francisco, of which he is now managing director. In the last thirty years, Salvadori has partipated in the structural design of important buildings around the world, working together with world famous architects such as Gropius, Breuer and Saarinen. As a result of his professional activity, he has received numerous awards from American engineering associations and has been named member honoris causa of the American Society of Civil Engineers and of the American Architects, and a member of the National Academy of Engineering.

For the last fourteen years, Mario Salvadori has dedicated his attention almost exclusively to the teaching of mathematics and science in elementary and secondary schools in the poor neighbourhoods of New York City, under the patronage of the Salvadori Educational Center on the Built Environment, for which he has written a manual for teachers, a textbook for children and prepared 10 films regularly broadcast on American television stations. The New York City educational authority has established 8 "Salvadori Schools" in which his educational methods are applied and has honoured him with various awards.

Eugenio Fubini was born in Turin in 1913, the son of Turinese mathematician Guido Fubini. He completed his first two years of physics at the University of Turin and the last two years in Rome, where he received his degree in 1933 with Enrico Fermi as his thesis supervisor. From 1935 to 1938, Fubini worked at the Istituto Nazionale di Elettrotecnica in Turin and after his move to the United States found a job as an engineer for the Columbia Broadcasting System in New York from 1939 to 1942.

From 1942 to 1945, Fubini was research associate at the Harvard University Radio Research Laboratory, working on the design, development and operation of radios and radar for countermeasures, identification and tracking.

As scientific consultant and technical observer for the US Army and Navy in the European theatre of operations from 1943 to 1944, he participated in the construction of the electronic identification and jamming systems for the invasion of Italy and southern France. During 1944 and 1945, he was in England with the Eighth US Air Wing as head of reconnaissance and electronic countermeasures. In 1945, Fubini also worked as special consultant for electronic countermeasures to the Air Communications Office of the Italian Dipartimento di Guerra.

Fubini's life was hectic in those days, taking him from northern Africa, to Corsica, central Italy and Great Britain. But upon his return, he married Jane Elisabeth (Betty) Mathmer, the daughter of the dean of the University of Massachusetts at Amherst, whom he had met in 1941. Betty and Eugenio had six children: five girls and a boy. They owned five acres of land and gave each of their children a horse when they turned twelve, which they had to care for all by themselves. Thus the family stables grew and the children rode on their parents' land as their grandchildren now do.

Eugenio Fubini entered the Airborne Instruments Laboratory (AIL) in Melville, New York, in 1945 as an engineer and worked on the development of microwave components as well as magnetic detectors of electronic test equipment, anti-jamming apparatuses, antennae, homing devices and reconnaissance systems. He held various positions in this company, which later became the Airborne Laboratory Division of the Cutler-Hammer Corporation. In 1960, he was named vice president of the Research and Systems Engineering Division of the AIL Division.

In March 1961, Fubini joined the Office of Defense Research and Engineering for Research and Information Systems of the Secretary of Defense, when President Kennedy named him Assistant Secretary of Defense. In this last capacity, he was in charge of directing the entire network of military research and development problems, supervising the National Security Agency and revising planning of all command, control and communication equipment of the Department of Defence.

Eugenio Fubini was Assistant Secretary of Defense and Assistant Head of Defense Research and Engineering until June 1965, when he left to become one of the vice presidents of the International Business Machines (IBM) Corporation. He was promoted to vice president and group executive in 1966, in charge of IBM's Research Division, Science Research Associates, Inc., the Production Systems Department and the Instructional Systems Development Department.

In April 1969, Fubini resigned from IBM to become a private consultant to industry and the government.

Fubini has about 40 technical publications and 11 patents to his credit. He has received honorary degrees from the Brooklyn Polytechnic, the Rensselaer Polytechnic Institute and the Pratt Institute and numerous decorations and medals from the Department of Defence. He has given courses at Harvard University (1956) and has been on numerous scientific and technical commissions. He is a member of the US National Academy of Engineering.

Roberto M. Fano, born in Turin in 1917, the younger brother of Ugo Fano, studied engineering at the Turin Politecnico from 1935 to 1939 and completed his studies at the Massachusetts Institute of Technology (MIT), Cambridge, Mass: B.Sc. in 1941 and PhD in 1947.

His university career was based mainly at MIT, where he started out and gradually advanced from a teaching assistant (1941-43), to an instructor, research associate, associate professor (1955-56), professor (1956-84) and professor emeritus (after 1984).

Fano started out as an electrical engineer but soon deviated into computer science, working on the design, construction and operation of large-scale projects such as the MAC Project, of which he was the founding director (1963-68).

Fano worked as a consultant for various foundations (Ford Foundation 1961-63, Agnelli Foundation 1972-74) and industrial companies (Radio Corporation 1961-63) and in 1974-75 was visiting scientist at the Zurich research lab of the IBM Department of Electrical Engineering.

He was nominated to the National Academy of Engineering in 1973 and the National Academy of Sciences in 1978.

Along with R. B. Adler and L. J. Chu, Fano has written two excellent texts: *Electromagnetic Fields, Energy and Forces* and *Electromagnetic Transmission and Radiation*, published by Wiley & Sons (1966). On his own, he has written *Transmissions and Informations*, brought out in 1961 by MIT Press and Wiley & Sons, which has been translated into Russian (1963), Japanese (1965) and German (1966).

7. A few concluding remarks

To conclude, I would like to point out that the racial laws, first, and the Second World War, later, with all its immediate and future consequences, brought about the total or almost total destruction of this group of Italian physicists, scattering them throughout the world. Something which certainly could not have been predicted at the time of Via Panisperna by even the most far-sighted person such as O. M. Corbino. Many of them were faced with decisions of unprecedented gravity and after much thought made up their minds. And each of the emigrant physicists unwittingly became an ambassador of Italian culture and style at the highest level. But what is even more evident looking back now is the enormous intellectual potential, the vast scientific competence and the invaluable ability for undergraduate and post-graduate teaching that the racial laws drove from the country. It is true that many of the people of whom I have spoken returned to teach in Italy after the war for periods of varying length. Fermi returned for the first time in September 1948 to take part in an international conference held in Como[39] and again in the summer of 1954 to hold the well known course of lectures at the International School in Varenna,[40] which was to be the last time he taught before his death only four months later.

After retiring at 65 years of age from the American universities in which they had taught, Bruno Rossi and Emilio Segrè returned to teach in Italian universities.

---

[39] ENRICO FERMI, *Conferenze di Fisica Atomica*, Accademia Nazionale dei Lincei, Fondazione Donegani (1950).

[40] ENRICO FERMI, *Lectures on Pions and Nucleons*, B. T. Field editor, "Nuovo Cimento", *2*, Suppl. (1955), 17-95.

Bruno Rossi taught a supplementary course to general physics at the University of Palermo in 1974-75 and 1975-76 and Emilio Segrè taught general nuclear physics at the University of Rome in 1973-74 and 1974-75. In addition to these courses, both Rossi and Segrè have returned to Italy almost every year, giving lectures and holding conferences in many Italian universities. Fano has also returned frequently to teach in Rome and elsewhere. Pincherle has been back to Varenna and Perugia, De Benedetti has been back to Turin and since 1978, Pontecorvo has come back to Italy almost every year for two or three months. We are very grateful to all of them.

Yet, it is impossible to forget nor must we ever forget what happened at the end of the thirties: the material, physical and moral injury inflicted on people, the upheaval caused in the families of all those who were directly or indirectly affected by the racial laws, but also the damage, which was much less painful and apparent on the personal plane but certainly no less serious from a social point of view, that those laws caused the younger generations above all, the very generations that they were allegedly supposed to defend but which they seriously wronged by depriving them of a great many teachers endowed with profound personal knowledge and extraordinary research and teaching abilities.

EDOARDO AMALDI

VICENDE DELL'ACCADEMIA NAZIONALE DEI LINCEI DURANTE IL FASCISMO

Credo doveroso ricordare sia pure brevemente le vicende dell'Accademia Nazionale dei Lincei durante il Fascismo, quali del resto si possono leggere anche nell'Annuario.

Nei primi anni del regime fascista l'Accademia aveva dato prova della sua indipendenza esaminando con spirito critico provvedimenti governativi che interessavano la cultura nazionale. Tale atteggiamento non era piaciuto al Capo del Governo, il quale, volendo disporre di un organismo più docile che esercitasse insieme funzioni culturali e politiche, istituì, con Regio Decreto Legge (n. 87) del 7 gennaio 1926 l'Accademia d'Italia; in un articolo (n. 9) del decreto si dichiarava però che nulla sarebbe stato innovato nei riguardi della nostra Accademia. Ma più tardi, con il Regio Decreto 11 ottobre 1934 (n. 2309), fu approvato un nuovo Statuto dell'Accademia dei Lincei, cui seguì un nuovo regolamento emanato con decreto ministeriale 11 marzo 1936. Lo Statuto del 1934 imponeva il giuramento di fedeltà al regime da parte degli Accademici, dava al Capo del Governo diritto di scelta entro terne proposte per le nomine di nuovi soci e disponeva che le nomine del Presidente e del Vicepresidente fossero attribuite al Capo del Governo di concerto con il Ministro dell'Educazione Nazionale. Finalmente, non sembrando ancora sufficienti queste misure restrittive dell'autonomia accademica, con Legge dell'8 giugno 1939 (n. 755) fu stabilita la fusione dell'Accademia dei Lincei con l'Accademia d'Italia, fusione che equivaleva in realtà alla soppressione della prima, ricca di gloriose tradizioni, a favore della seconda di recente fondazione.

Caduto nel luglio 1943 il regime fascista, Benedetto Croce, con un articolo dell'agosto di quell'anno, propose la soppressione dell'Accademia d'Italia e la ricostituzione dell'Accademia dei Lincei. Il provvedimento, che non poté allora essere preso, in conseguenza dell'armistizio dell'8 settembre 1943 e della occupazione dell'Italia da parte dell'esercito tedesco, fu adottato l'anno successivo dal Governo Bonomi, costituitosi dopo la liberazione di Roma. Lo stesso giorno 28 settembre 1944 furono promulgati due Decreti Legislativi Luogotenenziali, il primo (n. 359) riguardante la ricostituzione dell'Accademia dei Lincei, il secondo (n. 363) la soppressione dell'Accademia d'Italia, le funzioni e il patrimonio della quale dovevano essere devoluti a quella; il che, fra l'altro, fece sì che da allora l'Accademia dei Lincei venisse a disporre di un secondo splendido palazzo, cioè della rinascimentale Villa della Farnesina, capolavoro del Peruzzi.

Per compiere le relative pratiche fu nominato un Commissario, mentre la ricostituzione, per la parte scientifica, fu affidata ad un Comitato presieduto inizialmente da Benedetto Croce e composto di alcuni soci anziani dell'Accademia dei Lincei. I compiti assegnati al Comitato furono stabiliti nel D.Lv.Lgt. del 12 aprile 1945 (n. 178).

Con successivo D.Lv.Lgt. del 16 novembre 1945 (n. 801) fu affidato alla Presidenza del Comitato l'incarico di reggere, governare e amministrare l'Accademia durante l'anno accademico 1945-46 e di provvedere alle elezioni di nuovi soci nazionali, per coprire i numerosi posti vacanti. Con lo stesso D.Lv.Lgt. nella classe di scienze morali fu determinata la composizione della nuova Categoria di critica delle arti e delle lettere con 9 soci nazionali e 9 corrispondenti, istituita con il predetto D.Lv.Lgt. del 12 aprile 1945.

Così soltanto nell'ottobre del 1946 l'Accademia, in base allo Statuto del 1920 che era stato richiamato in vigore, poté procedere alla regolare costituzione del suo ufficio di Presidenza e, con l'anno accademico 1946-47, riprendere appieno il ritmo delle sue attività.

Questo quadro d'insieme non solo è importante come informazione di fondo; esso è anche indispensabile per poter comprendere i rapporti fra l'Accademia dei Lincei ed Albert Einstein, quali risultano da alcune lettere conservate nell'archivio di Palazzo Corsini (1).

Einstein era stato nominato socio straniero dei Lincei nel 1921. Venuto a conoscenza, probabilmente dalla stampa internazionale delle recenti leggi razziali del Governo Fascista, il 3 ottobre 1938, egli ne chiese conferma per lettera (*fig.1*) all'Accademia dei Lincei. Dall'archivio non sembra che l'Accademia abbia risposto. In data 15 dicembre 1938 Einstein scrisse una seconda lettera (*fig. 2*), con cui chiedeva che il suo nome venisse cancellato dalla lista dei Membri corrispondenti (in realtà stranieri) dell'Accademia; a questa rispose, in data 2 gennaio 1939, il Presidente dell'Accademia, Federico Millosevich, dicendo di prendere atto della sua richiesta (*fig. 3*). Millosevich era un professore di Mineralogia dell'Università di Roma che era stato nominato da Mussolini Presidente dell'Accademia dei Lincei il 14 febbraio 1938, e che rimase in tale posizione fino alla dissoluzione totale dei Lincei.

Questa storia, tuttavia, non terminò in maniera così sgradevole. Finita la guerra, a primo Presidente dell'Accademia Nazionale dei Lincei, ricostituita come ho spiegato sopra, fu eletto il matematico Guido Castelnuovo (1865-1952), che aveva subito le conseguenze delle leggi razziali italiane restando a Roma, aveva difeso nei limiti del possibile i giusti interessi degli studenti

(1) Sono grato al Professor John Stachel, direttore de «The Collected Papers of Albert Einstein»,745 Commonwealth Avenue, Boston University, MA 02215, e al Professor Tullio Regge dell'Università di Torino, per aver richiamato la mia attenzione su questo carteggio fra Einstein e l'Accademia dei Lincei.

— 45 —

THE INSTITUTE FOR ADVANCED STUDY
SCHOOL OF MATHEMATICS
FINE HALL
PRINCETON, NEW JERSEY

den 3.Oktober 1938

Reale Accademia Nazionale
dei Lincei
R o m a

Gemäss Zeitungsmeldungen soll das Ausscheiden jüdischer italienischer Gelehrter aus dortigen Akademien verfügt worden sein. Ich erlaube mir die höfliche Anfrage, ob diese Meldung auf Wahrheit beruht.

Mit ausgezeichneter Hochachtung
A. Einstein
Professor Albert Einstein

Einschreiben!

Fig. 1. - Lettera di Albert Einstein alla Reale Accademia Nazionale dei Lincei del 3 ottobre 1938 (Archivio Storico dell'Accademia).

36 102

den 15.Dezember 1938

Reale Accademia Nazionale
Dei Lincei
R o m

Sehr geehrte Herren!

Ich ersuche Sie hiermit, meinen Namen aus der Liste Ihrer korrespondierenden Mitglieder zu streichen!

Hochachtungsvoll

Professor Albert Einstein.

Registered!
Return Receipt requested!

Fig. 2. - Minuta della lettera di Albert Einstein del 15 dicembre 1938 messa a disposizione da «The Collected Papers of Albert Einstein».

— 47 —

36 104

R. ACCADEMIA NAZIONALE
DEI LINCEI
IL PRESIDENTE

Roma 2 Gennaio 1939 - XVII°

Prot. 835 S

Chiarissimo Professore,

prendo atto delle Vostre dimissioni da Socio Straniero di questa Reale Accademia.

Con ossequio

IL PRESIDENTE

Chiarissimo
Sig.Prof. ALBERT EINSTEIN
Princeton University
NEW JERSEY

Fig. 3. - Lettera del 2 gennaio 1939 di Federico Millosevich, Presidente nominato dal Governo Fascista (1938-1943) della Accademia dei Lincei, ad Albert Einstein in cui si prende atto delle sue dimissioni.

36 105

ACCADEMIA NAZIONALE DEI LINCEI

PRESIDENZA

Roma 26 aprile 1946

Illustre
Prof.Rodolfo EINSTEIN
Institute for advanced study
Princeton(U.S.A.)

L'Accademia Nazionale dei Lincei,soppressa dal Governo Fascista nel 1939,risorge oggi in regime di libertà e riprende,con la data odierna,la sua attività scientifica.

L'Accademia desidera vivamente riannodare le relazi
ni con tutte le Società consorelle,e riavere nel suo seno gli uomini illustri che ha eletto nel passato suoi Soci Stranieri.

La prego pertanto di inviarci con la Sua adesione il Suo esatto indirizzo odierno.

Nella fiducia di una Sua risposta cortesemente sollecita,Le inviamo i nostri saluti distinti

Il Presidente della Classe di
Scienze Fisiche,Matematiche e Naturali
(Prof.Guido Castelnuovo)

G. Castelnuovo

Valendomi delle mie antiche relazioni con Lei aggiungo la mia preghiera personale perchè Ella consenta a riprendere il posto di socio straniero nella rinnovata Accademia dei Lincei che si onorava di averla tra i suoi membri più illustri

G. C.

Fig. 4. - Lettera del 26 aprile di Guido Castelnuovo primo Presidente della Accademia dei Lincei ricostituita, con cui si invita Albert Einstein a ritirare le sue dimissioni. Il nome di Einstein è, purtroppo, errato per un errore di battitura.

— 49 —

C/2155 - h

THE INSTITUTE FOR ADVANCED STUDY
SCHOOL OF MATHEMATICS
PRINCETON, NEW JERSEY

Scienze Fisiche

June 26,1946

To the President of the
Academia Nationale dei Lincei
Rome, Italy

Sir:

With great pleasure I see from your letter of April 26th 1946 that the Academia Nationale dei Lincei has resumed its activities for the benefit of science, your country having been liberated from fascist oppression. My mailing-address (private) is:112 Mercer Str.Princeton N.J.

comunicato a Foglio e preso nota. M.L.

Faithfully yours,
A. Einstein.
Albert Einstein.

P.S. Dear Dr.Castelnuovo:
I shall be happy indeed to become again socio straniero of your Academy as I have been in the good times of the past.

Saluti affettuosi, A. E.

Fig. 5. - Lettera di Albert Einstein del 26 giugno 1946 con cui Einstein ritira le dimissioni.

universitari ebrei (2) ed era sempre stato un relativista convinto tanto da pubblicare nel 1923 il bel volumetto «*Spazio e tempo secondo le vedute di A. Einstein*» (3).

In data 26 aprile 1946 Castelnuovo scrisse ad Einstein per comunicargli che l'Accademia Nazionale dei Lincei era stata ricostituita e per invitarlo a riallacciare le sue vecchie relazioni (*fig. 4*).

La risposta positiva di Einstein, in data 26 giugno 1946, è mostrata nella *fig. 5*.

Edoardo Amaldi

THE ACCADEMIA NAZIONALE DEI LINCEI DURING THE FASCISM

I think it is fair to remind, although briefly, the vicissitudes of the Accademia Nazionale dei Lincei during the Fascism, such as can be read, moreover, also in the Annuario.

In the first years of the fascist regime the Academy had given proof of its independence by examining critically those actions of the government that were relevant to the national culture.Such critical stand had not been liked by the Chief of the Government, who, wishing to have at his disposal a more docile body exercising at the same time cultural and political functions, established, by virtue of a Regio Decreto Legge (n. 87), on January 7, 1926, the Accademia d'Italia; an article (n. 9) of the decree, however, stated that nothing would change regarding our Academy. But later on, with the Regio Decreto of October 11, 1934 (n. 2309), a new Statute of the Accademia dei Lincei was approved, followed by a new regulation issued by ministerial decree on March 11, 1936. The 1934 Statute required a loyalty oath to the regime to be sworen by members of the Academy, gave to the Chief of the Government the right of choice within terns of names proposed for the nomination of new members, and ordered that the nominations of the President and Vicepresident be granted to the Chief of the Government, acting in concert with the Minister of National Education. Finally, these restrictive measures of academic autonomy being considered not sufficient yet, a Law of June 8, 1939 (n. 755) ordered the fusion of the Accademia dei Lincei with the Accademia d'Italia, a fusion actually equivalent to the suppression of the former, rich of glorious traditions, to the benefit of the latter, of recent foundation.

After the fall of the fascist regime in July 1943, Benedetto Croce, in an article of August of the same year, proposed the suppression of the Accademia d'Italia and the re-establishment of the Accademia dei Lincei. The action, which could not be taken at that time, because of the armistice of September 8, 1943, and of the occupation of Italy by the German Army, was passed the following year by the Bonomi Government, installed after the liberation of Rome. On the same day September 28, 1944, two Decreti Legislativi Luogotenenziali were issued, the first (n. 359) re-establishing the Accademia dei Lincei, while the second (n. 363) abolished the Accademia d'Italia, whose functions and properties were to be devolved to the former: which had as a consequence, amongst others, that the Accademia dei Lincei had since then at its disposal a second splendid building, the Renaissance Villa della Farnesina, masterpiece of Peruzzi.

A Commissar was nominated in order to fulfil the related practices, while the reconstitution, on the scientific side, was entrusted to a Committee, initially chaired by Benedetto Croce, composed by a few senior members of the Accademia dei Lincei. The tasks assigned to the Committee were defined in the D.Lv.Lgt. of April 12, 1945 (n. 178).

A successive D.Lv.Lgt. of November 16, 1945 (n. 801) entrusted the Committee's Presidency with the task of managing, governing and administering the Academy during the academic year

1945-46, and of arranging for the elections of new national members, so as to fill the numerous vacancies. The same D.Lv.Lgt. determined the composition (nine national members and nine correspondants) of the new Category of criticism of arts and letters, which had been established, for the class of moral sciences, by the said D.Lv.Lgt. of April 12, 1945.

It was then only in October 1946 that the Academy, according to the Statute of 1920 which had been reinforced, could proceed to the regular constitution of its Presidency office and, during the academic year 1946-47, fully regain the rythm of its activities.

This general framework is important not only as background information; it is also indispensable in order to understand the relations between the Accademia dei Lincei and Albert Einstein, as evidenced by several letters held in the archive in Palazzo Corsini.[1]

Einstein had been elected foreign member of the Lincei in 1921. Having known, probably through the international press, of the recent racial laws issued by the Fascist Government, he wrote, on October 3, 1938, a letter to the Accademia dei Lincei, asking for confirmation (fig. 1). It does not appear, from the archive's content, that the Academy answered. On December 15, 1938, Einstein wrote a second letter (fig. 2), asking that his name be canceled from the list of corresponding (actually foreign) members of the Academy; to this letter was answered, on January 2, 1939, by the President of the Academy, Federico Millosevich, saying that the request had been accepted (fig. 3). Millosevich was a professor of Mineralogy from the University of Rome who had been nominated by Mussolini as President of the Accademia dei Lincei on February 14, 1938, and occupied that position until the total dissolution of the Lincei.

The end of this story, however, is not as sad. At the end of the war, as the first President of the Accademia Nazionale dei Lincei, re-established as I said before, was elected the mathematician Guido Castelnuovo (1865-1952), who had suffered the consequences of the racial laws while staying in Rome, had defended, as far as possible, the right interests of jewish university students, and had always been a convinced relativist, so that he had published, in 1923, the beautiful booklet "*Space and time according to the views of A. Einstein*".

On April 26, 1946 Castelnuovo wrote Einstein communicating that the Accademia Nazionale dei Lincei had been reconstituted and inviting him to resume the old relations (fig. 4).

Einstein's positive answer, dated June 26, 1946, is shown in fig. 5.

[1] I am grateful to Professor John Stachel, director of "The Collected Papers of Albert Einstein", 745 Commonwealth Avenue, Boston University, MA 02215, and to Professor Tullio Regge from the University of Turin, for directing my attention to this correspondence between Einstein and the Accademia dei Lincei.

Section B

Post-War Italian Physics

During the 1970s, Amaldi's interest in Italian physics extended to the period following Fermi's departure, that is, from the time just before the outbreak of the war to the period of post-war reconstruction. The papers selected in this section refer to that period. Though written on different occasions, they were all closely linked in that they were written in the 1970s while he was preparing the fifth chapter, entitled *The Collapse and the Reconstruction*, of his as yet unpublished book on the history of the Institute of Physics (which is currently being published along with selected documents and letters, and edited by G. Battimelli and M. De Maria).

The first four papers have a lot in common with each other and with two texts found in the first section (*Neutron Work in Rome in 1934–36 and the Discovery of Uranium Fission* and *Gian Carlo Wick During the Thirties*) of this book: not only are certain topics and opinions a recurrent feature in them, and in some cases entire passages are identical, but they are also taken from the above-mentioned manuscript. Nevertheless, there are differences in the emphasis due in part to the different circumstances under which each was written, and partly to Amaldi's desire to consider certain events from a different point of view in each paper. In some ways, they complement the other pieces that he had already published or was preparing. In the last two papers, which are devoted to people with whom he had very close ties with, Amaldi, in a style and scientific rigor which, in historiographic terms, go far beyond that of traditional academic commemorations, described events that took place at the end of the 1960s or later. The piecing together of the biographies of Persico and Pancini (along with others that are found in Part II of this volume) flanked Amaldi's historical research activities throughout the 1970s and, as part of them, marked a fundamental turning point in his thinking about the course of Italian physics.

The first paper (*Personal Notes on Neutron Work in Rome in the 30s and Post-War European Collaboration in High Energy Physics*) comprises the texts of three lectures given by Amaldi during the 57th course of the International School of Physics in Varenna. The subject of the course, held in 1972, was *History of Twentieth Century Physics* and the teachers, who hovered between recollection and history, were some of the most important names in post-war physics. The first two parts of the paper are devoted to research on artificial radioactivity and neutron physics. This was the first time Amaldi had turned his attention to this subject in a historical work. The narration and sources used (including documents and instruments) are far more complex than what he would take credit for when he modestly claims that "in these two lectures, unavoidably I will very often repeat in different words or, sometimes, practically the same words what can be found in these various books" (the biographies of Fermi published by Emilio Segrè and Laura Fermi and the introductions to Fermi's *Collected Papers*, published a few years before the course was held). The two lectures spanned a three-year period from 1934 to 1936 and ended on this thought: "The leading role that the group had maintained for about three academic years had finished. This was certainly due in part to the worsening political situation in Europe in general and, in particular, in Italy. It was also due, perhaps, to the fact that it was always becoming more difficult to compete with other groups which had, in the meantime, equipped their laboratories with various types of accelerators that provided neutron sources much better than those at our disposal". This only hints at Amaldi's deep contemplation on the reasons or the dispersal of Fermi's group. It was behind many of the choices he made as a research policymaker during the war and in its aftermath, and which went beyond the received wisdom concerning those events based on the (certainly tragic and indubitable) role played by the racial laws. The third lecture is dedicated to the first joint international work on high energy physics and, in particular, to the collaboration among Europeans which led to the founding of CERN. Here, Amaldi used his diary and files as sources for his reconstruction (from his point of view, as he himself points out) of the origins of that great laboratory in Geneva. One — and certainly not the least — of the merits of this text is the clarity with which the link between research strategy, context and results emerges in each situation considered.

The attentive reader will have noticed the absence of all references to the years between 1938 and 1944 in the preceding work. This is a subject which Amaldi devoted considerable attention to a few years later. This is seen in the first part of the second paper in this section, *The Years of Reconstruction*. This piece includes the texts of the valedictory lectures he gave on 7–9 September 1978 at a farewell meeting organized upon his departure from the teaching service. And his choice of topic on an occasion such as this is certainly significant. The text vividly recaptures the events that took place in Italian physics from a Roman point of view: the dispersion of Fermi's group; the decisions taken by those still working at the Institute of Physics in Rome during the war and, in particular, the decision to abandon the line of research that could be of military interest; the revival of activity after the war and the feeling of the huge gulf separating both sides of the Atlantic when relations were taken up once again; the beginnings of applied nuclear physics in Italy (at the CISE in Milan and Comitato Nazionale per le Ricerche Nucleari when it was founded subsequently); and the research on cosmic rays and the problems of accelerators. What was lacking in this work, which suggestively opens with Fermi's departure from Italy in December 1938, are those technical comments which are typically found in almost all of Amaldi's works. This is most certainly due to the fact that the journal *Scientia,* which had published it, is aimed at a well-educated though not specialist audience.

The third paper, dedicated to Niels Bohr and Italian physics, is Amaldi's welcoming address to the symposium organized by the research group on the history of physics at the University of Rome "La Sapienza" and the *Rivista di Storia della Scienza* as part of the celebrations on the centenary of Bohr's birth. Amaldi also spoke of Bohr on another occasion during those celebrations, by recalling the latter's work in the early years of CERN. In that speech, published in the second part of this book, he underlines Bohr's relations with Italian physicists and, in particular, with Gian Carlo Wick. Indeed, it was when he was talking about the relations between Bohr and Wick that Amaldi delved more deeply into the reasons that led the Rome group (made up at this point of Amaldi, Wick, Mario Ageno, Daria Bocciarelli and Giulio Cesare Trabacchi) to abandon — having perceived the possible military applications — the line of research on nuclear fission developed with Bohr's decisive help between 1939 and 1941.

The fourth and last paper in this first group, *Physics in Rome in the 40's and 50's*, complements the others. The text of a speech given during a symposium organized on the seventieth birthday of Marcello Conversi gives a strictly scientific reading (and considering the context in which it was presented) of events which occurred in the 1940s and 1950s , and which had already been examined from other perspectives elsewhere. The title should not be misleading: Amaldi attempts to give a general overview of the fundamental research carried out in physics in Italy in those years (albeit from a Roman point of view) by describing the lines of research and the development of certain institutions. The occasion itself is significant: Conversi came into contact with the Rome group when he was a student while Fermi was still the director. Moreover in the 1940s, he participated in important research (it suffice to recall his collaboration with Ettore Pancini and Oreste Piccioni) and became the director of the Pisa branch of the Istituto Nazionale di Fisica Nucleare in the 1950s, where some of the fundamental research activities which led to the design and construction of Italian computers was carried out (initially in collaboration with Olivetti).

The piece entitled *Ricordo di Enrico Persico*, co-authored with Franco Rasetti, was not written for any special occasion. Indeed, Persico had specifically asked not to be commemorated. A high school friend of Fermi who was shortlisted for the first chair in theoretical physics along with Fermi and Aldo Pontremoli, Persico first taught in Florence and Turin before moving to the Universitè Laval in Québec after the war, and settled down in Rome when he finally returned to Italy. During his career, Persico played a crucial role in the teaching of quantum mechanics. He also made important theoretical contributions in many fields and worked in the post-war era with the theoretical group of the Accelerator section of the Istituto Nazionale di Fisica Nucleare, which was charged with the task of constructing the electron synchrotron in Frascati. For this reason, Amaldi and Rasetti dedicated a biographical work to him ten years after his death from heart attack in Rome in 1969, which was published by the Accademia Nazionale dei Lincei and reprinted in the *Giornale di Fisica.* This short work is actually the result of the systematic collection of testimonials which Amaldi undertook over a five-year period after he had gained access to and begun an initial examination of Persico's personal archives, which are being maintained in the Department of Physics at the University of Rome "La Sapienza".

Once again, a commemorative speech, the *Ricordo di Ettore Pancini*, and the last work in this section, was given by Amaldi after the death of Pancini, who was an extremely generous man, both intellectually and in the human sphere. Amaldi had already begun work on a biography of the physicist during the last years of his life and the work is strictly linked (as the author states in a note) to the preparation of the manuscript on physics in Rome and Italy during the war. Amaldi describes not only Pancini's scientific work (from his collaboration with Gilberto Bernardini and work with Conversi and Piccioni to his later studies on synchrotron light), but also his political activities during the partisan war and in post-war Italy, which were strongly characterized by his extreme left-wing views (not infrequent among Italian physicists) and role as a promoter of research.

Personal Notes on Neutron Work in Rome in the 30s and Post-war European Collaboration in High-Energy Physics.

E. Amaldi

Istituto di Fisica dell'Università - Roma

1. – Recollections of research on artificial radioactivity.

As an introduction to an article on the production and slowing-down of neutrons [1] that I wrote about 15 years ago, I tried to give a rather detailed and accurate account of the succession of discoveries and contributions that opened the field of artificial radioactivity and neutron physics.

In the two lectures that I give on the same subjects today and tomorrow I will be much more biased since they are in great part based on my recollections of what happened in the thirties at the University of Rome, when I was working in the group led by FERMI.

Much information about the work done by FERMI and his collaborators in that period can be found in various publications: among these I should recall Fermi's *Collected Papers*, published a few years ago, jointly by the Accademia Nazionale dei Lincei and the University of Chicago Press. The first volume [2] contains the papers of the «Italian period» (1923-1938); the second the papers of the «American period» (1939-1954) and a number of unpublished reports declassified by the AEC on the occasion.

Each paper or group of papers on a specific subject is preceded by an introduction written by one of the members of the Editorial board (or exceptionally by some other scientist) who was to remember or reconstruct the circumstances, sometimes even of political nature, under which the work was done by FERMI.

These introductions provide a succession of flashes on Fermi's life, which have been in some way co-ordinated in a complete and excellent biography by SEGRÈ in his recent book: *Enrico Fermi: Physicist* [3].

Other information not only on FERMI, but, more in general, on the life at the Istituto di Fisica of Via Panisperna, can be found also in the book by Fermi's wife, Laura [4]. In these two lectures unavoidably I will very often repeat in different words or, sometimes, practically in the same words, what can be found in these various books.

1'1. – The discovery of the artificial radioactivity was announced by I. CURIE and F. JOLIOT in a note presented to *Comptes Rendus* (and *Nature*) in January 1934 [5]. They had observed that boron and aluminium, when bombarded with polonium α-particles, gave a positron emission which did not start immediately when the α-particle source was placed close to samples of these elements, but increased in intensity from that moment onwards reaching, some time later, a limiting value.

When the Po source was taken away, the positron emission did not cease immediately, but started to decrease exponentially with time as does the activity of a radioactive substance. The lifetime was 14 min for B, 15 s for Al and 2.5 min for Mg, which was found, slightly later [6], to show a similar behaviour with the exception that the emitted electrons were of both signs.

Some detail about how this discovery was made can be found in the speech that F. PERRIN pronounced in July 1964 on occasion of the 30th anniversary of the discovery of artificial radioactivity [7]. This discovery, made and interpreted correctly in a few days, was the result of about two years of experimental work of remarkable quality. Using all possible techniques for the investigation of the atomic radiations known at that time, the CURIE-JOLIOT had studied, about two years before, the penetrating radiations which had been discovered by BOTHE and BECKER to be emitted by beryllium bombarded with α-particles. Their experiments, as I will say in more detail tomorrow, had paved the way to the discovery of the neutron by CHADWICK.

Furthermore, about six months before the discovery of artificial radioactivity, they had found that aluminium bombarded by α-particles emits not only protons and neutrons [8] but also positrons [9]. Later it became clear that these positrons were due to artificial radioactivity induced in aluminium, but for six months their origin appeared rather mysterious and therefore was a matter of discussion everywhere. In particular this effect was the subject of long—and rather inconclusive—debates at the « Conference on Nuclear Physics » held at Leningrad, in September 1933, and at the 7th « Conseil de Physique Solvay » that took place in Bruxelles about one month later.

The discovery of artificial radioactivity was due to an accidental observation made by JOLIOT. In order to study the emission of positrons by aluminium bombarded with α-particles, he used a cloud chamber in a magnetic field with a window in the side wall closed by a thin aluminium foil. The source of α-particles (emitted by polonium with 5.3 MeV energy) was usually placed very close to the window, outside the chamber.

One day, in January 1934, JOLIOT noticed that the emission of positrons persisted when the Po source was taken away. As he later told PERRIN, he immediately understood the importance of his observation and the necessity of trying to study the new phenomenon from different points of view. Thus, he went to look for his wife Irène, who was working in a nearby laboratory, with the idea of associating her in all physical and chemical tests that he could foresee would be necessary in order to provide decisive proofs about the qualitative nature and the quantitative aspects of the new phenomenon.

According to F. PERRIN, the first observation was made on Friday morning and the note announcing the discovery of nuclei showing a new type of radioactivity, the radioactivity by positrons, was presented to the Academie des Sciences by J. PERRIN, the successive Monday. It contained the correct interpretation of the phenomenon and the results of a few physical and chemical tests which became classical examples that were followed by all other physicists and chemists working later in this same field. One year later, Irène and Frédéric JOLIOT-CURIE received the Nobel Prize for Chemistry for this discovery.

In Stockholm, on receiving the Nobel Prize, they gave two lectures by dividing the subject as one would not have expected. Irène, who was a pupil of her mother, was mainly a chemist: she treated the physical aspects, in particular the radioactivity by positrons. Frédéric had studied at the École de Physique et de Chimie de la Ville de Paris and was mainly a physicist and engineer; he discussed the chemical aspects, underlining the extraordinary consequences that were opened by the possibility of producing artificially a number of radioisotopes.

The last point should be emphasized also today. With the exception of elastic and inelastic scattering, all processes produced by the absorption of an incident α-particle give rise to a nuclide of atomic number different from that of the target element. Thus it becomes possible to separate it by applying the same classical procedures that had been used years before by Marie and Pierre CURIE for the discovery of radium.

In fact, by detecting the decay electrons, one can easily test the various chemical fractions and recognize in some of them the presence of exceedingly small amounts of transmutation products [8].

Although its basic principle had been used before in the case of natural radioactive substances, radiochemistry started to become an important branch of modern science with extraordinary applications in many fields of chemistry, biochemistry, biology, technology, etc., only with the discovery by the JOLIOT-CURIES of artificial radioactivity.

1·2. – This discovery gave to FERMI and, in general, to the group working at the University of Rome the occasion to initiate really important new experimental work. For some years there had been talk of the advisability of gradually abandoning atomic physics—the field in which every one had worked for some years—and concentrating the main research effort on nuclear physics. These ideas were beginning to take practical shape towards the end of 1931. In October, on my return from Leipzig, where I had spent about ten months working on X-ray diffraction by liquids under DEBYE, I was pleased to accept the task of systematically presenting, in a series of seminars, the content of the classical treatise by RUTHERFORD, CHADWICK and ELLIS [10], which I had begun to study a few months earlier. FERMI, RASETTI, SEGRÈ, MAJORANA and a few others attended the seminars. My lecture was often interrupted by observations of the most varied kind from members of the group, which gave rise to long discussions and gave FERMI the opportunity of developing *extempore* the theory of some of the phenomena mentioned. Exceptionally, MAJORANA made observations almost always very penetrating. Many of the ideas and approaches that emerged in these discussions were presented a few years later in the book on the nucleus by RASETTI that was published in Italy as well as the United States [11] and that, in the late thirties, became one of the most popular texts of the new nuclear physics.

The attendance at the seminar fell off in one or two months: in November SEGRÈ went to work in Hamburg at the Institute directed by STERN to learn the technique of molecular beams, and RASETTI went to the Kaiser Wilhelm Institut für Physik in Berlin-Dahlem, directed by Lise MEITNER, to work on the penetrating radiation emitted by beryllium bombarded with α-particles, already mentioned above. At the same time FERMI and I in Rome began to construct a cloud chamber of the Blackett type and average dimensions in order to become familiar with one of the most important techniques then in use for studying radioactivity and transmutations.

On the return of RASETTI from Berlin, in the Fall of 1932, FERMI and RASETTI organized a program of research in nuclear physics. A rather large cloud chamber, essentially designed after those in use in Berlin-Dahlem, was constructed and worked excellently as soon as it was assembled. A gamma-ray crystal spectrometer was built by FERMI and RASETTI, who also developed the technique of growing bismuth monocrystals of large dimensions [12].

Various types of counters were also put into operation and RASETTI separated a strong source of RaD from a radium solution. He further separated polonium and, mixing the latter with beryllium powder, prepared a neutron source comparable to the most powerful ones then in use elsewhere. These developments were made possible by a grant from the Consiglio Nazionale delle Ricerche, which had raised the research budget of the department to an amount of the order of \$ 2000 to \$ 3000 per year, corresponding to about ten times the average budget of the physics departments in Italian universities in those years.

During the Summer and Fall of 1933 SEGRÈ and I did not participate in these developments. We were occupied in finishing some spectroscopic work on a new phenomenon that we had recently observed and which had been correctly interpreted by FERMI only in November of that year. I will say a few words on this work in tomorrow's lecture since its theoretical interpretation turned out to be very useful in the description of the behaviour of slow neutrons.

The switching from atomic to nuclear physics was taking place gradually also in the theoretical activity of FERMI and others. In 1930 FERMI started to work on the hyperfine structure of spectral lines, a subject he further developed in collaboration with SEGRÈ in 1932 [13].

From this work as well as from his interventions in the discussions taking place at various conferences and seminars in Italy and abroad, Fermi's competence on the properties of nuclei had started to be recognized, so that he was asked to report on the status of the physics of the nucleus at a nuclear conference held in Paris in 1932 as part of a large international conference on electricity. In his report he mentioned Pauli's hypothesis on the existence of the neutrino in order to explain the apparent nonconservation of energy and momentum in beta-decay.

MAJORANA, after the discovery of the neutron, had proceeded to develop a nuclear model based on neutrons and protons [14] without electrons and FERMI towards the end of 1933 wrote his paper on the beta-decay, where he introduced a new type of force, the « weak interaction » described by a proper Hamiltonian [15]. I will not try to enter into a discussion of this fundamental paper because it would bring me too far away from my main subject. What I have said is enough to give an idea of the experimental and theoretical background that already existed at the University of Rome when the Joliot-Curies announced the discovery of artificial radioactivity produced by α-particle bombardment.

1'3. – Shortly after the first papers of these authors were read in Rome, FERMI, in March 1934, suggested to RASETTI that they try to observe similar effects with neutrons by using the Po+Be source prepared by RASETTI.

In a few weeks several elements were irradiated and tested for activity by means of a thin-walled Geiger-Müller counter, with a totally negative result, obviously due to lack of intensity. RASETTI left for a vacation in Morocco, while FERMI continued the experiments. The idea then occurred to FERMI that in order to observe a neutron-induced activity it was not necessary to use a Po+Be source. A much stronger Rn α+Be source could be employed, since its beta and gamma radiations (absent in the Po+Be sources) were no objection to the observation of a delayed effect. Radon sources were familiar to FERMI since they had been supplied previously by Professor TRABACCHI (head of the Laboratorio Fisico dell'Istituto di Sanità Pubblica) for use with the gamma-ray spectrometer mentioned above [12].

All one had to do was to prepare a similar source, consisting of a glass bulb filled with beryllium powder and radon (Fig. 1 and 2). When FERMI had his stronger neutron source (about 30 mCu) he systematically bombarded the elements in order of increasing atomic number, starting from hydrogen and following with lithium, beryllium, boron, carbon, nitrogen and oxygen, all with negative results. Finally, he was successful in obtaining a few counts on his Geiger-Müller counter when he bombarded fluorine and aluminium. These results and their interpretation in terms of (n, α) reactions were announced in

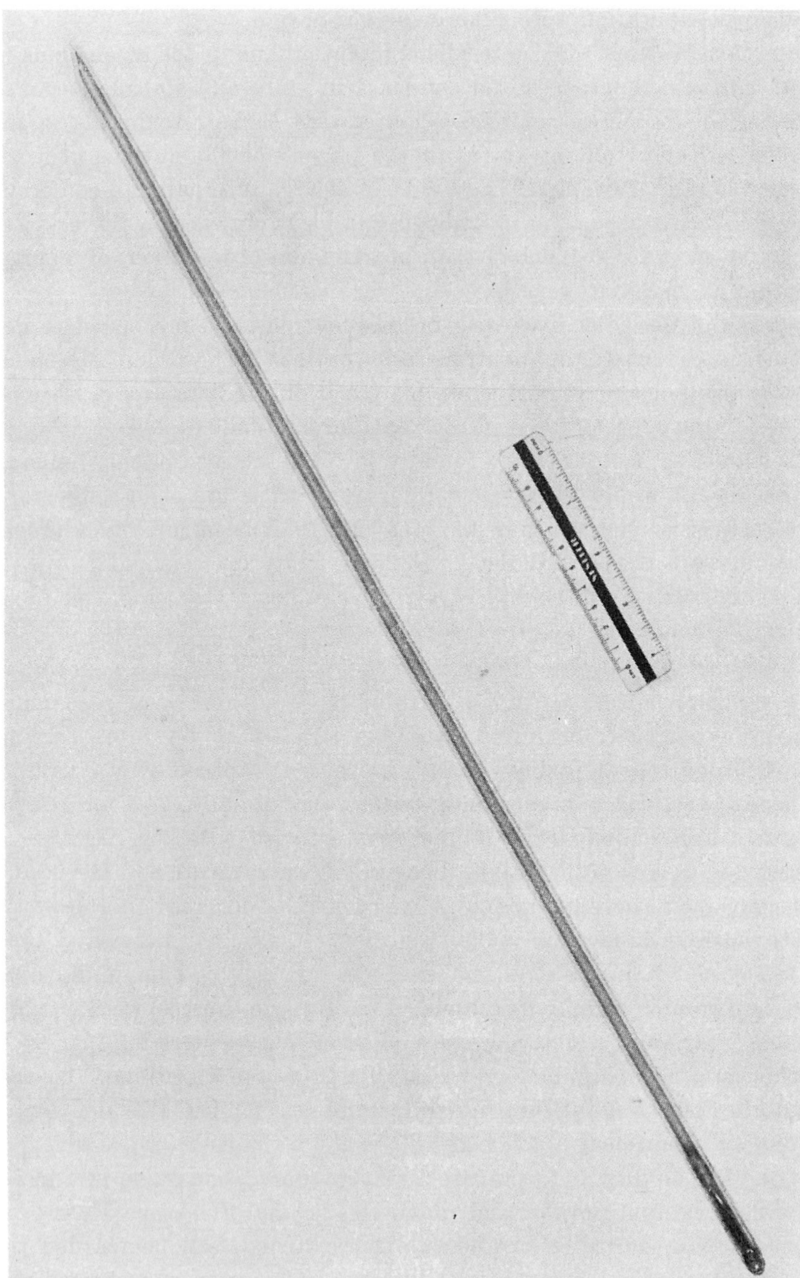

Fig. 1. – Photograph of one of the Rn α+Be sources used in Rome in 1934-36. The long glass tube is used only for handling the source without having the hand of the operator exposed too heavily to the gamma-radiation emitted by the decay products of Rn.

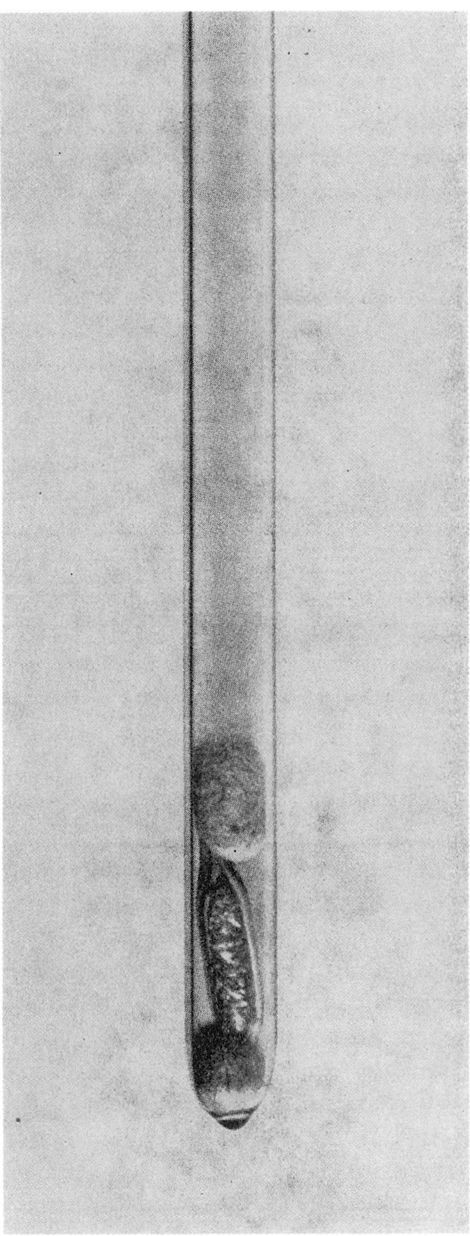

Fig. 2. – Enlarged photograph of the essential part of the Rn α+Be shown in Fig. 1. It consists of a glass bulb (1.0÷1.5) cm long filled with beryllium powder and radon, kept in a fixed position at the end of the handling tube.

a letter to *Ricerca Scientifica* on 25 March 1934 [16]. The title « Radioattività indotta da bombardamento di neutroni - I » indicated his intention to start a systematic study of the phenomenon which would have led to the publication of a long series of similar papers.

FERMI wanted to proceed with the work as quickly as possible and therefore asked SEGRÈ and me to help him with the experiments. A cable was sent to RASETTI asking him to come back from his vacation. The work immediately was organized in a very efficient way: FERMI did a good part of the measurements and calculations; SEGRÈ secured the substances to be irradiated, the sources and the necessary equipment, and later became involved in most of the chemical work. I took care of the construction of the Geiger-Müller counters (Fig. 3 and 4)

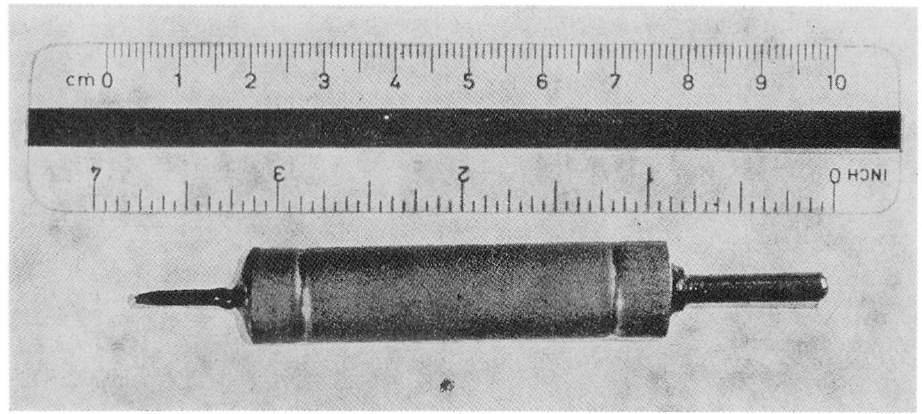

Fig. 3. – One of the Geiger-Müller counters used in 1934 by Fermi's group. The wall was of aluminium between 1 and 2 tenths of a millimetre thick; the small cylinder was obtained by cutting the bottom of a box of medicinal tablets.

and of what we now call electronics. This division of the activities, however, was not rigid at all and each of us participated in all phases of the work. We immediately realized that we needed the help of a professional chemist. Fortunately, we succeeded almost immediately in convincing D'AGOSTINO to work with us. He had been a chemist in the laboratory of Professor TRABACCHI and, at the time I am talking about, he held a fellowship in Paris at the laboratory of Madame CURIE where he was learning radiochemistry. He had come back to Rome for a few days during the Easter vacations but we showed him our work and, on request of FERMI, he remained with us and never went back to Paris.

During the succeeding months our group published in quick succession a long series of experimental results: about sixty elements were irradiated with neutrons and in about forty of them at least one new radioactive product was discovered and often identified.

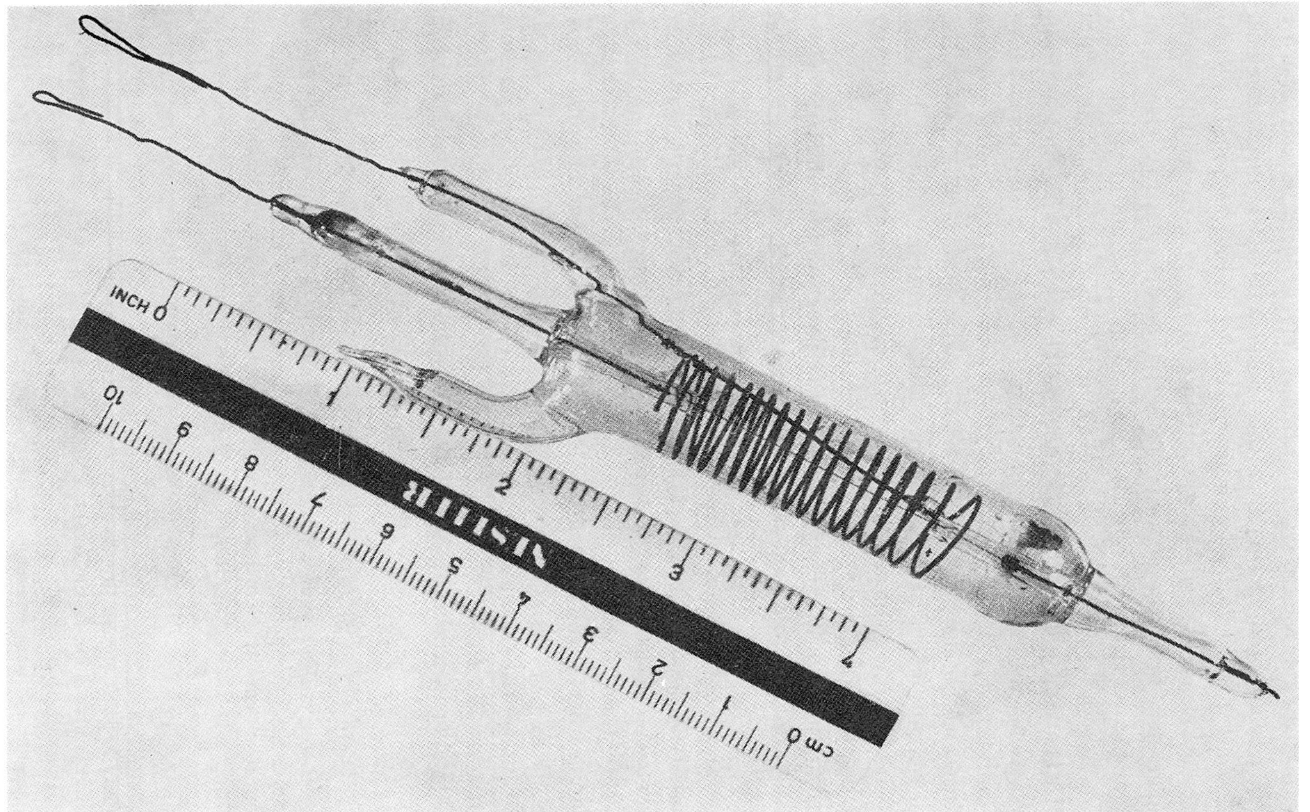

Fig. 4. – One of the Geiger-Müller counters constructed by RASETTI and that later replaced the counters of Fig. 3. Its glass walls were thin enough (about (2÷3) tenths of a millimetre) to be crossed by soft beta-rays. In order to reduce the effect of electric charges deposited on the glass, the cylindrical part of the glass tube was painted with a thin layer of colloidal graphite not shown in the photograph.

These results were sufficiently abundant to allow the beginning of a systematic classification of nuclear reactions produced by neutrons. We had found that all elements, whatever their atomic weight, could be activated by neutrons. The nuclide product was sometimes an isotope of the target nucleus, on other occasions it had an atomic number lower by one or two units. From this point of view a marked difference was found in the behaviour of the light and heavy elements. For light elements the active products had in general an atomic number smaller than that of the target nucleus, while for heavy elements the active product was always an isotope of the bombarded nucleus. The results obtained with the light elements can in general be explained as due to (n, p) and (n, α) reactions, in which the nuclear charge of the nucleus decreases by one or two units, respectively. In these processes it is the outgoing particle which has to cross the electrostatic potential barrier; the higher this is, the heavier the residual nucleus is. The energies of the neutrons emitted from Be bombarded with the α-particles from Po or Rn in equilibrium with its product (and all neutron sources used at that time were of this type) are a few MeV so that in the case of heavy elements the penetrability of the electrostatic barrier turns out to be so small that the corresponding cross-sections are negligible.

The same argument applied to the incident particle explained why the artificial radioactivity produced by α-particles could be observed by the JOLIOT-CURIES, as well as by others, only in the case of light elements.

The interpretation of the nuclear reactions in which Z does not change met, on the contrary, some difficulties, the complete solution of which took some time.

The importance of the work on artificial radioactivity produced by neutrons was obvious to us as well as to all nuclear physicists. It was only one month after the beginning of this work that Lord RUTHERFORD wrote the following letter to FERMI:

23rd April 1934

Dear FERMI,

I have to thank you for your kindness in sending me an account of your recent experiments in causing temporary radioactivity in a number of elements by means of neutrons. Your results are of great interest, and no doubt later we shall be able to obtain more information as to the actual mechanism of such transformations. It is by no means clear that in all cases the process is as simple as appears to be the case in the observations of the Joliots.

I congratulate you on your successful escape from the sphere of theoretical physics! You seem to have struck a good line to start with. You may be interested to hear that Professor DIRAC is also doing some experiments. This seems to be a good augury for the future of theoretical physics!

Congratulations and best wishes,

Yours sincerely,
(RUTHERFORD)

Send me along your publications on these questions.

Two remarks may be in order at this point: the first is that our group was probably the first large physicists' team working successfully for about two years in a very well organized way. The second that we were perhaps the first to introduce the use of preprints. In order to communicate rapidly our results to our colleagues we wrote almost weekly short letters in Italian to the *Ricerca Scientifica*, the journal of the Consiglio Nazionale delle Ricerche, and obtained what we would now call preprints of these letters that where mailed to a list of about forty of the most prominent and active nuclear physicists all over the world, and the letters appeared a couple of weeks later in the journal. This procedure was facilitated by the fact that my wife, Ginestra, was working at that time at the *Ricerca Scientifica*.

1·4. – Proceeding according to increasing atomic number, before Summer 1934 we irradiated finally thorium and uranium. We observed a number of new activities which were not easily interpreted. We thought that the irradiation of uranium should produce transuranic elements for which we expected properties similar to those of rhenium, osmium, iridium and platinum. This erroneous expectation was then common and, since we had proved that a few of the activities produced were not due to isotopes of elements with atomic number from 86 to 92, we concluded that these activities seemed to be due to elements with atomic number higher than 92. More precisely we thought we had succeeded in separating an ekaRe ($Z = 93$) and an ekaOs ($Z = 94$). The possibility of fission suggested by a German chemist, Ida NODDACK [17], was not considered seriously by us although later we were not able to understand the reason.

One should say that our results were confirmed during 1935 and 1936 by HAHN and MEITNER [18] who even extended them: they thought at that time they had identified two beta-radioactive families, originating by neutron capture in two different uranium isotopes and in each of which appeared, besides an ekaRe ($Z = 93$) and an ekaOs ($Z = 94$), also an ekaIr ($Z = 95$) and in one of the two an ekaAu (given as uncertain). Only after the discovery of fission by HAHN and STRASSMANN in 1939 was it shown by workers in various countries, independently of each other, that the transuranic elements would not behave like Re, Os, Ir and Pt, but would form a second family of rare earths indicated later as «actinides». This conclusion was reached in particular with semi-empirical arguments by ABELSON and MCMILLAN [19] in 1940 and was proved in 1941 by GÖPPERT-MAYER [20] who, at Fermi's instigation, calculated the energy of the $5f$ atomic orbits by the Fermi-Thomas statistical method, as FERMI had done for the $4f$ orbits of the rare earths in 1928 [21].

The work accomplished by Summer 1934 was summarized in a paper that was brought by SEGRÈ and me to Lord RUTHERFORD in Cambridge at the beginning of July 1934. When FERMI wrote to RUTHERFORD if we could spend the summer at the Cavendish Laboratory, he answered with the letter shown in

Fig. 5 [22]. The manuscript given by us personally to RUTHERFORD was presented by him to the Royal Society and was published very quickly in the *Proceedings of the Royal Society* [23].

2. – Recollection of early research on the properties of the neutron.

2·1. – When, at the beginning of July 1934, SEGRÈ and I arrived in Cambridge, the Cavendish Laboratory appeared to us as the very world capital for nuclear physics. RUTHERFORD was working with OLIPHANT on the D+D reactions but with his strong personality dominated the whole laboratory. CHADWICK was working with M. GOLDHABER; they had discovered shortly before the photodisintegration of the deuteron [24]. COCKCROFT, DEE, ELLIS and FEATHER were there and ASTON was going on improving the accuracy of his measurements of atomic masses. In his laboratory there was a young American, BAINBRIDGE, who had recently made an important step forward in this fundamental technique. From time to time, going around in the Cavendish, one could meet J. J. THOMSON who had retired only a few years before.

The only people working on artificial radioactivity produced by neutrons were T. BJERGE and C. H. WESTCOTT, the first from Copenhagen, the second from Canada. We visited systematically all research groups active at that time in the Cavendish, as well as KAPITZA who was in charge of the Mond Laboratory and had—not long before—constructed his famous rotating generator for producing extremely intensive magnetic fields.

The Cavendish was the birth place of the neutron and this was not purely accidental. The process of generation of this particle actually started in Charlottenburg, where in 1930 BOTHE and BECKER [25] had discovered the emission of a very penetrating radiation from beryllium under α-particle bombardment. During 1932 this radiation had been studied also by RASETTI while he was working in Lise Meitner's laboratory in Berlin-Dahlem [26].

I remember very clearly when, towards the end of January 1932, we began to receive in Rome the issues of the *Comptes Rendus* containing the notes by the JOLIOT-CURIES on the properties of this penetrating radiation.

In the first of these notes it was shown that the penetrating radiation emitted by Be under bombardment with polonium α-particles could transfer kinetic energies of about 5 million electronvolt to the protons present in small layers of various hydrogenated materials (such as water or cellophane). In order to interpret these observations the JOLIOT-CURIES at first [27] put forward the hypothesis that the phenomenon was similar to the Compton effect, namely that the incident photon undergoes an elastic collision with a proton. They had calculated, by applying the laws of energy and momentum conservation, that the incident photons must have an energy of at least 50 million electronvolt. Later, however, they realized that the Klein-Nishina formula applied

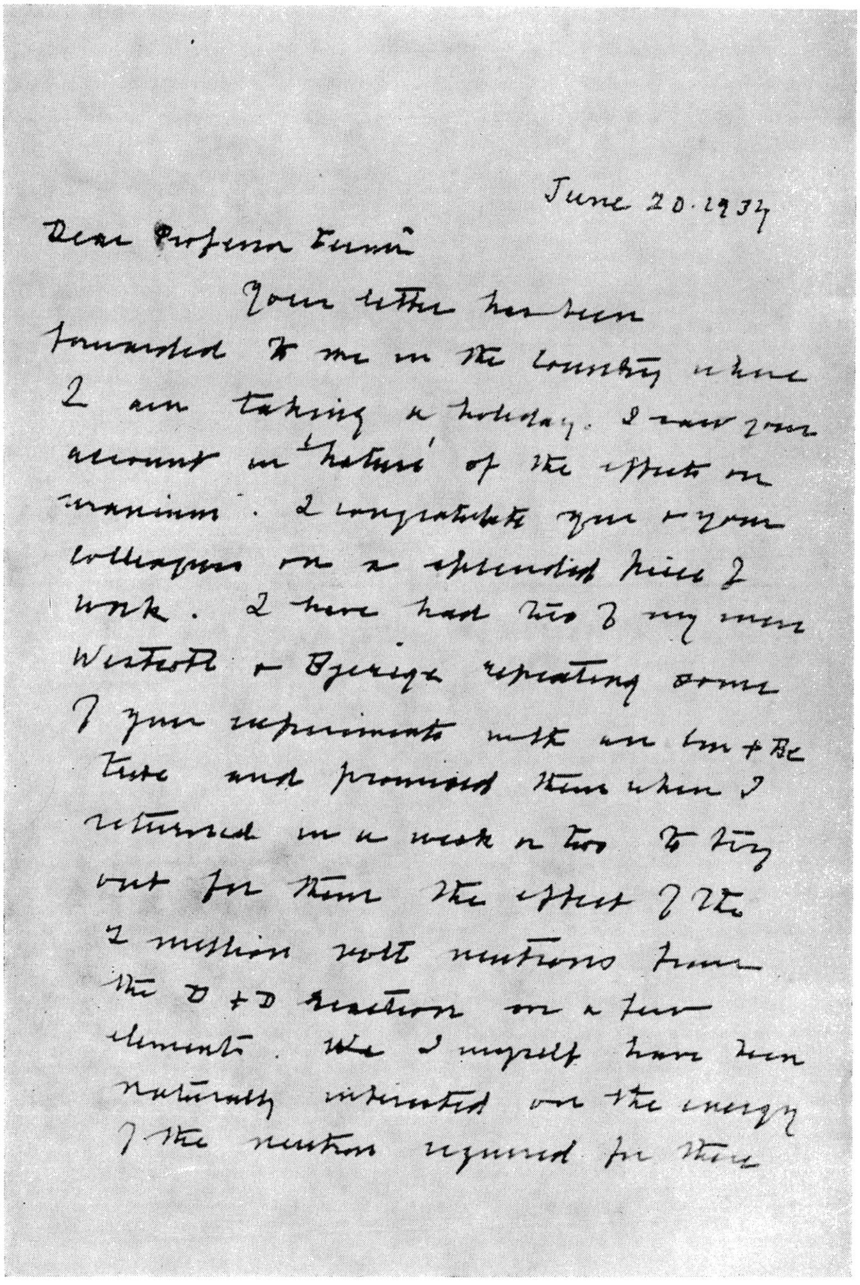

June 20. 1934

Dear Professor Fermi

Your letter has been forwarded to me in the country where I am taking a holiday. I saw your account in "Nature" of the effects on uranium. I congratulate you & your colleagues on a splendid piece of work. I have had two of my men Westcott & Bjerge repeating some of your experiments with an Em + Be tube and promised them when I returned in a week or two to try out for them the effect of the 2 million volt neutrons from the D + D reaction on a few elements. We & myself have been naturally interested on the energy of the neutron required for these

Fig. 5. – Letter of RUTHERFORD to FERMI [22].

transformation. We do not, however, propose at the moment to make systematic experiments in this direction as our installation is required for other purposes. I cannot at the moment give you a definite statement as to the output of neutrons from our tube but it should be of the same order as for an Em + Be tube containing 100 millicuries & may be pushed much higher.

If your assistants come to Cambridge say in the 2nd week of July, I shall be delighted to give them the benefit of our experience and to see the mode of operation of our installation for transmutation in general. I hope one or both speak English as the knowledge of Italian in the laboratory is very limited!. The two men to see are Dr Oliphant & Dr Cockcroft.

Excuse this hastily written letter. I do not take a secretary with me on holiday!

Yours —

Rutherford

to the protons gave a value for the cross-section too small by many orders of magnitude [28]. Consequently, they suggested that the observed effect was due to a new type of interaction between gamma-rays and protons, different from that responsible for the Compton effect [29].

I like to recall that when MAJORANA [30] read these notes he said, shaking his head, « They haven't understood a thing. They are probably observing recoil protons produced by a heavy neutral particle ». A few days later we got, in Rome, the issue of *Nature* containing the letter to the Editor from CHADWICK [31], dated 17 February 1932, entitled « Possible existence of a neutron », in which he demonstrated the existence of the neutron on the basis of a classical series of experiments, in which recoil nuclei of some light elements (such as nitrogen) were observed in addition to recoil protons.

In order to understand how MAJORANA could guess this discovery, which was suggested but certainly not demonstrated by the Joliot-Curies' results, it should be remembered that he was familiar—through a paper published a few years earlier by GENTILE jr.—with the nuclear model proposed in 1927 by Lord RUTHERFORD [32]. It attempted to make the value of the radius of the uranium nucleus, deduced from the deviations observed in the elastic scattering of particles with respect to the predictions based on Coulomb repulsion alone ($3.2 \cdot 10^{-12}$ cm), agree with the value deduced from the energy of α-particles emitted by radioactive nuclei ($\simeq 6 \cdot 10^{-12}$ cm). According to this model, the nucleus consisted of a central part with a positive charge Ze, around which turned neutral satellites kept in their circular orbits by an attractive force originating from the electric polarization to which the satellites were subjected under the effect of the central electric field.

RUTHERFORD had actually suggested already in 1920 that, if the nucleus consisted of protons and electrons, inside it there might be neutral particles consisting of a proton and an electron closely bound together. This idea had remained so much alive in Cambridge that some of Rutherford's students and colleagues, and particularly CHADWICK, had tried several times between 1920 and 1932 to show experimentally that there existed neutral particles with a mass of the same order as, or greater than, that of the proton. The story of these attempts has been given by CHADWICK himself in «Some Personal Notes on the Search for the Neutron » published in the proceedings of the 10th International Conference on the History of Science [33].

In 1928 GENTILE had shown [34] the inconsistency of the model suggested by RUTHERFORD the year before. However, the idea that there might exist in Nature neutral particles of subatomic dimensions had also remained in the air in Rome.

Soon after Chadwick's discovery, various authors understood that the neutron must be one of the components of the nucleus [35] and began to propose various models which included particles, protons, electrons and neutrons.

The first to publish the idea that the nucleus consists solely of protons and neutrons was probably IWANENKO [36]. Neither I nor his other friends remember whether MAJORANA came to this conclusion independently. What is certain is that before Easter of that same year MAJORANA had tried to work out a theory of light nuclei assuming that they consisted solely of protons and neutrons and that the former interacted with the latter through exchange forces. He also reached the conclusion that these exchange forces must act only on the space co-ordinates (and not on the spin) if one wanted the α-particle, and not the deuteron, to be the system saturated with respect to binding energies. He was, however, dissatisfied with his results and refused to publish them in spite of the insistence of FERMI and all his friends [37]. He published his paper on the nuclear forces, later called Majorana forces, only in late 1933 from Leipzig [38]. He was persuaded to do this by the sheer weight of the authority of HEISENBERG who had published his famous three papers on the nuclear forces already in 1932 [39] without knowing anything of Majorana's work.

Going back to the Summer of 1934, SEGRÈ and I brought with us the manuscript of the long paper in which our work on artificial radioactivity produced by neutrons was summarized. We handed it over to Lord RUTHERFORD who received us, commenting favourably on our work and making jokes one after the other while smoking his pipe. We could not understand what he was saying but anyhow he took care that the paper was published as soon as possible [23].

After a visit to most laboratories I devoted some time to learn from Wynn WILLIAMS how to construct linear amplifiers for measuring with good accuracy the energy lost by a single α-particle or proton in the gas of a small ionization chamber. SEGRÈ and I, however, spent most of the time discussing our common problems with BJERGE and WESTCOTT.

2·2. – One of the questions that was left unsolved in our paper in the *Proceedings of the Royal Society* [23] was whether the reactions that produced an isotope of the target were (n, γ), *i.e.* radiative capture, or (n, 2n). They could also have been processes of inelastic scattering of the incident neutron with formation of an isomer of the target nucleus. The last process however was not considered in Summer 1934 and in any case it would have been excluded by the same experiments and considerations reported below. The objections against the (n, 2n) reactions were based on the clearly endoenergetic character of this process. Therefore, the (n, γ) reaction would have remained as the most natural interpretation if it had not been for the following fact based on the « one-particle model » universally accepted at that time. The probability of emission of one or more photons of a few MeV energy during the short time that a neutron takes to cross a nucleus is so small that the calculated cross-section for a (n, γ) process turns out to be negligibly small.

The most important argument, and the decisive one in favour of radiative capture, consisted in testing whether the same unstable nuclide of atomic number Z could be produced by bombarding elements of the same atomic number Z, as well as elements of atomic number $Z+1$ and $Z+2$. For example if the *same* $_{11}$Na is produced in the three reactions

$$^{27}_{13}\mathrm{Al}(\mathrm{n},\alpha)^{24}_{11}\mathrm{Na}\,,\qquad _{12}\mathrm{Mg}(\mathrm{n},\mathrm{p})_{11}\mathrm{Na}\,,\qquad ^{23}_{11}\mathrm{Na}(\mathrm{n},\,?)_{11}\mathrm{Na}\,,$$

necessarily the last one should be a radiative capture.

The first step to prove that this was just what happened was made by BJERGE and WESTCOTT [40]. While we were in Cambridge they found that, after neutron bombardment, sodium shows a weak activity, the period of which (about 10 h) is, within the error of measurement, the same as that of the long periods produced in magnesium and aluminium, which were known [23] to to be due to an isotope of Na [41].

Back in Rome we completed the proof by showing, with the help of D'AGOSTINO, that this long activity produced in Na was actually due to an isotope of the same elements.

I have insisted on this particular problem because I believe that the proof that neutrons can undergo radiative capture with an appreciable cross-section was one of the first experimental evidences that the « one-particle model » was inadequate for describing many important properties of the nuclei.

We also found a second case of « proven » radiative capture, which was based on the discovery of a new radioisotope of Al with a lifetime of almost 3 minutes. But, when, a few days after our results were made known, I tried to repeat the measurements, I did not find this new activity anymore.

Unfortunately, we had communicated our result to FERMI who, on his way back from South America (where he had spent the Summer), was attending the International Conference on Physics in London. He had even mentioned our experiment at one of the meetings. The fact that we were not able to confirm our result was hurriedly communicated to FERMI who was angry and embarrassed at having presented an erroneous result at the Conference.

We were unhappy and confused because we were not able to understand the origin of our fault. This was the first hint of unexpected complications which were fully clarified in about one and a half months.

2·3. – In the paper published in the *Proceedings of the Royal Society* [23] the activity of the various artificial radioactive bodies had been classified only qualitatively as strong, medium and weak. This classification was clearly unsatisfactory and, therefore, at the beginning of the academic year 1934-35, we decided to try to establish a quantitative scale of activities which for the moment could be in arbitrary units. This work was assigned to me and PONTECORVO, one of our best students who had taken the degree (laurea) in July 1934 and after the summer vacations had joined the group. We started by studying the conditions of irradiation most convenient for obtaining well-reproducible results. For this type of work we used the activity of 2.3 min lifetime of silver [23].

We immediately found, however, some difficulty because it became apparent that the intensity of activation depended on the conditions of irradiation. In particular there were certain wood tables near a spectroscope in a dark

Riassunto – Em+Be 15. 30 mc

P.t. 39^5	NE.	A_1	A_2	B_1	B_2	C_1	C_2	D
110	7,5							
		153	162	38	22	42	25	34
		131	145	32	23	39	31	27
116	9,9							
		149	139	34	25	25	27	30
		132	114	38	25	21	23	24
109	9,1							
115	8,6							
		131	110	41	28	28	26	31
107	8,2							
114	11							
		150	143	37	19	31	21	33
112	8,0							
		846	813	220	142	186	153	179
		738	705	112	34	78	45	71

Fig. 6. – Photograph of page 3 of the notebook *B*1. It shows the summary of measurements taken under various conditions sketched in the lower part of the page.

room which had miraculous properties, since silver irradiated on those tables gained much more activity than when it was irradiated on a marble table in the same room.

In order to clarify the situation I started a systematic investigation. According to the notebook $B1$ where the measurements of that period are recorded [42] these measurements were started on 18th October 1934. Figure 6 reproduces page 3 containing the summary of a typical series of measurements made inside and outside a lead housing (castelletto), the walls of which were 5 cm thick. The Rn α+Be source was always in position A_1 (or A_2). The irradiation was made by placing the Ag cylinder inside the lead housing in the four positions A_1, B_1, C_1 and D. The distance from D to A_1 was the same (11 cm) as from B_1 to A_1. Outside the lead housing the measurements were made in the three positions, A_2, B_2 and C_2, which were at the same relative distances as A_1, B_1 and C_1. The results of the measurements, given in the lower part of the page, clearly show that the activation decreases with the distance from the source more slowly inside than outside the « castelletto ». In particular B_1 is about 4 times larger than B_2 and D is about equal to C_1 and twice C_2. The fact that D was equal to C_1 was an indication that in D the absorption of lead was roughly compensated by the « scattering in » of the neutrons moving from the source in directions different from that of the detector. It was then decided to test this conclusion in a better geometry, *i.e.* to compare the activation of the same Ag cylinder under the two conditions shown in Fig. 7. The lead wedge

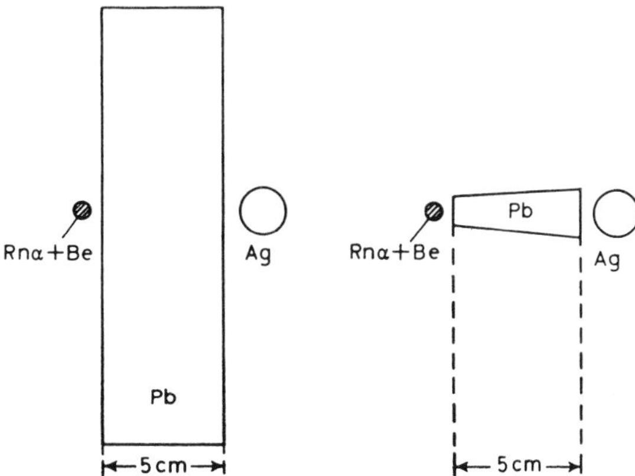

Fig. 7. – Arrangements used by FERMI and collaborators for studying the transmission of lead under conditions similar but « cleaner » than those shown in Fig. 1.

necessary for the arrangement shown on the right-hand side of Fig. 7 was ready the morning of October 20, but it was actually used only a few days later.

In the morning of October 22 most of us were busy doing examinations and FERMI decided to proceed in making the measurements. ROSSI from the University of Padua and PERSICO from the University of Turin were around in the Institute of Via Panisperna and PERSICO was, I believe, the only eye-witness of what happened. At the moment of using the lead wedge FERMI decided suddenly to try with a wedge of some light element, and paraffin was used first [43]. The results of the measurements are recorded on pages 8 and 9 of the same notebook *B*1 (Fig. 8). They are written at the beginning by FERMI and towards the end by PERSICO. Towards noon we were all summoned to watch the extraordinary effect of the filtration by paraffin: the activity was increased by an appreciable factor.

The work was, as usual, interrupted shortly before one o'clock and when we came back, as usual, at 3 p.m., FERMI had found the explanation of the strange behaviour of the filtered neutrons. The neutrons are slowed down by a large number of elastic collisions against the protons present in the paraffin and in this way become more effective. This last point, *i.e.* the increase of the reaction cross-secticn by reducing the velocity of the neutrons, was at that time still contrary to our expectation. The same afternoon the experiment was repeated in the pool of the fountain in the garden of the Institute and we also had succeeded in clarifying the reasons for the discrepancy of the two sets of measurements on the activation of Al that I mentioned above [44].

FERMI went on to hypothesize that the neutrons could be thermalized and that same day an experiment was devised to test the validity of this assumption.

The evening of the 22nd October all the group came to my house and a letter announcing our results was written to the *Ricerca Scientifica* [45].

2'4. – The discovery of the hydrogen effect obviously opened a number of problems and we had to modify our previous program. The first step was to measure what we called the « coefficient of aquaticity », *i.e.* how much immersion in water would increase the activity when a thin cylinder of the target material was placed around the neutron source. These measurements gave us confirmation that the (n, γ) reactions were the only ones sensitive to hydrogeneous substances and by early November we were convinced that slowing-down was the correct explanation of the phenomenon observed.

Then we concentrated our study in trying to understand the behaviour of slow neutrons rather than the nuclide they produced. Among other things we made an experiment aimed at establishing whether the neutrons were thermalized by comparing the activation of a piece of Rh surrounded by hydrocarbons at two different temperatures: room temperature and 200 °C [45]. We did not observe any effect; shortly afterwards MOON and TILLMAN, working in London, realized that such an effect can be observed only in the vicinity of the boundary between two media at different temperatures, and succeeded in obtaining a positive result [47].

FERMI, PONTECORVO and RASETTI [48] also found that some substances like B, Cl, Co, Y, Rh, Ir, Ag, Cd, etc. had very large capture cross-sections σ_c:

(8) 20 ottobre 34

Assorbimento neutroni Em + Be che attivano l' Ag in 4 cm di Paraffina.

N. E. $\frac{2050-1980}{9} = \frac{70}{9} = 8$

Rit 19 $\frac{2357-2070}{3} = \frac{287}{3} = 96$

Con 4 cm paraffina			senza		
0'	2483	81	0'	2610	49
1'	2564		1'	2659	
	2711	99	0'	2874	54
	2810		1'	2928	
0'	2961	90	0'	3074	58
1'	3051		1'	3132	
0'	3156	74	0'	3249	50
1'	3230		1'	3299	

Fig. 8. – Photograph of pages 8 and 9 of the notebook *B*1 of FERMI *et al.* showing the record of the first observation of the effect of hydrogeneous substances on the radioactivity induced by neutrons.

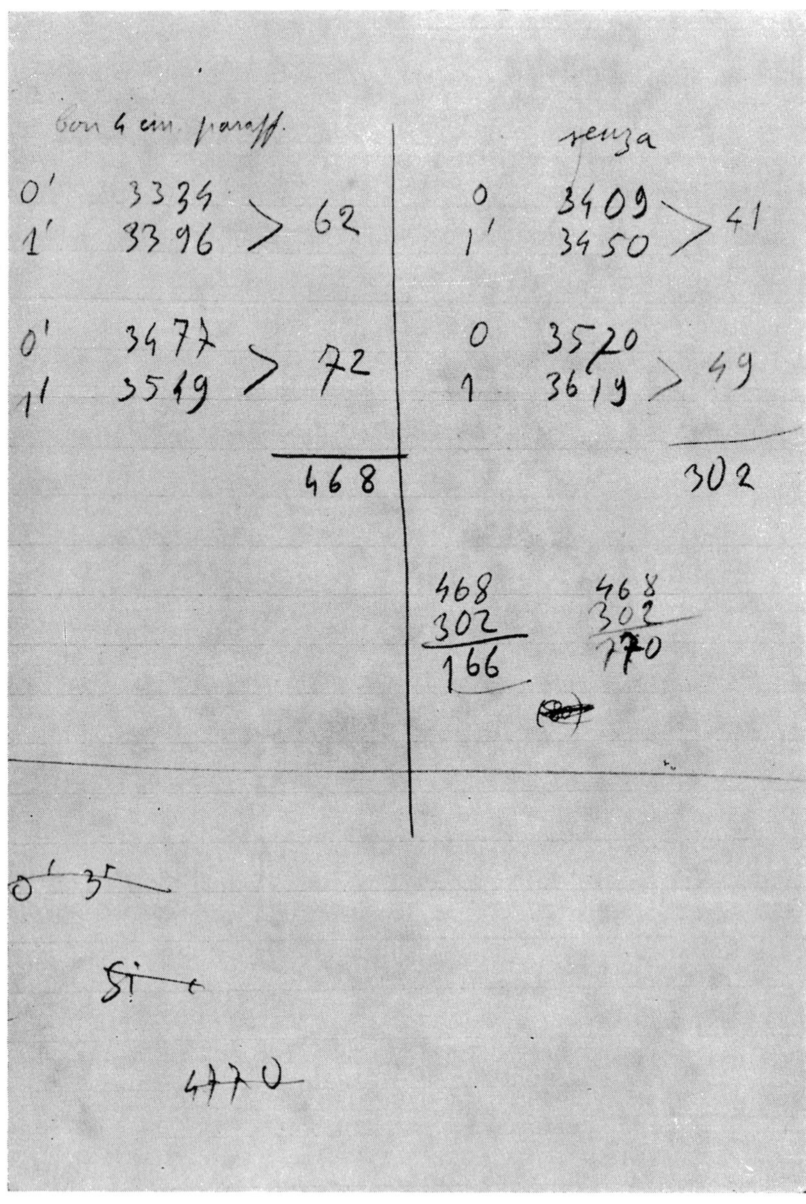

Con 4 cm. paraff.

0' 3334
1' 3396 > 62

0' 3477
1' 3549 > 72

468

senza

0 3409
1' 3450 > 41

0 3520
1 3619 > 49

302

468
302
166

468
302
770

0' 3'

Sì

4770

the measurements were still very rough but sufficient to establish that in some cases σ_c was by 10^3 or even 10^4 greater than the geometric cross-section of the nuclei.

The explanation of these anomalous capture cross-sections clearly required quantum mechanics: for particles of such a small velocity that their wavelength λ is much larger than the radius R of the target obstacles, the upper limit of the cross-section is not πR^2 but $\lambda^2/4\pi$ multiplied by a numerical factor which can be not much smaller than 1. The same type of reasoning brought FERMI as well as others to foresee a general law for the dependence of the capture cross-section σ_c on the velocity v of the neutrons. The probability P of capture of a neutron by a nucleus is given by $P = \sigma_c \cdot v$ and if one describes the process in a simple approach, as that provided by the « one-particle model », one easily finds that

$$\lim_{v \to 0} \sigma_c \cdot v = \text{constant}.$$

This means the so-called $1/v$-law, *i.e.* that for very low velocities σ_c is proportional to v^{-1}.

We also studied the gamma-rays emitted by various nuclei in the radiative capture of neutrons and tried to slow down neutrons by collisions with substances other than hydrogen. Thus we found some effect of inelastic collisions which explained our original observations of the effect of surrounding the source and the detector with lead. This was accomplished by December 1934, *i.e.* within about six weeks of the discovery of slow neutrons.

The large capture cross-sections for slow neutrons observed in boron and lithium did not correspond to any radioactivity nor to a prompt emission of gamma-rays (as we had observed, for example, in cadmium). The behaviour of boron (and less extensively of lithium) was then studied by means of an ionization chamber connected to an electrometer as well as by means of a small ionization chamber connected to a linear amplifier built following Wynn Williams' general guidelines. Thus we arrived at the correct conclusion that slow neutrons produce in ^{10}B an (α, n) reaction.

The results of all these experiments and of many others that I have no time to mention were published in a second paper in the *Proceedings of the Royal Society* [49] and in a series of more detailed articles in the *Nuovo Cimento* [50].

2'5. – I have not yet mentioned Prof. CORBINO, the head of the Department of Physics of the University of Rome and the person responsible more than any other for the creation, in his Institute, of a group of young physicists around FERMI. In 1926 he had succeeded in creating a chair of « Theoretical Physics » for FERMI and, a couple of years later, a chair of « Spectroscopy » for RASETTI. SEGRÈ and I were his assistants but he pushed us to work with FERMI. He protected all of us from the criticisms of the traditional university environment, which in many cases was not too favourable to us. He was an extremely intelligent person. He had been a very distinguished physicist, but from a certain moment on he had devoted most of his activity to politics and industry. He had become a member of the Italian Senate, and had been Minister of Public Education and later of Economics and Industry.

Shortly after the discovery of the effect of hydrogeneous substances he suggested that slow neutrons might have important practical applications and that it could be advisable to take out a patent on this work. This was done and resulted in Italian patent n. 324458 of October 26, 1935, which was later extended to other countries.

Coming back to our work, I like to recall that in January-February 1935 we made, however, an attempt in a different direction, which, although unsuccessful, is of some interest. In order to explain the great number of new activities induced in thorium and uranium which had been isolated by various groups working in Rome, Paris (I. CURIE and co-workers) and Berlin (HAHN and MEITNER) one had invoked a number of transuranic elements and their possible decay products. It was then rather natural to consider that besides beta activities there must be new radioactive alpha-emitters. Thus the decision was taken that I had to look for activities of this type by means of a small ionization chamber connected to the linear amplifier. Helped from time to time according to necessity by one or the other of my colleagues, I began to irradiate some foil of uranium (or thorium) in the form of oxides and put them immediately after irradiation in front of the thin-window ionization chamber. Since we could not observe any activity, we thought that this might be due to the fact that the corresponding lifetimes were too short, perhaps fractions of a second. Therefore, the uranium foil was placed in front of the ionization chamber and irradiated there with a neutron source surrounded by paraffin. A piece of lead was placed between the neutron source and the chamber in order to reduce the background due to gamma-rays.

We also thought that, if alpha-emitters existed, and they had a short lifetime, they had to emit (according to the Geiger-Nuttal law) alpha-particles of a range considerably longer than that of the particles emitted spontaneously by uranium and thorium. Thus all the experiments were carried out with uranium and thorium covered with an aluminium foil equivalent to 5 or 6 cm of air.

The experiments gave negative results, but, if we had occasionally forgotten the aluminium foil, we should have observed the recoiling nuclei due to fission already in January or February 1935. Some years later, talking with HAHN after he had discovered in collaboration with STRASSMANN the fission in 1939, I learned that a very similar experiment, also with negative results, was made at about the same time in Berlin-Dahlem by VON DROSTE. Looking backward it is difficult to say what would have been our reaction if we had observed fission fragments at the beginning of 1935.

2·6. – A very serious difficulty was clearly emerging from our experiments [49] as well as from those of other groups, in particular the group working at Columbia University: DUNNING, PEGRAM and MITCHELL [51]. There was no correlation between the scattering and the capture cross-section in cases where the capture cross-section was very large as, for instance, for cadmium. FERMI tried to explain the phenomenon but without success. As I will say later, it was explained by BOHR in January of the following year, 1936.

For interpreting the elastic cross-sections in the second paper in the *Proceedings of the Royal Society* [49] the concept of scattering length was used as a very convenient artifice for describing what happens in the limit of infinite wavelength of the neutron; here infinite means, of course, very large with respect to the dimensions of the nucleus. The concepts of scattering length and pseudopotential had been, however, developed by FERMI in 1933 [52] in order to explain some spectroscopic phenomena (pressure shift of the spectral lines) that had been found experimentally by SEGRÈ and myself [53].

Before summer 1936 we succeeded in extending our previous results and we made a mechanical experiment by which one could compare the velocity reached by the neutrons with the velocity of the edge of a fast rotating wheel [54, 55].

An equivalent, although different, experiment was made at about the same time in Copenhagen by FRISCH and SØRENSEN [56]. During the summer the Columbia University group with the addition of SEGRÈ, who was visiting there, constructed and operated the first mechanical velocity selector for slow neutrons [57].

After the summer vacation of 1935, FERMI and I found ourselves alone in Rome. Most of the group had dispersed by now. The general atmosphere in Italy was chiefly to blame for this as the country prepared for the Ethiopian war. RASETTI had gone to the U.S.A. and planned to stay at Columbia University for at least one year. SEGRÈ, too, had left for a Summer in the United States, and in the meantime had been appointed professor at the University of Palermo. Upon his return to Italy, he also left Rome to go to Sicily. D'AGOSTINO no longer worked with us; he had taken a position at the Istituto di Chimica del Consiglio Nazionale delle Ricerche. PONTECORVO had returned to Rome shortly after us and for a few months worked with WICK on the back-scattering of slow neutrons by various elements [58]. Later he won a Ministero dell'Educazione Nazionale scholarship for study abroad and left Rome in the spring of 1936 to work with the JOLIOTs at the Curie Laboratory in Paris. Thereafter, his visits to Rome were very brief and infrequent.

Upon resuming work, FERMI and I turned our attention to some results of BJERGE and WESTCOTT [59] and of MOON and TILLMAN [60], who had observed that the absorption of slow neutrons by various elements differed slightly depending on the element used as detector.

This fact was not explained by the current theory of the absorption of neutrons by nuclei. As I said before, this theory predicted for all nuclei a capture cross-section inversely proportional to the velocity of the neutrons. This energy dependence was supposed to be valid for such a large energy interval as to certainly cover the energy range of slow neutrons.

We went to work with even greater energy than in the past, as if by our own more intensive efforts we intended to compensate for the loss of manpower in our group. We had prepared a systematic plan of attack which we jokingly summarized by saying that we would measure the absorption coefficient of all 92 elements combined in all possible ways with the 92 elements used as detectors. In jest we added that after combining all the elements two by two, we would also combine them three by three. By this we meant that we would also study the absorption properties of the neutron radiation filtered in several ways.

Actually, after having measured the absorption coefficient of eleven different elements in all possible combinations with 7 detectors, we were convinced that the observations of the groups quoted above were correct, and that in general the rule was valid that the absorption coefficient of a given element was greater when the same element was used as detector. We began to study the particular cases of silver, rhodium and cadmium in great detail. The absorption properties of cadmium were investigated more thoroughly. We performed on the neutrons filtered by a cadmium layer absorption measurements of different elements with various detectors, as we had already done in studying the unfiltered neutrons. Thus, early in November 1935, we established that, if the neutrons were previously filtered with cadmium, the self-absorption effect mentioned above was considerably greater [61].

SZILARD [62] independently had the same idea of studying selective absorption of the neutrons filtered through cadmium and showed that In and I had a behaviour similar to that found by us for Ag and Rh. Indeed the idea of the existence of selective absorption was already contained in the papers of the two British groups [59, 60], but the experiments carried out by us and by SZILARD made this interpretation compelling.

I think it is interesting to note that on this occasion I, as many others, tended to make a simple picture of the phenomenon. I tried to interpret the different groups of neutrons as different bands of energy. FERMI, however, did not want to accept this description. He, too, was convinced that this was obviously the simplest hypothesis, but maintained that it was not strictly necessary, at least for the moment, and was therefore harmful if introduced into our mental picture. He insisted that one must proceed by reasoning with the observed experimental facts only. The correct interpretation of the nature of the neutron groups would finally emerge as a necessary consequence of the data. He was afraid that a preconceived interpretation, however plausible it sounded, would sidetrack us from an objective appraisal of the phenomenon that confronted us.

Therefore, we began a systematic study of the absorption and diffusion properties of the various neutron groups, labelling them with letters, both for brevity and in order to avoid any trace of interpretation. The expression « group C » was used for neutrons strongly absorbed by cadmium; « group D » for neutrons strongly absorbed by rhodium, but not by cadmium; « group A and B » for the two components which we believed we had characterized in the radiation strongly absorbed by silver, but not absorbed by cadmium.

In a second letter to the *Ricerca Scientifica*, dated December 12, 1935 [63], the groups of slow neutrons are clearly defined and their absorption and other properties are studied. In this paper experiments are reported which were performed to establish the number of neutrons belonging to each group (numerosity) among those that emerge from the surface of the moderator (paraffin or water), the reflection coefficient of the various group (albedo) [64] and their diffusion distance.

Thus we showed evidence that group C—the neutrons strongly adsorbed by cadmium—had properties very different from those of the neutrons that passed through cadmium. The albedo of the neutrons of group C was very high (0.83), while that of groups D, A, etc. was negligible. The diffusion length in paraffin was about 3 cm for group C, while it was about six times less for the other groups.

These latter results were further clarified by a more accurate experiment performed in January 1936, in which the diffusion length was determined from the escape probability of a neutron originally found at a depth x within a medium filling a half-space [65]. The expression of such a probability is derived from the diffusion equation which had been adopted for the description of the properties of group C. In this same paper, the interpretation of the neutron groups, as due to energy difference, is discussed as the most likely interpretation, without, however, precluding others. This work concludes with reference to an experiment well under way, but not yet completed, whose purpose was to clarify this point: if various groups differ only in energy, the neutrons which at a certain moment belong to a group, as a result of further slowing-down, must transform into neutrons of another group. At that time we had already learned from several experiments by other workers [66] that group C included thermal neutrons. Therefore, if the interpretation of the groups in terms of different bands of energy was correct, all the other groups ought to be transformable by slowing-down into group C. The definitive results of these experiments were given in a final long paper [67]. In the meantime an experiment of the same type had been published by Halban and Preiswerk [68].

In order to establish precisely the diffusion properties of group-C neutrons, it was necessary to supplement the measurements of the diffusion length mentioned above with a measurement of the mean free path. The result of a first measure of this quantity showed a clear difference between the values of the group-C neutrons mean free path and the mean free path of all other neutron groups [69, 67].

The explanation of the neutron groups in terms of differences in energy had in the meantime been imposed by various experiments in particular by those of the type referred to above: the transformability of various groups into group *C*.

2·7. – At the same time an important step towards solving the difficulties mentioned above (*i.e.* the existence of resonances and the fact that the very large capture cross-sections were not associated with appreciable scattering) was made in two independent papers by BREIT and WIGNER [70] and by BOHR [71].

The first authors, guided by the analogy with certain molecular resonances, assumed that the neutron, once it had penetrated the nucleus, could give part of its energy to one of the nuclear components and thus create an excited metastable state whose lifetime was long enough to produce the desired small width of the level. They derived an expression for the radiative capture cross-section which is valid when the neutron is captured in a single resonance level and which is usually indicated as the « one-level formula of BREIT and WIGNER ».

The paper by BOHR, presented at the Danish Academy on 27th January 1936, is also based on the idea of an intermediate level. But instead of postulating its existence and of deriving the expression of the corresponding cross-sections, BOHR developed a new conception of the mechanism of nuclear processes which justifies the existence of many excited levels in nuclei of intermediate and high atomic mass number, whose lifetime is sufficiently long to match the experimental results. The extraordinary stability of the intermediate level is explained by BOHR by noticing that as soon as the incident neutron has entered the nucleus it starts to collide with the constituent neutrons and protons. As a consequence its energy is rapidly shared among many particles no one of which acquires an energy large enough to leave the nucleus. This situation can last a long time, until one of the particles, through a momentary fluctuation which concentrates in it a sufficient energy, flies out of the nucleus.

According to this point of view a nuclear process can be described as taking place in two independent steps; the first is the formation on an « excited compound nucleus » as a consequence of the capture of the incident neutron, the second is the decay of the compound nucleus either by emission of a particle, or—more frequently—by irradiating a photon.

However, the difference in mean free path reported in ref. [69] was interpreted by FERMI as due to the chemical bond, since he could not imagine any nuclear phenomenon that could possibly give rise to a similar effect. The theory of this phenomenon was given by FERMI in a fundamental extensive paper containing the theory of a number of phenomena observed with slow neutrons [72].

At this point, perhaps, it is useful to remember that, while this work progressed, measurement techniques were becoming considerably more refined, During the first period of investigation by the group at the University of Rome.

the activity measurements were taken exclusively by means of Geiger-Müller counters. However, after the discovery of the effect of hydrogenous substances, the activity had become so high that it was frequently possible to use an ionization chamber connected to an electrometer (Fig. 9). This technique was then

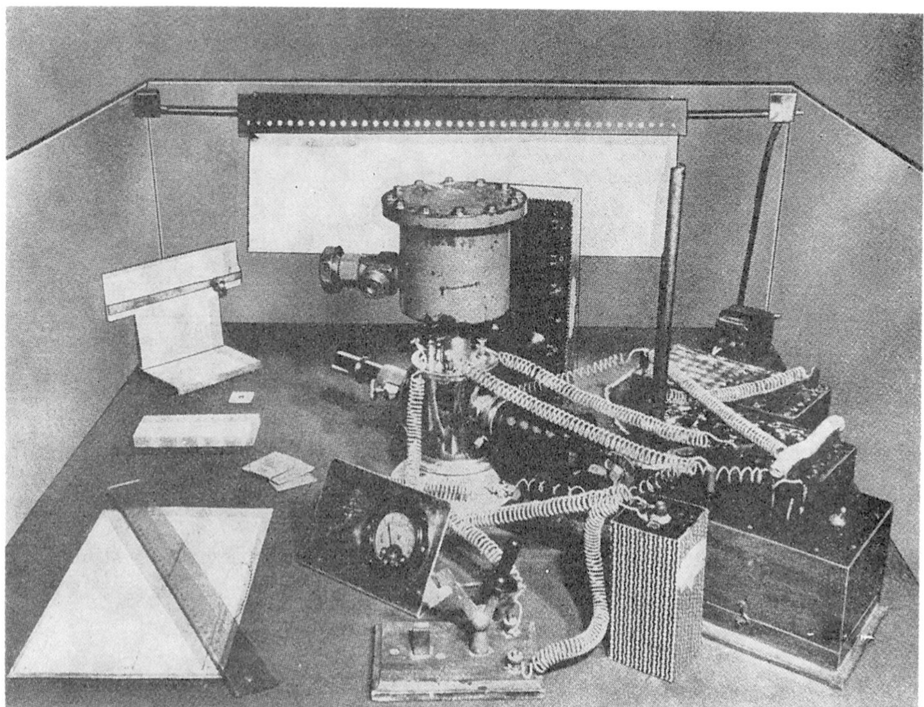

Fig. 9. – Ionization chamber and electrometer used by FERMI and collaborators to measure the activity induced by slow neutrons. Note on the left the scale on which the wire of the Edelman electrometer was projected and the nomogram for deduction of the activation of Rh to standard measuring conditions (Fig. 10). The white shield on the background was introduced, for a better illumination of the instrument, only when this photograph was taken (in 1962 on occasion of the celebration of the 20th anniversary of the operation of the first nuclear pile by FERMI *et al.* in Chicago).

developed and perfected by experimenting with new types of ionization chambers and new ways of using the electrometers. These had been calibrated with great care in order to know well their characteristics and utilize them to their maximum potentialities. The preparation of nomograms (Fig. 10) and graphs allowed a rapid computation, from the readings made on the electrometer scale, of the activity of the radioactive body being measured.

Once the interpretation of the phenomena observed on the basis of the « compound nucleus » resonance levels, according to Bohr's hypothesis, was

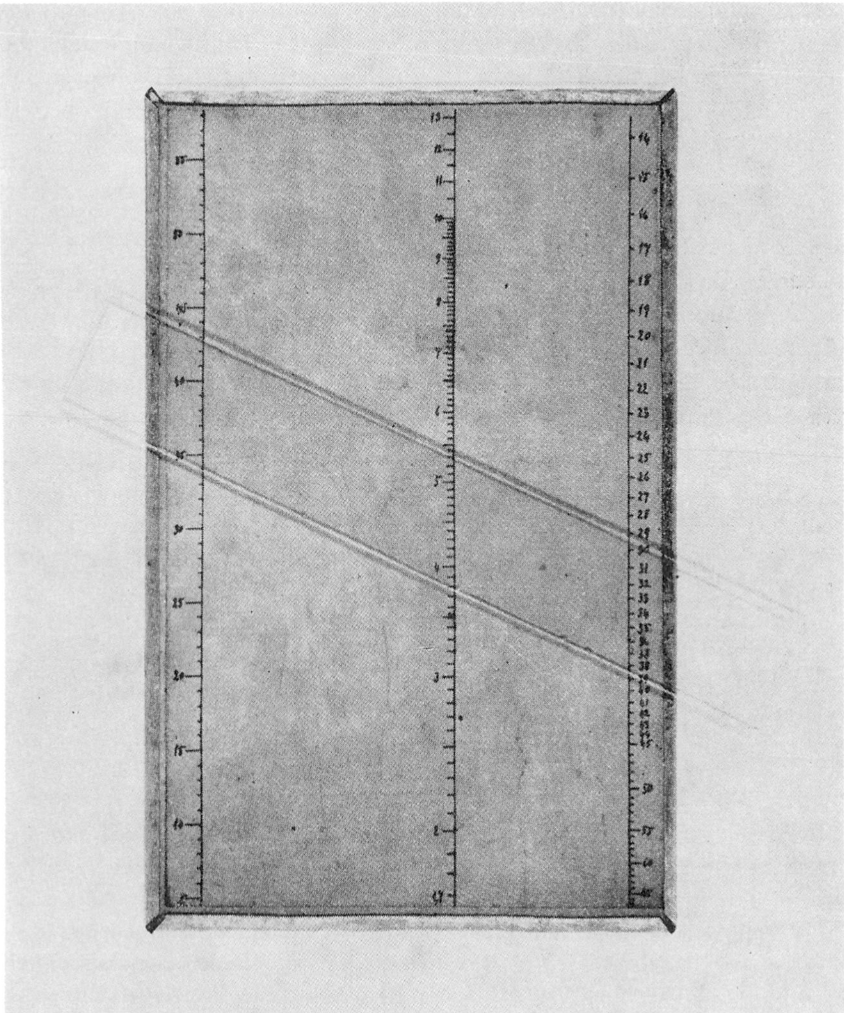

Fig. 10. – Photograph of the nomogram calculated and drawn by G. C. WICK to help the work of FERMI and AMALDI. It regards the activity of $T_{\frac{1}{2}} = 44$ s of Rh. On the central scale the total activity (*i.e.* the activity measured from the end of the irradiation, $t = 0$, to $t = \infty$) is read on the straight line determined by the time passed between the end of the irradiation and the beginning of the measurements (left-hand scale) and the duration of the measurement (right-hand scale).

accepted, the problem of determining the width and energy of these resonance lines naturally arose. This was done by FERMI and me in a series of experiments [67, 73] the theory of which was given by FERMI in the extensive theoretical paper quoted above [72]. This work demonstrates how the mean value of the square of the distance traveled by the neutrons, before reaching the resonance energy of the detector, increases as the resonance energy of the detector decreases. In this way a quantitative relation between spatial distribution of resonance neutrons and their energy was established.

2'8. – The academic year, which had slipped by in an atmosphere of frenzied work and isolation, was by now drawing to a close. RASETTI wrote to us every now and then about what was happening at Columbia University; reprints by HALBAN and PREISWERK kept us informed of the work in Paris and a correspondence with PLACZEK kept us in contact with Copenhagen. Through this latter correspondence we learned of Bohr's work, as well as of that of FRISCH and PLACZEK, concerning the $1/v$ absorption law in boron [74]. Through this latter correspondence the joke that, « as the captain's age can be determined from the length of the ship's mast, so the energy of a neutron group can be determined by the distance it travels as it slows down », spread from Rome. The expression « age », used later by FERMI to represent the quantity $\langle r\rangle^2/6$, might date back to this period. At first the expression « the age of the captain » was used to refer to experiments concerning the transformation of one group into another of lower energy (Fig. 11).

Besides FERMI and me, WICK and PONTECORVO were also working at the Institute, as I said before. There was also MAJORANA, and occasionally SEGRÈ came to visit us from Palermo.

We worked with incredible stubbornness. We would begin at eight o'clock in the morning and take measurements almost without a break until six or seven in the evening, and often later. The measurements were taken with a chronometric schedule as we had studied the minimum time necessary for all the operations. They were repeated every three or four minutes, according to need, for hours and hours and for as many successive days as were necessary to reach a conclusion on a particular point.

Having solved one problem we immediately attacked another without a break or feeling of uncertainty: « Physics as soma » [75] was the phrase we used to refer to our work performed while the general situation in Italy grew more and more bleak, first as a result of the Ethiopian campaign, and then as Italy took part in the Spanish Civil War.

This was more or less the end of the golden period for the investigation of the neutron properties at the University of Rome. The work went on for a few years and some interesting results were still obtained. But the leading role that the group had maintained for about three academic years had finished. This was due certainly in part to the political situation becoming worse and worse in Europe in general and in Italy in particular. It was due, perhaps, also to the fact that it was becoming always more difficult to compete with other groups that, in the meantime, had equipped their laboratories with accelerators of various types which provided neutron sources much better than those at our disposal.

114

ore 16.52 messo Pb I$_2$ A (in acqua)
ore 17.17 tolto " 0,27 Cd

Pb I$_2$ A

4,1	20/176,7	113 -38,4	74,6	83,8	84,0
7,8	20/189,8	105,5	66,9	82,7	
11,5	20/201,2	99,4	61,0	84,0	attiv. iniziale 256,2
15,5	20/214,5	93,2	54,8	84,1	83,2 38,4
19,8	20/229	87,3	48,9	84,5	5,95
24,2	20/252	79,4	41,0	80,3	84,1

Età del capitano argento
misure delle attività dal cubito
4 esperienze a, b, c, d

a) P(7) S P(2,37) 0,54 Cd Ag 0,3 Ag 0,54 Cd

b) P(7) S P(2,37) 0,54 Cd 0,057 Ag Ag 0,3 Ag 0,54 Cd

c) P(7) S P(0,5) 0,54 Cd Ag 0,3 Ag 0,54 Cd P(1,87)

d) P(7) S P(0,5) 0,54 Cd 0,057 Ag Ag 0,3 Ag 0,54 Cd P(1,87)

Ag (2,74)

18.15 a) Ag β	58,8 14,3 73,1 (-2,3)	35,0		84,3
18.20 b) Ag γ	60,0 8,8 68,8	18,9	25,7	84,3
18.25 c) Ag δ	59,1 21,2 80,3	55,1		84,3
18.30 d) Ag α	59,8 13,9 73,7	33,8	38,2	84,3

Fig. 11. – Photograph of p. 114 of the notebook *B*3 handwritten by FERMI.

3. – First international collaborations between Western European countries after World War II in the field of high-energy physics.

3·1. – In this Section, I will try to outline the stage of development of the study of subnuclear particles in Europe immediately after the Second Wordl War and to summarize tbe most important steps that brought the creation of CERN. The two subjects are closely related and the case of CERN is of particular interest for a number of reasons. Among the various international research bodies created in Western Europe after the Second World War, CERN was the first one in order of time. It entered into operation at a very early date and acted from the beginning with the full satisfaction of the governments of all participating countries as well as of the involved scientific circles.

It would be very interesting but it is not easy to find out and disentangle all the elements that made such an early and—at the same time—lasting success possible. Among these, certainly one should keep in mind the general idea—held in many political circles of Western European countries—of the necessity of moving towards some form of political unification of at least a considerable part of the old continent.

A second favourable element was the scientific, technical and administrative experience in various countries, during and immediately after the war, of wide and complex organizations operating in the field of the nuclear sciences and their applications. I will not say more on this point since it will be treated by KOWARSKI. I should mention, however, that this experience had brought about the creation in the U.S.A. of a few big research laboratories such as the Argonne National Laboratory and the Brookhaven National Laboratory. The latter had been created and run—very successfully—by the Associated Universities Inc. Furthermore the dimensions of the geographic region involved and of the laboratory were very similar to those of a possible future European research establishment. In the latter case a program of research had to be chosen completely free from any limitation or restriction originating from military, political or even industrial secrecy.

To these one should add a further element of paramount importance: immediately after the Second World War cosmic-ray research had a very high level in Western Europe and was in part carried out through successful international collaborations.

Today we are used to considering cosmic rays and high-energy physics as two separated chapters of physics, but originally it was not so. For many years the only available source of high-energy particles was the cosmic radiation and high-energy physics was the part of cosmic-ray research regarding the study of the behaviour of the component particles. Even today, for the investigation of particles of energies greater than 10^{13} eV, the only thing that one can do is to use cosmic rays.

3·2. – It would be very interesting to review in detail the development of cosmic rays from the point of view of high-energy physics, starting from the discovery of the positron and the mu-meson in the thirties. It would be a very exciting story but too long, at least in the frame of my lecture. In the first part of which I will try to underline—mainly from my personal direct knowledge and experience—the transition undergone in Europe by high-energy physics from being a part of cosmic-ray research to becoming an independent and strong activity going on in many European universities and national laboratories, but mainly in the Meyrin Laboratories of CERN.

The whole story can be seen in the right perspective only if one realizes that, in spite of the enormous destruction suffered by most European countries, research in the field of cosmic rays was still at a very high level at the end of the Second World War.

The opportunity for a first encounter between physicists was provided by the conference organized as a joint effort by the British Physical Society and the Cavendish Laboratory. The Conference took place in Cambridge on 22-27 July 1946 and its subject was «Fundamental Particles and Low Temperature». Contact between physicists in different parts of the world had been impossible for years and this conference provided a welcome opportunity to renew old friendships and to hear what others had been doing [76]. Theoretical papers were presented by BOHR on problems of elementary-particle physics, PAULI on difficulties of field theories and field quantization, DIRAC on difficulties in quantum electrodynamics, BORN on relativistic quantum mechanics and the principle of reciprocity and by BHABHA on the relativistic wave equations for elementary particles. The experimental contributions referred to classical problems. For example, JANOSSI presented results of the investigation into the production of mesons, LEPRINCE RINGUET and WILSON discussed measurements of the mass of mesons, CLAY and WATAGHIN cosmic-rays showers, bursts and showers of penetrating particles, BERNARDINI the meson decay and its secondary electrons. There was also a number of papers on nuclear physics and slow-neutron physics. Among these one should recall a few papers by FERMI and others on the diffraction and reflection of slow neutrons.

Shortly after the conference, however, a number of papers started to appear which opened completely new perspectives. A few examples can be recalled here. For instance, towards the end of 1946 CONVERSI, PANCINI and PICCIONI [77], in the wake of the cosmic-ray work started in Rome by BERNARDINI, found the unexpected result [78] that negative mesons reduced to rest in carbon are captured by the nuclei at a very low rate so that they undergo spontaneous decay. The natural interpretation of this result was that the mesons constituting the hard component of cosmic rays had an interaction with nuclei much weaker than expected at that time. There was, however, a possible different interpretation based on the suspicion that for some reason the time required by a meson to be reduced to rest in carbon was longer than its lifetime. This possibility was pretty soon excluded by FERMI, TELLER and WEISSKOPF [79].

Thus it was definitely established that the « mesons » constituting the hard component of cosmic rays were not the particles hypothesized by YUKAWA, as quanta of the field responsible for nuclear forces. The experiment of CONVERSI *et al.* has been frequently quoted by many authors, like BETHE, SCHWEBER and DE HOFFMANN [80] and ALVAREZ [81], as the one marking the origin of modern high-energy physics.

At the beginning of October 1947, LATTES, OCCHIALINI and POWELL—working in Bristol with the newly developed nuclear research emulsions—observed a new particle that they called « π-meson » [82]. It decayed into a lighter particle correctly identified by the same authors as the weak interacting particle [77, 79] constituting the hard component of cosmic rays, and that they called « μ-meson ».

Already at the beginning of the same years, PERKINS [83] and OCCHIALINI and POWELL [84] had observed in nuclear emulsions a few tracks of slow mesons ending in a « star », *i.e.* a group of divergent tracks due to protons and other charged nuclear fragments, and had interpreted them as due to mesons.

One week after the appearance of the discovery of the $\pi \rightarrow \mu$ decay, the Bristol group published a second paper [85] in which they pointed out that the number of mesons giving rise to the $\pi \rightarrow \mu$ decay was of the same order of magnitude as the number of mesons giving rise to nuclear disintegrations at the end of their track. They identified the star-generating mesons as negative π-mesons, and from this as well as from other considerations they arrived at the conclusion that the π-meson had all the main properties of the Yukawa particle.

Almost at the same time ROCHESTER and BUTLER [86] at the Manchester University observed in a cloud chamber, triggered by a number of counters, two « V events » identified later as decays of a $\theta^0(\equiv K^0)$-meson and a Λ^0-hyperon.

These are only a few outstanding examples. Many more papers could be quoted, the quality of which was a clear proof that the European tradition in the investigation of the structure of matter and its tiniest components was still alive in many places, in spite of the incredible material and moral destruction that had taken place throughout a great part of Europe during the recent war. An important international encounter was the « Symposium on Cosmic Rays » organized in Krakow on October 1947 by the International Union of Pure and Applied Physics [87]. Besides many problems already discussed during conferences in previous years, the evidence for the existence of the π-meson was presented.

A few groups operating cloud chambers at mountain altitudes made many remarkable contributions to the study of the new particles. Among these, three should be mentioned because of the extent and importance of their work. Two groups were led by BLACKETT from Manchester University and later from

Imperial College, when he moved to London in 1953. The first group started to operate in 1950 the same cloud chamber used in 1947 by ROCHESTER and BUTLER [86] at the Laboratoire du Pic du Midi in the French Pyrénées (2867 m) [88]. The other group started to operate in 1951 a much larger cloud chamber [89] at the Hochalpine Forschungs-Station at the Jungfraujoch (3460 m) in Switzerland [90]. The third group was led by LEPRINCE-RINGUET from the École Polytechnique. It started to operate at the Laboratoire du Pic de Midi two large cloud chambers [91] placed one on top of the other in an arrangement that allowed determination of the momentum as well as of the range of the tracks with good accuracy [92, 93].

At the Laboratorio della Testa Grigia (3500 m) [94] near the Teodule Pass and, later, at the Osservatorio della Marmolada (2030 m) [95] in the Italian Alps other investigations with nuclear emulsions, counters, ionization chambers and cloud chambers were carried out [96, 97].

Some of the problems tackled in those years were discussed at the 8th Solvay Conference on Elementary Particles that took place in Bruxelles in September 1948 and in a rather large International Conference on Cosmic Rays that was held in Como about one year later [98].

All the work with cloud chambers and counters at mountain altitudes was rather expensive and required the physicists involved to remain for rather long periods far away from their institutions. From both points of view, the nuclear-emulsion technique—developed mainly at the Bristol University and later also at the Université Libre de Bruxelles—was the essential vehicle providing physicists in many European Universities the possibility of participating in the most important developments of high-energy physics in spite of the poverty of the budgets of the corresponding laboratories and without imposing long absences from their teaching duties. We should be grateful to OCCHIALINI and POWELL, who played a fundamental role in promoting the development of this technique, which during the early fifties allowed an extensive study of the particles produced by collisions of cosmic-ray primaries in the upper atmosphere and, shortly afterwards, the participation of many European groups in interesting research on the new particles produced by high-energy accelerators built in the U.S.A. [99].

This participation was due mainly to the generosity of a few U.S.A. laboratories, among which particularly the Lawrence Radiation Laboratory in Berkeley, and the Brookhaven National Laboratory, L.I., should be mentioned. But without the emulsion technique such participation would not have been possible at all.

The mountain laboratories mentioned above as well as the nuclear-emulsion laboratories of the Universities of Bristol, Bruxelles (where OCCHIALINI was working since 1948) and a few other Western European cultural centres had become in those years points of encounter of young physicists originating from many different countries. The life in common in mountain huts and the co-ordination of the experiments planned by different groups were elements that paved the way to the idea of wider and more ambitious collaborations which—as I will say in a few minutes—were popping out in various places before and around 1950.

3'3. – The exploration of high altitudes by means of balloons to study cosmic radiation had been tried extensively with success in the U.K., particularly by the Bristol Group, around 1950. This group thus arrived at the discovery of the τ- [100] and the charged K-meson [101].

In 1951 it was suggested that a study of high-energy events produced by cosmic radiation could be more conveniently carried out at lower latitudes. At a latitude of 40°, for example, only that part of the primary radiation which has an energy above 7 GeV per nucleon is allowed by the magnetic field of the Earth to enter the atmosphere. Thus, at these latitudes the magnetic field of the Earth removes a large number of primaries which are not very efficient in producing the new secondary particles because of their relatively low energy. The nearer one goes to the equator the stronger this effect is.

Thus a first international expedition was planned in 1952 [102] under the sponsorship of CERN that recently had entered its «Planning Stage» as I will explain below. Naples and Cagliari were chosen as the most convenient of the available bases in the Mediterranean. Thirteen universities took part in the expedition which met with some success. In particular it was the first successful attempt to recover balloons at sea. Furthermore, the expedition made a survey of the winds at high altitudes which was of great importance for the expedition of the subsequent year. The results of this first expedition were briefly reported to the Third Session of the Council of CERN held on 4-7 October 1952 in Amsterdam [103].

A second expedition took place, also to Sardinia, in June-July 1953. This was much on the same lines but on a larger scale than the first one [104]. Eighteen laboratories from European countries in addition to one from Australia took part in it. 25 balloons were launched and over 1000 emulsions were exposed 7 hours at altitudes between 25 and 30 km (Fig. 12). This corresponds to 9.27 litres of nuclear emulsions weighing about 37 kg. They were successfully recovered at sea on account of the employment of a seaplane and a corvette (Pomona) generously placed at disposal of the expedition by the Italian Airforce and the Italian Navy (Fig. 13).

While in the 1952 expedition only glass-backed emulsions had been used, in the 1953 expedition «stripped emulsions» were introduced. Thus, the tracks were followed almost always from one emulsion to the adjacent one, a point of great importance when high-energy events are studied. The development required special care and in particular the construction of special developing systems at the Universities of Bristol, Padua and Rome, where the emulsions of the whole expedition were processed.

In October 1953, a meeting was held at the Department of Physics of the University of Bern (Switzerland) for distributing the packages of exposed and developed emulsion among the participating universities and, in April 1954, an international conference was organized in Padua [105] to discuss the first results and plan jointly the most efficient methods for the investigation.

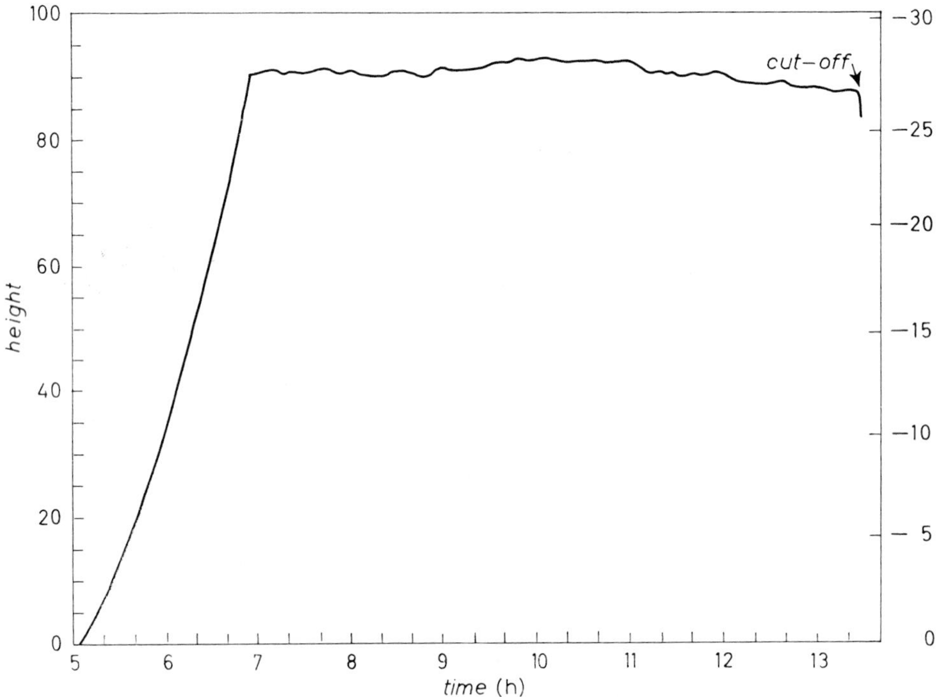

Fig. 12. – Example of balloon record of the altitude as a function of time: flight No. 15.

Further results appeared in the normal scientific literature and in lectures given by various participants at the second Course of the Varenna International Summer School that took place during the Summer 1954 [106]. At the same course, FERMI gave a series of « Lectures on pions and nucleons » which were his last contribution to the teaching of physics before his death in Chicago on November 29 of the same year.

Among the main results of the expedition one can recall a number of determinations of the values of the masses of heavy mesons giving rise to different decays which brought suspicion that all these processes could be alternative decays of the same particle. Also the identification of the various hyperons and the determination of their masses and decay modes were considerably improved [107].

In July of the same year, 1953, a very important conference had been held

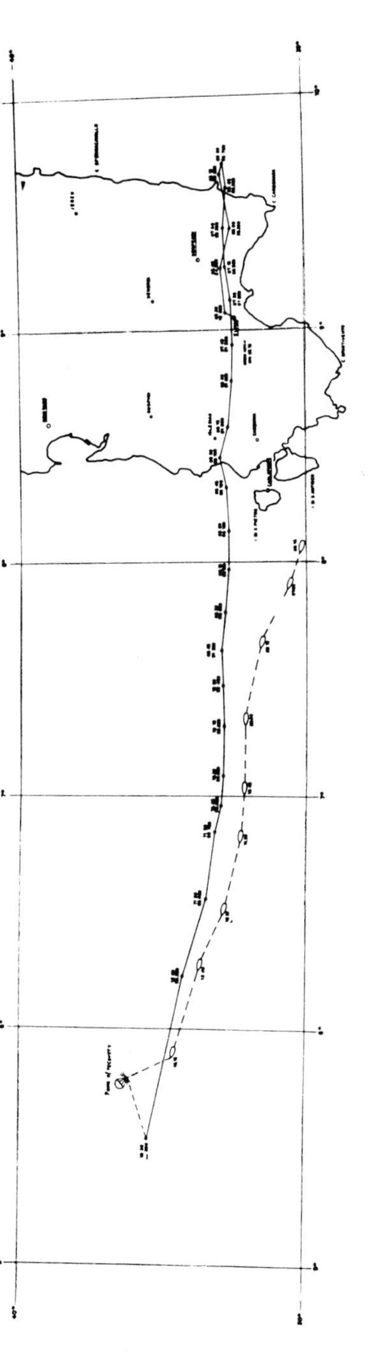

path of the balloon
path of the ship

scale: nautical miles
km

Fig. 13. – Example of « flight map »: flight No. 15.

at Bagnères de Bigorre [108] in the French Pyrénées where the new heavy particles were discussed at length. At this meeting, for example, the Dalitz-Fabri plot for the τ-meson was discussed by DALITZ, MICHEL presented the selection rules for decay processes and the Rome Group introduced the logarithm of $\text{tg}\,\theta$ as a very appropriate random variable for the description of high-energy events in which many secondary particles are emitted.

A third expedition was organized by the Universities of Bristol, Milan and Padua in October 1954 from Northern Italy (Novi Ligure). It consisted in the launching of a single stack of 15 litres of nuclear emulsions (corresponding to a weight of 63 kg) which is indicated in the literature as « G-stack » [109].

In the landing, which took place in the Apennines, the stack was partly damaged. The delicate operation of recovery of the damaged emulsions was made by the Milan group, while its processing, involving an exercise in small-scale chemical engineering, was made at the Universities of Bristol and Padua.

The great advantage in employing a very large stack is evident for studying the different modes of decay of the heavy mesons produced in the collision of a high-energy particle with a nucleus of the emulsions. A substantial fraction of the secondaries come to rest inside the stack so that their modes of decay can be observed and the energy determined with high accuracy from their range.

A great part of the results [110] was presented at the International Conference on Elementary Particles held in Pisa in June 1955 to celebrate the Centenary of the *Nuovo Cimento* [111].

One of the most important results of the G-stack study was the final recognition that the values of the masses of heavy mesons giving rise to different decay processes were identical within rather small experimental errors. Thus the interpretation of all these processes as alternative decays of the same particle was strengthened, contributing in an essential way to the general recognition of the so-called θ-τ puzzle. This found its explanation only in 1956 with the discovery by LEE and YANG [112] of the nonconservation of parity by weak interactions.

A striking fact that emerged in Pisa was that the time for important contributions to subnuclear particle physics from the study of cosmic rays was very close to an end. A few papers presented by physicists from the U.S.A. showed clearly the advantage for the study of these particles presented by the Cosmotron of the Brookhaven National Laboratory (3 GeV) but even more by the Bevatron of the Lawrence Laboratory in Berkeley (6.3 GeV).

In order to complete this outline of the status of elementary-particle research in Europe the creation of International Summer Schools should also be mentioned. The idea was of Mrs. C. MORETTE, who married Dr. B. S. DE WITT shortly before the beginning of the first course of the International Summer School she founded in 1951 at Les Houches [113]. This example was followed by the International Summer School founded in Varenna in 1953 by POLVANI, President of the Società Italiana di Fisica. As mentioned by C. DE WITT in the ad-

Fig. 14. – The preparation of the launching of the G-stack (by courtesy of FRANZINETTI).

dress she gave at the ceremony of opening of the first course at Varenna [93], the program of the Italian School was in some way complementary to that of Les Houches since it regarded mainly the experimental aspects of modern physics. The subjects of the first two courses (1953 and 1954) both regarded elementary particles.

At this point it appears in order to add the following remarks. The nuclear-emulsion technique was used in those years for the study of subnuclear particles in cosmic rays more extensively in Europe than in the U.S.A. where the group of SCHEIN, working at the Chicago University, was, however, one of the most efficient and outstanding.

The cloud chamber technique was employed by an even larger number of groups in the U.S.A. Among these one should recall ANDERSON *et al.* at Pasadena, BRODE, FRETTER *et al.* at Berkeley, THOMPSON at the Indiana University and ROSSI, BRIDGE *et al.* at M.I.T.

Many very important results were obtained by all these groups which are not mentioned at length here only because the lecture refers mainly to the work done in Europe. Furthermore, the collaboration was always very active between the groups working in the two continents. In particular, the collaboration was very tight between the groups of M.I.T. and the École Polytechnique.

3·4. – This very superficial review of the state of development of high-energy physics in Europe in the period 1945-1950, with its extension to later times, is sufficient, I believe, to explain how in different places—in a more or less independent way—the idea of a wide European collaboration started to be considered with the aim of tackling and solving in common those problems that no single European country would have had the technical, organizational and financial strength to face and solve successfully. But how did one arrive at the creation of an organization of the type of CERN and what were the procedures followed and the problems faced in order to reach such a goal?

If one looks at the chronicle of the time one finds that the prehistory of CERN can be divided in three periods [114-116]. The first one includes the first initiatives and extends from about 1948 to 15th February 1952. On that date the representatives of eleven European governments [117] signed in Geneva the so-called *Agreement* establishing a provisional organization with the aim of planning an International Laboratory and organizing other forms of co-operation in nuclear research.

The second period, the so-called *planning stage* extends from February 1952 to 1st July 1953 when the *Convention* establishing the final organization was signed by the representatives of twelve European governments.

Finally, the third period, usually called *interim stage*, runs from July 1953 to 29th September 1954, when a prescribed point of the ratification procedure by the Parliaments of the Member States was reached and the Convention entered into force.

It is impossible to establish the beginning of the first stage since around 1948 the idea of starting an international collaboration among European countries in the field of nuclear physics and elementary particles was making its first appearance, in more or less nebulous forms, in many places. Many scientists were aware of the continually increasing gap between the means available in Europe for research in general and in particular for research in the field of nuclear physics and elementary particles, and the means available in the United States where a few high-energy accelerators had started to produce results, while others had already reached advanced stages of construction or design [99]. It was becoming more and more evident that such a situation would be changed only by a considerable effort made in common by many European nations.

Most of the historical remarks that I will give in the following are based mainly on my diary of that period and on my personal recollections, with the result that the presentation may have some advantage of liveliness but the obvious disadvantage of being onesided.

3'5. – I remember that in the years 1948-1950 the various aspects of the problem including energy and cost of machines were examined in Rome in frequent discussion between FERRETTI and myself and in letters exchanged with BERNARDINI, who in those years was at Columbia University and thus had the opportunity of contributing to stimulating the interest of RABI in this subject.

I remember that I became aware that similar problems were discussed in other European countries, in particular in France, when I heard of the « European Cultural Conference » held in Lausanne in December 1949. At the meeting a message from DE BROGLIE was read by DAUTRY, Administrator of the French Commissariat à l'Energie Atomique. In the message the proposal was made to create in Europe an international research institution without mentioning, however, nuclear physics or fundamental particles [115].

At that time the opinions were still rather divided about the type of research to be tackled by the new organization and the nature of the collaboration to be established.

In June 1950 the General Assembly of UNESCO was held in Florence and RABI, who was a member of the delegation from the U.S.A., made a very important speech about the urgency of creating regional centres and laboratories in order to increase and make more fruitful the international collaboration of scientists in fields where the effort of any one country in the region was insufficient for the task. In the official statement approved unanimously by the General Assembly along the same lines, neither Europe nor high-energy physics were mentioned. But this specific case was clearly intended by many people taking part in the Assembly, in particular by AUGER, who was Director of Natural Sciences of UNESCO, and by RABI, with whom I had discussed the subject at length a few days before.

A further endorsement came from IUPAP which was at that time under the presidency of KRAMERS. I was one of the vice-presidents and I had asked KRAMERS, at the beginning of the Summer 1950, to include the discussion of Rabi's proposal, with specific reference to Europe and high-energy physics, in the agenda of the meeting of the Executive Committee of IUPAP that was to take place at the beginning of September of the same year in Cambridge, Mass. Although KRAMERS could not preside at the meeting because of bad health, the problem was discussed at length under the chairmanship of Sir DARWIN [118] assisted by the Secretary General FLEURY. As a conclusion I was asked to get in contact with RABI and with physicists from various European countries in order to clarify the aims and structure of the new organization and to help in the co-ordination of the different efforts. My first step was to write to AUGER [119] who in the meantime had presented the problem to the conference on nuclear physics held in Oxford during the month of September, where he had found enthusiastic support from many parts, in particular from the young physicists.

AUGER had now the authority to act but there was no money appropriated on the scale required for a detailed expert study of such a project.

In December 1950, however, the European Cultural Centre (which was founded at the already mentioned Lausanne meeting of 1949) called at Geneva a commission for scientific co-operation. AUGER, KRAMERS, FERRETTI, PREISWERK and RANDERS were present. As a result of the meeting funds were made available, immediately by the President of the Italian Council of Research Prof. COLONNETTI, and soon afterwards by the French and Belgian Governments. The total sum collected by AUGER was very modest, about $ 10 000; it was, however, sufficient to initiate the first steps for arriving at the planning and construction of a large particle accelerator.

At the beginning of 1951 AUGER established a small office at UNESCO and invited me to Paris at the end of April to discuss the constitution of a working group of European physicists interested in the problem.

The first meeting of this « Board of Consultants » was held at UNESCO, in Paris, at the end of May 1951. Two goals were immediately established: a long-range, very ambitious, project of an accelerator second to none in the world, and in addition the construction of a less powerful and more standard machine which could allow at an earlier date experimentation in high-energy physics by European teams.

Many other features of the new organization were also examined, as preparatory work for a *conference of delegates of governments* that was convened by UNESCO in Paris, in December 1951. This conference, which took place under the chairmanship of DE ROSE, a French diplomat who a few years later was elected President of CERN, led to the signing of the Agreement, which took place, as I said before, in Geneva in February 1952.

The first problem tackled by the Council of the provisional organization created by the agreement was the nomination of the officers responsible for the appointment of the remainder of the staff and for planning the laboratory.

BAKKER from the Netherlands was nominated Director of the Synchro-Cyclotron Group, DAHL from Norway (with GOWARD from the U.K. as deputy) Director of the Proton-Synchrotron Group, KOWARSKI from France (with PREISWERK from Switzerland as deputy) Director of the Laboratory Group, which had to take care of site, buildings, workshops, administrative forms, financial rules, etc., BOHR Director of the Theoretical Group and finally myself as Secretary General with the task of maintaining cohesion between the four groups of which the provisional organization was composed.

Almost all these people had worked on the Board of Experts, nominated by UNESCO, during the first stage and contributed later to the creation of the new organization and to the development of its activities.

In July of the same year, 1952, an international nuclear-physics conference was held in Copenhagen; on that occasion the type of accelerator to be built as the main goal of the new European organization was amply discussed. A report of the conclusions reached by the participants was presented by HEISENBERG to the Council which held its Second Meeting in Copenhagen immediately after the Conference. Thus the decision was taken that the Proton-Synchrotron Group should explore the possibility of constructing a 10 GeV proton-synchrotron which, at that time, represented the biggest machine in the world.

During the month of August DAHL and GOWARD went to Brookhaven in order to study in detail the *Cosmotron* (with a maximum energy of about 3 GeV) that was very close to completion [99]. During their two-week visit, and in some way in connection with the discussions going on in relation to the European project, COURANT, LIVINGSTONE and SNYDER came out with the « strong-focusing principle » [120].

This important discovery came soon enough to allow a change of the plans of the provisional organization: with the approval of the Council, the PS Group embarked on the study of a strong-focusing PS of $(20 \div 30)$ GeV instead of the weak-focusing 10 GeV machine considered until then.

During the summer of 1952 four sites were offered for the construction of the new laboratory: one near Copenhagen, one near Paris, one in Arnhem in the Netherlands and one in Geneva.

After long and lively discussions the site in Meyrin near Geneva was unanimously selected [103].

Another point I would like to touch about those times concerns the participation of the European nations.

All the European members of UNESCO had been invited to the Conference opened in Paris in December 1951 and closed in Geneva on 15th February 1952, but no response came from the countries of Eastern Europe. Furthermore, you have probably noticed that while the agreement was signed at that date by the representatives of eleven countries, the Convention establishing the permanent organization was signed by twelve countries.

The difference was due to the U.K. which, at the beginning, was rather cautious in committing itself to take part in the new organization. The U.K. government preferred to remain in the formal position of an observer during the first two stages, while, by signing the Convention, it became a full-right member of the permanent organization.

I remember that in the Autumn of 1952 it was decided that BAKKER, DAHL and myself should go to Brookhaven to take part in the dedication of the cosmotron that was foreseen for December 15th.

Our trip was already arranged when I was called on the telephone from London by Sir John COCKCROFT who had been from the start very much in favour of the participation of the U.K. in the new venture.

As a consequence of our conversation on the telephone I decided to leave earlier and to pass through London on my way to Brookhaven.

In London I went to the D.S.I.R. where I met its chairman Sir Ben LOCKSPEISER, who a few years later was elected president of the permanent CERN.

After rather long discussions about various organizational and financial aspects of the project as a whole, Sir John brought me to Lord CHERWELL [121] who, at that time, was a member of the Churchill government.

Lord CHERWELL appeared to be very clearly against the participation of the U.K. in the new organization. As soon as I was introduced in his office he said that the European laboratory was to be one more of the many international bodies consuming money and producing a lot of papers of no practical use. I was annoyed and answered rather sharply that it was a great pity that the U.K. was not ready to join such a venture which, without any doubt, was destined to full success, and I went on by explaining the reasons for my convictions. Lord CHERWELL concluded the meeting by saying that the problem had to be reconsidered by His Majesty's Government.

When we left the Ministry of Defence, where the meeting had taken place, I was rather unhappy about my lack of self-control, but Sir John and Sir Ben were rather satisfied and tried to cheer me up.

A few weeks later Sir Ben LOCKSPEISER wrote an official letter asking that the status of observer be given to the U.K. in the provisional organization, and the D.S.I.R. started to regularly pay «gifts», as they were called, corresponding exactly to the U.K. share calculated according to the scale adopted by the other eleven countries.

In spite of its very particular legal position, the U.K. gave in practice fundamental support to the provisional organization and was the first among the European countries to ratify the Convention.

3·6. – The four groups at the beginning had started to work at the institutions of the corresponding directors. In October 1953 the PS staff was assembled in Geneva, partly at the Institute of Physics of the University and partly in temporary huts built in its vicinity. At that time the transition took place from mainly theoretical work to experimentation and technical designing. In the same month of October an International Conference on protonsynchrotrons was organized at the Institute of Physics by the PS staff. In March 1954 GOWARD died after a short and tragic illness. He was succeeded by ADAMS who became Director of the PS Division in 1955 when DAHL went back to Norway to direct the design and construction of the first Norwegian nuclear research reactor.

During October 1953 an administrative nucleus began to function in a temporary Geneva office and, from January 1954 on, in the Villa Cointrin at the Geneva airport.

First instrumentation workshops, then a library and a few laboratories were gradually set up also at the airport in the Summer 1954. On August 13 of that same year the bulldozers started to break the ground at Meyrin and the construction of the European laboratory began.

The SC group remained centered in Amsterdam and the theoretical study group in Copenhagen, where the training of young theoreticians recruited as CERN Fellows went on. The training of experimentalists went on at Uppsala [122], Liverpool [99] and the Jungfraujoch. Through an agreement between CERN and Imperial College the cloud chamber at this high-altitude station was operated jointly by the two institutions starting from 1955 [123].

At the end of the interim period (September 1954) the total staff of CERN, not counting the holders of fellowships and the consultants on a part-time fraction, amounted to 120, of which about one half belonged to the PS group. The total floor area at disposal of the staff was some 2500 square metres.

On September 29, 1954, when a prescribed point in the ratification procedure was reached all the assets of provisional CERN became suddenly masterless. For eight days I had the honour—as Secretary General—of sole responsibility of ownership on behalf of a new-born permanent organization. Then the first meeting of the permanent Council assembled on the 7th October in Geneva, the Secretary General presented his final report, BLOCH was nominated Director General and CERN entered its final permanent form. The SC was operated for the first time in 1958 and the proton beam circulated in the PS on November 29, 1959, *i.e.* almost one year before the time schedule.

REFERENCES

[1] E. AMALDI: *The production and slowing-down of neutrons*, in *Handbuch der Physik*, Vol. **38**/2, edited by S. FLÜGGE (Berlin, 1959), p. 1.

[2] E. FERMI: *Collected Papers*, Vol. **1**, Italy 1921-1938, Accademia Nazionale dei Lincei and the University of Chicago Press (1962), edited by E. AMALDI, H. L. ANDERSON, E. PERSICO, C. S. SMITH, A. WATTENBERG and E. SEGRÈ.

[3] E. Segrè: *Enrico Fermi, Physicist* (Chicago, 1970).

[4] L. Fermi: *Atoms in the Family* (Chicago, 1954).

[5] I. Curie and F. Joliot: *Compt. Rend.*, **198**, 254 (1934); *Nature*, **133**, 201 (1934).

[6] I. Curie and F. Joliot: *Compt. Rend.*, **198**, 559 (1934); *Journ. Phys. Rad.*, **5**, 153 (1934).

[7] Allocution de M. F. Perrin, presenté à la *Séance Commémorative du XXX Anniversaire de la Découverte de la Radioactivité Artificielle, Grand Amphitéâtre de la Sorbonne, Vendredi 3 Juillet 1964*, published at page 21 of Vol. 1 of the proceedings of the *Congrès International de Physique Nucléaire, Paris, 2-8 Juillet 1964*.

[8] The Joliot-Curies proved, for example, that the radioactive nuclide produced by bombarding aluminium with α-particles has the chemical properties of phosphorous. Therefore the reaction produced necessarily involves the absorption of the incident α without emission of charged particles. Furthermore aluminium consists of a single isotope, $^{27}_{13}\mathrm{Al}$; therefore one can conclude that the observed transformation is probably due to an (α, n) reaction:

a) $$^{27}_{13}\mathrm{Al}(\alpha, \mathrm{n})^{30}_{15}\mathrm{P}\,.$$

Since the nuclide $^{30}_{15}\mathrm{P}$ does not exist in Nature, it is very reasonable to consider that it is unstable and, therefore, decays with the emission of positrons

b) $$^{27}_{13}\mathrm{P} \to {}^{30}_{14}\mathrm{Si} + \mathrm{e}^{+}\,.$$

$^{30}\mathrm{Si}$ is stable and was already known to be the final product of the reaction

c) $$^{27}_{13}\mathrm{Al}(\alpha, \mathrm{p})^{30}_{41}\mathrm{Si}\,.$$

[9] I. Curie and F. Joliot: *Journ. Phys. Rad.*, **4**, 494 (1933).

[10] Sir E. Rutherford, J. Chadwick and C. D. Ellis: *Radiations from Radioactive Substances* (Cambridge, 1930).

[11] F. Rasetti: *Il nucleo atomico* (Bologna, 1936); *Elements of Nuclear Physics* (New York, 1936).

[12] E. Fermi and F. Rasetti: *Ric. Scient.*, **4** (2), 299 (1933).

[13] E. Fermi: *Nature*, **125**, 16 (1930), and p. 328 of ref. [2]; *Zeits. Phys.*, **60**, 320 (1930) and p. 336 of ref. [2]; E. Fermi and E. Segrè: *Zeits. Phys.*, **82**, 11 (1932) and p. 514 of ref. [2].

[14] For more detail on this point see the second Section.

[15] E. Fermi: *Nuovo Cimento*, **11**, 1 (1934); *Zeits. Phys.*, **88**, 161 (1934).

[16] E. Fermi: *Ric. Scient.*, **5** (1), 283 (1934) and p. 645 of ref. [2].

[17] I. Noddack: *Ang. Chem.*, **47**, 653 (1934).

[18] O. Hahn and L. Meitner: *Naturwiss.*, **24**, 158 (1936); *Ber. Dtsch. Chem. Ges.*, **69**, 905 (1936); L. Meitner: *Kernphysik*, edited by E. Bretscher (Berlin, 1936), p. 24.

[19] E. McMillan and P. H. Abelson: *Phys. Rev.*, **57**, 1185 (1940).

[20] M. Mayer: *Phys. Rev.*, **60**, 84 (1941). The same problem has been discussed by Y. Suguira and H. Urey (*Kgl. Danske Vid. Selsk. Mat.-fys. Medd.*, **7**, 3 (1926)) on the basis of the old quantum theory.

[21] E. Fermi: *Zeits. Phys.*, **48**, 73 (1928).

[22] The letter of Rutherford shown in Fig. 5 reads as follows:

June 20th, 1934

Dear Professor Fermi,

Your letter has been forwarded to me in the country where I am taking a holiday. I saw your account in « Nature » of the effects on uranium. I congratulate you and your colleagues on a splendid piece of work. I have had two of my men WESTCOTT and BJERGE repeating some of your experiments with an Em+Be tube and promised them when I returned in a week or two to try out for them the effect of the 2 million volt neutrons from the D+D reaction on a few elements. I myself have been naturally interested in the energy of the neutron required to start transmutations. We do not, however, propose at the moment to make systematic experiments in this direction as our installation is required for other purposes. I cannot at the moment give you a definite statement as to the output of neutrons from our tube but it should be of the same order as from an Em+Be tube containing 100 millicurie and may be pushed much higher.

If your assistants come to Cambridge say in the first week of July, I shall be delighted to give them the benefit of our experience and to see the mode of operation of our installations for transmutations in general.

I hope one or both speak English as the knowledge of Italian in the laboratory is very modest. The two men to see are Dr. OLIPHANT and Dr. COCKCROFT. Excuse the hand written letter, I do not take a secretary with me in holiday!

Yours sincerely,
RUTHERFORD

[23] E. FERMI, E. AMALDI, O. D'AGOSTINO, F. RASETTI and E. SEGRÈ: *Proc. Roy. Soc.*, A **146**, 486 (1934).

[24] J. CHADWICK and M. GOLDHABER: *Nature*, **134**, 237 (1934).

[25] W. BOTHE and H. BECKER: *Zeits. Phys.*, **66**, 289 (1930). See also the report by BOTHE on page 153 of the *Proceedings of the International Conference*, held in Rome in 1931 as well as W. BOTHE and H. BECKER: *Naturwiss.*, **20**, 349 (1932); *Zeits. Phys.*, **76**, 421 (1932).

[26] F. RASETTI: *Naturwiss.*, **20**, 252 (1932).

[27] I. CURIE and F. JOLIOT: *Compt. Rend.*, **193**, 1412 (1932); F. JOLIOT: *Compt. Rend.*, **193**, 1415 (1932).

[28] According to this formula the cross-section is inversely proportional to the square of the mass of the target particle.

[29] I. CURIE and F. JOLIOT: *Compt. Rend.*, **194**, 708 (1932).

[30] E. AMALDI: *La vita e l'opera di Ettore Majorana*, Accademia Nazionale dei Lincei (1966): translated into English as: *Ettore Majorana, man and scientist*, in *Strong and Weak Interactions. Present Problems*, edited by A. ZICHICHI (New York, 1966).

[31] J. CHADWICK: *Nature*, **129**, 322 (1932); see also by the same author: *Proc. Roy. Soc.*, A **136**, 692 (1932).

[32] Sir E. RUTHERFORD: *Phil. Mag.*, **4**, 580 (1927).

[33] J. CHADWICK: *Actes du X Congrès International d'Histoire des Sciences* (Ithaca, N. Y., 1962), (Paris, 1964), Vol. **1**, p. 159.

[34] G. GENTILE: *Rend. Acad. Lincei*, **7**, 346 (1928).

[35] F. PERRIN: *Compt. Rend.*, **195**, 236 (1932); see also W. HEISENBERG: *Rapports et Discussions du VII Congrès de l'Institut International de Physique Solvay* (1934), p. 289.

[36] D. IWANENKO: *Nature*, **129**, 798 (1932).

[37] More detail about the refusal by MAJORANA to publish his results can be found in ref. [30, 2, 3].

[38] E. MAJORANA: *Zeits. Phys.*, **82**, 137 (1933); *Ric. Scient.*, **4** (1), 559 (1933).

[39] W. HEISENBERG: *Zeits. Phys.*, **77**, 1 (1932); **80**, 156, 587 (1932).

[40] T. BJERGE and C. H. WESTCOTT: *Nature*, **134**, 286 (1943).

[41] In this case the proof is very strong, because aluminium is formed by the single

isotope ^{27}Al and $^{22}_{11}Na$ was already known to emit positrons instead of electrons, as would be expected for nuclides having an excess of neutrons with respect to the corresponding stable isotopes. In the case of the (n, 2n) processes one would expect, at least in general, an emission of positrons.

[42] This as well as most notebooks of that period have been collected by Mr. Lodovico ZANCHI and me and later have been deposited at the Domus Galilaeana in Pisa: E. AMALDI: *The Fermi Manuscripts at the Domus Galilaeana*: *Physics*, **1** (2), 69 (1959); T. DERENZINI: *Analisi dei manoscritti di Enrico Fermi alla Domus Galilaeana*, *Physics*, **6** (1), 75 (1954).

[43] In the introductory remarks to Fermi's paper on *The Magnetic Fields in Spiral Arms* (*Fermi's Collected Papers*, Vol. **2**, p. 927) CHANDRASEKHAR wrote that Fermi told him once more or less what follows:

« I will tell you how I came to make the discovery which I suppose is the most important one I have made. We were working very hard on the neutron-induced radioactivity and the results we were obtaining made no sense. One day, as I came to the laboratory, it occurred to me that I should examine the effect of placing a piece of lead before the incident neutrons. Instead of my usual custom, I took great pains to have the piece of lead precisely machined. I was clearly dissatisfied with something: I tried every excuse to postpone putting the piece of lead in its place. I said to myself: "No, I do not want this piece of lead here; what I want is a piece of paraffin". It was just like that with no advance warning, no conscious prior reasoning. I immediately took some odd piece of paraffin and placed it where the piece of lead was to have been ».

In a short note dedicated to Norman FEATHER—to mark his twenty-five-year tenure of the chair of Natural Philosophy at the University of Edimburgh—Maurice GOLDHABER has circulated at the end of 1971 a few *Remarks on the Prehistory of the Discovery of Slow Neutrons*. Having read in Segrè's book (ref. [3]) the conversation of FERMI with CHANDRASEKHAR reported above, GOLDHABER apparently had the impression that the idea of using paraffin instead of lead could suddenly have come in the mind of FERMI as a consequence of what had been written by CHADWICK and GOLDHABER at the end of the paper on the photodisintegration of the deuteron (ref. [24]). What actually CHADWICK and GOLDHABER state is that the value obtained for the cross-section of the inverse process—*i.e.* the radiative capture of fast neutrons by protons to form deuterons—calculated by applying the detailed balance to their experimental results was too small to provide an explanation for the gamma-ray emission by hydrogeneous substances observed by D. E. LEA. The idea that neutrons could lose energy through a number of collisions is not mentioned although it is possible that CHADWICK and GOLDHABER had thought about it.

In order to logically arrive at recognizing the high efficiency of slow neutrons starting from the photodisintegration of the deuteron, a rather long chain of assumptions of unknown facts was necessary in 1934:

a) The extrapolation of the cross-section of photoeffect to very low energy gives a negligible value, as stated by GOLDHABER in his *Remarks*. Only the photomagnetic effect is important at very low energy and this was introduced only in 1936 by FERMI who proved that the corresponding cross-section follows the $1/v$ law and reaches the value of ~ 0.3 barns at energies around $kT \simeq 0.025$ eV.

b) The slowing-down is determined by the *elastic cross-section* σ_e which is not related in a simple way to that of the previous process. At energies around

1 eV σ_e is about 20 barn, as it was shown by us in Rome—as well as by other authors—in the course of 1935-1936.

c) From 1 eV to thermal energy the elastic cross-section increases by another large factor due to chemical bond of hydrogen in molecules. This effect was found experimentally by FERMI and myself only in 1936 and explained correctly by FERMI the same year.

In conclusion it appears inconceivable that FERMI arrived at the idea of the slowing-down through this complicated reasoning. On the contrary it seems to me that the fact that in the mind of FERMI came suddenly the idea of using paraffin instead of lead is remarkable but not so extraordinary if one considers the many experiments made by our group on irradiation under various conditions. These were discussed every day with all the group: at the beginning RASETTI tried to explain our strange results as due to some mistake made by PONTECORVO and me. We protested and proved we were not wrong and this gave rise to long and vivacious discussions in which FERMI took always an active part.

[44] E. AMALDI, O. D'AGOSTINO and E. SEGRÈ: *Ric. Scient.*, **5** (2), 381 (1934).

[45] E. FERMI, E. AMALDI, B. PONTECORVO, F. RASETTI and E. SEGRÈ: *Ric. Scient.*, **5** (2), 282 (1934) and p. 757 of ref. [2].

[46] O. D'AGOSTINO, E. AMALDI, F. RASETTI and E. SEGRÈ: *Ric. Scient.*, **5** (2), 467 (1934) and p. 655 of ref. [2].

[47] P. B. MOON and J. R. TILLMAN: *Nature*, **135**, 904 (1935); *Proc. Roy. Soc.*, A **153**, 476 (1936). The indication of a small effect was found independently by J. R. DUNNING, G. B. PEGRAM, D. P. MITCHELL and G. A. FINK: *Phys. Rev.*, **47**, 888 (1935).

[48] E. FERMI, B. PONTECORVO and F. RASETTI: *Ric. Scient.*, **5** (2), 380 (1934) and p. 759 of ref. [2].

[49] E. AMALDI, O. D'AGOSTINO, E. FERMI, B. PONTECORVO, F. RASETTI and E. SEGRÈ: *Proc. Roy. Soc.*, A **149**, 522 (1935) and p. 765 of ref. [2].

[50] E. FERMI and F. RASETTI: *Nuovo Cimento*, **12**, 201 (1935) and p. 795 of ref. [2]; B. PONTECORVO: *Nuovo Cimento*, **12**, 211 (1935); E. AMALDI: *Nuovo Cimento*, **12**, 233 (1935); E. SEGRÈ: *Nuovo Cimento*, **12**, 232 (1935).

[51] J. R. DUNNING, G. B. PEGRAM, G. A. FINK and D. P. MITCHELL: *Phys. Rev.*, **48**, 265 (1935).

[52] E. FERMI: *Nuovo Cimento*, **2**, 157 (1934) and p. 706 of ref. [2].

[53] E. AMALDI and E. SEGRÈ: *Nature*, **133**, 141 (1934); *Nuovo Cimento*, **11**, 145 (1934).

[54] E. AMALDI, O. D'AGOSTINO, E. FERMI, B. PONTECORVO and E. SEGRÈ: *Ric. Scient.*, **6** (1), 581 (1935) and p. 669 of ref. [2].

[55] E. AMALDI: *Phys. Zeits.*, **38**, 692 (1937).

[56] O. FRISCH and E. T. SØRENSEN: *Nature*, **136**, 258 (1935).

[57] J. DUNNING, G. B. PEGRAM, G. A. FINK, D. P. MITCHELL and E. SEGRÈ: *Phys. Rev.*, **48**, 704 (1935).

[58] B. PONTECORVO and G. C. WICK: *Ric. Scient.*, **7** (1), 134, 220 (1936).

[59] T. BJERGE and C. H. WESTCOTT: *Proc. Roy. Soc.*, A **150**, 709 (1935).

[60] P. B. MOON and J. R. TILLMANN: *Nature*, **135**, 904 (1935); **136**, 106 (1935). See also L. ARTSIMOVICH, I. KOURTCHATOV, G. LATYSCHEV and W. CROMOW: *Zeits. Sovjet.*, **8**, 472 (1935); B. PONTECORVO: *Ric. Scient.*, **6** (2), 145 (1935); D. P. MITCHELL, J. R. DUNNING, E. SEGRÈ and G. B. PEGRAM: *Phys. Rev.*, **48**, 175 (1935).

[61] E. FERMI and E. AMALDI: *Ric. Scient.*, **6** (2), 344 (1935) and p. 808 of ref. [2].

[62] L. SZILARD: *Nature*, **136**, 951 (1935). Some time later a similar behaviour was

found also for Au by O. R. FRISH, G. HEVESY and H. A. C. MC KAY: *Nature*, **137**, 149 (1936).

[63] E. AMALDI and E. FERMI: *Ric. Scient.*, **6** (2), 443 (1935) and p. 816 of ref. [2].

[64] The expression « albedo » is used by astronomers for the fraction of incident light diffusely reflected from the surface of a planet or satellite. The most common use refers to the Moon. It is, however, also used for other surfaces, for instance, for snow.

[65] E. AMALDI and E. FERMI: *Ric. Scient.*, **7** (1), 56 (1936).

[66] The first indication was provided by the fact that a velocity distribution roughly Maxwellian was observed (ref. [55]) using a velocity selector in which the « selection » was made by means of Cd absorbers. This argument was confirmed and strengthened by the observation of P. PREISWERK and H. VON HALBAN (*Nature*, **136**, 951 (1936)) in the case of Ag and of F. RASETTI and G. A. FINK (*Phys. Rev.*, **49**, 642 (1936)) in the case of Ag, Rh, In and I, that while all these elements show a thermal effect of the type of that investigated by MOON and TILLMAN (ref. [47]), no effect at all is observed if the same elements are screened with Cd. This means that the absorption lines of these elements are above the thermal region, which, on the contrary, is practically all included in the region absorbed by Cd.

[67] E. AMALDI and E. FERMI: *Ric. Scient.*, **7** (1), 454 (1936) and p. 841 of ref. [2]; *Ric. Scient.*, *Phys. Rev.*, **50**, 899 (1936) and p. 892 of ref. [2].

[68] H. VON HALBAN and P. PREISWERK: *Nature*, **136**, 951, 1027 (1935); *Compt. Rend.*, **202**, 840 (1936).

[69] E. AMALDI and E. FERMI: *Ric. Scient.*, **7** (1), 223 (1936) and p. 828 of ref. [2]; **7** (1) 393 (1936) and p. 837 of ref. [2].

[70] G. BREIT and E. WIGNER: *Phys. Rev.*, **49**, 519 (1936).

[71] N. BOHR: *Nature*, **137**, 344 (1936); *Naturwiss.*, **24**, 241 (1936).

[72] E. FERMI: *Ric. Scient.*, **7** (2), 13 (1936) and p. 943 of ref. [2]. The English translation by G. N. TEMMER of this fundamental paper is given at p. 990 of ref. [2].

[73] E. AMALDI and E. FERMI: *Ric. Scient.*, **7** (1), 310 (1936).

[74] O. R. FRISH and G. PLACZEK: *Nature*, **137**, 357 (1936). The same arguments were developed independently by D. F. WEEKS, M. S. LIVINGSTON and H. A. BETHE: *Phys. Rev.*, **49**, 471 (1936).

[75] This expression comes from Aldous Huxley's novel « Brave New World » and refers to a pill with sexual hormones base used by men in the year 2000 to combat spleen.

[76] Report on an *International Conference on Fundamental Particles and Low Temperature*, held at the Cavendish Laboratory, Cambridge on 22-27 July 1946, Vol. **1**, *Fundamental Particles* (London, 1947).

[77] M. CONVERSI, E. PANCINI and O. PICCIONI: *Nuovo Cimento*, **3**, 372 (1945); *Phys. Rev.*, **71**, 209 (1947).

[78] S. TOMONAGA and G. ARAKI: *Phys. Rev.*, **58**, 90 (1940). These authors had foreseen that only positive mesons at rest should undergo spontaneous decay, whereas the negative ones should be captured in a time far shorter than the meson lifetime.

[79] E. FERMI, E. TELLER and V. WEISSKOPF: *Phys. Rev.*, **71**, 314 (1947); E. FERMI and E. TELLER: *Phys. Rev.*, **72**, 399 (1947).

[80] S. S. SCHWEBER, H. A. BETHE and F. DE HOFFMANN: *Mesons and Fields* (Evanston, Ill., 1956).

[81] L. ALVAREZ: *Recent Developments in Particle Physics* of « Les Prix Nobel of 1968 » (Stockholm, 1969), p. 125.

[82] C. M. C. LATTES, G. P. OCCHIALINI and C. F. POWELL: *Nature*, **160**, 453 (1947).

[83] D. H. PERKINS: *Nature*, **159**, 126 (1947).

[84] G. P. OCCHIALINI and C. F. POWELL: *Nature*, **159**, 186 (1947).

[85] C. M. C. LATTES, G. P. OCCHIALINI and C. F. POWELL: *Nature*, **160**, 486 (1947).

[86] C. D. ROCHESTER and C. C. BUTLER: *Nature*, **160**, 855 (1947). The chamber, of rather modest dimensions (28 cm in diameter and about 7 cm deep), was placed in a rather intensive magnetic field (7500 G). It was crossed by a lead plate about five cm thick. Already in 1944 L. LEPRINCE RINGUET and M. LHÉRITIER (*Compt. Rend.*, **219**, 618 (1944)) had otained at the École Polytechnique in Paris a mass value very close to that established a few years later for the K-meson, from the application of the principles of conservation of energy and momentum to a cloud chamber picture of a collision of a positive particle with an electron. The result was, however, affected by such a large experimental error that it could not be considered as a proof of the existence of new particles of mass about three times greater than that of the π-meson.

[87] *Symposium on Cosmic Rays*, Krakow, October 1947, IUPAP Document RC 48-1.

[88] R. ARMENTEROS, K. H. BARKER, C. C. BUTLER, A. CACHON and A. H. CHAPMAN (*Decay of V-particles*, *Nature*, **167**, 501 (1951)) identified protons in the decay of the V^0; R. ARMENTEROS, K. H. BARKER, C. C. BUTLER and A. CACHON (*The properties of neutral V-particles*, *Phil. Mag.*, **42**, 1113 (1951)) separated the $V_0^1(\equiv \Lambda^0)$ from the $V_0^2(\equiv K^0 \to \pi^+ + \pi^-)$; R. ARMENTEROS, K. H. BARKER C. C. BUTLER, A. CACHON and C. M. YORK (*The properties of charged V-particles*, *Phil. Mag.*, **43**, 597 (1952)) observed the first Ξ.

[89] The dimensions of the chamber were $(55\times 55)\ \text{cm}^2\times 16$ cm. It was placed in a 5000 G magnetic field.

[90] J. P. ASTBURY, J. S. BUCHANAN, P. CHIPPENDALE, D. D. MILLAR, J. A. NEWTH, D. I. PAGE, A. RITZ and A. B. SAHIAR: *The mean life of charged V-particles*, *Phil. Mag.*, **44**, 242 (1953).

[91] The dimensions of the two chambers were $(68\times 64)\ \text{cm}^2\times 30$ cm. The upper chamber had a 3600 G magnetic field, the lower one contained 15 copper plates 1 cm thick.

[92] B. GREGORY, A. LAGARRIGUE, L. LEPRINCE RINGUET, F. MULLER and CH. PEYROU: *Nuovo Cimento*, **11**, 292 (1954). This paper contains the suggestion of the $K_{\mu 2}$ decay mode and mass measurements by momentum and range. R. ARMENTEROS, B. GREGORY, A. HENDEL, A. LAGARRIGUE, L. LEPRINCE RINGUET, F. MULLER and CH. PEYROU: *Nuovo Cimento*, **1**, 915 (1955). It contains the final proof of the $K_{\mu 2}$ decay mode and the recognition that the K_μ-particle is essentially positive.

[93] A review of the cloud chambers operated in the various laboratories with a discussion of the possibilities and limits of this technique is given in P. M. S. BLACKETT: *Lectures* (p. 264) *to the First International Summer School in Varenna*: *Rendiconti del Corso tenuto nella Villa Monastero a Varenna 19 Agosto-12 Settembre 1953*, Vol XI (Serie IX), *Suppl. Nuovo Cimento*, **1**, 141 (1953).

[94] G. BERNARDINI, C. LONGO and E. PANCINI: *Ric. Scient.*, **18**, 91 (1948); G. BERNARDINI and E. PANCINI: *Ric. Scient.*, **20**, 966 (1950).

[95] A. ROSTAGNI: *L'Energia Elettrica*, **28**, 211 (1951). A better laboratory was rebuild in 1953. For more detail on all these laboratories see: *The World's High Altitude Research Stations*, Research Division, College of Engineering, New York University.

[96] The following researches were made at the Laboratorio della Testa Grigia-Nuclear disintegrations produced by cosmic rays were studied by G. BERNAR:

DINI *et al.* with nuclear emulsions (*Phys. Rev.*, **74**, 845, 1878 (1948); **76**, 1792 (1949); **79**, 952 (1950); *Nature*, **163**, 981 (1949)), by I. F. QUERCIA *et al.* (*Nuovo Cimento*, **7**, 457 (1950)) and E. AMALDI *et al.* (*Nuovo Cimento*, **7**, 697 (1950)) with ionization chambers. Extensive showers by E. AMALDI *et al.* (*Nuovo Cimento*, **7**, 401, 816 (1950)) and C. BALLARIO *et al.* (*Phys. Rev.*, **83**, 666 (1951)); the electromagnetic component produced in nuclear explosions by G. SALVINI *et al.* with a cloud chamber (*Nuovo Cimento*, **6**, 207 (1949); **7**, 36, 943 (1950); *Phys. Rev.*, **77**, 284 (1950)); penetrating showers from hydrogen and other elements by M. CONVERSI, G. FIDECARO *et al.* (*Nuovo Cimento*, **1**, 330 (1955)); the determination of the mean lifetime of the new strange particles by means of a cloud chamber by P. ASTBURRY, C. BALLARIO, R. BIZZARRI, A. MICHELINI, E. ZAVATTINI and A. ZICHICHI *et al.* (*Nuovo Cimento*, **2**, 365 (1955); **6**, 994 (1957)).

[97] The following researches were made at the Laboratorio della Marmolada: P. BASSI *et al.* studied with counter techniques the positive excess of mesons (*Nuovo Cimento*, **8**, 469 (1957)), with ionization chambers the zenith distribution of the nucleonic component (*Nuovo Cimento*, **9**, 722 (1952)) and of the particles of extensive showers (*Nuovo Cimento*, **10**, 779 (1953)). The last problem was investigated also by M. CRESTI *et al.* (*Nuovo Cimento*, **10**, 779 (1953)). M. DEUTSCHMANN, M. CRESTI *et al.* investigated by means of a cloud chamber the decay of V^0 (*Nuovo Cimento*, **3**, 180, 566 (1956); **4**, 747 (1956)). H. MASSEY, E. H. S. BURHOP *et al.* the production of secondary particles in hydrogen by means of a high-pressure (~ 100 atmospheres) cloud chamber (*Nature*, **175**, 445 (1955); *Suppl. Nuovo Cimento*, **4**, 272 (1956)).

[98] *Congresso Internazionale di Fisica sui Raggi Cosmici, Como 11-16 Settembre 1949*; *Suppl. Nuovo Cimento*, **6**, 309 (1949).

[99] The accelerators capable of producing at least π-mesons that entered into operation before 1956 are the following:

Institution	Type of machine	Energy (MeV)	Date first operated
In the U.S.A.			
Univ. of Calif. Berkeley	synchrocyclotron	350	1946
Univ. Rochester	synchrocyclotron	240	1948
Columbia Univ.	synchrocyclotron	400	1950
Univ. of Chicago	synchrocyclotron	460	1951
Carnegie Inst. Tech.	synchrocyclotron	450	1952
Brookhaven Nat. Laboratory (Cosmotron)	proton synchrotron	3200	1952
Univ. of Calif. Berkeley (Bevatron)	proton synchrotron	5700 6200	1954 1955
Univ. of Calif. Berkeley	electronsynchrotron	320	1949
Cornell Univ.	electronsynchrotron	300	1951
M.I.T.	electronsynchrotron	300	1952
Univ. of Mich.	electronsynchrotron	300	1953
Purdue Univ.	electronsynchrotron	300	1954

Institution	Type of machine	Energy (MeV)	Date first operated
Univ. of Ill. (Urbana)	betatron	300	1950
In Europe			
Univ. of Liverpool	synchrocyclotron	400	1954
Univ. of Glasgow	electronsynchrotron	350	1954

For more information, see, for example, M. S. LIVINGSTONE and J. P. BLEWETT: *Particle Accelerators* (New York, 1962).

[100] R. M. BROWN, U. CAMERINI, P. H. FOWLER, H. MUIRHEAD, C. F. POWELL and D. M. RITSON: *Nature*, **163**, 47 (1948).

[101] G. O'CEALLAIGH: *Phil. Mag.*, **42**, 1032 (1951); M. G. K. MENON and C. O'CEALLAIGH: *Proc. Roy. Soc.*, A **221**, 2941 (1954).

[102] *Report on the Expedition to the Central Mediterranean for the Study of Cosmic Radiation: Napoli 18-28 May, Cagliari 30 May-13 July*, CERN/16, Rome, 30 September 1952. As much as I know a single copy of this document exists in the files of CERN. It contains also a number of photographs (in original) of the two ships (A. S. Altair and cannoniera Bracco) and of the airplane that the Italian Navy and Airforce had put at disposal of the expedition in Naples and Cagliari. The Universities that took part in the expedition were Bristol, Bruxelles, Cagliari, Genova, Glasgow, Göttingen (Max Planck Institut), London (Imperial College), Lund, Milano, Padova, Paris (École Polytechnique), Roma, Torino. The expedition started in Naples and shifted to Cagliari because of the strength and direction of the winds at high altitude. A total of thirteen balloons were launched. The document includes a report on the technical details by J. H. DAVIS from Bristol and C. FRANZINETTI from Rome.

[103] *Minutes of the Third Session, Amsterdam 4-7 October 1952*, CERN/GEN/4, Rome, 15 February 1953. At this meeting, that took place under the chairmanship of P. SCHERRER, the Council decided to propose Geneva as a site for the European Laboratory.

[104] J. DAVIS and C. FRANZINETTI: *Suppl. Nuovo Cimento*, **2**, 480 (1954). The same report, in a slightly abbreviated form, appeared also as a CERN document: CERN/GEN/11, *Report on the Expedition to the Central Mediterranean for the Study of Cosmic Radiation.* The Physics Laboratories of the following Universities participated in this expedition: Bern, Bristol, Bruxelles (Université Libre), Catania, Copenhagen, Dublin, Genova, Göttingen (Max Planck Institut), London (Imperial College), Lund, Milano, Oslo, Padova, Paris (École Polytechnique), Roma, Sidney, Torino, Trondheim, Uppsala. The balloons were constructed at Bristol and Padova under the direction of Dr. H. HEITLER and the gondolas and the radio equipment were built in Milano and Roma. Altogether the corvette Pomona covered over 5000 nautical miles for the recovery of the equipment. Until it was recovered it was kept underwater—at a depth convenient for assuring a low and constant temperature—by a buoy, whose position was tracked by a radio-wind transmitter that was heard by the receiver on the ship about 5 nautical miles away. Furthermore a radio cut-off apparatus was developed which allowed the cut-off of the « gondola » from the balloon to be operated at any time from the ship. Among the technicians who made an essential contribution to the success of the expedition A. PELLIZZONI from Roma and M. ROBERTS from Bristol should be particularly mentioned.

[105] *Rendiconti del Congresso Internazionale sulle Particelle Elementari Instabili Pesanti e sugli Eventi di Alta Energia nei Raggi Cosmici, Padova 12-15 Aprile 1954, Suppl. Nuovo Cimento*, **2**, 163 (1954).

[106] *Rendiconti del Corso che fu tenuto, nella Villa Monastero, a Varenna dal 18 Luglio al 7 Agosto 1954 a cura della Scuola Internazionale di Fisica, Suppl. Nuovo Cimento*, **2**, 1 (1955).

[107] It may be interesting to recall that also two stars connected by a heavy track were observed and tentatively interpreted as due to the production and annihilation of an antiproton (E. AMALDI, C. CASTAGNOLI, G. CORTINI, C. FRANZINETTI and A. MANFREDINI: *Nuovo Cimento*, **1**, 492 (1955)). When, about one year later, the properties of artificial antiprotons became known, doubts were raised about such an interpretation mainly because of a large-angle scattering undergone by the presumed antiproton (see for example at p. 419 of C. POWELL, P. H. FOWLER and D. H. PERKINS: *The Study of Elementary Particles by the Photograph Method* (London, 1959)). The observation, however, was at the time of some interest and is still today in the frame of the history of antiparticle hunting.

[108] *Congrès Internationale sur le Rayonnement Cosmique*, organisé par l'Université de Toulouse, sous le patronage de l'UIPPA, avec l'appui de l'UNESCO, *Bagnères de Bigorre, 6-11 Juillet 1953.*

[109] A short information on the expedition can be found in the « Introduction » by C. F. POWELL to the extensive paper of ref. [110]. This Introduction has been included at p. 318 in *Selected Papers of C. F. Powell*, edited by E. H. S. BURHOP, W. O. LOCK and M. G. K. MENON (Amsterdam, 1972). As stated by C. F. POWELL if anybody has played a distinctive and leading part in this joint effort, it was Dr. A. MERLIN from Padua.

[110] *Observations on Heavy Meson Secondaries* (*G*-stack collaboration), p. 398 of ref. [110]. Some emulsions were distributed also to the Institute for Advanced Studies and the University College in Dublin, and to the University of Genova.

[111] *Conferenza Internazionale sulle Particelle Elementari, Pisa, 12-18 Giugno 1955, Suppl. Nuovo Cimento*, **2**, 135 (1956).

[112] T. D. LEE and C. N. YANG: *Phys. Rev.*, **104**, 254 (1956).

[113] École d'Eté de Physique Theorique de l'Université de Grenoble.

[114] E. AMALDI: *Suppl. Nuovo Cimento*, **2**, 339 (1955).

[115] L. KOWARSKI: *An account of the origin and beginning of CERN*; CERN 61-10, 10 April 1961.

[116] E. AMALDI: *CERN, past and future*, p. 415 of *Topical Conference on High-Energy Collisions of Hadrons, CERN, Geneva 15-18 January 1968.*

[117] Belgium, Denmark, France, Greece, Italy, Netherlands, Norway, Sweden, Switzerland, West Germany, Yugoslavia.

[118] C. G. DARWIN, professor of Theoretical Physics at the University of Cambridge, is well known for many important contributions among which one should recall the treatment of the anomalous Zeeman effect by means of the Dirac equation and in general the study of solutions of the Dirac equation in the case of low velocities, so that the four-component Dirac spinors can be replaced by two-component spinors, of which one is large, the other small.

[119] Letter of E. AMALDI to P. AUGER: Rome, le 3 Octobre 1950

Prof. P. AUGER
12 Rue Emile Faguet
Paris XIV

Cher Professeur AUGER,

À l'occasion de la réunion du Comité Exécutif de l'Union Internationale de Physique Pure et Appliquée qui a eu lieu à Cambridge Massachusetts (U.S.A.) les jours 7-8-9 septembre, on a parlé de la proposition présentée par le Prof. I. I. Rabi à l'Assemblée Génerale de l'U.N.E.S.C.O. à Florence, au sujet de la construction d'un Laboratoire Européen de Physique Nucléaire. Après quelques discussions, l'Exécutif a décidé de faire préparer deux rapports sur cet argument: l'un devrait être rédigé par Rabi, qui devrait préciser aussi bien que possible sa pensée, l'autre par moi, qu'on a chargé de prendre contact avec les physiciens européens, de manière à rejoindre, si possible, un accord sur plusieurs points fondamentaux.
Je savais déjà que Vous vous interessés de l'argument pour l'U.N.E.S.C.O.; à présent je viens d'apprendre de Ferretti qu'à l'occasion du Congrès d'Oxford Vous avez provoqué et dirigé une discussion intéressante à ce sujet.
Jusqu'ici j'ai écrit seulment à Rabi. Avant de m'adresser à d'autres physiciens européens, j'ai pensé vous écrire afin d'éviter que mon action puisse paraître en désaccord avec celle que vous êtes en train de développer.
Je Vous serais trés reconnaissant si Vous vouliez me communiquer aussitôt que possible, quels sont les moyens que Vous suggérez pour obtenir que l'action de l'U.N.E.S.C.O. et de l'U.I.P.P.A. résultent s'additioner efficacement.
J'ai appris de Ferretti les lignes générales que Vous avez données à la discussion à Oxford et je suis complètement d'accord avec Vous. Surtout je suis convaincu de l'importance spécifique et générale de ce qu'un tel projét puisse se réaliser.
Je penserais d'écrire aus physiciens suivants:

Angleterre	—	J. D. Cockcroft
France	—	?
Suisse	—	P. Scherrer
Allemagne	—	W. Heisenberg
Belgique	—	M. Cosyns
Hollande	—	C. J. Bakker
Danemark	—	N. Bohr
Suède	—	K. Siegbahn
Norvège	—	?
Espagne	—	?
Autriche	—	?

en présentant la chose et en posant quelques problèmes, par exemple:

1) lieu
2) direction
3) financement.

Quant au premier point, il me semble essentiel que le lieu soit beau et assez central, et il me sembre qu'on pourrait choisir une localité située en Suisse. Pour ce qui se rapporte au deuxième point je n'ais aucune idée particulière. Pour le troisième, je désirerais connaître votre avis et celui de Rabi, pour pouvoir prèsenter des proposition concrètes. En effet je pense que, si on écrit d'une façon générique, sans proposer des solutions raisonnables, les sceptiques vont faire mourir la chose avant qu'elle naisse.

Votre avis à ce sujet me sera très précieux afin que je puisse accomplir la tâche qu'on m'a donnée aussi bien que possible, surtout pour l'effective réalisation de ce projet.

Avec mes meilleures amitiés,

(Prof. EDOARDO AMALDI)

[120] E. D. COURANT, M. S. LIVINGSTONE and H. S. SNYDER: *Phys. Rev.*, **88**, 1190 (1952).

[121] Before becoming an eminent personality in public life, Lord CHERWELL was well known as a physicist under the name of F. A. LINDEMANN. Among other contributions, his work on the theory of melting is now classical.

[122] A 200 MeV cyclotron was in operation at the Werner Institute since 1953.

[123] W. A. COOPER, H. FILTHUTH, J. A. NEWTH, G. PETRUCCI, R. A. SALMERON and A. ZICHICHI: *Nuovo Cimento*, **4**, 1433 (1956); **5**, 1388 (1957).

SCIENTIA

Reprint from the 5-6-7-8 1979 issue

Annus LXXIII

Volume N. 114

The Years of Reconstruction

Part I

EDOARDO AMALDI
Università di Roma

INTRODUCTION - I

The organizers of this meeting asked me to speak of « *The Years of Reconstruction* »*, but for reasons of continuity with Emilio Segré's speech, I can not avoid beginning from autumn 1938, when Fermi and a few other collegues were forced to leave Italy. In other words I will start with a short account of what I call 'The disaster of physics in Italy', and then, for lack of time, I will jump to the years of reconstruction, i.e. to the period that opens with the arrival of the allied troops in Rome and ends at a time that can not easily be defined but I would place between the middle and the end of 1954. In this account I will omit the war years which are an essential part of the life of the Istituto di Fisica Guglielmo Marconi. Only here and there I will mention some wartime episodes because they are essential to understand what happened later.
What I will say is based essentially on my personal recollections so that, in many parts my account will have an autobiographic character. Initially I tried to avoid this form, but I immediately realized that my speech was becoming even more boring. Thus, I left it in its initial form; I apologize for this and I ask you to take the parts of more personal nature as belonging to 'the story of a poor physicist' in Italy during the years 1938-1954. Mine is just one of the many stories of other physicists or intellectuals that lived in our country in those years and witnessed the many typhoons that passed sufficiently close to allow them a clear view of their dramatic consequences, but were fortunate enough to be sufficiently distant from their center to remain practically undamaged.

THE DISASTER OF PHYSICS IN ITALY - II

The train with Fermi's family had left the Termini station for Stockholm the evening of December 6, 1938[1,2]. Franco Rasetti, my wife Ginestra, I and a few of their relatives were remained on the platform to say goodbye to them and went home.
I looked at the people in the streets, who clearly could not be aware of what was going on, but I knew, or rather all of us knew, that a period, very short indeed, of the history of culture in Italy had definitely ended that evening. A period which could have been extended and developed and, perhaps, have had a wider influence on the university system, and, eventually, perhaps even on the whole country.
Our little world had been upset, almost certainly destroyed by forces and circumstances completely foreign to our field of action. An attentive observer could have told us that it had been naive to think we could build such a delicate and frail construction on the slopes of a volcano which showed such clear signs of increasing activity. But we were born and had grown up on those slopes and we had always thought that what we were doing was much more

lasting than the political phase the country was passing through.

The departure of Fermi had been determined by a number of causes, the last and more direct of which were the « racial laws » promulgated by the Fascist government the July 14, 1938 and which hit his wife Laura Capon and their children. This was only one of the many steps of the years of long degradation produced by the Fascist countries in Europe. I may recall here two of the more dramatic of these events. The *Anschluss* of Austria by Germany in March 1938 and the Munich agreement of September 29-30, 1938, where Great Britain and France had accepted the dismembrement of Czechoslovakia with the transfer of a few of its border provinces to Germany, Poland and Hungary. Fermi was not the only physicist leaving our country.

Bruno Rossi, who had been nominated professor in Padua in autumn 1933, had left with Nora for Copenhagen on October 12, 1938. They could afford a sudden departure by using the residue of a previous fellowship he had used for a few months stay at the W. Bothe laboratory at the Reichsanstahlt in Charlottenburg near Berlin.

Emilio Segré, who had been nominated professor in Palermo in 1935, at the beginning of summer 1938, had gone to Berkeley, to work on the short lived isotopes of element 43, technetium, he had discovered with Carlo Perrier in Palermo in 1937[3].

As a consequence of the trend of the political situation in Italy and in Europe in general, he had decided to remain there and asked his wife, Elfride, to join him with their son Claudio, not yet two years old.

Giulio Racah, Ugo Fano, Eugenio Fubini, Sergio De Benedetti, Leo Pincherle and a few others had already left or were ready to leave the country.

THE SITUATION OF THE PHYSICISTS IN ITALY AT THE BEGINNING OF THE WAR - III

The remainder of our group at the Institute Guglielmo Marconi, where we had moved at the end of summer 1936 from the old Institute of Via Panisperna, consisted of Franco Rasetti, professor of Spectroscopy, I, professor of Experimental Physics, Bruno Ferretti, Fermi's assistant and Mario Ageno, who had arrived in 1934-35 from Genoa, as a student. He had taken his laurea with Fermi in 1936, and in 1938 had a position as an assistant. There was also Oreste Piccioni, who had taken his laurea a few months before, and a very clever student, Marcello Conversi.

There was also, of course, Antonino Lo Surdo, professor of Fisica Superiore, who at the death of Orso Mario Corbino in 1937 had succeded him as director of the Institute. In 1938 he had also succeded Fermi in the annual appointment to give the course of Fisica Terrestre, a subject that he was cultivating with his pupils, Enrico Medi, first assistant (*aiuto*), and Antonio Bolle, assistant. Following Fermi's suggestion, the Faculty had appointed Ferretti to replace him for the academic year 1938-39 in the course of Theoretical Physics. In fact Fermi had left Italy with a regular 'leave of absence' for one year teaching at Columbia University. These bureaucratic steps, legally unexceptionable, had been carefully prepared by Fermi. When a few weeks before, he had communicated to Rasetti and me his decision to leave Italy, he also had told us his intention to avoid any action or declaration that could damage those remaining in the country.

Here and there scattered throughout Italy in the various universities, there were a few individual active researchers and in some places groups, which, however, had lost their leader just in their first infancy.

Gilberto Bernardini, who in 1938 was passed from the chair of Physics of the University of Camerino, to the chair of Fisica Superiore of the University of Bologna, went on devoting a few days per week to research work done in Rome, since in Bologna, as before in Camerino, he had not much possibility for serious experimentation[4].

G. C. Wick had been very recently called to the chair of Theoretical Physics in Padua from Palermo where he had the same chair from autumn 1937. Of the experimental group in Padua only Ettore Pancini, a very young 'extraordinary' assistant, remained. Soon the faculty called Antonio Rostagni from Messina to the chair left by Rossi.

Also Giuseppe Cocconi appeared in Rome from time to time in 1938-39. After his 'laurea' taken in Milan in 1937, he had come to Rome and in January 1938 had started to work with Fermi, Rasetti and Bernardini for a direct observation of the decay of the mesons constituting the hard component of cosmic rays, discovered by Rossi in Florence in 1932. During summer 1938 Cocconi returned to Milan, and started to work independently, always in cosmic rays, with the full support of G. Polvani, who was the Director of the Istituto per le Scienze Fisiche Aldo Pontremoli. At the University of Milan the chair of Theoretical Physics had been covered, since autumn 1937, by G. Gentile jr., who, however died prematurely in 1942[5], not long after the publication of an important paper in which he had introduced and discussed the intermediate statistics[6].

Gentile and Caldirola, who came regularly to Milan from Pavia to lecture on Statistical Mechanics, gave moral and, occasionally, scientific support to A. Borsellino of the Milan Polytechnic. He had studied at the Scuola Normale of Pisa where, at the beginning of 1937-38, had started to work on his thesis under the supervision of Racah, who, in autumn 1937 had been appointed professor of Theoretical Physics at the University of Pisa.

Borsellino had taken his laurea after Racah's departure, and, shortly afterwards, had accepted the position of assistant to A. Amerio, professor of Fisica Sperimentale at the Milan Polytechnic.

The situation in other Universities was the following. E. Persico had the chair of Theoretical Physics in Turin and was extremely shocked and depressed by the general trend of political events. He devoted those years mainly to writing excellent physics books. A few assistants of Segré remained in Palermo. B. N. Cacciapuoti from the Scuola Normale di Pisa who had published with Segré an important paper on the long lived isotopes of technetium[7]. M. Mandò from Florence, who went back to his University after the departure of Segré and M. Santangelo from Palermo.

In Naples A. Carrelli, a spectroscopist, had the chair of Experimental Physics, and in Pavia P. Caldirola was assistant professor* of Theoretical Physics. After his laurea, taken in 1937, he had come to Rome to work with Fermi in theoretical physics, with a grant from his college, the Collegio Ghisleri of Pavia. A few months after his arrival, Fermi had left Rome and Caldirola had moved to Padua, to work under the guidance of Wick.

To complete the picture I should recall also a few departures of Italian physicists which took place a few years before or shortly after the departure of Fermi.

At the end of winter 1935-36 B. Pontecorvo had moved from Rome to Paris where he had started to work in the laboratory of the Joliot-Curie with a scholarship of the Ministero della Educazione Nazionale. He never came back to Italy from Paris, except for short visits that, before as well as after the war, took place almost regularly at Christmas, Easter and for the summer vacations. E. Majorana, nominated professor of Theoretical Physics in Naples in 1937, had disappeared in spring 1938[8], G. Wataghin and G. Occhialini had moved in 1937 to the University of S. Paulo, from Turin and Florence respectively, through a cultural agreement between Italy and Brasil.

At the beginning of summer 1939, F. Rasetti left Italy for Quebec (Canada), where he had been nominated professor of physics at Laval University. The Ministero della Educazione Nazionale had put him at disposal of the Ministero degli Affari Esteri, which had authorized him to take this job for a few years.

* I use the following equivalence between Italian and American levels in the career of a university professor: assistente = assistant; aiuto = first assistant; professore incaricato = assistant professor; professore straordinario = associated professor; professore in cattedra = full professor.

A few more names must be mentioned here. Among these I should recall Eligio Perucca of the Polytechnic of Turin, a remarkable physicist and man, whose interests always were oriented towards classical physics[9].

At the end of 1938 it was clear that in order to succeed in surviving from a scientific point of view, it was necessary to try to concentrate, in a smaller number of places, at least a part of those who remained scattered here and there. This idea, amply discussed with Bernardini, Wick and Ferretti, for years remained a basic point which often guided our actions. Thus, in autumn 1939, I succeded in obtaining Wick's move to the chair of Theoretical Physics at the University of Rome. In fact, Fermi's leave of absence had expired at the end of October of that year, and since he was not returned to Italy, the Ministero della Educazione Nazionale had considered him 'resigner' and thus it had become possible to proceed to the nomination of his successor. This, after all, conformed to Fermi's desire.

At the beginning of 1940 I succeeded in obtaining Cacciapuoti's move to Rome from Palermo and Pancini's from Padua, the first as an assistant to the chair of Fisica Sperimentale, the second as a researcher at the Istituto Nazionale di Geofisica (ING), founded, in 1939, in the frame of CNR by Antonino Lo Surdo who was also its director. A similar move was possible also for Santangelo about one year later. Bernardini had succeeded in convincing Lo Surdo to include at least a part of cosmic ray work in the research program of ING and therefore to provide some financial means for instrumentation and staff.

THE RECONSTITUTION OF THE INSTITUTE - IV

With the arrival in Rome of the Allied troops, the 5th June 1944, the war was very far from its end. It was still going on in all the world, even in central and north Italy. It was necessary to wait about one year for the liberation of Milan and all north Italy, the 25th April 1945, and for the unconditioned surrender of Germany, the 7th May of the same year.

But in June 1944, for Rome and its inhabitants, a period of many months of very scarce food supply finally ended. The German occupation which had been marked by the deportation of Jews and political prisoners, that had seen the beginning and growth of the partisans fight and the Fascist and Nazi retaliations and had culminated in the tortures of Via Tasso and the massacre of the Fosse Ardeatine had finished.

Rome had entered a new phase of still uncertain, perhaps even obscure, issue. But whatever this might be, it was clear in which direction everybody, out of necessity, had to move. The material damages undergone by the country had to be repaired, one had to surpass the levels reached in the past and contribute to the construction of a society which would retain and develop only some of the features that just passed, refusing and eliminating superficial and profound deteriorative aspects left by Fascism.

The first rule clearly consisted of trying to work seriously, without ridicolous nationalistic arrogances, without prosopopoeia or rhetoric but also without false modesties nor inferiority complexes.

This general state of mind was particularly strong at the Institute, where all the staff was aware that a long night was over and that the new day required a great effort.

The more delicate scientific material was brought back to our building. About one month after the armistice of September 8, 1943 — i.e. about ten months before the beginning of our story — the German Command in Rome had promulgated the order not to remove any scientific material from the Institutes and research Laboratories, in particular from those of the University, so that they could requisition it, if they wished to do so. The idea did not please us and so, with a truck, loaded a couple of times in the internal court of the Institute by Cacciapuoti, Conversi, our technician Renato Berardo and with the help of Cavaliere L. Zanchi, for years our administrator, I brought all or almost all the material easily removable and that we considered important to rescue to a safer place known only to few people.

Immediately after the arrival of the Allied, we brought back to the Institute from the Liceo Virgilio also the equipment that Conversi and Piccioni had built and used for measuring the meanlife of mesons. The measurements had been carried on almost without interruption also during the months of occupation. To keep this experiment in operation, at any cost, had become, for all of us, a kind of symbol of our will of cultural and scientific continuity.

The counters and the electronics of this experiment had been taken away from the Institute a few days after the first bombardment of Rome by the American air force. The goal of the incursion had been the goods station San Lorenzo, but more than eighty bombs had fallen within the perimeter of the Città Universitaria, damaging various buildings. It was the 19th of July 1943. I remember that I was with Gian Carlo Wick in my office when we heard the air-raid warning and that while we were moving rapidly towards the stairs to reach the basement, we clearly saw through the windows the bombs falling on the Chemistry building in front of our Institute.

At that moment we were more or less, all in the building. I remember that there was G. Bernardini, B. Ferretti, M. Conversi, C. Ballario, A. Lo Surdo, E. Medi, R. Cialdea, a student and later assistant of Lo Surdo, R. Berardo, M. Berardo, L. Zanchi and many others.

All the window glasses of our building had gone to pieces since four bombs had fallen within a few meters from each of the corners of the building, but the structure had not been damaged.

We were afraid however, that other bombings could follow the first one, making the continuation of the work impossible. For various reasons, the interruption was by now unavoidable for all researches except for the measurements of Conversi and Piccioni, whose experimental set-up with some spare electronic material, formed in that moment a kind of little closed system.

It was necessary to find a place sufficiently close to the Vatican City, to be within an area reasonably protected from the air-raids. These conditions appeared to be fulfilled by the Liceo Virgilio, whose deputy-headmaster, Prof. L. Fagiolo was an acquaintance of Conversi's. With the authorization of the Headmaster Prof. A. Bandini, it was transported towards the end of July 1943 by a hand cart.

I remember that besides Piccioni there were three new students: C. Franzinetti, F. Lepri and L. Mezzetti. I accompanied them along Via Nazionale and Corso Vittorio Emanuele on my bicycle. Sometimes I preceded, sometime I followed the cart trying to avoid road accidents and asking to the policemen to give us the right of way.

After the liberation of Rome, besides the instruments and this apparatus fully operational, the young people returned to the Institute. All or almost all had been involved in this or that phase of the war: many on various fronts in Africa, Greece, Russia and Yugoslavia, many in the partisans fight although at different levels of unvolvement and danger.

A few came back later. For example Mandò who had been mobilized and sent to North Africa in 1939 as a lieutenant of the 204th Regiment of Divisional Artillery, had been taken prisoner at Sidi El Barrani the 10th December 1940 on the occasion of the first British offensive, and came back from India in June 1946.

M. Ageno also had been mobilized in 1939 and sent to North Africa as lieutenant in the Artillery Corps. When in April 1941 the German Armoured Corps and a few Italian Divisions, under the command of General Rommel, had broken the British lines, reconquering Cirenaica, part of the English troops remained besieged in the stronghold of Tobruk.

Ageno took part in the siege, in the advanced

observatory of his Grouping, from which he could see and communicate the result of the fires. After one hundred days of siege, the 14th December 1941, he was urgently moved to a military hospital, first in Bengasi and later in Italy. He remained in hospitals and sanatories until 1948, i.e. eight years after his mobilization and only in 1960 he could definitively abandon the stick he used for moving within the laboratory.

The research activity was immediately resumed with extraordinary energy at the Istituto Guglielmo Marconi. There were still essentially two main lines of research as in 1938: the study of cosmic rays, guided by G. Bernardini, and nuclear physics, which was at the center of my interests. The latter activity was carried out in collaboration with Daria Bocciarelli and Giulio Cesare Trabacchi of the Istituto Superiore di Sanità (ISS) using, as neutron source the 1.1 million volt accelerator we had constructed with F. Rasetti in the Laboratorio Fisico of that Institute in 1937-38[10]. The idea of building this machine had been fostered by Fermi, supported by Trabacchi, director of the Laboratorio di Fisica dell'ISS and favoured by the Director General of that Institute, D. Marotta[11].

The machine had a performance not inferior to that of the other similar devices already at the disposal of physicists of other European and oversea laboratories.

It might have allowed the continuation of research on neutrons and, at the same time, the production of considerable quantities of artificial radioactive substances of great interest for their therapeutic applications and their use as tracers in many biological, chemical and industrial processes. I have used the conditional form because, first, the dispersion of the group of Roman physicists, and later the war, enormously reduced its immediate utilization and, later, after the war, the nuclear reactors became sources of neutron greatly superior to similar installations for producing artificial radioactive substances. As we shall see in the following passage this machine was very usefully employed for many years in research work, but its exploitation was certainly inferior to what we had hoped when we began designing it.

A SHORT DETOUR DEVOTED TO LITTLE DOMESTIC BUROCRATIC STEPS AND A GREAT DRAMA FOR MANKIND - V

September 7th, 1944, the Allied Government in Italy nominated the mathematician Guido Castelnuovo (1865 - 1952) Extraordinary Commissioner for the CNR. Almost immediately Castelnuovo nominated his deputy the mathematician Francesco Tricomi (1897-1978) who had arrived shortly before in Rome from Turin, where he had participated in the partisans fight. In that period, which lasted only to the end of the year, I was entrusted with the function of President of the Committee for Physics and Astronomy of the CNR.

At the beginning of 1945, the Italian Government, anxious to bring the CNR back to normal conditions, nominated President of CNR Gustavo Colonnetti (1886-1968), professor of Construction Sciences at the Turin Polytechnic, who was recently returned to Italy from Switzerland, where he had fled for political reasons and where he had organized, in a camp for Italian political refugies, students university courses. I was appointed member of the Committee for Physics and Astronomy, whose president was Eligio Perucca.

At the beginning of August 1945 the United States dropped two atomic bombs on the Japanese cities of Hiroshima and Nagasaki. Thus, suddenly and dramatically the entire world learned that the applications of nuclear knowledge for pacific as well as for military purposes, were no more only a possible development dimly foreseen by experts, but by now had become part of reality. We were all on vacation at Rocca di Mezzo in Abruzzi. The Bernardinis, the Wicks, the Ferretti swere there as well as a few people belonging to the new generations like G. Careri, I. F. Quercia, B. Rispoli, L. Mezzetti and C. Franzinetti. All of us were extremely shocked by the gravity of the problems that mankind had to face from now on in the future. At the

same time we began to realize that the scientific and technical developments made overseas had clearly established a much greater gap than we had imagined until then, between the technological levels reached on the opposite sides of.the Atlantic.
A few weeks later all of us returned to Rome. The research activities at the Istituto Marconi and the Laboratorio Fisico of the ISS had already been resumed with a new impetus before summer. It was generally recognized that these two laboratories were among the very few of the whole country that had maintained alive research activity at a good level during all the war. Thus the CNR approved the proposal already presented before summer of the creation at the Istituto di Fisica Guglielmo Marconi, of a 'Centro di Studio della Fisica Nucleare e delle Particelle Elementari' of which I was nominated director and Bernardini vice-director. The research program consisted in the natural development of previous activities.
The creation of the 'Center' ensured a continuity in the appropriation of financial means by the CNR, which until then had been extremely limited and irregular.
A few months after the liberation of Milan, Polvani organized at Como a Physics Conference for celebrating the second centenary of Alessandro Volta's birth[12]. From Rome A. Giacomini, director of the Istituto di Elettroacustica O. M. Corbino, and I took part in the meeting.
We reached our destination after a 36 hours trip with our rucksacks full of supplies, having crossed on foot pontoon bridges over a few rivers, on the banks of which the trains mainly composed of cattle carriages stopped. This was the first reunion of the physicists of Central and South Italy with those of North Italy. Besides Polvani, Persico, Perucca, Carrelli and Somigliana there were Rostagni, Caldirola and a few younger physicists including Carlo Salvetti and Giorgio Salvini, who had been formed during the difficult war years by studying when on leave from their military duties, the first under the influence of Giovanni Gentile jr., the other in the wake of Giuseppe Cocconi who had been mobilized at the beginning of 1941 and moved, a few months later to Rome, as I will say below.

THE BEGINNING OF APPLIED NUCLEAR RESEARCH IN ITALY - VI

Passing through Milan, after the Como meeting, Livio Gratton and I met the chemist Luigi Morandi, brother of the socialist leader Rodolfo Morandi.
Gratton was 'first astronomer' at the Merate Observatory (not far from Como). He had been one of our first students at the Institute of Via Panisperna and had prepared the meeting. Dr. Luigi Morandi, who had been nominated Commissioner to the chemical industry Montecatini S.p.A. after the liberation of Milan, was interested in the problem of the relations between University research and industry. Following our long and interesting conversation, which among others gave rise to a lasting personal friendship, I prepared during the following month a thirty odd typed page report on « Physics in Italy ».
It contained a synthetic presentation of my views about what had to be done immediately in Italy to acquire the necessary scientific equipment and to obtain qualified staff in view of also developing peaceful applications of nuclear physics[13].
I also sent a copy of the same report to Valletta, director of FIAT, who, according to Morandi, was also interested in the same problems. More or less at the same time G. Bernardini also came in contact with Valletta through completely different channels.
At the beginning of 1946, I believe it was February, Giuseppe Bolla came to visit me at the Institute. Bolla had been the successor of Segré in Palermo and later had moved to Milan where he had the chair of Fisica Superiore.
He had come to talk to me about a scheme he had worked out in collaboration with Salvetti, Salvini and M. Silvestri, a young en-

gineer of the electrical utility Edison S.p.A. The essence of the scheme was the creation, in Milan, of a laboratory devoted to the development of applied nuclear physics with funds provided by a few industries. The program of the Milanese collegues was much more specific although partial with respect to the picture I had sketched in my report about one month before. Furthermore, it was endowed with a considerable concretness since Bolla and his collaborators were already in contact with a few industrialists of North Italy who had declared that they were ready, in principle, to finance the undertaking. The only governmental body which, institutionally, could have been interested in this fundamental problem, was CNR, which, however, was absolutely not in the condition, at that moment nor for many years to come, to take upon itself such an heavy burden. Thus it appeared to me right and my duty accept to collaborate with the Milanese collegues. The same answer was given by Bernardini and Ferretti when asked on the same question a little later. Thus, together with our Milanese collegues, we were among the founders of the body created a few months later under the name 'Centro Informazioni, Studi ed Esperienze' (CISE)[14].
An agreement was established from the beginning between the physicists of CISE and those of the 'Centro per lo studio della Fisica Nucleare e delle Particelle Elementari'. The research of applied nuclear physics was the specific subject of CISE, and that of fundamental nuclear physics the institutional subject of research of the Centre founded in Rome, the separation at the borderline being left to common sense in a spirit of reciprocal understanding.
My contribution to CISE during the subsequent years consisted in going to Milan, for a few days at least once every month, where besides discussing with Bolla, director of CISE, the general organization of the laboratory and the development of its programmes, I lectured on neutron physics, trying to transfer what I knew in the field of nuclear physics, to the young people mentioned above and a few others, even younger very gifted physicists that Bolla had succeeded in collecting together in the new laboratory. Among these young people it is enough to recall Sergio Barabaschi, Alberto Cacciari, Laura Colli, Ugo Facchini, Sergio Gallone, Emilia Gatti, Enrico Germagnoli, Cesare Marchetti and the chemist Enrico Cerrai.
Bolla has been an excellent director and to him, more than anybody else, are due the very good results obtained by CISE, from the beginning, the most important of which was certainly the formation of highly qualified personnel.
The contributions of Bernardini and Ferretti to the CISE's life had the same form and were similar to mine, but Bernardini taught the young Milaneses mainly the techniques for the detection of particles and Ferretti was an extraordinary general consultant for all theoretical and computational problems involved in the treatment of any subject of pure and applied physics.
My activity for CISE started to decrease gradually as my interest in the creation of the European laboratories of CERN started to find a favourable and increasing response from the outside world. I had also begun to realize that my contribution to CISE was becoming less important. I had already taught the researchers of CISE what was really useful for their work, and most of them could now procede independently.
My relationship with CISE came to a complete end the 26th of June 1952, when I was nominated a member of the Comitato Nazionale per le Ricerche Nucleari (CNRN), a body that in 1960 was transformed in Comitato Nazionale per l'Energia Nucleare (CNEN).
A few initiatives, non competitive with CISE, aiming at transferring the knowledge of the Institute's researchers to the applied field took place also in Rome. The first of these was made by Bernardini, Cacciapuoti, Ferretti, F. Lepri, Pancini and Piccioni who devoted part of their time to a laboratory financed by FIAT, and called 'Laboratorio Beta'. During the years 1946-1949 the Laboratorio Beta tackled a number of problems

mainly of electronics but sometimes of a more complex nature, such as, for example, the elimination of the dazzling due to car lights.

A lasting success was obtained by a young graduated of the Istituto Marconi, Renato Casale. In July 1947 he founded, under the name « Italelettronica », an industry that initially used the competence of some members of the Istituto Marconi, but soon created an independent laboratory. The Italelettronica gradually extended its activity from the initial construction of various types of counters, mass spectrometers, demultiplication systems, amplifiers and uranium ore detectors to a wide range of electronic instruments.

Among these initiatives of an applied nature, I should also recall the study and development of radioactive lodging started by I. F. Quercia with the support of AGIP in the years 1949-51.

THE REVIVAL OF RESEARCH - VII

The building of our Institute, during the winter 1945-46, still had a large part of the window glasses broken and replaced by sheets of cardboard or masonite, which let in terrible draughts. Furthermore there was no heating at all in the whole university since the bombing has destroyed the pipes connecting the central thermal station to the buildings of the various departments.

During the winter the temperature inside our Institute was so low that, after the night, the rotating pumps could not turn because of the excessive viscosity of the oil. Therefore when the people whose work required the use of vacuum, arrived at the Institute in the morning, they had to devote at least one hour to heating, gradually and uniformily, the body of their rotating pumps with a gas flame.

Careri recalls that this was still the situation during the winter 1947-48. But I can state that the measurements made by him with his mass spectrometer in such a tiring way, were certainly not inferior to those made, at the same time, in British or American laboratories under normal conditions.

In that period we remained with only one tube of 'Apiezon' vacuum grease. It had been placed in a cabinet, the key of which was in an open drawer in my office, where any researcher came to take it when necessary for his work.

The tube was placed on a paper on which was written: « This is the last vacuum grease tube we have. When it is finished it will not be possible to make vacuum any more ». Everybody used it with great care until, almost one year later, we received from a friend from overseas some new grease tubes.

Our instrumentation and in particular our electronic apparatus was obsolete and insufficient. Miniaturized tubes had appeared in United States during the war and therefore were unknown to us until the arrival of the allied troops. From my trip to United States that I will mention later (§ VIII) I had brought back some valuable and necessary material, part of which, I had bought, and a larger part received as gifts from various American friends (of Italian origin and not).

But this certainly was not enough. The continuation and development of our work was made possible by using materials acquired from the camps of ARAR (Azienda Rilievo e Alienazione Residuati) where the residual war materials had been collected. A few of our young people had specialized in this type of recovery operation which sometime was rather adventurous. Once S. Sciuti and F. Lepri, another time Pancini and Quercia, brought to the Institute from the ARAR camp near Capua a truck of electrical (oscilloscopes, amplifiers) and optical (cameras, theodolites, etc.) materials.

Considerable amounts of equipment were brought in a number of expeditions, from a camp near Rome by Pancini and Quercia, from another near Bari by Pancini, Quercia and Rispoli and from the camp of Tombolo, near Pisa, by Cacciapuoti, Lepri and Pancini, who arrived at the Institute with a six truck caravan.

The fact that we had at our disposal for

the first time, thanks to those expeditions, hundreds of electronics tubes of well defined types, had also a considerable influence on our approach to the planning and designing of new experimental setups, in which soon we started to include, for example, even counters hodoscope of large dimensions.

In the recovery operations, but even more in the work of reparation and/or adaptation of the recovered equipment, a number of technicians of our Institute gave a remarkable contribution. Besides Renato Berardo (1902-1976) and Mario Berardo (b. 1915) should be mentioned: Azelio Mancini (b. 1911), Vincenzo Bettini (b. 1913), Alceste Macrino (b. 1920), Francesco Ocello (b. 1904), Serafino Generosi (b. 1914) and Eugenio Zattoni (1924-1972).

As it appears from these few examples, there were plenty of difficulties, but in spite of all of them, the work had been fully taken up again with a great dedication by all the staff.

The main scientific lines tackled during 1945 by the Centro were essentially the same that had been fostered during the war[15].

The Study of Cosmic Rays - VII.1

In the field of cosmic rays, that as I said before, was guided mainly by G. Bernardini, the research subjects were: the study of the equilibrium between the electronic and the penetrating components at various altitudes above sea level[16], the study of the positive excess of the penetrating component, by means of the magnetic deflection in iron cores[17], and, finally, the decay of the meson tackled by two methods: in the years 1939-43, by the indirect method based on the anomalous absorption shown by air[18], and, beginning from 1943, by the direct method of the delayed coincidences, developed in those years by Conversi and Piccioni[19]. This represented a considerable step forward with respect to the measurements of the mean life of mesons made for the first time in 1941: by Rasetti at Quebec and by Auger and co-workers in Paris[20].

Furthermore the results of young Romans turned out to be in excellent agreement with those obtained by similar methods at about the same time by Rossi and Nereson at Cornell University (Ithaca)[21].

In Rome, at the Istituto Guglielmo Marconi, a number of younger people had entered the production phase[22]: E. Corinaldesi, R. Querzoli, M. Verde (1946), I. F. Quercia, R. Rispoli, S. Sciuti (1947), C. Ballario, G. Cortini, A. Manfredini, L. Mezzetti and G. Morpurgo (1948).

The Study of a Few Nuclear Processes - VII.2

As I said before the work in the field of nuclear physics was made at the ISS in collaboration with D. Bocciarelli and G. C. Trabacchi (1884-1959).

During the war years we had measured the cross section of fast neutrons of various energies against nuclei of different atomic number[23]. Later we had studied the dependence on the neutron energy of the fission cross section[24], a subject, however, that we decided to abandon after the entrance of Italy into the war and my return from six months service on the North African front. Ageno[25], Cacciapuoti, Bocciarelli, Trabacchi and I, i.e. all the people involved in this research, were afraid that being active and acknowledged experts in this subject could expose us to the danger of being obliged to work for the Powers of the Axis for the development of military applications of fission. The decision was taken after a deep discussion in which also Bernardini, Ferretti and Wick took part.

Thus we turned to the investigation of the collision of fast neutrons against protons[26] and deuterons[27] in energy regions still scarcely explored. Later, in 1943, our attention had been attracted by the typed copy, received from Placzek after Italy's entrance into the war, of a theoretical paper by Bohr, Peierls and Placzek[28] on the optical theorem and the relationship between the absorption and scattering cross section and the diffractive nature of the latter, for neutrons of sufficiently large energy against medium and heavy nuclei. The experiments on this subject kept us occupied during the years 1944 and 1945[29].

In those years we also committed a few mistakes, consisting in starting new programmes that we soon had to abandon because they were beyond our possibilities. One of these was the construction of an installation for isotopic separation by thermal diffusion[30], the other the design of a 20 MeV betatron, in connection with which Cacciapuoti spent few months during 1945 at Kerst laboratory in Urbana (Illinois)[31,32].

A Clusius tube for thermal diffusion was constructed successfully, but it could only serve as a prototype for the construction of the few hundred similar devices necessary for a significant installation for isotopic separation. Such a plant was clearly beyond our financial and organizational capability.

Towards the end of winter 1945-46, the Institute in Rome suffered a serious loss: G. C. Wick accepted the offer of a chair at the University of Notre Dame in South Bend (Indiana). Everybody was very sorry about his departure, but I more than anybody else. Not only we were losing an excellent theoretical physicist, but I was also losing a close friend. Furthermore, during the more recent years we had started the preparation of a book on neutrons, of which only a few parts were written. These, however, were used, years later, as a part of an extensive article I wrote on this subject[33].

Other Cosmic Rays Investigations in Milan and Catania - VII.3

From 1938 on, Cocconi had started in Milan a considerable research activity on cosmic rays, done in part alone in part in collaboration with Vanna Tongiorgi. The main problems were: the secondaries of the hard component, the neutrons present in cosmic rays, the dependence of the zenital effect on the mean life of the meson[34]. In these experiments he used, besides Geiger counters, also borontrifluoroside counters (for neutrons) and a vertical cloud chamber[35].

Towards the end of 1942, Cocconi was put in the 'tern' of a national competion[36] and nominated associated professor of Experimental Physics at the University of Catania, where he transferred a part of his research activity, in spite of the fact that he had been mobilized as 'car driver' at the end of February 1941. Some time later from Milan he had been moved to the 'Centro Studi dell'Aeronautica Militare di Guidonia', in the outskirt of Rome, in the quality of « aviere di governo » and attached to the Istituto di Fisica Guglielmo Marconi to work, in collaboration with lieutenant Giorgio Fea[37] of the Metheorological Service of the Italian Airforce, on the development of infrared detectors sensitive enough to detect airplain motors at a far distance in the dark.

Towards the end of the war, in collaboration with A. Loverdo and Vanna Tongiorgi[38], Cocconi initiated a study of the structure of extensive showers, which lasted many years, even after he moved, to Cornell University at the end of 1947. The whole set of these papers, and even their initial part done in Italy, even today should be considered as an example, also for text books, of a particularly elegant and significant experimental study of a very complex phenomenon[39].

After the departure of Cocconi from Milan, the investigation of cosmic rays and the instruments for their detection, with particular regard to extensive showers, was continued at the Università Statale by G. Salvini, who, with a few collaborators (A. Lovati, A. Mura e G. Tagliaferri), tackled the study of the penetrating component present in extensive showers, and the local production of penetrating particles by extensive showers[40].

In 1949 Salvini moved to Princeton, on an invitation from that University, where he studied, with success, the production ratio of charged and neutral mesons in cosmic rays[41], until his return to Italy in 1951.

As I said before, in Milan there were also a few theoreticians, in particular C. Salvetti

(1946) at the Università Statale and A. Borsellino (1947) at the Polytechnic.

Also in Padua, A. Rostagni could count on a number of clever young physicists, whose scientific products started to appear on *Il Nuovo Cimento* and other scientific journals between 1946 and 1948[42]. Among these I recall: G. Puppi (1946)[22], who, after a short period devoted to the theory of polar biatomic molecules, passed to the study of cosmic rays, which, in those years, was also the central subject of the work of the other Padua's physicists: N. Dallaporta, E. Clementel (1946), I. Filosofo (1947), P. Bassi, A. Kind and A. Loria (1948).

The revival of the interest in physics was, however, even more general. In Turin, G. Montalenti and L. Radicati (1947) joined C. M. Garelli and G. Lovera (1943), in Pavia A. Gigli (1947) joined L. Giulotto (1943), A. Ricamo (1946) started to work in Bologna, in Florence, in the field of Optics, G. Toraldo Di Francia (1944) joined F. Scandone (1943) and in that of cosmic rays, M. Della Corte (1946), T. Fazzini and S. Franchetti (1948) started to work, in Rome at the National Institute of Electroacustics O. M. Corbino, P. G. Bordoni (1947) and D. Sette (1948), in Catania G. Milone (1947).

THE RENEWAL OF CONTACTS WITH THE INTERNATIONAL CIRCLES - VIII

A circumstance of considerable importance for us in Rome, but, I believe, of relief for all physicists in Europe, was the first international conference after the war, organized by the British Physical Society and the Cavendish Laboratory. The subject of the meeting, that took place in Cambridge on 22-27 July 1946, was « Fundamental Particles and Low Temperatures »[43]. Contacts between physicists in different parts of the world had been impossible for years and this conference provided a welcome opportunity to renew old friendships and to hear, what others had been doing. Theoretical papers were presented by Bohr, Pauli, Dirac, Born and Bhabha, experimental results by Janossi, Leprince Ringuet, Clay, G. Wataghin (still in S. Paulo, Brazil), H. Anderson, L. Marshall and D. H. Wilkinson. I presented a report on the experimental work made at the Laboratorio fisico of the ISS on the diffraction of neutron and the optical theorem, Bernardini presented two reports, one on meson decay and its secondary electrons, the other on the use of low efficiency counters in coincidence experiments. Finally Ferretti gave a paper on the relationship between the absorption by nuclei of mesons at rest, in which he discussed the fact, first observed by Rasetti at Quebec, and carefully investigated only in the previous months by Conversi and Piccioni, that in iron only positive mesons decayed, as foreseen in 1940, by Tomanaga and Araki[44], in the frame of Yukawa's theory.

In those days, in Cambridge, I met again John Cockcroft, that I knew from the time of my first visit to the Cavendish Laboratory in summer 1934. This short renewal of contact and his unespected interest for physics in Italy, were for me indications of the possibility of initiatives on a European scale, that I had not yet imagined.

I had closer contact with the American collegues in the period September - December 1946, during which I held, on their invitation, a series of seminars in seven or eight United States' universities[32]. In the seminars I presented not only the work done in Rome on cosmic radiation and neutron physics[29], but also the work done in Milan by Cocconi and co-workers on extensive showers[38]. They had been prepared with the help of all the authors who had provided me with tables of data and slides of results. My seminars certainly contributed to make the work done in Italy during the war known, and, perhaps, contributed also to consolidate the invitation of Piccioni to M.I.T., starting from the winter 1946-47 and to set the premises for the invitations to Cocconi by Cornell University, starting from autumn 1947, and to G. Bernardini to give a course on cosmic radiation at

Columbia University starting from October 1948.

I began my tour in USA by participating in the International Conference on « The Future of Nuclear Science » that was held in Princeton (N. J.) at the beginning of September to celebrate the bicentenary of the foundation of that University. It was there that I reencountered for the first time since the war a number of old friends, like G. Breit, E. U. Condon, I. I. Rabi, V. Weisskopf, R. R. Wilson. It was there also that I met for the first time many people that I had known for a long time only from their scientific work: P. M. S. Blackett, Irène Curie, F. Joliot, L. Kowarski and H. A. Kramers.

In particular I met Fermi who told me that the University of Chicago was ready to offer me a chair. He also told me that work had already started on the design of an electron accelerator and a proton accelerator which were expected to make the Institute for Nuclear Studies one of the better equipped centers in the world.

Such an agreable and attractive offer strongly shook my resolve to remain in Italy, where I felt certain responsibility towards the younger researchers and a duty to contribute to the reconstruction in the changed political climate.

A contributing factor influencing my decision to return to Italy was the clear, although unexpressed attitude of Ginestra who accompained me on the trip and had no doubt as to where our duties being.

I do not know how the news of Fermi's offer reached Rome so quickly: nor why, despite my written reassurances, it had created uneasiness at the Institute. However, when we arrived in Rome from Le Havre, where we had disembarked shortly before Christmas 1946, Ginestra and I were warmly received at the Termini railroad station by a group or researchers and technicians.

UNESCO (United Nations Educational and Cultural Organization) was founded in London the 16th November 1945 and its headquarters was established in Paris in 1946. The principal conceiver of UNESCO had been the British biologist Julian Huxley, brother of the writer Aldous Huxley.

Initially the countries of the Tripartite Pact i.e. Germany, Italy and Japan, were not accepted in the new organization, and the Italian diplomacy of the De Gasperi Government had initiated the steps for obtaining the admission of our country. In the course of spring 1947 the Minister of Foreign Affairs, Count Carlo Sforza, succeded in obtaining the unofficial assurance that Italy would be admitted on occasion of the General Conference foreseen to be held at Mexico City from November 6 to December 4 of the same year.

The Italian delegation nominated by Count Sforza consisted of Guido De Ruggero, philosopher and historian, head of the delegation, Ranuccio Bianchi Bandinelli, archeologist, and I. Shortly before the departure the delegation was received at Palazzo Chigi by Count Sforza, who gave us a few general instructions.

In Mexico City we had to wait until the official request by the Italian Government, set at one of the first points of the Agenda was discussed by the General Assembly.

Immediately after the largely positive result of the votation, we entered the Assembly and started to partecipate in its work that lasted almost one month. In that period not only had I the great opportunity to get to know my two colleagues well, but also to discover their remarkable qualities as men of culture and to establish with both of them a lasting friendship. It was also a great occasion for encounters with first class personalities of other countries. Thus, for example, we had reiterated contacts with the French delegation, in particular with its president, Jean Maritain, who, shortly after the admission of Italy to UNESCO, cordially invited us to an official lunch. I, in particular, met for the first time P. Auger, that I knew very well from scientific literature and from conversation with B. Pontecorvo and Paul Ehrenfest junior, one of Auger's collaborators in the work on extensive showers. We had met on the occasion of a ski-vacation

we from Italy (E. Fubini, G. Racah, F. Rasetti, G. C. Wick and I) had organized with them, at Fent, in Austria between Christmas 1938 and Epiphany 1939.
Auger had accepted to be director of the Department of Natural Sciences of UNESCO and thus he had become one of J. Huxley's principal collaborators.
I first met him in the airplain between Paris and Mexico City, I saw him again a few times, but only very briefly, during the Conference. This, however, was enough to establish such a relationship between us that later, on a few occasions, he wrote me for scientific questions or the selection of people, and I, naturally, contacted him, when a few years later one started to consider the possibility of establishing new international organizations such as CERN[45] and ESRO[46].
Already a few months before I had come into direct contact with physicists beyond the 'iron curtain'. I partecipated to the 'Mission of the Italian Government for the Reestablishment of Cultural Relations with Poland', i.e. to the visit that, on invitation of the Polish Government, about twenty Italian professors of the various disciplines paid, between 25 May and 10 June 1947, to various Universities and cultural centers in Poland, such as Warsaw, Danzig, Gdinia, Breslau, and Cracow. From this last city we visited also the extermination camps of Auschwitz and Birkenau and had ascertained *de visu* the organization of industrial type for the extermination of Jews and political prisoners[47].
Following my visit to Poland during 1948, the Centro of CNR in Rome had, as scientific guests, from the University of Warsaw Prof. A. Soltan, a well known nuclear physicist, for about one month, and from the University of Cracow, Prof. Miensowicz, a recognized experimentalist in cosmic rays, for about two months[48].
The contacts with Polish colleagues were stregthned by G. Bernardini who participated, from 5 to 12 October of the same year, to the Symposium on Cosmic Rays, that took place in Cracow under the sponsorship of the International Union of Pure and Applied Physics (IUPAP)[49] and where he reviewed the work made in Rome on the decay of the muon. Scientific connection were extended a few years later to Indian and Japanese colleagues on occasion of other conferences organized in the frame of IUPAP of which I had become vicepresident in 1948: the International Conference on Elementary Particles held in Bombay from 14 to 22 December 1950[50] and the International Conferences on Cosmic Rays and Theoretical Physics held in Tokyo and Kyoto in September 1953, and to all of which I participated reviewing the work made recently in Italy on the interaction of high energy muons with nuclei.

A FEW PARTICULARLY IMPORTANT RESULTS - IX

Towards the end of my trip in the United States, while I was in Washington D.C., perhaps at the beginning of December 1946, I received a letter from Piccioni in which he informed me that he had finished, with Conversi and Pancini, the measurements on the decay of positive and negative mesons, brought to rest in carbon. The experiment had been proposed by Pancini, who, having learned about the paper of Tomanaga and Araki[44], immediately understood the possibility of checking their theoretical conclusions by combining the delayed coincidence technique, developed by Conversi and Piccioni[19], with that of the magnetized iron cores, of which he had become an expert during the last few years[17].
Conversi, Pancini and Piccioni, had observed that in carbon, at variance with what they had found in iron, the number of decay electrons of negative mesons was roughly equal to the number of decay electrons of positive mesons[51]. It was natural to conclude that the absorption of negative mesons by carbon nuclei was negligible.
This result appeared to indicate that the mesons constituting the penetrating component of cosmic rays had an interaction with nuclei much weaker than that foreseen

for the particles hypotized by Yukawa. The result was in clear contradiction with the theoretical conclusions of Tomanaga and Araki, who had foreseen that the negative mesons had to be absorbed even in light nuclei in a much shorter time than their mean life. I immediately wrote these results of Conversi, Pancini and Piccioni, to Fermi in Chicago. Fermi, in collaboration with E. Teller and V. Weisskopf[52], examined an alternative interpretation, i.e. the possibility that for some reason not yet clear, the time required by a meson to be reduced to rest in carbon, were much longer than its mean life. Since the result of such a theoretical study was negative, it was definitively demonstrated that the mesons of the penetrating radiation discovered by Anderson and Neddermayer in 1936[53] could not be identified with the particles hypotized by Yukawa. The enigma, as is well known, was clarified about one year later by Lattes, Occhialini and Powell[54], who at Bristol in Great Britain discovered, with the technique of nuclear photographic plates, the π-meson or pion and its decay in a neutrino and one of the particles constituting the hard component of cosmic rays, that they called μ-meson or muon.

The experiment of Conversi, Pancini and Piccioni is often quoted in international literature as the beginning of high energy physics[55].

Among many other results obtained in those years in Italy, I like to recall two more of considerable importance. In 1948, at Padua, G. Puppi[56] was the first to suggest, and anyhow independently of other authors, the universality of weak interactions, a law similar to the universality of electromagnetic interactions, expressed for any pair of charged particles by the unique constant $e^2/\hbar c$. In trying to interpret the weak decays of the nucleon, muon and pion (or the equivalent weak absorption process $\mu^- + p \to n + \nu_\mu$), Puppi pointed out that the weak interaction between the three pairs of particles (p, n), (ν_μ, e) and (ν_μ, μ) had to be the same. He illustrated the situation with a triangular graphic representation, that became known in literature as Puppi's triangle. The successive year, P. G. Bordoni, of the Istituto di Elettroacustica O. M. Corbino, during a visit to M.I.T. discovered a rather wide maximum of the curve of internal dissipation of a single copper crystal as a function of temperature, in the region around 90 kelvin and at frequencies of about 50 kHz[57]. The phenomenon, completely general is indicated in literature as 'Bordoni effect'. It provides one of the first experimental indications of the existence of dislocation in crystals.

E. A.

* *Edoardo Amaldi farewell lecture delivered at the Meeting organized by Istituto di Fisica, Facoltà di Scienze, Università di Roma, held in Rome, 7-9 September 1978.*

BIBLIOGRAPHY AND NOTES

1 L. Fermi, *Atoms in the Family*, The University of Chicago Press, Chicago, 1954; *Atomi in Famiglia*, Mondadori, Milano, 1965.

2 E. Segrè, *Enrico Fermi Physicist*, The University of Chicago Press, Chicago, 1970; *Enrico Fermi Fisico*, Zanichelli, Bologna, 1971.

3 E. Segrè, C. Perrier, *Radioactive Isotopes of Element 43*, Nature (London) *140*, 193 (1937); Journal of Chem. Phys. *5*, 712 (1937); *7*, 155 (1938).

4 The Institute of Physics of Bologna (Via Irnerio, 46) was under the direction of Quirino Majorana. It was relatively well equipped with instruments; however, due to its antiquated organization, in practice, only its director was allowed to carry out experimental research.

5 G. Polvani, *Giovanni Gentile junior*, Reale Istituto Lombardo di Scienze e Lettere, *75*, fasc. 11 (1941-42); C. Salvetti, *Giovanni Gentile junior*, Rendiconti del Seminario Matematico e Fisico di Milano, *16* (1942); A. Sommerfeld, *Zum Gedächtniss an Giovanni Gentile jun.*, Nuovo Cimento *1*, 151 (1943); G. Polvani, *Rievocazione di Giovanni Gentile jun.*, Nuovo Cimento *1*, 155 (1943).

6 G. Gentile jr., Nuovo Cimento *17*, 493 (1940); Rend. R. Ist. Lombardo, Scienze e Lettere *74*, I (1940-41), Rend. Seminari Mat. Fis., Milano (1941).
The problem was later reconsidered by A. Sommerfeld (Ber. Dtsch. Chem. Ges. *75*, 1988 [1942]) and others (H. Wergeland, Kgl. Norske Vid. Selsk. Froh. *17*, 51 [1944], G. Schubert, Zeit. Naturf. *1*, 113 [1946], H. Muller, Ann. Phys. Lpz. *7*, 420 [1950], D. ter Haar: Physica *18*, 199 [1952]); and interest in it has been revived in recent years after the discovery of particles with half integral spin greater than ½ and/or more than one internal degree of freedom: see, for instance, H. S. Green, Phys. Rev. *90*, 270 (1953); O. W. Greenberg, A. M. L. Messiah, Phys. Rev. B, *138*, 1155 (1965).

7 B. N. Cacciapuoti, E. Segrè, Phys. Rev. *52*, 1252 (1937).

8 E. AMALDI, *La vita e l'opera di Ettore Majorana*, Acc. Naz. Lincei, Roma, 1966; *Ettore Majorana, Man and Scientist*, in A. ZICHICHI (ed.), *Strong and Weak Interactions - Present Problems*, Academic Press Inc. New York (1966). *Ricordo di Ettore Majorana*, Giornale di Fisica *9*, 300, 1968.

9 G. WATAGHIN, *Eligio Perucca* (b. 28-3-1890 Potenza, d. 5-1-1965 Roma), Problemi Attuali di Scienza e Cultura, Quaderno N. 77, Acc. Naz. Lincei, Roma, 1966; R. DEAGLIO, *L'opera di Eligio Perucca*, Accad. delle Scienze di Torino, Vol. 100, 1966.

10 E. AMALDI, D. BOCCIARELLI, F. RASETTI, G. C. TRABACCHI, *Generatore di neutroni da 1000 kV*, Ric. Scientifica *10*, 623 (1939).

11 The decision to build this apparatus originates from an exchange of letters between D. Marotta, Director General of the Istituto Superiore di Sanità and E. Fermi. Carbon copies of some of those letters still exist and are kept in the Archives of the Istituto Superiore di Sanità. In a letter, of November 16th 1936, Prof. Marotta asked Fermi his opinion about the construction of a apparatus for the artificial production of radioactive substances, which had been proposed by Prof. Trabacchi, Director of the Physics Laboratory of the I.S.S. He also asked Fermi to collaborate in the running of the machine once it was built. Fermi's reply to that letter has not been found, but there is a large section of it in « Appunti per l'On. Gabinetto », Prot. R/211 of December 20th 1936 - XV, which Prof. Marotta sent to the Ministry of the Interior, to which the I.S.S. was affiliated. The introductory section of it reads: « Given the scarsity of Radium and the difficulty encountered in extracting it from minerals, recently scientists have tried to obtain artificial — so to speak — radioactive substances, i.e. substances which have properties similar to Radium and can be used in its place in most of its applications. Given the relatively easy way of obtaining them and their low cost, there could be an extension of their application and further new applications attempted. 'Artificial' radioactive substances are obtained by bombarding with neutrons some elements such as jodium, manganese, arsenic, cobalt, iridium; their production has today acquired a considerable importance. In Italy, the production of 'artificial' radioactive substances has been highly successful, due to the research of Academician Professor Enrico Fermi, whose opinion I have asked regarding the possibility and the convenience of preparing such substances in our Laboratories. He has replied as follows. 'In reply to your letter of 16th of this month, I am including the data regarding the possibility of producing artificial radioactive substances for medical purposes, using neutrons. If a 1-million volt machine were avalaible, the most convenient way of producing neutrons would be that of bombarding a beryllium or lithium target with heavy hydrogen nuclei. With a beryllium target and a 1-million volt accelerating potential differences, a yield of 1.5×10^{11} neutron per second is obtained using a current of deuterium ions of one milliamps. With lithium, the yield attainable is probably slightly higher: however the use of beryllium is more convenient because of its better chemical and mechanical features. The slowing down of neutrons allows one to use up to 50% of the neutron flux, in the most favourable cases; that means that using a one milliamp current one can obtain up to 2 Curie of artificial radioactive matter. Assuming that one could build a 1-million volt tube for a 1-milliamp deuterium ion current, one could prepare artificial radioactive substances in a sufficient quantity for medical applications. Of the many elements which can be activated using this process, one could select some with a convenient mean life according to the duration of the irradiations. Thus, for short irradiations one might advise the use of iodium (25 min period); for irradiations of intermediate lengths, manganese (2.5 hrs period); for long irradiations, arsenic (26 hrs period) or other substances such as, for example, cobalt or iridium which have periods of several months. In addition to the applications for the therapy of malignant tumors, considerable quantities of artificial radioactive substances could be used as indicators in chemical and biochemical research. Finally, I should like to state that I shall be very pleased to collaborate with the Physical Laboratory of the Istituto di Sanità Pubblica to build and make use of such an apparatus, if the necessary funds are granted to you' ».
What follows contains an evaluation of the cost (Lit. 300.000 at the start and Lit. 100.000 - yearly for the running of it) and a summary of the advantages to be derived from the construction of such a project. Moreover we were quite ready to deal with such a problem. Already, in Via Panisperna, Fermi Rasetti and I had built and run a hydrogen ion source; and during summer 1936 I had been in the United States in order to study the construction of the accelerating tube of the Department of Terrestrial Magnetism of the Carnegie Institution of Washington D. C. (E. AMALDI, *Istituti di Fisica negli Stati Uniti di America*, luglio-ottobre 1936, Vol. IV of *Viaggi di studio promossi dalla Fondazione Volta*). Furthermore, in 1936, Fermi, Rasetti and I had built a small 200 kV pilot machine at the Guglielmo Marconi Institute (Ric. Scient., *8*, 40 [1937]).

12 Reference to this meeting can be found in G. POLVANI, *Commemorazione dell'Ing. Eugenio Somaini*, Suppl. Nuovo Cimento *6*, 514 (1968); in the rubric *Notizie varie* of Ric. Scient. *15*, 674 (1945); in the journal *Il Popolo Comasco* of Friday 16 November 1945.

13 The last page of this report (p. 35), par. 5.3, includes the « Summary of proposals » where it is mentioned that one should: 1) send seven experimenters and three theorists abroad every year for three years; 2) give financial support to researchers who remain in Italy; 3) build in Italy: one cyclotron, one betatron, one atomic pile and one plant for the separation of isotopes; 4) start, as soon as possible, the production of deuterium on a semi-industrial scale. A 66-pages Appendix included the biographical data and the list of publications of twentynine physicists who were in a position to partecipate usefully in such a program.

14 Unfortunately no written and documented history of CISE exists; only a report on its activity is available, *CISE, Centro Informazioni, Studi, Esperienze, Programmi, Competenze, Attrezzature, 1977*, Segrate, Milano, 1977.

15 E. AMALDI, *Sulle ricerche di fisica nucleare eseguite a Roma nel quadriennio di guerra*, Ric. Scient. *16*, 61 (1946).

16 G. BERNARDINI, B. N. CACCIAPUOTI, B. FERRETTI, Ric. Scient. *10*, 731 (1939); G. BERNARDINI, B. N. CACCIAPUOTI, B, FERRETTI, O. PICCIONI, Phys. Rev. *58*, 1017 (1940); G. BERNARDINI, B. N. CACCIAPUOTI, B. FERRETTI, O. PICCIONI, G. C. WICK, Ric. Scient. *10*, 1010 (1939), Nuovo Cimento *17*, 317 (1940); G. BERNARDINI, B. N. CACCIAPUOTI, Ric. Scient. *12*, 981 (1941); G. BERNARDINI, M. SANTANGELO, E. SCROCCO, Ric. Scient. *12*, 321 (1941). The ratio between the electron component and the penetrating (muon) component remained unexplained until the pion was discovered and one understood that the electron component in the high atmosphere is prevalently due to gamma rays from the decay of neutral pions, and that the muons originate from the decay of charged pions.

17 G. BERNARDINI, M. CONVERSI, Ric. Scient. *11*, 840 (1940); G. BERNARDINI, M. CONVERSI, E. PANCINI, G. C. WICK, Phys. Rev. *60*, 535 (1941); G. BERNARDINI, M. CONVERSI, E. PANCINI, E. SCROCCO, G. C. WICK, Phys. Rev. *68*, 109 (1945).

The idea of using magnetized iron blocks to deflect muons had been proposed by B. Rossi, but his first experiment had been unsuccessful (Nature *128*, 300, 1931).

18 M. Ageno, G. Bernardini, B. N. Cacciapuoti, B. Ferretti, G. C. Wick, Ric. Scient. *10*, 1073 (1939); Phys. Rev. *57*, 945 (1940); G. Bernardini, B. N. Cacciapuoti, E. Pancini, O. Piccioni, G. C. Wick, Phys. Rev. *60*, 910 (1941); G. Bernardini, B. N. Cacciapuoti, E. Pancini, O. Piccioni, Nuovo Cimento *19*, 69 (1942); G. Bernardini, Zeit. f. Phys. *120*, 413 (1942); G. Bernardini, G. Festa, Rend. Acc. d'Italia *4*, 166 (1943).

19 M. Conversi, O. Piccioni, Nuovo Cimento *2*, 40, 71 (1944).

20 F. Rasetti, Phys. Rev. *59*, 613 (1941); *60*, 198 (1941); P. Auger, R. Maze, R. Chaminade, Compt. Rend. *213*, 381 (1941); R. Maze, R. Chaminade, Compt. Rend. *214*, 266 (1942).

21 B. Rossi, N. Nereson, Phys. Rev. *62*, 417 (1942); *64*, 199 (1943).

22 The year, indicated in brackets, after one or more names is hereafter that of the year in which the people quoted have appeared as authors of articles in *Il Nuovo Cimento*.

23 E. Amaldi, D. Bocciarelli, F. Rasetti, G. C. Trabacchi, Ric. Scient. *10*, 633 (1939); Phys. Rev. *56*, 881 (1939).

24 M. Ageno, E. Amaldi, D. Bocciarelli, B. N. Cacciapuoti, G. C. Trabacchi, Ric. Scient. *11*, 302, 413 (1940), *12*, 134 (1941); Phys. Rev. *60*, 67 (1941).

25 Ageno was not present at the final discussions but he had previously examined this problem with me.

26 E. Amaldi, D. Bocciarelli, G. C. Trabacchi, Ric. Scient. *11*, 121 (1940); M. Ageno, E. Amaldi, D. Bocciarelli, G. C. Trabacchi, Ric. Scient. *12*, 830 (1941); E. Amaldi, D. Bocciarelli, B. Ferretti, G. C. Trabacchi, Ric. Scient. *13*, 502 (1942); E. Amaldi, D. Bocciarelli, G. C. Trabacchi, Naturwiss. *30*, 582 (1942).

27 M. Ageno, E. Amaldi, D. Bocciarelli, G. C. Trabacchi, Nuovo Cimento *1*, 253 (1943); M. Ageno, E. Amaldi, D. Bocciarelli, B. N. Cacciapuoti, G. C. Trabacchi, Naturwiss. *31*, 231 (1943); M. Ageno, E. Amaldi, D. Bocciarelli, G. C. Trabacchi, Phys. Rev. *71*, 30 (1947).

28 N. Bohr, R. Peierls, G. Placzek, Nature (London) *144*, 200 (1939).

29 E. Amaldi, D. Bocciarelli, B. N. Cacciapuoti, G. C. Trabacchi, Nuovo Cimento *3*, 15, 203 (1946).

30 B. N. Cacciapuoti, Nuovo Cimento *18*, 114 (1941); *1*, 126 (1943). G. Boato also worked on this subject. In 1948 he went to Zürich for six months to work under Prof. Clusius.

31 In order to study the constructional features of the betatron built by D. W. Kerst, B. N. Cacciapuoti went to Urbana (Illinois) from Nov. 1945 to February 1946 with a subsidy from the CNR. Prof. G. M. Pestarini and his assistant Dr. Salmi, of the Faculty of Engineering of Rome University contributed very effectively in the designing of the magnet. News of this work is to be found in Ref. 32.

32 E. Amaldi, *Centro di Studio per la Fisica Nucleare, Attività svolta durante l'anno 1946*, Ric. Scient. *17*, 391 (1947).

33 E. Amaldi, *The Production and Slowing down of Neutrons*, pp. 1-659 of Vol. 38/2, Handbuck der Physik, Springer Verlag, Berlin, 1958.

34 G. Cocconi, Ric. Scient. *10*, 733 (1939); Naturwiss. *27*, 740 (1939); Nuovo Cimento *16*, 78, 447 (1939); Phys. Rev. *57*, 61 (1939); Ric. Scient. *11*, 788 (1940); Naturwiss. *29*, 335 (1941); Zeit. f. Phys. *118*, 88 (1941); Phys. Rev. *60*, 532, 533 (1941); Naturwiss. *30*, 328 (1942).

35 G. Polvani, G. Cocconi, Ric. Scient. *12*, 410 (1941).

36 The board of examiners for the selection of the Professor for the Chair of Experimental Physics at the University of Palermo, was formed by Professor A. Lo Surdo (President), L. Tieri, E. Persico, G. Polvani, A. Carrelli (Secr.) and it concluded its work on Nov. 18th, 1942. The selected candidates were: 1st E. Medi, 2nd G. Cocconi, 3rd V. Polara, who were called respectively by the University of Palermo, Catania and Messina. Among the candidates who were declared *maturi* (eligible) the following were included: B. N. Cacciapuoti (who had a vote for the 1st place, one for the 2nd and two for the third) and G. Occhialini (who was favourably judged although he did not receive any votes).

37 Giorgio Fea had graduated with Rasetti in 1934 discussing a Thesis on the construction and operation of a boron-trifluoride cloud chamber. In 1935 he joined the Metheorologic Office of the Italian Air Force, but continued to partecipate in scientific research in the Institute as an « assistente volontario » with Rasetti and, later, with Lo Surdo.

38 G. Cocconi, A. Loverdo, V. Tongiorgi, Nuovo Cimento *1*, 49, 314 (1943); *2*, 14, 28 (1944).

39 G. Cocconi, Rev. Mod. Phys. *21*, 26 (1949).

40 G. Salvini, Nuovo Cimento *3*, 283 (1946); Ric. Scient. *17*, 914 (1947); A. Mura, G. Salvini, G. Tagliaferri, Nuovo Cimento *4*, 10, 102, 279 (1947); Nature (London) *159*, 367 (1947); Rend. Lincei *2*, 437 (1947); Phys. Rev. *73*, 261 (1948); *75*, 1112 (1949); A. Lovati, A. Mura, G. Salvini, G. Tagliaferri, Nuovo Cimento *6*, 207, 291 (1949); Nature (London) *163*, 1004 (1949); Nuovo Cimento *7*, 36, 943 (1950); Phys. Rev. *77*, 284 (1950).

41 Y. Kim, G. Salvini, Phys. Rev. *85*, 921 (1952); *88*, 40 (1952); G. Salvini, Nuovo Cimento *10*, 1018 (1953).

42 It is not possible to cite here all the papers published by these authors in that period - not even if one could limit oneself to the main works; a complete picture can be obtained from Vol. 3 (1946), Vol. 4 (1947) and Vol. 5 (1948) of the Ninth Series of the *Nuovo Cimento*.

43 Report on an *International Conference on Fundamental Particles and Low Temperature*, held at the Cavendish Laboratory on 22-27 July 1946, Vol 1. Fundamental Particles, London, 1947.

44 S. Tomanaga, G. Araki, Phys. Rev. *58*, 90 (1940).

45 E. Amaldi, *First International Collaborations between Western European Countries after World War II in the Field of High Energy Physics*, p. 326 of *Rendiconti della Scuola Internazionale di Fisica Enrico Fermi*, LVII Corso, Varenna, 31 luglio - 12 agosto 1972, on ***History of Twentieth Century Physics***, C. Weiner (ed.), Academic Press, New York, 1977.

46 W. B. Walsh, *Science and International Public Affairs*, International Relations Program, The Maxwell School of Syracuse University, 1967.

47 For more detail on this 'Mission' to a Poland almost completely destroyed but with a very high dignity and strong will of reconstruction, see: G. Devoto, *Polonia*, Edizioni di Letteratura, Vallecchi, Firenze, 1948; P. Bilinski, *Biblioteca e Centro di Studi a Roma dell'Accademia Polacca delle Scienze, nel 50° Anniversario della Fondazione*, 1927-1977, Conferenze e Studi 70, p. 103. The participants were: M. Aloisi, E. Amaldi, R. Cacciapuoti, V. Caglioti, G. Calogero, G. Devoto, G. Fiocco, G. Mayer, G. Natta, M. Picone, N. Sapegno, G. Supino, E. Volterra and others.

48 E. Amaldi, *Centro di Studio della Fisica Nucleare e delle Particelle Elementari, Attività svolta durante l'anno 1948*, Ric. Scient. *20*, 269 (1950); . . . ,

Attività svolta durante l'anno 1949, Ric. Scient. *20*, 927 (1950); ..., *Attività svolta durante l'anno 1950*, Ric. Scient. *21*, 1149 (1951); ..., *Attività svolta durante l'anno 1951*, Ric. Scient. *22*, 1175 (1952).

49 Symposium on Cosmic Rays, Cracow, October 1947, IUPAP Document RC48-1.

50 E. Amaldi, *La Ricerca Scientifica in India*, Ric. Scient. *21*, 1305 (1951).

51 M. Conversi, E. Pancini, O. Piccioni, *3*, 372 (1945); Phys. Rev. *71*, 209 (1947).

52 E. Fermi, E. Teller, V. Weisskopf, Phys. Rev. *71*, 314 (1947); E. Fermi, E. Teller, Phys. Rev. *72*, 399 (1947).

53 C. D. Anderson, S. H. Neddermeyer, Phys. Rev. *50*, 263 (1936).

54 C. M. Lattes, G. P. Occhialini, C. F. Powell, Nature (London) *160*, 453 (1947).

55 S. S. Schweber, H. A. Bethe, F. de Hoffmann, *Mesons and Fields*, Row, Peterson & Co., Evanston (Ill.), 1956; L. W. Alvarez, *Recent Developments in Particle Physics* in *Les Prix Nobel of 1968*, Imprimerie Royale P.A. Norstedt & Söner, Stockholm, 1969, p. 125.

56 G. Puppi, Nuovo Cimento, *5*, 587 (1948); *6*, 194 (1949).

57 P. G. Bordoni, Ric. Scient. *19*, 1951 (1949); J. Acoust. Soc. Am. *26*, 495 (1954).

The Years of Reconstruction

Part II

EDOARDO AMALDI
Università di Roma

THE RE-EXAMINATION OF THE RESEARCH PROGRAM - X

I returned to Rome from the United States at the end of December 1946 and with Bernardini and Ferretti and all young people of the 'Center' we re-examined in detail our research program.

From the Cambridge Conference and from my trip to the United States, it appeared evident that in the field of fundamental research we had maintained a good level during all the war[58]. On the contrary we had remained behind in many experimental techniques, not to speak of the applications of neutron physics.

Therefore it was necessary to make an effort to bring our experimental techniques up-to-date[59]. In carrying out this program in the following years we were helped by various people, in particular by P. M. S. Blackett and B. Rossi. A. Loria worked for about three years (1947-48, 1948-49, and 1950-51) and S. Sciuti for one year (1952) in Blackett's Laboratory at the University of Manchester. After Blackett moved to Imperial College in London in 1953, B. Brunelli went there to work with C. C. Butler during 1954-55. D. Broadbent came to work at the Istituto Guglielmo Marconi in Rome from Manchester, in the period 1950 to 1952, and P. Astbury from London from 1954 to 1956.

Our greatest help came from B. Rossi who, first at Los Alamos during the war and later at M.I.T., had developed various techniques for the fast detection of particles, largely described in the nice book he published with Staub[60]. In those years the following physicists from Italy went to work in Rossi's laboratory at M.I.T.: O. Piccioni (in 1947), P. Bassi (in 1951-52), R. Giacconi (in 1954), L. Scarsi (in 1957-60), O. Occhialini and C. Dilworth (in 1959-60), A. M. Conforto (in 1963), A. Egidi (in 1963-64). B. Rossi, however, did not only help young Italian physicists. He also offered hospitality to a number of young physicists of other countries, in particular to a few belonging to the group of the Ecole Polytechnique guided by Louis Leprince Rinquet.

I like to recall, among these, Bernard Gregory, who worked with Rossi from 1947 to 1950, and took a Ph. D., and three other also remarkable young French physicists who spent about one year each at M.I.T.: Francis Muller at about the same time as Gregory, Charles Peyrou in 1952-53 and, some time later, the theoretician R. Stora.

Let us go back to the decisions taken at the Istituto di Fisica Guglielmo Marconi. Besides the effort to bring our experimental techniques up-to-date, it was clearly necessary to restrict the spectrum of researches tackled by the 'Centro'. We thus decided to abandon the idea of constructing a 20 MeV betatron, not only because our means were inadequate, but also because we could not count on the support of Italian industry, which was totally engaged in the work inherent to the general reconstruction of the country[59].

We also decided to abandon, for the moment, the investigation of neutron physics, since the CISE had already started to tackle in a promising way its applications, while our instrumentation for years to come would not be in condition to compete with that at the disposal of American colleagues for most fundamental problems[59].

Thus we concentrated our whole effort in the study of cosmic radiation. In 1947 Bernardini conceived and, with the help of Pancini and Conversi, directed the construction of the Laboratorio della Testa Grigia (L.T.G. altitude: 3,500 m) on the Swiss - Italian border near the Theodul Pass[61]. The site was well known to the Roman physicists, who had made two expeditions of a few months each there, the first one during the winter 1940-41[62], the second during the winter 1942-43, using as a laboratory a small room of the upper station of the aerial cableway. Also Cocconi and Tongiorgi had spent a few months there to take measurements in the same place, but the construction of an *ad hoc* laboratory allowed an activity of much greater amplitude during the whole solar year. Many physicists from the Universities of Rome, Bologna, Milan and Turin worked at LTG for about ten years. Among these I like to recall, besides Bernardini, Pancini, Conversi and Mezzetti, a large number of younger people such as R. Bizzarri, B. Brunelli, C. Castagnoli, M. Cervasi, M. Cini (who later started to work in theoretical physics), G. Fidecaro, A. Gigli, A. Michelini, S. Sciuti, G. Stoppini, E. Zavattini and A. Zichichi. These last, of course, at later times[63]. The LTG is still used today for cosmic ray research by the Laboratorio di Cosmo-Geofisica of the CNR, in Turin.

At the same time, in Rome, a series of experiments were started to test if muons also behaved in the same way, at high energy, as demonstrated for muons at rest by the Conversi, Pancini and Piccioni experiment, i.e. a complete absence of strong interactions. The first of these experiments, made by Fidecaro and me[64] with an hodoscope of more than one hundred Geiger counters, showed that the measured elastic cross section of a few GeV muons against lead nuclei was in agreement with that computed by taking into account only the electromagnetic interaction.

The same conclusion was obtained shortly after by Castagnoli, Gigli, Sciuti and me for the inelastic cross section measured at such a depth underground, that the mean energy of the muons is about 10 GeV[65]. Similar results were obtained by means of a cloud chamber by Lovati, Mura, Succi and Tagliaferri[66] of the University of Milan. Some time later my young collaborators repeated the experiment at a greater depth, and confirmed the absence of strong interactions for muons of energy as large as about 50 GeV[67].

IN SEARCH OF YOUNG STAFF OUT OF ROME - XI

During the academic year 1946-47 very few universities, besides Rome, Padua and Milan, were in condition to provide the means for experimental research work for the students' thesis. This was in particular true for Pisa which had been badly damaged by the allied air-raids. In order to remedy this inconvenience, at least partially, the 'Centro' of Rome constituted four scholarships of Lit. 100.000 each, for students preparing an experimental thesis, with funds given by a few industries (Snia Viscosa, Ente Nazionale Metano) and persons (Signor F. Solbiati). Three of these scholarships were assigned to students of the Scuola Normale Superiore di Pisa, i.e. to Carlo Castagnoli, Alfonso Merlini, and Bruno Rumi, the fourth scholarship was given to a young chemist of the University of Genoa, Giovanni Boato, who had expressed the intention also to take the laurea in Physics[59].

A similar situation was encountered in 1951 but for theoretical physics. The Scuola Normale Superiore of Pisa sent to Rome, with one of its scholarship, the undergraduate student R. Gatto, who prepared his thesis under the guidance of Ferretti and after his laurea remained at the Istituto Guglielmo Marconi, first with a scholarship of the Fondazione Della Riccia, and later with funds of the 'Centro' and INFN.

In 1953 it occurred to me that the number

of young Sicilian physicists working at our Institute was relatively small. Fidecaro, Gatto and Beneventano were all of Sicilian origin, but most probably, as we knew from past experience, there were other capable young people trying, in spite of many difficulties, to work in the Institutes of Physics of the Universities in Sicily.

I discussed the matter with a few colleagues, in particular with Ballario, who had the position of first assistant and was of great help in dealing with all possible problems. He contacted Mariano Santangelo from Palermo, who a little later told us of the existence in Palermo of a young Trapanese, « slightly wild but intelligent ».

Ballario sent him an invitation to come to Rome for an interview. The young man, Antonino Zichichi, accepted the invitation, came to the Institute, and in the conversation that Ballario and I had with him, the judgement of Santangelo was fully confirmed. We proposed that he should take a written examination under Ballario's supervision. Zichichi accepted our proposal and solved the problem we had assigned to him quite well. Consequently he was appointed a research worker at our Institute and put to work in Ballario's group which had constructed a cloud chamber and installed it at LTG[63].

A FEW NOTABLE PERSONAGES - XII

Besides G. Bernardini there was, in those years, another key figure, Bruno Ferretti. He had been Fermi's assistant, in 1947 he had been nominated associate professor of Theoretical Physics at the University of Milan and in 1948 he had moved to the chair left vacant in Rome by G. C. Wick. In those years not only had he already published many important papers on various aspects of cosmic radiation, on the properties of muons and pions, on field theory and quantum electrodynamics, but he had also an exceptional influence during and after the war in Rome, Milan and more in general in all Italian Universities on the formation of young theoreticians. Among these I like to recall E. Corinaldesi, M. Verde, B. Zumino, G. Morpurgo, R. Gatto and G. Bernardini[48].

Ferretti also had a remarkable influence on many experimentalists, to whom he frequently suggested significant experiments, or the most clever way of designing them. Furthermore he used to discussing with them, always in great details, their experimental results with a quite exceptional competence, insight and generosity. He contributed considerably to establishing a close relationship with a few active British groups, by going, with a fellowship from the British Council, for ten months to Manchester, in the laboratory of P. M. S. Blackett in 1946, and giving, in 1947, a course on cosmic radiation at the University of Birmingham. Here he established a close relationship with Rudolf Peierls, with whom he published an interesting paper on the theory of radiation damping and the propagation of light[68].

Later, as a member of the directorate, first of CNRN and later of CNEN, Ferretti contributed considerably to the beginning of applied nuclear research in our country.

It was mainly Ferretti's personality that, in 1952, attracted and permanently fixed in Rome an occasional visitor, Bruno Touschek, so different from all of us and that it was so pleasant and useful to have in our country for sixteen years[69].

Another fact of considerable importance for the future of physics in Italy and for the enlargement of the fields of research seriously studied in our country was the appearance on the scene of the Institute, at the beginning of 1945, of Ing. Giorgio Careri. Shortly before he had taken his laurea in chemical engineering, but he did not want to work as an engineer and even less as chemist; he wished to learn physics at the Institute, but not the type of physics we were doing. We discussed his problems together: I was very much in favour of his general frame of mind since I had always felt uneasy about the disharmonious development of physics in Italy. But it is not easy to teach what one does not know, at a level of serious and productive research.

The solution of the problem was found by him when he designed and constructed a mass spectrometer, not planned for nuclear measurements, but for the quantitative analysis of the isotopic composition of various elements, in view of a few applications he made later to the abundance of deuterium and argon 40 (produced in the decay of potassium 40) in vapour of endogenous origin [70]. Later he passed to the study of simple liquids, such as liquid metals, and finally to that of liquid helium and of its superfluid phase.

Careri discovered his true vocation, thermodynamics and statistical mechanics, on the occasion of the International Conference organized by the SIF in Florence, in 1949. Besides Bohr, Heisenberg, Kramers and Pauli, J. G. Kirkwood, J. E. Mayer and L. Onsager[71] were also present. Careri was in some way followed by Giovanni Boato, who arrived in Rome in 1947 took his laurea in physics at the end of the same year, and remained to work at the Istituto Guglielmo Marconi.

Careri and Boato, together with Luigi Giulotto of Pavia, a spectroscopist and expert in nuclear and paramagnetic resonances, and Adriano Gozzini of Pisa, a light and microwave spectroscopist, are, perhaps, the people that more than anybody else share the great merit of having always cultivated, and always at a very good level, the researches on the structure of matter.

G. Toraldo Di Francia in Florence played a similar role for Optics. All these were followed immediately or later by many others, in Pavia: A. Loinger (1949), G. Chiarotti, G. Gulmanelli (1951), F. Bassani (1954); in Milan: R. Fieschi, F. G. Fumi (1951), A. Vaciago (1954), F. Tosi (1955); in Pisa: E. Polacco (1953); in Rome: D. Sette (1948), A. Paoletti (1954), G. Caglioti (1955); in Palermo: U. Palma (1955), and many others that I can not quote for lack of space, but to all of whom the development reached today in many important and highly diversified fields of physics is due.

THE ITALIAN PHYSICAL SOCIETY (SIF) AND THE INTERNATIONAL SCHOOL OF VARENNA - XIII

Quirino Majorana (1871-1957)[72], director of the Istituto di Fisica Augusto Righi at the University of Bologna, had been President of SIF and director of *Il Nuovo Cimento* since 1925. The years of war had weighted heavily on the life of SIF and on its journal, *Il Nuovo Cimento*, which now was reduced to a pitiful condition. Only during the last years Bernardini had become vice-President of SIF, had taken the direction of its journal and started to reverse the course of its decay. A remarkable desire for revival had already appeared in Como on the occasion of the meeting promoted and organized by Polvani in November 1945 for the bicentenary of the birth of Alessandro Volta, that I mentioned above (§ V)[12]. In public discussions, as well as private conversations held in Como, the need for meeting again and giving a periodic character to similar physicists' encounters clearly appeared. Thus, a few months later, in spring 1946, the Italian physicists met again in Bologna, at the Institute of Via Irnerio, and decided to give new life to SIF. They elected president of SIF Giovanni Polvani (b. in Spoleto December 17, 1892; d. in Milan August 11, 1970).

He fulfilled this function for 15 years, from 1947 to 1962, when he was nominated President of C.N.R.[73].

In Bologna, at the same time, I was elected vice-President, a position that I kept until October 1950, when Ageno was elected in my stead. This position gave me the opportunity of closely following what Polvani was doing.

He used to regularly convene the Directive Council of SIF and discuss, with the vice-President and members of the Council, all matters concerning the social life and the management of *Il Nuovo Cimento*, but when one passed to carrying out the decisions, it was he, Polvani alone, who personally saw to everything. For me, vice-President, it was only possible to express my approval of the steps he had made and, in most cases, my admiration for the way in which he had made them. I believe that my successor, Ageno, succeeded better than I, in contributing to the life of SIF.

In those years Polvani rapidly brought *Il Nuovo Cimento*, to take the role of a great international scientific journal, every year he organized a national conference of SIF in a different city; he created a series of prizes for young researchers and secondary school physics teachers with funds that he succeeded in obtaining from public and private bodies, and which were confered on the occasion of the national conferences. Polvani also organized a number of international conferences such as that on Statistical Mechanics held in Florence in May 1949[71] and the other on cosmic rays held in Como in September of the same year[74]. Enrico Fermi returned to Italy for the first time for the latter conference, which closed with a visit of all participants to the L.T.G.

At the national conferences, quite often, the enthusiasm of the participants, in great part young, expressed itself in cheerfully boister ous forms, especially after the speech that Polvani used to pronounce at the social banquet or at the end of the general lecture on the status of advancement of physics that often was given by Ferretti with uncommon vivacity and depth.

The work, as President of SIF, of which Polvani rightly was most proud, was the creation of the International School of Physics at Villa Monastero, in Varenna on the Como Lake.

The person that conceived and created the first summer school of physics has been the French physicist Cécile Morette, who married Dr. B. S. De Witt shortly before the first course, in 1951, of the summer school in Theoretical Physics at Les Houches. Her example was followed by Polvani who, in 1953, founded the Scuola Internazionale di Fisica di Varenna. As mentioned in the address that C. Morette De Witt pronounced at the cerimony of opening of the first course at Varenna[75], the program of the Italian School was in some way complementary to that of Les Houches since it regarded mainly the experimental aspects of modern physics while the French School was devoted to theoretical physics.

The subjects of the two first courses of the Varenna School (1953[75] and 1954[76]) both regarded elementary particles. But, already in 1954, Polvani started to include among the activities taking place at Villa Monastero, not only courses in mathematics and all natural sciences but also in humanities, history and social sciences. For this aim he worked unremittingly for the creation of the 'Ente Villa Monastero', which, in collaboration with the Administration of Como's Province, even today regulates and coordinates the multifold activities that take place in the beautiful Villa Monastero.

A FEW REMOVALS AND NOMINATIONS - XIV

After the war, in 1946, it was my unpleasant duty to have to write to ask Rasetti to take a decision about his university position in Italy. Since he had been put at disposal of the Ministery of Foreign Affairs, he still formally occupied the chair of Spectroscopy. Since 1959, the lectures had been given by Cacciapuoti. Rasetti immediately resigned and at the end of 1947 the Faculty of Natural Sciences called Gilberto Bernardini to the chair of Spectroscopy. However, from October 1948 onwards, Bernardini began to spend a large fraction of the year in United States, first at Columbia University (New York) and later at the University of Illinois (Urbana) and the spectroscopy course was given again by a young (annually appointed) assistant professor, Carlo Franzinetti.

On the 7th June 1949 Antonino Lo Surdo died in Rome. Not only did the chair of Fisica Superiore became vacant, but also the course of Fisica Terrestre, the direction of the Istituto di Fisica Guglielmo Marconi, and that of the Istituto Nazionale di Fisica required new appointments. The INC initially created within the CNR in 1939, was later made into an autonomous body, dependent on the Minister of Public Education, who proceeded to nominate as successor of Lo Surdo, Enrico Medi. Since 1942 Medi had been professor of Fisica Sperimentale at the University of Palermo[36] but, as I said before, he had been for many years assistant of Lo Surdo. For the chair of Fisica Superiore of the University of Rome, I enquired, very discretely, from the various colleagues that had been forced to leave Italy years before, in particular Segré and Pontecorvo, who, judging from their frequent visits to our country, appeared to be the most inclined to return. Finally I got a positive answer from Enrico Persico who had been at the Université Laval of Quebec (Canada) since 1947, when he had left the University of Turin for the chair left by Rasetti[78], who had moved to Johns Hopkins University in Philadelphia (Pa.).

At about the same time the Ministry of Public Education assigned to the Faculty of Natural Sciences a chair of Fisica Terrestre to which, about one year later, E. Medi was called.

Finally the Faculty of Natural Sciences appointed me director of the Istituto di Fisica Guglielmo Marconi. Mario Ageno had the position of first assistant, having succeeded Cacciapuoti who had occupied the same position since 1943 and in 1948 had been asked by UNESCO, first to prepare the « Conference of Scientific Experts of UNESCO for Latin America » and later for the creation in Montevideo of the 'Center for Scientific Cooperation of UNESCO in Latin America'.

When, at the end of 1949, Ageno, with Occhialini and Cacciapuoti, was put in the 'tern of winners' of the national competition for the chair of Fisica Superiore of the University of Cagliari[79], his successor as first assistant was E. Pancini, who, at the end of 1954, in his turn was put with Conversi and Lovera in the 'tern of winners' of the national competition for the chair of Physics of the University of Sassari.

They were called to the University of Sassari, Pisa and Modena[80] and the position of first assistant passed to Carlo Ballario. Ageno, Pancini and Ballario, in successive periods of time, were of extraordinary help in the organization of the Institute and in looking for, and finding the solution of an infinity of little problems that continuously crop up in a department in rapid expansion.

THE GROWTH OF PHYSICS IN ITALY - XV

In those days physics was expanding everywhere in the world, but mainly in the countries technologically more advanced. The reasons of such a rapid growth are multifold, complex and not easy to analyse[81]. Certainly, among these, not least was the impact on the imagination of the man of the street, and therefore of the political circles, of the success of a few technological applications the roots of which were in advanced fields of physics, such as radar and nuclear energy in their peaceful and military forms. Irrespective of their ideological orientation, the governments had become more inclined than in the past, not to grudge funds for fundamental research.

This general tendency had penetrated, although with difficulty, even in Italy and the existence in the country of a few oases of rather high levels of activity in advanced research, constituted an element of great attraction for the new generations.

In Rome, as well as in other universities, the number of students anxious to work seriously was steadily increasing, but the university population was far from such numbers as to create problems for the staff or for the library and laboratory facilities. This was also because teaching duties were assigned to young researchers, who, in most cases, fulfilled them with enthusiasm and care.

An outline of the researches tackled by the 'Centro per lo Studio della Fisica Nucleare e delle Particelle Elementari' is provided by the list of scientific papers published each year by the research staff which is given for the first five years of the Center's life at the end of the Annual Report for 1950, and for the successive years at the end of the corresponding annual Reports[48].

Bet-veen 1949 and 1954 many new names start to appear in addition to those mentioned above: G. Baroni, D. Cunsolo, F. Mariani, G. Nencini (1950), A. Alberigi Quaranta, F. Bachelet, E. R. Caianello, A. M. Conforto, R. S. Liotta, G. Martelli, G. Stoppini (1951), G. Cortellessa (1952), F. Bolle, L. Mezzetti (1953), M. Beneventano and A. Paoletti (1954).

Some of the lines of research, already quoted above and that constituted the leitmotiv of the Center's activity, started to be tackled with new techniques. For example the nuclear disintegrations produced by cosmic rays were investigated using two new techniques: that of nuclear photographic plates, initially introduced in Italy by G. Bernardini, and later developed by his closest collaborators and by C. Franzinetti who, from September 1947 to the end of 1950, had been working at Bristol (G. B.) in the laboratory where C. F. Powell and G. Occhialini had given an extraordinary impetus to the development of this technique. In a series of experiments a few packets of nuclear plates had been exposed to cosmic rays at high altitude by G. Cortini, A. Manfredini and A. Persano, by means of balloons given as gift to the Center, partly by M. A. Turve, director of the Department of Terrestrial Magnetism of the Carnegie Institution of Washington, D.C., and partly by S. K. Allison, director of the Institute for Nuclear Studies of Chicago.
The same phenomenon was studied at the L.T.G. with the technique of ionization chambers, developed by the group of B. Rossi at M.I.T., in order to establish the main features of the generating radiation (I. F. Quercia, B. Rispoli and S. Sciuti) and in relation to this production in different materials as well as in association with extensive showers (E. Amaldi, C. Castagnoli, A. Gigli and S. Sciuti).

The experimental determination of the positive excess was extended up to altitudes of 5,100 and 7,300 meters by means of counters telescopes with magnetized iron cores, placed on board airplains kindly put at their disposal by the Aeronautica Militare (I. F. Quercia, B. Rispoli and S. Sciuti).
The decay processes and more generally the nature of secondary particles generated locally by cosmic rays were studied by means of cloud chambers (C. Ballario, G. Bernardini, A. De Marco and A. Loreti, A. Mura, G. Salvini, G. Tagliaferri).
In Rome, at the Laboratorio Fisico of the ISS, a number of experiments was carried out, such as, for example, a precision test of the Klein-Nishina formula (M. Ageno, G. Chiazzotto and R. Querzoli) and an extended study of scintillations in liquids and of the corresponding processes of molecular excitation (M. Ageno).
Furthermore in 1950, at the Istituto Guglielmo Marconi, Ballario and Beneventano started a new activity that became permanent, i.e. the datation, by means of carbon 14, of samples of archeological and geological interest, which, for the chemical part, was, and still is today, based on a collaboration with the Institute of Geochemistry of the University of Rome (Cortesi *et al.*).
Another research of considerable interest, also because it was a forerunner of spark chambers, was the construction during 1950 by F. Bella and C. Franzinetti of plane counters, about one hundred times faster than Geiger counters which, at that time, still were the most commonly used particle detectors.

A similar expansion was also taking place in other universities. The following names were appearing for the first time on *Il Nuovo Cimento*: R. Levi Setti (1949) of Pavia; P. Budini (1949) and C. Cernigoi (1954) of Trieste; G. Quareni (1951), D. Brini, O. Rimondi and P. Veronesi (1952), M. Galli (1953) of Bologna, A. Bonetti and G. Tomasini (1951) of Genoa.
The research activity was specially intense in Padua, and so the CNR on the 1st January 1947 created at the Institute of Physics of that University a 'Centro per lo studio degli ioni veloci', similar to that created in Rome fifteen months before. The expediency of such a decision was confirmed by the

large number of capable young researchers that shortly afterwards entered their production phase: M. Ceccarelli, A. Merlin (1949), F. Ferrari and C. Villi (1952), M. Baldo, M. Cresti, B. Vitale, G. Zago and T. Zorn (1953).
Also in Turin things were moving, at the Polytechnic as well as at the University.

G. Wataghin's return to Italy from Brasil in 1949, certainly had an influence on the young people, among which I recall: F. De Michelis, A. Gamba, R. Malvano (1949), M. Cini, S. Fubini, M. Panetti (1951), P. Brovetto, S. Ferroni (1952), A. Debenedetti, T. Regge, M. Vigone (1953), B. Bosco, G. Ghigo and A. Tallone (1954). Consequently the C.N.R. created, the 1st July 1951, a 'Centro sperimentale e teorico di Fisica Nucleare' with its seat in the Institute of Physics of the University of Turin.
During the same year, 1949, also G. Occhialini returned to Italy from Belgium. He had remained in S. Paulo until 1945 and then had moved to Bristol where he had worked with C. F. Powell until 1948. In that year he moved to the 'Centre de Physique Nucléaire' in Brussells and returned to Italy after he had been selected, as I said above, in the competition for the chair of Fisica Superiore of the University of Cagliari[79].
Nominated professor at the University of Genoa he had a strong influence on the research activity in that University. In 1952 he was called to the University of Milan, where, in August 1951, a fourth Center of CNR had been created, where, in 1951 C. Succi and in 1952 L. Scarsi joined the physicists already mentioned above.
Under the influence of Occhialini, and of his wife Constance Dilworth, not only were groups of experts in nuclear emulsion technique formed in Genoa and Milan, but also the already existing groups using this same technique in other Universities were appreciably reinforced.
At the same time, A. Bisi, G. Bertolini (1952) and L. Zappa (1953) began working with G. Bolla who had moved to the chair of Fisica Generale of Milan Polytechnic in 1950.

A NEW FORM OF ADMINISTRATION OF RESEARCH - XVI

It was at that time that the necessity for a close coordination of the research activities of the four 'Centers' of the CNR emerged very clearly and, under the parallel action of G. Bernardini, E. Perucca, President of the Physics Committee of C.N.R., and mine, the Istituto Nazionale di Fisica Nucleare (INFN) was created the 8th of August 1951. This body has a new structure, the aim of which is a grouping and coordination of most research activities of similar nature carried on in different universities, completing and integrating, whenever necessary and/or possible, the staff and the financial means assigned by the Ministry of Education.
According to its statute, the INFN has the following aims[82]:
a) to promote, coordinate and carry on theoretical and experimental researches in the field of fundamental nuclear physics; i.e. researches aiming to enlarge the knowledge of facts and laws providing the foundations of the physics of nucleus and elementary particles, irrespective of their importance or interest for practical applications;
b) to promote and carry out researches in those fields of pure science, that could turn out to be of interest for the progress of nuclear science;
c) to promote and support initiatives capable of increasing and deepening the scientific culture and instruction in the fields mentioned in points (*a*) and (*b*) above; in particular to deal with the training of an always larger number of scholars of a high standard in those fields;
d) to maintain the relationships and develop the collaboration with Organizations, Bodies and Institutes that, in Italy and abroad, deal with the same problems, in agreement with general lines established by CNRN (later CNEN).
The first President of INFN was G. Bernardini who, in that period, spent about one half of his time in the USA, first at Columbia University and, later, at the University of

Illinois. Thus he ensured automatically a close contact with many American physicists who already had particle accelerators at their disposal. In those years Bernardini, in collaboration with a few researchers of the Rome « Center » (M. Beneventano, G. Stoppini, L. Tau) studied the photoproduction of pions by means of nuclear emulsions exposed to gamma rays, produced by means of the betatron of the University of Illinois. In the meantime Rostagni had constructed the Observatory of Marmolada (altitude 2,030 m)[83], where besides the physicists from Padua, also E. H. S. Burhop and H. Massey from the University of London and Deutschmann from the University of Munich worked[84].

The launching of balloons for the exposition of nuclear plates to cosmic rays at altitudes of about 30 km, made by the University of Padua in collaboration with that of Bristol in 1947-48[85], was the prelude to a series of international expeditions, promoted by C. F. Powell of the University of Bristol and based on the support of the Milan, Padua and Rome Sections of INFN. The first expedition took place in 1953[86], under the auspices of CERN, which was entered in its 'Planning Stage' as I will say below.

Naples and Cagliari were chosen as the most convenient of the available bases in the Mediterranean. Thirteen Universities participated in this expedition, which was the first successful attempt of recoving packets of nuclear plates deposited in the sea after the balloon flight.

It also allowed the collection of information about winds at high altitude which was of great importance for the expedition of the following year.

This second expedition also took place in Sardinia, during June and July 1953. It was very similar to the previous one, except for its size: 18 laboratories belonging to different European Universities, plus an Australian laboratory took part in the expedition. 25 balloons were launched and more than 1000 emulsions, corresponding to a volume of 9.3 liters and a weight of 37 kg, were exposed for at least 7 hours at altitudes between 25 and 30 km[87].

The packets of exposed emulsions were recovered from the sea by the joint use of a seaplane and a corvette (Pomona) generously placed at disposal of the expedition by the Aeronautica Militare and the Marina Militare Italiana.

While nuclear plates (i.e. glass-backed nuclear emulsions) were used in the 1952 expedition, in the 1953 expedition, for the first time, 'stripped emulsions' i.e. photographic emulsions without any inactive support were employed on a large scale. With this new technique the particle's tracks can be followed from one emulsion to the adjacent one with a great advantage for the study of the events produced by cosmic rays. The development, fixing and washing of 1000 nuclear emulsions required special care and the construction of special processing plants at the Universities of Bristol, Padua and Rome where the exposed emulsions of the whole expedition were processed.

An important technical and organizational role was played by Ing. H. Heitler from Bristol, who directed, among others, the construction of all balloons made, part in Bristol and part in Padua, with the help of Ing. I. Scotoni of the Istituto di Fisica of Padua.

The base of the expedition was established at Elmas airport near Cagliari. J. Davis from Bristol and C. Franzinetti from Rome played a central role during the launching and recovery operation. They were helped by many physicists and technicians. Among the latter I should like to recall Max Roberts (1923-1976) from Bristol and Alfredo Pellizzoni (n. 1923) from Rome, helped by another very young Roman technician, Romolo Diotallevi (b. 1935). Among the many things constructed in Bristol by Roberts I may mention the automatic system for ballast release, and among those made by Pellizzoni in Rome, the clock cut-off which also could be triggered by means of a radio signal from the boat for the release of the emulsion packet with its parachute, buoy, etc.[87].

During this second expedition, from the 6th to the 12th of July 1953, an important international conference was organized by the University of Toulouse, in Bagnère de Bigorre in the French Pirénées[88]. Some very important aspects of particle physics were discussed in detail at the Conference. Dalitz, for example, discussed the so-called 'Dalitz-Fabri plot' for the representation of the decay processes of the τ mesons, Michel presented the selection rules for the decay processes, and the Rome group the use of the logaritm of tg $\vartheta/2$ as stochastic variable particularly appropriate for the description of high energy events in which many secondary particles are emitted and which many years later was called pseudorapidity.

In October 1953 a meeting took place at the Institute of Physics of the University of Bern (Switzerland) to distribute among the universities participating in the expedition the packets of emulsion completely processed. In April 1954 an International Conference was held in Padua to discuss the first results of the expedition and to establish together the most appropriate methods for the data analysis and a deeper study of the results[89].

Later results were made known through the normal scientific press and the lectures held by a number of the participants, at the second course of the International School of Varenna that took place during summer 1954[76].

At the same course, Enrico Fermi gave a series of « Lectures on Pions and Nucleons » which represent the last contribution he gave to physics teaching before his death, that took place in Chicago the 29th November of the same year 1954.

Among the main results of the expedition I should like to recall a few determinations of the values of the masses of heavy mesons undergoing different decay processes. These results suggested that all these processes could be due to alternative decay modes of the same particle. Also the identification of various hyperons and the determination of their masses and decay modes were considerably improved.

A third expedition was organized by the Universities of Bristol, Milan and Padua in October 1954, during which a single balloon was launched from Novi Ligure[90].

The balloon carried a package or emulsions of 15 liters, corresponding to 63 kg, which in the literature is often indicated as 'G-stack' (G for giant). The use of a single large stack has clearly the great advantage of allowing one to follow the tracks of the great majority of the secondary particles to their end so that their decay mode is observed and their initial energy can be measured with considerable precision by using the well known range - energy relations.

One of the most important results of the G-stack study was the final recognition that values of the masses of heavy mesons giving rise to different decay processes were equal to within rather small experimental errors[91]. Thus the interpretation of all these processes as alternative decays of the same particle was strengthened, contributing in an essential way to the general recognition of the so-called 'ϑ-τ puzzle'. This was only explained in 1956 with the discovery by Lee and Yang of the nonconservation of parity by weak interactions[92].

The greatest merit for the success of this expedition rightly was attributed by C. F. Powell to A. Merlin from Padua[90].

Let us go back, for a moment, to the Istituto Guglielmo Marconi. At the beginning of the fifties we started to realize that the effort made from 1947 onwards to bring our experimental techniques up-to-date, had, in some way, a negative influence on the cultural scientific training of the young physicists, especially when compared with their oversea coevals. Thus, after studying the Post-graduate Schools in Physics at the University of Chicago and M.I.T., I established, in 1952, the 'Scuola di perfezionamento in Fisica' of the University of Rome. This initiative followed that taken one year before by Bolla at the Milan Polytechnic, where he had started a 'Scuola di perfezionamento' planned for training young people with laureas in physics, chemistry and engineering, for the applications of nuclear sciences.

During the following years a few other

schools similar to that of the University of Rome were created by other 'Sections' of the INFN. All of them contributed considerably to the improvement of the scientific training of young researchers.

During the same year 1952, as I already said above, the arrival of Bruno Touschek, contributed to render even more brilliant the already active theoretical group of the University of Rome. The spectrum of problems investigated was also very wide, going from isobaric states of nucleons and pion-nucleon collisions (B. Ferretti) to the binding energy of ^{3}H and ^{4}He (G. Morpurgo), from the stability of the proton orbits in strong focusing machines (E. R. Caianello, A. Turin) to the tolerance limits in the alignment of the magnets in machines of this type (M. Sands, B. Touschek), from the quadrupole moments of heavy nuclei (R. Gatto), to the decay of the τ-meson (E. Fabri, B. Touschek)[93].

THE PROBLEM OF PARTICLE ACCELERATORS IN ITALY AND EUROPE - XVII

The idea of a wide scale European collaboration with the aim of tackling and solving together those problems that no single European country would have had the technical, organizational and financial strength to face and solve alone, started to appear in various universities, laboratories and cultural centers in Europe already in the years 1948-50. In particular the opportunity of creating a European laboratory for research on elementary particles, endowed with such equipment that no single country could have afforded alone was considered. The prehistory of CERN (Centre Européen pour la Recherche Nucléaire) was thus beginning and the Istituto Guglielmo Marconi was involved from the start, since I had been nominated Secretary General of the Organization already in its provisional phase[45,94]. This came to an end the 29th September 1954 when a prearranged point was reached in the procedure of ratification by the Parliaments of the Member States of the 'Convention' establishing definitively the existence of this new Organization and fixing its main features.

In the meantime, in Italy, the national structure of INFN and the larger funds it had received, led us to consider the opportunity of an enlargement of its research activities by facing problems of greater dimensions. Thus, at the beginning of 1953, under the chairmanship of Bernardini, it was decided to construct an electrosynchrotron of 1000 MeV and Giorgio Salvini was appointed to direct the design and construction of this machine and of the corresponding laboratory.

Salvini completed this task, by creating the Laboratori Nazionali di Frascati del CNRN (later CNEN) around a machine of excellent performance, within 1958, i.e. within the time schedule and with the funds foreseen from the start.

The Frascati machine accelerates electrons and, therefore, not accidentally, is complementary to the two machines of the initial program of CERN, which both accelerate protons: the synchrocyclotron of 600 MeV, designed and built under the direction of C. J. Bakker, which entered into operation in 1957, and the protosynchrotron of 30 GeV, designed and constructed under the direction of J. Adams, and that started to operate the 29th of November 1959.

Italy was a Member State of CERN from the beginning and its physicists, in a few years time, would have at disposal not only the Laboratori Nazionali di Frascati, equipped for the investigation of electron and photon physics, but could participate, in collaboration with those of the other countries, also in the researches carried on in the Geneva laboratories on the lepton and hadron physics.

In the course of 1954, clearly the years of the reconstruction ended as a result of a collective work not very frequent in our country for numerical participation, quality of people and length of time (about ten years). Even the organizational structures were really new and could have served as an example for other activities, not only in pure and applied physics, but in research in general.

A new phase was beginning in Italy, nay in Europe, not only for the study of elementary particles, but in general for all branches of research.

E. A.

Edoardo Amaldi farewell lecture delivered at the Meeting organized by Istituto di Fisica, Facoltà di Scienze, Università di Roma, held in Rome, 7-9 September 1978.

BIBLIOGRAPHY AND NOTES

58 These opinions and the decisions which followed are reported in Ref. 59.

59 E. Amaldi, *Centro di Studio per la Fisica nucleare e delle particelle elementari, Attività svolta durante l'anno 1947*, Ric. Scient. *18*, 54 (1948).

60 B. Rossi, H. H. Staub, *Ionization Chambers and Counters, Experimental Techniques*, Mc Graw-Hill Book Co., New York, 1949.

61 G. Bernardini, C. Longo, E. Pancini, *Relazione sulla costruzione del L.T.G.*, Ric. Scient. *18*, 91 (1948). Architect C. Longo had designed the building: due to its aluminium structure, it could very well stand the gusts of wind and snow which were often extraordinary strong. G. Bernardini, E. Pancini, *Ampliamento del L. T. G.*, Ric. Scient. *20*, 966 (1950). The direction of L. T. G. passed to Pancini and later to G. Fidecaro: *Il Laboratorio della Testa Grigia: attività svolta dal 1°-1-1953 al 30-9-1954.*

62 During this expedition, on January 5th, 1941, B. N. Cacciapuoti fell and badly damaged his right eye with the point of one of his skis, while returning to Breuil, towards evening. The accident is described in some detail by G. C. Wick, the senior physicist of the group, in a letter of 7/1/41 to the Director of the Guglielmo Marconi Institute.

63 The following researches were made at the Laboratorio della Testa Grigia. Nuclear disintegrations produced by cosmic rays were studied by G. Bernardini *et al.*, with nuclear emulsions (Phys. Rev. *74*, 845, 1878 [1948]; *76*, 1792 [1949]; *79*, 952 [1950]; Nature *163*, 981 [1949]), by I. F. Quercia *et al.* (Nuovo Cimento 7, 437 [1950]) and E. Amaldi *et al.* (Nuovo Cimento 7, 697 [1950]) with ionization chambers. Extensive showers by E. Amaldi *et al.* (Nuovo Cimento 7, 401, 816 [1950] and C. Ballario *et al.* (Phys. Rev. *83*, 666 [1951]); the electromagnetic component produced in nuclear explosions by G. Salvini *et al.*, with a cloud chamber (Nuovo Cimento *6*, 207 [1949]; *7*, 36, 943 [1950]; Phys Rev. *77*, 284 [1950]; penetrating showers from hydrogen and other elements by M. Conversi, G. Fidecaro *et al.* (Nuovo Cimento *1*, 336 [1955]); the determination of the mean lifetime of the new strange particles by means of a cloud chamber by P. Astbury, C. Ballario, R. Bizzarri, A. Michelini, E. Zavattini and A. Zichichi *et al.* (Nuovo Cimento, *2*, 365 [1955]; *6*, 994 [1957]).

64 E. Amaldi, G. Fidecaro, Helv. Phys. Acta *23*, 93 (1950); Nuovo Cimento 7, 535 (1950); Phys. Rev. *81*, 338 (1951).

65 E. Amaldi, C. Castagnoli, A. Gigli, S. Sciuti, Nuovo Cimento *9*, 453, 969 (1952); Proc. Phys. Soc. *A65*, 556 (1952); E. Amaldi, Proc. Int. Conf. of Theoretical Physics, Kyoto-Tokyo, Sept. 1953, p. 106.

66 A. Lovati, A. Mura, C. Succi, G. Tagliaferri, Nuovo Cimento *10*, 105, 1201 (1953); *11*, 92 (1954) and p. 215 of Ref. 88.

67 P. E. Argan, A. Gigli, S. Sciuti, Nuovo Cimento *11*, 530 (1954).

68 B. Ferretti, R. Peierls, Nature (London) *160*, 531 (1947).

69 E. Amaldi, *Una grande perdita per la fisica: Bruno Touschek* (b. Vienna 3/2/1921, d. Innsbruck 25/5 1978), to be published by Acc. Naz. Lincei e il Giornale di Fisica.

70 G. Boato, G. Careri, M. Santangelo, Nuovo Cimento *9*, 44 (1952); G. Boato, G. Careri, G. Volpi, Nuovo Cimento *9*, 538 (1952); G. Boato, A. Cimino, G. Careri, E. Molinari, G. Volpi, Nuovo Cimento *9*, 993 (1952); Naturwiss. *42*, 388 (1955).

71 Convegno Internazionale di Meccanica Statistica, Firenze 17-20 maggio 1949, Nuovo Cimento *6*, 146 (1949).

72 E. Perucca, *Commemorazione di Quirino Majorana*, Rend. Acc. Lincei *25*, 354 (1958).

73 V. Caglioti, *Commemorazione di Giovanni Polvani*, Genova, Palazzo Tursi, 3 ottobre 1970, Pubbl. dell'Istituto Intern. delle Comunicazioni; C. Salvetti, *Ricordo di Giovanni Polvani*, Istituto Lombardo, Rendiconti *104*, 115 (1970); P. Caldirola, *Commemorazione di Giovanni Polvani*, Bollettino SIF n° 84, 15 luglio 1971; A. Carrelli, *Giovanni Polvani*, Accad. Naz. Lincei, Celebrazioni Lincee n° *57* (1972).

74 Congresso Internazionale di Fisica sui Raggi Cosmici, Como 11-16 settembre 1949, Suppl. Nuovo Cimento *6*, 309 (1949).

75 Rendiconti del Corso tenuto nella Villa Monastero di Varenna, 19 agosto - 12 settembre 1953, Suppl. Nuovo Cimento *1*, 264 (1954).

76 Rendiconti del Corso che fu tenuto, nella Villa Monastero, a Varenna dal 18 luglio al 7 agosto 1954 a cura della Scuola Inter. di Fisica, Suppl. Nuovo Cimento *2*, 1 (1955).

77 E. Medi, *Antonino Lo Surdo* (b. Messina 4-2-1880, d. 7-6-1949), Annali di Geofisica *2*, 159 (1949).

78 E. Amaldi, F. Rasetti, *Ricordo di Enrico Persico*, in printing in *Celebrazioni Lincee*, Acc. Naz. Lincei, Roma, 1979.

79 The board of exminers concluded its work on November 14th 1949, with the following result: 1st G. Occhialini (five votes), 2nd B. N. Cacciapuoti (three votes), 3rd M. Ageno (four votes). Ageno (was also awarded two votes for the second place. The selected professors went to Genoa, Trieste and Sassari respectively. However Ageno, due to his poor health, resigned and accepted a position at the I.S.S.

80 The board of examiners, formed by Professors G. Valle (president), G. Polvani, A. Rostagni, R. Deaglio, E. Amaldi (secretary) concluded its work on Nov. 11th, 1950, with the following result: 1st E. Pancini (three votes), 2nd M. Conversi (five votes), 3d G. Lovera (three votes). Conversi also received two votes for the first place; A. Drigo and G. Salvini each one vote for the third place. The winners went to Sassari, Pisa and Modena respectively.

81 See, for example, P. Auger, *Tendences Actuelles de la Recherche Scientifique*, UNESCO, Paris, Nov. 1961.

82 E. Amaldi, *L'Istituto Nazionale di Fisica Nucleare*, Notiziario del CNEN, *9*, n° 1, gennaio (1963).

83 A. Rostagni, *L'energia elettrica 28*, 211 (1951). An ampler Laboratory was built in 1953. For further details on these high altitude laboratories, see: *The World's High Altitude Research Stations*, Research Division, College of Engineering, New York University.

84 The following researches were made at the Laboratorio della Marmolada: P. BASSI *et al.* studied with counter techniques the positive excess of mesons (Nuovo Cimento *8*, 469 [1957], with ionization chambers the zenith distribution of the nucleonic component (Nuovo Cimento *9*, 722 [1952]) and of the particles of extensive showers (Nuovo Cimento *10*, 779 [1953]). The last problem was investigated also by M. CRESTI *et al.* (Nuovo Cimento *10*, 779 [1953]). M. DEUTSCHMANN, M. CRESTI *et al.* investigated by means of a cloud chamber the decay of V^0 (Nuovo Cimento *3*, 180, 556 [1956]; *4*, 747 [1956]). H. MASSEY, E. H. S. BURHOP *et al.*, the production of secondary particles in hydrogen by means of a high-pressure ($\sim$ 100 atmospheres) cloud chamber (Nature *175*, 445 [1955]; Suppl. Nuovo Cimento *4*, 272 [1956]).

85 *Report on the activity of the INFN, by G. Bernardini, president*, p. 374 of *Un piano quinquennale per lo sviluppo delle ricerche nucleari in Italia*, published by the Comitato Nazionale per le Ricerche Nucleari, Roma, 1958.

86 *Report on the Expedition to the Central Mediterranean for the Study of Cosmic Radiation: Napoli 18-28 May, Cagliari 30 May - 13 July*, CERN/16, Rome, September 30, 1952.
To my knowledge only one copy of this report exists, now in the CERN Archives. For further details see Ref. 45.

87 J. DAVIES, C. FRANZINETTI, Suppl. Vol. 12th Nuovo Cimento, *2*, 480, 1954. The Physical Laboratories of the following Universities took part in the expedition: Bern, Bristol, Brusseles (Université Libre), Catania, Copenhagen, Dublin, Genoa, Göttingen (Max Planck Institut), London (Imperial College), Lund, Milan, Oslo, Padua, Paris (Ecole Politechnique), Rome, Sydney, Turin, Trondheim, Uppsala. The balloons were built in Bristol and Padua under the direction of Dr. H. Heitler; the gondolas and the transmitters in Milan and Rome. During the expedition the Corvette 'Pomona' covered more than 5,000 nautical miles for the recovery of the emulsion from the sea. These were attached to a buoy and kept under water until retrieved, in order to have them at a sufficiently low temperature and avoid damage from the Mediterranean sun. The buoy was located, thanks to a small radio-transmitter on it, its signal could be heard by the 'Pomona' within a range of five n. miles. A radio controlled release mechanism which allowed one to trigger the release of the gondola from the ship, whenever wanted, was also built and tested successfully but never used.

88 Congrès International sur le Rayonnement Cosmique, Organisé par l'Université de Toulouse, sous le patronage de l'UIPPA, avec l'appui de l'UNESCO, Bagnères de Bigorre, Juillet 1953.

89 *Rendiconti del Congresso Internazionale sulle Particelle Elementari Instabili Pesanti e sugli Eventi di Alta Energia nei Raggi Cosmici*, Padova 12-15 aprile 1954, Suppl. Vol. 9 Nuovo Cimento *2*, 163 (1954).

90 Some information on this expedition can be found in the *Introduction* by C. F. POWELL to the extensive paper (Ref. 91). This introduction has been included on p. 318 of *Selected Papers of C. F. Powell*, edited by E. H. S. BURHOP, W. O. LOCK, M. G. K. MENON, Amsterdam, 1972. As Powell wrote: « If anybody has played a distinctive and leading part in this effort, it was Dr. M. Merlin from Padua. He played a very important role in the early days during the discussion of the feasibility of flying a very large stack: throughout the expedition, his enthusiasm and his confidence in a successful outcome were of the greatest importance; and finally his drive and enthusiasm in the period of the examination of the plates were largely responsible of the fact that the whole enterprise was brought to a successful conclusion . . . ».

91 *Observations on Heavy Meson Secondaries* (G-stack collaboration), Suppl. Vol. 4 Nuovo Cimento, Vol. 2, 398 (1956).

92 T. D. LEE, C. N. YANG, Phys. Rev. *104*, 254 (1956).

93 E. AMALDI, *Istituto Nazionale di Fisica Nucleare, Sezione di Roma, Attività svolta durante l'anno 1952-53*, Ric. Scient. *25-1*, 1353 (1955); *Istituto Nazionale di Fisica Nucleare, Sezione di Roma, Attività svolta durante l'anno 1953-54*, Ric. Scient. *25-2*, 1977 (1955).

94 E. AMALDI, *CERN, the European Council for Nuclear Research*, Suppl. Nuovo Cimento *2*, 339 (1955); L. KOWARSKI, *An Account of the Origin and Beginning of CERN*, CERN 61-10, 10 April 1961; E. AMALDI, *CERN, Past and Future*, in *Topical Conference on High Energy Collisions of Hadrons*, CERN, Geneva 15-18 January 1968, p. 415; L. KOWARSKI, *New Forms of Organization in Physical Research after 1945*, p. 370 of *Rendiconti della Scuola Internazionale di Fisica Enrico Fermi*, LVII Corso, Varenna 31 luglio, 12 agosto, in *History of Twentieth Century Physics*, C. WEINER (ed.), Academic Press, New York, 1977.

INTRODUCTION

EDOARDO AMALDI*

The International Symposium on Niels Bohr, held in Rome from 25 to 27 February, 1985, was promoted by the Faculty of Natural Sciences and organized by the Group for the History of Physics of the Department of Physics of the University "La Sapienza". It had been conceived as a part of the celebrations of the centenary of Niels Bohr's birthday which, during 1985, were organized by many Universities, Academies and other organizations all over the world. The format of the Rome Symposium shows a dual nature: some of the contributions refer to the historical aspects of the life and scientific activity of Niels Bohr, while many others concern the foundations of quantum mechanics with special regards to the ideas of Niels Bohr and their evolution, also in connection with the views of some of his contemporaries that worked under his more or less direct influence or developed the seeds that he had sowed with his writings, lectures or discussions.

Although the human and cultural contacts of Bohr, for obvious geographic reasons, were in great part concentrated in Scandinavian countries and in the English and German speaking scientific circles, as a consequence of his central position in the development of atomic and nuclear physics during the first half of our century and his humanitarian views, he had a considerable influence also on physicists of far away countries, such as the south European countries, USSR and Japan.

It appears to me in order to take this occasion for trying to summarize the direct contacts of Niels Bohr with Italian physicists, at list in the limit of my knowledge.

1. The first contact, not very fortunate, took place in 1924-25 when the 23 years old Enrico Fermi published in *Zeitschrift für Physik* a paper on the theory of the collisions of a charged fast moving particle against

* E. Amaldi, Dipartimento di Fisica, Università di Roma "La Sapienza".

the atoms of a medium [1]. The basic idea was that the electric field of the fast particle can be replaced by its Fourier integral and treat this as the frequency spectrum of the electric field of light. This approach was applied by Fermi to a few problems such as the excitation of the atoms of mercury's vapour by low energy electrons, and the calculations of the number of ion's pairs per path unit produced in helium by the α particles of RaC and of the range of these same α particles in helium. Each problem is dealt with the help of mathematical methods of sufficient approximation but not better than warranted by the underlying physical hypothesis. The results agree with experiment, but only to an order of magnitude, and Fermi, typically, remarks that "in one of the applications the agreement with experiment is better than should be expected" [2].

Soon after the publication of Fermi's paper, appeared the paper by Bohr "On the Behaviour of Atoms in Collisions" [4] conceived in the general frame of the theory of Bohr-Kramers-Slater.

In a footpage note Bohr observed that Fermi's method, when applied to the energy distribution of the electrons ejected, disagrees with experiments and concluded:

> Under these circumstances, it can hardly be regarded as a support of the assumptions made by Fermi that an estimate of the stopping power based on the requirement of energy balance gives results in approximate agreement with experiments.

As pointed out by Persico [2] "Later on, when quantum mechanics was developed, Fermi's method for the study of collisions with electrically charged particles found adequate theoretical justification through a theorem by Dirac [6] as remarked by Williams [7]. This author discussed carefully the validity limits of Fermi's method and, using the quantum mechanical values of the matrix elements, calculated the excitation and ionization probability in distant collisions. A little later Weizsäcker used the same method for the calculation of *Bremsstrahlung* [8]. Williams and Weizsäcker were working at the time at Copenhagen. The method under name of Weizsäcker-Williams has found ever wider applications to atomic and nuclear problems and Fermi himself used it on many occasions. Indeed its fundamental idea was a favorite one with him for his whole life" [2].

The next important contact of Niels Bohr with Italian physicists was provided by his partecipation in the International Conference of the Physicists held in Como, Pavia and Rome, on September 11 to 20, 1927 for celebrating the first centenary of the death of Alessandro Volta [9]. Important speaches were given by E. Rutherford, J. Franck, W. Gerlach, O. Stern, A.H. Compton, H.A. Lorentz, R.A. Millikan, O.W. Richardson, F. Paschen, P. Zeeman, M. Born, A. Sommerfeld, T. Levi-Civita, P. Debye, A.S. Eddington, H.A. Kramers and Niels Bohr. The title of Bohr's speech was "The Quantum Postulate and the Recent Development of Atomic Theory" [10].

The "Como Lecture" was actually delivered twice: first on September 16, at the Como Conference, and again at the Fifth Solvay Conference, held in Bruxelles from 24 to 29 October. The later occasion marked the beginning of the second phase of the Bohr-Einstein dialogue on the fundamentals of the quantum-mechanical description of nature [11]. In this lecture, published in *Nature* in 1928 [12], Bohr dealt with the following topics: the quantum postulate and causality, quantum of action and kinematics, measurements in quantum theory, the correspondence principle and matrix theory, wave mechanics and the quantum postulate, the reality of stationary states, and the problem of elementary particles.

As Jørgen Kalckar tell us [5b]: in Como

> [...] The reception of Bohr's presentation of his new ideas by the distinguished audience was remarkably cool. The discussion [...] hardly touches on the fundamental issues brought forward by Bohr [...] the main reason is without doubt to be found in the character of the lecture, in highly non-technical nature, with only a few elementary formulae, but an abundance of subtle comments on various aspects of quantum theory. It was of course not the last time that this characteristic style of Bohr created difficulties for the comprehension of his views [...].

In the discussion that followed a number of people highly competent in at least some of the problems dealt with by Bohr in his wide presentation took the floor: Born, Kramers, Heisenberg, Fermi and Pauli, but all of them developed their proper line of approach to some specific problem with very little or none connection with Bohr presentation.

For example, Fermi's remarks concern the following points:
1) a procedure very clever but, admittedly, rather arbitrary, for introducing in the frame of quantum mechanics the radiation reaction on the emitting atom [13], which was very soon outdated by Dirac's radiation theory [14];
2) the necessity and advantages of revising, in the frame of quantum mechanics, the statistical laws of gases, with special regard to the two alternative solutions proposed by Einstein and Fermi himself of such a problem, and the clarification of the relation between these two statistical laws provided by the works of Heisenberg, Dirac and Winter; and finally
3) the application of the antisymmetric statistics to various specific cases: by Sommerfeld to the conduction electrons in metals and by Fermi (and Thomas) to the electron's cloud of the atom.

The International Conference on Nuclear Physics held in Rome on October, 1931 [15] was another important encounter between Niels Bohr and Italian physicists. Among the many Italian physicists and the few Italian chemists partecipating in the Conference [16] I may recall in particular G. Marconi, Orso Mario Corbino, Director of the Institute of Physics of Via Panisperna, Fermi, Persico, Rasetti and Bruno Rossi. E. Segrè and I had contributed, under the supervision of Fermi and Rasetti, to the preparation of the Conference and during the Conference we fulfilled the function of scientific secretaries.

Among the foreigners there was R.A. Millikan, E. Rutherford, Maria Shlodowska-Curie, N. Bohr, W. Heisenberg, W. Pauli, G. Gamow, A. Sommerfeld, A.H. Compton, J. Perrin, P.M.S. Blackett, and many others.

Of the many problems discussed one of the more controversial was the explanation of the continuity of the β-decay spectra. Pauli presented, for the third or fourth time since December 1930, his suggestion that in β-decay the electron is emitted simultaneously with a neutrino. As he wrote years later [17]: at the Rome Conference

> I met in particular Fermi – who showed immediately a great interest for my idea and a very positive attitude towards my new neutral particles – and Bohr, who on the contrary maintained his idea that in β-decay the energy is conserved only statistically [...].

In his contribution entitled "Atomic Stability and Conservation Laws" [18] Bohr also casted doubts about the applicability of quantum mechanics to systems as small as a nucleus and justified that by "the failure of the fundamental quantum mechanical rules of statistics when applied to nuclei". This sentence clearly refers to the experimental result of Rasetti [19] who a few year before had shown that the nucleus of ${}_7N^{14}$ obeys the Bose-Einstein statistics and not the Fermi-Dirac statistics as expected according to the then universally accepted model of the nucleus as a system composed of protons and electrons.

It is well known that, a few months after the 1931 Rome Conference, nuclear physics went through a period of exceptionally rapid evolution. In February 1932, in Cambridge, James Chadwick discovered the neutron, in April of the same year, also in Cambridge, J.D. Cockcroft and E.T.S. Walton succeded in observing, for the first time, nuclear disintegrations produced in light elements by accelerated particles, and during summer C.D. Anderson in Pasadena provided the first evidence for the existence of positive electrons in cosmic rays. Anderson's work was followed by a number of papers by Blackett and Occhialini, F. Joliot and I. Curie, L. Meitner and a few others that established the local production of showers by cosmic rays, and clarified their electromagnetic nature by observing, among others, the production of e^+e^- pairs by γ-rays and their annihilation.

Before summer 1933 Heisenberg started to develope the theory of nuclei as systems composed of neutrons and protons interacting with exchange nuclear forces and in December of the same year Fermi published a first short presentation of his theory of β-decay based on Pauli assumption.

In January 1934 I. Curie and F. Joliot discovered the artificial radioactivity produced by α-particles and in March E. Fermi that produced by neutrons, which opened the way to the exploration of the interaction of neutrons with nuclei, to the discovery of slow neutrons and their peculiar properties.

Thus, when Niels Bohr and his colleagues organized the International Conference on "Probleme der Atomkernphysik" that took place in Copenhagen from 14 to 20 June, 1936, the situation was completely new from the point of view of the experimental results as well as of the theoretical picture. All the groups active in the field were well represented at the Conference.

Heisenberg spoke on the neutron-proton nuclear model and of the nuclear forces, which were discussed also by G.C. Wick, M. Goldhaber talked about the neutron-proton collision and Meitner gave a detailed presentation of the uranium business. I presented the work carried out in Rome on the slow neutron-resonances, and Wick discussed the theory of diffusion of slow neutrons through an hydrogeneous medium, developed in great part by Fermi, but to which Wick himself had given several important contributions. This admittedly approximate approach was slightly extended and generalized by Fermi during the war in USA, and became known as the "slow neutron age theory".

The most important contribution to the Conference was that given by Niels Bohr, who illustrated the compound nucleus model of nuclear reactions.

Bohr's ideas were not new to the audience since he had presented them as an address to the Danish Academy of Sciences on January 24, 1936 and had appeared in the issue of *Nature* of February 29 [20]. But it was a completely different experience and a great excitement to lissen Bohr, a few months later, during which various authors, like H. Bethe, G. Placzek, R. Oppenheimer and R. Serber has employed them, implemented by the more formal approach introduced independently in February 1936 by Breit and Wigner [21], for explaining on interpreting quantitatively a number of experimental results.

The subject was of special interest for the Rome group which had provided the most extensive set of experiments in favour of the existence of narrow slow neutron resonances, arriving, before the end of March 1936, to a fairly good estimate of their width: about one tenth of electronvolt for nuclei of atomic mass number slightly higher than one hundred (Rh, Ag and I)[22].

Besides G.C. Wick and me, two more Italian physicists partecipated in the Copenhagen Conference: Giulio Racah and Ugo Fano. Unfortunately the proceedings of the conference have not been published and almost no written material can be found about this conference, even in the archives of the Copenhagen Institute for Theoretical Physics.

About one year later another international conference was held in Bologna for celebrating the second centenary of the birth of Luigi Galvani [23]. Among the participants there were ten Nobel laureates, nine physicists (F.W. Aston, N. Bohr, L. de Broglie, P. Debye, W. Heisenberg, C.V. Raman, O.W. Richardson, E. Schrödinger and M. Siegbahn), and one physiologist (Lord O.M. Adrian).

Interesting speeches were presented by Aston on the determination of the atomic weights of isotopes by the "doublet method", by Bethe on resonances observed in (n,d) and (n,p) nuclear reactions, and Segrè on the discovery he had recently made, with Perrier, of element Z=43.

During the morning of October 20, Bohr was informed by telegram of the sudden death of Lord Rutherford. He was profoundly struck. I accompanied him to the central telephon office, where I succeded in helping him to get in touch with Rutherford's widow. He was accompanied by his son Aage, who was 15 years old. Before learning Bologna for Cambridge the same day, Bohr pronounced in front of the Conference a short but deeply felt commemoration of Rutherford.

2. The first Italian physicist that went to Copenhagen for spending a few months at the Institute for Theoretical Physics directed by Niels Bohr, was Ettore Majorana [24].

Fermi had persuaded him to go abroad, first to Leipzig and then to Copenhagen, and had obtained a grant from the Consiglio Nazionale delle Ricerche for his journey, which began at the end of January, 1933 and lasted about seven months. From Rome he went to Leipzig where he remained until the beginning of summer; from Leipzig he went to Copenhagen where he spent about three months, and from there back to Leipzig where he stayed for about a month before finally returning to Italy at the beginning of autumn 1933. Both in Leipzig and Copenhagen he remained very much alone, partly because of his natural avversion to making new acquaintances, but also because, although he understood German well, he spoke it rather badly. In Leipzig Majorana became friendly with Heisenberg, for who he had a great admiration and a feeling of friendship. It was from Leipzig that Majorana sent for publication his paper on the exchange forces [25], which, however, he had already conceived in spring 1932 [24].

In Copenhagen he met, of course, Niels Bohr, C. Møller, L. Rosenfeld and many others. At that time George Placzek [26] was also there, and Majorana was friendly with him, as he had already known him in Rome in 1931-32. According to Rosenfeld, who, on my request, wrote me a letter in 1964 [27], in Copenhagen he never saw Majorana without Placzek and he heard his voice only once on occasion of a conversation in a café [24].

The Italian physicist that, after the June 1936 Conference, visited more regularly the Bohr Institute, is Gian Carlo Wick. On occasion of a short staying in Copenhagen from 10 to 16 September 1937, they started to talk also of the political situation in Italy. Thus they discovered to have in common a definite dislike for fascism, which contributed to establish and reinforce their friendship.

In 1938 Wick got a fellowship which allowed him a two months stay in Copenhagen, starting from August 29. For Gian Carlo this was the most interesting visit, although it was in someway upset by what was going on in Europe as we shall see below.

Shortly after his arrival Wick mentioned to Bohr some considerations he had recently developed about the meaning of the expression $\rho = h/mc$ given by Yukawa for the range of the nuclear forces in terms of the mass m of the meson enchanged between the two interacting nucleons. Bohr liked them and when, around the middle of september he went for a few days to Stockholm, Wick accompained him. In Stockolm Bohr gave a seminar in which, among other things, mentioned Gian Carlo idea. Bruno Pontecorvo was also in Stockholm, went to lissen Bohr lecture and immediately after told Gian Carlo, in his typical jocular style: "Now I can be proud of knowing you!" [28]. Encouraged by Bohr, Wick send a letter to the Editor of *Nature* [29], in which he explains that his approach is in someway inspired by the arguments used, many years before, by Bohr in connection with Gamow theory of α-decay. Then he shows that the expression of ρ given above can be interpreted as the upper limit for the distance to which the virtual transitions involved in the emission and absorption of mesons by the two interacting nucleons can make themselves felt without contradiction with the energy conservation principle.

During this period Wick was invited several times to Bohr's house and on these, as well as on other occasions, they discussed again and again the political situation in Europe, where the tension was extreme especially during and after the Munich Meeting (September 28-30, 1938) where Great Britain, represented by the Prime Minister Chaimberlain, and France, represented by the President of the Ministers Council Daladier, under the pressure of Hitler, Führer of the Third Reich, supported by Mussolini, Duce of Italy, accepted the dismembrement of Czechoslovakia, with the transfer of a few of its border promices to Germany,

Poland and Hungary. This meeting was only one of the many steps of the years of long degradation produced by the Fascist countries in Europe.

Less than two weeks after the Munich Meeting, Bruno Rossi and his wife, Nora Lombroso, also arrived in Copenhagen. Bruno was professor of physics at the University of Padua since 1933, but as a consequence of the (antisemitic) "racial laws" promulgated by the Fascist Government the June 14, 1938, had lost his job. He had to leave Italy as many others: E. Segrè, G. Racah, U. Fano, E. Fubini, S. De Benedetti, L. Pincherle and even Fermi whose wife, Laura Capon, was hit by the "antisemitic" trend taken by Mussolini's Government [30]. Bruno had written to Niels Bohr, "who, most graciously" invited him to visit his Institute [31].

The Rossi could affront a sudden departure by using the residue of a previous fellowship he had used for a few months stay at the W. Bothe laboratory at the Reichsanstahl in Charlottenburg, near Berlin [30]. They remained in Copenhagen two months during which Niels Bohr and his wife Margrethe made everything possible to make their stay as pleasant as possible. Niels Bohr even organized a small international conference that brought to Copenhagen many scientists, among them Fermi and a number of cosmic ray physicists. One of his motives was to give Rossi the opportunity of meeting people who might be able to help him find a job. In the case, that is what happened, because, shortly thereafter Patrick Blackett invited Rossi to Manchester on a fellowship from the Society for the Protection of Science and Learning [31].

Bohr was also trying to find a good excuse for giving some finantial support to the Rossi, but was afraid of offending them. The solution he found was the following. One day, shortly before the conference mentioned above, Bohr went to see Rossi who was reading in the library and asked him to pass by his secretary. When Bruno went to the secretary, she gave him an envelope containing a substantial sum of money. Then Bruno, surprised, asked explanations and the secretary answered: "It is for the conference". "But I am already here, and therefore I do not have any expenditure" replied Bruno. "It is true – said the secretary – but if you had not been here, we would have reimbursed your traveling expenses" [32].

As I said above, also Fermi arrived in Copenhagen for the conference organized by Bohr. In a letter from Copenhagen to his mother in Turin Gian Carlo wrote: "F. has appeared to me rather depressed". In his recent letter to me [28] he notices that perhaps it would have been more appropriate to say "preoccupied". A few weeks later Fermi came back to Rome and informed, in great secrecy, Rasetti and me, that he had decided to leave the country for good. This actually happened at the beginning of December of the same year [30]. Towards the end of October also Ugo Fano arrived in Copenhagen.

The immediate and far away consequences of the Munich Meeting, were the subject of many discussions also during the little conference held in Copenhagen in autumn 1938. One of these took place, I believe in Bohr's house, with the partecipation of Patrick Blackett, Max von Laue and others. Blackett was furious and forsaw the most cathastrophic consequences. After a long exchange of views, von Laue, well known for his antinazi position, perhaps for introducing an element of hope, said more or less "The Munich Meeting has at least avoided that we are shooting to each other among us. Don't you think so?" Blackett remained silent for a few seconds and then said sharply "I dont-know!". The next day Niels Bohr with discretion told the episode to Wick and commented: "Blackett has not lost his fighting spirit!".

Later Wick visited Bohr two more times. In 1938, from Pittsburg (USA), where he was professor at the Carnegie Institute of Technology, Wick partecipated in the "Conference on Quantum Physics", organized by Niels Bohr and Stephan Rozental and that took place in Copenhagen from 6 to 10 July. Other Italian partecipants were Bruno Ferretti, and Ugo Fano, who, however, remained at the Bohr Institute for about six weeks also for refining the presentation of quantum physics according to the Copenhagen approach, in the book, directed primarly to research workers in the natural sciences, he had written in those years with his wife Lilla [33].

I also was there. This was the meeting where the physicists started to realize that in addition to pions and muons, also several other particles, first discovered by Rochester and Butler as V-events, were of paramount importance. Cecil Powell, as experimentalist, and Hans Bethe, as theoretician, were the two people that with their speeches and interventions in the discussions, remained at the center of the attention of the audience during those days.

After the conference I remained in Copenhagen a few more days for partecipating in the 7th General Assembly of IUPAP that took place, from 11 to 13 of the same month, and which is even to day of considerable interest in the prehistory of CERN [34].

The last encounter of Wick with Niels Bohr took place eight years later. On his way from Brookhaven to USSR for partecipating in the Ninth International Annual Conference on High Energy Physics (Kiev, 15-25 July, 1959) Gian Carlo stopped in Danemark and visited the Bohrs in their country house at Tisvilde. On that occasion Niels told Wick with some detail and great humor his encounter in May 1944 with Churchill when he – without success – tried to convene to the Prime Minister of Great Britain his ideas about "the open world" without military secrets [35].

When Bohr, on June 9, 1950, had send his "Open Letter to the United Nations" in which these ideas were further developed and recast in a more complete and elaborate form, he had send a copy also to Wick. In his letter of thanks Gian Carlo had written of being in full agreement with the desirable developments envisaged by Niels Bohr, but of being less optimist than him about the possibility of finding a hearing from people and governements.

3. During winter 1939-1940. with a few collegues of the University of Rome, I had started a series of experiments aiming to verify one of the many results given by Bohr and Wheeler in their two famous papers on the theory of nuclear fission [36]. They had shown that the compound nuclei formed by neutron capture in ${}_{92}U^{238}$, ${}_{91}Pa^{231}$ and ${}_{90}Th^{232}$ can undergo fission only if the energy of the incident neutron is superior to certain thresholds ($E_{thr} \approx 0.7$; 0.1 and $1.7 MeV$), above which the fission cross section very rapidly becomes energy independent.

Our experiments on uranium bombarded with fast neutrons (from 0.2 to $\leq 15 MeV$) had shown that the fission cross section was constant from threshold ($E_{thr} = 0.7 MeV$) to energies not far to $10 MeV$, but that around and above this energy, an unespected clear increase was appearing [37]. Since our theroretical collegues did not give us a satisfactory answer, on April 4, 1940 I wrote a letter to Niels Bohr for informing him of our results and asking his advice for their interpretation. He send me a detailed answer in a kind letter, dated May 11, 1940 [38]. He showed that our results could be understood since a compound nucleus, formed by capture of a neutron of such a high energy ($10 MeV$ or more), may, even after emission of an evaporation neutron, remain sufficiently highly excited to undergo fission subsequently.

I wrote him again on May 26, 1940 for expressing our thanks and announcing the dispach of the reprints of our two papers on U, in the second of which his interpretation had been used and confirmed. I mentioned also a set of measurements carried on more recently with a new technique and a minor quantitative difficulty we had found in comparing some of the new experimental data with the numerical values deduced by Bohr and Wheeler from their theory [36].

Towards the end of the letter I added:

> I am leaving Rome tomorrow and also two of the young fellows working with me left a few days ago; therefore I am afraid that our research will be stopped for some time.

Ageno had been drafted already at the end of August 1939 and had partecipated in the work on the fission cross section only during a few months licence, Cacciapuoti had been mobilized a few days before,

I had got the order of mobilization on May 24, and, at the beginning of June, left from the arbor of Naples for North Africa as a lieutenant of artillery. The Second World War had begun on September 1, 1939 when the Hitler troops, from one side, and the Stalin troops, from the other, had crossed the Polish border, and on September 3, Great Britain and France had declared war to Germany in spite of their inadequate military preparation. Mussolini was remained out of the conflict but, almost one year later, when he saw Great Britain in difficulties and the French army collapsing under Hitler attack, thought that the war was close to the end, became hasty in trying to gain low-cost titles for sitting at the table of the negotiations on the winner's side and the 10 June, 1940 declared war to the Allied.

Niels Bohr answer to my second letter is dated July 19, 1940, and was accompanied by a copy of the manuscript of the article he had sent for publication to *Physical Review* [39]. A great part of this paper is devoted to a detailed discussion of the increase of the fission cross section we had observed for neutrons of sufficiently high energy. In his letter Bohr suggested to do experiments with thorium similar to those we had made with uranium. We had already began measurements with this element, but the work had been interrupted by the entrance in war of Italy.

I saw this letter of Bohr when, at the beginning of December 1940, I come back from North Africa for taking up again my teaching duties. This had been possible on the request of the Faculty of Sciences of the University of Rome, which had asked the application to my case of the recent decree-law regarding full professors belonging to the class 1908 or senior. Back to Italy, at the beginning of 1944, I carried out a few more measurements on thorium and protoactonium with the only two colleagues still in Rome (a lady, D.B., and a senior colleague G.C.T.) and send a third short paper to *La Ricerca Scientifica* [40].

By now we had understood that the United States were close to enter in the war. Therefore we summarized all our results on the fast neutron fission cross section in a paper written in English and sent it to *Physical Review* where it appeared after the Japonese attach to Pearl Harbour [41]. We also decided to stop any work on fission and to employ our limited manpower in completely different problems. We were arrived to the conviction that almost any problem related to the investigation of fission could become of interest for the possible construction of weapons and we did not want to be involved in this type of work [42].

On 22 February, 1941, a letter signed by G.C. Wick and me, was sent again to Niels Bohr. We informed him that I was back from military duties and that Wick had moved from the University of Padua to that of Rome. Almost all the letter regarded new results obtained on thorium fission and their comparison with Bohr and Wheeler theory. The agreement was as good as could have been expected. Towards the end of the letter we mention of having read somewhere that the extensive paper by Bohr, Peierls and Placzek was expected to appear pretty soon and we expressed our great interest. Finally we asked about him and his family. At that time we had started to be warried about the possibility that something unpleasant could happens to some of them.

Danemark had been occupied by Hitler's troops around the 10th of April, 1940, but recently from the press and other information sources we had got the impression that the situation in Copenhagen was becomes worse. This was one of the reasons why we had wrote this last letter in German instead of English, as were all the others. Perhaps the German censorship could be less alarmed. Bohr answer, dated March 18, 1941, also in German, contained a rather detailed discussion of our results and at the end information about his recent work on the energy loss and scattering of the nuclei produced in fission recoiling at high velocity in matter.

We wrote him again on 8 August, 1942, from Siusi (Bolzano) where we were for summer vacations. The letter refers to results of measurements on the scattering of protons of 10-14 MeV by protons, we had carried on during the last months. The answer of Bohr, dated September 12, 1942, and a letter from Wick and me, dated Rome 13 October 1942 refers to the same subject, that had become the center of our interest, followed shortly later by the problem of the experimental test of some of the results contained in the paper of Bohr, Peierls and Placzek.

4. A number of young Italian theoreticians later when to work at the Bohr Institute, after the decision of the Council of the Provisional CERN (May 2, 1952) [34] to create the Group of Theoretical Studies at Copenhagen under the directorship of N. Bohr. The first to go over there was E. Caianello from Naples, at that time at the University of Rome, who spent in Copenhagen one semester in 1953, working on the stability in the orbits of strong focusing accelerators, on Fermi's interaction and quantum field theory and one semester in 1955 working on the problem of renormalization in field theory. At that time he started to get interested in cybernetics. A. Loinger from Pavia passed a long period during 1954-55 at the Bohr Institute. His main interest was the theory of measure. He had interesting interactions with L. Rosenfeld which lasted for years after the departure of Loinger from Copenhagen. M. Cini, from Turin, was there for about two months starting from the middle of September 1954. He worked on the pion nucleon scattering. V. Wataghin, also from Turin, was in Copenhagen from September 1957 to August 1959 and worked on the proton electromagnetic form-factors. A few others,

like A. Molinari, T. Regge and G. Pollarolo, that were also for rather long period overthere, arrived after the departure of Niels Bohr and the creation of Nordita [34].

My personal contact with Niels Bohr during the early period of CERN were very frequent. From the paper I have presented to the conference [34] they may appear very formal, but it was not so at all. In all, or almost all the occasions I enjoyed enormously to see him again and I always felt that he also was in some way pleased to see me. We talked a bit of physics; he usually asked me, kindly, what I was doing or what was going on in general at our Institute in Rome. And we always discussed the Italian situation, the political situation, with special regard to the state of the Universities and of research. He always knew, from me, from Wick, or some other Italian physicists, which were our problems and he always took a great interest in their possible solution. I still have a vivid memory of the evening of 16 April 1959. Bohr was arrived in Geneva for partecipating the next day in the 12th meeting of the SPC [34]. I was already there for a small symposium on the future programme of research of the 30 GeV PS. Bohr invited me at dinner at his hotel, the Hotel du Rhone, where I spent that evening in an unforgottable conservation with Niels and Margrethe Bohr, a conversation that touched only slightly upon physics and biology. The main interest was on the problems of south Italy, modern Danish artists and recent Italian books and films, in a relaxed atmosphere on deep and warm consideration of all possible aspect of human life.

Rome October 8, 1986

References

[1] E. FERMI, "Über die Theorie des Stösses Zwischen Atomen und electrisch geladenen Teilchen", *Zeit. f. Phys.*, *29*, 1924, 315-327, received on October 20, 1924; "Sopra l'urto tra atomi e corpuscoli elettrici", *Nuovo Cimento, 2*, 1925, 143-158.

[2] E. Persico comment on p. 142 of Ref. [3].

[3] E. FERMI, *Collected Papers*, Vol. 1 (Italia 1921-38), Accademia Nazionale dei Lincei, The University of Chicago Press, 1962.

[4] N. BOHR, "Über die Wirkung von Atomen bei Stössen", *Zeit. f. Phys.*, *34*, 1925, 142-157, received on March 30, 1925. See also Ref. [5a], 177-193, 1nd for his Engl. trans. Ref. [5a], 194-206.

[5] N. BOHR, *Collected Papers*, E. Rüdinger (gen. ed.), North-Holland, Amsterdam: a) Vol. 5, "The Emergence of Quantum Mechanics (Mainly 1924-1926)", edited by Klaus Stolzenburg, 1984; b) Vol. 6, "Foundations of Quantum Physics (1926-1932)", edited by Jørgen

Kalckar, 1985; c) Vol. 9, "Nuclear Physics (1929-1952)", edited by Sir Rudolf Peierls, 1986.

[6] P.A.M. DIRAC, *Principles of Quantum Mechanics*, p.167 of 1st Edition, p. 176 of 5th Edition.

[7] E.J. WILLIAMS, "Application of the Method of Impact Parameter in Collision", *Proc. Roy. Soc., 139*, 1933, 163-186.

[8] C.F. v. WEIZSÄCKER, "Ausstrahlung bei Stössen seher Schneller Elektronen", *Zeit. f. Phys., 88*, 1934, 612-625.

[9] *Atti del Congresso Internazionale dei Fisici, tenuto a Como, Pavia e Roma dall'11 al 20 Settembre 1927 per celebrare il primo centenario della morte di Alessandro Volta*: a) Vol. 1, cap. 1, "Inaugurazione del Congresso", cap. 2, "Esperienze sulla struttura della materia", cap. 3, "L'elettricità e le sue applicazioni"; b) Vol. 2, cap. 4, "Elettrologia", cap. 5, "Ottica Fisica", cap. 6, "La teoria sulla struttura della materia e sulle radiazioni", cap. 7, "Seduta di Pavia. Sintesi dei lavori compiuti e dei lavori attuali", cap. 8, "Commemorazione Ufficiale di Alessandro Volta in Campidoglio e chiusura del Congresso".

[10] Ref. [9b], 565-588; Ref. [5b], 109-145.

[11] *Niels Bohr, A Centenary Volume*, edited by A.P. French and P.J. Kennedy, Harvard University Press, Cambridge Mass. 1985.

[12] N. BOHR, "The Quantum Postulate and the Recent Development of Atomic Theory", *Nature, 121*, 1928, 580-590, *Supplement to Nature*, April 14, 1928.

[13] See also E. FERMI, "Sul meccanismo nell'emissione nella meccanica quantistica", *Rend. Lincei, 5*, 1927, 795-800.

[14] For the remarks of E. Persico see Ref. [3], 271.

[15] *Atti del Convegno di Fisica Nucleare, 1931*, Reale Accademia d'Italia, Fondazione Alessandro Volta, *Atti dei Convegni, 1*, 1932.

[16] The Italian physicists partecipating in the Rome Conference were, in alphabetic order: M. Cantone, A. Carrelli, O.M. Corbino, E. Fermi, A. Garbasso, G. Giorgi, Q. Majorana, E. Persico, F. Rasetti, B. Rossi. W. Wataghin. The chemists were: F. Giordani, N. Parravano.

[17] W. PAULI, see the last chapter of *Aufsätze und Vorträge über Physik und Erkenntnistheorie*, 1933-1938, entitled "Zur älteren und neuen Geshichte des Neutrinos"; it is reproduced in Vol. 2, 1313-1337 of *Collected Scientific Papers by Wolfgang Pauli*, eds. R. Kronig and V.F. Weisskopf, Interscience Publ., New York 1964.

[18] Ref. [15], 119-130.

[19] F. RASETTI, "Selection Rules in the Raman Effect", *Nature, 123*, 1929, 757-759; "On the Raman Effect in Diatomic Gases", *Nat. Acad. Sci., 15*, 1929, 234-237, 515-519; "Über die Rotations-Raman

Spektren von Stickstoff und Sauerstoff", *Zeit. f. Phys., 61*, 1970, 598-601.

[20] N. BOHR, "Neutron Capture and Nuclear Constitution", *Nature, 137*, 1936, 344-348; see also the first of the "News and Views" reported in the same issue at p. 351.

[21] G. BREIT, E. WIGNER, "Capture of Slow Neutrons", *Phys. Rev., 49*, 1936, 519-531, received February 15, 1936.

[22] E. AMALDI, E. FERMI, "Sopra l'assorbimento e la diffusione dei neutroni lenti", *Ric. Scient., 7* (1), 1936, 454-503; "On the Absorption and Diffusion of Slow Neutrons", *Phys. Rev., 50*, 1936, 899-928.

[23] "Celebrazione del secondo centenario della nascita di Luigi Galvani", XXIX Riunione della Società Italiana di Fisica (Bologna 18-21 ottobre 1937): a) "Account of the Official Celebration", *Nuovo Cimento, 14*, 1937, 389-391; b) "Atti della XXIX Riunione della Società Italiana di Fisica e del Congresso di Fisica", *ibidem*, 492-533.

[24] E. AMALDI, *La vita e l'opera di Ettore Majorana*, Accademia Nazionale dei Lincei, Roma 1966, translated in English as "Ettore Majorana, Man and Scientist", pp. 10-77 of *Strong and Weak Interations. Present Problems*, edited by A. Zichichi, Academic Press, New York 1966.

[25] E. MAJORANA, "Über die Kerntheorie", *Zeit. f. Phys., 82*, 1933, 137-145; "Sulla teoria dei nuclei", *Ric. Scient., 4* (1) 1933, 559-565.

[26] E. AMALDI, "G. Placzek (1905-1955)", *Ric. Scient., 26*, 1956, 2037-2041; L. van HOVE, "George Placzek (1905-1955)", *Nucl. Phys., 1*, 1956, 623-626.

[27] The letter of Léon Rosenfeld to E. Amaldi dated 10 August, 1964 contained a memo, three quarter of a typewritten page long, entitled "Recollections about Majorana", in which the "rather meagre information" mentioned in Ref. [22] are given.

[28] I thank my dear friend G.C. Wick for the letter he has written to me on September 27, 1986, after I had called him on the telephon one day before, asking for precise information about his contacts with N. Bohr. He gave me further detail, not contained in the letter quoted above, when I called him again on the telephon a few days later.

[29] G.C. WICK, "Range of the Nuclear Forces in Yukawa's Theory", *Nature, 142*, 1938, 993-994.

[30] E. AMALDI, "The Years of Recontruction", pp. 379-461 of *Perspectives of Fundamental Physics*, edited by Carlo Schaerf, Harwood Academic Publ., Amsterdam 1979, and *Scientia, 114*, 1979, 51-68, 439-451.

[31] B. ROSSI, "The Decay of 'Mesotrons' (1939-1943): Experimental

Particle Physics in the Age of Innocence", pp. 183-205 of *The Birth of Particle Physics*, edited by L.M. Brown and L. Hoddeson, Cambridge University Press, Cambridge 1983.

[32] This typical "Niels Bohr story" will appear in a book that Bruno Rossi is now writing for the publisher Zanichelli (Bologna). A xerox copy of the page of the manuscript containing it has been sent to me by Bruno with a letter of September 29, 1986.

[33] U. FANO, L. FANO, *Basic Physics of Atoms and Molecules*, John Wiley & Sons, New York 1959.

[34] E. AMALDI, "Niels Bohr and the Early History of CERN", this volume.

[35] See, for example: a) Margaret Gowing, "Niels Bohr and Nuclear Weapons", invited paper presented at the "Niels Bohr Centenary Symposium", Copenhagen, October 4-7, 1985; b) Martin Sherwin, "Niels Bohr and the Origin of Nuclear Arms Control", invited paper presented at the "Niels Bohr Symposium", organized by the American Academy of Arts and Sciences, November 12-14, 1985.

[36] N. BOHR, J.A. WHEELER, "The Mechanism of Nuclear Fission", *Phys. Rev., 56*, 1939, 426-450, received June 28, 1939, and pp. 363-389 of Ref. [5c]; "The Fission of Protoactinium", *Phys. Rev., 56*, 1939, 1065-1066, "letter" dated October 20, 1939, and pp. 403-404 of Ref. [5c].

[37] M. AGENO, E. AMALDI, D. BOCCIARELLI, B.N. CACCIAPUOTI, G.C. TRABACCHI, "Sulla scissione degli elementi pesanti", *Ric. Scient., 11*, 1940, 302-311, (May 1940); M. AGENO, E. AMALDI, D. BOCCIARELLI, G.C. TRABACCHI, "Scissione dell'uranio con neutroni veloci", *ibidem*, 413-417, (June 1940).

[38] The originals of all Bohr's letters to me (or Wick and me) mentioned here are deposited in the Archives of Accademia Nazionale delle Scienze in Rome. Copies of these same letters are in the Archives of the Niels Bohr Institute, together with the originals of our letters.

[39] N. BOHR, "Successive Transformations in Nuclear Fission", *Phys. Rev., 58*, 1940, 864-866, and pp. 475-479 of Ref. [5c].

[40] E. AMALDI, D. BOCCIARELLI, G.C. TRABACCHI, "Sulla fissione del Torio e del Protoattinio", *Ric. Scient., 12*, 1941, 134-138, dated 10 February 1941.

[41] M. AGENO, E. AMALDI, D. BOCCIARELLI, B.N. CACCIAPUOTI, G.C. TRABACCHI, "Fission Yield by Fast Neutrons", *Phys. Rev., 60*, 1941, 67-75, received April 30, 1941.

[42] For some detail on this point see my intervention at pp. C8-315-316 of "Colloque International sur l'Histoire de la Physique des Particules", Paris, 21-23 Juillet 1982, *Journal de Physique*, Colloque C8, Supplement au n. 12, Tome 43, Décembre 1982.

Physics in Rome in the 40's and 50's

by
Edoardo Amaldi
Dipartimento di Fisica, Università 'La Sapienza' Roma

1. - Introduction -

I am very grateful to the organizers of this Symposium in honour of Marcello Conversi 70's birthday for the invitation to contribute a lecture that gives me the opportunity of expressing to Marcello my friendship and admiration for his work and for his human qualities. I remember him when he was a student during the last period of Fermi's life in Italy. He was a nicelooking, brilliant, intelligent and hard working young man, who, with the passing of time, added to his natural very remarkable qualities, an unusual balance of judgement and clarity of vision on any scientific, technological or human matter he was brought to face during his very active life.

The subject suggested for my speech by the organizers of this symposium, in part overlaps what I said on two previous occasions:on September 10th, 1978 in the talk on *"The years of reconstruction"* (1) and on 2nd October, 1985, in the talk *"Mario Ageno, ricercatore e trattatista"*(2) (*"Mario Ageno, research worker and treatis' writer"*).

In all parts of the speach of today, where overlaps are unavoidable with previous presentations of my recollations of old times, I will try to stress more the purely scientific aspects of the story, and leave in the background as much as possible the political and historical scenarios.

2. - The research lines followed during the 40's in Rome -

All the activities of the Istituto di Fisica 'Guglielmo Marconi' were heavily perturbed by the entrance of Italy into the Second World War when, on 10 June, 1940, Mussolini's Government declared war to France and Great Britain. Several young research workers, and even some less young people (like me, for example), were mobilized at that time, or even earlier, so that all activities had to be carried on with reduced staff until life conditions gradually returned to normality through a few phases marked by the following historical events: the entrance of the Allied troops in Rome on 5 June, 1944, the liberation of Milan and all North Italy on 25 April, 1945, and the unconditioned surrender of Germany, and the end of the Second World War in Europe, on 7 May of the same year.

Even during those years of war the research activity was never stopped completely. A few people could remain all the time in Rome because either they had not been drafted in the armed forces or their military duties kept them in Rome leaving them some free time. Others contributed to various researches during the leaves they obtained for different reasons from their military commands.

During this whole period the main research lines were essentially the same as immediately before the war: the study of cosmic rays, guided by G.Bernardini and nuclear physics, which was at the centre of my interest.

2.1 - The researches in nuclear physics.

The researches in the field of nuclear physics were carried out in collaboration with Daria Bocciarelli and Giulio Cesare Trabacchi of the Istituto Superiore di Sanità (I.S.S.) using as neutron source the 1.1 milion volt Cockcroft and Walton voltage multiplier, we had constructed with Franco Rasetti at the Laboratorio Fisico of that Institute in 1937-38(3).

During 1939-41 Ageno, Bocciarelli, Cacciapuoti, Trabacchi and Amaldi published a few papers(4) on the dependence of the fission cross section σ_f of $_{92}U^{238}$ and $_{90}Th^{232}$ on the energy of the incident neutron. The problem was quite new. The paper by Hahn and Strassmann announcing the discovery cf fission, had appeared in the first issue of January 1939 of Naturwissenschaften(5) and the papers by L.Meitner and O.Frisch(6) and by S.Flügge and G.v.Droste(7) providing the explanation of fission in terms of mass defect considerations, carry respectively the dates of 16 and 22 January 1939.

After the observation of the single fission fragments by means of a small ionization chamber and proportional amplifier made by Ageno and myself in winter 1939(2), and the recognition that the number of people working on fission induced in U^{235} by slow neutron was incredibly large, we had decided to work on U^{238}, a subject much less appealing from the point of view of possible applications but certainly a subject of research much less crowded and of comparable scientific interest. The results of our first measurements showed, in agreement with the theory of fission published shrotly before by Bohr and Wheeler(8), that σ_f remains practically constant from threshold (0.7 and 1.7 MeV) up to about 10 MeV. Above this energy, however, σ_f increases rapidly. This was, at the time, an unespected result.

After consultation with our theoreticians, Gian Carlo Wick and Bruno Ferretti, it was decided to submit the problem to Niels Bohr. My first letter to Bohr is dated 22 April, 1940, and the third and last one of our correspondece on this subject, is signed also by G.C.Wick and dated 22 February, 1941. To all these letters Niels Bohr answered, in his usual very kind style, giving an explanation of what we had observed in U. When the energy of the incident neutron is large enough, the compound nucleus U^{239}, produced by capture of the incident neutron, is so excited that the evaporation of a neutron becomes competitive with fission and the residual nucleus U^{238} is still sufficiently excited to undergo fission itself. The theoretical estimates made by Bohr for the case of Thorium were in excellent agreement with the measurements in the mean time we had carried out on this element. N.Bohr wrote a paper on this effect with the title "Successive transformations in nuclear fission", which appeared in Physical Review of November 1940(9).

After my return from six months service on the North Africa front, and having matured the conviction that almost any problem related to the investigation of fission could become of interest for the construction of nuclear weapons, we summarized our results in a paper written in English, sent it to Physical Review, where it appeared after the Japanese attack at Pearl Harbor(4d), and stopped working in this field. These decisions were taken in a discussion between the people directly involved in this work, plus G.Bernardini, B.Ferretti and G.C.Wick(10).

Immediately after we concentrated our activity on the investigation of the collisions of fast neutrons against protons(11) and deuterons(12) in energy regions still scarcely explored.

Later, in 1943, our attention was attracted by the typed copy, received from G.Placzek after Italy's entrance into the war, of a theoretical paper of Bohr, Peierls and Placzek(13) on the optical theorem and the relationship between the absorption and scattering cross section and the diffractive nature of the latter, for neutrons of sufficiently large energy against medium and heavy nuclei. The experiments on this subject kept us occupied during the years 1945 and 1946(14).

2.2 - The researches in cosmic rays

The research subjects in the field of cosmic rays, that as I said before, were guided by G.Bernardini, were:

- the study of the equilibrium between the electronic and the penetrating components at various altitude above sea level(15);
- the study of the positive excess of the penetrating component, by means of the magnetic deflection in magnetized iron cores(16), and, finally,
- the decay of the mesotron, tackled during the years 1939-43, by the indirect method, suggested independently by Kulemkampff(17) and Euler and Heisenberg(18), of comparing the absorption undergone by the penetrating component of cosmic rays in a layer of air with that undergone in a layer of the same mass per unit area of a condensed material like carbon or lead(19).

For what concerns the first line of research, I should remind that the ratio between the electronic component and the penetrating component remained unexplained until 1947 when the pion was discovered and one understood that the electron component in the high atmosphere is prevalently due to gamma rays from the decay of neutral pions, and that the muons originate from the decay of charged pions.

The people involved in this research, in addition to Gilberto Bernardini, were Bernardo Nestore Cacciapuoti, Oreste Piccioni, Bruno Ferretti and Gian Carlo Wick and, in the last experiment, Mariano Santangelo and Eolo Scrocco.

The first paper on the second line of research which concerns the positive excess of the mesotrons is signed by Bernardini and Marcello Conversi.Later also G.C.Wick, E.Pancini and E. Scrocco were involved.

The people that contributed to the third line, in addition to Gilberto, were Ageno, Cacciapuoti, Ferretti, Pancini, Piccioni and Wick.

I like to stress that the two theoreticians Gian Carlo Wick and Bruno Ferretti contributed to this world not only by developing detailed theories of the instruments and suggesting appropriate refinements of the data analysis, but also by helping the experimentalists in collecting the data at mountains altitudes.

By 1940 this kind of experiments, carried out by various groups in the world, in particular the new group that B.Rossi had set up in United States(20) and the group of G.Bernardini that I mentioned above, had provided conclusive evidence that the mesotron is an unstable particle decaying with a lifetime of a few microseconds. A direct measurement of its halftime, however, was desirable.

The first successful measurement was carried out by Rasetti at the University Laval in Quebec(21) where he had moved in summer 1939. At about the same time a similar experiment was made also by Auger et al. in Paris(22).

Already late in 1941 or perhaps early in 1942, Piccioni proposed to Conversi to work together for measuring the mean life of the mesotron with an experiment superior to those of the previous authors. All of them had recorded the number of retarded decay electrons in wide time intervals, assuming a priori that the decay law of the mesotron is exponential. By developing an appropriate electronic technique, Piccioni thought they could obtain a precise determination of this mean life, through the detection of several points of the mesotron exponential decay curve. This was the beginning of a very fruitful collaboration which lasted until 1946. In spite of the war conditions, Conversi and Piccioni obtained a good determination of the mesotron mean life(23), a measurement that had been already carried out, about one year before, in USA by Rossi and Nereson(24) but was not known in Italy because of the war, and was made at about the same time also in Paris by Chaminade et al.(25).

Immediately after Conversi and Piccioni started to study the behaviour of positive and negative mesotrons reduced at rest, in a first experiment in iron and later in a lighter element, carbon. By now was 1946. At this point of the work they were joined by E.Pancini, who had been first in military service and later strongly involved in the partisan fighting in North Italy(26).

I will not give detail about the important discovery of Conversi, Pancini and Piccioni(27) of the leptonic nature of the mesotron which has been put in full light by the excellent papers that O.Piccioni and M.Conversi have presented at the International Symposium on the History of Particle Physics, held at Fermilab in May 1980(28) and by the paper that Conversi has presented at the conference to celebrate the "40th Anniversary of the Discovery of the Pi Meson", held in Bristol in July, 1987(29).

In addition, the next contribution to this Symposium is by Oreste Piccioni, who will also treat this same important subject.

In the mean time a few groups had started to do research on cosmic rays also in other Italian universities. Giuseppe Cocconi, after his "laurea" examination at the University of Milan, towards the end of 1937, had moved to Rome, where at the beginning of January 1938 he had worked with E.Fermi and G.Bernardini on the products of the mesotron decay. Returned to Milan at the end of the academic year, he had started a considerable activity in cosmic rays, done in part alone, in part in collaboration with Vanna Tomgiorgi, who, a few years later, became his wife.

The main problems he tackled were: the study of the secondaries of the hard component, that of the neutrons present in cosmic rays and of the dependence of the zenithal effect on the mean life of the meson(30). In these experiments he used, besides Geiger counters, borontrifluoroside counters (for neutrons) and a vertical cloud chambers, the first to be constructed and used in Italy for the investigation of cosmic rays(31).

Towards the end of 1942, Cocconi was put in the 'tern' of a national competition(32) and nominated associated professor (professore straordinario) of Experimental Physics at the University of Catania, where he transferred a part of his research activity, in spite of the fact that he had been mobilized as 'car driver' at the end of February 1941. Some time later from Milan he was moved to the 'Centro Studi dell'Aeronautica Militare di Guidonia', in the outskirt of Rome, in the quality of "aviere di governo" and attached to the Istituto di Fisica Guglielmo Marconi to work in collaboration with the lieutenant Giorgio Fea(33) of the Metheorological Service of the Italian Airforce, on the development of infrared detectors sensitive enough to detect even in the dark airplain motors at a far distance.

Towards the end of the war, in collaboration with A.Loverdo and Vanna Tongiorgi(34), Cocconi initiated a study of the structure of extensive showers, which lasted many years, even after he moved, to Cornell University at the end of 1947. The whole set of these papers, in particular their initial part done in Italy, even today should be considered as an example, also for text books, of a particularly elegant and significant experimental study of a very complex phenomenon(35).

After the departure of Cocconi from Milan, the investigation of cosmic rays and the instruments for their detection, with particular regard to extensive showers, was continued at the Università Statale by G.Salvini, who, with a few collaborators (A.Lovati, A.Mura e G.Tagliaferri), tackled the study of the penetrating component present in extensive showers, and the local production of penetrating particles by extensive showers(36).

In 1949 Salvini moved to Princeton, on invitation from that University, where he studied, with success, the production ratio of neutral and charged pions in cosmic rays(37), a work that kept him busy even after his return to Italy in 1951.

In Milan there were also a few theoreticians, in particular C.Salvetti (1946)(38) at the Università Statale and A.Borsellino (1947) at the Polytechnic.

Also in Padua, A.Rostagni - who had succeeded to B.Rossi - could count on a number of clever young physicists, whose scientific production started to appear in *Il Nuovo Cimento* and other scientific journals between 1946 and 1948. Among these I recall: G.Puppi (1946), who, after a short period devoted to the theory of polar biatomic molecules, passed to the study of cosmic rays, which, in those years, was also the central subject of the work of the other Padua's physicists: N.Dallaporta, E.Clementel (1946), I.Filosofo (1947), P.Bassi, A.Kind and A.Loria (1948).

A result of particular importance was obtained in 1949 by G.Puppi who was one of several physicists who independently from each other, arrived to formulate the universality of weak interaction(39), i.e. the fact that the coupling constants are the same for all processes due to weak interaction among the three pairs of particles (p, n), (ν_e, e) and (ν_μ, μ). This relationship appears particularly evident from the Puppi triangle, which is clearly described in words but not shown as a figure in Puppi's paper.

The revival of the interest in physics was, however, even more general. In Turin, G.Montalenti and L.Radicati (1947) joined C.M.Garelli and G.Lovera (1943), in Pavia A.Gigli (1947) joined L.Giulotto (1943), A.Ricamo (1946) started to work in Bologna; in Florence, in the field of Optics, G.Toraldo Di Francia (1944) joined F.Scandone (1943) and in that of cosmic rays, M.Della Corte (1946) started to work, in Rome at the National Institute of Electroacustics O.M.Corbino, P.G.Bordoni (1947) and D.Sette (1948), in Catania G.Milone (1947).

2.3 - The researches in other fields of physics

Until now I have presented only researches and results in the field of nuclear physics and cosmic rays, the latter seen, almost exclusively, as the source of subatomic particles.

Among the young people that I mentioned above as new forces entering research during the 40's, a number, however, were devoting their activity to completely different fields. And among them a few had pretty soon a great success. This was, in particular, the case of Pier Giorgio Bordoni of the Istituto Nazionale di Elettroacustica O.M.Corbino (now Istituto di Acustica O.M.Corbino) of the CNR (Consiglio Nazionale delle Ricerche), who, in 1949, using statistical mechanics methods and the recent theory of anelasticity of Zener, forsaw a new relaxation effect to be looked for in plastically deformed crystals at low temperature. This effect is due to preexisting dislocations that move through the crystal because of the simultaneous action of small elastic stresses and thermal fluctuations[(40)]. Shortly after Bordoni discovered this effect in a few mono- and poli-crystalline metals face-centered-cubic (fcc) lattices (Pb, Cu, Al, Ag) tested in experiments carried out at the low-temperature laboratory of M.I.T. with the equipment he had completely designed and constructed at the Istituto Nazionale di Elettroacustica[(41)].

About 130 original papers before 1960 and more than 300 papers until 1987 appeared on this subject, which is currently indicated as "Bordoni effect". It takes place not only in all elements with fcc lattices, but also in those with bcc and ec lattices.

This relaxation effect, due to dislocations has provided one of the first experimental proves of the existence of dislocations in crystals and preceded their visualization, which was carried out in 1953 in the case of transparent crystals by Hedges and Mitchell.

3. - The parallel development of the research institutions -

Let me go back for a moment to 1945, when the II World War came to an end. It was generally recognized that the Istituto Guglielmo Marconi was one of the few laboratories of the whole country which had maintained alive a few research activities at a good level during the war period. Thus, at the end of summer 1945, the CNR approved the proposal presented a few months before, of creating at the Istituto di Fisica Guglielmo Marconi a "Centro di Studio della Fisica Nucleare e delle Particelle Elementari" of which I was nominated director and Bernardini vice-director. The research programme obviously consisted in the continuation and natural development of previous activities. The creation of the Centre, however, ensured a continuity in the appropriation of financial means by CNR, which until then had been extremely limited and irregular.

During summer and autumn 1946 Bernardini, Ferretti and I renewed our contacts with the international scientific circles on occasion of a very important international conference held in July at the Cavendish Laboratory in Cambridge (U.K.), and a tour of seminars I made in United States from the beginning of September to the end of the same year.

These renewed contacts confirmed the conviction, that, during those difficult years, in the field of fundamental research we had succeeded in maintaining a good level, while we had remained behind in many experimental techniques, not to speak of the applications of neutron physics(1).

Therefore it was necessary to make an effort for bringing our experimental techniques up-to-date and to abandon, for the moment, the investigation of neutron physics since our instrumentation for years to come would not be in conditions to compete even for the most simple fundamental problems, with that at disposal of United States physicists(42). As a consequence of these decisions a few programmes still in the initial phase, in the field of nuclear physics, were abandoned(42), my research group passed to the study of cosmic rays, and a few new techniques for their investigation were introduced in Italy(43).

In the same year 1947 Bernardini conceived and, with the help of Pancini and Conversi, directed the construction of the Laboratorio della Testa Grigia (L.T.G. altitude: 3,500 m.s.l.) on the Swiss -Italian border near the Theodul Pass(44). The site was well known to the Roman physicists, who had made two expeditions of a few months each there, the first one during the winter 1940-41, the second during the winter 1942-43, using as a laboratory a small room of the upper station of the aerial cableway. Also Cocconi and Tongiorgi had spent a few months there to take measurements in the same place, but the construction of an *ad hoc* laboratory allowed an activity of much greater amplitude during the whole solar year. Many physicists from the Universities of Rome, Bologna, Milan and Turin worked at LTG for about ten years. Among these I like to recall, besides Bernardini, Pancini, Conversi and Mezzetti, a large number of younger people such as R.Bizzarri, B.Brunelli, C.Castagnoli, M.Cervasi, M.Cini (who later started to work in theoretical physics), G.Fidecaro, A.Gigli, A.Michelini, S.Sciuti, G.Stoppini, E.Zavattini and A.Zichichi. These last, of course, at later times. The LTG is still used today for cosmic ray research by the Laboratorio Nazionale di Cosmo-Geofisica of the CNR, in Turin.

Research activities were appearing or reappearing and expanding also in other universities. Since they were particularly intense in Padua, the CNR on the 1st January 1947 created at the Institute of Physics Galileo Galilei of that University a "Centro per lo Studio degli ioni veloci" similar to that created in Rome fifteen months before. The expedency of such a decision was confirmed by the large number of capable young researchers that shortly afterwards entered their production phase: M.Ceccarelli, M.Merlin, (1949), F.Ferrari and C.Villi (1952), M.Baldo, M.Cresti, B.Vitale, G.Zago and T.Zorn (1963).

Also in Turin things were moving, at the Polytechnic as well as at the University. G.Wataghin's return to Italy from Brasil in 1949, certainly had an influence on the young people, among which I recall: F. De Michelis, A.Gamba, R.Malvano (1949), M.Cini, S.Fubini, M.Panetti (1951), P.Brovetto, S.Ferroni (1952), A.Debenedetti, T.Regge, M.Vigone (1953), B.Bosco, G.Ghigo and A.Tallone (1954). Consequently the CNR created, the 1st July 1951, a 'Centro sperimentale e teorico di Fisica Nucleare' with its seat in the Institute of Physics of the University of Turin.

During the same year, 1949, also G.Occhialini returned to Italy from Belgium. He had remained in S.Paulo in Brasil until 1945 and then had moved to Bristol where he had worked with C.F.Powell until 1948. In that year he moved to the 'Centre de Physique Nucléaire' in Brussels and returned to Italy after he had been nominated professor at the University of Genoa, where he had a strong and lasting influence on the research activity. In 1952 he was called to the University of Milan, where, in August 1951, a fourth Center of CNR had been created, and where, in 1951 C.Succi, in 1952 L.Scarsi joined the physicists already mentioned above.

Under the influence of Occhialini, and of his wife Constance Dilworth, not only were groups of experts in nuclear emulsion technique formed in Genoa and Milan, but also the already existing groups using this same technique in other Universities were appreciably reinforced.

At the same time, A.Bisi, G.Bertolini (1952) and L.Zappa (1953) began working with G.Bolla (1901-1980) who had moved to the chair of Fisica Generale of Milan Polytechnic in 1950(45).

It was at that time that the necessity for a close coordination of the research activities of the four 'Centers' of the CNR emerged very clearly and, under the parallel action of G.Bernardini, E.Perucca, President of the Physics Committe of CNR , and me as a member of the same committee, the Istituto Nazionale di Fisica Nucleare (INFN) was created on 8 August , 1951. This body had a new structure, the aim of which was a grouping and coordination of most research activities of similar nature carried on in different universities, completing and integrating, whenever necessary and/or possible, the staff and the financial means assigned by the Ministry of Education(46).

The first President of INFN was G.Bernardini who, in that period, spent about one half of his time in the USA, first at Columbia University and, later, at the University of Illinois. Thus he ensured automatically a close contact with many American physicists who already had particle accelerators at their disposal.

In the meantime Rostagni had constructed (1950) the Observatory of Marmolada (altitude 2,030 m.s.l.)(47), where besides the physicists from Padua, also E.H.S.Burhop and H.Massey from the University of London and Deutschmann from the University of Munich worked for rather long periods.

Two more typical developments that characterize the 50s should be mentioned here.

The idea of a wide scale European collaboration with the aim of tackling and solving together those problems that no single European country would have had the technical, organizational and financial strength to face and solve alone, started to appear in various universities, laboratories and cultural centers in Europe already in the years 1948-50. In particular the opportunity of creating a European laboratory for research on elementary particles, endowed with such equipment that no single country could have afforded alone was considered. The prehistory of CERN (Centre Européen pour la Recherche Nucléaire) was thus beginning and the Istituto Guglielmo Marconi was involved from the start, since I was nominated Secretary General of the Organization in its provisional phase. This came to an end on September 29th, 1954, when a prearranged point was reached in the procedure of ratification by the Parliaments of the Member States of the 'Convention' establishing definitively the existence of this new Organization and fixing its main features[(48)].

In the meantime, in Italy, the national structure of INFN and the larger funds it had received, led to consider the opportunity of an enlargement of its research activities by facing problems of greater dimensions. Thus, at the beginning of 1953, under the chairmanship of Bernardini, it was decided to construct an electronsynchrotron of 1000 MeV and Giorgio Salvini was appointed to direct the design and construction of this machine and of the corresponding laboratory.

Salvini completed his task, by creating the Laboratori Nazionali di Frascati del CNRN (later CNEN, now ENEA) around a machine of excellent performance, within 1958, i.e. within the time schedule and with the funds foreseen from the start[(49)].

The Frascati machine accelerated electrons and, therefore, not accidentally, was complementary to the two machines of the initial program of CERN, which both accelerated protons: the synchrocyclotron of 600 MeV, designed and built under the direction of C.J.Bakker from the Netherlands, which entered into operation in 1957, and the protosynchrotron of 30 GeV, designed and constructed under the direction of J.Adams from U.K., and that started to operate on 25 November, 1959.

Italy was a Member State of CERN from the beginning and now the Italian physicists had at their disposal not only the Laboratori Nazionali di Frascati, equipped for the investigation of electron and photon physics, but could participate, in collaboration with those of the other European countries, also in the researches carried on in the Geneva laboratories in the field of hadron and lepton physics.

4. - The researches during the 50's -

While the 40's were the years of the Second World War and of the beginning of the reconstruction, the 50's include the end of the reconstruction period and the transition to a new era in natural sciences in general, and in physics in particular.

It is in this decade that, for example, high energy physics is transformed from being a branch of cosmic rays investigation carried on by observing the secondaries of primaries of not well defined momentum and sometimes even dobtful identification into a science constructed piece after piece, by detailed studies of the rich phenomenology produced by beams of particles provided by accelerators.

It is during the 50's that visual techniques, like nuclear emulsions and buble chambers, become current instruments extremely appropriate for tackling the problems that the physicists start to face immediately after the II World War: here it is enough to recall the discovery of the pion and of its decay made by Lattes, Muihead, Occhialini and Powell(50) a few months after the Conversi, Pancini, Piccioni experiment and the discovery of the two first examples of strange particles made, shortly later, by Rochester and Buthler with a cloud chamber(51).

Passing from cosmic rays to accelerators the cloud chambers were replaced by bubble chambers because of a number of excellent reasons, the first of which is the higher density of the medium in which any charged particle leaves its trace. But also other chapters of physics seem to have undergone a development at an accelerated pace in those years: the discovery of nuclear magnetic resonances was made in 1946, independently, by the groups of Purcell at Harvard and Bloch at Stanford, the investigation of semiconductors brought to the discovery of the transistors on the Christmas' eve of 1947, the idea of quantizing the circulation in superfluid helium was for the first time formulated by Onsager at the International Conference on Statistical Mechanics held in Fluorence in 1949(52), but passed unnoticed and was rediscovered and applied to a few specific cases (such as turbulent flux and rotating helium) by Feymann in 1955, and the theory of superconductivity appeared in 1957.

4.1 - Particle physics

Shortly after the publication of the results of the Conversi, Pancini and Piccioni experiment, some cosmic rays experimentalists as well as theoreticians raised, on completely different grounds, the question whether the muons could not have an energy-dependent interaction with nucleons, which would become apparent only at very high energy. In order to clarify this point a series of experiments was carried out at the University of Rome by means of hodoscopes of about one hundred or more Geiger counters. The first experiment by G.Fidecaro and E.Amaldi, showed that the measured elastic cross section of a few GeV energy muons against lead nuclei was in agreement with that computed by taking into account only the electromagnetic interaction(53).

The same conclusion was obtained shortly after by Castagnoli, Gigli, Sciuti and me for the inelastic cross section measured at such a depth undergound, that the mean energy of the muons is about 10 GeV(54). Similar results were obtained by means of a cloud chamber by Lovati, Mura, Succi and Tagliaferri(55) of the University of Milan. Some time later my young collaborators repeated our experiment with counters at a greater depth, and confirmed the absence of strong interactions for muons of energy as large as about 50 GeV(56).

A number of researches were carried out in those years at the two high altitude laboratory in the Alps. The main subjects studied at the L.T.G. (3.500 m.s.l.) were the following: the nuclear disintegration produced by cosmic rays were studied with nuclear emulsions(57) as well as with fast ionization chambers(58), extensive showers(59), the electromagnetic component produced in nuclear explosions with a cloud chamber of the University of Milan(60), the penetrating showers from hydrogen and other elements by means of an hodoscope of Geiger counters(61), and a determination of the mean life of the new strange particles by means of a cloud chamber(62) both by groups of the University of Rome.

The researches made at the Observatory of the Marmolada were carried on mainly by Padua physicists and concerned: the positive excess of mesons(63), the zenithal distribution of the nucleonic component(64) and of the particles of extensive showers(65). Furthermore Deutschmann, Cresti et al investigated by means of a cloud chamber the decay of V°(66) and Massey, Burhop et al(67) the production of secondary particles produced in hydrogen by means of a cloud chamber at 100 atmosphere.

In autumn 1951 Marcello Conversi and about one year later Giorgio Salvini became professors at the University of Pisa. In the mean time Conversi had been for about two years in Chicago, where Fermi was leading a group of outstanding students. In that period Marcello extended the experimental study of cosmic rays mesons and protons at several latitudes and altitudes (up 9.000 m.s.l.) by means of counter arrays carried in airplane flights(68).

As I mentioned before, Salvini had spent very fruitfully those years in Princeton, where he extended, in part in collaboration with Y.B.Kim, in part alone(37) the investigation of the themes he had started to tackle at the L.T.G., leading a group from the University of Milan, i.e. the study of the production of neutral pions by means of a cloud chamber controlled by an array of counters. In the experiment carried out in United States, the mountain laboratory was that at Echo Lake (3.200 m.s.l.).

The more interesting results of these new experiments were the determination of the value of the pion production cross section and of the ratio of the number of neutral pion to that of the charged pions produced per event with multiplicity ranging from 1 to 5 or more. The value found in these experiments, for this ratio, was 0.5; it corrected previous results obtained by other authors, and settled a point of relevance in connection with the problem of extending to the interaction intervening in pion production, the concept of charge independence, which had been established in 1936 for low energy nucleon-nucleon interaction.

Starting from 1953 Salvini was busy with the direction of the design and construction of the electrosynchrotron. In 1955 he moved from the University of Pisa to that of Rome, from where he could more easily fullfil his function which included the transformation of a few hectars in the Colli Albani, from a vignard into the Frascati Laboratory, in full operation.

Conversi remained at the University of Pisa until 1958, i.e. a time interval long enough for exerting a remarkable influence on the young people grown up in those years at the University and the Scuola Normale.

For brevity reasons I will recall here only three of the many research activities he developed in those years in collaboration of a few colleagues and a number of very gifted young pupils.

The first is the conception and construction, in collaboration with Adriano Gozzini, of the flash chamber(69); a new instrument which allows the electrical reconstruction of the tracks of charged particles, still used today in cosmic rays experiments and in the proton decay underground experiment in the Frejus Laboratory, and provided the point of departure for the later development of spark chambers.

The second problem tackled by Conversi with a Pisa-Milan group was the search at mountains altitudes, by means of a cloud chamber triggered by an array of counters, of particles of 550 electron masses suggested by Alicanian et al.

The experiments gave negative result(70) and thus closed a problem that, in the middle of the 50's, was for some time at the centre of the attention of high energy physicsts.

The third activity that I like to mention is a research carried out at Brookhaven by a group lead by Jack Steinberg and in which Conversi and Puppi entered with a few of their young collaborators from the Universities of Pisa and Bologna.

Following the suggestion advanced by T.D.Lee and C.N.Yang in their famous paper on non-conservation of parity , this composite group studied with a bubble chamber the correlation between production and decay angle of the processes initiated by a pi-minus plus proton collision and ending either in a $\Lambda^\circ + \theta^\circ$ or in a $\Sigma^- + K^+$(71).

This experiment followed shortly later the first experiments on this fundamental problem: the paper by Mrs. Wu et al on Co^{60} beta decay(72) and the papers by Garwing, Lederman et al(73) and by Telegdi et al(74) on the $\pi \rightarrow \mu \rightarrow e$ decay chain.

In parallel with these research activities Conversi founded in Pisa *"Il Centro di Studi sulle Calcolatrici Elettriche "* which eventually became a national institute for the elaboration of information of the CNR.

In the post war situation prevailing in many European countries, and in particular in Italy, the nuclear emulsion technique developed mainly at the Bristol University and later also at the Université Libre de Bruxelles - was the essential tool providing the physicists the possibility of participating in the most important developments of high energy physics in spite of the modesty of the budjets of the corresponding laboratories. Occhialini and Powell played a fundamental role in promoting the development of this technique, which during the early fifties allowed an extensive study of the particles produced by collisions of cosmic rays primaries in the upper atmosphere and, shortly afterwards, the participation of many European groups in interesting researches on the new particles produced by high energy accelerators built in USA. This participation was due mainly to the generosity of a few USA laboratories, among which particularly the Lawrence Radiation Laboratory in Berkeley, and the Brookhaven National Laboratory, L.I., should be mentioned. But without the emulsion technique such a participation would have not been possible at all.

As I said before Bernardini was the first to use this technique in Italy for studying, in collaboration with G.Cortini and A.Manfredini the nuclear evaporations produced by the nucleonic component of cosmic rays already in 1948-50(57).

When later - in 1953-55 - he was at the University of Urbana, Bernardini maintained a collaboration with M.Beneventano, G.Stoppini and L.Tau of the University of Rome, carrying on researches on photoproduction of pions by means of nuclear emulsions exposed to gamma rays produced by means of the betatron constructed by Kerst et al at the University of Illinois. The main theme of these papers was the photoproduction of charged pions in deuterium and hydrogen near threshold(75)(76). Their more specific and interesting result was a fairly good determination of the value of the coupling constant of strong interactions.

The exploration of high altitude by means of balloons to study cosmic radiation had been tried extensively with success in the U.K., particularly by the Bristol Goup, around 1950. This group thus arrived at the discovery of the τ^-(77) and the charged K-meson(78).

In 1951 it was suggested that a study of high energy events produced by cosmic radiation could be more conveniently carried out at lower latitudes. At a latitude of 40°, for example, only that part of the primary radiation which has an energy above 7 GeV per nucleon is allowed by the magnetic field of the Earth to enter the atmosphere. Thus, at these latitudes the magnetic field of the Earth removes a large number of primaries which are not very efficient in producing the new secondary particles because of their relatively low energy. The nearer one goes to the equator the stronger this effect is.

Thus a first international expedition was planned in 1952 under the sponsorship of CERN that recently had entered its "Planning Stage". Naples and Cagliari were chosen as the most convenient of the available bases in the Mediterranean. Thirteen universities took part in the expedition which met some success. In particular was the first successful attempt to recover balloons at sea. Furthermore, the expedition made a survey of the winds at high altitudes which was of great importance for the expedition of the subsequent year. The results of this first expedition were briefly reported to the Third Session of the Council of CERN held on 4-7 October 1952 in Amsterdam.

A second expedition took place, also to Sardinia, in June-July 1953. This was much on the same lines but on a larger scale than the first one. Eighteen laboratories from European countries in addition to one from Australia took part in it. 25 balloons were launched and over 1000 emulsions were exposed 7 hours at altitudes between 25 and 30 km. This corresponds to 9.27 litres of nuclear emulsions weighing 37 kg. They were successfully recovered at sea on account of the employment of a seaplane and a corvette (Pomona) generously placed at disposal of the expedition by the Italian Airforce and the Italian Navy[(79)].

While in the 1952 expedition only glass-backed emulsions had been used, in 1953 expedition "stripped emulsions" were introduced. Thus, the tracks were followed almost always from one emulsion to the adjacent one, a point of great importance when high-energy events are studied. The development required special care and in particular the construction of special developing systems at the Universities of Bristol, Padua and Rome, where the emulsions of the whole expedition were processed.

In October 1953, a meeting was held at the Department of Physics of the University of Bern (Switzerland) for distributing the packages of exposed and developed emulsion among the participating universities and, in April 1954, an international conference was organized in Padua to discuss the first results and plan jointly the most efficient methods for the investigation[(80)].

Further results appeared in the normal scientific literature and in lectures given by various participants at the second Course of the Varenna International Summer School that took place during the summer 1954. At the same course, Fermi gave a series of "Lectures on pions and nucleons" which were his last contribution to the teaching of physics before his death in Chicago on November 29 of the same year[(81)].

Among the main results of the expedition I can recall a number of determinations of the values of the masses of heavy mesons giving rise to different decays which brought suspicion that all these processes could be alternative decays of the same particle. Also the identification of the various hyperons and the determination of their masses and decay modes were considerably improved.

In July of the same year, 1953, a very important conference was held at Bagnéres de Bigorre in the French Pyrénées where the new heavy particles were discussed at length(82). At this meeting, for example, the Dalitz plot for the τ-meson was discussed by Dalitz, Michel presented the selection rules for decay processes and the Rome Group introduced the logarithm of tgθ (later called rapidity) as an appropriate variable for the description of high-energy events in which many secondary particles are emitted.

The conference dinner speaker was Cecil Powell who, in brilliant terms, pressed for the urgency of constructing big European accelerators for avoiding that in the old continent subnuclear particle research could be drawn by the mounting level of the sea of machine results already noticeable in the United States.

A third expedition was organized by the Universities of Bristol, Milan and Padua in October 1954 from Northern Italy (Novi Ligure). It consisted in the launghing of a single stack of 15 litres of nuclear emulsions (corresponding to a weight of 63 kg) which is indicated in the literature as "G-stack"(83).

The great advantage in employing a very large stack is evident for studying the different modes of decay of the heavy mesons produced in the collision of a high energy particle with a nucleus of the emulsions. A substantial fraction of the secondaries come to rest inside the stack so that their modes of decay can be observed and the energy determined with high accuracy from their range.

A great part of the results was presented at the International Conference on Elementary Particles held in Pisa in June 1955 to celebrate the Centenary of Il Nuovo Cimento(84).

One of most important results of the G-stack study was the final recognition that the values of the masses of heavy mesons giving rise to different decay processes were identical within rather small experimental errors. Thus the interpretation of all these processes as alternative decays of the same particle was strengthened, contributing in an essential way to the general recognition of the so-called $\theta-\tau$ puzzle. This found its explanation only in 1956 with the discovery by Lee and Yang of the nonconservation of parity by weak interactions.

A striking fact that emerged in Pisa was that the time for important contributions to subnuclear particle physics from the study of cosmic rays was very close to an end. A few papers presented by physicists from the U.S.A. showed clearly the advantage for the study of these particles presented by the Cosmotron of the Brookhaven National Laboratory (3 GeV) but even more by the Bevatron of the Lawrence Radiation Laboratory in Berkeley (6.3 GeV).

The emulsion group of the University of Rome took a very active part in two of the three baloon expeditions I mentioned above but not in the third one. We had observed in the emulsions exposed on occasion of the second baloon expedition two stars connected by a track the main features of which suggested a rather slow particle of protonic mass, which we thought could be due to an antiproton[85]. The observation was made at the beginning of 1955 and the Bevatron in Berkeley was rather close to enter into operation. An agreement was reached with the Berkeley physicists so that, in autumn of the same year, shortly after the successful experiment of Chainberlain, Segrè, Wiegand and Ypsilantis[86] a stack of emulsion was exposed in the same beam of negative protons and the emulsions were divided between the Berkeley and the Rome laboratory.

The first annihilation star was observed in December in Rome, the second one shortly later in Berkeley, and many more in due time in both places. This "antiproton collaboration" between Berkeley and Rome lasted from autumn 1955 to the end of 1959 and brought to a rather extensive study of the main features of antiproton annihilation in emulsions[87].

These results were, in part, superceded a few years later, when liquid hydrogen bubble chambers of large dimensions started to provide a great number of clean proton-antiproton data.

There are many other important results obtained in those years by groups working with the emulsion technique in different Italian Universities that I would like to remind. But for time reasons I will recall very briefly only one more of them: I refer to the discovery made at Padua by Massimilla Baldo-Ceolin and Prouse of the first example of an antilambda appeared in Il Nuovo Cimento of November 1958[88].

4.2 - A few hints about the investigation of optics and the structure of condensed matter

This short review of what happened in physics in Rome, and more in general in Italy, during the 40's and the 50's, would be unilateral and too much distorted if I would not touch upon a few of the many results obtained in those years in other fields of physics like Optics and in structure of condensed matter.

In Italy the more relevant contributions to Optics, since the years of war, are due to Giuliano Toraldo di Francia, who, in Florence, in 1961, introduced the principle of inverse interference, and, a few years later forsaw the existence of evanescent waves. He provided the experimental evidence of the existences of these waves by means of microwaves in 1949. At about the same time Toraldo provided also the prove of the directionality of the retina's cones by means of the concept of retinic microwaves, an interpretation later largely accepted. In later periods Toraldo di Francia developed an interpretation of the acoustic propagation in the deep ocean (1950-56), introduced the concept of superesolution of optical instruments (1952), founded the theory of optical resolution from the point of view of the information theory (1957) and developed a complete theory of all effects of

"Cerenkov type", decomposing the relativistic field of the electron in evanescent waves (1960).

The development of the investigation of the structure of condensed matter in Italy has been dealt with from the historical point of view at least twice: a first time by the late Luigi Giulotto (1911-1986)(89) of the University of Pavia, in an excellent lecture he gave, on occasion of the 20th anniversary of the creation of the Gruppo Nazionale di Struttura della Materia(90) and a second time at the international conference on *"The Origin of Solid State Physics in Italy: 1945-1960"* held in Pavia on 21-24 September 1987(91).

Let me start from the Istituto di Elettroacustica 'Orso Mario Corbino' (now Istituto di Acustica O.M.Corbino(92)) where in 1939, Amedeo Giacomini (1905-1976)(93) created a section devoted to the investigation of ultrasounds. At the beginning of the 40's Giacomini started a series of measurements of the velocity of ultrasounds in liquids by an optical technique based on the diffraction of light by the ultrasound field. This method is clearly limited to transparent liquids.

In solids the velocity of the different types of waves (longitudinal, transversal, extensional, etc.) can be determined in the most convenient way from the measurement of the corresponding eigenfrequencies of bodies of simple shape (cylinder, plate, etc) and known dimensions. This kind of measurement was initiated in 1947 by Bordoni, who had developed an electrostatic method for measuring vibration amplitudes appreciably smaller than the lattice constant of solids and which in addition permits the determination of the energy dissipation.

The measurement of the energy dissipation associated to ultrasound propagation involves greater difficulties in liquids than in solids. This problem was tackled with considerable amplitude and depth starting from 1949 by Daniele Sette, whose many remarkable results have been amply illustrated by Pier Giorgio Bordoni in the speech he gave on the activity of the Institute O.M.Corbino on occasion of the international symposium held in Rome in April 1987, for celebrating the 50th Anniversary of the foundation of the Institute and of the death of its founder(94).

In his speech Bordoni gave the due credit to the many other members of that Institute who, over the years, contributed interesting scientific results.

At the beginning of 1945, Doctor Engineer Giorgio Careri appeared at the Institute 'Guglielmo Marconi'. Shortly before he had taken the 'laurea' in chemical engineering, but he did not want to work as an engineer and even less as a chemist. He wished to learn physics at the Institute but not the type of physics we were doing. We discussed his problems together. I was very much in favour of his general frame of mind since I had always felt uneasy about the disharmonious development of physics in Italy. But it is not easy to teach what one does not know at a level of serious and productive research. The solution of the problem was found pretty soon: Careri designed and constructed a mass spectrometer, not planned for nuclear measurement, but for the quantitative analysis of the isotopic composition of various elements, in view of a few applications he made later to the abundance of deuterium and argon 40 (produced in the beta decay of potassium 40) in vapour of endogeneous origin(95). Later he passed to the study of simple liquids, such as liquid metals(96) and finally to that of liquid helium amd its superfluid phase.

This work was started in the cryogenic laboratory set up in Frascati under the responsability of Careri for providing liquid hydrogen and liquid helium targets for the synchrotron as soon as it would be ready to accelerate electrons for the experimental programme. The technique invented by Careri and developed by him and his collaborators(97) was based on the investigation of the movement of an He ion inside liquid Helium. This work brought a few years later to the experimental evidence of the quantization of the circulation by Careri and collaborators(98).

A very nice account of this long series of experiments has been presented by Careri and Scaramuzzi to the 1987 Conference on History of Solid State Physics in Italy I mentioned before(99).

Careri was in some way followed by Giovanni Boato who arrived in Rome in 1947 from Genova, where he had taken his laurea in Chemistry with Carlo Perrier(100). In Rome Boato took his laurea in physics at the end of the same year and remained to work at the Institute Guglielmo Marconi. In the middle of the 50.s he became professor at the University of Genova where, with a number of students and collaborators, investigated various isotopic effect under condition of phases equilibrium, arriving to formulate a law which regulates this kind of equilibria.

When, in 1955, the CNRN (Comitato Nazionale per le Ricerche Nucleari, later CNEN, still later ENEA) obtained the financial means for installing at Ispra, under the direction of Carlo Salvetti of the University of Milan, a CP-5 type research reactor, immediately a small group of young physicists was formed in Rome for the development of the instrumentation necessary for the exploration of the structure and properties of condensed matter by slow neutron diffraction.

Initially the group consisted of Giuseppe Caglioti, who had previously worked on the secondaries produced by high energy muons(101), Antonio Paoletti and Francesco Paolo Ricci, who had both worked with Careri, on self-diffusion of liquid metals.

A very nice account of the first period of this activity in Italy has been given by G.Caglioti(102). Caglioti learned slow neutron techniques first at Argonne National Laboratory and later at Chalk River, where he started to work with Bert Brockhouse on the dynamics of atoms in crystals; Paoletti went to Brookhaven National Laboratory where he started to work with Robert Nathans in the field of magnetism and F.P.Ricci to M.I.T. where he worked with C.G.Shull on theoretical problems and experimental methodologies.

The scientific fruits produced at the European Laboratories of Ispra, Casaccia and Grenoble started to appear only after 1960, and therefore do not enter inside the period covered by my review.

In the middle of the 50's, Ugo Palma and Beatrice Palma Vittorelli, of the University of Palermo, set up, after a period spent at M.I.T., a research activity carried on by means of a "home made" spectrometer for the electronic paramagnetic resonance and auxiliary equipment which was used, by them and several collaborators, for dealing with a number of problems such as the vibronic spectra of magnetic crystals, the collective motion of protons in non-ferromagentic crystals, the cascade of ionic and electronic processes triggered at various temperatures in photosensitive crystals.

Adriano Gozzini, immediately after his return from war imprisonment, started an activity in the field of microwaves spectroscopy in Pisa. Already during the second part of the 40's and the 50's he obtained a number of important results and created at the same time a remarkable school still very active today. Both these aspects have been amply illustrated on occasion of the Conference held in Pisa on 5 September 1987 for celebrating his 70th birthday. The nice volume collecting the contributions to this meeting and also a few article sent by people unable to be present at the ceremony(103), illustrates not only the work carried on by Gozzini, and more in general by the physicists working in Pisa on various problems of structure of matter, but also, the relations between Gozzini and his collaborators and other groups, in particular the group of Alfred Kastler of the École Normale Superieur of Paris.

A considerable activity on ferromagnetism was developed in Turin and Ferrara. At the Istituto Elettrotecnico Nazionale 'Galielo Ferraris' in Turin Biorci, Ferro and Giorgio Montalenti carried on a systematic investigation on magnetic alloys which brought to the clarification of various until then unsolved problems, in particular the relationships between cohercitive force, permeability and magnetic viscosity.

The radiation damage produced by neutron bombardment of high permeability alloys was investigated later by Ferro and Montalenti. The last author carried on similar investigations also about the dielectric constant.

In Ferrara A.Drigo has studied the magnetic properties of thin films. He was the first to confirm the theory by Kittel about the disappearance of the Bloch walls in thin films and the decrease undergone by the magnetic moments when the films tend to become two-dimensional structures.

Finally I come to Pavia where Luigi Giulotto led directly and undirectly, a number of activities without interruption for more than 45 years. A paper by Giulotto, initiated in 1942 and concluded only in 1947 because of the war conditions(104), concerns the fine structure of the spectral line H_α. In this paper Giulotto provided the first clear experimental proof of a small separation between the terms n =2, $\ell = 0$ and n = 2, $\ell = 1$ of the hydrogen atom, which shortly later was correctly interpreted by Lamb and became known as Lambshift. Giulotto's paper is quoted by Lamb and in a comment on this important effect appeared in the Encyclopedia Britannica(105). According to Giulotto the subject had been proposed to him by Piero Caldirola (1914-1984)(106) who had in those years a considerable influence also on the research programme of the experimentalists of the University of Pavia.

Shortly later the nuclear magnetic resonance (NMR) became a central subject of activity of the Alessandro Volta Institute. The first NMR apparatus was working in Pavia already in 1947, only one year after the discovery by Bloch and coworkers and Purcell and coworkers. The first to enter this field as a collaborator of Giulotto was Alberto Gigli, followed by Chiarotti and Lanzi who contributed to the development of a new technique for measuring the nuclear spin-lattice relaxation time based on the inversion of the nuclear magnetization i.e. on a negative temperature of the spin system(107).

The technique was employed in a number of problems like: accurate measurements of the relaxation times in liquids of high purity or containing paramagnetic impurities, the investigation of the molecular association and polimerization, of the different mobilities of water molecules adsorbed in colloidal solutions of organic substances (Bonera, Lanzi, Rigamonti) and the phase transitions (Rigamonti; Bonera and Tosca).

Other interesting researches carried on in Pavia were, in 1948, the investigation of the Raman spectrum of crystals, in some way inspired by the preceeding work of Giuseppe Bolla (1901-1980) of the University of Milan(108). This study brought, among others, to a satisfactory interpretation of two low frequency Raman lines observed in calcite and other similar crystals and the beginning between 1951 and 1955 of a lasting activity, catalized by Caldirola and Fumi on the lattice defects. This line began with a work by Chiarotti, and Camagni and was pursued for a number of years by Chiarotti and coworkers who studied under many aspects lattice deflects, colour centres and surface phenomena in semiconductors(109).

These remarks bring me to mention the work carried on by the Gruppo di Fisica dello Stato Solido of the University of Milan, founded in 1951-52, by initiative of Piero Caldirola which, during the fifties, devoted its activity, mainly of theoretical nature, to the investigation of lattice defects in ionic crystals as well as metals and provided a place of encounter and formation of a number of young theoreticians dedicated to the physics of solids. In addition to Fausto G.Fumi, who was its initial more enthusiastic member, the group included Franco Bassani, Roberto Fieschi and M.Tosi and many others. All these people had a remarkable cultural influence on their experimental colleagues which, at the beginning was concentrated on the people working in Milan and Pavia, but later, when they became professors in universities far away from their original centre, was spread all over our country.

- Notes -

(1) E.Amaldi: "The years of reconstruction", pp.379-455 of "Perspectives of Fundamental Physics", ed. by C.Schaerf, Harwood Academic Publisher, New York (1979); and Scientia 23 (1979) 51-68 and 439-451; in Italian: ibidem 29-50 and 421-439; Giornale di Fisica 20 (1979) 186-225.

(2) E.Amaldi: "Mario Ageno, ricercatore e trattatista", in course of publication.

(3) E.Amaldi, D.Bocciarelli, F.Rasetti, G.C.Trabacchi: "Generatore di neutroni di 1000 kV", Ric.Scient. 10 (1939) 623-632.

(4) (a) M.Ageno, E.Amaldi, D.Bocciarelli, G.C.Trabacchi: "Sulla scissione degli elementi pesanti", Ric.Scient. 11 (1940) 302-311.
(b) M.Ageno, E.Amaldi, D.Bocciarelli, B.N.Cacciapuoti, G.C.Trabacchi: "Scissione dell'uranio con neutroni veloci" Ric.Scient. 11 (1940) 413-417.
(c) E.Amaldi, D.Bocciarelli, G.C.Trabacchi: "Sulla scissione del torio e del protoattinio", Ric.Scient. 12 (1941) 134-138.
(d) M.Ageno, E.Amaldi, D.Bocciarelli, B.N.Cacciapuoti, G.C.Trabacchi: "Fission Yield by Fast Neutrons", Phys.Rev. 60 (1941) 67-75.

(5) O.Hahn, F.Strassmann: Über den Nachweiss und das Verhalten der bei Bestrahlung des Urans mittels Neutronen entstehenden Erdalkalimetallen", Naturwissen. 27 (1939) 11-15, received December 22, 1938.

(6) L.Meitner, O.Frisch: "Disintegration of Uranium by Neutrons: A New Type of Nuclear Reaction", Nature (London) 43 (1939) 239-240, dated January 16, 1939.

(7) S.Flügge, G.v.Droste: "Energetische Betrachtungen zu der Entstehung von Barium bei der Neutronenbestrahlung von Uran", Zeit.f.Phys.Chemie, Abteil B42 (1939) 274-280, received on 22 January, 1939.

(8) N.Bohr, J.A.Wheeler: "The mechanism of nuclear fission", Phys.Rev. 56 (1939) 426-450, dated June 28, 1939; "The fission of Protoactinium", ibidem 1065-1066, dated October 20, 1939.

(9) N.Bohr: "Successive Transformations in Nuclear Fission", Phys.Rev. 58 (1940) 864-866, received on August 12, 1940, and pp.475-479 of: "Niels Bohr Collected Papers", E.Rüdiger (Gen.ed.) North Holland, Amsterdam: Vol 9, Nuclear Physics (1929-1952), ed. Sir Rudolf Peierls (1986).

(10) As I explained elsewhere, we did not take this decision for humanitarian reasons, but only because we did not like become recognized experts in fission so that some Italian authority, directly or perhaps on request of one of Italy's allies, could invite us to participate in this kind of development. See Ref(1) as well as my intervention at pp. C8-315-316 of the: Comptes Rendus du Colloque International sur l'Historie de la Physique des Particules", Paris, France, 21-23 juillet 1982; Journal de Physique, Colloque C8, supplement au n° 12, Dec. 1982.

(11) E.Amaldi, D.Bocciarelli, G.C.Trabacchi: "Misura della sezione d'urto elastico fra neutroni e protoni", Ric.Scient. 11 (1940) 121-127;
M.Ageno, E.Amaldi, D.Bocciarelli, G.C.Trabacchi: "Sull'urto fra protoni e neutroni I", Ric.Scient. 12 (1941) 830-842; E.Amaldi, D.Bocciarelli, B.Ferretti, G.C.Trabacchi: "Sull'urto fra protoni e neutroni II", Ric.Scient. 13 (1942) 502-531; E.Amaldi, D.Bocciarelli, G.C.Trabacchi: "Streuung von 14-MV-Neutronen an Protonen", Naturwiss. 30 (1942) 582-583.

(12) M.Ageno, E.Amaldi, D.Bocciarelli, G.C.Trabacchi: "Sull'urto di neutroni contro protoni e deuteroni", Nuovo Cimento 1 (1943) 253-278;
M.Ageno, E.Amaldi, D.Bocciarelli, B.N.Cacciapuoti, G.C.Trabacchi: "Streuung von schnellen Neutronen an Protonen und Deuteronen", Naturwiss. 31 (1943) 231-232; M.Ageno, E.Amaldi, D.Bocciarelli, G.C.Trabacchi: "On the scattering of fast neutrons by protons and deuterons", Phys.Rev. 71 (1947) 30-31.

(13) Niels Bohr, R.Peierls, G.Placzeck: "Nuclear Reactions in the Continuous Eenergy region", Nature (London) 144 (1939) 200-201.

(14) E.Amaldi, D.Bocciarelli, B.N.Cacciapuoti, G.C.Trabacchi: "Effetti di diffusione nello sparpagliamento di neutroni veloci", Nuovo Cimento 3 (1946) 15-21; "Sullo sparpagliamento elastico dei neutroni veloci da parte dei nuclei medi e pesanti", ibidem 203-234; "On elastic scattering of fast neutrons by medium and heavy nuclei", Phys.Soc.Cambr.Conf. Report (1947) 97-113.

(15) G.Bernardini, B.N.Cacciapuoti, B.Ferretti: "Misura del rapporto fra l'intensità della componente dura della radiazione cosmica penetrante sotto uno strato equivalente a 4 metri di acqua al livello del mare", Ric.Scient. 10 (1939) 731-733;
G.Bernardini, B.N.Cacciapuoti, B.Ferretti, O.Piccioni: "The Genetic Relation between the Electronic and Mesonic Components of Cosmic Rays Near and Above Sea Level", Phys.Rev. 58 (1940) 1017-1026;
G.Bernardini, B.N.Cacciapuoti, B.Ferretti, O.Piccioni, G.C.Wick: "Sulle Condizioni di Equilibrio delle Componenti Elettronica e Mesonica in Mezzi Diversi e a Varie Altezze sul Livello del Mare", Ric.Scient. 10 (1939) 1010-1017; "Sulle condizioni di equilibrio delle Componenti Elettronica e Mesonica intorno al Livello del Mare", Nuovo Cimento 17 (1940) 317-344;
G.Bernardini, B.N.Cacciapuoti: "Sulla Componente Elettronica della Radiazione Cosmica e la Teoria dei Processi Moltiplicativi", Ric.Scient. 12 (1942) 981-1019;
G.Bernardini, E.Pancini, M.Santangelo, E.Scrocco: "Sulla produzione di radiazione secondaria elettronica da parte dei mesoni", Ric.Scient. 12 (1942) 321-340.

(16) G.Bernardini, M.Conversi: "Sulla deflessione dei corpuscoli cosmici in un nucleo di ferro magnetizzato", Ric.Scient. 11 (1940) 840-848;
G.Bernardini, G.C.Wick, M.Conversi, E.Pancini: "Positive Excess in Mesotron Spectrum", Phys.Rev. 60 (1941) 535-536;
B.Bernardini, M.Conversi, F.Pancini, E.Scrocco, G.C.Wick: "Researches on the Magnetic Deflection of the Hard Component of Cosmic Rays", Phys.Rev. 68 (1945) 109-120.
The idea of using magnetized iron blocks to deflect muons had been proposed by B.Rossi, but his first experiment had been unsuccessful (Nature 128 (1931) 300).

(17) H.Kulenkamff: "Bemerkung über die durchdringenden Komponente der Ultrastrahlung", Verh.Deutsch.Phys.Ges. 19 (1938) 92.

(18) H.Euler, W.Heisenberg: "Theoretische Gesichtspunkte zur Deutung der kosmischen Strahlung", Erg.der. Exakten Naturwiss. 17 (1938) 1-69.

(19) M.Ageno, G.Bernardini, B.N.Cacciapuoti, B.Ferretti: "Sulla instabilità del mesone", Ric.Scient. 10 (1939) 1073-1081; "The Anomalous Absorption of the Hard Component of Cosmic Rays in Air", Phys.Rev. 57 (1940) 945-950;
G.Bernardini, B.N.Cacciapuoti, E.Pancini, O.Piccioni, G.C.Wick: "Differential Measurements of the Meson's Lifetime at Different Elevations", Phys.Rev. 60 (1941) 910-911;
G.Bernardini, B.N.Cacciapuoti, E.Pancini, O.Piccioni: "Sulla vita media del mesotrone" Nuovo Cimento 19 (1942) 69-99;
G.Bernardini: "Über die anomale Absorption in Luft und die Lebendauer des Mesons", Zeit.f.Phys. 120 (1942) 413-436;
G.Bernardini, C.Festa: "Su di un metodo per la determinazione della vita media del

mesone basato sugli effetti integrali di assorbimento", Rend.Acc. d'Italia 4 (1943) 166-181.

(20) B.Rossi, N.H.Hilberry, J.D.Hoag: "The Variation of the Hard Component of Cosmic Rays with Height and the Disintegration of the Mesotron", Phys.Rev. 57 (1940) 461-468.

(21) F.Rasetti: "Disintegration of Slow Mesotrons", Phys.Rev. 59 (1941) 613; ibidem 60 (1942) 198-204.

(22) P.Auger, R.Maze, R.Chaminade: "Une misure directe de la vie moyenne du méson au repos", C.R.Acad.Sci. Paris 214 (1942) 266-269.

(23) M.Conversi, O.Piccioni: "Sulla disintegrazione dei mesoni lenti", Nuovo Cimento 2 (1944) 71-87; "On the Disintegration of the Mesons", Phys.Rev. 70 (1946) 874-881.

(24) B.Rossi, N.C.Nereson: "Experimental Determination of the Disintegration Curve of Mesotrons", Phys.Rev. 62 (1942) 417-422; Further Measurements on the Disintegration Curve of Mesotrons", Phys.Rev. 64 (1943) 199-201.

(25) R.Chaminade, R.Fréon, R.Maze: "Nouvelle mesure directe de la vie moyenne du méson au repos", C.R.Acad.Sci. Paris 218 (1944) 402-404.

(26) E.Amaldi: "Ricordo di Ettore Pancini", Il Saggiatore (1988), in press.

(27) M.Conversi, E.Pancini, E.Piccioni: "On the Decay Process of Positive and Negative Mesons", Phys.Rev. 68 (1945) 232; "On the Disintegration of Negative Mesons", Phys.Rev. 71 (1947) 209-210.

(28) O.Piccioni: "The observation of the leptonic nature of the mesotron by Conversi, Pancini and Piccioni", pp.222-241;
M.Conversi: "The period that led to the 1946 discovery of the leptonic nature of the mesotron", pp. 242-250 of "The Birth of Particle Physics", ed.by L.M.Brown and L.Hoddeson, Cambridge University Press, Cambridge (1983).

(29) M.Conversi: "From the discovery of the Mesotron to that of its Leptonic Nature", pp. 1-20 of "40 Years of Particle Physics", Proceedings of the International Conference to celebrate the 40th Anniversary of the Discovery of the π Meson, 22-24 July 1987, University of Bristol, ed. by B.Foster and P.H.Fowler, Adam Hilger, Bristol and Philadelphia (1987).

(30) G.Cocconi, V.Tongiorgi: "Sui secondari della componente mesonica nella radiazione cosmica", Ric.Scient. 19 (1939) 733-736;
G.Cocconi: "Über die Neutronen der kosmischen Ultra-Strahlung", Naturwiss. 27 (1939) 740-741; "Sulla Produzione degli Jukoni", Nuovo Cimento 16 (1949) 78-85;
G.Cocconi, V.Tongiorgi: "Sulla Radiazione Secondaria dei Raggi Cosmici", Nuovo Cimento 16 (1949) 447-455;
G.Cocconi: "A New Proof of the Instability of the Mesotron", Phys.Rev. 57 (1939) 61-62;
G.Cocconi, V.Tongiorgi: "Misure sugli sciami estesi di raggi cosmici a 2200 metri sul livello del mare", Ric.Scient. 11 (1940) 788-790;
G.Cocconi: "Determinazione della vita media del mesotrone", Ric.Scient. 12 (1941) 421-425;
G.Cocconi, V.Tongiorgi: "Die Elektronenkomponente der Ultrastrahlung und die Intensität des Mesotrons", Zeit.f.Phys. 118 (1941) 88-103;
G.Cocconi: "On the protonic nature of the Primary Cosmic Radiation", Phys.Rev. 60 (1941) 532-533;
G.Cocconi: "On the Presence of Strongly Ionizing Particles in Cosmic Ray

Showers", Phys.Rev. 60 (1941) 533;
G.Cocconi, V.Tongiorgi: "Über das Sperktrum der Ultrastrahlung in 2200 Meter Höhe ü.d.M.", Naturwiss. 30 (1942) 328-329.

(31) G.Polvani, G.Cocconi: "La camera di Wilson dell'Istituto di Fisica della Università di Milano", Ric.Scient.12 (1941) 410-420.

(32) The board of examiners for the selection of the Professor for the Chair of Experimental Physics at the University of Palermo, was formed by Professor A. Lo Surdo (President), L.Tieri, E.Persico, G.Polvani, A.Carrelli (Secr.), and it concluded its work on Nov. 18th, 1942. The selected candidates were: 1st E.Medi, 2nd G.Cocconi, 3rd V.Polara, who were called respectively by the University of Palermo, Catania and Messina. Among the candidates who were declared maturi (eligible) the following were included: B.N.Cacciapuoti (who had a vote for the 1st place, one for the 2nd and two for the 3rd) and G.Occhialini who was favourably judged although he did not receive any vote).

(33) Giorgio Fea had gratuated with Rasetti in 1934 discussing a Thesis on the construction and operation of a boron-trifluoride cloud chamber. In 1934 he joined the Metheorologic Office of the Italian Air Force, but continued to participate in scientific research in the Institute as an "assistente volontario" with Rasetti and later, with Lo Surdo.

(34) G.Cocconi, A.Loverdo, V.Tongiorgi: "Sulla presenza di sciami estesi di mesoni negli sciami estesi dell'aria", Nuovo Cimento 1 (1943) 49-55; "Sugli sciami estesi dell'aria", ibidem 314-324; "Spettro di densità degli sciami estesi dell'aria", ibidem 2 (1943) 14-27; "Sulla costituzione degli sciami estesi dell'aria", ibidem 28-34.

(35) G.Cocconi: "Results and Problems Concerning the Extensive Air Showers", Rev.Mod.Phys. 21 (1949) 26-30.

(36) G.Salvini: "Sull'assorbimento della radiazione cosmica a 2100 m.", Nuovo Cimento 3 (1946) 283-284; "Un contatore di Geiger Müller di forma sferica", Ric.Scient. 17 (1947) 914-916;
A.Mura, G.Salvini, G..Tagliaferri: "Osservazioni in camera di Wilson sullo sparpagliamento delle particelle degli sciami estesi", Nuovo Cimento 4 (1947) 102; "Sulla componente penetrante degli sciami dell'aria", ibidem 279.
G.Salvini, G.Tagliaferri: "On the Penetrating Compound of Air Showers", Phys.Rev. 73 (1948) 261-262;
A.Lovati, A.Mura, G.Salvini, G.Tagliaferri: "Alcune proprietà delle esplosioni nucleari nella radiazione cosmica", Nuovo Cimento 6 (1949) 207-216; "Sulla natura e sul numero delle particelle penetranti nelle esplosioni nucleari prodotte nel piombo della radiazione cosmica", ibidem 291-293; "Nuclear Interactions of the Particles Produced in Cosmic Ray Burst", Nature (London) 163 (1949) 1004-1006; "Proprietà delle particelle emesse nelle esplosioni nucleari e confronto fra esplosioni in C e in Pb", Nuovo Cimento 7 (1950) 36-47; "Cloud Chamber Observations on the Electromagnetic Component for Nuclear Explosions and the Development of the Nuclear Cascade", ibidem 943-953; "Mean Free Path of the Particles Produced in Nuclear Emulsions and Comparison between Explosions in C and in Pb", Phys. Rev. 77 (1950) 284-285.

(37) G.Salvini, Y.B.Kim: "Production Cross Section and Frequency of Neutral Mesons in Cosmic Rays", Phys.Rev. 88 (1952) 40-50;
G.Salvini: "Interaction mean free path and charge exchange of the π-meson", Nuovo Cimento 10 (1953) 1018-1033.

(38) The year, indicated in brackets, after one or more names is hereafter that of the year in which the people quoted have appeared for the first time as authors of articles in Il Nuovo Cimento.

(39) B.Pontecorvo: "Nuclear capture of mesons and the meson decays", Phys.Rev. 72 (1947) 246-247;
O.Klein: "Mesons and Nucleons", Nature (london) 161 (1948) 897-899;
G.Puppi: "Mesons of Cosmic Rays", Nuovo Cimento 5 (1948) 587.
T.D.Lee, M.Rosenbluth, C.N.Yang: "Interactions of mesons and nucleons with light particles", Phys.Rev. 75 (1949) 905.
J.Tiommo, J.A.Wheeler: "Charge Exchhange Reaction of the μ-Meson with the Nucleus", Rev.Mod.Phys. 21 (1949) 144-152.

(40) P.G.Bordon: "Teoria della dissipazione elastica nei monocristalli secondo la meccanica quantistica: un nuovo effetto di rilassamento", Ric.Scient. 19 (1949) 851-862.

(41) P.G.Bordoni: "Assorbimento di ultrasuoni nei solidi", Suppl.Nuovo Cimento 7 (1950) 144-160; "Elastic and Anelastic Behaviour of Some Metals at Very Low Temperatures", J.Ac.Soc.Am. 26 (1954) 495-502; "Comportamento elastico ed anelastico di alcuni metalli a bassissima temperatura", Ric.Scient. 23 (1953) 1193-1202.

(42) E.Amaldi: "Centro di Studio per la Fisica nucleare e delle particelle elementari, Attività svolta durante l'anno 1947", Ric.Scient. 18, (1948) 54-60.

(43) As it will appear more clear below I refer here to the emulsion technique and to the fast ionization chambers of the type developed by B.Rossi and H.H.Staub ("Ionization Chambers and Counters, Experimental Techinques", New York, N.Y. (1949).

(44) G.Bernardini, C.Longo, E.Pancini: "Relazione sulla costruzione del L.T.G.", Ric.Scient. 18 (1948) 91-98. Architect C.Longo had designed the building: due to its aluminium structure, it could very well stand the gusts of wind and snow which were often extraordinary strong;
G.Bernardini, E.Pancini: "Ampliamento del L.T.G.", Ric.Scient. 20 (1950) 966-968;
The direction of L.T.G. passed to Pancini and later to G.Fidecaro: "Il Laboratorio della Testa Grigia: attivit`svolta dal 1°-1-1953 al 30-9-54.

(45) E.Gatti: "Commemorazione di Giuseppe Bolla", Seduta del 25-11-1982 dell'Istituto Lombardo di Scienze e Lettere.

(46) E.Amaldi: "L'Istituto Nazionale di Fisica Nucleare", Notiziario del CNEN, 8, n°1, gennaio (1963).

(47) A.Rostagni: L'energia elettrica 28 (1951) 211;
An ampler Laboratory was built in 1953. For further details on these high altitude laboratories, see: "The World's High Altitude Research Stations", Research Division, College of Engineering, New York University.

(48) A.Hermann, J.Krige, U.Mersits, D.Pestre: "History of CERN Vol 1", North Holland, Amsterdam (1987).

(49) "L'elettrosincrotrone e i Laboratori Nazionali di frascati", a cura di G.Salvini, Suppl. Nuovo Cimento 24 (1962) 3-388.

(50) M.C.Lattes, H.Muirhead, G.P.Occhialini, C.F.Powell: "Processes involving charged mesons", Nature (London) 159 (1947) 694-697;
M.C.Latter, G.P.Occhialini, C.F.Powell: "Observations of the tracks of slow mesons in nuclear emulsions", Nature (London) 160 (1947) 453-456.

(51) C.D.Rochester, C.S.Butler: "Evidence for the Existence of New Unstable Elementary Particles", Nature (London) 160 (1947) 855-857.

(52) "Convegno Internazionale di Meccanica Statistica", Firenze 17-20 maggio 1949, Suppl. Nuovo Cimento 6 (1949) 146-308.

(53) E.Amaldi, F.Fidecaro: "A research for anomalons scattering of μ-mesons by nucleons", Helv.Phys.Acta 23 (1950) 93-102; Nuovo Cimento 7 (1950) 535-552; "An experiment on the anomalous scattering of μ-mesons by nucleons", Phys.Rev. 81 (1951) 338-341.

(54) E.Amaldi, C.Castagnoli, A.Gigli, S.Sciuti: Nuovo Cimento 9 (1952) 453, 969; Proc.Phys.Soc. A65 (1952) 556;
E.Amaldi: "Underground Experiments in Europe", Proc.Int.Conf. of Theoretical Physics, Kyoto-Tokyo, Sept. 1953, p.106-110.

(55) A.Lovati, A.Mura. C.Succi, G.Tagliaferri: "A Search for the Production of Penetrating Secondaries by μ-Mesons Underground", Nuovo Cimento 10 (1953) 105-107; "Further Results on the Interactions of Cosmic Rays Underground", ibidem 1201-1204.

(56) P.E.Argan, A.Gigli, S.Sciuti: "On the Interaction of μ-Mesons with Matter at High Energies", Nuovo Cimento 11 (1954) 530-538.

(57) G.Bernardini, G.Cortini, A.Manfredini: "Nuclear Evaporations produced by cosmic rays", Phys.Rev.74 (1948) 1878-1879; "Sulle evaporazioni nucleari nei raggi cosmici e l'assorbimento della componente nucleonica", Nuovo Cimento 6 (1949) 456-469; "On the nuclear evaporation in cosmic rays and the absorption of the nucleonic component II", Phys.Rev. 79 (1950) 952-963.

(58) J.Buschmann, I.F.Quercia, B.Rispoli: "Sulle disintegrazioni nucleari prodotte dalla radiazione cosmica a 3500 m. s.l.d.m" Nuovo Cimento 7 (457-469;
E.Amaldi, C.Castagnoli, A.Gigli, S.Sciuti: "Sull'effetto di transizione nel fenomeno di produzione di stelle da parte della radiazione cosmica", Nuovo Cimento 7 (1950) 697-699.

(59) E.Amaldi, C.Castagnoli, A.Gigli, S.Sciuti: "Contributi allo studio degli sciami estesi", Nuovo Cimento 17 (1950) 401-456 and 816-834.
C.Ballario, B.Brunelli, A. De Marco, G.Martelli: "On the Diurnal Variation of Extensive Air Showers of High Density at 3500 m above Sea Level", Phys.Rev. 83 (1951) 666-667.

(60) A.Lovati, A.Mura, G.Salvini, G.Tagliaferri: "Alcune proprietà delle esplosioni nucleari nella radiazione cosmica", Nuovo Cimento 6 (1949) 207-216.

(61) M.Cervasi, G.Fidecaro, L.Mezzetti: "Penetrating Showers from Hydrogen and Other Elements", Nuovo Cimento 1 (1955) 300-313.

(62) G.Alexander, P.Astbury, C.Ballario, R.Bizzarri, B.Brunelli, A.De Marco, A.Michelini, E.Zavattini, A.Zichichi: "A Cloud Chamber Observation of a Singly Charged Unstable Fragment", Nuovo Cimento 2 (1955) 365-369;
C.Ballario, R.Bizzarri, B.Brunelli, A.De Marco, E.Di Capua, A.Michelini, G.C.Moneti, E.Zavattini, A.Zichichi: "Life Time Estimate of Λ° and θ° Particles", Nuovo Cimento 6 (1957) 994-996.

(63) P.Bassi, F.Filosofo, C.Manduchi, L.Prinzi: "Eccesso positivo dei mesoni a 2000 metri", Nuovo Cimento 8 (1951) 469-474.

(64) P.Bassi, C.Manduchi, P.Veronesi: "Sulla distribuzione zenitale della componente nucleonica di media energia", Nuovo Cimento 9 (1952) 722-725.

(65) P.Bassi, A.M.Bianchi, D.Cadorni, C.Manduchi: "Sulla distribuzione zenitale delle particelle degli sciami estesi", Nuovo Cimento 9 (1952) 1037-1043;
M.Cresti, A.Loria, G.Zago: "Sulla distribuzione zenitale delle particelle degli sciami estesi", ibidem 10 (1953) 779-783.

(66) M.Deutschmann, M.Cresti, W.D.B.Greening, L.Guerriero, A.Loria, G.Zago: "An AnomalousV° Event", Nuovo Cimento 3 (1956) 180-183.

(67) E.H.S.Burhop, H.S.W.Massey et al.: "The High Pressure Cloud-Chamber: Cosmic-Ray Programme at La Marmolada", Nature (London) 175 (1955) 445-448. "The Use of the High Pressure Cloud Chamber in the Study of the Unstable Particles of the Cosmic Radiation", Suppl. Nuovo Cimento 4 (1956) 272-285

(68) M.Conversi: "Positive excess of mesons at 30.000 Feet", Phys.Rev. 76 (1949) 311-312; "Latitude Dependence at 30.000 Feet of Penetrating Particles Slowed Down Other Traversing 15 cm of Lead", Phys.Rev. 76 (1949) 444-445; "Experiments on Cosmic Mesons and Protons at Several Altitude and Latitude", Phys.Rev. 79 (1950) 749-767.

(69) M.Conversi, A.Gozzini: "The Hodoscope Chamber: A New Instrument for Nuclear Research", Nuovo Cimento 2 (1955) 189-191;
G.Barsanti, M.Conversi, S.Focardi, G.P.Murtas, C.Rubbia, G.Torelli: "On the Hodoscope Chamber", Proceedings of CERN Symposium 1956, p.56, Geneva (1956).

(70) M.Conversi, E.Fiorini, S.Ratti, C.Rubbia, S.Succi, G.Torelli: "A Search of Particles of 550 m_e", Nuovo Cimento 9 (1958) 740-744;
M.Conversi, G.M.De Munari, A.Egidi, E.Fiorini, S.Ratti, C.Rubbia, C.Succi, G.Torelli: "Mass 550 Particle m_e", Phys.Rev. 114 (1959) 1150-1151.

(71) F.Eisler, P.Plano, A.Prodell, N.Samios, M.Schwartz, J.Steinberger, P.Bassi, V.Borelli, G.Puppi, H.Tanaka, P.Waloschek, V.Zoboli, M.Conversi, P.Franzini, I.Mannelli, R.Santamgelo, P.Silvestrini, D.A.Glaser, C.Graves, M.L.Perl: "Demostration of Parity Nonconservation Hyperon Decay", Phys.Rev. 108 (1957) 1353-1355;
F.Eisler, P.Plano, A.Prodell, N.Samios, M.Schwartz, J.Steinberger, M.Conversi, P.Franzini, I.Mannelli, R.Santamgelo, P.Silvestrini: "Leptonic Decay Mode of Hyperons", Phys.Rev. 112 (1958) 979-981.

(72) S.Wu, E.Amber, R.W.Hayward, D.D.Hoppes, R.P.Hudson: "Experimental test of parity conservation in beta decay of Co^{60}", Phys.Rev. 106 (1957) 386-387.

(73) R.L.Garwin, L.Lederma, M.Weinrich: "Observation of the failure of conservation of parity and charge conjugation in meson decay", Phys.Rev. 105 (1957) 1415-1417.

(74) J.I.Friedman, V.L.Telegdi: "Nuclear emulsion evidence for parity non conservation in the decay $\pi^+ \to \mu^+ \to e^+$", Phys.Rev. 105 (1957) 1681-1682.

(75) M.Beneventano, G.Bernardini, D.Carson-Lee, E.L.Goldwasser, G.Stoppini: "The pi^+/pi^- ratio from deuterium near photo-pion threshold", Nuovo Cimento 12 (1954) 156-159.

(76) M.Beneventano, G.Bernardini, D.Carson-Lee, G.Stoppini, L.Tau: "Differential cross section for photoproduction of positive pions in Hydrogen", Nuovo Cimento 4 (1956) 323-356;
M.Beneventano, G.Bernardini, G.Stoppini, L.Tau: "Photoproduction of charged pions in Deuterium", Nuovo Cimento 10 (1958) 1109-1142.

(77) R.M.Brown, V.Camerini, P.H.Fowler, H.Muirhead, C.F.Powell, D.M.Ritson: "Observations with Electron-Sensitive Plates Exposed to Cosmic Radiation", Nature (London) 163 (1949) 47-51.

(78) O'Ceallaigh: "Masses and Modes of Decay of Heavy Mesons X-Particles. Part I", Phil.Mag. 42 (1951) 1032-1039;M.G.Menon, O'Callaigh: "Observations on the Decay of Heavy Mesons in Photographic Emulsions, Proc.Roy.Soc. A221 (1954) 292-318.

(79) J.Davis, C.Franzinetti: "Report on the Expedition to the Central Mediterranean for the Study of Cosmic Radiation", Suppl.Nuovo Cimento 2 (1954) 480-497 and CERN/GEN/11.

(80) "Rendiconti del Congresso Internazionale sulle Particelle Elementari e sugli Eventi di Alte Energie nei Raggi Cosmici", Padova 12-15 aprile 1954, Suppl. Nuovo Cimento 2 (1954) 163-497.

(81) "Rendiconti del Corso che fu tenuto, nella Villa Monastero, a Varenna dal 18 luglio al 7 agosto 1954, a cura della Scuola Internazionale di Fisica", Suppl. Nuovo Cimento 2 (1955) 1-469.

(82) "Congrés International sur le Rayonnement Cosmique Organizé par l'Université de Tolouse, sous le patronage de l'UIPPA, avec l'appui de l'UNESCO", Begnére de Bigorre, Juillet 1953.

(83) "G-stack Collaboration: Observations on Heavy Meson Secondaries", Suppl. Nuovo Cimento 4 (1956) 398-424.

(84) "Conferenza Internazionale sulle particelle elementari", Pisa 12-18 giugno 1955, Suppl. Nuovo Cimento 4 (1956 135-1078.

(85) E.Amaldi, C.Castagnoli, G.Cortini, C.Franzinetti, A.Manfredini: "Measured event produced by cosmic rays", Nuovo Cimento 1 (1955) 492-500.

(86) O.Chamberlain, E.Segrè, C.Wiegand, T.Ypsilantis: "Observation of Antiprotons", Phys.Rev. 100 (1955) 947-950.

(87) O.Chamberlain, W.W.Chupp, G.Goldhaber, E.Segrè, C.Wiegand, E.Amaldi, G.Baroni, C.Castagnoli, C.Franzinetti, A.Manfredini: "Antiproton star observed in emulsion", Phys.Rev. 101 (1956) 909;
O.Chamberlain, W.W.Chupp, A.G.Ekspong, G.Goldhaber, S.Goldhaber, E.J.Lofgren, C.Wiegand, E.Amaldi, G.Baroni, C.Castagnoli, C.Franzinetti, A.Manfredini: "Example of anti-nucleon annihilation, Phys.Rev. 102 (1956) 921-923;
W.H.Barkas, W.W.Chupp, A.G.Ekspong, G.Goldhaber, S.Goldhaber, H.H.Heckman, D.H.Perkins, J.Sandweiss, E.Segrè, F.M.Smith, D.H.Stork, L. Van Rossum, E.Amaldi, G.Baroni, C.Castagnoli, C.Franzinetti, A.Manfredini: "Antiproton-nucleon annihilation process", Phys.Rev. 105 (1957) 1037-1058;
E.Amaldi, C.Castagnoli, M.Ferro-Luzzi, C.Franzinetti, M.Manfredini: "Further

Results on Antiproton Annihilation", Nuovo Cimento 3 (1957) 1797-1800; E.Amaldi, G.Baroni, G.Bellettini, C.Castagnoli, M.Ferro-Luzzi, A.Manfredini: "Study of antiproton with emulsion technique", Nuovo Cimento 14 (1959) 977-1026.

(88) D.J.Prowse, M.Baldo Ceolin: "A Decay of an Antihyperon, $\bar{\Lambda}°$, in Nuclear Emulsion", Nuovo Cimento 10,2 (1958) 635-645.

(89) A.Rigamonti: "Luigi Giulotto", Rendiconti dell'Istituto Lombardo di Scienze e Lettere 121 (1987).

(90) L.Giulotto: "L'avvio della ricerca fisica in struttura della materia", Convegno Nazionale di Struttura della Materia in occasione del 20° anniversario della costituzione del Gruppo Nazionale per la Struttura della Materia, Pavia, giugno 1982.

(91) "The Origins of Solid State Physics in Italy: 1945-1960", Pavia 21 to 24 September, 1987 (in course of publication).

(92) See the publication by CNR on the "Istituto di Acustica Orso Mario Corbino, 50° anniversario", Roma (1947), by P.E.Giua, Director of the Istituto O.M.Corbino.

(93) A.Alippi: "In Memoriam di Amedeo Giacomini", Rivista Italiana di Acustica 3, N°3 (1979) 191-192.

(94) "The Proceedings" of this Conference, which took place at the Accademia Nazionale dei Lincei on 23-24 April, 1987, are in course of publication.

(95) G.Boato, G.Careri, M.Santangelo: "Argon Isotopes in Natural Gases", Nuovo Cimento 9 (1952) 44-49;
G.Boato, G.Careri, G.Volpi: "Hydrogen Isotopes in Steam Wells", ibidem 539-540;
G.Boato, A.Cimino, G.Careri, E.Molinari, G.G.Volpi: "A Perturbing Factor on the Kinetics of the Homogeneous Hydrogen Deuterium Exchange Reaction", Nuovo Cimento 10 (1953) 993-994; "Neue Messungen der homogenen Austauschgeschwindigkeit zwischen Wasserstoff und Deuterium", Naturwiss. 42 (1955) 388.

(96) G.Careri, A.Paoletti: "A Direct Interchange Mechanism in Liquid Tin and Indium self-Diffusion", Nuovo Cimento 1 (1955) 517-518; "Self-Diffusion in Liquid Metals", Suppl. Nuovo Cimento 1 (1955) 161-165; "Self-Diffusion in Liquid Indium and Tin", Nuovo Cimento 2 (1955) 574-591.

(97) The first of a long list of papers by Careri et al on this subject is:
G.Careri, J.Reuss, F.Scaramuzzi, J.O.Thomson: "Movement of ions and ^{3}He in Liquid Helium", Proc. Fifth Intern. Conf. on Low Temperature Physics and Chemistry (Madison, 1957) ed. by J.F.Dillinger, The University of Wisconsin Press, Madison 155 (1958).

(98) G.Careri, S.Cunsolo, P.Mazzoldi, M.Santini: "Experiments on the creation of Charged Quantized Vortex Rings in Liquid Helium at 1 K", Phys.Rev.Lett. 15 (1965) 392-396.

(99) G.Careri, F.Scaramuzzi: "Low Temperature Physics in Frascati, Padua and Rome", presented at the 1987 Conference mentioned in Ref.(91).

(100) F.Fumi: "Carlo Perrier (Torino 1886 - Genova 1948)", Atti dell'Accademia Ligure di Scienze e Lettere 5 (1948) 1-8.

(101) G.Caglioti, A.Gigli, S.Sciuti: "On the Production of Secondary Electrons by High Energy Mesons", Nuovo Cimento 1 (1955) 851-852.

(102) G.Caglioti: "Start-up in Italy", p.81-85 of: "Fifty Years of Neutron Diffraction, the Advent of Neutron Scattering", ed. by G.E.Bacon, Adam Hilger, Bristol.

(103) "Interaction of Radiation with Matter, A Volume in Honour of Adriano Gozzini", Scuola Normale Superiore di Pisa, Pisa, 1987.

(104) L.Giulotto: "Fine Structure of H_α", Phys.Rev. 71 (1947) 562; "Struttura fine della H_α", Ric.Scient. 17 (1947) 209-216.

(105) The comment given in the Britannic Encylopedia under the heading "Quantum Mechanics" is the following: "Rather there was a real discrepancy for a number of years, a matter of controversy, but painstaking spectroscopic measurements by L.Giulotto and others in 1947 and 1948, made it clear the discrepancy was real".

(106) F.Bassani: "Ricordo di Piero Caldirola", Il Nuovo Saggiatore 1, fasc. 4 (1985).

(107) L.Giulotto, G.Chiarotti: "Proton Relaxation in Water", Phys.Rev. 93 (1954) 1241; G.Chiarotti, G.Cristiani, L.Giulotto, G.Lanzi: "A Nuclear Inductor for Measurements of Thermal Relaxation Time in Liquids", Nuovo Cimento 12 (1954) 519-525.

(108) E.Gatti: "Commemorazione di Giuseppe Bolla", Istituto Lombardo di Scienze e Lettere, Seduta del 25 novembre (1960) 1226.

(109) P.Camagni, G.Chiarotti: "A Study of the $F \rightarrow Z_1$ Conversion in KCl Crystals with Divalent Impurities", Nuovo Cimento 11 (1954) 1-10;
P.Camagni, S.Ceresara, G.Chiarotti: "Thermal Equilibrium of Color Centers in Doped KCl Crystals", Phys.Rev. 118 (1960) 1226-1228.

Estratto dal *Giornale di Fisica* - Vol. XX - N. 4 - Ottobre-Dicembre 1979

Ricordo di Enrico Persico (9 agosto 1900 - 17 giugno 1969) (*).

E. Amaldi e F. Rasetti

È passato ormai parecchio tempo dal 17 giugno 1969, quando nelle prime ore del mattino Enrico Persico spirò nella sua abitazione in Roma in sèguito ad un'affezione cardiaca.

Sia l'Accademia Nazionale dei Lincei che la Facoltà di Scienze dell'Università di Roma ritennero di dover rispettare il desiderio, da lui espresso fra le sue ultime volontà, di non essere commemorato.

Ma a distanza di tempo è doveroso ricordare, almeno per iscritto e nella forma piú semplice, la figura di Enrico Persico, per il contributo che egli diede alla ricerca e all'insegnamento universitario e per le sue qualità umane. Il non raccogliere i dati piú importanti sulla sua persona e il suo lavoro, il non riordinarli, significherebbe lasciare una grave lacuna nella storia dello sviluppo storico della Fisica in Italia durante il periodo 1925-1970 (1).

1. Infanzia e giovinezza.

Enrico Persico nacque a Roma il 9 agosto 1900, figlio unico di Gennaro e Rosa Massaruti. Il padre, di origine napoletana al pari della madre, era cassiere della Banca d'Italia, e morí nel 1920 o poco prima.

Enrico frequentò il Ginnasio-Liceo Umberto, ove fu compagno di classe di Giulio Fermi, fratello, maggiore di un anno, di Enrico Fermi, che era di un anno indietro a loro. Nell'inverno 1915 Giulio Fermi morí in sèguito ad una piccola operazione ritenuta da tutti di secondaria importanza, e suo fratello Enrico ed Enrico Persico cominciarono a frequentersi e divennero amici (2). I due ragazzi, Persico di 15 e Fermi di 14 anni, erano accomunati da un grande interesse per le scienze, come testimonia Persico stesso (2-23) nella mezza pagina ove ricorda le loro « lunghe passeggiate da un capo all'altro di Roma, parlando

(*) L'Accademia Nazionale dei Lincei ha gentilmente autorizzato la pubblicazione anche sul *Giornale di Fisica* di questa biografia, preparata per le *Celebrazioni Lincee*, **115** (1979).

(1) Questa è l'ultima stesura del Ricordo di Enrico Persico. La stesura precedente, completata durante la primavera 1977, fu spedita alla seguente lista di amici e conoscenti di Persico con la preghiera d'inviare suggerimenti di correzioni o aggiunte. Tutti hanno risposto alla richiesta, chi oralmente e chi per iscritto (l). Le lettere inviate in tale occasione vengono depositate presso l'Accademia dei Lincei in quanto costituiscono esse tesse un insieme d'importanti testimonianze sulla figura e l'opera di Persico: M. Ageno, C. Bernardini (l), G. Bernardini (l), R. Bizzarri, B. Brunelli (l), G. Careri, M. Cini, M. Conversi, E. Denina (l), B. De Tollis, L. Fermi (l), E. Ferrari, C. Guazzaroni Ferrari (l), A. Gamba (l), M. Magistrelli (l), G. Occhialini (l), L. Radicati (l), B. Rossi, A. Rostagni (l), G. Salvini (l), E. Segrè (l), S. Segre, F. Tricomi (l), M. Verde, G. C. Wick (l), T. Zeuli. A tutti questi amici e colleghi esprimiamo i nostri vivi ringraziamenti.

(2) E. Fermi: *Note e Memorie*, Accademia Nazionale dei Lincei e The University of Chicago Press, Vol. **1** (Italia, 1921-1933), Roma (1961). Vedi anche l'edizione inglese degli stessi editori e dello stesso anno: E. Fermi: *Collected papers*.

Fig. 1. – Enrico Persico a Roccaraso nel 1923.

di argomenti di ogni genere » ma soprattutto di problemi di matematica e di fisica. Questo comune interesse portò i due giovani a decidere di studiare Fisica quando l'uno nel luglio 1917, l'altro nel luglio dell'anno successivo conseguirono la licenza liceale avendo entrambi saltato la 3ª Liceo.

Persico s'iscrisse cosí al 1º anno di Fisica all'Università di Roma ove insegnavano il fisico Orso Mario Corbino e il fisico matematico Vito Volterra, ai quali si aggiunse, nel 1919, il matematico Tullio Levi-Civita.

Nell'autunno del 1918 Fermi partecipò al concorso per l'ammissione alla Scuola Normale di Pisa e, avendolo vinto, si trasferí in quella città. Ebbe cosí inizio una corrispondenza abbastanza sistematica fra i due amici, buona parte della quale (lettere di Fermi a Persico) è stata conservata da Persico e messa a disposizione di Emilio Segrè per il suo libro biografico su Enrico Fermi ove figura come Appendice I ([3]). Queste lettere possono servire molto bene a mettere in evidenza gl'interessi dei due giovani e la loro evoluzione con il procedere degli studi universitari, anche se la mancanza delle lettere di risposta di Persico a Fermi costituisce una considerevole lacuna. Le prime due lettere di Fermi a Persico (una del 1917, l'altra del 1918) sono precedenti all'andata di Fermi a Pisa, l'ultima è del settembre 1926, cioè alla vigilia del concorso alla cattedra di Fisica Teorica bandito su richiesta dell'Università di Roma.

Il 22 novembre 1921 Persico si laureò con una tesi svolta sotto la guida di O. M. Corbino sull'effetto Hall, argomento su cui pubblicò anche un lavoro con L. Tieri.

Nominato, súbito dopo la laurea, assistente all'Osservatorio Astronomico, diretto dall'ing. Alfonso Dilegge, passò nel 1922, sempre come assistente, all'Istituto di Fisica dell'Università di Roma, diretto da O. M. Corbino, e vi rimase fino al 1927, salvo un periodo di circa un anno (1925) trascorso all'Università di Cambridge (Inghilterra) ove ebbe contatti personali con A. S. Eddington e P. A. M. Dirac.

Gli altri assistenti di O. M. Corbino e quell'epoca erano Laureto Tieri (1879-1952), Nella Mortara (n. 1893), Aldo Pontremoli (1896-1928) e Giulio Bisconcini (1880-1969). La Mortara, donna straordinariamente sportiva, di modi semplici e diretti, si occupava delle Esercitazioni di laboratorio del 1° e 2° anno di Fisica e Ingegneria, funzione che seguitò a svolgere fino al 1948, ben dopo il trasferimento dell'Istituto alla Città Universitaria (1936).

Gli studenti (maschi e femmine e soprattutto queste ultime) venivano trascinati dalla Mortara a sciare d'inverno, a remare sul Tevere nelle stagioni intermedie e a nuotare sulle spiagge vicine a Roma durante tutte le stagioni dell'anno.

Fu cosí che essa divenne una figura caratteristica dell'Istituto di Fisica indicata da tutti come « zia Nella ». Persico negli anni venti s'innamorò di lei e le propose di sposarla, ma Nella Mortara non ne volle sapere perché non si sentiva in alcun modo portata per il matrimonio. Essi rimasero tuttavia ottimi amici per tutta la vita, ma certamente questo episodio non fu di secondaria importanza nella vita di Persico.

Nel periodo 1922-26, oltre che con O. M. Corbino di cui era assistente, Persico

([3]) E. Segrè: *Enrico Fermi, fisico*, Bologna (1971). Vedi anche l'edizione inglese: *Enrico Fermi, physicist*, Chicago (1970).

Fig. 2. – Da sinistra: Enrico Persico, Maria Fermi, sorella di Enrico Fermi, ed Enrico Fermi, Monte Cavo (Roma), 1923.

collaborò con Vito Volterra e Tullio Levi-Civita, di cui raccolse e mise in forma adatta per la pubblicazione (1925) un classico corso di lezioni sul *Calcolo differenziale assoluto* (3.1). L'interesse destato dal volume fu tale che, per iniziativa del matematico E. T. Whittaker, ne fu fatta poco dopo un'edizione in lingua inglese (1927) (3.2).

In quagli anni Persico e Fermi, e piú tardi anche gli altri giovani fisici romani, frequentavano le case dei matematici Tullio Levi-Civita, Federigo Enriques e Guido Castelnuovo, in particolare la casa di quest'ultimo, che, tutti i sabati sera, era aperta con grande semplicità e cordialità ai colleghi e ai giovani matematici e fisici che vivevano a Roma o che vi si trovavano occasionalmente di passaggio (4).

Persico conseguí la libera docenza nel 1924 e nel 1925 prese parte al primo concorso bandito in Italia per la cattedra di Fisica Teorica dell'Università di Roma. La terna risultò cosí composta: 1° Enrico Fermi, 2° Enrico Persico, 3° Aldo Pontremoli. Essi furono chiamati rispettivamente alle Università di Roma, Firenze e Milano.

Fra i suoi lavori di questo primo periodo (1.1–1.17) si debbono ricordare due ricerche sperimentali sull'effetto Hall, una con O. M. Corbino, l'altra con L. Tieri, una ricerca sull'effetto Ettingshausen, altre sul funzionamento del triodo, sulla teoria della relatività, sulla teoria cinetica dei gas altamente ionizzati e sulle oscillazioni delle Cefeidi. Questi ultimi due lavori derivano dal

(4) L. Fermi: *Atoms in the family*, Chicago (1954). Vedi anche la traduzione in italiano: *Atomi in famiglia*, Milano (1954).

suo interesse per l'astrofisica, iniziato nel periodo di assistentato all'Osservatorio Astronomico di Roma, e rafforzato dai suoi contatti con gli astrofisici di Cambridge.

Fra i suoi lavori di questo periodo sono particolarmente interessanti le due note sulla polarizzazione rotatoria magnetica in un campo magnetico alternato (1.14,1.15), che rappresentano una sistemazione definitiva del complesso problema, la nota, in collaborazione con E. Fermi (1.17), sul principio delle adiabatiche e la nozione di forza viva nella meccanica ondulatoria e quella sulla teoria cinetica dei gas ionizzati (1.13) Dalla nota con Fermi del novembre 1926 traspare lo sforzo fatto in quell'epoca dai due amici per penetrare le idee di Schrödinger sulla meccanica ondulatoria apparsi poco prima sugli *Annalen der Physik*. Nel loro lavoro, basato sull'analogia tra meccanica del punto materiale e ottica, essi deducono, senza specificarne il significato fisico, certe espressioni per l'energia cinetica e l'energia potenziale. Queste, quando qualche mese dopo fu sviluppata l'interpretazione probabilistica del meccanica ondulatoria, risultarono essere proprio le espressioni dei valori probabili (*expectation values*) di quelle grandezze.

Il lavoro piú importante di questo periodo è probabilmente quello sulla teoria cinetica dei gas ionizzati fatto nella scia di Eddington e spesso citato accanto a quelli dei pionieri nel campo della fisica del plasma (5).

In questo lavoro Persico sostituisce il potenziale coulombiano degli ioni con un potenziale di Debye-Hückel che è essenzialmente coulombiano per distanze inferiori alla distanza di Debye, oltre la quale decresce esponenzialmente.

Persico fu sempre di un'onestà completa, esente da qualsiasi ombra, incapace di fare anche la minima concessione o di accettare anche piccoli compromessi che lo facessero allontanare dalla piú rigida moralità di vita o intellettuale. Egli ebbe fin da giovane idee chiarissime e ferme sul fascismo, che gli era estraneo, inaccettabile, ancor prima che per il suo contenuto politico, per la retorica, l'autoincensamento e il trionfalismo di cui era impastato.

La sua riservatezza, la ritrosia a parlare di sé, la cautela con cui esprimeva giudizi potevano facilmente essere presi per timidezza. In realtà, forse era anche un poco timido, ma soprattutto non voleva esprimere opinioni affrettate o manifestare, anche momentaneamente e occasionalmente, giudizi superficiali.

2. Il periodo trascorso a Firenze (1927-31).

Quando all'inizio del 1927, Persico si trasferí a Firenze, Franco Rasetti, che era già a Firenze dal 1922, prima come assistente e poi come aiuto di Garbasso, si trasferí a Roma, ove assunse il posto di aiuto di O. M. Corbino, posto che era stato lasciato libero da Persico. Persico e Rasetti si conoscevano piuttosto poco, per quanto ciascuno fosse informato degl'interessi e della personalità dell'altro dal comune amico Enrico Fermi. In quell'epoca essi s'incontrarono piú volte e súbito divennero grandi amici.

Giunto a Firenze, Persico dedicò uno sforzo assai considerevole all'inse-

(5) Vedi, per esempio, L. Oster: *Emission, absorption and conductivity of a fully ionized gas at radio frequency*, in *Rev. Mod. Phys.*, **33**, 525 (1961); S. Chapman e T. G. Cowling: *The Mathematical Theory of Nonuniform Gases*, Cambridge (1960).

Fig. 3. – Da sinistra: Antonio Rostagni, Gleb Wataghin, Enrico Persico, Enrico Fermi, Matilde Rostagni. Sopra Gressoney La Trinité, dicembre 1932 (grazie alla cortesia di A. Rostagni).

gnamento dell'allora nuova meccanica quantistica, influendo notevolmente sui giovani fisici che si andavano raccogliendo all'Istituto di Fisica attorno al suo direttore Antonio Garbasso (1871-1933) che ebbe a Firenze un ruolo simile a quello che O. M. Corbino (1876-1937) esercitò a Roma e su tutto il paese.

Fra questi giovani si debbono ricordare Bruno Rossi, Gilberto Bernardini, Giuseppe Occhialini e Giulio Racah, a cui si aggiunsero in sèguito Daria Bocciarelli e Lorenzo Emo Capodilista.

Con tutti loro Persico stabilí immediatamente rapporti molto semplici e amichevoli. Come ricorda Occhialini, Persico possedeva una piccola automobile (una Fiat Balilla) con cui spesso li portava dal centro della città all'Istituto di Fisica ad Arcetri e, la domenica, a Vallombrosa o all'Abetone a sciare.

Daria Bocciarelli ricorda in modo particolare il seminario organizzato ad Arcetri da Persico in modo estremamente semplice ma efficace. Persico stesso o uno dei giovani, a turno, esponeva un lavoro apparso recentemente in una rivista scientifica e la discussione che ne seguiva era una straordinaria occasione di approfondimento critico dell'argomento.

L'influenza di Persico sull'insegnamento della nuova meccanica quantistica si estese ben presto da Firenze a tutta Italia tramite un ben noto volume di dispense *Lezioni di meccanica ondulatoria* [3,4], che fu usato per anni nella maggior parte delle università italiane, e sulle cui vicende torneremo fra poco.

A quell'epoca all'Istituto di Fisica

dell'Università di Roma il gruppo di giovani fisici che si era formato attorno ad Enrico Fermi, con l'aiuto di Franco Rasetti e sotto la vigile e lungimirante protezione del direttore dell'Istituto O. M. Corbino, aveva preso l'abitudine di attribuire scherzosamente titoli ecclesiastici ai componenti dell'Istituto ([2,3]). Cosí Corbino veniva chiamato il Padre Eterno, Fermi il Papa, Rasetti il Cardinal Vicario, Segrè e Amaldi gli abati e cosí via. Nel quadro di questo scherzo Enrico Persico, che era andato a diffondere la fisica teorica moderna in un'altra università, era stato nominato Prefetto di Propaganda Fide. Persico, che partecipava allo scherzo, inviò agli amici di Roma, probabilmente nel corso del 1929, la seguente poesia:

Padre Enrico il missionario
Se ne andò tra gl'infedel
In un'isola selvaggia
A spiegare l'evangel.

Dei cannibali il gran capo
Non appena egli approdò
La sua pentola piú grossa
Sopra il fuoco preparò,

I sentieri di sua terra
Tutti fé fortificar
E la caccia all'uomo bianco
Ordinò per terra e mar.

Ma a piantare le sue tende
Il buon padre pur riuscí
E dei negri il sacerdote
Alla fede convertí.

D'alto sdegno allora invaso
L'antropofago signor
Lo segnò nel libro nero
Per tremila sacchi d'or.

Ma il capitolo ambrosiano
Premiò invece tanto zel
Con l'invito a predicare
A quel popolo fedel.

A tal nuova i convertiti
Calde lacrime versar
E il buon padre, assai commosso,
Non li volle abbandonar.

Ma quest'anno il negro sire
Trasferito ha regno e tron
In un'isola vicina
Conquistata agli scorpion.

Nella terra ch'egli lascia
Regna ovunque ora la fè
E di padri missionari
Piú bisogno ormai non v'è.

Per ciò dunque Padre Enrico
Ora accingesi a partir
E, varcato l'Appennino,
Altre genti a convertir.

Il gran capo dei cannibali era il prof. Vasco Ronchi, noto ottico, i cui interessi erano rivolti all'ottica strumentale e fisiologica e non alla nascente meccanica ondulatoria. Il sacerdote dei negri era Gilberto Bernardini, ritornato nella sua città natale dopo aver conseguito la laurea all'Università di Pisa e il diploma della Scuola Normale Superiore di Pisa.

Per sopravvivere Bernardini per qualche tempo aveva costruito cannocchiali per la Marina presso una piccola industria (Officina Cipriani e Baccani a Rifredi) ma al tempo stesso aveva seguíto le lezioni di Persico, nonostante le proteste dei suoi datori di lavoro. In breve tempo Persico, essendosi reso conto del valore di Bernardini, lo invitò a fargli da assistente e a svolgere per lui le esercitazioni. Date le dimissioni dalla Cipriani e Baccani, Bernardini divenne cosí assistente straordinario in soprannumero di Persico, ma poiché le 270 lire che riceveva come stipendio non gli permettevano di vivere insieme alla giovane moglie, prese con il consenso di Persico una supplenza nelle scuole medie. Ma qualche mese dopo Ronchi gli propose di diventare suo assistente all'Istituto di Ottica ([6]) e Bernardini accettò sperando di

([6]) L'Istituto di Ottica nacque, ad opera di V. Ronchi, nel 1927 presso l'Istituto di Fisica di Arcetri, come trasformazione del « Laboratorio di Ottica e Meccanica di Precisione » istituito nel 1917 sotto la spinta congiunta delle Forze Armate e dei vari Enti industriali e diretto per anni da R. Occhialini. La villetta, originariamente di sei stanze, era stata costruita per accogliere la Fisica

potersi cosí inserire nelle attività dell'Istituto di Fisica di Arcetri, dove già c'era anche Bruno Rossi. Ma Ronchi vedeva con ostilità questi interessi di Bernardini che lo distraevano dall'ottica e cercava d'impedirgli di andare a sentire le lezioni di Persico, tanto da giungere, in qualche caso, a chiuderlo a chiave in laboratorio. L'anno successivo, Bernardini ottenne di essere nominato assistente di Garbasso e poté cosí inserirsi completamente nel nuovo ambiente che stava formandosi in Arcetri.

« L'isola vicina conquistata agli scorpion », nominata nella poesia di Persico, era una villetta, costruita in Arcetri, a poca distanza dall'Istituto di Fisica dell'Università, che era stata lasciata in completo abbandono, e in cui si trasferí l'Istituto di Ottica a quell'epoca (6). Ampliata successivamente, essa è ancora oggi la sede dell'Istituto Nazionale di Ottica del Ministero della Pubblica Istruzione.

Nella poesia si accenna infine alla offerta della cattedra di Fisica Teorica fatta a Persico dall'Università di Milano dopo la morte di Aldo Pontremoli, che nel 1928 aveva partecipato come fisico alla seconda spedizione Nobile al Polo Nord ed era morto drammaticamente sul pack polare (7).

Terrestre di Lo Surdo, ma in sèguito al trasferimento di Lo Surdo a Roma (1919) era stata destinata al Laboratorio di Ottica e Meccanica di Precisione di R. Occhialini. Nel corso degli anni trenta, attraverso successivi ingrandimenti, sotto la direzione di Ronchi, fu ampliata a circa 40 stanze. Ringraziamo F. T. Arecchi, attuale direttore dell'Istituto Nazionale di Ottica, per la lettera in proposito e per la corrispondente indicazione bibliografica: V. Ronchi: *Una istituzione caratteristica: l'Istituto Nazionale di Ottica di Arcetri*, in *Dialogues*, luglio-settembre 1964. Vedi anche dello stesso stesso autore gli *Atti della Fondazione Giorgio Ronchi*, **11**, n. 6 (1976); n. 1, 2, 3, ecc. (1977).

(7) Nella seconda spedizione al Polo Nord con dirigibile, organizzata e diretta da U. Nobile (nella prima, che aveva avuto luogo nel 1926 sotto la direzione di R. Amudsen, Nobile aveva avuto il còmpito di pilotare l'aeronave), al ritorno dal secondo volo sul Polo, il dirigibile Italia, che si era appesantito troppo e troppo rapidamente a causa della formazione d'incrostazioni di ghiaccio, era precipitato sulla banchisa, lasciando sul ghiaccio la quasi totalità dell'equipaggio. Nelle tragiche vicende che ne seguirono, trovarono la morte otto membri dell'equipaggio fra cui Aldo Pontremoli Persero la vita anche altre sei persone, fra cui R. Amudsen, che tentarono di portare soccorso ai naufraghi.

Occhialini ricorda che durante una gita a Larderello, organizzata in occasione di un congresso a Firenze, Pugno Vanoni, professore di Elettrotecnica all'Università di Milano, voleva parlare tranquillamente con Persico per convincerlo ad accettare l'offerta, ma che i giovani fisici fiorentini, secondo un piano orchestrato da Racah, fecero di tutto per impedire tale colloquio, trovando il modo d'intromettersi fra i due e non lascinadoli mai soli. Essi consideravano la partenza di Persico come un fatto molto grave per le attività scientifiche e didattiche che si stavano sviluppando in modo cosí promettente a Firenze.

Sotto le amichevoli ma ferme pressioni dei giovani fisici fiorentini, Persico rinunciò all'invito, ma quando, non molto tempo dopo, ricevette un'analoga proposta dall'Università di Torino accettò e, nell'autunno del 1930, si trasferí nella nuova sede.

I lavori di Persico del periodo fiorentino (1.18-1.22) riguardano questioni di statistica molecolare, applicazioni della meccanica ondulatoria e l'effetto Hall, argomento già da lui trattato nel periodo romano.

3. A Torino (1930-47).

Il suo passaggio dall'Università di Firenze a quella di Torino fu in parte determinato dal fatto che le università

Fig. 4. – Da sinistra: A. Caracciolo di Forino, Giacomo Morpurgo, Giorgio Careri, Enrico Persico, D. Cunsolo, A. Gamba (appena visibile), R. Stroffolini, B. Bosco. Dopo il banchetto sociale tenuto in occasione del Congresso Nazionale della SIF a Cagliari nel 1953, Persico di forza fu posto a sedere sopra il tetto di un'automobile e portato in trionfo.

italiane erano divise in due categorie: quelle di categoria A erano finanziate completamente dallo Stato, mentre quelle di categoria B avevano un carattere locale anche se ricevevano un considerevole contributo finanziario dallo Stato. L'Università di Torino era di categoria A, mentre quella di Firenze era di categoria B, circostanza questa che dava a Persico un senso di non completa sicurezza.

La notizia che Persico lasciava Firenze per Torino fu accolta dai suoi amici romani con qualche dispiacere, perché a loro sembrava difficile che nella nuova sede si potese formare un gruppo di giovani, sia sperimentali che teorici, di tanto valore quali erano quelli che già c'erano o si stavano formando ad Arcetri.

Persico tuttavia anche nel periodo trascorso a Torino riuscí a contribuire alla nascita di un'importante scuola di Fisica Teorica moderna. Gian Carlo Wick si laureò nel 1931 con Carlo Somigliana (1860-1955) ed ebbe con Persico contatti non molto approfonditi, solo durate l'anno accademico 1931-32 ([8]). Egli ricorda, tuttavia, di essere andato a sciare piú volte durante quell'inverno con Enrico Persico e Maria Daviso di Charvensod, insegnante di Scienze al Liceo Massimo d'Azeglio.

Fra i vari giovani teorici che subirono l'influenza di Persico ricordiamo: Nicolò Dallaporta, Luigi Radicati di

([8]) Dopo la laurea G. C. Wick andò, con una borsa di studio, in Germania, prima a Gottinga e poi a Lipsia ove trascorse, alle scuole di Max Born e Werner Heisenberg, praticamente tutto l'anno accademico 1930-31. Tornato in Italia passò a Torino gli anni accademici 1931-32 e 1932-33 e nell'autunno del 1933 si trasferí a Roma ove aveva avuto il posto di assistente di E. Fermi.

Brozolo, Marcello Cini, Augusto Gamba e Tino Zeuli.

Dallaporta, che da Bologna si era trasferito a Catania seguendo O. Specchia, accettò nel 1938 un posto di assistente di A. Pochettino con il preciso scopo di studiare con Persico, il quale lo indirizzò a trattare problemi di urto di ioni e atomi neutri di bassa energia, in connessione con le esperienze di Rostagni.

Radicati ricorda che fu « il bellissimo corso di fisica matematica di Persico il primo vero incontro con la fisica, una vera rivelazione, ... » e che piú tardi, dopo la guerra, fu Persico, anche se sfiduciato sulle possibilità per un giovane fisico di trovare un lavoro in Italia, ad aiutarlo a ritornare alla fisica.

Qualche tempo dopo il suo arrivo a Torino, Enrico Persico inviò agli amici romani un'altra poesia di cui riproduciamo il testo:

Al di là dell'Appennino
Padre Enrico giunto è
Ed insegna altrui il cammino
Sulla strada della fè.

Incomincia la lezione
Col precetto elementar
Ch'è cattiva educazione
Carne umana divorar.

Narra poscia ch'oltre i monti
Vivon popoli fedel
Che del ver le sacre fonti
Ricevuto hanno dal ciel.

Essi han d'h il sacro culto
Han nei quanti piena fè
E per loro è grave insulto
Dir che l'atomo non c'è.

Sono pur bestemmie orrende
Il negar che v'è la *psi*,
Che un valor non nullo prende
$\Delta q \times \Delta p$,

Che dell'orbite ai momenti
S'addizionano gli spin
E elettroni equivalenti
Son vietati dal destin.

Credon poi, con fè profonda
Cui s'inchina la ragion
Che la luce è corpo ed onda
Onda e corpo è l'elettron.

Sono questi i dogmi santi
Ch'egli insegna agli infedel
Con esempi edificanti
Appogiandosi al Vangel.

In essa si parla del Vangelo. Questo, nel gergo di allora, era il volumetto di dispense del corso di meccanica quantistica tenuto da Persico appena giunto a Firenze e raccolto da Bruno Rossi e Giulio Racah, di cui si è parlato sopra. La prima edizione ([3.4a]) era risultata tipograficamente quasi illeggibile, tanto che i fisici romani la avevano chiamata « Vangelo copto ». Ciò non ostante essa era la migliore presentazione che esistesse a quell'epoca, e in qualunque lingua, delle idee fondamentali della meccanica quantistica e delle sue applizioni. Corbino, che seguiva abbastanza da vicino queste vicende scientifiche e umane, grandi e piccole, ne fece fare una nuova edizione ben stampata a spese del suo Istituto ([3.4b]) e Persico ne regalò una copia ai giovani fisici di Roma con la seguente dedica:

Coptum hoc evangelium
Italicis tipis transcriptum
Auctumque et nonnullis locis emendatum
Patris Aeterni munificientia editum
Doctis Abbatibus
In limine solii pontificii assidentibus
Offerit dicatque
Enricus Persicus de Propaganda Fide Cardinalis.

Da Torino Persico faceva, di quando in quando, un viaggio a Roma, ove veniva sempre accolto con simpatia e piacere da Corbino, da Fermi e in generale da tutto il gruppo dei fisici romani. Fermi era sempre particolarmente lieto di rivedere il vecchio amico con cui passava ore a esporre quello che lui stesso o altri stavano facendo in Istituto. Una conversazione scientifica con Persico era sempre piacevole e utile per il suo interesse a qualsiasi problema scientifico e per la chiarezza delle sue idee che si manifestava nelle osservazioni e domande come del resto in ogni suo scritto scientifico.

Fu proprio in occasione di una di queste visite all'Istituto di Roma che Persico assistette alla scoperta del fenomeno del rallentamento dei neutroni e scrisse sul Quaderno di misure i risultati delle letture dei conteggi dei contatori usati da Fermi per misurare l'attività indotta dai neutroni in un cilindretto di argento in presenza o assenza di paraffina (9).

A Torino Persico riallacciò l'amicizia con F. G. Tricomi e Gino Castelnuovo, che già conosceva da Roma. L'ingegner Castelnuovo, figlio del matematico Guido Castelnuovo, era diventato uno dei dirigenti della RAI e come tale viveva a Torino. Tricomi che conosceva Persico dal 1922, quando era assistente del matematico Francesco Severi all'Università di Roma, era diventato professore di Analisi all'Università di Torino e aveva contribuito in modo determinante alla chiamata di Persico a Torino, vincendo l'ostilità (non alla persona ma alla materia, cioè alla Fisica Teorica) di Somigliana.

Persico entrò in un gruppo composto allora prevalentemente di assistenti universitari come E. Denina, piú tardi professore di Chimica-Fisica ed Elettrochimica al Politecnico di Torino, R. Margaria, piú tardi professore di Fisiologia all'Università di Milano, O. M. Olivo, piú tardi professore di Anatomia all'Università di Bologna, e A. Rostagni, in sèguito professore di Fisica Generale all'Università di Padova. Questo gruppo, che comprendeva l'architetto D. Morelli, faceva spesso gite in montagna a cui talvolta partecipavano anche Alessandro e Benvenuto Terracini e la figlia Eva di quest'ultimo.

Con Rostagni che allora era assistente di A. Pochettino, titolare di Fisica Sperimentale all'Università, e si occupava di esperienze sui raggi neutrali, faceva anche giri automobilistici. Ad un giro sui laghi dell'Alta Italia attorno a Pasqua del 1934 parteciparono anche Enrico e Laura Fermi e la sorella di questa, Paola Capon; a un altro estivo, in Svizzera, nello stesso anno, partecipò anche Franco Rasetti. Anche con Denina faceva passeggiate, in macchina o a piedi, o passava le vacanze estive, il piú delle volte nelle zone alpine. Persico amava moltissimo la montagna, e seguitò ad andarvi fino al 1961, quando cominciò ad accusare qualche disturbo di pressione.

Oltre ad alcuni lavori su varie applicazioni della meccanica quantistica e a qualche altro sui raggi neutrali (in parte in collaborazione con Rostagni (1.25-1.26)) Persico pubblicò in quegli anni numerosi libri che ebbero una influenza molto considerevole sulla cultura scientifica italiana. Basti qui ricordare *L'Ottica*, stampata da Vallardi nel 1932 (3.6), che è un rifacimento completo di un parte del trattato di Fisica di Battelli e Cardani, nella cui stesura fu aiutato da B. Rossi, come è ricordato nella prefazione; *L'introduzione alla Fisica Matematica*, redatta da T. Zeuli, pubblicata nel 1936 da Zanichelli (3.11) e i *Fondamenti della Meccanica Ondulatoria*, pubblicati ancora da Zanichelli nel 1939 (3.12).

(9) L'episodio è descritto a p. 642 del rif. (2) e a p. 83 del rif. (3) (p. 80 del testo inglese) e, in maggior dettaglio, da E. AMALDI: *Personal notes on neutron work ...*, p. 294 del LVII Corso della Scuola Internazionale di Fisica «Enrico Fermi» della S.I.F., Varenna, 31 Luglio - 12 agosto 1972; a cura di C. WEINER, Academic Press, New York (1977).

Fig. 5. – Enrico Persico negli ultimi anni di vita.

Questo ultimo libro rappresenta un sostanziale rifacimento delle dispense del corso tenuto da Persico a Firenze. Esso ebbe un'importanza notevolissima perché rese accessibile lo studio della meccanica ondulatoria a giovani che conoscevano solo l'italiano. Il testo, però, era ottimo in senso assoluto, tanto che oltre a varie edizioni successive italiane, ne apparve nel 1950 un'edizione ampliata in inglese per i tipi di Prentice Hall, New York: *Fundamentals of Quantum Mechanics* (3.14). Questo testo, la cui traduzione in inglese era stata fatta da G. M. Temmer, ebbe larga diffusione non solo negli Stati Uniti, ma anche in vari altri paesi.

I *Fondamenti della Meccanica Ondulatoria* facevano parte di un progettato *Trattato di Fisica* sovvenzionato dal CNR ([10]), di cui uscirono solo altri due volumi: *Molecole e Cristalli* di E. Fermi ([11]) e *Fisica Nucleare* di F. Rasetti ([12]). Un quarto volume sulla *Spettroscopia atomica* di E. Segrè rimase allo stato di manoscritto in sèguito alle leggi razziali ([13]). Il contenuto di ciascun volume fu discusso dall'autore con Fermi che aveva assunto il còmpito di coordinatore dell'intera opera ed ebbe cosí, inizialmente, una notevole influenza sulle sue caratteristiche e portata.

Negli ultimi anni Persico parlava spesso di una nuova stesura dei *Fondamenti della Meccanica Ondulatoria*, alla quale aveva cominciato a lavorare, principalmente con l'idea di aggiungere un capitolo dedicato ai problemi d'urto, la cui importanza era grandemente aumentata rispetto agli anni trenta. Per il momento, tuttavia, non abbiamo trovato, tra i suoi manoscritti, traccia evidente di questo lavoro.

Va ancora ricordato che fra i fisici della sua generazione Persico fu quello che manifestò piú chiaramene un interesse per problemi epistemologici, come risulta da varie sue pubblicazioni, in particolare dall'articolo *Analisi del determinismo fisico* (2.21).

4. Uno sguardo d'insieme attraverso la lettera di un amico.

La figura di Persico e alcune sue caratteristiche particolari possono essere colte da chi non lo conobbe personalmente dalla lettera che Tricomi scrisse a Rasetti, su richiesta di quest'ultimo, nel novembre 1974.

Torino, 4 novembre 1974

Caro Rasetti,

effettivamente io ho conosciuto molto bene il nostro comune amico Persico, non solo perché fu mio collega qui a Torino dal 1929 al 47, ma anche perché coabitò con me e la mia famiglia durante lo « sfollamento » degli anni di guerra (1942-45) e perché anche dopo il 47 c'incontrammo abbastanza spesso, in America e a Roma. Tuttavia non posso riferire molti « fatti caratteristici » relativi al nostro amico, giusto perché una delle sue peculiarità piú salienti fu proprio di non avere speciali caratteristiche, a meno che si consideri come tale quella di avere molto buon senso e molto equilibrio. E anche molta prudenza che l'induceva a tenersi un po' in disparte nei turbinosi tempi in cui ci è toccato di vivere.

Per quel che concerne la figura

([10]) Il CNR versava a ciascun autore 10 000 lire, a totale compenso dei diritti di autore.

([11]) E. Fermi: *Molecole e cristalli*, Bologna (1934), di cui apparve anche una edizione in inglese: *Molecules, Crystals and Quantum Statistics*, New York (1966).

([12]) F. Rasetti: *Il nucleo atomico*, Bologna (1936), di cui apparve anche una edizione in inglese: *Elements of Nuclear Physics*, New York (1936).

([13]) Su richiesta di E. Segrè, Persico lesse molto coscienziosamente il manoscritto del libro di *Spettroscopia atomica* e quando glielo restituí, con varie osservazioni, lo accompagnò con un disegnino di strumenti di tortura, raffiguranti simbolicamente le leggi razziali.

scientifica di Persico credo di potere asserire che le sue ben note doti didattiche e di brillante espositore derivavano non tanto da doti, diciamo cosí, letterarie (che pur non gli mancavano) quanto da un'approfondita rielaborazione personale dei concetti fondamentali della Fisica, che non si arrestava se non quando riusciva a vedere le cose con piena chiarezza, oppure a constatare che — nello stato delle cose — questa non era raggiungibile. Mai perciò tentava d'impasticciare le cose per nascondere, ai suoi stessi occhi o a quelli degli altri, le reali difficoltà, per esempio, della teoria della costituzione della materia.

Riguardo la vita privata di Persico, il fatto piú saliente di cui sono a conoscenza è la profonda e lunga depressione psichica in cui cadde dopo l'estate del 40, in sèguito alla subitanea morte della vecchia e amata madre, per effetto di un lieve incidente automobilistico capitatogli sulla via Emilia, mentre la conduceva quasi coattivamente (negli ultimi suoi tempi la madre era pressocché demente) ad un convento in Toscana per sottrarla al pericolo dei bombardamenti aerei che cominciavano ad imperversare su Torino. Tale depressione fu cosí prolungata che allorché, nel novembre del 42, i bombardamenti divennero davvero minacciosi, egli era ancora cosí apatico e senza volontà che dei vicini di casa lo strapparono quasi a forza dal suo alloggio, conducendolo provvisoriamente con loro in delle campagne a sud della città. E poco dopo fui io stesso a trascinarmelo meco a Torre Pellice, mettendogli a disposizione una camera che era libera, nell'appartamento in cui ero sfollato assieme coi miei, dando cosí origine ad una simpatica convivenza prolungatasi fino all'armistizio del settembre 43. Ma anche dopo che io dovetti darmi alla montagna (e successivamente occultarmi a Roma) perché troppo compromesso contro il fascismo, Persico continuò a vivere a Torre Pellice e fu di valido aiuto ai miei, nonostante che il coraggio fisico non fosse la sua qualità piú spiccata.

Ricordo che piú tardi ci trovammo insieme (io ero tornato a Torre nel maggio del 45) quando giunse la notizia della bomba atomica di Hiroshima. Non ne fummo troppo stupiti perché precedentemente avevamo discusso piú volte assieme della possibilità di qualcosa del genere, concludendo che tutto dipendeva dal valore numerico, allora sconosciuto, di una costante: il numero medio di neutroni liberati da un nucleo di uranio quando ne assorbiva uno. Perciò il nostro quasi simultaneo commento alla notizia fu: « si vede che quella costante era sufficientemente grande ».

Credo superfluo soffermarmi sull'amore di Persico per lo sci e per i gatti, perché sono cose abbastanza note nell'ambiente dei fisici. Quanto al secondo, credo che molto dipendesse dal fatto che egli era assai solo, tanto che, a quanto mi disse, una delle principali considerazioni che lo indussero ad accettare il trasferimento a Roma dopo Quebec, fu che costà aveva dei cugini che gli avrebbero fatto un po' di compagnia.

Una delle cose che mi sorprese un po' quando, nel 50, andai a fargli visita a Quebec, fu di trovarlo tanto mangiapreti quanto non era mai stato nel suo sereno ateismo. Ma era una comprensibile reazione all'ambiente ultra clericale dell'Università Laval, che tu ben conosci. Mi raccontò che fra l'altro, quando si annunciava al telefono come « Persico », gli interlocutori, abituati a quell'ambiente, talvolta intendevano « Pére Sicò », donde ripetuti: « Oui mon Pére » ecc. da lui non eccessivamente graditi. Ma all'Università Laval lo apprezzavano tanto che, dopo che lui volle andarsene, cercarono di mantenere con lui qualche legame.

In quegli anni Persico ebbe dei ripetuti attacchi di uno strano male alle dita delle mani — forse conseguenza di esposizioni a radiazioni — che gli costò la perdita di diverse falangi. L'ultima volta fu a Quebec ove, grazie alla penicillina, poté essere evitata un'ulteriore mutilazione. E mi raccontò, ridendo, delle comiche manovre delle suore-infermiere dell'ospedale per fargli delle inizioni senza « offesa al pudore ». ...

Con i piú cordiali saluti,

tuo F. Tricomi

La lettera di Tricomi mette bene in evidenza alcune delle vicende familiari e ambientali del periodo trascorso da Persico a Torino e che contribuirono alla sua decisione di accettare una cattedra all'Università Laval di Quebec.

Nella stessa lettera figura però una frase sul coraggio fisico di Persico che non ci trova d'accordo e tanto meno è d'accordo A. Gamba che negli anni 1943-45 fu studente di Fisica a Torino, fece con Persico gli esami di Fisica Matematica e Fisica Teorica e, dopo la liberazione, prese la tesi con Persico, laureandosi nel 1946. All'inizio della primavera 1944, Gamba, allora partigiano, aveva preso un appuntamento con Persico all'Istituto di via Giuria per discutere qualche questione di esami. Dopo mezz'ora di colloquio disse che doveva andarsene e gli chiese di accompagnarlo per continuare la conversazione. Gamba gli rispose di essere armato e gli mostrò la pistola facendogli presente che, se qualche milite fascista lo avesse fermato, sarebbe stato costretto ad usarla. Persico non fu menomamente preoccupato di ciò e andò con Gamba a piedi dall'Istituto fino a casa sua per una distanza di circa tre chilometri.

Un ultimo particolare non molto rilevante che figura nella lettera di Tricomi è l'accenno alla possibilità che il male alle dita di cui sofferse Persico potesse essere conseguenza di esposizioni a radiazioni. È vero che nel periodo torinese Persico preparò alcune sorgenti di neutroni di radon piú berillio, simili a quelle usate a Roma, e che con queste fece alcune esperienze sui neutroni. Ma poiché noi a Roma preparammo settimanalmente per anni sorgenti molto piú intense, che usammo sistematicamente ogni giorno, senza avere nessun inconveniente del genere, siamo propensi a escludere tale interpretazione.

5. A Quebec, in Canada (1947-50).

Nel 1938 l'Università Laval di Quebec era giunta nella determinazione di dare un impulso alle sue attività nel campo della fisica che fino ad allora erano state assai poco curate e il suo decano, Monseigneur Alexandre Vachon, tramite l'Accademia Pontificia, di cui Rasetti era stato nominato membro, si mise in contatto con lui poco dopo la partenza di Fermi dall'Italia. Rasetti accettò l'invito e si trasferí a Quebec nell'estate 1939. Egli praticamente creò il Dipartimento di Fisica e riuscí, durante la guerra, a fare varie esperienze di notevole rilievo. Nel 1947 Rasetti accettò l'offerta di una cattedra di Fisica della Johns Hopkins University di Baltimora (USA) e scrisse a Persico per sapere se era interessato a succedergli. Persico accettò l'offerta anche perché nel 1947 era molto sfiduciato sulle possibilità di ripresa della fisica in Italia e, in generale, di un rapido ritorno del nostro Paese a condizioni di vita accettabili.

A Quebec, come ricorda Tricomi nella sua lettera, Persico fu nominato direttore del Dipartimento di Fisica. Egli cominciò in quel periodo ad occuparsi della teoria degli spettrometri per raggi beta, argomento a cui diede vari importanti contributi (1.31-1.35) e su cui indirizzò alcuni allievi ([14]).

Nel 1949, essendosi fatto visitare in sèguito a qualche disturbo, un noto medico gli diagnosticò un male incurabile al colon. Purtroppo Persico non pensò subito a farsi esaminare da un secondo specialista e passò cosí circa un anno attendendosi il peggio da un momento all'altro. In questo periodo egli fece lunghi viaggi in macchina attraverso il Canada e gli Stati Uniti. Dopo circa un anno, anche sotto la spinta di qualche amico, si fece visitare da un altro specialista che attribuí ad altra causa la lieve opacità che appariva nelle radiografie. Per quanto Persico avesse accettato la siuazione con notevole freddezza, la nuova diagnosi ebbe senza dubbio un'influenza positiva sul sul suo umore e forse sul suo atteggiamento verso la vita negli anni successivi.

6. Di nuovo a Roma (1950-69).

La morte di Antonio Lo Surdo (n. 1880) avvenuta il 7 giugno 1949 pose tutta una serie di problemi. Egli era professore di Fisica Superiore all'Università di Roma fin dal 1919 e alla morte di O. M. Corbino, avvenuta all'inizio del 1937, gli era succeduto nella direzione dell'Istituto di Fisica. Quando, nell'autunno 1938, Fermi aveva lasciato l'Italia, Lo Surdo gli era succeduto nell'incarico del corso di Fisica Terrestre, campo di cui Lo Surdo si era interessato fin dal 1917.

Nel 1939 egli aveva fondato l'Istituto Nazionale di Geofisica (I.N.G.) inizialmente sostenuto dal Coniglio Nazionale delle Ricerche, e piú tardi eretto in ente autonomo dipendente dal Ministero dell'Educazione Nazionale. Questo Istituto, di cui Lo Surdo era direttore, aveva sede presso l'Istituto di Fisica alla Città Universitaria.

Alla morte di Lo Surdo, il ministro della Pubblica Istruzione aveva nominato direttore dell'I.N.G. Enrico Medi, professore di Fisica Sperimentale all'Università di Palermo ma che proveniva da Roma, essendo stato per molti anni assistente di Lo Surdo.

La chiamata del professore di Fisica Superiore da parte della Facoltà di Scienze dell'Università di Roma presentò all'inizio qualche difficoltà. Alcuni membri della Facoltà erano in favore della chiamata di Medi, mentre altri sostenevano l'opportunità di chiamare Persico, per le sue straordinarie doti di maestro, ed anche perché cosí facendo si otteneva i rientro in Italia di un fisico la cui perdita definitiva sarebbe stata di notevole danno per il nostro Paese. La soluzione fu trovata con la creazione da parte del ministro di una ulteriore cattedra ([15]) per la

([14]) A. Boivin: *Journ. Opt. Soc. Am.*, **42**, 60 (1952).

([15]) Le cattedre di Fisica all'Università di Roma erano ancora quattro, come nel 1931: Fisica Sperimentale (E. Amaldi), Fisica Superiore (vacante), Fisica Teorica (B. Ferretti), che nel 1947 era succeduto a G. C. Wick quando questi aveva accettato l'offerta di una cattedra di Fisica da parte dell'Uni-

Fisica Terrestre a cui circa un anno dopo fu chiamato Enrico Medi, mentre Persico fu chiamato súbito alla Fisica Superiore. Egli tenne questo insegnamento fino al 1958, anno in cui passò, all'interno della Facoltà stessa, alla cattedra di Fisica Teorica che ricoprí fino alla morte.

Il successore di Persico a Fisica Superiore fu Marcello Conversi, che, dopo aver studiato e lavorato a Roma, Chicago (1947-49) e Portorico (1950), aveva vinto, nel 1950, il concorso di Fisica Sperimentale ed era stato chiamato all'Università di Pisa. Giunto a Roma alla fine del 1958, diventò subito amico di Persico. Negli anni successivi Persico, Conversi, Ageno e Nella Mortara (spesso chiamati gli scapoloni dell'Istituto) presero l'abitudine di fare gite domenicali nelle piccole bellissime città nei dintorni di Roma, gite che tutti ricordarono per anni con gran piacere.

Durante il periodo romano Persico svolse anche la funzione di supervisore della biblioteca dell'Istituto, che aveva sempre piú acquistato il carattere di un istituto policattedra. La direttrice della biblioteca, dr.ssa C. Guazzaroni Ferrari, si rivolgeva a Persico per la classificazione dei libri, e soprattutto per i nuovi acquisti, non sempre facili data la limitatezza dei fondi e la necessità di mantenere aggiornata la biblioteca. Persico fece questo lavoro per diciotto anni con una assiduità e continuità eccezionali, durante i quali la consistenza scientifica della biblioteca fu pressoché triplicata, il numero degli abbonamenti a pubblicazioni scientifiche periodiche fu portato da diciotto a centocinquanta, fu introdotta la classificazione decimale, fu ampliata la sala di lettura per gli studenti e creato il servizio di fotocopia a disposizione anche degli studenti.

A Roma Persico seguitò per qualche tempo ad occuparsi di problemi di ottica elettronica. In particolare ideò due diversi tipi di rete bidimensionale di resistenze (1.36, 1.37) che permettono di affrontare il problema della simulazione di campi elettrici (o magnetici) generati da elettrodi (o poli) di forma qualsiasi. Le reti di Persico sono assai vantaggiose rispetto ai metodi, già in uso in molti laboratori, basati sull'impiego di vasche contenenti un elettrolita, oppure sull'uso di fogli di alluminio. Questa linea di ricerca di concluse con la costruzione di una rete eseguita con l'aiuto di F. Magistrelli (1.36).

Anche a Roma il suo contributo didattico fu del tutto straordinario: basti qui ricordare il suo corso di Fisica Superiore di cui pubblicò vari volumi di dispense, via via migliorate e ampliate, raccolte e redatte da L. Tau (3.15) e da R. Bizzarri (3.16, 3.18), e vari corsi di perfezionamento: presso la Scuola di Perfezionamento dell'Università di Roma il corso di « Macchine acceleratrici e reattori », tenuto dal 1953-54 al 1959-60 e trasformato piú tardi (dal 1960-61 al 1963-64) in corso sui « Reattori nucleari ». Le lezioni sulle macchine acceleratrici raccolte in un primo tempo da G. Caglioti, N. D'Angelo e A. Reale (3.17) poi da E. Ferrari e S. E. Segre, furono piú tardi ampliate e pubblicate da Benjamin nel 1968 (3.23). Le lezioni sui « Reattori nucleari » furono invece raccolte da G. Riguzzi (3.19) e costituiscono un altro notevole contributo di Persico alla didattica ad alto livello. In quegli stessi anni egli tenne anche le

versità di Notre Dame nell'Illinois (USA), e Spettroscopia (ancora coperta formalmente da Rasetti che quando era andato a Quebec era stato messo a disposizione dal Ministero degli Affari Esteri).

« Lezioni teoriche di fisica del reattore » nel quadro del « Corso di Fisica Nucleare Applicata e Chimica Nucleare » del CNEN (dal 1956-57 al 1962-63) e, con diversa accentuazione, nel quadro del « Corso di Perfezionamento in Ingegneria Nucleare » del CNEN (dal 1959-60 al 1963-64).

Nel 1953 il Consiglio Direttivo dell'Istituto Nazionale di Fisica Nucleare (INFN), presieduto in quegli anni da Gilberto Bernardini, aveva deciso di procedere alla costruzione di un elettrosincrotone nazionale da 1.1 GeV. I fondi a disposizione dell'INFN erano ormai abbastanza consistenti e pertanto sembrava doveroso concentrarli in una singola impresa di notevole impegno piuttosto che suddividerli in tanti rivoletti, il piú delle volte non molto significativi. La direzione di tale progetto fu affidata a Giorgio Salvini, allora professore di Fisica Generale all'Università di Pisa.

Persico, invitato da Salvini a partecipare alle prime riunioni del gruppo che doveva eseguire la progettazione e la costruzione della macchina, accettò fin dall'inizio la direzione della Sezione Teorica, ma non la direzione generale offertagli da Salvini stesso.

La sua attività iniziò súbito con una serie di lezioni sulla dinamica ultrarelativistica, sulle distribuzioni di potenziale elettrico o di campo magnetico, determinate da elettrodi o poli di forme non usuali, sul modo di risolvere questi problemi con metodi di simulazione, sulle trasformazioni di Schwarz-Christoffel e sulla tecnica del filo percorso da corrente per la determinazione delle linee di forza di un campo magnetico. Tutti problemi connessi con alcuni aspetti della progettazone del sincrotrone, che coinvolgeva la costruzione, per esempio, di magneti non usuali (dotati, magari, di dentini correttivi o « tip »).

Nel 1955 il gruppo di progetto si trasferí da Pisa all'Università di Roma e di qui, nel 1956, a Frascati, ove era iniziata l'effettiva costruzione dei Laboratori Nazionali del CNEN e dell'elettrosincrotrone nazionale. Nel frattempo Salvini era stato chiamato (nel 1955) a ricoprire la seconda cattedra di Fisica Sperimentale dell'Università di Roma.

Fin dall'inizio di questa attività Persico ebbe come allievi e collaboratori Carlo Bernardini, Pier Giorgio Sona e Angelo Turrin, i quali erano appena laureati e si formarono seguendo il corso di macchine acceleratrici tenuto da Persico alla Scuola di Perfezionamento dell'Università di Roma.

Un'esposizione completa dei problemi teorici affrontati dalla Sezione Teorica si trova nella Parte II del volume del Supplemento al Nuovo Cimento dedicato alla progettazione e costruzione del sincrotrone ([16]).

Alle riunioni dell'intero gruppo di progetto presiedute da Salvini, nascevano spesso vivaci discussioni su quale soluzione adottare per questo o quel problema. Di solito l'accordo veniva raggiunto piuttosto facilmente. Ma su di un punto essenziale la discussione si protrasse a lungo e fu assai vivace. Si trattava di fissare l'altezza del traferro del magnete e quindi le dimensioni verticali della ciambella per la quale Persico sosteneva un valore piuttosto elevato, per tenere conto con prudenza di tutte le possibili cause di errore nella localizzazione del fascio elettronico, mentre Salvini riteneva che i calcoli teorici (di altri autori e, soprattutto, del gruppo di

([16]) *L'elettrosincrotrone e i Laboratori di Frascati*, a cura di G. Salvini, Bologna (1962).

Persico) fossero ormai degni della massima confidenza e che pertanto si potesse accettare un'ampiezza totale della ciambella pari a quella strettamente prevista dal gruppo teorico di Persico. Il mediatore dell'importante problema fu in quella occasione Mario Ageno, insieme al quale furono concordate dimensioni di ragionevole compromesso tra le esigenze della prudenza e dell'economia.

Oltre a vari studi e calcoli originali inerenti alla progettazione dell'elettrosincrotone ([1.44]), Persico sviluppò in quel periodo la teoria generale dell'iniezione di particelle cariche in macchine acceleratrici ([1.39]), teoria che viene oggi usata correntemente dai progettisti e costruttori di tali macchine in tutto il mondo.

Tutti coloro che presero parte alla progettazione dell'elettrosincrotrone nazionale ricordano con affetto e ammirazione il contributo di Persico sempre estremamente preciso e puntualizzato.

La partecipazione di Persico a queste attività continuò fino al 1957, anno in cui la parte di progettazione e di calcolo era giunta al suo termine. In quel momento egli cessò la sua collaborazione continua e ufficiale, ma rimase disponibile, e con precedenza assoluta su altri suoi impegni, per consigli, verifiche e calcoli. Salvini ricorda, per esempio, di aver discusso con Persico, anche a sincrotrone ormai funzionante, i problemi umani, non indifferenti, nella collocazione e attività futura dei fisici e degl'ingegneri che avevano partecipato all'impresa. Comunque, Persico non volle comparire, anche se sollecitato, nel momento del successo, cioè quando ebbe inizio il funzionamento del sincrotrone, non ostante che tale successo fosse dovuto, in parte notevole, anche al suo contributo. E ciò non per ritrosia o disinteresse, ma perché veramente superiore (come osservò Salvini) a « certo caldo e baccano scientifico in un'epoca di gran baccano ».

Quando si staccò dai Laboratori Nazionali di Frascati del CNEN Persico cominciava ormai ad avere altri interessi. Buona parte del suo tempo veniva assorbita dai vari corsi di perfezionamento sulla fisica del « reattore nucleare » a cui si è sopra accennato. Egli, inoltre, era attratto da alcuni problemi di fisica del plasma, di cui si era già occupato nel 1926. Nel 1958 si interessò delle perdite di particelle da una bottiglia magnetica, e pubblicò il primo lavoro significativo su questo argomento in collaborazione con L. G. Linhart ([1.40]) che allora lavorava al CERN e aveva incontrato Persico in occasione di alcune lezioni che era stato invitato a tenere a Roma.

7. Qualche altro ricordo personale.

Persico condusse la sua vita di scapolo senza il minimo lusso né spreco, ma senza farsi mancare nulla. Il suo appartamento al terzo piano del numero 34 di via di Villa Emiliani era confortevole e tenuto in maniera perfetta da una governante che, nella sua serena serietà e rispettoso riserbo, provvedeva a tutto il necessario senza mai apparire. Lo studio di Persico era gradevole e ben fornito di libri e di un buon radiogrammofono con cui sentiva spesso musica classica. In qualche occasione invitava a casa sua qualche amico a prendere un buon tè, sempre con ottimi pasticcini.

Andava molto spesso al cinematografo e talvolta ai concerti o a teatro e gradiva molto gl'inviti degli amici a passare qualche ora in conversazione, nella quale inseriva osservazioni spiritose sui fatti del giorno, che facevano trapelare la sua « natura creativamente arguta e piena di dry humor », per usare un'espressiva frase di G. Occhialini.

I figli degli amici famigliarizzavano con lui, qualunque fosse la loro età, purché avessero raggiunto l'età della ragione. La domenica, o durante le vacanze primaverili o autunnali, faceva spesso gite in automobile, da solo o con qualche amico, aventi per meta il piú delle volte bellezze naturali ma non di rado anche opere d'arte.

G. C. Wick ricorda che durante l'anno accademico 1931-32, trascorso all'Università di Torino, era stato spesso terzo membro della commissione di esami di Fisica Matematica e Fisica Teorica con Somigliana e Persico.

Somigliana, notevole matematico ma uomo molto testardo, era talvolta difficile da ricondurre al buon senso, ma Persico riusciva ad ammansirlo con fine diplomazia. In quelle occasioni egli dava a Wick un'occhiata significativa carica di compiacimento e sollievo.

M. Ageno ricorda sempre una conversazione che ebbe luogo su di un vaporetto a Venezia nel settembre 1937 fra Fermi, Persico, Rasetti, Amaldi e altri sul significato della meccanica quantistica. Eravamo tutti a Venezia per il Congresso della Società per il Progresso delle Scienze. Il punto di vista espresso da Persico rivelò fin d'allora una comunità d'interessi e un'affinità di punti di vista fra Persico ed Ageno, che era appena laureato, che si mantenne per tutta la vita e che si rinsaldò dopo il ritorno di Persico a Roma.

Fino al 1961 Persico seguitò ad andare in montagna d'estate e a sciare d'inverno, associandosi il piú delle volte ad amici, i cui figli trascinavano con sé numerosi altri ragazzi e bambini tanto da formare spesso compagnie di oltre venti o trenta persone. Nonostante il chiasso, spesso assordante, Persico sembrava gradire queste situazioni e verso la fine della decina di giorni trascorsi fra Natale e la Befana, a Madonna di Campiglio o a Selva in Val Gardena, egli offriva un gran vassoio di paste a tutta la compagnia, la componente piú giovane della quale si abbandonava a rumorose e ripetute acclamazioni allo « zio Persico ».

Negli ultimi anni sui campi di sci andava ancora quasi dappertutto, sia pure piano piano. Qualche amico diceva scherzosamente che Persico aveva inventato lo « sci adiabatico »: guardandolo per un istante sembrava fermo e solo con due osservazioni successive, separate da un lungo intervallo di tempo, era possibile stabilire se stava salendo o scendendo il pendio.

Questo tipo di scherzi non era nuovo. Da giovane, Persico aveva una motocicletta con cui andava cosí piano che, secondo qualche amico, una volta era caduto in curva, ma dalla parte interna.

Un'altra caratteristica di Persico era l'amore per i gatti. Sia le pareti della sua casa che quelle del suo studio in Istituto erano rallegrate da numerose bellissime fotografie di gatti. In casa non aveva nessun gatto, ma egli provvedeva sistematicamente il cibo ai gatti del vicinato, che in breve tempo imparavano ad apprezzare l'esistenza di un cosí benefico individuo. C. Bernardini ricorda che, allo scopo di fornir loro il cibo, Persico aveva costruito un « orologio a trippa », come lui diceva, che faceva cadere dall'alto il mangiare per i gatti a intervalli di varie ore durante la sua assenza da casa per lavoro.

Fra i molti episodi su « Persico e i gatti » due forse vanno ricordati perché caratteristici. Una volta, da Quebec, andò a Chicago, ospite dei Fermi, e Laura Fermi, « per conservare intatta la sua reputazione di padrona di casa, si fece prestare uno dei gatti degli Urey ([17]) » ma Persico « rimase imbarazzato di questa iniziativa », forse anche perché si trattava di un gatto cosí « altolocato ».

Un'altra volta E. Amaldi, di passaggio per Copenhagen, vide in una vetrina, molto bella e moderna, un gatto dipinto su stoffa, molto originale e vivace e, consultatosi con sua moglie, concluse che fra tutte le fotografie di gatti possedute da Persico non ve ne era nessuna di quel tipo. Forse sarebbe piaciuta a Persico in quanto arricchiva la sua collezione di una nuova e diversa concezione del « gatto ». Ma quando, tornato a Roma, portò a Persico il suo regalo, questi lo ringraziò con grande cortesia ma fu súbito chiaro che quella raffigurazione del gatto lo aveva disturbato. Il gatto danese stava eretto sulle zampe posteriori, con la coda ritta e con le zampe anteriori sosteneva una lunga tromba dinnanzi alla bocca in una mossa sfacciatamente allegra e chiassosa che nulla aveva a che fare con l'espressione di tutti i gatti che guardavano dalle pareti dello studio di Persico, tutti pieni di grazia e di riservatezza, che non disturbavano nessuno con i loro lazzi e che ancor meno desideravano essere disturbati.

Persico era profondamente e risolutamente contro ogni forma di retorica, iperbole o esibizionismo non solo nei riguardi dei fatti quotidiani, ma anche negli apprezzamenti e giudizi d'insieme o, per cosí dire, di natura storica. Un esempio di questo atteggiamento è dato dalla sua bella prolusione al 37º Congresso della Società Italiana di Fisica, tenuto a Bergamo nel 1952, intitolata « Leonardo e la fisica » ([2.22a+b]). In questa egli smonta, con la sua limpida e calma razionalità, alcune esagerazioni e deformazioni dell'informazione effettivamente disponibile che vengono di solito ripetute dalla stampa dei piú diversi livelli.

Chi ha avuto il piacere di conoscere Persico a fondo è convinto che in realtà egli fosse molto di piú di quanto non appaia dai suoi lavori di ricerca e perfino dai suoi trattati.

Tutto sembra indicare che egli non fosse un solitario per natura, ma che lo fosse diventato per necessità, forse perché non era abbastanza deciso di fronte alle circostanze della vita da fare lo sforzo necessario per mutare il corso degli eventi nel suo intorno immediato.

La sua coerenza, la sua indipendenza di giudizio, la sua assoluta non disponibilità al benché minimo compromesso erano caratteristiche dominanti della sua personalità.

Politicamente era piuttosto conservatore essendo portato, come osserva G. Careri, ad un certo scetticismo di fronte a cambiamenti improvvisi, non di rado piú appariscenti che sostanziali. Gamba ricorda il suo scetticismo sui presunti benefici della nazionalizzazione dell'energia e, dopo il 1968, la sua vivace avversione per « il salto alla quaglia a sinistra » diventato di moda in molti ambienti universitari e culturali.

8. Il maestro.

Gl'interessi scientifici di Persico si estendono su di un arco di tempo di quasi quarant'anni e riguardano pro-

([17]) M. H. C. Urey ebbe nel 1934 il Premio Nobel per la Chimica per la scoperta dell'idrogeno pesante o deuterio.

blemi che vanno dall'astrofisica a varie applicazioni della meccanica quantistica, dalla conduzione dei metalli e nei gas alla fisica dei plasmi, dagli spettrometri a raggi beta alla dinamica ultrarelativistica coinvolta nella progettazione di macchine acceleratrici. Alcuni di questi lavori sono certamente importanti: qui basti ricordare quello giovanile sulla teoria cinetica dei gas ionizzati ([1.13]), l'altro sulla teoria dell'iniezione nelle macchine acceleratrici ([1.39]) o quello sulle perdite delle bottiglie magnetiche ([1.40]).

Ciò nonostante la sua caratteristica piú peculiare è stata e rimane quella del didatta e maestro. Non solo egli ha pubblicato numerosi articoli di rassegna e alta divulgazione (elenco n. 2), ma un numero di volumi, di dispense, libri e trattati (elenco n. 3) come ben pochi altri fisici hanno saputo fare in quegli stessi anni. La maggior parte di questi libri si rivolgono a studenti di livello universitario o postuniversitario, altri come *Gli atomi e la loro energia* ([3.20]) non a fisici ma a persone di formazione scientifica e tecnica diversa, come chimici, biologi e ingegneri, altri ancora sono testi per le scuole medie inferiori ([3.5]) e superiori ([3.9, 3.10]). Da tutti questi scritti appaiono evidenti le sue capacità didattiche, la sua continua preoccupazione di spianare la strada al lettore, in particolare agli studenti, per aiutare i quali giunse a tradurre un ben noto libro di Max Planck a carattere epistemologico ([3.22]) e perfino a pubblicare una *Guida all'inglese tecnico* ([3.24]), un vocabolarietto di grande utilità per gli studenti che cominciano a studiare su testi o articoli di fisica in lingua inglese.

Il suo interesse per l'insegnamento della fisica nelle scuole secondarie non solo lo spinse a scrivere i libri sopra citati e a tenere conferenze di vario argomento, ma anche a collaborare con il *Giornale di Fisica* su cui scrisse, nel periodo 1956-65, una serie di osservazioni e consigli che ancor oggi possono essere letti e meditati con grande profitto (elenco n. 4).

Chi ha avuto la fortuna di conoscerlo e frequentarlo è sempre restato colpito dalla sua straordinaria chiarezza di idee, dall'enorme numero di problemi classici (teorici e sperimentali) che conosceva profondamente, conoscenza che quasi sempre gli permetteva di suggerire a chi si rivolgeva a lui il procedimento piú opportuno per cercare, e trovare, la soluzione di problemi nuovi, anche difficili e di tipo strano. E questo era sempre vero — come osserva Carlo Bernardini — sia che si trattasse di problemi nel campo della meccanica classica o relativistica, o in quello dell'ottica o dell'elettricità.

Un'altra qualità che colpiva tutti coloro che lo avvicinavano era la sua straordinaria disponibilità. Se uno studente o un collega gli poneva un problema, Persico immediatamente interrompeva quello che stava facendo e si metteva con serietà, pazienza e impegno a cercare di aiutare chi si era rivolto a lui. Egli metteva una vera passione nello svolgimento del suo lavoro didattico, avente origine da un'altissima considerazione del problema della formazione dei giovani.

Come ricorda Giorgio Careri, all'inizio degli anni sessanta, Persico cominciò a preparare un questionario che verso la fine delle lezioni distribuiva ogni anno ai suoi studenti. Nel questionario veniva posta una serie di domande sulle sue lezioni, come, per esempio, l'interesse destato nell'allievo da ciascun argomento particolare, quali erano state le difficoltà incontrate, e infine un invito a esprimere critiche all'esposizione dei vari argomenti, alle esercitazioni e cosí via. Persico si serviva delle risposte degli studenti per cercare di migliorare il suo corso di anno in anno.

Non può quindi far meraviglia che Persico abbia chiaramente sofferto della contestazione studentesca, iniziata all'Università di Roma all'inizio del 1968. Per lui sempre cosí disponibile per tutti, in particolare per gli studenti, non era facile comprendere le ragioni di una contestazione riguardante problemi interni all'Istituto su cui non vi era stato alcun tentativo né di amichevole dialogo né di acceso dibattito.

9. La morte.

Le prime avvisaglie del male che lo avrebbe portato alla morte vari anni dopo si manifestarono al principio di agosto 1961 a Verbier nel Vallese, ove si era recato per le vacanze estive con gli Amaldi, Mario Ageno e i Conversi. I medici gli consigliarono di non andare piú in alta montagna, cosa che egli fece, anche se a malincuore. Negli anni successivi ebbe in piú occasioni periodi di qualche settimana o mese in cui la pressione arteriosa subiva sbalzi notevoli.

Cosa si dovesse o potesse fare per curare il male non fu mai ben chiaro, finché il 17 giugno 1969, dopo alcuni giorni di letto imposti da un nuovo attacco dello stesso male, concluse la sua vita limpida e appartata.

Era Socio Nazionale della'Accademia delle Scienze di Torino dal 1943 e dell'Accademia Nazionale dei Lincei dal 1952. Nell'ottobre 1969, in occasione del 55º Congresso della Società Italiana di Fisica tenuto a Bari, il Consiglio della SIF, che aveva deliberato non molto prima della sua morte di assegnargli una delle medaglie d'oro della SIF, procedette, in via eccezionale, all'assegnazione di tale medaglia alla memoria di Enrico Persico.

A parte i libri lasciati alla biblioteca dell'Istituto di Fisica dell'Università di Roma e qualche proprietà lasciata alla sua governante, Persico, nel suo testamento, nominò l'Accademia Nazionale dei Lincei suo erede universale. Questa, nel 1970, istituí la Fondazione Persico avente la finalità di assegnare borse di studio a studenti di fisica universitari o anche liceali, purché interessati alla fisica, tratte dal reddito del patrimonio Persico comprendente tre appartamenti oltre a un non grande capitale in titoli.

Ogni autunno la commissione nominata dall'Accademia per gli esami dei candidati e l'assegnazione delle Borse Persico non fa altro che contribuire a perpetuare, *post mortem*, la totale disponibilità di Enrico Persico al problema della formazione dei giovani.

ELENCO DELLE PUBBLICAZIONI DI ENRICO PERSICO (*)

1. *Note scientifiche originali.*

1. *L'effetto Hall nel bismuto solidificato nel campo magnetico* (E. Persico e L. Tieri), in *Rend. Lincei*, **30**, II, 464 (1921).
2. *Sul moto lento e quasi stazionario di un sistema rigido di cariche elettriche*, in *Nuovo Cimento*, **23**, 239 (1922).
3. *L'effetto Hall nelle lamine anisotrope e l'interpretazione di talune esperienze*, in *Rend. Lincei*, **31**, II, 500 (1922).
4. *Sul principio di equivalenza in relatività*, in *Rend. Lincei*, **31**, II, 98 (1922).

(*) Siamo grati al prof. M. Verde per l'aiuto che ci ha dato nella compilazione di questo elenco di pubblicazioni. Egli stesso è autore di un *Cenno commemorativo* di E. Persico ([18]).

([18]) M. Verde: *Enrico Persico* (1900-1970), *Cenni commemorativi*, in *Accad. Scienze, Torino*, **105**, 313 (1970-71).

5. *Sulla massa mutua di due elettroni*, in *Rend. Lincei*, **32**, II, 280 (1923).
6. *Effetto von Ettingshausen « apparente »*, in *Nuovo Cimento*, **26**, 123 (1923).
7. *Sul significato fisico della seconda forma fondamentale in relatività*, in *Rend. Lincei*, **32**, II, 208 (1923).
8. *Sulle correnti rotanti*, in *Nuovo Cimento*, **26**, 41 (1923).
9. *Sui criteri per la caratterizzazione concreta dello spazio e del tempo*, in *Rend. Lincei*, **32**, I, 524 (1923).
10. *Sul diagramma corrente oscillatoria - corrente di placca in un oscillatore a lampada* (O. M. Corbino ed E. Persico), in *Rend. Lincei*, **1**, 412 (1925).
11. *Oscillazioni secondarie in un generatore con lampade a tre elettrodi* (O. M. Corbino ed E. Persico), in *Rend. Lincei*, **1**, 538 (1925).
12. *Sull'ampiezza delle oscillazioni prodotte da una lampada a tre elettrodi*, in *Rend. Lincei*, **1**, 723 (1925).
13. *Kinetic theory of an ionized gas*, in *Month. Not. Roy. Astron. Soc.*, **86**, 93 (1926).
14. *La polarizzazione rotatoria magnetica in campo alternato*, in *Rend. Lincei*, **3**, 561 (1926).
15. *Ancora sulla polarizzazione rotatoria magnetica in campo magnetico*, in *Rend. Lincei*, **3**, 603 (1926).
16. *Effect of viscosity on a pulsating star*, in *Mont. Not. Roy. Astron. Soc.*, **86**, 98 (1926).
17. *Il principio delle adiabatiche e la nozione di forza viva nella nuova meccanica ondulatoria* (E. Fermi ed E. Persico), in *Rend. Lincei*, **4**, 452 (1926).
18. *Velocità molecolari, stati d'eccitazione e probabilità di transizione in un gas degenere*, Nota I, in *Rend. Lincei*, **7**, 137 (1928).
19. *Velocità molecolari, stati d'eccitazione e probabilità di transizione in un gas degenere*, Nota II, in *Rend. Lincei*, **7**, 235 (1928).
20. *La risonanza ottica secondo la meccanica ondulatoria*, Nota I, in *Rend. Lincei*, **8**, 55 (1928).
21. *La risonanza ottica secondo la meccanica ondulatoria*, Nota II, in *Rend. Lincei*, **8**, 160 (1928).
22. *L'effetto Hall con elettrodi estesi* (E. Persico e F. Scandone), in *Rend. Lincei*, **10**, 238, 361, 437 (1929).
23. *Sulla relazione $E = h\nu$ nella meccanica ondulatoria* in *Rend. Lincei*, **11**, 985 (1930).
24. *Un problema di meccanica ondulatoria unidimensionale*, in *Nuovo Cimento*, **9**, 284 (1932).
25. *Teoria del dispositivo a campo trasversale per lo studio dei raggi positivi e neutrali*, in *Accad. Sci. Torino*, Atti, **73**, Disp. 1ª, 161 (1937).
26*a*. *Über die Anwendung von Transversalfeldmethoden zur Messung von Ionisations- und Umladungsquerschnitten* (E. Persico e A. Rostagni), in *Ann. Physik*, **32**, **3**, 245 (1938).
26*b*. *Sui dispositivi a campo trasversale per lo studio dei raggi positivi e neutrali* (E. Persico e A. Rostagni), in *Accad. Sci. Torino*, Atti, **73**, Disp. 3ª, 363 (1938).
27. *Dimostrazione elementare del metodo di Wentzel e Brillouin*, in *Nuovo Cimento*, **15**, 133 (1938).
28. *Sulle collisioni atomiche a parametro d'urto definito*, in *Accad. Sci. Torino*, Atti, **74**, Disp. 2ª, 164 (1939).
29. *Derivation of the principle of electromagnetic equivalence from Laplace's law*, in *Rev. Univ. Nac. Tucuman*, A **1**, **1** and **2**, 63 (1940).
30. *Sugli stati elettronici legati alle irregolarità dei cristalli*, in *Ricerca Scientifica*, **11**, 419 (1940)
31. *Optimum conditions for beta-ray solenoid spectrometer*, in *Phys. Rev. Lett.*, **73**, 1475 (1948).
32. *A theory of the solenoid beta-ray spectrometer*, in *Rev. Sci. Instrum.*, **20**, 191 (1949).
33. *Il dipolo come lente magnetica divergente*, in *Rend. Lincei*, **8**, 191 (1949).
34. *Sulla disposizione dei diaframmi negli spettrometri beta elicoidali*, in *Rend. Lincei*, **10**, 344 (1951).
35. *Beta-ray spectroscopes* (E. Persico e C. Geoffrion), in *Rev. Sci. Instrum.*, **21**, 945 (1950).
36. *A new resistor network for the integration of Laplace's equation*, in *Nuovo Cimento*, **9**, 74 (1952).
37. *Le reti di resistenze come strumenti di calcolo*, in *Atti del Convegno di Venezia su « I modelli della tecnica »* (1-4 Ottobre 1955), Vol. II, 247 (1955).
38. *Le synchrotron et ses problèmes*, in *Journ. Phys. et Rad.*, **16**, 360 (1955).
39. *A theory of the capture in high injected synchrotron*, in *Suppl. Nuovo Cimento*, **2**, 459 (1955).
40. *Plasma loss from magnetic bottles* (E. Persico e J. G. Linhart), in *Nuovo Cimento*, **8**, 740 (1958).
41. *Operation at* 1000 MeV *of the Frascati electrosynchrotron* (A. Alberigi Quaranta, F. Amman, C. Bernardini, U. Bizzarri, G. Bologna, G. Corazza, G. Diambrini, G. Ghigo, A. Massarotti, G. P. Murtas, M. Puglisi, I. F. Quercia, R. Querzoli, G. Sacerdoti, G. Salvini, G. Sanna, P. G. Sona, R. Toschi, A. Turrin, M. Ageno e E. Persico), in *Nuovo Cimento*, **11**, 311 (1959).
42. *Sulla situazione dei lavori per l'elettrosincrotrone da* 1200 MeV *e sul programma di ricerche* (M. Ageno, A. Alberigi Quaranta, F. Amman, C. Bernardini, U. Bizzarri, G. Bologna, G. Corazza, G. Cortellessa, G. Diambrini, G. Ghigo, A. Massarotti, G. Moneti, G. P. Murtas, E. Persico, M. Puglisi, I. F. Quercia, R. Querzoli, G. Sacerdote, G. Salvini, G. Sanna, R. Toschi e A. Turrin), in *Suppl. Nuovo Cimento*, **11**, 324 (1959).
43. Parte I: *Descrizione generale. L'elettrosincrotrone* (M. Ageno, A. Alberigi Quaranta, F. Amman, C. Bernardini, U. Bizzarri, G. Bologna, G. Corazza, G. Diambrini, G. Ghigo, R. Habel, C. Infante, A. Massarotti, G. Moneti, G. P. Murtas, E. Persico, M. Puglisi, I. F. Quercia, R. Querzoli, G. Sacerdoti, G. Salvini, G. Sanna, S. Sircana, P. G. Sona, R. Toschi e A. Turrin), in *Suppl. Nuovo Cimento*, **24**, 17 (1962).
44. Parte II: *Il progetto teorico* (C. Bernardini, E. Persico, P. G. Sona e A. Turrin), in *Suppl. Nuovo Cimento*, **24**, 64 (1962).
45. Parte VII: *Comandi e controlli* (M. Ageno, A. Alberigi Quaranta, F. Amman, C. Bernardini, U. Bizzarri, G. Bologna, G. Corazza, G. Diambrini, G. Ghigo, R. Habel, C. Infante, A. Massarotti, G. Moneti, G. P. Murtas, E. Persico, M. Puglisi, I. F. Quercia, R. Querzoli, G. Sacerdoti, G. Salvini, G. Sanna, S. Sircana, P. G. Sona, R. Toschi e A. Turrin): in *Suppl.*

Nuovo Cimento, **24**, 312 (1962).

46. Parte VIII: *Il funzionamento dell'elettrosincrotrone* (M. Ageno, A. Alberigi Quaranta, F. Amman, C. Bernardini, U. Bizzarri, G. Bologna, G. Corazza, G. Diambrini, G. Ghigo, R. Habel, C. Infante, A. Massarotti, G. Moneti, G. P. Murtas, E. Persico, M. Puglisi, I. F. Quercia, R. Querzoli, G. Sacerdoti, G. Salvini, G. Sanna, S. Sircana, P. G. Sona, R. Toschi e A. Turrin), in *Suppl. Nuovo Cimento*, **24**, 334 (1962).
47. *Sviluppi e risultati della meccanica ondulatoria*, p. 452 degli *Atti del Convegno Lagrangiano*, tenuto presso l'Accademia delle Scienze di Torino (22-25 Ottobre 1963) (Torino, 1963).

2. *Conferenze, articoli di rassegna e alta divulgazione.*

1. *I principi della teoria dei quanti*, in *Elettrotecnica* **10**, 432 (1923).
2. *Fenomeni galvano e termo-magnetici*, in *Elettrotecnica*, fascicolo speciale nel 1° centenario della morte di A. Volta, Uniel, Roma, 1927.
3. *Meccanica ondulatoria*, in *Elettrotecnica*, **15**, 197 (1928).
4. *La teoria del magnetismo*, in *Conferenze di Fisica e Matematica della R. Univ. e della R. Scuola Ing., Torino*, **1**, 21 (1931).
5. *Prolusione al corso di Fisica Teorica*, in *Conferenze di Fisica e Matematica della R. Univ. e della R. Scuola Ing., Torino*, **2**, 3 (1932).
6. *Geofisica e Cosmogonia*, in *Conferenze di Fisica e Matematica della R. Univ. e della R. Scuola Ing., Torino*, **2**, 77 (1932).
7. *Il vuoto*, in *Conferenze di Fisica e Matematica della R. Univ. e della R. Scuola Ing., Torino*, **4**, 51 (1934).
8. *Questioni di aggiornamento nella fisica atomica*, in *Conferenze di Fisica e Matematica della R. Univ. e della R. Scuola Ing., Torino*, **4**, 173 (1934).
9. *I raggi molecolari*, in *Nuovo Cimento*, **11**, 118 (1934).
10. *Questioni di matematica applicata*, trattate nel II Congresso di Matematica Applicata, Roma, 1939, da E. Persico, G. C. Wick, E. Pistolesi, A. Eula, A. Ghizzetti e W. Gröbner, Zanichelli, Bologna (1939).

11*a*. *Une fenètre ouvert sur le monde atomique*, in *Scientia*, gennaio 1965.

11*b*. *Una finestra sul mondo atomico: la camera di Wilson*, in *Nuovo Cimento*, **11**, 725 (1934).

12. *Le statistiche di Bose-Einstein e di Fermi*, in *Conferenze di Fisica e Matematica della R. Univ. e della R. Scuola Ing., Torino*, **5**, 89 (1937).
13. *Marconi e la scienza*, in *Conferenze di Fisica e Matematica della R. Univ. e della R. Scuola Ing., Torino*, **6**, 105 (1938).
14. *Sulla teoria cinetica del gas di neutroni*, in *Conferenze di Fisica e Matematica della R. Univ. e della R. Scuola Ing., Torino*, tenute negli anni 1936-37 e 1937-38, p. 29.
15. *Osservazioni statistiche sulla fisica contemporanea*, in *Saggiatore*, n. 4, 178 (1940).
16. *Collisioni atomiche*, in *Saggiatore*, n. 3, 65 (1941).
17. *L'esperienza mentale nel metodo galileiano*, in *Scienza e Tecnica*, n. 12, 871 (1941).
18. *L'idea di probabilità nella fisica classica*, in *Rend. Seminario Mat. e Fis. dell'Univ. e Politec. Torino*, **7**, 25 (1941).
19. *Galileo e la fisica*, in *Nel III centenario della morte di Galileo Galilei*, Pubblic. dell'Università Cattolica del S. Cuore, Serie quinta. *Scienze Storiche*, Vol. XX, Milano (1942).
20. *Il nuovo fuoco*, discorso inaugurale tenuto in occasione dell'inizio dell'anno accademico 1945-46, Università di Torino.
21. *Analisi del determinismo fisico*, p. 25 di *Fondamenti logici della Scienza*, Francesco de Silva, Torino (1947).

22*a*. *Léonard et la physique*, in *Scientia*, dicembre 1952.

22*b*. *Leonardo e la Fisica*, in *Suppl. Nuovo Cimento*, **10**, 201 (1953).

23. *Souvenir de Enrico Fermi*, in *Scientia*, ottobre 1955.
24. *L'ottica elettronica*, in *Giornale di Fisica*, **1**, n. 1 (1956).
25. *Che cos'è che non va?*, in *Giornale di Fisica*, **1**, n. 1 (1956).
26. *I fondamenti della fisica nucleare*, in occasione della 3ª Rassegna di Elettronica Nucleare, Roma, EUR, 2 Luglio 1956.
27. *Che cosa non va (e che possiamo farci)*, in *Giornale di Fisica*, **1**, n. 2 (1957).
28. *Il valore educativo della Fisica*, in *Giornale di Fisica*, **1**, n. 4 (1957).
29. *Come presentare la fisica a chi non dovrà usarla*, in *Giornale di Fisica*, **4**, n. 4 (1965).
30. *Le vie della certezza in fisica*, a p. 265 di *Conferenze di Fisica*, Feltrinelli, Milano (1963).
31. *La cultura in Fisica*, in *Giornale di Fisica*, **4**, n. 3 (1963).
32. *Voci dell'Enciclopedia Italiana: Materia, Quanti, Quantistica Meccanica.*

3. *Dispense di corsi universitari, libri e trattati.*

1. *Lezioni di calcolo differenziale assoluto*, di T. Levi-Civita, redatte da E. Persico, Stock, Roma (1925).
2. *The absolute differential calculus*, by T. Levi-Civita, edited by Dr. E. Persico, authorized translation by Miss M. Long, Blackie & Son Ltd., London and Glasgow (1927).
3. *Lezioni di meccanica razionale*, redatte per uso degli studenti, a.a. 1928-29. Dispense dell'Università di Firenze.

4*a*. *Lezioni di meccanica ondulatoria*, redatte dai Dr. B. Rossi e G. Racah, Tipo-Lito Filippi, Firenze (1929).

4*b*. *Lezioni di meccanica ondulatoria*, redatte dai Dr. B. Rossi e G. Racah, seconda edizione migliorata ed accresciuta, pubblicata sotto gli auspici dell'Istituto di Fisica di Roma, CEDAM, Padova (1929-30, 1935).

5. *Elementi di Fisica e Chimica ad uso delle scuole di avviamento professionale*, Zanichelli, Bologna (1931 e 1942).

6. *Ottica*, Vallardi, Milano (1932).
7. *Lezioni di meccanica razionale*, redatte per uso degli studenti, a.a. 1933-34. Dispense dell'Università di Torino.
8. *Lezioni di fisica matematica*, a.a. 1936-37. Dispense dell'Università di Torino.
9. *Fisica ad uso degli Istituti Magistrali*, Zanichelli, Bologna (1937).
10. *Fisica per le scuole medie e superiori* (E. Fermi ed E. Persico), Zanichelli, Bologna (1938).
11. *Introduzione alla fisica matematica*, redatta da T. Zeuli, Zanichelli, Bologna (1936), di cui sono state pubblicate nuove edizioni nel 1941, 1947 e 1952.
12. *Fondamenti di meccanica atomica*, Zanichelli, Bologna (1939), di cui sono state pubblicate varie successive edizioni.
13. *Lineamenti della struttura della materia. Sguardo d'insieme e sviluppo storico*, V. Giorgi, Torino (1944).
14. *Fundamentals of Quantum Mechanics*, translated by G. M. Temmer, Prentice Hall, New York (1950).
15. *Lezioni di ottica elettronica*, raccolte da L. Tau, Tipo-Litografia Marves, Roma (1950-51).
16. *Metodi di analisi delle radiazioni*, dispense dell'Università di Roma, redatte da R. Bizzarri, La Goliardica, Roma (1951-52).
17. *Lezioni sugli acceleratori di particelle*, raccolte da G. Caglioti, N. D'Angelo e A. Reale, Tipo-Litogafia Conti e Prioda, Roma (1954).
18. *Le radiazioni*, dispense dell'Università di Roma, redatte da R. Bizzarri, La Goliardica, Roma (1957).
19. *Lezioni di fisica del reattore*, dispense dell'Università di Roma, redatte da G. Riguzzi, La Goliardica, Roma (1958 e 1960).
20. *Gli atomi e la loro energia*, Zanichelli, Bologna (1959).
21. *Corso di fisica teorica*, appunti per le lezioni, raccolti da R. Rosei e P. Rotoloni, Ilardi, Roma (1960-61).
22. *Max Planck: La conoscenza del mondo fisico*, traduzione dal tedesco di E. Persico e A. Gamba di *Wege zur Physikalischen Erkenntis* (1908-1933) e *Wissenschaftliche Selbstbiographie* (1936-1947), Boringhieri, Torino (1964).
23. *Principles of particles accelerators*, in collaborazione con E. Ferrari e S. E. Segre, Benjamin, New York (1968).
24. *Guida alla lettura dell'inglese tecnico*, Zanichelli, Bologna (1965).

4. *Osservazioni e consigli di natura didattica.*

su: *Il Giornale di Fisica*

1. *Esperienze di lezione*, **1**, n. 1 (1956).
2. *Struttura della materia*, **1**, n. 2 (1957).
3. *Ottica*, **1**, n. 3 (1957).
4. *Relatività*, **1**, n. 4 (1957).
5. *Elettricità e magnetismo*, **2**, n. 1 (1958).
6. *Storia della fisica*, **3**, n. 1 (1962).
7. *Fisica nucleare*, **3**, n. 2 (1962).
8. *Sullo spettro continuo di raggi* X, **4**, n. 1 (1963).

In memory of Enrico Persico (August 9, 1900 - June 17, 1969)(*)

E. Amaldi and F. Rasetti

A long time has passed since Enrico Persico died of heart disease in his home in Rome on June 17, 1969. At the time, both the Accademia Nazionale dei Lincei and the Faculty of Science of the University of Rome felt obliged to respect his desire, expressly set down in his will, not to be commemorated.

Time has passed, however, and a very simple written commemoration is now due Enrico Persico for his contribution to research and university teaching and for his qualities as a human being. Failure to gather and order the most important facts of his life and work would mean leaving a serious gap in the history of Italian physics in the period between 1925 and 1970.(1)

1. Childhood and youth

Enrico Persico was born in Rome on August 9, 1900, the only son of Gennaro Persico and Rosa Massaruti. His father, of Neapolitan origin like his mother, worked as a cashier at the Banca d'Italia and died in late 1919.

Enrico went to the classical liceo Umberto, where he was a classmate of Giulio Fermi, brother of Enrico, who was one year younger and therefore a year behind them at school. In the winter of 1915, Giulio Fermi died following a minor operation that all had considered of little importance. That was when the friendship between his brother Enrico and Enrico Persico began.(2) The two young boys, Persico 15 and Fermi 14, shared a strong interest in the sciences, as attested to by the half page (2-23) in which Persico remembers "long walks from one side of Rome to the other, talking about all kinds of things", but above all problems of mathematics and physics. This common interest led the two young men to decide to study physics when they graduated, the one in July 1917, the other the next year, as they both skipped the last year of liceo.

Thus, Persico registered for first year physics at the University of Rome, where the faculty

(*)The Accademia Nazionale dei Lincei has kindly authorized publication in *Giornale di Fisica* of this biography, prepared for the *Celebrazioni Lincee*, **115** (1979).

(1) This is the last version of the "Ricordo di Enrico Persico". The previous version, completed during the summer of 1977, was sent to the following list of Persico's friends and acquaintances with the request to send back suggestions for corrections or additions. All answered either verbally or in writing (l). The letters sent on that occasion have been deposited at the Accademia dei Lincei as they represent an important testimonial to Persico's life and works: M. Ageno, C. Bernardini (l), G. Bernardini (l), R. Bizzarri, B. Brunelli (l), G. Careri, M. Cini, M. Conversi, E. Denina (l), B. De Tollis, L. Fermi (l), E. Ferrari, C. Guazzaroni Ferrari (l), A. Gamba (l), M. Magistrelli (l), G. Occhialini (l), L. Radicati (l), B. Rossi, A. Rostagni (l), G. Salvini (l), E. Segrè (l), S. Segre, F. Tricomi (l), M. Verde, G. C. Wick (l), T. Zeuli. We express our most heartfelt thanks to all these friends and colleagues.

(2) E. FERMI: *Note e Memorie*, Accademia Nazionale dei Lincei e The University of Chicago Press, Vol. **1** (Italy, 1921-1933), Rome (1961). See the English edition by the same publishers and in the same year: E. FERMI: *Collected papers*.

included physicist Orso Mario Corbino, physicist-mathematician Vito Volterra and--most recent addition in 1919--mathematician Tullio Levi-Civita.

In the fall of 1918, after taking and passing the examination for admission to the Scuola Normale, Fermi moved to Pisa. There ensued a rather systematic correspondence between the two friends, much of which (Fermi's letters to Persico) was preserved by Persico and later made available to Emilio Segrè for his biography of Enrico Fermi, in which it appears as Appendix I.([3]) These letters are quite useful in pointing out the interests of the two young men and their development through university, even though the lack of Persico's letters of response create a gap. Fermi's first two letters to Persico (one in 1917, the other in 1918) were written before he left for Pisa. The last letter was written in September 1926, that is, on the eve of the competitive examination for the chair of theoretical physics at the University of Rome.

On November 22, 1921, Persico received his degree with a thesis supervised by O. M. Corbino on the Hall effect, a subject on which he later published another paper together with L. Tieri.

Upon graduation, Persico was immediately given the position of assistant to Alfonso Di Legge, director of the Astronomical Observatory and in 1922 became assistant lecturer at the Physics Institute of the University of Rome under O. M. Corbino. There he remained until 1927, with the exception of approximately one year (1925) at Cambridge University (England), during which time he met A. S. Eddington and P. A. M. Dirac.

Corbino's other assistants at the time were Laureto Tieri (1879-1952), Nella Mortara (born 1893), Aldo Pontremoli (1896-1928) and Giulio Bisconcini (1880-1969). Nella Mortara, an extraordinary sportswoman, with simple and straighforward ways, was in charge of first and second year physics and engineering lab courses, a job she continued to carry out until 1948, long after the Institute had moved to the new univerity campus (Città Universitaria) (1936).

Students (both male and female, but above all female) were swept along by Mortara's enthusiasm for skiing in the winter, rowing on the Tiber in spring and autumn and swimming at the seaside near Rome in all seasons.

Thus she became a characteristic figure of the Physics Institute, known to all as Zia Nella (Aunt Nella). In the twenties, Persico fell in love with her and proposed to her, but she turned him down, saying she was not made for marriage. They remained good friends throughout their lives, but this episode was to be of considerable importance in Persico's life.

From 1922 to 1926, in addition to his work as assistant to O. M. Corbino, Persico also worked with Vito Volterra and Tullio Levi-Civita, whose classic course on *Calcolo differenziale assoluto* (Absolute differential calculus) (3-1) he compiled and adapted for publication (1925). The volume raised so much interest that an English version was prepared shortly afterward at the urgings of mathematician E. T. Whittaker (1927)(3-2).

In those years, Persico and Fermi, and later also other young Roman physicists frequented

([3]) E. SEGRE': *Enrico Fermi, fisico,* Bologna (1971). See also the English edition: *Enrico Fermi, physicist,* Chicago (1970).

the homes of mathematicians Tullio Levi-Civita, Federigo Enriques and Guido Castelnuovo. In particular, the home of the latter was open with great simplicity and cordiality every Saturday evening to colleagues and young mathematicians and physicists living in or passing through Rome.(4)

In 1924, Persico became university teacher and in 1925 sat the first competitive examination in Italy for the chair of theoretical physics at the University of Rome. The short list was composed of Enrico Fermi, first, Enrico Persico, second, and Aldo Pontremoli, third, who were called to the universities of Rome, Florence and Milan, respectively.

Works worthy of mention from this early period (1-1, 1-17) are two experimental studies of the Hall effect, one carried out with O. M. Corbino, the other with L. Tieri; a study on the Ettingshausen effect, as well as studies on the functioning of the triode, the theory of relativity, the kinetic theory of highly ionized gases and cepheid oscillations. The last two originate in his interest in astrophysics which arose during the time he spent as assistant at the Rome Astronomical Observatory and intensified after his contacts with astrophysicists at Cambridge.

Particularly interesting works from this period are two notes on magnetic rotating polarization in an alternating magnetic field (1-14, 1-15), which represent the final word on the complex problem; a paper written in collaboration with E. Fermi (1-17) on the principle of adiabatics and the notion of vis viva in wave mechanics; and one on the kinetic theory of ionized gases (1-13). The note written with Fermi in November 1926 reveals the effort made by the two friends to comprehend Schrödinger's ideas on wave mechanics published a short time earlier in *Annalen der Physik*. In this note, based on the analogy between mechanics of a mass point and optics, they work out certain expressions for kinetic and potential energy, without however specifying their physical significance. When the probabilistic interpretation of wave mechanics was developed only a few months later, these expressions turned out to be the expectation values of those parameters.

Persico's most important paper from this period is probably the one related to his work on the kinetic theory of ionized gases carried out in Eddington's wake and often mentioned alongside those of the pioneers in plasma physics.(5)

In this paper, Persico replaces the Coulomb potential of ions with a Debye-Hückel potential which is essentially Coulombian for distances shorter than the Debye distance, while it decreases exponentially for longer distances.

Persico was always a scrupulously honest person, incapable of making the slightest concession or compromise that would deviate him from his strict ethical and intellectual code. From youth, he had clear and firm ideas on Fascism, which was alien and unacceptable to him, not only for its political content, but also for the rhetoric, self-aggrandizement and triumphalism that characterized it.

(4) L. FERMI: *Atoms in the family*, Chicago (1954). See also the Italian translation: *Atomi in famiglia*, Milano (1954).

(5) See, for example, L. OYSTER: *Emission, absorption and conductivity of a fully ionized gas at radio frequency*, in *Rev. Mof. Phys.*, **33**, 525 (1961); S. CHAPMAN and T. G. COWLING: *The Mathematical Theory of Nonuniform Gases*, Cambridge (1960).

His reserve and unwillingness to speak about himself and the caution with which he would express his views could easily have been taken for shyness. But although he may indeed have been a bit timid, this attitude was mainly dictated by his desire not to utter, even momentarily or occasionally, hasty judgements or superficial opinions.

2. The period in Florence (1927-31)

When Persico moved to Florence in early 1927, Franco Rasetti, who had been there since 1922 first as an assistant lecturer (*assistente*) and later as associate lecturer (*aiuto*) to A. Garbasso, was transferred to Rome, where he took the position of associate to O.M. Corbino left open by Persico. Persico and Rasetti did not really know each other, although they had both been informed of the other's interests and personality by their mutual friend Enrico Fermi. Given the occasion to see a lot of each other in this period, they soon became good friends.

After his arrival in Florence, Persico dedicated much time to the teaching of quantum mechanics, a new subject at the time, thereby greatly influencing the young physicists at the Physics Institute. Most of these had been drawn to the Institute by its director Antonio Garbasso (1871-1933), who played the same role in Florence as O. M. Corbino (1876-1937) played in Rome and the rest of Italy.

Among these young students were Bruno Rossi, Gilberto Bernardini, Giuseppe Occhialini and Giulio Racah and, later, Daria Bocciarelli and Lorenzo Emo Capodilista.

Persico immediately struck up easy and friendly relations with all of them. As Occhialini recalls, Persico had a small car (a Fiat Balilla) and often gave them a lift from the city center to the Physics Institute in Arcetri or took them skiing in Vallombrosa or Mount Abetone on Sundays.

Daria Bocciarelli still remembers the simple but particularly effective way in which Persico organized seminars at Arcetri. Either Persico personally or one of the students would in turn report on an article that had recently appeared in some scientific journal. The discussion that ensued was always an extraordinary opportunity for a thorough critical examination of the subject at hand.

Persico's influence on the teaching of the new quantum mechanics soon spread beyond Florence to the rest of Italy by means of his well known volume containing the texts of a course of lectures on wave mechanics, *Lezioni di meccanica ondulatoria* (3-4), used for years in most Italian universities. We will return to this matter shortly.

The group of young physicists that had formed around Enrico Fermi at the Physics Institute in Rome at that time with the help of Franco Rasetti and under the attentive and far-sighted protection of the Institute's director, O. M. Corbino, had the habit of jokingly giving the members of the Institute ecclesiastical nicknames ([2, 3]). Thus Corbino was called the Eternal Father, Fermi the Pope, Rasetti the Cardinal Vicar, Segrè and Amaldi the abbots and so on. And Enrico Persico, who had gone to spread the teaching of modern theoretical physics to other universities was known as the Prefect of Propaganda Fide. Persico, in keeping with this joke, sent his friends the following poem, probably some time in 1929:

Father Enrico the missionary

Went out among the infidels
To a savage island
To explain them the gospel.

The great chief of the cannibals
When he saw him appear
Brought out his biggest cauldron
And set it on the fire.

The paths on his territory
He had fortified
And the hunt for the white man
He ordered on sea and land.

But the good father nevertheless succeeded
In setting up his camp
And the priest of the black men
He converted to his faith.

Overcome with indignation
The anthropofagous lord
Marked him in his black book
For three thousand sacks of gold.

But the Ambrosian chapter
Rewarded instead his zeal
And invited him to preach
To the believers there.

At the news, the converts
Wept hot tears
And the good father, deeply moved,
Did not want to abandon them.

But this year, the black sire
Has moved his realm and throne
To a nearby island
Conquered from the scorpions.

In the land he has abandoned
Faith now reigns everywhere
And missionary fathers
Are no longer needed there.

That is why Father Enrico
Is now about to leave
to cross the Appennines,
and make other convertees.

The chief of the cannibals was Prof. Vasco Ronchi, a well known optician, whose interests were directed at instrumental and physiological optics rather than the nascent wave mechanics. The priest of the blacks was Gilberto Bernardini, who had returned to his city of birth after having received his degree from the University of Pisa and finished his specialization at the Scuola Normale Superiore in Pisa.

To make enough money to live, Bernardini had for some time worked for a small company (Officina Cipriani e Baccani in Rifredi) which built Navy telescopes while following Persico's lectures on the side, despite his employers' protests. Realizing his value, Persico soon asked him to be his assistant and to direct lab procedure for him. Bernardini resigned from Cipriani e Baccani, and became a special assistant (*assistente straordinario in soprannumero*) to Persico. But since the salary offered (270 lire per month) was not enough to support himself and his young wife, he took on, with Persico's approval, a job as supply teacher at the lower secondary school. A few months later, however, when Ronchi offered him a job as assistant lecturer at the Optics Institute,(6) Bernardini accepted in the hopes of being able to continue to be involved in the activities of the Physics Institute in Arcetri, which now included Bruno Rossi. But Ronchi was hostile to Bernardini's interests which he felt distracted him from optics and tried to prevent him from going to listen to Persico's lectures, to the point that he sometimes locked him in his laboratory. The following year, Bernardini was named assistant lecturer to Garbasso and was thus able to become an integral part of the new group that was forming in Arcetri.

"The nearby island conquered from the scorpions" mentioned in Persico's poem was an abandoned building not far from the Physics Institute in Arcetri to which the Optics Institute had been transferred.(6) Subsequently enlarged, it still houses the Ministry of Education's Istituto Nazionale di Ottica.

---

(6) In 1927, V. Ronchi transformed the Laboratory of Optics and Precision Mechanics of the Physics Institute in Arcetri into the Institute of Optics. The former institute, had been set up in 1917 under joint pressure from the armed forces and various industries and had been directed for years by R. Occhialini. In 1919, after Lo Surdo's transfer to Rome, the Laboratory of Optics and Precision Mechanics moved to a six-room building originally constructed by Lo Surdo to house the Institute for Terrestrial Physics. During the thirties, Ronchi undertook successive enlargements of the Optics Laboratory until it consisted of approximately 40 rooms. We thank F. T. Arecchi, the current director of the Istituto Nazionale di Ottica, for his letter on this subject and for the bibliographic information provided: V. RONCHI: *Una istituzione caratteristica: l'Istituto Nazionale di Ottica di Arcetri*, in *Dialogues*, July-September 1964. See also by the same author, *Atti della Fondazione Giorgio Ronchi*, **11**, no. 6 (1976); no. 1, 2, 3, etc. (1977).

Finally, the poem mentions the offer made to Persico of the chair of Theoretical Physics at the University of Milan after the death of Aldo Pontremoli, who died on the ice pack while participating as a physicist in Nobile's second expedition to the North Pole in 1928.(7)

The young Florentine physicists felt that Persico's departure would have had serious consequences on the scientific and didactic activity that was developing in such a promising way in Florence. Thus, Occhialini remembers that during an excursion to Larderello, organized as part of a conference in Florence, they implemented a plan worked out by Racah to prevent Persico from talking to Pugno Vanoni, professor of electrotechnology at the University of Milan, who wanted to speak to Persico to convince him to take the position. The plan called for them to do everything in their power to keep the two from talking, including interrupting them and never leaving them alone.

Under the friendly but firm insistence of the young Florentine physicists, Persico turned down the offer, but when he received a similar offer from the University of Turin only a short time later, he accepted, leaving Florence in 1930.

While in Florence, Persico's attention was focussed on problems of molecular statistics, and applications of wave mechanics and the Hall effect (1-18, 1-22), a subject he had already investigated during his period in Rome.

3. Turin (1930-47)

Persico's move from the University of Florence to the University of Turin was partly a result of the fact that Italian universities were divided into two categories at that time: those in category A were completely state-financed, while those in category B were of a local nature, although they did receive considerable financing from the state. The University of Turin belonged to category A, while Florence belonged to category B, and this gave Persico a slight feeling of instability.

Persico's friends in Rome were unhappy to hear that he was leaving Florence for Turin. They felt it was difficult that a group of young people including both experimental and theoretical physicists of the quality of the one that had developed and was developing at Arcetri could be formed in the new location.

Persico nevertheless managed to contribute to the formation of an important school of modern theoretical physics in Turin, too. Gian Carlo Wick, who received his degree under Carlo Somigliana (1860-1955), only had rather superficial contacts with Persico during the 1931-32 academic year.(8) Yet he remembers having gone skiing a number of times with Persico and Maria Daviso di Charvensod, a science teacher at the Massimo d'Azeglio liceo.

Of the many young people who were influenced by Persico, we can name Nicolò Dallaporta,

(7) During the return from the second dirigible expedition to the North Pole, organized and headed by U. Nobile (in the first 1926 Amudsen expedition, Nobile had piloted the aircraft), the dirigible "Italia" gained too much weight too quickly as a result of ice formation and crashed, leaving almost the entire crew on the ice pack. Eight people died during the tragic events that followed, among them was Aldo Pontremoli. Six more, including R. Amudsen, were killed while attempting to save the survivors.

(8) After his degree, G. C. Wick received a scholarship and spent almost the entire academic year 1930-31 in Germany, first in Göttingen under Max Born and later in Leipzig under Werner Heisenberg. On his return to Italy, he spent the academic years 1931-32 and 1932-33 in Turin, before moving to Rome in 1933 when he was named assistant to E. Fermi.

Luigi Radicati di Brozolo, Marcello Cini, Augusto Gamba and Tino Zeuli.

Dallaporta, who had followed O. Specchia from Bologna to Catania, accepted the job of assistant to A. Pochettino in 1938 for the precise purpose of studying under Persico, who guided him towards study of the collision of low energy neutral atoms and ions in connection with experiments by A. Rostagni.

Radicati remembers that "Persico's wonderful course in mathematical physics was his first real encounter with physics, a real revelation, . . . " and that later, after the war, even though Persico was skeptical about the possibilities of finding a job in physics in Italy, he helped him return to the field.

Shortly after his arrival in Turin, Enrico Persico sent his friends in Rome another poem:

Beyond the Appennines
Father Enrico has arrived
And is now guiding others
Along the road to faith.

The lecture begins
With the elementary rule
That it is not polite
To eat human flesh.

He then recounts that beyond the hills
Live pious people who
have received the sacred sources of the truth
Directly from heaven.

For them, the *h* is a sacred cult
They have complete faith in quantums
And for them it is a grave insult
To say that the atom does not exist.

It is equally blasphemous
To negate the existence of *psi*
Or the non null value of
Δq xΔp,

Or that the spin of orbits and moments
Must be added together
And equivalent electrons

Are forbidden.

They also believe, with a profound faith
To which reason bows
That light is both body and wave
Body and wave is the electron.

These are the holy dogmas
That he teaches to the infidels
With edifying examples
Taken from the Gospel.

This poem mentions the Gospel. In slang at the time, the Gospel was the volume containing the course of lectures on quantum mechanics that Persico gave when he first arrived in Florence and which had been compiled into a volume by Bruno Rossi and Giulio Racah. The first edition (3-4a) was printed so poorly that it was almost illegible and became known among Roman physicists as the "Copt Gospel". It was nevertheless the best presentation that existed at that time and in any language of the fundamental ideas of quantum mechanics and its applications. Corbino, who took considerable interest in scientific and human events, both large and small, had a new edition printed at the Institute's expense (3-4b) and Persico gave a copy to the young physicists in Rome with the following dedication:

Coptum hoc evangelicum
Italicis tipis transcriptum
Auctumque et nonnullis locis emendatum
Patris Aeterni munificientia editum
Doctis Abbatibus
In limine solii pontificii assidentibus
Offerit dicatque
Enricus Persicus de Propaganda Fide Cardinalis.

Persico often travelled from Turin to Rome, where he was always warmly greeted by Corbino, Fermi and the entire group of Roman physicists. Fermi was always particularly happy to see his old friend and would spend hours telling him what he and others were doing at the Institute. A scientific conversation with Persico was always pleasant and worthwhile, given his interest in all scientific problems and the lucidity of his ideas, which were as manifest in his observations and questions as they were in his scientific works.

It was during one of these visits to the Institute in Rome that Persico witnessed the discovery of the phenomenon of neutron slowdown, attested to by the fact that the results of the readings of the counters used by Fermi to measure the activity induced by the neutrons in a small silver cylinder in the presence or absence of paraffin are jotted down in the lab measurements notebook in Persico's handwriting.(9)

In Turin, Persico renewed his friendship with F. G. Tricomi and Gino Castelnuovo, whom he had known in Rome. Castelnuovo, an engineer and the son of mathematician Guido Castelnuovo, was living in Turin because he had become an executive of the RAI, the Italian Broadcasting Corporation. Tricomi, who had met Persico in 1922, when he was assistant lecturer in mathematics to Francesco Severi at the University of Rome, had become professor of calculus at the University of Turin and had been decisive in bringing Persico to Turin, overcoming Somigliana's hostility (not to the person, but to the subject, that is theoretical physics).

Persico became part of a group made up mainly of assistant lecturers such as E. Denina, later professor of Chemistry for Physics and Electrochemistry at the Turin Polytechnic; R. Margaria, later professor of physiology at the University of Milan; O. M. Olivo, later professor of anatomy at the University of Bologna; and A. Rostagni, later professor of general physics at the University of Padua. This group, which also included architect D. Morelli, frequently went for walks in the mountains. Alessandro and Benvenuto Terracini and the daughter of the latter, Eva, sometimes came along on these hikes.

Persico also used to go on trips in his car with Rostagni, assistant at the time to A. Pochettino, professor of experimental physics at the University and involved in experiments on neutral rays. Enrico and Laura Fermi and her sister, Paola Capon, took part in one of these trips to the lakes in northern Italy around Easter 1934. Franco Rasetti came along on a car trip to Switzerland in the summer of the same year. Persico also toured with Denina, either in car or on foot, and sometimes went on summer vacation with him, mainly in the Alpine region. Persico loved the mountains and continued to go walking until 1961, when high blood pressure started to become a problem.

In addition to several papers on various applications of quantum mechanics and on neutral rays (partly in collaboration with Rostagni (1-25, 1-26)), Persico published a number of books in those years which were to have a strong influence on Italian scientific culture. It is enough to recall *L'Ottica*, brought out by Vallardi in 1932(3-6), a total review of a part of the physics text by Battelli and Cardani, in which Persico was aided by Bruno Rossi, as mentioned in the preface; *L'Introduzione alla Fisica Matematica*, edited by T. Zeuli, and published in 1936 by Zanichelli (3-11) and *I Fondamenti della Meccanica Ondulatoria*, also published by Zanichelli in 1939 (3-12). The latter was basically a revision of the course of lectures that Persico had given in Florence. It was of great importance in that it opened up the study of wave mechanics to young people who spoke only Italian. But the text was of such excellent quality that, besides various later editions in Italian, an expanded English version was also put out by Prentice Hall, New York, in 1950: *Fundamentals of*

(9) This episode is described on p. 642 of the reference at footnote 2 and on p. 83 of the reference at footnote 3 (p. 80 of the English text) and, in greater detail, by E. AMALDI: *Personal notes on neutron work . . .* , p. 294 of the SIF's LVII Corso della Scuola Internazionale di Fisica "Enrico Fermi", Varenna, July 31 - August 12, 1972; edited by C. WEINER, Academic Press, New York (1977).

Quantum Mechanics (3-14). This text, which was translated by G. M. Temmer, was widely used in both the United States and various other countries.

The *Fondamenti della Meccanica Ondulatoria* was part of a planned *Trattato di Fisica* funded by the CNR,(10) of which only two other volumes appeared however: *Molecole e Cristalli* by E. Fermi(11) e *Fisica Nucleare* by F. Rasetti.(12) The fourth volume, *Spettroscopia atomica* by E. Segrè never got beyond the manuscript stage because of the racial laws.(13) The author of each volume discussed the contents with Fermi, who was the coordinator of the entire work and therefore exerted considerable initial influence on its characteristics and scope.

In his later years, Persico often spoke of a redraft of the *Fondamenti della Meccanica Ondulatoria* on which he had already begun work. The idea was mainly to add a chapter on collision problems, the importance of which had greatly increased with respect to the thirties. Yet, no evident trace of this work has been found among his papers.

It should also be pointed out that of the physicists of his generation, Persico was the one who showed the most evident interest in epistemology, as is attested to by various publications, in particular, the article *Analisi del determinismo fisico* (2-21).

4. An overall view through the letter of a friend

For those who did not know Persico personally, the letter that Tricomi wrote to Rasetti in November 1974 at the latter's request may help to clarify the figure and personality of Enrico Persico.

Turin, November 4, 1974

Dear Rasetti,

I did indeed know our common friend Persico well. Not only was he my colleague here in Turin from 1929 to 1947, but he also lived with me and my family during the "evacuation" of the war years (1942-45), and even after 1947 we often met in the United States and Rome. Nevertheless, I cannot recall any "characteristic facts" concerning our friend, for the very reason that one of his most outstanding peculiarities was that he did not have special characteristics, unless having common sense and being well balanced can be considered such. He was also very prudent and this kept him slightly removed from the turbulent times in which we lived.

As for Persico the scientist, I believe that his well known gift as a teacher and brilliant commentator derived not so much from, let us say, literary talents (which he certainly did not lack) as from a thorough personal reelaboration of the fundamental concepts of physics. This process

(10) The CNR paid each author 10,000 lire for all copyrights.

(11) E. FERMI: *Molecole e Cristalli*, Bologna (1934), of which an English edition, *Molecules, Crystals and Quantum Statistics*, New York, appeared in 1966.

(12) F. RASETTI: *Il nucleo atomico*, Bologna (1936), of which the English edition, *Elements of Nuclear Physics*, New York, came out in 1936.

(13) At the request of E. Segrè, Persico read the manuscript of the book *Spettroscopia atomica* very scrupulously. When he returned it with his observations, he added a drawing of instruments of torture, symbolically representing the racial laws.

only came to a halt when he was able to see things absolutely clearly or when he realized that that was not possible in the current state of knowledge. Therefore, he never attempted to confuse things to hide from himself or others the real difficulties, for example, in the theory of the make-up of matter.

The most important fact that I know about Persico's private life is the deep and long depression from which he suffered after the death of his aging and beloved mother in the summer of 1940. His mother died during a minor automobile accident which he had on the via Emilia while driving her almost forcibly (she was almost completely senile in her last days) to a convent in Tuscany to remove her from the dangers of the bombings that were starting to rain down on Turin. The depression lasted so long that when the bombings really became dangerous in November 1942, he was so apathetic and indifferent that his neighbours had to move him from his apartment and almost carry him off to the country south of the city. A short time later, I dragged him to Torre Pellice, where I gave him a room that was free in the apartment to which I had been evacuated with my family. That was the beginning of our pleasant coexistence which lasted until the armistice in September 1943. Afterward, when I had to take to the mountains (and later hide out in Rome) because of my anti-Fascist activity, he continued to live in Torre Pellice and was a valid help to my family, even though physical courage was never one of his outstanding qualities.

I remember that we were together (I had returned to Torre in May 1945) when news of the atomic bomb in Hiroshima reached us. We were not too surprised because we had frequently discussed the possibility of something like that happening and had concluded that everything depended on the numerical value of a certain constant, unknown at the time: the average number of neutrons freed by one nucleus of uranium when it absorbs another. Therefore when we heard the news we both commented almost simultaneously: "I guess the constant was sufficiently large."

I think it superfluous to dwell on Persico's love of skiing and cats, both well known in physics circles. His love for cats, I believe, depended on the fact that he was very solitary, so much so that he once told me that one of the main reasons he accepted the position in Rome after Quebec was that he had some cousins there who would provide a little company.

Something which surprised me when I visited him in Quebec in 1950 was that he had become far more anti-clerical than he had ever been in his serene atheism. But it was a comprehensible reaction to the ultra-clerical environment at Laval University, which you yourself know only too well. In fact, he told me that sometimes when he identified himself on the phone as Persico, the persons on the other end of the line, who were used to that environment, took him for "Père Sicò", repeatedly replying "Oui mon Père", etc. which he did not appreciate too much. But Persico was so greatly esteemed at Laval that they went to great lengths to keep up contact with him after his departure.

In those years, Persico had repeated attacks of a strange ailment affecting his fingers--perhaps the consequence of radiation--which cost him the loss of several phalanges. The last attack occurred in Quebec, where penicillin saved him from further mutilation. And he told me laughingly of the comic maneouvres of the nun-nurses in their attempts to give him an injection without "offending his decency". . . .

Most cordially,

Your F. Tricomi

Tricomi's letter may clarify some of the events that took place in Persico's family and working life during the period he spent in Turin and which contributed to persuade him to accept the chair at Laval University in Quebec.

But the letter also contains a phrase about Persico's physical courage with which I do not agree at all. Even less in agreement is A. Gamba, who studied physics in Turin from 1943-45. He sat his mathematical physics and theoretical physics exams with Persico and after the liberation had Persico supervise his thesis for graduation in 1946. In early spring 1944, Gamba, then fighting in the resistance movement, had made an appointment with Persico at the Institute in via Giuria to discuss examinations. After they had talked for half an hour, Persico said that he had to go, but asked Gamba to accompany him so that they could continue their discussion. Gamba told him that he was armed and showed him his pistol. He told him that if they were stopped by militiamen, he would have to use it. Persico was not worried in the least by this and walked with Gamba all the way from the Institute to his home, a distance of approximately three kilometers.

Another particular in Tricomi's letter is his suggestion that the ailment Persico suffered from may have been the consequence of exposure to radiation. It is true that in Turin Persico prepared some neutron sources from radon and beryllium, like those used in Rome, and that he carried out some experiments with these, but since we prepared much more intense sources every week for years in Rome and used them every day without any reaction of the kind, I would tend to exclude that interpretation.

5. Quebec, Canada (1947-50)

In 1938, Laval University, Quebec, decided to step up its activities in the field of physics, which had been rather neglected up to that time. After Fermi had left Italy, the dean, Monseigneur Alexandre Vachon, contacted Rasetti through the Accademia Pontificia, to which he had been named a member. Rasetti accepted the invitation and moved to Quebec in the summer of 1939. There, he practically set up the physics department and managed to carry out several experiments of considerable importance during the war. In 1947, Rasetti accepted the chair of physics at Johns Hopkins University in Baltimore (USA) and wrote to Persico asking him whether he was interested in taking his place at Laval. Persico accepted the offer, also because he had little confidence in the possibilities of a recovery of physics in Italy and more generally of a rapid return to acceptable living levels in the country.

As Tricomi points out in his letter, Persico was named director of the Department of Physics at Laval. At that time, he was starting to turn his attention to the spectrometer theory for beta rays, a subject on which he wrote various important papers (1-31, 1-35) and towards which he directed several students.[14]

In 1949, Persico had some trouble with his health and went to a well known physician who diagnosed an uncurable disease. Unfortunately, Persico did not think of getting a second opinion and spent approximately one year expecting the worst at any time. In this period, he made long car trips around Canada and the United States. After about a year, however, urged by his friends, he went to another specialist who attributed the shadows on the x-rays to another cause. Although Persico had accepted the verdict with great outward calm, the new diagnosis had a positive effect on his mood and perhaps even on his attitude towards life in subsequent years.

6. In Rome again (1950-69)

The death of Antonio Lo Surdo on June 7, 1949 (born 1880) posed a number of problems. He had been professor of superior physics at the University of Rome from 1919 and after the death of O. M. Corbino in 1937, had taken his place as director of the Physics Institute. When Fermi left Italy in autumn 1938, Lo Surdo replaced him in geophysics, a field that had interested him since 1917.

In 1939, he founded the Istituto Nazionale di Geofisica (ING), initially supported by the Consiglio Nazionale delle Ricerche, and later turned into an autonomous agency under the Ministry of Education. This institute, which Lo Surdo directed, was located in the Physics Institute on the university campus (Città Universitaria).

Upon Lo Surdo's death, the Minister of Education named Enrico Medi head of the ING. Medi was professor of experimental physics at the University of Palermo at the time, but he came from Rome, where he had been assistant to Lo Surdo for many years.

Therefore, there was some disagreement in the beginning about who should take the chair for superior physics at the Rome Faculty of Science. Some members of the faculty were in favour of Medi, while others wanted to call in Persico, not only for his exceptional talents as a teacher but also because it would be a way of bringing home a physicist whose continued stay abroad would have been a loss for the country. The ministry solved the dilemma by creating another chair for geophysics[15] approximately a year later and assigning it to Enrico Medi, while Enrico Persico immediately took over the chair of superior physics. He kept this position until 1958, when he transferred to theoretical physics, where he stayed until his death.

Persico's successor in superior physics, Marcello Conversi, who had studied and worked in Rome, Chicago (1947-49) and Puerto Rico (1950), had won a competitive exam in experimental physics in 1950 and had been offered a position at the University of Pisa. When he came to Rome in 1958, he immediately made friends with Persico. In the following years, Persico, Conversi, Ageno and Nella Mortara (often called the singles of the Institute) got into the habit of taking Sunday trips

---

(14) A. BOIVIN: *Journ. Opt. Soc. Am.*, **42**, 60 (1952).

(15) There were still four physics chairs at the University of Rome, as in 1931: experimental physics (E. Amaldi), superior physics (vacant), theoretical physics (B. Ferretti, who was succeeded in 1947 by G. C. Wick, when the former accepted the offer of the physics chair at the University of Notre Dame in Illinois), and spectroscopy (still formally held by Rasetti, who made himself available to the Ministry of Foreign Affairs when he left for Quebec).

to the beautiful villages lying around Rome, trips which all remembered with pleasure for years.

During his stay in Rome, Persico also took over the job of supervising the Institute's library, which was now to serve several chairs. The chief librarian, Dr. C. Guazzaroni Ferrari, turned to Persico for the classification of books and above all, for help in deciding on new acquisitions, which was not always easy, given the limited resources and the need to keep the library up to date. Persico carried out this job with exceptional diligence and continuity for eighteen years, during which time the number of scientific works in the library almost tripled, the number of subscriptions to scientific periodicals rose from eighteen to 150, the decimal system was introduced, the reading room expanded and a photocopy service for students set up.

In Rome, Persico continued to investigate problems of optical electronics for some time. In particular, he invented two different kinds of two-dimensional resistance grids (1-36, 1-37), which made it possible to deal with the problem of simulating electric (or magnetic) fields generated by electrodes (or poles) of any kind. Persico's grids were a considerable improvement on the methods already used by many laboratories involving tanks containing electrolytes or aluminum sheets. This line of research led to the construction of a grid with the help of F. Magistrelli (1-36).

In Rome, Persico made some extraordinary contributions to teaching: suffice it to recall his course in superior physics, of which the text was published in various volumes that were gradually improved and expanded, compiled and edited by L. Tau (3-15) and R. Bizzarri (3-16, 3-18), and various advanced courses. Among the latter was the course on "Accelerating machines and reactors" given in 1953-54 and 1959-60 at the School of Advanced Studies of the University of Rome and later transformed (from 1960-61 to 1963-64) into the course on "Nuclear reactors". The lectures on accelerating machines were first assembled by G. Caglioti, N. D'Angelo and A. Reale (3-17), and later reedited by E. Ferrari and S. E. Segre. In 1968, they were expanded and published by Benjamin (3-23). Persico's lectures on "Nuclear reactors", on the other hand, were compiled by G. Riguzzi (3-19) and constitute another remarkable contribution to high quality teaching. In those years, Persico also gave "Theoretical lessons on reactor physics" in the framework of the CNEN's "Course on Applied Nuclear Physics and Nuclear Chemistry" (from 1956-57 to 1962-63) and, with a different accent, of the CNEN's "Advanced Course in Nuclear Engineering" (from 1959-60 to 1963-64).

In 1953, the Executive Council of the Istituto Nazionale di Fisica Nucleare (INFN), headed at the time by Gilberto Bernardini, decided to undertake the construction of an Italian 1.1 GeV electron synchrotron. This decision stemmed from the fact that there were enough funds available to the INFN to allow them to be concentrated in one important effort rather than divided in a number of minor endeavours, often much less significant. The project was headed by Giorgio Salvini, professor of general physics at the University of Pisa at the time.

Salvini invited Persico to take part in the meetings of the group that was to design and construct the machine, offering him the position of director. Persico readily accepted the job of director of the theoretical section, but not that of director of the entire project.

He immediately began activity with a series of lectures on ultrarelativistic dynamics, the distribution of electric potential or magnetic field determined by unusually shaped electrodes or poles, ways to solve these problems using simulation, Schwarz-Christoffel transformations and the electric wire technique for determining the lines of force of a magnetic field--all problems connected to the design of the synchrotron, which involved the construction, for example, of unusually-shaped magnets (equipped with corrective teeth or "tips").

In 1955, the group in charge of design moved from Pisa to the University of Rome and in 1956 from there to Frascati, where construction of the CNEN's National Laboratories and of the electron synchrotron was already underway. In the meantime (1955), Salvini had taken over the second chair of experimental physics at the University of Rome.

From the beginning of this activity, Persico's students and collaborators included Carlo Bernardini, Pier Giorgio Sona and Angelo Turrin, neo-graduates taking his course on accelerating machines at the School of Advanced Studies of the University of Rome.

A complete review of the theoretical problems dealt with by the Theoretical Section can be found in Part II of the supplementary volume to *Nuovo Cimento* dedicated to the design and construction of the synchrotron.(16)

During the meetings of the design group chaired by Salvini, lively discussions often arose about the solutions to be adopted. Agreement was generally reached quite easily, but on one basic point, discussion dragged on and became very heated. The point in question was the height of the magnetic gap and, therefore, the vertical dimensions of the doughnut. Persico maintained that it had to be rather high, to take into account all possible causes of error in localizing the electron beam. Salvini felt that the theoretical calculations (of others, but above all of Persico's group) could be relied upon and that the total width of the doughnut should therefore be kept exactly equal to the theoretical width calculated. In the end, the dispute was mediated by Mario Ageno, who helped to work out dimensions that represented a reasonable compromise between caution and economy.

In addition to the various original studies and calculations related to design of the electron synchrotron (1-44), Persico also developed the general theory of injection of charged particles into accelerators in that period, a theory which is now commonly used around the world in the design and construction of such machines.

All those who participated in the design of Italy's electron synchrotron remember Persico's extremely precise and detailed contribution with affection and admiration.

Persico continued his participation in this activity until 1957, the year in which design and calculation were completed. Although his official and permanent collaboration came to an end at that time, he remained available--giving it absolute priority over other commitments--for advice, verifications and calculations. For example, Salvini remembers having discussed with Persico once the synchrotron was already in operation the rather notable human problems involved in the placement and future activity of the physicists and engineers who had worked on the project. Despite invitations to do so, Persico refused to receive recognition at the moment of success, that is,

(16) *L'elettrosincrotrone e i Laboratori di Frascati*, edited by G. SALVINI, Bologna (1962).

when the synchrotron came into operation, even though that success was to a considerable extent due to his work. And the reason for this was not shyness or lack of interest, but that he really was above (as Salvini put it) a "certain scientific fervour and hubbub, in an age of great hubbub".

When he left the CNEN's National Laboratories in Frascati, Persico started to be drawn by other interests. Much of his time was taken up by advanced courses in the physics of nuclear reactors mentioned previously. He was also interested in some problems of plasma physics which he had already worked on in 1926. In 1958, he started work on the loss of particles by a magnetic bottle and published his first important paper on the subject with L. G. Linhart (1-40), who was working at CERN and had met Persico during some lectures he was invited to give in Rome.

7. A few other personal memories

Persico lived a bachelor's life of neither luxury nor waste, but without depriving himself of anything either. His apartment on the third floor of 34 via di Villa Emiliani was comfortable and perfectly kept by a housekeeper who with serene trustworthiness and respectful reserve saw to everything without ever appearing. Persico's study was pleasant and well supplied with books. He also had a good radio-gramophone on which he often listened to classical music. He would sometimes invite a friend over for a cup of tea, which was always accompanied by excellent pastries.

He frequently went to the movies and sometimes to the theatre or concerts and enjoyed being invited over by friends for a chat, during which he would make witty comments about current events, betraying a "creatively keen spirit full of dry humour", to use the words of G. Occhialini.

He was on excellent terms with the children of all his friends, regardless of their age as long as they were old enough to reason. On Sundays or during the spring and autumn holidays, he would regularly go for drives alone or with friends to visit natural sights, but sometime also works of art.

G. C. Wick remembers that during the academic year 1931-32, which he spent at the University of Turin, he was often the third member of the examining committee for mathematical physics and theoretical physics composed of Somigliana and Persico. It was often difficult to get Somigliana, a notable mathematician but very stubborn man, to see reason, but Persico always managed to bring him around with his subtle diplomacy. He would then throw Wick a quick glance full of satisfaction and relief.

M. Ageno will always remember the conversation that took place aboard a seabus in Venice in September 1937 among Fermi, Persico, Rasetti, Amaldi and others on the meaning of quantum mechanics. We were all in Venice for the Congress of the Società per il Progresso delle Scienze. The view expressed by Persico at that time revealed an affinity of interests and perspective with Ageno, who had just received his degree, which continued throughout their lives and became even stronger after Persico's return to Rome.

Until 1961, Persico continued to go hiking in the summer and skiing in the winter, generally with friends, whose children brought along other young people and children forming groups of twenty, sometimes thirty people. Despite the noise, which was often deafening, Persico seemed to enjoy these occasions and at the end of the ten days between Christmas and the Epiphany spent in

Madonna di Campiglio or Selva di Val Gardena, he would offer the whole group a huge assortment of pastries leading the youngest members of the group to repeated acclamations for "Uncle Persico".

Even in his last years, Persico was still able to ski almost any route, albeit slowly. A friend used to quip that Persico was the inventor of "adiabatic skiing": looking at him for an instant he seemed to be standing still; only by means of two subsequent observations separated by long intervals of time could it be made out whether he was ascending or descending the slope.

These kinds of jokes were not new. As a young man, Persico had a motorcycle which friends used to say he drove so slowly that he once fell--inwards--while taking a curve.

Another characteristic was Persico's love for cats. The walls of both his home and his office in the Institute were covered with beautiful photographs of cats. He didn't have any cats of his own but regularly fed the cats in the neighbourhood which soon learned to appreciate the existence of their benefactor. C. Bernardini recalls that Persico built what he called a "tripe clock", that is, a mechanical device that dropped food for the cats at regular intervals of several hours while he was at work.

Of the many known episodes concerning Persico and his cats, two must be recalled as characteristic. While in Quebec, he once took a trip to Chicago to visit the Fermis. Laura Fermi, in order "to uphold her reputation as hostess, borrowed one of the Urey's cats" to make him feel at home([17]), but Persico "was embarrassed by her undertaking"--maybe because the cat came from such a "good" family.

Another time, while strolling through the streets of Copenhagen with his wife, E. Amaldi saw a very original and lively fabric painting of a cat displayed in a beautiful, modern show window. After consultation with his wife, he concluded that Persico did not have a cat like that among his many cat photographs and thought that Persico might like to enrich his collection with this new and completely different conception of a cat. But when he gave him the gift back in Rome, Persico thanked him with great courtesy but it was immediately obvious that this cat somehow bothered him. Standing on its hind legs, its tail extended, the Danish cat was playing the trombone with its front paws in an openly cheerful and boisterous way that had nothing to do with the gracious and reserved expression of all the other cats looking down from the walls of Persico's study. Those cats certainly did not bother anyone with their clowning; much less did they want to be bothered.

Persico was deeply and resolutely opposed to any kind of rhetoric, hyperbole or exhibitionism with regard to daily events or comments and overall opinions on history. An example of this attitude was his elegant inaugural speech entitled "Leonardo e la fisica" (2-22 a+b) to the 37th Congress of the Società Italiana di Fisica, held in Bergamo in 1952. In it, he used his lucid and tranquil rationality to cut down some of the exaggerations and distortions found in the information generally repeated by the press at all levels.

Those who had the pleasure of knowing Persico well believe that he was much more than

([17]) M. H. Urey won the Nobel Prize for Chemistry in 1934 for his discovery of heavy hydrogen or deuterium.

would appear from his research papers and even from his textbooks. It seems he was not a solitary man by nature, but that he became one by necessity, perhaps because he was not determined enough to make the effort required to change the course of events around him in certain circumstances.

His coherence, his independent opinion and his absolute unwillingness to compromise were the dominant characteristics of his personality.

Politically, he was rather conservative, since, as G. Careri points out, he was skeptical of sudden changes, often more apparent than substantial. Gamba recalls his scepticism about the supposed benefits of the nationalization of electrical utilities and, after 1968, his aversion for the "leapfrogging to the left", which had become fashionable in many university and cultural circles.

8. The teacher

Persico's scientific interests extended over a period of almost forty years and ranged from astrophysics to various applications of quantum mechanics, from electrical conduction in metals and gases to plasma physics, from beta ray spectrometers to the ultrarelativistic dynamics involved in the design of accelerators. Some of these works are certainly important: suffice it to recall the paper written in his youth on the kinetic theory of gases (1-13), the one on the theory of injection into accelerators (1-39) or the one on the losses of magnetic bottles (1-40).

Nevertheless his most outstanding characteristic was and remains his ability as a teacher. Not only did he publish numerous review articles and excellent popular science articles (list 2), he also brought out a number of books, handouts, volumes and textbooks (list 3) that was unequalled by most physicists of his time. The majority of these publications was aimed at university or post-graduate students, others, such as *Gli atomi e la loro energia*(3-20) were directed at people with a scientific or technical education, such as chemists, biologists and engineers, still others were textbooks for lower (3-5) and upper (3-9, 3-10) secondary schools. All these publications attest to his didactic abilities and his constant attempts to iron out the difficulties for his readers, in particular, students. Indeed, in order to help his students, he even translated a well known epistemological volume by Max Planck (3-22) and published a *Guida all'inglese tecnico*(3-24), a small English-Italian dictionary of great use to students beginning to study English textbooks or articles on physics.

His interest in the teaching of physics in secondary schools not only motivated him to write the textbooks mentioned above and to hold conferences on various subjects, but also to work (from 1956 to 1965) with the *Giornale di Fisica*, for which he wrote a series of observations and suggestions that can still be read and pondered with benefit today (list 4).

Those who were fortunate enough to know and frequent Persico were always impressed by his extraordinary clear-mindedness and his profound knowledge of an enormous number of classic (theoretical and experimental) problems, which almost always allowed him to suggest to his interlocutors the best procedure for seeking and finding the solution to new and even difficult and unusual problems. And this was true, as Carlo Bernardini pointed out, not only in the field of classic or relativistic mechanics, but also in optics and electricity.

Another quality which struck people was his exceeding availability. If a student or a colleague presented him with a problem, Persico would immediately drop what he was doing and dedicate himself seriously, patiently, and conscientiously to trying to help that person. He had a passion for his work as a teacher and this was based on the importance he attributed to the problem of education of the young.

Giorgio Careri remembers that Persico prepared a questionnaire at the beginning of the sixties which he would hand out to his students at the end of the lectures. It contained a number of questions about his lectures, such as the interest that each subject aroused in the student, what difficulties the student encountered, and finally an invitation to criticize the exposition of the various subjects, the lab course, etc. Persico used the students' replies to try to improve his course year after year.

It is not surprising, therefore, that Persico was hurt by the student protests that began at the University of Rome in 1968. He, who had always been so open to everyone, especially his students, found it difficult to understand the reasons for disputes about problems within the Institute when no attempt had been made to solve them by means of either friendly dialogue or even heated debate.

9. Death

The first signs of the disease that was to lead to Persico's death some years later manifested themselves in early August 1961, during a summer vacation at Verbier in the Valais with the Amaldis, Mario Ageno and the Conversis. The physicians advised him to give up hiking, which he did with great reluctance. But in the following years, he nevertheless had week or month-long spells in which he suffered remarkably high blood pressure.

It was never quite clear what he could or should do to treat this disorder. On June 17, 1969, after a few days in bed following another attack, his transparent and secluded life came to an end.

He was national member of the Accademia delle Scienze in Turin from 1943 and of the Accademia Nazionale dei Lincei from 1952. In October 1969, at the 55th Congress of the Società Italiana di Fisica held in Bari, the Council of the SIF, which had not long before his death deliberated to honour Persico, posthumously awarded him the SIF gold medal.

Except for his books which were left to the library of the Physics Institute of the University of Rome and a few personal belongings which went to his housekeeper, Persico named the Accademia Nazionale dei Lincei his universal heir. In 1970, with the income from Persico's estate, consisting of three apartments and a small quantity of bonds, the Accademia set up the Persico Foundation, aimed at granting scholarships to upper secondary and university students interested in physics.

Once a year in autumn, the commission nominated by the Accademia to examine the candidates and select the winner of the Persico scholarship contributes to perpetuating even beyond his death Enrico Persico's total availability to the problems of the education of the young.

FISICA E ...

RICORDO DI ETTORE PANCINI (1915-1981)

Edoardo Amaldi
Dipartimento di Fisica
Università «La Sapienza» di Roma

Il 20 aprile 1982, nell'Aula Magna della Società Nazionale di Scienze, Lettere e Arti, a Via Mezzocannone 8, in Napoli, si è svolta una cerimonia commemorativa di Ettore Pancini, indetta congiuntamente dall'Università di Napoli, dalla Società Italiana di Fisica e dall'Istituto Nazionale di Fisica Nucleare. Dopo gli interventi del Magnifico Rettore, professor Carlo Ciliberto, del Presidente della Società Nazionale di Scienze, Lettere e Arti, professor Paolo Corradini, del Presidente della Società Italiana di Fisica, professor Renato Ricci, Edoardo Amaldi ha espresso alla signora Elda Rupil, vedova di Ettore, ai loro figli, Barbara, Alessandra, Giulio e Alice, e congiunti, e a tutti coloro che, nelle varie fasi della vita di Ettore, gli sono stati affettivamente e intellettualmente vicini, il cordoglio di tutti i presenti, e suo personale, per la scomparsa di questo indimenticabile amico, e
34 ne ha quindi rievocato la figura con il seguente discorso.

1. – Un rapido sguardo alla sua vita

Ettore Pancini nacque a Stanghella, allora provincia di Padova, oggi provincia di Rovigo, il 10 agosto 1915, secondo figlio di Giulio e Maria Galeazzi. Il padre, nato a Varmo in provincia di Udine, discendeva da una vecchia famiglia friulana di quella città; era ingegnere del comune di Venezia, e, nel corso della sua carriera, divenne ispettore per la provincia di Venezia della Magistratura delle Acque.

Il primo fratello di Ettore, Mario, di circa un anno più vecchio di lui, diventò ingegnere dell'ENEL, e la sorella Irene, di 14 anni più giovane, sposò il dr. Franco Franco, medico e ora Primario dell'Ospedale per le Malattie Infettive di Venezia.

Ettore fece gli studi elementari e medi a Venezia, ma in considerazione della sua vivacità e intraprendenza giovanile, i genitori lo mandarono a fare il liceo nel Collegio Militare di Napoli.

Superato l'esame di maturità classica si iscrisse al corso per la laurea in Matematica all'Università di Padova, ma un anno dopo passò al corso di laurea in Fisica ove fu allievo e laureando di Bruno Rossi che, a partire dall'autunno 1933, ricopriva la cattedra di Fisica Sperimentale di quella Università. In quegli anni Bruno Rossi non solo aveva iniziato, e praticamente portato a termine, la costruzione di un nuovo edificio, come sede

Fig. 1. – Ettore Pancini attorno al 1940.

dell'Istituto di Fisica «Galileo Galilei», ma aveva anche creato un centro di ricerche sulla radiazione cosmica tra i più avanzati e vivaci in Europa e nel mondo. Ma il 12 ottobre 1938 aveva lasciato l'Italia in seguito alle leggi razziali promulgate dal governo fascista il 14 luglio 1938. Cosicché, quando alcune settimane dopo Ettore si presentò all'esame di laurea, il prof. A. Drigo dovette svolgere la funzione di relatore in sostituzione di Bruno Rossi.

Gli anni di studio all'Università di Padova furono per Ettore di importanza determinante sotto varii punti di vista (fig. 1). Sul treno, con cui quotidianamente andava e tornava da Venezia a Padova, conobbe una studentessa di Lettere, la signorina Elda Rupil, con cui si sposò nel 1941. All'Università di Padova si legò d'amicizia con varii studenti che mal tolleravano il regime fascista e fra i quali debbo ricordare lo studente in Matematica, Giorgio Trevisan, più tardi professore di Matematica, in quella stessa Università([1]), che certamente ebbe un'influenza su di lui e subì a sua volta l'influenza di Ettore per quanto riguardava le loro idee politiche e sociali. In quella stessa epoca, o poco dopo, venne in contatto anche con Eugenio Curiel([2]), esponente del Partito Comunista clandestino di Venezia. Queste sue amicizie determinarono qualche anno dopo il passaggio di Ettore dal Partito d'Azione, a cui apparteneva inizialmente, al Partito Comunista Italiano.

Dopo la laurea, Ettore fu subito nominato assistente straordinario dell'Istituto di Fisica di Padova, ma, in seguito alla partenza di Bruno Rossi, egli si sentiva in qualche modo abbandonato, cosicché cominciò a cercare una nuova sistemazione in un'altra sede, che gli permettesse di lavorare seriamente, più o meno nello stesso campo in cui lo aveva indirizzato Bruno Rossi.

Fu così che venne a Roma ad informarsi della situazione alla fine del 1939 ove io lo incontrai per la prima volta. A Roma infatti vi era una molto vivace attività di ricerca sulla radiazione cosmica, diretta da Gilberto Bernardini, allora professore di Fisica Superiore all'Università di Bologna, ma che dedicava alcuni giorni alla settimana a queste ricerche svolte a Roma, presso l'Istituto «Guglielmo Marconi». Bernardini era anche riuscito a convincere il prof. Antonino Lo Surdo, direttore dell'Istituto «Guglielmo Marconi» e dell'Istituto Nazionale di Geofisica (ING), a far rientrare, una parte almeno, delle ricerche sui raggi cosmici nell'ambito dell'ING e di conseguenza a provvedere mezzi per materiali e personale.

Pancini in un primo momento (gennaio-novembre 1940) fu assunto come ricercatore dell'ING, ma, a partire dal primo dicembre 1940, fu nominato, in seguito a concorso, assistente di ruolo presso l'Istituto di Fisica dell'Università di Roma, e più precisamente come assistente dell'«Ufficio del Corista Uniforme», un ufficio creato da Blaserna([3]) nel 1887 per il controllo degli strumenti musicali prodotti da varii costruttori e la verifica della scala musicale usata dagli enti musicali.

Ma il primo febbraio 1941 Ettore veniva chiamato alle armi e la sua attività di insegnante universitario e di ricercatore veniva interrotta fino alla fine della guerra, salvo un periodo di licenza, durante l'inverno 1942-43, durante il quale riusciva a lavorare per qualche mese a Plateau Rosà, insieme a Bernardini, Cacciapuoti e Piccioni, come preciserò nel seguito.

Egli era sottotenente dell'artiglieria contraerea e seguì le sorti del suo Gruppo che fu assegnato a diverse sedi, nelle quali, a partire dal 1941, lo seguiva la moglie Elda: a Cormons e Montegliano in Friuli, alla Cecchignola, allora alla periferia di Roma, a Riccione e S. Remo, anche dopo la nascita, nel 1943, della prima figlia Barbara.

Essendo stato ferito mentre si trovava alla Stazione Termini in partenza per Venezia, in occasione del bombardamento di Roma del 19 luglio 1943, l'8 settembre dello stesso anno, giorno dell'armistizio fra le Forze Alleate 35
e l'Italia, Ettore si trovava in licenza di convalescenza a Venezia. Barbara aveva 6 o 7 mesi. All'ordine del governo di Mussolini a tutti gli ufficiali di complemento di presentarsi per riprendere servizio, Ettore Pancini entrava immediatamente nel movimento partigiano e diveniva in breve tempo il comandante dei Gruppi di Azione Partigiana (GAP) di Venezia alle dipendenze di quel Comitato di Liberazione Nazionale, con il ben noto nome di comandante Achille([4]). Nel 1944 gli veniva affidata la responsabilità di Comandante della Zona Militare di Venezia e in tale veste organizzava un notevole numero di azioni militari di assalto, a tutte le quali prendeva parte personalmente. Tra queste rimasero famose l'assalto di Ca' Giustinian, sede del Comando Tedesco in Venezia, e pochi giorni prima della liberazione della città l'interruzione di uno spettacolo al teatro Goldoni, ove «Achille»

lesse agli spettatori un proclama in cui si annunciava l'imminente liberazione di Venezia e la fine dell'oppressione nazista e fascista[5].

Un episodio che vorrei ricordare del periodo della sua attività partigiana risale al febbraio 1945, epoca in cui Ettore era sulle montagne nei pressi di Novara. La moglie Elda, a Venezia, era in attesa di un secondo figlio ed Ettore, come se d'istinto sapesse esattamente la data della nascita, si spostò rapidamente a Venezia ove poté assistere la moglie, in assenza dell'ostetrica per circostanze casuali, facendo tutto da solo, fino al taglio del cordone ombelicale, e alla sistemazione dell'ombelico di una nuova bambina, Alessandra, nata il 19 febbraio 1945. Ettore parlava di questo episodio con evidente soddisfazione solo con pochi intimi, a cui diceva anche di esserne molto fiero. Dopo il parto, Ettore si affrettò a ripartire per il suo posto di combattimento, anche perché vi erano segni evidenti che la presenza del comandante Achille a Venezia era stata segnalata e cominciavano i preparativi per la sua cattura.

Gli altri due figli, Giulio e Alice, sono nati in epoche successive, e precisamente nel 1949 e 1950, in condizioni certamente meno drammatiche.

Tornando alla sua attività partigiana debbo ricordare che Ettore fu arrestato dai tedeschi all'inizio di aprile 1945, in una casa di campagna, in seguito ad una spiata, ma circa 20 giorni dopo riuscì ad evadere con l'aiuto di alcuni militi di reparti fascisti che speravano di farsi qualche amico in vista dell'ormai pre-
36 vedibile esito della guerra, e tornò immediatamente al suo posto di combattimento. Fu poco dopo che, alla testa di un gruppo di partigiani operanti nella zona, il comandante Achille ingaggiò, nelle vicinanze di Portogruaro, una vera e propria battaglia con un reparto tedesco, che, in previsione della inevitabile sconfitta, cercava di ritirarsi verso il Brennero.

Fino all'aprile 1945 fu il rappresentante del Corpo Volontari della Libertà nel Comitato di Liberazione Nazionale di Venezia. Fu poi impegnato nella smobilitazione del Corpo e rientrò a Roma, ove riprese la sua attività presso l'Istituto «Guglielmo Marconi», nel settembre 1945.

Il suo stipendio di assistente all'Istituto di Fisica dell'Università ammontava a L. 14000 al mese ed era del tutto inadeguato per il mantenimento suo e della sua giovane famiglia. Per qualche mese fu gentilmente ospitato da Gilberto Bernardini che viveva con la famiglia in un appartamento preso in affitto a Via Dalmazia (fig. 2). Quindi insieme ad un altro giovane dell'Istituto, Ruggero Querzoli, misero due letti in una stanza dell'Istituto, normalmente adibita a laboratorio. E quando questa si rese necessaria per il lavoro, Ettore sistemò un lettino da campo in uno dei gabinetti dell'Istituto di Fisica, ove rimase per lungo tempo, coabitandovi, per periodi più o meno lunghi, con altri giovani, fra i quali debbo ricordare J. Buschmann, un ricercatore tedesco, e successivamente D. Broadbent, un ricercatore inglese, entrambi venuti a lavorare al nostro Istituto con borse di studio estremamente modeste.

Fig. 2. – Ettore Pancini attorno al 1945.

Per mesi in quell'epoca Ettore Pancini e Carlo Ballario vissero nutrendosi prevalentemente di farina di castagne che Ballario aveva portato dall'Appennino bolognese.

Quando nel 1947 Gilberto Bernardini ideò la costruzione di un laboratorio per lo studio della radiazione cosmica in alta quota, Ettore Pancini divenne il suo più stretto e permanente collaboratore nella costruzione e messa in marcia del Laboratorio della Testa Grigia (a 3500 m s.l.m.) sul confine italo-svizzero vicino al Passo del Teodulo [13]. La direzione del Laboratorio, per espresso desiderio di Bernardini, fu affidata ad Ettore Pancini che la tenne fino al 1953, provvedendo nel 1950 anche al suo ampliamento [14].

Alla fine degli anni 40, Pancini, insieme ad altri giovani dell'Istituto «Guglielmo Marconi», fu molto attivo anche nel recuperare materiale elettrico dai campi A.R.A.R. (Azienda Rilievo e Alienazione Residuati), ove erano stati raccolti i residuati di guerra. Questi materiali furono estremamente importanti per la ripresa e lo sviluppo dell'attività di ricerca dell'Istituto di Fisica dell'Università di Roma, la cui modesta dotazione non permetteva l'acquisizione di considerevoli quantitativi di materiali nuovi (6).

Nel novembre 1949 Ettore Pancini fu nominato aiuto del direttore dell'istituto, funzione che io avevo assunto da alcuni mesi, in seguito alla morte di Antonino Lo Surdo avvenuta il 7 giugno di quello stesso anno. Ma circa un anno dopo risultava primo nella terna del concorso alla cattedra di Fisica dell'Università di Sassari, ove veniva nominato professore straordinario a partire dal 1 novembre 1950.

Nel 1952 veniva chiamato dalla Facoltà di Scienze dell'Università di Genova a ricoprire la cattedra di Fisica Sperimentale, lasciata da Giuseppe Occhialini trasferitosi all'Università di Milano. Ettore rimaneva in questa sede fino al 1961, anno in cui veniva chiamato dalla Facoltà di Scienze dell'Università di Napoli, ove rimaneva fino alla sua scomparsa il 1 settembre 1981.

2. – L'attività di ricerca del periodo giovanile

Le sue prime ricerche sulla radiazione cosmica risalgono al 1940. Esse erano svolte in collaborazione con G. Bernardini, M. Santangelo e E. Scrocco [1,2] e riguardavano le condizioni di equilibrio fra la componente penetrante e i suoi secondari elettronici, i quali, come mostrarono gli autori, sono in notevole eccesso rispetto a quanto calcolato sulla base dei soli processi di urto, irraggiamento e creazione di coppie. Questi lavori sono fra i primi ad indicare la presenza di elettroni secondari generati nel decadimento delle particelle costituenti la componente penetrante della radiazione cosmica, chiamate, a partire dal 1936, «mesotroni» ma la cui natura a quell'epoca non era ancora chiarita.

Nello stesso periodo Pancini prendeva parte alla preparazione della spedizione, organizzata da Bernardini nell'ambito dell'ING, per estendere lo studio della radiazione cosmica ad alte quote (Plateau Rosà; 3500 m s.l.m.), spedizione a cui egli riuscì a partecipare grazie a qualche mese di licenza. Le tematiche affrontate in quel periodo sono sostanzialmente due.

La prima, in collaborazione con Bernardini, Conversi, Scrocco e Wick [3,4,7] riguarda la misura dell'eccesso di «mesotroni» positivi rispetto a quelli negativi che gli autori affrontano con l'uso di «lenti magnetiche» che permettono di concentrare, nei contatori del telescopio, i mesotroni dell'uno oppure dell'altro segno, a piacere. Questa tecnica, proposta inizialmente da B. Rossi (7), era già stata impiegata, e con successo, da G. Bernardini e M. Conversi (8).

La seconda tematica, studiata in collaborazione con Bernardini, Cacciapuoti, Conversi, Scrocco e Wick [5,6], riguarda l'assorbimento anomalo dei «mesotroni» nei mezzi di bassa densità rispetto a quelli ad alta densità. L'effetto dovuto al decadimento dei mesotroni e la sua misura permette di ricavare il rapporto fra la vita media di queste particelle e la loro energia di quiete. I valori dedotti per tale rapporto dal gruppo italiano e quelli misurati indipendentemente, circa alla stessa epoca, da un gruppo guidato da Bruno Rossi negli Stati Uniti (9), risultarono in ottimo accordo fra loro e costituirono un importante risultato 37 scientifico all'inizio degli anni 40.

Rientrato a Roma dal servizio militare nel 1945, egli riprese a lavorare con Conversi e Piccioni, che nel frattempo avevano sviluppato una tecnica molto raffinata per la misura delle «coincidenze ritardate» in vista del suo impiego nella misura diretta della vita media del mesotrone (10). In particolare Piccioni aveva inventato un nuovo tipo di circuito per le coincidenze rapide, basato sull'uso di tubi a emissione secondaria (11), che risultò una parte essenziale dell'apparecchiatura elettronica sviluppata e usata successivamente da lui e da Conversi (12) per la misura diretta della vita media del mesotrone che avevano pensato di fare fin dalla fine del 1941, quando avevano deciso di lavorare insieme. Con questa apparecchiatura gli autori riuscirono a conseguire risultati assai migliori di quelli ottenuti dai pionieri in questa misura: F. Rasetti a Quebec (13) e Auger e collaboratori a Parigi (14), e paragonabili con quelli ottenuti, a loro insaputa già prima in U.S.A., da Rossi e Nereson (15).

In un successivo esperimento (16) basato sulla stessa tecnica (10,11), Conversi e Piccioni dimostrarono, in accordo con una misura preliminare di Rasetti in alluminio, che solo circa la metà dei mesotroni a fine percorso in ferro subisce il decadimento spontaneo; un risultato dunque in apparente accordo con la congettura che solo i mesotroni positivi si disintegrino una volta ridotti in quiete, mentre quelli negativi subiscono la cattura da parte del nucleo atomico, come previsto da Tomonaga e Araki, per le particelle di Yukawa.

La tecnica impiegata da Conversi, Pancini e Piccioni in questo gruppo di lavori particolarmente importanti, consisteva nel combinare le lenti magnetiche, già usate nei lavori sull'eccesso positivo (8)[3,4,7], con l'impiego delle coincidenze ritardate sviluppate da Piccioni e Conversi (10,11).

In un primo lavoro, in cui i mesotroni cosmici venivano ridotti in quiete nel ferro (elemento di numero atomico $Z=26$) Conversi, Pancini e Piccioni [8] mostrarono che, in apparente accordo con le previsioni teoriche di Tomonaga e Araki, basate sulla teoria di Yukawa dei mesotroni come mediatori delle forze nucleari, queste particelle, se positive, una volta ridotte in quiete, decadono spontaneamente, mentre, se negative, vengono assorbite dai nuclei degli atomi del mezzo in tempi as-
38 sai più brevi della loro vita media.

Sulla base di varie congetture, assai ben escogitate, gli stessi autori procedevano poco dopo a ripetere l'esperienza, questa volta, però, riducendo in quiete i mesotroni in un materiale costituito di atomi di numero atomico molto più piccolo di quello del ferro. A tale scopo essi scelsero il carbonio.

Il risultato trovato in questo caso era però del tutto imprevisto: l'assorbimento dei mesotroni negativi da parte dei nuclei di carbonio è trascurabile: anche quelli negativi, al pari di quelli positivi, decadono spontaneamente prima di venire catturati dai nuclei di carbonio [9,10,11] in pieno contrasto con le previsioni di Tomonaga e Araki.

Non è questa la sede per discutere in dettaglio la storia di questa esperienza e tutte le sue conseguenze e implicazioni. Ciò è stato fatto in modo molto completo e affascinante, nei contributi presentati da Oreste Piccioni e Marcello Conversi al Simposio sulla storia della fisica delle particelle elementari, tenuto al Fermilab nel maggio 1980 (17). Qui mi limiterò a ricordare che l'esperienza di Conversi, Pancini e Piccioni dimostrò che le particelle costituenti la componente penetrante della radiazione cosmica, e fino ad allora chiamate mesotroni, non sono le particelle ipotizzate da Yukawa, ma particelle dotate *solo* di interazioni deboli (e, se cariche, anche di interazioni elettromagnetiche), ossia particelle che appartengono alla categoria (o famiglia) indicata oggi come quella dei *leptoni*. Fino ad allora si conosceva un solo leptone carico, l'elettrone; dopo l'esperienza di Conversi, Pancini e Piccioni la famiglia dei leptoni si era allargata con l'aggiunta di un nuovo importante membro, a cui, in seguito, fu dato il nome di *muone*. Inoltre l'esperienza diede l'avvio a numerosi lavori teorici fra i quali ricorderò uno di Fermi, Teller e Weisskopf (18) che dedussero il fattore di discrepanza fra il risultato sperimentale di Conversi, Pancini e Piccioni e quello teorico (ben 10^{12}) e uno di Wheeler (19), che derivò la legge di dipendenza della probabilità di cattura di un muone negativo da parte di un nucleo, dal valore del suo numero atomico Z (la cosiddetta legge Z^4). Infine l'esperimento di Conversi, Pancini e Piccioni, aprì la via allo studio degli atomi mesici, divenuto oggi un campo della fisica su cui sono stati scritti moltissimi lavori, articoli di rassegna e libri (20).

È quindi ben comprensibile che autori come Bethe e collaboratori (21) e Louis Alvarez (22) abbiano indicato, in importanti pubblicazioni, l'esperimento di Conversi, Pancini e Piccioni come l'inizio della fisica delle alte energie.

Pochi mesi dopo questa importante scoperta, prima Piccioni e poi Conversi andarono a lavorare negli Stati Uniti, il primo per rimanervi permanentemente, il secondo per un periodo di studio.

Pancini, a Roma, mandò avanti, insieme ad A. Alberigi Quaranta, lo studio del comportamento dei muoni ridotti in quiete nel carbonio [19] e pubblicò, con lo stesso giovane collaboratore, L. Mezzetti e G. Stoppini anche un lavoro tecnico sul comportamento degli elettroni nelle miscele usate per il riempimento dei contatori di Geiger [17].

Già precedentemente insieme a Lucio Mezzetti e Gherardo Stoppini [16] aveva sviluppato la tecnica di misura di intervalli di tempo molto brevi ($\sim 10^{-8}$ s) in vista della sua applicazione allo studio dei tempi di arrivo di particelle di diversa massa facenti parte di uno stesso sciame esteso.

Parallelamente a questa attività, strettamente di ricerca scientifica, Pancini si dedicò, negli anni passati a Roma, alla preparazione e messa a punto di alcune apparecchiature elettroniche alquanto avanzate. Fra queste vorrei ricordare un generatore di impulsi per lo studio degli stimoli e della propagazione di segnali elettrici nei tessuti nervosi, ricerca condotta in collaborazione con il ben noto fisiologo del sistema nervoso dell'Università di Pisa, Prof. G. Moruzzi e l'amico Franco Lepri [[15]].

Negli anni 1946-1949 Pancini, insieme a G. Bernardini, B. N. Cacciapuoti, B. Ferretti, F. Lepri e O. Piccioni dedicò parte del suo tempo anche alla nascita in Roma di un laboratorio finanziato dalla FIAT e chiamato «Laboratorio Beta», ove essi affrontarono numerosi problemi applicativi, prevalentemente di natura elettronica.

3. – L'attività di costruttore e promotore di ricerca

Quando nell'autunno 1950 Ettore lasciò Roma per Sassari egli si trovò di fronte ad una situazione in un certo senso nuova. Lo stato dell'Istituto di Fisica di quell'Università era disastroso e i mezzi a sua disposizione erano così modesti da non permettere, praticamente, alcun intervento sostanziale. La sua principale azione, nel breve periodo trascorso in quella Università, consistette nell'investire il poco danaro di cui disponeva nel miglioramento della biblioteca, che era quasi priva di trattati moderni e delle riviste scientifiche essenziali per l'inizio di qualsiasi attività di ricerca.

In quel periodo egli seguitò a frequentare, sia pure a tempo parziale, l'Istituto di Fisica dell'Università di Roma, ove lavorò, per qualche tempo, insieme a Franco Bonaudi e Lorenzo Resegotti, alla messa a punto di scintillatori liquidi.

Un piccolo episodio del periodo trascorso a Sassari, che vorrei ricordare per mostrare lo spirito giovanile di Ettore anche dopo la sua nomina a professore, è il seguente. Una sera, passeggiando con alcuni amici per la città, in stato di euforia, a quanto pare per aver bevuto un bicchiere di buon vino, Ettore, ad un certo punto per scommessa, diede la scalata, sfruttando il bugnato della facciata, al palazzo del Rettorato, fino a raggiungere la finestra del Magnifico Rettore, fatto che, il giorno dopo, diede luogo a commenti dei più svariati toni.

Giunto a Genova nel 1952, trovò un Istituto che già aveva avuto un primo risveglio grazie alla presenza, per un paio di anni, di Giuseppe Occhialini. Questi aveva dato l'avvio a un gruppo di ricerca sulla radiazione cosmica a mezzo della tecnica delle emulsioni nucleari, a cui lui stesso, come ben pochi altri, aveva dato un contributo straordinario, trasformandola da un metodo di osservazione piuttosto grossolana, in un metodo di misura di notevole precisione, in grado di fornire dettagli praticamente non osservabili con tutte le altre tecniche.

Ettore era stato chiamato alla cattedra di Fisica Sperimentale dell'Università di Genova soprattutto per l'azione svolta in suo favore da Antonio Borsellino, che già da qualche anno era professore di Fisica Teorica presso quell'Università. «L'arrivo di Ettore a Genova — come ha scritto recentemente Borsellino ([23]) — ... segnò una svolta decisiva nello sviluppo della fisica presso l'Università di Genova. In quegli anni la fisica italiana si stava dando una organizzazione che mai aveva avuto negli anni precedenti, trasformando il fervore, lo spirito di sacrificio e di collaborazione dei primi anni post-bellici in solide strutture, capaci di assicurare possibilità di lavoro più efficaci, aperte a schiere crescenti di giovani ricercatori.

La Società Italiana di Fisica (SIF) da un 39 lato, con il prestigio della propria rivista *Il Nuovo Cimento* e della Scuola Internazionale Enrico Fermi di Varenna, l'Istituto Nazionale di Fisica Nucleare (INFN) dall'altro, furono gli strumenti operativi che consentirono ai fisici italiani di poter produrre ricerca scientifica competitiva e significativa a livello internazionale. Ettore Pancini fu attivo sia come consigliere della Società Italiana di Fisica, sia nello sviluppare, nell'Istituto di Fisica di Genova, le strutture nuove indispensabili».

Alberto Gigli Berzolari, che fu assistente ed aiuto di Pancini nel periodo genovese, mi ha inviato la seguente testimonianza ([24]).

«L'impegno di Ettore nel periodo in cui diresse l'Istituto di Fisica dell'Università di Genova fu caratterizzato da uno spirito di liberalità sia nei rapporti con le varie componenti interne dell'Istituto sia nei rapporti con quelle esterne che raramente ho riscontrato, così alto e nobile, in altri Istituti Universitari. Spirito di liberalità che derivava certamente dalla Sua educazione familiare e che si rafforzò prima nell'ambito dell'Istituto di Fisica di Padova ma soprattutto — in tempi successivi — nell'ambito dell'Istituto di Fisica di Roma.

Costruì a Genova, in un vecchio edificio e circondandosi di persone attive e spregiudicate, un autentico e moderno Istituto Universitario, operando scelte oculate in tutti quei settori che a Suo giudizio, e correttamente, Egli riteneva bisognosi di strutture di base essenziali per una solida prospettiva culturale e di ricerca: biblioteca, strutture didattiche e scientifiche, officina meccanica, officina elettronica, falegnameria, personale di alta professionalità che Egli, con il nostro aiuto, seppe scegliere anche approfittando della crisi momentanea di alcune industrie locali ad alto contenuto tecnologico. Fu facile per noi ricercatori impostare programmi di attività complessi che senza tali strutture sarebbe stato certamente impossibile realizzare. È mia opinione che gran parte delle fortune successive dell'Istituto di Genova siano da collegare anche a tale sforzo operativo e promozionale.

Accattivante e trascinatore, apparentemente burbero ma di infinita generosità e bontà, Egli riuscì a creare una comunità scientifica di autentici amici. E questo fu uno dei Suoi più grandi meriti.

A quell'epoca mi occupavo della realizzazione di "camere a diffusione" nell'ambito di un programma suggerito da Giorgio Salvini durante la mia permanenza a Roma e spostato
40 to poi a Genova in seguito al mio trasferimento in quella sede. Ettore ci impegnò severamente nella realizzazione di una "camera a diffusione" di grandi dimensioni (credo la più grande mai realizzata) mosso dall'idea che in quel momento bisognava fattivamente cooperare — patriotticamente, Egli mi ammoniva; e non scherzava — allo sforzo nazionale volto alla realizzazione in Frascati dell'Elettrosincrotrone ed alla necessaria strumentazione per la sperimentazione successiva; era, il Suo, vero e generoso spirito di servizio. E aveva ragione!

Nello stesso tempo egli ci stimolava ad ideare programmi nuovi ed accettò con entusiasmo l'idea che i miei collaboratori ed io ci impegnassimo su strade originali (era l'anno 1954!) nel settore dei rivelatori di particelle ionizzanti. Riuscimmo così a realizzare la prima "camera a bolle a gas disciolto" che differiva, per la sua concezione, da quelle ormai tradizionali a "liquido surriscaldato" già in fase di studio da parte di Pietro Bassi a Padova e di Giuseppe Martelli a Pisa. Grande fu la gioia di Ettore quando gli annunciai che i nostri tentativi erano stati coronati da successo e che l'effetto delle radiazioni ionizzanti sul comportamento delle soluzioni gas-liquido — da noi solo empiricamente ipotizzato — era una realtà ([25]).

A quell'epoca si frequentava l'Istituto anche di sera, di sabato e di domenica. Abbandonando il laboratorio dopo il lavoro quotidiano, lasciavo sempre alcune note su quanto fatto nel corso della giornata perché sapevo che Ettore, prima di lasciare l'Istituto, girava per i laboratori per rendersi conto dei progressi del nostro lavoro. Lo sorpresi una volta, a notte fonda, nel nostro laboratorio che esaminava le prime fotografie (particolarmente ben riuscite) ottenute lo stesso giorno con un prototipo di camera a bolle a gas disciolto. Non potrò mai dimenticare l'entusiasmo che Egli mi manifestò nel commentare il buon livello del nostro lavoro e la lunga e stimolante discussione, piena di osservazioni acute e ricca di suggerimenti, che ne seguì.

Infine, è per me doveroso ricordare che fu grazie all'interessamento e al sostegno di Ettore che l'Istituto Nazionale di Fisica Nucleare intervenne nella sede di Pavia. Poiché, diceva Ettore, con i vostri Collegi Universitari, avete notevoli possibilità di fornire quadri giovanili qualificati per le fortune della Fisica Nazionale e l'INFN non può ignorare tale realtà».

In un periodo successivo, sempre grazie all'esistenza di una buona officina meccanica e di personale tecnico qualifico, fu costruito, sotto la direzione di Giovanni Boato, uno spettrometro di massa per l'analisi delle abbondanze isotopiche ([26]), un secondo esemplare del quale, per iniziativa di Pancini, fu ceduto a Tongiorgi di Pisa per le sue ricerche di geochimica degli isotopi.

In parallelo con l'officina meccanica, Pancini diede vita anche ad un laboratorio di elettronica capace di produrre prototipi di strumenti, alcuni dei quali furono adottati anche da altri laboratori. Questa officina servì inoltre da base tecnica indispensabile per le ben note attività di Augusto Gamba ([27]) e di Guido Palmieri ([28]) rivolte al riconoscimento delle figure o immagini da punti di vista e con finalità totalmente diverse da quelle convenzionali.

Infine si adoperò per l'acquisizione, da

Napoli 13 marzo 74

Cari ragazzi, rispondo un po' in ritardo alla vostra lettera perchè ho aspettato che la vostra maestra tornasse a scuola.

Vi devo dire sinceramente che, mentre ho letto con molto piacere quello che mi avete scritto perchè una lettera è sempre una manifestazione di affetto, quello che la maggioranza di voi mi ha voluto dire non mi è piaciuto proprio.

Si tratta della gelosia che vien fuori da molte delle vostre lettere: ora – secondo me – la gelosia è, fra tutti, il sentimento più sbagliato. Infatti ci sono delle cose che devono essere divise: una torta o che so io. Altre invece sono tali che più si distribuiscono a diverse persone e più diventano grandi e fra queste c'è l'affetto, l'amicizia, la cultura.

I legami che ci sono fra voi e me sono appunto legami di affetto, di amicizia, di cultura e perciò voi sbagliate se pensate di doverli dividere con altri.

molto affettuosamente vostro Ettore P.

41

Fig. 3. – Lettera di Ettore ai bambini della IV classe della scuola elementare «De Luca», nel rione Traiano di Napoli, conservata e gentilmente messa a disposizione dalla maestra signora Piera Nazzaro.

parte della Sezione di Genova dell'INFN, del betatrone da 30 MeV, costruito dalla Brown Boveri, e di proprietà della Sezione INFN di Torino, quando questa riuscì ad acquistare una macchina di energia più elevata. Il betatrone da 30 MeV rese possibile la formazione in Genova di una scuola di fisica nucleare con particolare riguardo all'uso delle radiazioni gamma.

Nel periodo trascorso a Genova Ettore dedicò una notevole parte della sua intelligenza ed energia a sviluppare i programmi a carattere scientifico svolti presso la «Casa della Cultura», allora diretta da Enrica Basevi di cui divenne collaboratore e caro amico.

Chiamato alla cattedra di Fisica Superiore dell'Università di Napoli, soprattutto ad opera di Giulio Cortini, professore di Fisica Sperimentale in quella Università già dal 1959, fu membro della Giunta Esecutiva dell'INFN dal febbraio 1959 al 30 giugno 1962 e direttore della Sezione dell'INFN di Napoli dal 1 luglio 1962 al 30 giugno 1965, dell'Istituto di Fisica Superiore dal 1965 al 1970 e di quello di Fisica Sperimentale dal 1970 al 1972. Con lo sdoppiamento delle cattedre Ettore era infatti passato dalla cattedra di Fisica Superiore a quella di Fisica Generale I nel 1962, insegnamento che svolse fino al 1978, quando decise un nuovo cambiamento, passando alla cattedra di Ottica.

Oltre a queste attività ufficiali, Ettore si interessò fortemente, dal 1961 al 1967, prima della fondazione, da parte di Adriano Buzzati Traverso, e poi alla conduzione del Laborato-
42 rio Internazionale di Genetica e Biofisica del CNR ([29]); dal 1972 al 1975 egli fu autorevole membro del corrispondente Consiglio Scientifico.

In alcune delle sue attività Pancini ebbe il pieno appoggio oltre che di Giulio Cortini anche di Ruggero Querzoli che fu titolare della cattedra di Fisica Superiore a Napoli dal 1965 al 1967. Cortini, fin dal 1962 aveva creato, come Istituto della Facoltà di Scienze, e aveva successivamente diretto, un «Seminario Didattico» mirante all'aggiornamento degli insegnanti di Fisica delle Scuole medie e allo svolgimento di ricerche didattiche.

Ettore fu autorevole membro del Comitato che dirigeva il Seminario e si impegnò personalmente in un gruppo di ricerca sull'insegnamento delle scienze al livello di scuola elementare, con la collaborazione della psicanalista Jacqueline Mehler Amati, cui era legato da viva amicizia. La sua partecipazione diretta in numerose classi della scuola elementare «De Luca», nel rione Traiano, in una zona povera di Napoli, ha lasciato ricordi indimenticabili, come appare dalla lettera manoscritta riportata in fig. 3.

Quando, nel 1975, Cortini lasciò l'Università di Napoli per quella di Roma, Ettore assunse anche la direzione del Seminario Didattico.

Nel periodo 1968-69, insieme a Giulio Spadaccini e Vittoria Santoro, con la quale aveva lavorato già in precedenza su altri argomenti, e alla quale era legato da una lunga amicizia, si mise a studiare ciò che accade sotto l'azione dei raggi gamma all'interfaccia fra un idrocarburo e una soluzione salina, ricerca in qualche modo collegata con il problema della formazione di molecole organiche all'inizio dell'evoluzione prebiologica. In realtà Pancini e collaboratori trovarono che ben poco si sapeva anche su un problema molto più semplice, la gammaradiolisi di un singolo idrocarburo e pertanto rivolsero le loro ricerche in questa direzione. I primi risultati furono pubblicati in un lavoro relativo alla gammaradiolisi del pentano e dell'esano [[20]].

L'iniziativa, presa da Pancini nel 1970, di acquistare, con fondi dell'INFN, un acceleratore Tandem da 3 + 3 MeV, ebbe purtroppo un iter burocratico e realizzativo molto lungo e cominciò a dar frutti solo recentemente.

4. – Il suo ritorno alla ricerca attiva

Quando, nel 1978, Ettore lasciò l'insegnamento di Fisica Generale per passare a quello di Ottica, egli creò, a Napoli, un gruppo di ricercatori nel campo dell'ottica con luce coerente di cui facevano parte Solimeno, Arimondo e Santamato, l'ultimo dei quali proveniva da Roma, ove era stato allievo e collaboratore di Francesco de Martini che nel frattempo era diventato professore a Napoli di Struttura della Materia.

Ma già qualche tempo prima Ettore aveva ripreso un grande interesse, dovrei anzi dire un interesse giovanile, per la ricerca attiva. Quando nel periodo 1976-77 presso i Laboratori Nazionali di Frascati dell'INFN, era stato avviato il «Progetto Utilizzazione Luce di Sincrotrone» (PULS) come impresa congiunta CNR-INFN, Ettore aveva cominciato a tenere contatti frequenti con Franco Bassani, presidente della Commissione Scientifica preposta al progetto PULS, e con alcuni sperimentatori coinvolti nella sua realizzazione, come Emilio Burattini e Mario Iannuzzi, il primo dei quali era stato, nel 1961, suo studente a Genova, durante l'ultimo anno di attività di Pancini presso quella Università.

Successivamente il suo interesse cominciò a prendere forme più concrete in connessione con lo sviluppo di magneti «wiggler» ed ondulatori ([30]) per la produzione di luce di sincrotrone che ebbe luogo presso i Laboratori Nazionali di Frascati dell'INFN.

Partendo da una proposta originale di Ginsburg ([31]) H. Motz aveva realizzato un wiggler con cui riuscì a produrre, nel 1951, luce visibile usando un fascio di elettroni di 100 MeV ([32]).

Successivamente magneti wiggler furono usati su alcune macchine circolari per scopi vari (per esempio: al CEA intorno al 1965 per ottenere lo smorzamento delle oscillazioni di betatrone).

Il primo wiggler progettato e utilizzato come sorgente intensa di radiazione di sincrotrone dura fu quello (a 6 poli interi equivalenti) dello SSRL (Stanford Synchrotron Radiation Laboratory), messo in funzione nel marzo 1979, sull'anello di accumulazione SPEAR, da un gruppo guidato da H. Winick ([33]).

Il wiggler di Adone, progettato e realizzato dal gruppo Adone ([34]), fu provato con successo sulla macchina nel settembre 1979. Il gruppo PWA (Napoli, Frascati, Trento, Daresbury), nato per la realizzazione del Laboratorio di utilizzo della radiazione di sincrotrone, ebbe un notevole apporto da Napoli (E. Burattini, N. Cavallo, M. Foresti, C. Mencuccini, P. Patteri, R. Rinzivillo e U. Troya) fin dal 1978. Pancini entrò a farne parte alla fine del 1979, quando la strumentazione costruita a Napoli fu trasferita ad Adone.

Nel 1980 il gruppo osservò i primi intensi fasci di raggi X e iniziò un primo ciclo di sperimentazione [21].

I wiggler di Brookhaven erano e sono ancora in costruzione, mentre sono attualmente in funzione magneti wiggler in URSS, in Inghilterra (Daresbury) e in Francia (Orsay).

Ettore si entusiasmò della proposta di fare alcune prove sull'emissione di luce coerente nel visibile (ossia dell'uso del wiggler come ondulatore) avanzata da Barbini, Cattoni e Vignola del progetto LELA (Laser ad Elettroni Liberi in Adone), e ne seguì tutti gli

Fig. 4. – Ettore Pancini a Napoli nel 1963 (fotografia del dr. Mario Muchnik).

sviluppi molto da vicino. Il risultato fu molto appariscente permettendo di osservare rilevanti effetti cromatici [21].

A partire dal giugno 1980 il PWA è stato ampiamente utilizzato dallo stesso e da altri gruppi di ricercatori che tuttora se ne servono per varie esperienze di spettroscopia di assorbimento fra 10 e 30 keV (ossia fra 1 e 43
0.4 Å).

Successivamente, quando Barbini e Vignola fecero la proposta di fare un esperimento di fattibilità di un laser a elettroni liberi su Adone (LELA) ([35]) e si formò, nel 1979, un gruppo Frascati-Napoli [21], Ettore entrò a farne parte portando un notevole contributo di idee ed entusiasmo.

Come osserva Cortini ([36]) «... la sua partecipazione all'impostazione scientifica di molte delle ricerche che i colleghi più giovani facevano attorno a lui e l'aiuto che egli continuamente dava loro, anche sul piano delle tecniche sperimentali, gli avrebbero senz'altro dato titolo per firmare, insieme agli altri, molte pubblicazioni. Ma Ettore era fatto così. Mi diceva l'altro giorno Salvini, venuto a Venezia per il suo funerale, che non ha mai conosciuto un ricercatore tanto disinteressato alla notorietà e al potere».

5. – La sua sensibilità ai problemi politici, sociali e umani

Tornando per un momento alla sua attività politica debbo ricordare che dopo il Decimo Congresso del Partito Comunista Sovietico, svoltosi a Mosca nel novembre 1956, e in occasione del quale Kruschev fece una severa critica del periodo staliniano, Ettore, come del resto altri, si allontanò dal PCI, ma molto silenziosamente.

In seguito a pressanti inviti da parte di varii amici, accettò di presentarsi, come candidato di Democrazia Proletaria per la Camera dei Deputati, alle elezioni politiche del 1979 nella circoscrizione di Napoli-Caserta, e nelle amministrative, comunali (N. 2) e regionali (N. 3) di Napoli del 1980.

Chi non ha conosciuto Ettore Pancini di persona forse non riuscirà a cogliere da quanto ho detto fino ad ora alcuni aspetti fondamentali della sua personalità e del suo carattere: la sua spregiudicata intelligenza, e la sua contenuta, ma profonda, affettuosità verso parenti ed amici, il suo distacco da ogni potere personale e la sua tolleranza e rispetto per le opinioni degli altri.

44

Se ben ricordo doveva essere l'inverno 1947 o 1948, quando un giorno Ettore venne nel mio studio all'Istituto «Guglielmo Marconi» e mi disse di aver incontrato per strada, in uno stato di grande tristezza e depressione, Lamberto Allegretti, fisico pisano, che era stato durante la guerra la persona maggiormente coinvolta negli aspetti tecnici dei mezzi di assalto usati dalla Marina Militare e che era uscito da poco da una penosa azione giudiziaria e dalle sue conseguenze, determinate dalla sua appartenenza alla «X Mas» del Governo di Salò.

Ettore mi chiese se io avevo qualcosa in contrario ad aiutare Allegretti a uscire da quello stato, invitandolo a venire a lavorare in Istituto, e fu molto lieto della mia risposta. Fu così, che a partire dal giorno dopo, anche Allegretti cominciò a frequentare l'Istituto e a contribuire alla ricostruzione e alla ripresa del lavoro in un'atmosfera indimenticabile di fervida attività.

Talvolta nelle discussioni tra colleghi, o con amici, il desiderio e il gusto del contraddittorio lo portavano a posizioni quasi paradossali, che non di rado erano bene al di là di quello che era il suo meditato limpido pensiero.

Nei quarant'anni di amicizia che ci ha legato ci siamo trovati d'accordo, o per lo meno dalla stessa parte, in un grandissimo numero di occasioni. In alcuni casi, tuttavia, siamo stati di opinione diversa, in due o tre casi di opinione nettamente opposta. Ma anche in questi casi il rispetto e la stima reciproca non sono mai venuti meno, e ciascuno di noi ha vissuto con profondo dispiacere questi distacchi derivanti da profonde e meditate convinzioni irrinunciabili.

In questi casi io ho percepito meglio che in altre circostanze la sua personalità, sempre complessa, talvolta tormentata, ma sempre guidata da un'aspirazione alla giustizia nel più alto senso della parola.

La sua tolleranza verso gli altri era accompagnata da un'estrema rigidezza di principi che guidavano il suo comportamento di uomo politico e sociale; il suo fascino personale, la sua capacità di allegria e di dedizione ad una persona o a una causa, erano per tutti quelli che lo hanno conosciuto e amato, indimenticabili sorgenti di ammirazione e, al tempo stesso, di intimo piacere.

Elenco delle pubblicazioni di Ettore Pancini

[1] *Il rapporto fra l'intensità della componente elettronica e della componente mesotronica a 10 e 70 metri di acqua equivalente sotto il livello del mare*, Ric. Sci., **11**, 952 (1940) (in collaborazione con M. Santangelo ed E. Scrocco).

[2] *Sulla produzione della radiazione secondaria elettronica da parte dei mesotroni*, Ric. Sci., **12**, 321 (1941) (in collaborazione con G. Bernardini, M. Santangelo ed E. Scrocco).

[3] *Positive excess in mesotron spectrum*, Phys. Rev., **60**, 535 (1941) (in collaborazione con G. Bernardini, M. Conversi and G. C. Wick).

[4] *Sull'eccesso positivo della radiazione cosmica*, Ric. Sci., **12**, 1227 (1941) (in collaborazione con G. Bernardini, M. Conversi e G. C. Wick).

[5] *Differential measurements of the meson's lifetime at different elevations*, Phys. Rev., **60**, 910 (1941) (in collaborazione con G. Bernardini, B. N. Cacciapuoti, O. Piccioni e G. C. Wick).

[6] *Sulla vita media del mesotrone*, Nuovo Cimento, **19**, 69 (1942) (in collaborazione con G. Bernardini, B. N. Cacciapuoti e O. Piccioni).

[7] *Researches on the magnetic deflection of the hard component of cosmic rays*, Phys. Rev., **68**, 109 (1945) (in collaborazione con G. Bernardini, M. Conversi, E. Scrocco e G. C. Wick).

[8] *On the decay process of positive and negative mesons*, Phys. Rev., **68**, 232 (1945) (in collaborazione con M. Conversi e O. Piccioni).

[9] *On the disintegration of negative mesons*, Phys. Rev., **71**, 209 (1947) (in collaborazione con M. Conversi e O. Piccioni).

[10] *Sul comportamento dei mesoni positivi e negativi alla fine del loro percorso*, Rend. Acc. Lincei, Serie VII, **2**, 54 (1947) (in collaborazione con M. Conversi e O. Piccioni).

[11] *Sull'assorbimento e sulla disintegrazione dei mesoni alla fine del loro percorso*, Nuovo Cimento, **3**, 372 (1946) (in collaborazione con M. Conversi e O. Piccioni).

[12] *Sulla tecnica costruttiva dei contatori di Geiger interamente metallici*, Nuovo Cimento, **5**, 370 (1948) (in collaborazione con R. Berardo e L. Mezzetti).

[13] *Relazione sulla costruzione del «Laboratorio della Testa Grigia»*, Ric. Sci., **18**, 91 (1948) (in collaborazione con G. Bernardini e C. Longo (architetto).

[14] *Ampliamento del Laboratorio della Testa Grigia*, Ric. Sci., **20**,

966 (1950) (in collaborazione con G. BERNARDINI).

[15] *Un generatore di impulsi specialmente adatto per ricerche di elettrofisiologia, Rend. Acc. Lincei*, **9**, 84 (1950) (in collaborazione con F. LEPRI e G. MORUZZI).

[16] *Delays of penetrating particles in atmospheric showers, Phys. Rev.*, **81**, 629 (1951) (in collaborazione con L. MEZZETTI e G. STOPPINI).

[17] *Sulla velocità di migrazione degli elettroni nelle miscele di argon e alcool, Nuovo Cimento*, **9**, 618 (1952) (in collaborazione con A. ALBERIGI QUARANTA, L. MEZZETTI e G. STOPPINI).

[18] *The disintegration of mesons in carbon, Nuovo Cimento*, **9**, 959 (1952) (in collaborazione con A. ALBERIGI QUARANTA).

[19] *Sulla cattura dei mesoni μ da parte dei nuclei leggeri, Nuovo Cimento*, **11**, 607 (1954) (in collaborazione con A. Alberigi Quaranta).

[20] *γ-Radiolysis of alkanes I. Liquid n-pentane and n-hexane, Int. J. Radiat. Phys. Chem.*, **2**, 147 (1970) (in collaborazione con V. SANTORO e G. SPADACCINI).

[21] *Adone wiggler facility, Riv. Nuovo Cimento*, **4**, n. 8, 2 (1981) (in collaborazione con R. BARBINI, M. BASSETTI, M. E. BIGNAMI, R. BONI, A. CATTONI, V. CHIMENTI, S. DE SIMONE, R. DULACH, S. FAINI, S. GUIDUCCI, A. V. LUCCIO, M. A. PREGER, C. SANELLI, M. SERIO, S. TAZZARI, F. TAZZIOLI, M. VESCOVI, G. VIGNOLA e A. VITALI (Gruppo INFN e Lab. Naz. di Frascati per il magnete); E. BURATTINI, N. CAVALLO, M. FORESTI, C. MENCUCCINI, P. PATTERI, R. RINZIVILLO e U. TROYA (Gruppo Utilizzazione luce: INFN, Napoli); G. DALBA, F. FERRARI, P. FORNASINI (Università di Trento e IRST); A. JACKSON e J. WERGAN (Daresbury Laboratory, Daresbury, G.B.).

Bibliografia

(1) Giorgio Trevisan (1916-1974), ha fatto gli studi universitari a Padova, ove, nel 1939, è diventato assistente di Analisi Matematica e libero docente nel 1954. Nel 1962, in seguito a concorso, fu nominato professore di Analisi Matematica a Palermo ove rimase fino al 1965, anno in cui tornò a Padova come professore della stessa materia.

(2) Eugenio Curiel (nato a Trieste nel 1912, morto a Milano nel 1945) autore di numerosi scritti raccolti nel volume: *Eugenio Curiel, Scritti 1935-1945* (Editori Riuniti, Roma, 1973). La prefazione di Giorgio Amendola fornisce molte informazioni sulla vita di politico e di fisico, legato a Bruno Rossi nell'Università di Padova, dal 1933 al 1939, anno in cui, in data 23 giugno, fu arrestato. Condannato a 5 anni di confino, fu trasferito nell'isola di Ventotene nel gennaio 1940. Caduto il Governo Mussolini, il 25 luglio 1943, poté lasciare Ventotene il 21 agosto 1943 e andò a Milano, ove riprese la sua attività politica in clandestinità; ma il 24 febbraio 1945, mentre camminava per la strada, fu indicato da un delatore ad una squadra fascista, che, al suo disperato tentativo di fuga, lo abbatteva con una scarica di mitra.

(3) Pietro Blaserna, nato a Fiumicello (Gorizia) nel 1836, studiò fisica a Vienna e Parigi. Fu professore di fisica all'Università di Palermo dal 1863 al 1872 e a quella di Roma a partire dal 1872; egli progettò e fece costruire (1877-1880) l'edificio dell'Istituto di Fisica di Via Panisperna, che diresse fino alla morte avvenuta in Roma nel 1918. Nel 1890 fu nominato senatore del Regno d'Italia e, nel 1896, fu eletto Vice Presidente del Senato, carica che ricoprì fino alla morte.

(4) Le notizie qui riportate sull'attività partigiana di Ettore Pancini sono basate sugli appunti che io raccolsi in una conversazione, di oltre due ore, che, su mia richiesta, ebbi con lui nel mio appartamento, in Roma, il 10 novembre 1978, in vista del mio proposito di scrivere la storia de «Gli anni di guerra» dell'Istituto Guglielmo Marconi. Tali notizie sono state poi controllate e completate con l'aiuto della signore Elda Rupil, vedova Pancini, e di Giulio Cortini. Alcune di esse si trovano nel volume: *Venezia nella Resistenza, 1943-1945*, a cura di Giuseppe Turcato e Agostino Zanon Dal Bo, pubblicato dal Comune di Venezia (1975-76), oggi purtroppo introvabile.

(5) Si veda: *Il secondo Risorgimento d'Italia* (Centro Editoriale di Iniziativa, 1975) ove però non viene nominato il partigiano che lo lesse, ma che si trattasse di Achille mi fu detto da Ettore Pancini il 10 novembre 1978(4).

(6) E. AMALDI: *Gli anni della Ricostruzione, G. Fis.*, **20**, 186 (1979).

(7) B. ROSSI: *Magnetic experiments on the cosmic rays, Nature* (London), **128**, 300 (1931).

(8) G. BERNARDINI e M. CONVERSI: *Sulla diffusione dei corpuscoli cosmici in un nucleo di ferro magnetizzato, Ric. Sci.*, **11**, 840 (1940).

(9) B. ROSSI, N. H. HILBERRY and J. D. HOAG: *The variation of the hard component of cosmic rays with height and the disintegration of the mesotron, Phys. Rev.*, **57**, 461 (1940); B. ROSSI and D. B. HALL: *Variation of the rate of decay of mesotrons with momentum, Phys. Rev.*, **59**, 223 (1944).

(10) M. CONVERSI e O. PICCIONI: *Sulla registrazione di coincidenze e piccoli tempi di separazione, Nuovo Cimento*, **1**, 279 (1943).

(11) O. PICCIONI: *Un nuovo circuito di registrazione di coincidenze, Nuovo Cimento*, **1**, 56 (1943).

(12) M. CONVERSI e O. PICCIONI: *Misura diretta della vita media dei mesoni frenati, Nuovo Cimento*, **2**, 40 (1944); la cui versione in inglese apparve in *Phys. Rev.*, **70**, 859 (1946).

(13) F. RASETTI: *Disintegration of slow mesotrons, Phys. Rev.*, **59**, 613 (1941); *Disintegration of slow mesotrons*, **60**, 198 (1942).

(14) P. AUGER, R. MAZE et R. CHAMINADE: *Une mesure directe de la vie moyenne du méson au repos*, C.R. **214**, 266 (1942); R. CHAMINADE, A. FRÉON et R. MAZE: *Nouvelle mesure directe de la vie moyenne du méson au repos*, C.R. **218**, 402 (1944).

(15) B. ROSSI and N. C. NERESON: *Experimental determination of the disintegration curve of mesotrons, Phys. Rev.*, **62**, 417 (1942); *Further measurements on the disintegration curve of mesotrons*, **64**, 199 (1943). Questi risultati giunsero in Italia solo a guerra finita, ossia nell'estate 1945.

(16) M. CONVERSI e O. PICCIONI: *Sulla disintegrazione dei mesoni lenti, Nuovo Cimento*, **2**, 71 (1944); la cui versione in inglese apparve in *Phys. Rev.*, **70**, 874 (1946).

(17) O. PICCIONI: *The observation of the leptonic nature of the «mesotron» by Conversi, Pancini and Piccioni*, p. 222; M. CONVERSI: *The period that preceded and led to the 1946 discovery of the leptonic nature of the «mesotron»: some personal recollections*, p. 242 in *Proc. Intern. Symp. on the History of Particles Physics*, held at Fermi National Accelerator Laboratory on May 28-31, 1980, edited by L. M. BROWN and L. HODDESON (Cambridge University Press, Cambridge, 1983).

(18) E. FERMI, E. TELLER and V. C. WEISSKOPF: *The decay of negative mesotrons in matter, Phys. Rev.*, **71**, 314 (1947). In un lavoro successivo Fermi e Teller (*The capture of negative mesotrons in nature, Phys. Rev.*, **72**, 399 (1947)) dimostrarono, con un calcolo dettagliato, che la remota possibilità che il tempo totale di rallentamento e cattura del mesotrone negativo potesse essere molto più lungo in carbone che in ferro è trascurabilmente piccola.

(19) J. A. WHELLER: *Mechanism of capture of slow mesons, Phys. Rev.*, **71**, 320 (1947).

(20) S. DEVONS and I. DUERDOTH: *Muonic Atoms*, in *Advances in Nuclear Physics*, edited by M. BARANGER and E. VOGT, vol. II, p. 295 (1969); C. S. WU and L. WILETS: *Muonic Atoms and Nuclear Structure*, in *Annual Review of Nuclear Science*, edited by E. SEGRÈ, vol. 19, p. 527 (1969); A. O. WEISSENBERG: *Muons* (North Holland Publ. Co., Amsterdam, 1967).

(21) S. S. SCHWEBER, H. A. BETHE and F. DE HOFFMANN: *Mesons and Fields* (Row, Peterson and Co., Evanston, Ill., 1956).

(22) L. W. ALVAREZ: *Recent Developments in Particle Physics*, in *Les Prix Nobel of 1968* (Stockholm, 1969) p. 125. 45

(23) La pagina 3 del numero de *L'Unità* del 1 ottobre 1987, sotto il titolo generale: *Un fisico di frontiera* contiene una serie di articoli in cui viene ricordata la figura di Ettore Pancini: a) Giorgio Napolitano: *Quattro ricordi di Ettore Pancini*; b) Antonio Borsellino: *Aprì l'Istituto al vento del luglio 1960*; c) Edoardo Amaldi: *La nostra vita negli anni della ricostruzione*; d) Marcello Conversi: *Come scoprimmo l'inganno del mesone*. Immediatamente dopo la scomparsa di Ettore erano già apparsi i seguenti articoli: e) su *L'Unità* del 3 settembre *In ricordo al caro Pancini* (non firmato); f) R. Fieschi: *Tra scienza e passione politica*; g) su *Il Contemporaneo* del 3 settembre: M. Cini: *Ettore Pancini, scienziato e politico*; h) sul *Manifesto* del 5 settembre: *Addio ad Ettore Pancini* (non firmato); i) su *L'Unità* dell'11 settembre: V. SANTORO e F. VANDI: *L'impegno di Pancini senza ripiegamento*; j) Il Notiziario mensile del Comune di Genova dedicò una pagina a *Ettore Pancini: un uomo che ha lasciato una traccia anche nella nostra città* (VII, 7, ottobre 1981, p. 14); k) *Ettore Pancini*, non firmato, *CERN Courrier*, vol. 21, p. 409 (1981).

(24) Allegata alla lettera del 21 maggio 1982 di Alberto Gigli a E. Amaldi.

(25) P. E. ARGAN and A. GIGLI BERZOLARI: *A new detector of ionizing radiation. The gas bubble chamber, Nuovo Cimento*, **3**, 1171 (1956); P. E. ARGAN and A. GIGLI BERZOLARI: *On the bubbles formation in supersaturated gas-liquid solutions, Nuovo Cimento*, **4**, 953 (1956); P. E. ARGAN, A. GIGLI, E. PICASSO, V. BISI, G. PIRAGINO, G. BENDISCIOLI e A. PIAZZOLI: *Due camere a diffusione per esperienze con gli elettrosincrotroni da 1100 MeV di Frascati e da 100 MeV di Torino, Suppl. Nuovo Cimento*, **17**, 215 (1960).

(26) G. Boato, R. Sanna, M. E. Vallauri e M. Reinharz: *Uno spettrometro di massa di elevata sensibilità, Suppl. Nuovo Cimento,* **16**, 215 (1960).
(27) A. Gamba, *Optimum Performance of Learning Machines, Proc. IRE,* **49**, 349 (1961).
(28) G. Palmieri and R. Sanna: *Automatic probabilistic programmer analizer for pattern recognition, Methods,* **48**, 1 (1960).
(29) Il Laboratorio Internazionale di Genetica e Biofisica del Consiglio Nazionale delle Ricerche, 1962/1967, cinque anni di attività, Dedalo litostampa in Bari 20-1-1969.
(30) Nei magneti ondulatori e «wiggler» un campo magnetico alternato, con andamento approssimativamente sinusoidale (e ad integrale nullo sulla lunghezza del magnete stesso) costringe un fascio di elettroni su di una traiettoria anch'essa approssimativamente sinusoidale. Percorrendo la traiettoria gli elettroni irraggiano intensi fasci di radiazione, dall'infrarosso fino alla regione dei raggi X duri. La distinzione fra wiggler e ondulatori è quantitativa piuttosto che qualitativa. Un magnete wiggler ha un campo magnetico intenso che oscilla spazialmente con grande lunghezza d'onda ($>$ 10 cm). Viene usato per produrre fasci intensi di radiazione dura, distribuita in frequenza secondo uno spettro continuo. Un ondulatore ha un campo più debole e lunghezza d'onda corta ($>$ 1 cm), e produce fasci intensissimi di radiazione più molle, in uno spettro di righe (accordabili).
(31) V. L. Ginsburg: *Izv. Akad. Nauk. Arm. SSR, Fiz.,* **11**, 165 (1947).
(32) H. Motz: *Applications of the radiation from fast electron beams, J. Appl. Phys.,* **22**, 527 (1951); Errata 1217.
(33) M. Berndt, W. Brunk, R. Cronin, D. Jensen, A. King, J. Spencer, T. Taylor and H. Winick: *Initial operation of SSRL wiggler in SPEAR, Proceedings of the 1979 Particle Accelerator Conference, IEEE Trans. Nucl. Sci.,* **26**, 3812 (1979).
(34) M. Bassetti, A. Cattoni, A. Luccio, M. Preger and S. Tazzari: *A transverse wiggler magnet for Adone, Proceedings of the X International Conference on High Energy Accelerators, Protvino, URSS* (1977).
(35) R. Barbini e G. Vignola: *Proposta preliminare di un esperimento di laser a elettroni liberi (FEL) da effettuarsi con Adone,* Memo G-30 (1979); R. Barbini, M. E. Biagini, R. Boni, A. Cattoni, V. Chimenti, S. Guiducci, M. Preger, A. Reale, C. Sanelli, M. Serio, S. Tazzari, F. Tazzioli, C. Vignola, M. Castellano, F. Cevenini, R. Rinzivillo, E. Sassi, U. Troya, T. Clauser e V. Stagno: *Proposta di esperimento per la realizzazione di un laser ad elettroni liberi con il fascio ricircolato di Adone (LELA = Laser ad Elettroni Liberi in Adone), 27 novembre 1979*; R. Barbini and G. Vignola: *LELA: A free electron laser experiment in Adone,* LNF-80/12(R) (1980).
(36) Giulio Cortini: *Il compagno, lo scienziato,* discorso funebre pronunciato il 4 settembre a Venezia in occasione del funerale di Ettore Pancini e pubblicato su *Rinascita* dell'11 settembre 1981, p. 40.

46

IN COMMEMORATION OF ETTORE PANCINI (1915 - 1981)

Edoardo Amaldi
Department of Physics
University of Rome "La Sapienza"

On April 20, 1982, a ceremony in commemoration of Ettore Pancini organized jointly by the University of Naples, the Società Italiana di Fisica and the Istituto Nazionale di Fisica Nucleare, was held in the Aula Magna of the Società Nazionale di Scienze, Lettere e Arti (Via Mezzocannone 8, Naples). Following speeches by the rector, Prof. Carlo Ciliberto, the president of the Società Nazionale di Scienze, Lettere e Arti, Prof. Paolo Corradini, the president of the Società Italiana di Fisica, Prof. Renato Ricci, Edoardo Amaldi expressed his personal condolences and the condolences of all present for the passing away of this unforgettable friend to Mrs. Elda Rupil, Ettore's widow, their children, Barbara, Alessandra, Giulio and Alice, relatives and all those who had been emotionally or intellectually close to Ettore during the various periods of his life. Amaldi then went on to portray the man with the following words.

1. A quick look at his life

Ettore Pancini was born in Stanghella, then in the province of Padua, on August 10, 1915, the second child of Giulio and Maria Galeazzi. His father, who came from an old family in Varmo (Udine province), where he was born, worked as municipal engineer for the city of Venice and during the course of his career became inspector of the Water Authority for the province of Venice.

Ettore's brother Mario, approximately a year and a half older, became an engineer for the Italian state electricity utility ENEL, while his sister, fourteen years his junior, married Dr. Franco Franco, physician and now head of the Hospital for Infectious Diseases in Venice.

Ettore attended elementary and lower secondary school in Venice, but in consideration of his youthful vivacity and initiative, was sent by his parents to the Military College in Naples to complete his secondary education.

After graduation from classical studies, Ettore registered with the faculty of mathematics at the University of Padua. One year later, however, he switched to the faculty of physics, where he studied and took his degree under Bruno Rossi, who chaired the Department of Experimental Physics at that university from 1933. In those years, Bruno Rossi had not only begun and practically completed construction of a new building as the seat of the "Galileo Galilei" Physics Institute, but had set up one of the most advanced and active centers of research on cosmic rays in Europe and the world. Nevertheless, as a result of the racial laws promulgated by the Fascist government on July 14, 1938, Bruno Rossi was forced to leave the country on October 12, 1938, and was replaced by Prof. A. Drigo as Ettore Pancini's supervisor when he sat his degree examination only a few weeks later.

Those years at the University of Padua were of decisive importance for Ettore from several points of view (Fig. 1). Travelling to and from Padua everyday from Venice, he met a young woman studying Italian literature, Elda Rupil, whom he married in 1941. At the University of Padua he became friends with a number of students who were intolerant of the Fascist regime. One of them was Giorgio Trevisan, a mathematics student and later professor of mathematics at the same university.[1] The two were to have a great effect on each other's political and social ideas. At that time or only a short time later, Ettore met Eugenio Curiel,[2] member of the clandestine Communist Party in Venice. These friendships were decisive in his shift a few years later from the Partito d'Azione, to which he initially belonged, to the Italian Communist Party.

After receiving his degree, Ettore was immediately given the position of special assistant (*assistente straordinario*) at the Physics Institute of the University of Padua. But Ettore somehow felt abandoned after the departure of Bruno Rossi and started to look for another place in which to undertake work seriously in more or less the same field in which he had started out with Rossi.

This led him to Rome, where I met him for the first time at the end of 1939 while he was making inquiries about the situation there. Much interesting research on cosmic rays was being carried out at that time under Gilberto Bernardini, professor of superior physics at the University of Bologna, who dedicated a few days a week to the research being carried out in Rome at the "Guglielmo Marconi" Institute. Bernardini had managed to convince Prof. Antonio Lo Surdo, director of the Guglielmo Marconi Institute and the Istituto Nazionale di Geofisica (ING) to bring at least some research on cosmic rays into the ING framework, thus providing a source of human and material resources.

Pancini was first (January-November 1940) hired as a researcher at the ING. But after winning a competitive examination, he was named permanent assistant lecturer at the Physics Institute of the University of Rome on December 1, 1940, and, more precisely, assistant to the "Ufficio del Corista Uniforme", an office created by Blaserna[3] in 1887 to control manufactured musical instruments and to check the musical scale used by musical associations.

On February 1, 1941, however, Ettore was called to arms and had to give up his teaching and research activity until the end of the war. The only exception was a brief leave in the winter of 1942-43, during which he worked for a few months at Plateau Rosà with Bernardini, Cacciapuoti and Piccioni, as I will recount in more detail later.

As second lieutenant, Pancini was stationed in various places with his anti-aircraft artillery group. As of 1941, his wife was with him to Cormons and Montegliano in Friuli; Cecchignola on the outskirts of Rome; Riccione and S. Remo, even after the birth of their first daughter, Barbara, in 1943.

On July 19, 1943, Ettore was wounded during the bombing of Rome while leaving for Venice from Termini Station. Therefore, on September 8, the day the armistice was signed between Italy and the Allied Forces, Pancini was on convalescent leave in Venice. In response to Mussolini's orders to all reserve officers to return to service, Ettore immediately joined the resistance movement. Taking on the well known name of Achille, he soon became the commander of the Gruppi di Azione Partigiana (GAP) in Venice headed by the Comitato di Liberazione Nazionale.[4] In 1944, he was named commander of the Venice Military District. In that capacity, he organized numerous military attacks in which he always participated personally. Famous among these are the attack on Ca' Giustinian, headquarters of the German Command in Venice, and his incursion, only a few days before the liberation of Venice, into Goldoni Theatre, where "Achille" read to the spectators the proclamation of the imminent liberation of Venice and the end of Nazi and Fascist oppression.[5]

One episode which I would like to recall from the period in which he took part in the resistance movement dates back to February 1945. He was in the mountains near Novara at the time, while his wife was in Venice, expecting their second child. As if by instinct, Ettore returned to Venice just in time not only to witness the birth but to deliver the child himself, as the obstetrician had been held up for some reason. Doing everything himself, he even cut and tied the umbilical cord of his second daughter, Alessandra, born on February 19, 1945. Ettore only spoke of this adventure with his most intimate friends, but he spoke of it with obvious satisfaction and pride. After the birth, Ettore had to leave immediately to return to his post in the mountains, also because word of his presence in Venice had spread and plans for his capture were already being prepared.

His other two children, Giulio and Alice, were born later--in 1949 and 1950, to be exact--in no doubt much less dramatic conditions.

In early April 1945, the Germans were informed that Ettore was hiding out in a farmhouse. He was arrested, but only 20 days later managed to escape with the help of Fascist soldiers who hoped to gain his favour in view of the already predictable outcome of the war. He returned to his combat post and only a short time later led a group of resistance fighters in a real battle in the Portogruaro area against a German contingent that was attempting to withdraw to the Brenner Pass.

Ettore was the representative of the Corpo Volontari della Libertà in the Venice Comitato di Liberazione Nazionale until the end of April 1945, when he helped to dismantle the corps and returned to Rome. He took up his research at the Guglielmo Marconi Institute in September 1945.

His salary as assistant lecturer at the Physics Institute at the university came to 14,000 lire per month, certainly not enough to keep a wife and family. For a few months he was put up by Gilberto Bernardini and his family in the apartment they rented in via Dalmazia (Fig. 2). Then, he and another young man from the Institute, Ruggero Querzoli, set up two camp beds in one of the lab rooms. When this room was finally needed for research, Ettore moved his camp bed to one of the restrooms of the Physics Institute, where it remained a long time, joined from time to time by those of other young people such as J. Buschmann, a German researcher, and later D. Broadbent, an English researcher, both of whom came to the institute on extremely meager scholarships.

At that time, Ettore and Carlo Ballario basically lived for months on the chestnut flour that Ballario brought back from the Appennines near Bologna.

When Gilberto Bernardini came up with the idea of building a high altitude lab for the study of cosmic rays, Ettore Pancini became his permanent and closest collaborator both during construction and while bringing the lab into operation [13]. It was Bernardini's express desire that Ettore Pancini should become director of the Testa Grigia Laboratory (altitude 3500 m), located on the Italian-Swiss border near the Teodulo Pass. Ettore held this position until 1953 and directed the enlargement of the facility in 1950 [14].

At the end of the forties, Ettore and other young people from the Guglielmo Marconi Institute were very active in recovering electrical material from the ARAR (Azienda Rilievo e Alienazione Residuati) camps, where war surplus materials were stored. This material was extremely important for the recovery and development of research activity at Rome University's Physics Institute, which had such scant resources that it could not afford to purchase large quantities of new materials.[6]

In November 1949, Ettore Pancini was appointed assistant to the director (*aiuto del direttore*) of the Institute, a position which I had been filling for several months after the death of Antonino Lo Surdo on June 7th of that year. But approximately one year later, he ranked first of three candidates in the competitive exam for the chair of physics at the University of Sassari, where he took up the position of visiting professor on November 1, 1950.

In 1952, he was called to the Faculty of Sciences of the University of Genoa to take over the chair of experimental physics left by Giuseppe Occhialini, who had moved to the University of Milan. Ettore remained there until 1961, when he moved to the Faculty of Science of the University of Naples, where he remained until his death on September 1, 1981.

2. Research activity in his early years

Ettore Pancini's first studies on cosmic rays date back to 1940. Carried out with G. Bernardini, M. Santangelo and E. Scrocco [1, 2], they deal with equilibrium conditions between the hard component and its secondary electrons, which, as the authors demonstrate, are in far greater number than calculations based solely on collision processes, radiation and the creation of pairs predict. These studies were among the first to point to the presence of secondary electrons generated during decay of the particles making up the hard component of cosmic rays. They had been named "mesotrons" in 1936, but their nature was far from clear at that time.

During the same period, Pancini helped to prepare the expedition that Bernardini was organizing under the auspices of the ING to extend study of cosmic rays to high altitudes (Plateau Rosà, 3500 m). He was even able to take part in the expedition itself, thanks to a few months' leave from military service.

Pancini's interest was focussed on two major questions at the time. The first, in collaboration with Bernardini, Conversi, Scrocco and Wick [3, 4, 7] concerned measurement of the surplus of positive mesotrons with respect to negative ones. The authors used "magnetic lenses" which allowed them to set the telescope counter so as to pick up either positive or negative mesotrons, whichever they wished. This technique, first proposed by B. Rossi,[7] had already been used successfully by G. Bernardini and M. Conversi.[8]

The second, studied together with Bernardini, Cacciapuoti, Conversi, Scrocco and Wick [5, 6], dealt with the anomalous absorption of mesotrons in low density media as compared to high density media. They managed to calculate the relation between the mean life of these particles and their rest energy by measuring the effect produced by the decay of mesotrons. The values found by this Italian group and those measured independently at more or less the same time by an American group headed by Bruno Rossi,[9] turned out to be in excellent agreement and constituted an important scientific success in the early forties.

After his return from military service in 1945, Ettore took up his work with Conversi and Piccioni, who had in the meantime developed a very sophisticated technique for measuring "delayed coincidences" for use in direct measurement of the mean life of the mesotron.[10] In particular, Piccioni had invented a new kind of circuit for fast coincidences based on the use of secondary-emission tubes,[11] which became an essential part of the electronic apparatus later developed and used by himself and Conversi for direct measurement of the mean life of the mesotron,[12] something they had considered doing since the end of 1941, when they first decided to work together. This apparatus allowed the authors to achieve results that were considerably superior to those obtained by the pioneers in such measurements (Rasetti in Quebec[13] and Auger and his colleagues in Paris,[14]) and were comparable to those obtained only shortly before them--but unknown to them--in the United States by Rossi and Nereson.[15]

In a subsequent experiment[16] based on the same technique,([10,11]) Conversi and Piccioni demonstrated, in accordance with Rasetti's preliminary measurement in aluminum, that in iron only about half of the mesotrons undergo spontaneous decay at the end of the range: a result which is in apparent agreement with the hypothesis that only positive mesotrons disintegrate once they have been brought to rest, while negative ones are captured by the atomic nucleus, as predicted by Tomonaga and Araki for Yukawa particles.

The technique employed by Conversi, Pancini and Piccioni in this group of particularly important papers combined use of magnetic lenses, employed previously in the work on the positive surplus([8]) [3, 4, 7], with that of delayed coincidences developed by Piccioni and Conversi.([10,11])

In an initial study [8], in which the cosmic mesotrons were reduced to rest in iron (atomic number Z = 26), Conversi, Pancini and Piccioni[8] showed that, in apparent agreement with Tomonaga and Araki's theoretical predictions based on Yukawa's theory that mesotrons are mediators of nuclear forces, these particles, if positive, decay spontaneously once reduced to rest, while if negative, they are absorbed by the nuclei of the atoms of the medium in a much shorter time than their mean life.

On the basis of some well worked out hypotheses, the same authors then repeated the experiment, reducing the mesotrons to rest in a material with a much smaller atomic number than iron, in this case, carbon.

The outcome of the experiment was totally unexpected: the absorption of negative mesotrons by the carbon nuclei was negligible; contrary to the predictions of Tomonaga and Araki, both negative and positive mesotrons decay spontaneously before being captured by the carbon nucleus [9, 10, 11].

This is not the place to enter into a detailed discussion of this experiment and its consequences and implications. This has been done in a much more thorough and fascinating way in the papers presented by Oreste Piccioni and Marcello Conversi to the Symposium on the History of Particles Physics held at the Fermi Laboratory in May 1980.[17] I will merely say that Conversi, Pancini and Piccioni demonstrated that the particles making up the hard component of cosmic rays, until that time called mesotrons, were not the particles hypothesized by Yukawa, but rather particles endowed *only* with weak interactions (and if charged, with electromagnetic interactions), that is, particles that belong to the category (or family) now known as *leptons*. Until that time, only one lepton with charge was known: the electron. After the experiment by Conversi, Pancini and Piccioni, the family of leptons gained a new, important member, later named *muon*.

The experiment also gave rise to a number of theoretical works. Let me mention only one by Fermi, Teller and Weisskopf,[18]which worked out the discrepancy factor (in the order of 10^{12}), between theoretical results and Conversi, Pancini and Piccioni's experimental results, and one by Wheeler,[19] which established a law of dependence of the probability of capture of a negative muon by a nucleus on its atomic Z (the so-called Z^4 law). Finally, the experiment by Conversi, Pancini and Piccioni paved the way for study of mesic atoms, a field in which many papers, articles and books have now been written.[20]

It is therefore understandable that authors such as Bethe and colleagues[21] and Louis Alvarez[22] referred in their important publications to the experiment carried out by Conversi, Pancini and Piccioni as the beginning of high-energy physics.

A few months after this important discovery, first Piccioni and then Conversi went to work in the United States: the former to remain there permanently, the latter to complete a period of study.

In Rome, Pancini continued the study of the behaviour of muons reduced to rest in carbon with A. Alberigi Quaranta [19] and published, together with the same young colleague and L. Mezzetti and G. Stoppini, a technical paper on the behaviour of electrons in the mixtures used to fill Geiger counters [17].

He had already worked out a technique for measurement of very short time intervals (10^{-8}) with the latter two researchers [16], in view of its application to the study of the arrival times of particles of different mass belonging to the same extended shower.

In parallel to this scientific research, Pancini dedicated his time in Rome to the development and improvement of some very advanced electronic equipment. Of these, I would like to recall an impulse generator for the study of the stimuli and the propagation of electric impulses in nervous tissue, a project carried out in collaboration with the well known physiologist of the nervous system from the University of Pisa, Prof. G. Moruzzi, and his friend, Franco Lepri [15].

From 1946 to 1949, Pancini also dedicated part of his time, along with G. Bernardini, B. N. Cacciapuoti, B. Ferretti, F. Lepri and O. Piccioni, to setting up the so-called "Beta laboratory" in Rome. Financed by FIAT, the Beta lab was aimed at dealing with problems of application, mainly in the field of electronics.

3. Activity in research construction and promotion

When Ettore moved from Rome to Sassari in 1950, he found himself in a rather new situation: the Physics Institute of the university was in such a state and the resources so scarce that no real work could be undertaken. In the short time at the university, Ettore's main efforts were aimed at investing the little money at his disposal in improving the library, which was almost totally devoid of the modern publications and scientific journals essential for any kind of research activity.

In that period, he continued to frequent, although on a part-time basis, the Physics Institute of the University of Rome, where he worked for some time together with Franco Bonaudi and Lorenzo Resegotti in perfecting liquid scintillators.

I would like to tell the following anecdote, which took place while Ettore was in Sassari, to underline Ettore's youthful spirit even after he had become professor. One evening, while walking in the city with friends, in a state of bliss apparently brought on by a few glasses of good wine, Ettore took up a bet and climbed the stone face of the rector's building up to the window of the rector's office. It seems the feat gave rise to a vast range of comments the next day.

When Ettore reached Genoa in 1952, he found an institute that had already been revived to some extent by the two year presence of Giuseppe Occhialini. Occhialini had started up a research group on cosmic rays using the technique of nuclear emulsions. A technique to which he himself had made an extraordinary contribution, transforming it from a rather rough method of observation into a very precise method of measurement, able to provide details that were practically imperceptible by any other means.

Ettore's appointment to the chair of experimental physics at the University of Genoa was largely the result of the influence of Antonio Borsellino, who was professor of theoretical physics at the university. "Ettore's arrival in Genoa," Borsellino recently wrote,[23] "marked a turning point for the development of physics at the University of Genoa. In those years, physics in Italy was being organized in an unprecedented way, transforming the fervour and spirit of sacrifice and cooperation of the first postwar years into solid structures making it possible to work more efficiently and to open up to growing numbers of young researchers.

The Società Italiana di Fisica (SIF) on the one hand, with its prestigious journal *Il Nuovo Cimento* and the Enrico Fermi International School in Varenna, and the Istituto Nazionale di Fisica Nucleare (INFN), on the other, were the operational instruments that allowed Italian physicists to produce internationally competitive and significant scientific research. Ettore Pancini was active both as a counsellor of the Società Italiana di Fisica and in setting up the new indispensable structures of the Physics Institute in Genoa."

Alberto Gigli Berzolari, who was assistant to Pancini during his time in Genoa, sent me the following testimonial.[24]

"Ettore's commitment while he was director of the Physics Institute of the University of Genoa was characterized by a high and noble sense of freedom in his relations with both the people working in the Institute and those outside it that I have rarely seen in other university institutes. An open spirit which certainly derived from his upbringing and which was strengthened by the milieu of the Physics Institutes in Padua and, above all, Rome.

In Genoa, he managed to turn an old building into a modern university institute by surrounding himself with active and broad-minded people and by making wise choices in all the sectors that he, rightfully, felt required essential basic structures on which to build solid future prospects for research and culture: the library, teaching and scientific structures, the mechanics workshop, the electronics workshop, the carpenters workshop, staff. With our help, he hired professional personnel, taking advantage of the momentary crisis in some local high-tech industries. It was easy for us researchers to plan projects involving complex activities that would have been impossible without those structures. In my opinion, much of the subsequent success achieved by the Institute in Genoa can be traced back to that operational and promotional effort. Charming and charismatic, apparently gruff, but boundlessly generous and kind, he managed to create a scientific community of real friends. And this was one of his greatest merits.

I was working on "diffusion chambers" at that time as part of a project proposed by Giorgio Salvini during my stay in Rome and relocated to Genoa after my transfer there. Ettore seriously undertook construction of a large diffusion chamber (the largest ever built, I believe), spurred by the idea that we all had to cooperate--patriotically he would say, and it wasn't meant as a joke--in the national effort aimed at constructing the electron synchrotron in Frascati and the instrumentation needed for subsequent experimentation. This was his real and generous sense of service for the public good. And he was right!

At the same time, he stimulated us to come up with new projects and was enthusiastic when my colleagues and I would try out new and original ideas (that was 1954!) in the sector of ionized particle detectors. Thus, we built the first "dissolved gas bubble chamber" which differed in its conception from the traditional "overheated liquid" chambers already under study by Pietro Bassi in Padua and by Giuseppe Martelli in Pisa. Ettore was overjoyed when we announced to him that our efforts had been rewarded and that the effect of ionizing radiation on the behaviour of the gas-liquid solutions was as we had empirically hypothesized.[25]

In those days, we used to stay on at the Institute in the evenings and on Saturdays and Sundays. Leaving the lab after a day's work, we would leave a note about what we had done during the day because we knew that Ettore would make the round of the labs to see what progress we were making before going home. Late one night, I came upon him in our laboratory while he was examining the first photographs (which were particularly good) taken that same day with a prototype of our gas bubble chamber. I will never forget the enthusiasm with which he commented upon the quality of our work and the long and stimulating discussion full of acute observations and suggestions that followed.

Finally, I must point out that it was thanks to Ettore's concern and support that the Istituto Nazionale di Fisica Nucleare intervened in Pavia. Because, as Ettore said, the universities have considerable opportunities to supply young and qualified personnel for the advancement of physics in Italy and the INFN cannot ignore this fact."

Later, thanks to the university's excellent mechanical workshop and qualified technical personnel, a mass spectrometer for analysis of isotope abundances was constructed under the direction of Giovanni Boato.[26] Pancini insisted that a second one be given Tongiorgi in Pisa for his research on isotope geochemistry.

Along with the mechanics workshop, Pancini also set up an electronics laboratory capable of producing instrument prototypes, of which some were adopted by other laboratories. The laboratory also served as an indispensable technical base for the well known activities of Augusto Gamba[27] and Guido Palmieri,[28] aimed at pattern and image recognition from points of view and with objectives that were totally unconventional.

Finally, he had the Genoa Section of the INFN purchase the 30 MeV betatron built by Brown Boveri and owned by the Turin Section of the INFN when the latter managed to buy a higher energy machine. This betatron laid the basis for the formation of a Genoese school of nuclear physics particularly interested in the use of gamma rays.

While in Genoa, Ettore dedicated a large part of his intelligence and energies to working out scientific programmes for the "Casa della Cultura", then directed by Enrica Basevi, of whom he became a collaborator and close friend.

When called to the chair of Superior Physics at the University of Naples, especially upon the urgings of Giulio Cortini, professor of experimental physics at that university since 1959, he became a member of the executive committee of the INFN from February 1959 to June 30, 1962. He was also director of the Naples Section of the INFN from July 1, 1962 to June 30, 1965, of the Institute of Superior Physics from 1965 to 1970 and of the Institute of Experimental Physics from 1970 to 1972. When the chairs were divided in 1962, Ettore passed from Superior Physics to General Physics I, where he taught until 1978, when he decided to change once again and move to Optics.

In addition to these official activities, Ettore was also strongly involved from 1961 to 1967 in the founding (by Adriano Buzzati Traverso) and running of the CNR's International Genetics and Biophysics Laboratory;[29] from 1972 to 1975, he was an authoritative member of the relative scientific committee.

Ettore was fully supported in some of his activities not only by Giulio Cortini, but also by Ruggero Querzoli, professor of Superior Physics in Naples from 1965 to 1967. In 1962, Cortini organized a "didactic seminar" in the Faculty of Science, aimed at updating high school physics teachers and doing didactic research. Cortini later directed this seminar, while Ettore was a member of the seminar's steering committee. But Ettore also personally committed himself to work in a research group on the teaching of sciences at the elementary level, in collaboration with psychoanalyst and friend Jacqueline Mehler Amati. His visits to a number of classes of the "De Luca" elementary school, located in Traiano, an extremely poor neighbourhood of Naples, left some unforgettable memories, as attested to by the letter shown in Fig. 3.

When Cortini transferred from the University of Naples to the University of Rome in 1975, Ettore took over as director of the didactic seminars.

Together with Giulio Spadaccini and Vittoria Santoro, an old friend and co-worker, Ettore

Pancini undertook study in 1968-69 of what happens at the interface between a hydrocarbon and a saline solution under the action of gamma rays. This research was in some ways linked to the problem of the formation of organic molecules at the beginning of prebiological evolution. Pancini and his colleagues soon realized that very little was known about an even simpler problem--gamma radiolysis of a single hydrocarbon--and therefore redirected their investigations in that direction. Initial results were published in a paper on gamma radiolysis of pentane and hexane [20].

Unfortunately, redtape and construction took such a long time that the fruits of the Tandem 3 + 3 MeV accelerator Pancini decided to buy in 1970 with INFN funds have only started to be seen recently.

4. His return to active research

When Ettore left General Physics for Optics in 1978, he set up a research group in the field of coherent light optics, which included Solimeno, Arimondo and Santamato. The latter came from Rome where he had studied under and worked with Francesco de Martini who had, in the meantime, taken over the chair of the structure of matter in Naples.

But Ettore had already returned to an interest--I might almost say a juvenile interest--in active research. While the joint "Progetto Utilizzazione Luce di Sincrotrone" (PULS) was being set up by the CNR-INFN at the National Laboratories in Frascati in 1976-77, Ettore took up frequent contact with Franco Bassani, president of the scientific commission of the PULS project, and with some of the researchers involved in the project, such as Emilio Burattini, a student of his in Genoa in 1961, Pancini's last year at that university, and Mario Iannuzzi.

Later, his interest turned more concretely towards development of "wiggler" and undulatory magnets[30] for the production of synchrotron light at the INFN National Laboratories in Frascati.

Starting out from an original proposal by Ginsburg,[31] H. Motz had constructed a wiggler with which he managed to produce visible light in 1951 using a 100 MeV electron beam.[32]

Wiggler magnets were subsequently used for various purposes in some circular machines (for example, around 1965, they were used at the CEA to dampen betatron oscillations).

The first wiggler designed and used as an intense source of synchrotron radiation was the one (with 6 equivalent poles) at the Stanford Synchrotron Radiation Laboratory (SSRL). It was brought into operation in March 1979 on the SPEAR storage ring by a group led by H. Winick.[33]

The Adone wiggler, designed and built by the Adone group,[34] was tested successfully in September 1979. Several members of the PWA group (Naples, Frascati, Trent, Daresbury), formed to set up a laboratory for use of synchrotron radiation, came from Naples (E. Burattini, N. Cavallo, M. Foresti, C. Mencuccini, P. Patteri, R. Rinzivillo and U. Troya). Pancini entered the group in late 1979, when the instrumentation built in Naples was transferred to Adone.

In 1980, the group observed the first intense X-ray beams and started the first cycle of experiments [21].

While the Brookhaven wigglers were and still are under construction, wiggler magnets are currently in operation in the USSR, England (Daresbury) and France (Orsay).

Ettore was enthusiastic about the proposal put forward by Barbini, Cattoni and Vignola from the LELA (Laser ad Elettroni Liberi in Adone) group to run some tests on the emission of visible coherent light (that is, to try to use the wiggler as an undulator) and followed all developments closely. The result was very spectacular and produced remarkable chromatic effects [21].

Since June 1980, the PWA has been intensively used by that and other research groups for experiments in absorption spectroscopy between 10 and 30 keV (that is, between 1 and 0.4 Å).

When Barbini and Vignola suggested performing an experiment to test the feasibility of a free electron laser in Adone (LELA)[35] and later, in 1979, when the Frascati-Naples group was formed [21], Ettore joined, enhancing the group with his ideas and enthusiasm.

As Cortini recalls,[36] ". . . his participation in working out the scientific approach to much of the research of his younger colleagues and the help he continuously provided, also with regard to experimental techniques, would have given him the right to put his name alongside others to many publications. But Ettore was like that. While in Venice the other day for Ettore's funeral, Salvini told me that he had never known a researcher so totally disinterested in power and recognition."

5. His sensitivity to political, social and human problems

Returning to Ettore's political activity, I must recall that after the Tenth Congress of the Soviet Communist Party held in Moscow in November 1956, in which Khrushchev severely criticized Stalin's rule, Ettore, like others, left the Italian Communist Party, but did so very quietly.

Upon the urging of various friends, he agreed to run for parliament as a candidate for Democrazia Proletaria in 1979 in the Napoli-Caserta riding and in the municipal (second on the list) and regional (third) elections in Naples in 1980.

For those who did not know Ettore Pancini personally, what I have said up to now may not have transmitted some of the most fundamental aspects of his personality and character: his open-minded intelligence, his reserved but deep affection for family and friends, his detachment from all forms of personal power and his tolerance and respect for the opinions of others.

If I remember correctly, it must have been the winter of 1947 or 1948 when Ettore came into my office at the Guglielmo Marconi Institute and told me that he had met Lamberto Allegretti on the street, in a very sad and depressed state. During the war, Allegretti, a physicist from Pisa, had been heavily involved in developing some technical aspects of the assault weapons used by the Italian Navy. When Ettore met him, he had just lived through a distressing trial and its consequences brought on by his participation in "X Mas" in the Salò Republic.

Ettore asked me if I had anything against helping Allegretti out of this state by inviting him to come and work at the Institute and was pleased to hear my answer. That was how, from the next day onward, Allegretti started to work for the Institute and contributed to reconstruction and recovery in an unforgettable atmosphere of fitful activity.

Sometimes, in discussions with colleagues and friends, his desire and taste for contradiction led him to take up positions that verged on the paradoxical and were frequently far from his own well-pondered and lucid thoughts. But in forty years of friendship, we were in agreement or at least on the same side on a great many occasions. There were nevertheless some cases in which our opinions differed and in two or three we had opposite stances. But even on those occasions, our reciprocal respect and esteem never abated and both of us were genuinely aggrieved by these divergences deriving from profound, well meditated and unrelinquishable convictions.

It was on these occasions, that I was able to perceive his personality even more clearly, a personality that was always complex, restless at times, but always guided by an aspiration for justice in the most noble sense of the word.

His tolerance for others was accompanied by strict principles in his behaviour as a political and social being. His personal charm, his cheerfulness and his dedication to a person or a cause were, for all those who knew and loved him, unforgettable sources of admiration and, at the same time, intimate pleasure.

---

[1] Giorgio Trevisan (1916-1974) studied in Padua where he became assistant lecturer in calculus in 1939 and university teacher in 1954. In 1962, Trevisan was named professor of calculus in Palermo, where he remained until 1965, when he returned to Padua in the same capacity.

[2] Eugenio Curiel (Trieste 1912 - Milan 1945) was the author of a number of works published in a single volume: *Eugenio Curiel, Scritti 1935-1945* (Editori Riuniti, Rome, 1973). The preface by Giorgio Amendola provides much information on his life as a politician and physicist, linked to Bruno Rossi at the University of Padua from 1933 to 1939, the year in which he was arrested on June 23. Sentenced to five years internment, he was transferred to the island of Ventotene in January 1940. After the fall of Mussolini's government on July 25, 1943, he left Ventotene on August 21, 1943 and went to Milan, where he took up clandestine political activity. While walking down the street on February 24, 1945, he was pointed out by an informer to a Fascist action squad, which, despite his desperate attempt to escape, machine-gunned him down.

[3] Pietro Blaserna, born in Fiumicello (Gorizia) in 1836, studied physics in Vienna and Paris. Professor of physics at the University of Palermo from 1863 to 1872 and at the University of Rome after 1872, he designed and had constructed (1877-1880) the Via Panisperna building of the Physics Institute, which he directed until his death in Rome in 1918. In 1890, he was named Senator of the Kingdom of Italy and in 1896 was elected Vice President of the Senate, a position which he held until his death.

[4] This information on the resistance activity of Ettore Pancini is based on notes that I took during a more than two-hour long conversation that I had with him, upon my request, in my apartment in Rome on November 10, 1978, in view of my intention to write the history of the "war years" of the Guglielmo Marconi Institute. The information was then reviewed and integrated by Mrs. Elda Rupil, Pancini's widow, and Giulio Cortini. Some of it can be found in the volume *Venezia nella Resistenza, 1943-1945*, edited by Giuseppe Turcato and Agostino Zanon Dal Bo and published by the Municipal Administration of Venice (1975-76), but unfortunately no longer available today.

[5] See *Il secondo Risorgimento d'Italia* (Centro Editoriale di Iniziativa, 1975) which does not, however, give the name of the resistance fighter who read the proclamation. Ettore Pancini told me that it was Achille during our conversation on November 10, 1978 (see footnote 4).

[6] E. AMALDI: *Gli anni della Ricostruzione, G. Fis.*, **20**, 186 (1979).

[7] B. ROSSI: *Magnetic experiments on cosmic rays. Nature* (London), **128**, 300 (1931).

[8] G. BERNARDINI and M. CONVERSI: *Sulla diffusione dei corpuscoli cosmici in un nucleo di ferro magnetizzato, Ric. Sci.*, **11**, 840 (1940).

[9] B. ROSSI, N. H. HILBERRY and J. D. HOAG: *The variation of the hard component of cosmic rays with height and the disintegration of the mesotron, Phys. Rev.*, **57**, 461 (1940); B. ROSSI and D. B. HALL: *Variation of the rate of decay of mesotrons with momentum, Phys. Rev.*, **59**, 223 (1944).

[10] M. CONVERSI and O. PICCIONI: *Sulla registrazione di coincidenze e piccoli tempi di separazione, Nuovo Cimento*, **1**, 279 (1943).

[11] O. PICCIONI: *Un nuovo circuito di registrazione di coincidenze, Nuovo Cimento*, **1**, 56 (1943).

[12] M. CONVERSI and O. PICCIONI: *Misura diretta della vita media dei mesoni fenati, Nuovo Cimento*, **2**, 40 (1944); the English version appeared in *Phys. Rev.*, **70**, 859 (1946).

[13] F. RASETTI: *Disintegration of slow mesotrons, Phys. Rev.*, **59**, 613 (1941); *Disintegration of slow mesotrons*, **60**, 198 (1942).

[14] P. AUGER, R. MAZE and R. CHAMINADE: *Une mesure directe de la vie moyenne du méson au repos*, C.R., **214**, 266 (1942); R. CHAMINADE, A. FRÉON and R. MAZE: *Nouvelle mesure directed de la vie moyenne du méson au repos*, C.R., **218**, 402 (1944).

[15] B. ROSSI and N. C. NERESON: *Experimental determination of the disintegration curve of mesotrons, Phys. Rev.*, **62**, 417 (1942); *Further measurements of the disintegration curve of mesotrons*, **64**, 199 (1943). These results only reached Italy at the end of the war, that is, in the summer of 1945.

[16] M. CONVERSI and O. PICCIONI: *Sulla disintegrazione dei mesoni lenti, Nuovo Cimento*, **2**, 71 (1944); the English version appeared in *Phys. Rev.*, **70**, 874 (1946).

[17] O. PICCIONI: *The observation of the leptonic nature of the "mesotron" by Conversi, Pancini and Piccioni*, p. 222; M. CONVERSI: *The period that preceded and led to the 1946 discovery of the leptonic nature of the "mesotron": some personal recollections*, p. 242 in *Proc. Intern. Symp. on the History of Particles Physics*, held at Fermi National Accelerator Laboratory on May 28-31, 1980, edited by L. M. BROWN and L. HODDESON (Cambridge University Press, Cambridge, 1983).

[18] E. FERMI, E. TELLER and V. C. WEISSKOPF: *The decay of negative mesotrons in matter, Phys. Rev.*, **71**, 314 (1947). In a subsequent paper (*The capture of negative mesotrons in nature, Phys. Rev.*, **72**, 399 (1947)), Fermi and Teller used a detailed calculation to demonstrate that the possibility of the total time of slowdown and capture of a negative mesotron being much longer in carbon than in iron was remote.

[19] J. A. WHEELER: *Mechanisms of capture of slow mesons, Phys. Rev.*, **71**, 320 (1947).

[20] S. DEVONS and I DUERDOTH: *Muonic Atoms*, in *Advances in Nuclear Physics*, edited by M. BARANGER and E. VOGT, vol. II, p. 295 (1969); C. S. WU and L. WILETS: *Muonic Atoms and Nuclear Structure*, in *Annual Review of Nuclear Science*, edited by E. SEGRE', vol. 19, p. 527 (1969); A. O. WEISSENBERG: *Muons* (North Holland, Publ. Co. Amsterdam, 1967).

[21] S. S. SCHWEBER, H. A. BETHE and F. DE HOFFMANN: *Mesons and Fields* (Row, Peterson and Co., Evanston, Ill., 1956).

[22] L. W. ALVAREZ: *Recent Developments in Particle Physics*, in *Les Prix Nobel de 1968* (Stockholm, 1969) p. 125.

[23] Page 3 of the October 1, 1987 edition of *L'Unità* contains a series of articles recalling Ettore Pancini under the general headline: *Un fisico di frontiera*: a) Giorgio Napolitano: *Quattro ricordi di Ettore Pancini*; b) Antonio Borsellino: *Aprì l'Istituto al vento del luglio 1960*; c) Edoardo Amaldi: *La nostra vita negli anni della ricostruzione*; d) Marcello Conversi: *Come scoprimmo l'inganno del mesone*. The following articles appeared immediately after Ettore's death: e) in the September 3 issue of *L'Unità*, *In ricordo al caro Pancini* (unsigned); f) R. Fieschi: *Tra scienza e passione politica*; g) in the September 3 issue of *Il Contemporaneo*: M. Cini: *Ettore Pancini, scienzato e politico*; h)in the September 5 issue of *Il Manifesto*, *Addio ad Ettore Pancini* (unsigned); i) in the September 11 edition of *L'Unità*: V. SANTORO and F. VANDI: *L'impegno di Pancini senza ripiegamento;* j) The monthly bulletin of the Genoa Municipal Administration dedicated a page to *Ettore Pancini: un uomo che ha lasciato una traccia anche nella nostra città* (VII, October 7, 1981, p. 14) k) *Ettore Pancini*, unsigned, *CERN Courier*, vol. 21, p. 409 (1981).

[24] Annex to the letter of May 21, 1982 by Alberto Gigli and E. Amaldi.

[25] P. E. ARGAN and GIGLI BERZOLARI: *A new detector of ionizing radiation. The gas bubble chamber, Nuovo Cimento*, **3**, 1171 (1956); P. E. ARGAN and A. GIGLI BERZOLARI: *On the bubbles formation in supersaturated gas-liquid solutions, Nuovo Cimento*, **4**, 953 (1956); P. E. ARGAN, A. GIGLI, E. PICASSO, V. BISI, G. PIRAGINO, G. BENDISCIOLI e A. PIAZZOLI: *Due camere a diffusione per esperienze con gli elettrosincrotroni da 1100 MeV di Frascati e da 100 MeV di Torino, Suppl. Nuovo Cimento*, **17**, 215 (1960).

[26] G. BOATO, R. SANNA, M. E. VALLAURI and M. REINHARZ: *Uno spettrometro di massa di elevata sensibilità, Suppl. Nuovo Cimento*, **16**, 215 (1960).

[27] A. GAMBA, *Optimum Performance of Learning Machines, Proc. IRE*, **49**, 349 (1961).

[28] G. PALMIERI and R. SANNA: *Automatic probabilistic programme analyzer for pattern recognition, Methods*, **48**, 1 (1960).

[29] *Il Laboratorio Internazionale di Genetica e Biofisica del Consiglio Nazionale delle Ricerche, 1962/1967, cinque anni di attività*, Dedalo litostampa in Bari 20-1-1969.

[30] In "wiggler" and "undulatory" magnets, an alternating magnetic field with an approximately sinusoidal pattern (and with integral zero over the length of the magnet) forces an electron beam into an also approximately sinusoidal trajectory. Along that trajectory, the electrons send out intense beams of radiation, from infrared to hard x-ray. The distinction between wiggler and undulatory magnets is quantitative rather than qualitative: a wiggler magnet has an

intense magnetic field that oscillates spatially with a long wavelength (>10 cm) and is used to produce intense beams of hard radiation distributed in a continuous frequency spectrum; an undulatory magnet has a weaker field and shorter wavelength (> 1 cm) and produces very intense beams of softer radiation in a spectrum of (tunable) lines.

[31] V. L. GINSBURG: *Izv. Akad. Nauk. Arm. SSR, Fiz.*, **11**, 165 (1947).

[32] H. MOTZ: *Applications of the radiation from fast electron beams, J. Appl. Phys.*, **22**, 527 (1951); Errata 1217.

[33] M. BERNDT, W. BRUNK, R. CRONIN, D. JENSEN. A. KING, J. SPENCER, T. TAYLOR and H. WINICK: *Initial operation of SSRL wiggler in SPEAR, Proceedings of the 1979 Particle Accelerator Conference, IEEE Trans. Nucl. Sci.*, **26**, 3812 (1979).

[34] M. BASSETTI, A. CATTONI, A. LUCCIO, M. PREGER and S. TAZZARI: *A transverse wiggler magnet for Adone, Proceedings of the X International Conference on High Energy Accelerators, Protvino, USSR* (1977).

[35] R. BARBINI and G. VIGNOLA: *Proposta preliminare di un esperimento di laser a elettroni liberi (FEL) da effettuarsi con Adone*, Memo G-30 (1979); R. BARBINI, M. E. BIAGINI, R. BONI, A. CATTONI, V. CHIMENTI, S. GUIDUCCI, M. PREGER, A. REALE, C. SANELLI, M. SERIO, S. TAZZARI, F. TAZZIOLI, C. VIGNOLA, M. CASTELLANO, F. CEVENINI, R. RINZIVILLO, E. SASSI, U. TROYA, T. CLAUSER and V. STAGNO: *Proposta di esperimento per la realizzazione di un laser ad elettroni liberi con il fascio ricircolato di adone (LELA= Laser ad Elettroni Liberi in Adone), 27 novembre 1979*; R. BARBINI and G. VIGNOLA: LELA: A free electron laser experiment in Adone, LNF-80/12(r) (1980).

[36] G. CORTINI: *Il compagno, lo scienzato*, funeral oration pronounced at the funeral of Ettore Pancini in Venice, September 4, 1981, published in *Rinascita* on September 11, 1981, p. 40.

Part II

European Physicists and Their Institutions

Section A
Physics At the Beginning of the Century

The two papers presented in this section share an unusual nature when compared with the rest of the texts collected in this volume. This is because they do not refer to events in which Amaldi was involved directly as a protagonist or as a mere witness, but instead describe events which took place in the early years of this century.

The first essay, *The Solvay Conferences in Physics*, was prepared by Amaldi in the spring of 1980. It was read at the XVII Solvay Conference in Chemistry which was held on 23–24 April of that year in Washington, D.C., jointly with the 117th annual meeting of the National Academy of Sciences on the occasion of the 150th anniversary of the independence of Belgium. Half of the first day of the conference was devoted to the history of the Solvay Institutes and Amaldi, who was a member of the Solvay Scientific Committee in Physics at that time, had been invited by the Director of the Solvay Institutes, Ilya Prigogine, to give a lecture on the history of the Solvay conferences in physics. This was followed by a lecture on the conferences in chemistry by A. Ubbelohde. The papers appeared in 1984 in the proceedings of the conference, edited by G. Nicolis, under the title *Aspects of Chemical Evolution*.

Radioactivity, a Pragmatic Pillar of Probabilistic Conceptions originated as a lecture delivered by Amaldi at the LXXII Summer Course of the International School of Physics "Enrico Fermi" held in Varenna from 25 July 1977 to 6 August 1977. The course, directed by G. Toraldo di Francia, who also edited the proceedings which was published in 1979, was devoted to *Problems in the Foundations of Physics*. Here, for the first time, Amaldi explicitly addressed an open historiographic question, namely, the roots of the "repudiation of causality" in quantum mechanics. The paper is an eloquent testimony of Amaldi's more than cursory acquaintance with the recent literature on the history of physics: in the introductory pages, reference is made to the contributions of M. Jammer, S. Brush and P. Forman to the subject. Actually, Amaldi had wanted to intervene in the debate to correct what was, in his opinion, an excessive emphasis on the "external influences" behind the rejection of strict determinism in physics. This was found, namely, in Paul Forman's controversial essay on Weimar physicists and causality. His aim was, in his characteristic "pragmatic" approach to history, to redress the balance by focusing on the more strictly "intrinsic influences" that lay, according to him, at the core of the physicist's acceptance of an inherent a-causal description of natural phenomena, which are rooted in the development of research on radioactivity.

Offprints From
Aspects of Chemical Evolution
Edited by Dr. G. Nicolis

THE SOLVAY CONFERENCES IN PHYSICS

EDOARDO AMALDI

Istituto di Fisica "Guglielmo Marconi," Universita degli Studi, Rome, Italy

I. INTRODUCTION

The number of Solvay Conferences in Physics, held in Brussels every few years, starting from 1911 until 1982, has been 17.

As it has been explained by Mr. Jacques Solvay, the first one had been conceived by Ernest Solvay in consultation with Walther Nernst, as a tool to help directly in solving a specific problem of unusual difficulty impending over the whole of physics: was the quantum structure of nature really unavoidable? Such a structure had been suggested by Max Planck[1] in 1900, for interpreting the observed spectrum of the blackbody radiation. Five years later Einstein[2] had again considered it in connection with the fluctuations of the energy density of the electromagnetic field and, as an example, he had applied it in an extended form, to the photoelectric effect. Einstein, had used it once more a few years later for the computation of the specific heat of solid matter.[3] But the majority of physicists disliked the introduction of such a procedure, which appeared in contrast with the whole conception of nature prevailing in those days.

In 1911 the idea of national and international conferences dealing with scientific matters, was about a century old. The first scientific conference, probably, was the one organized in 1815 in Geneva by the chemist H. A. Gosse on the "Physical and Natural Sciences." A number of scientific conferences had been held in many places during the following 96 years.

The first Solvay Conference in Physics, however, set the style for a new type of scientific meeting: as Mr. Jacques Solvay told us, the participants in the meeting formed a select group of the most informed experts in a given field which met to discuss one or a few related problems of fundamental importance and seek to define the steps for their solution.

This style was kept for many years although the extraordinary expansion undergone with the passing of time by the effort invested in the development of all the sciences, in particular of physics, has provided an

TABLE I. The Solvay Conferences in Physics

Inquire after laws in complex systems	Search for the last constituents of matter	Exploration of our environment at large
	1 1911 Radiation Theory and the Quanta	
	2 1913 The Structure of Matter	
	First World War	
	3 1921 Atoms and Electrons	
4 1924 The Electrical Conductivity of Metals		
	5 1927 Electrons and Photons	
6 1930 Magnetism		
	7 1933 The Structure of Atomic Nucleus	
	Second World War	
	8 1948 Spectrum of Elementary Particles	
9 1951 Solid State		
10 1954 Electrons in Metals		
		11 1958 The Structure and Evolution of the Universe
	12 1961 Quantum Field Theory	
		13 1964 The Structure and Evolution of Galaxies
	14 1967 Fundamental Problems in Elementary Particles	
15 1970 Symmetry Properties of Nuclei		
		16 1973 Astrophysics and Gravitation
17 1978 Order and Fluctuations in Equilibrium and Nonequilibrium Statistical Mechanics		

increasing number of other occasions for exchanges of views and discussions between experts in any subject.

The proceedings of the old Solvay Conferences in Physics forever will remain sources of information of a unique kind about the historical development of our present views. It is much more difficult, perhaps impossible, to make a similar statement about the more recent Solvay Conferences. Only in 10 or 20 years from now will they be seen in the right perspective. What we know with certainty, from now, is that the 17 Solvay Conferences in Physics provide in their ensemble a succession of pictures of the state of our knowledge of the physical world as it was changing at intervals of a few years, all taken with the "same camera" from essentially the "same point of view."

A global view of the Solvay Conferences in Physics can be grasped from Table I, where I have indicated the year in which each conference took place and its general theme. The conferences have been ordered in a rather oversimplified way into three columns. The central one contains the Solvay meetings aiming to search for the last constituents of matter and their properties, the column on the left those devoted to inquires about simple laws in complex systems, and that on the right the meetings concerned with the exploration of our environment at large.

Such a classification has been adopted here only to give an order to the presentation of the rather complex subject. The succession of these 17 conferences shows two interruptions due to the two World Wars, which lasted from 1914 to 1918 and from 1939 to 1945.

A volume by Jagdish Mehra, which appeared in 1975,[4] treats the subjects of these conferences and tries to indicate their scientific significance. It was of help in the preparation of my discussion.

II. THE CONFERENCES ON THE SEARCH OF THE LAST CONSTITUENTS OF MATTER

The participants in the 1st Solvay Conference in Physics are shown in Fig. 1, which reproduces the, perhaps, most famous photograph of physicists of any time.

Hendrik Antoon Lorentz, from Leyden (Holland), presided the conference, whose general theme was the "Theory of Radiation and the Quanta." The conference[5] was opened with speeches by Lorentz and Jeans, one on "Applications of the Energy Equipartition Theorem to Radiation," the other on the "Kinetic Theory of Specific Heat according to Maxwell and Boltzmann." In their talks, the authors explored the possibility of reconciling radiation theory with the principles of statistical mechanics within the classical frame. Lord Rayleigh, in a letter read to the

Photo Couprie, Bruxelles

GOLDSCHMIDT PLANCK RUBENS LINDEMANN HASENOHRL
NERNST BRILLOUIN SOMMERFELD DE BROGLIE HOSTELET
SOLVAY KNUDSEN HERZEN JEANS RUTHERFORD
LORENTZ WARBURG WIEN EINSTEIN LANGEVIN
PERRIN Madame CURIE POINCARÉ KAMERLINGH ONNES

Fig. 1. Photograph of the participants in the First Solvay Conference in Physics (1911).

conference, stressed again the difficulty he had brought out in his masterly analysis[6] of 1900 and added: "Perhaps one could invoke this unsuccess as an argument in favour of the opinion of Planck and his school, that the laws of Dynamics (in their usual form) can not be applied to the last constituents of bodies. But I should confess that I do not like this solution of the difficulty. I do not see any inconvenience, of course, in trying to follow the consequences of the theory of the elements of energy (i.e., quanta). This method has already brought interesting consequences, due to the ability of those who have applied it. But it is difficult for me to consider it as providing an image of reality."

Two papers, one by E. Warburg and the other by H. Rubens, summarized the experimental measurements of the blackbody radiation for values of λT (λ in μm) smaller and larger than 3000.

These contributions were followed by an extensive presentation of the "Law of the Black Radiation" by Max Planck, who discussed, among other aspects, the physical nature of the constant h. Does this "quantum of action," he said, possess a physical meaning for the propagation of electromagnetic radiation in vacuum, or does it intervene only in the emission and absorption processes of radiation by matter?

The first point of view had been adopted by Einstein in the frame of his hypothesis of the "light quanta."[2]

Views of the second kind had been adopted by Larmor and Debye,[7] who conceived the quantum of action h as an elementary domain of finite extension in the space of phases intervening in the computation of the probability $W(E)$ for the energy density to have the value E.

In his contribution Nernst dealt with "the application of the Theory of Quanta to a few Physico-Chemical Problems," in particular the connection between Nernst theorem[8] and the quantization of energy.

Sommerfeld applied the theory of quanta to the emission of X- and γ-rays, to the photoelectric effect, and sketched the theory of the ionization potential. At the beginning of his paper he discussed in some detail the relationship observed in the emission of X- (or γ-) rays by cathode (or β) rays and arrived at the conclusion that "large quantities of energy are emitted in shorter times and small quantities of energy in larger times."[9] According to Sommerfeld this empirical result speaks in favor of the central role played in atomic and molecular phenomena by the quantum of action h introduced by Planck, the dimensions of which are energy multiplied by time.

The problem of specific heats, treated by Jeans from the classical point of view, as I said above, was discussed by Einstein in the case of solids, with special regard to the discrepancy observed at low temperature between the measured values and those deduced from the theory he had constructed in 1907 by quantizing the mechanical oscillators[3] as Planck had quantized the radiation oscillators.

Knudsen reported on the experimental properties of ideal gases, Kamerlingh Onnes on the electrical resistance of metals at low temperature, in particular on superconductivity he had discovered in Leiden in 1911,[10] and Langevin on the kinetic theory of magnetism and the central role played by the *magneton*, that is the magnetic moment of the elementary magnets which had been introduced in different approaches by Weiss and Langevin himself.

Finally Jean Perrin presented an extensive (97 pages) "Rapport sur les Preuves de la Réalité Moleculaire" in which he summarized his famous experiments on the Brownian motion of emulsion droplets suspended in a liquid and discussed the fluctuations, the determination of the elementary charge, the α decay of some radioactive nuclei, and the corresponding production of helium. The last section of the paper contains a comparison of the values of Avogadro's number deduced by completely different methods. The very satisfactory agreement between all these values provides the proof of molecular reality announced in the title of the paper.[11]

Nothing was said during the conference about the structure of the atom except in a short remark to Jeans' report by Rutherford in which he pointed out that the atom can be divided into two parts, an external part and an interior part and that the generalized coordinates, which according to Jeans appear not to contribute to the specific heat, could be those connected with the internal part of the atom.

The 2nd Solvay Conference (Fig. 2) took place in 1913 and its theme was "The Structure of Matter."[12] The meeting was opened with a long contribution (44 pages) by J. J. Thomson on the "Structure of the Atom" in which he tried to explain in a qualitative way from the classical point of view many general properties of matter.

In his long paper, however, there is no mention at all of the Rutherford model which appeared in the *Philosophical Magazine* of 1911[13] or of the papers by Geiger published in 1908 and 1910[14] and by Geiger and Marsden,[15] which appeared shortly before the meeting, on the scattering of α particles by atoms, or of the theoretical paper in which Niels Bohr had quantized the circular orbits of the electron of the Rutherford model.[16] Only in the discussion that followed J. J. Thomson's paper, did Rutherford mention the recent results of Geiger and Marsden which "bring to the conclusion that the atom consists of a positive nucleus, surrounded by a

Photographie Benjamin Couprie.

VERSCHAFFELT LAUE RUBENS GOLDSCHMIDT HERZEN LINDEMANN de BROGLIE POPE GRUNEISEN HOSTELET
HASENOHRL JEANS BRAGG Mme CURIE SOMMERFELD EINSTEIN KNUDSEN LANGEVIN
NERNST RUTHERFORD WIEN J.J. THOMSON WARBURG LORENTZ BRILLOUIN BARLOW KAMERLINGH ONNES WOOD GOUY WEISS

Fig. 2. Photograph of the participants in the Second Solvay Conference in Physics (1913).

collection of electrons whose number is equal to one half the atomic weight. . . ." In a second intervention, in the same discussion, Rutherford added some more detail and said: "An accurate comparison of the theory with the experiments has been made by Geiger and Marsden and the conclusions of the theory have been found in perfect agreement with the experimental results."

Then Langevin pointed out that the central nucleus mentioned by Rutherford had to contain electrons in order to explain the emission of β-rays by radioactive atoms and Marie Curie elaborated the idea that within the atom there should be "two kinds" of electrons. The peripheral electrons responsible for the processes of absorption and emission of radiation and for conductivity of metals and, in addition, the electrons emitted in the β decay of some radioactive nuclei.

The successive report to the conference was presented by Marie Curie who, in 5 short pages, dealt with "The Fundamental Law of Radioactive Transformations." She discussed the exponential law of decay and its interpretation in terms of a probability for an atom to decay independently from its previous life. This point of view had been already recognized at the beginning of the century. However, on this occasion she added a short but adequate review of the idea of her pupil and collaborator Debierne[17] insisting on the existence of a "disorder inside the central part of the atom (the nucleus) where its constituents should move with very high velocity judging from that of the emitted particles." Madame Curie examined even the possibility of defining a kind of temperature internal to the nucleus, much higher than the external temperature.

The discussion that followed, in which Nernst, Rubens, Brillouin, Wien, Lundman, and Langevin took the floor, will remain forever one of the moments of highest interest in the history of the gradual infiltration of the probability law into physical sciences which foreran the advent of quantum mechanics and its statistical interpretation.[18]

The rest of the conference was dominated by the recent discovery of Friedrich, Knipping, and Laue[19] of the diffraction of X-rays in crystals and reviews of the first steps in X-ray spectroscopy and in the investigation of crystal properties.

The 3rd Solvay Conference in Physics took place in 1921, after a long interruption due to the First World War. Its theme was "Atoms and Electrons."[20] It was centered on the Rutherford model of the atom and Niels Bohr's atomic theory. Bohr, however, was not able to attend the conference because of illness.

In a speech, 20 pages long, Rutherford discussed "The structure of the Atoms." After a detailed presentation of the results of Geiger and Marsden on the scattering of α particles by atoms, Rutherford discussed the simple relation obtained by Moseley between the frequency of the X-ray lines of the different elements and their number of order in the periodic system,[21] which was then recognized to be identical with the ratio of the electric charge of the nucleus and the absolute value of the charge of the electron.

Rutherford gave an estimate of the dimensions of the nucleus and then discussed the passage of β and α particles through matter, the transmutation of nitrogen into oxygen he had observed 2 years before,[22] the existence of isotopes of both radioactive and stable elements, their separation, and finally the structure of nuclei for which a reduction of the mass had been established with respect to the sum of the masses of their constituents, which were assumed to be protons and electrons, in appropriate number.

Maurice de Broglie discussed the relation $E = h\nu$ in various phenomena, such as the photoelectric effect, the production of light and X-rays in collisions of electrons against atoms; Kamerlingh Onnes, the paramagnetism at low temperature and the superconductivity; and de Haas, the angular moment of a magnetized body. At the end of the conference Paul Ehrenfest read a paper sent by Niels Bohr on "The Application of the Theory of Quanta to Atomic Problems" and added a survey on "The Correspondence Principle." Also, a paper by Millikan on "The Arrangement and Movements of the Electrons Inside the Atoms" not presented to the conference was added at the end of the proceedings.

The next Solvay Conference along the same line of thinking was the 5th held in October 1927 on "Electrons and Photons."[23] Later Langevin said that this was the occasion in which "la confusion des idées atteignit son maximum!" Sir William L. Bragg commented "I think it has been the most memorable one which I have attended. . . ." Heisenberg and Bohr also expressed their highest satisfaction.[24]

Quantum mechanics had exploded between 1923 and 1927. A. H. Compton, in 1923, had discovered the change in frequency of X-rays scattered from the electrons (the Compton effect).[25] Compton and, independently, Debye had underlined the importance of this discovery in support of the Einstein conception of light-quanta or photon propagation in space.[26]

The paper of Louis de Broglie, associating a wavelength to any particle, had appeared in 1925.[27] The five notes of Schrödinger on wave mechanics appeared in 1926.[28] The various papers on the representation of physical quantities by matrices by Born, Heisenberg, Jordon, Dirac, and Pauli were published between 1925 and 1927.[29] The results of diffraction experiments by Davisson and Germer[30] of electrons scattered by a single crystal of nickel and by G. P. Thomsom and Reid[31] of electrons transmitted by celluloid films were in 1927.

The paper by Max Born on "Quantummechanik des Stossvorgänge," in which he had proposed the statistical interpretation of the wave function, had appeared in 1926.[32] Niels Bohr had presented his principle of complementarity at the Como Conference in September 1927[33] and Heisenberg had formulated the uncertainty principle shortly before the Solvay Conference.[34]

The Conference was opened on October 24 with reports by Lawrence Bragg and Arthur Compton about the new experimental evidence regarding scattering of X-rays by electrons exhibiting widely different features when firmly bound in crystalline structures of heavy substances and when practically free in atoms of light gases.

All other contributions regarded various aspects of the new quantum mechanics.

Louis de Broglie discussed the relations between energy, momentum, and wavelength for photons as well as for electrons and examined the results of the Davisson–Germer experiment on the diffraction of electrons by crystals, which were in perfect agreement with theory.

Born and Heisenberg presented the foundations of the quantum mechanics in which the classical kinematic and dynamical variables are replaced by operators obeying a noncommutative algebra involving Planck's constant and showed that the introduction of these operators in the Hamiltonian gives rise to the Schrödinger equation. Schrödinger summarized the main features of wave mechanics and applied it to radiation, showing that the electric moment deduced in this approach is equivalent to a matrix of the theory of Born and Heisenberg.

Finally, Niels Bohr gave a report on the epistemological problems involved in the new quantum mechanics and stressed the viewpoint of complementarity. In the very lively discussion that followed, differences in terminology gave rise to great difficulties for agreement between the participants. The situation was humorously expressed by Ehrenfest, who wrote on the blackboard the sentence from the Bible, describing the confusion of languages that disturbed the building of the Babel tower.

The discussions, started at the sessions, were continued during the evenings within smaller groups, in particular, among Bohr, Ehrenfest, and Einstein, who, as is well known, was reluctant to renounce the deterministic description.

The discussion on these matters between Einstein and Bohr went on for years. It was taken up again at the 1930 and 1933 Solvay Conferences as well as in other places. It has been summarized by Niels Bohr in a chapter of the well-known book *Albert Einstein: Philosopher-Scientist*, which appeared for the first time in 1949.[35]

At any occasion Einstein proposed a new argument or a new "thought experiment" designed to illuminate some paradoxical consequence of quantum mechanics. On some occasions immediately, on others some time later, Bohr always was able to give an answer fully satisfactory in the laguage of quantum mechanics. Bohr's arguments were accepted by the majority of the physicists but did not satisfy Einstein, or a few others[36] who still were reluctant to renounce a view of the world they considered obvious and natural. These world views, called by some authors "local realistic theories of nature,"[37] are based on three assumptions or premises which, in their opinion have the status of world established truths, or even self-evident truths.[38] This definition of "reality" is what we are used to ascribing to systems of macroscopic particles and differs from the "reality" of quantum states of atomic or subatomic systems.

The progress undergone in recent years toward a solution of the problem raised by Einstein 1927 has become possible because of two developments: (1) A few experimental techniques, in particular, the methods for measuring very short times with good accuracy, have permitted in recent years the execution of several experiments which in their essence are practical versions of the "thought experiment" proposed and discussed in 1935 by Einstein, Podolski, and Rosen.[39] (2) A procedure of analysis of their results has been made possible by the work of Bell[40] who has derived in the frame of "local realism" a relation (the Bell inequality) obeyed by "local realistic theories" but violated by quantum mechanics.

Although not all the results of the experiments carried on until now are consistent with one another, most of them not only violate the Bell inequality but are also in good agreement with the prediction of quantum mechanics. The experimental errors are still rather large but in a few years this matter will be definitely settled by the results of further, more accurate experiments. It is rewarding that the final word even in this philosophical debate started in 1927, when quantum mechanics was in its infancy, will be definitely settled by accurate results of experiments designed and carried on with the most refined techniques of half a century later.

The 7th Solvay Conference in Physics was held in 1933, and was entitled the "Structure and Properties of Atomic Nuclei."[41] It was chaired by Langevin who, after the death of Lorentz (4 February 1928), had become President of the Scientific Committee. It took place at a time when the field had recently undergone extremely rapid developments. The conference was really marking the beginning of modern nuclear physics.

Cockcroft presented an extensive paper (56 pages) on the "Disintegration of Elements by Accelerated Particles" in which he reviewed the various types of accelerators available in those days (Greinacher or Cockcroft and Walton voltage multiplier, van de Graaf electrostatic accelerator, cyclotron, etc.) and the reactions that he had discovered, in collaboration with Walton, not long before in a few light elements (Li, F, Be, C, N) irradiated with protons and deuterons accelerated up to energies of the order of a few hundred kiloelectronvolts.[42]

In the discussion that followed, Rutherford added information about the work he was carrying on, in collaboration with Oliphant, on various reactions produced in lithium by bombardment with protons and deuterons,[43] and Ernst Lawrence described in more detail the cyclotron he had invented and first constructed with a few collaborators in 1932.[44]

The next speech was by Chadwick who treated (in 32 pages) the anomalous scattering of the α particles, the transmutation of the elements, and the evidence for the existence of the "neutron" he had discovered in 1932.[45]

Frederick Joliot and Irene Curie discussed the γ-rays emitted in association with neutrons by berillium irradiated with α particles and reported to have observed under the same conditions also the emission of fast positrons. The origin of these particles was not yet clear at the time of the Solvay Conference. It was understood by the same authors a few months later when they discovered the artificial radioactivity induced by α-particle bombardment which normally takes place by emission of positrons.

In the discussion that followed a number of participants took the floor: Lise Meitner, Werner Heisenberg, Enrico Fermi, Maurice de Broglie, Wolfgang Bothe, and Patrick Blackett.

In his intervention Blackett treated the discovery of the positron in cosmic rays by C. D. Anderson in 1932[46] and its confirmation by Blackett and Occhialini,[47] who had introduced, for the first time, the technique of triggering a vertical cloud chamber by means of the coincidence between two Geiger counters, one placed above, the other below the chamber. Blackett also discussed a number of papers by Meitner and Philipp, Curie-Joliot, Blackett, Chadwick and Occhialini, and Anderson and Neddermeyer,[48] all appearing almost at the same time, on the production of positrons in various elements irradiated with the γ-rays of 2.62 MeV energy of Thc. These were the first observations of electron–positron pair production. He also pointed out that the observed production of positrons has a cross section larger than the nuclear dimensions, and therefore, most probably, does not originate from a nuclear process.

Then Blackett examined the nature of showers observed in cosmic rays starting from Bruno Rossi's experiment with three nonaligned counters in coincidence and Rossi's curve showing the dependence of the number of observed showers as a function of the thickness of the layer of lead in which they are produced.[49] Then he presented the results he had recently obtained in collaboration with Occhialini by means of their new experimental technique mentioned above.[47]

The next paper was by Dirac on the "Theory of the Positron." In the following discussion Niels Bohr made a long intervention on the correspondence principle in connection with the relation between the classical theory of the electron and the new theory of Dirac.

Then Gamow talked about the "Origin of γ-Rays and the Nuclear Levels" and Heisenberg, on "The Structure of the Nucleus," discussed the exchange forces of the two types that Heisenberg[50] himself and Majorana[51] had proposed not long before.

In the discussion that followed, Pauli again brought up the suggestion he had already made in June 1931 on the occasion of a Conference in Pasadena. In order to explain the continuous spectrum of the β-rays emitted in the decay of many radioactive nuclei the emission of the electron had to be accompanied by the emission of a neutrino, that is, a neutral particle of a very small mass, possibly zero mass, and spin $\frac{1}{2}$.

Before the end of 1933 Fermi developed his theory of β decay,[52] in January 1934 Joliot-Curie discovered the radioactivity induced by α particles,[53] and Fermi that induced by neutrons.[54] Neutron physics was at the beginning of a new unexpected fast development.

Rutherford's comment to the 7th Solvay Conference was: "The last conference was the best of its kind that I have attended."

For a first time the theme selected for the 8th Conference was "Cosmic Ray and Nuclear Physics," but a long period of illness of the President of the Scientific Committee, Paul Langevin, imposed a first adjournment. Later it was decided that the conference would deal with the problems of elementary particles and their mutual interactions and that it would be held in October 1939. Even the list of speakers was prepared but World War II started on 3 September 1939 and the conference was postponed to an indefinite date.

The next Solvay Conference in Physics, the 8th from the beginning, took place in October 1948, that is, 15 years since the previous one. The Scientific Committee was now chaired by Sir Lawrence Bragg. The general theme of the conference was "Elementary Particles."[55]

Already in 1936 Anderson and Neddermeyer[56] had shown that the penetrating component of cosmic rays, first discovered in 1932 by Bruno Rossi,[57] consisted of charged particle of mass intermediate between those of the electron and the proton. Shortly later it was shown that this particle was unstable with a mean life on the order of 10^{-6} sec.

Both the mass and the mean life of this particle were in qualitative agreement with those suggested about 1 year before by Yukawa for the mediator of the nuclear forces.[58]

The picture, however, had changed rapidly after the end of the Second World War. The experiment of Conversi, Pancini, and Piccioni[59] had shown that this particle had an interaction with nuclei much weaker than that expected for the Yukawa mediator. At the beginning of October of the same year 1947, Lattes, Occhialini, and Powell[60] in Bristol had discovered in cosmic rays a new particle, that they called π-meson. It is unstable and decays, with a mean life of $\sim 10^{-8}$ sec, into a neutrino and the particle of Anderson and Neddermeyer that was called μ-meson or muon.

Almost at the same time Rochester and Butler[61] at Manchester observed in a cloud chamber triggered by counters two "V events" identified later as decays of a θ^0 ($\equiv K^0$)-meson and a Λ^0-hyperon.

These were the first examples of "strange particles," the adjective "strange" referring to the following anomalous property. Their decay takes a time of the order of 10^{-10} sec instead of the much shorter time ($\sim 10^{-23}$ sec) expected from the time observed to be involved in their production ($\sim 10^{-23}$ sec).

In Brussels the 8th Solvay Conference was opened with two speeches on mesons, one by Cecil F. Powell who reported on the results of work made in cosmic rays, and the other by R. Serber who presented the results on the production of "artificial mesons" obtained in Berkeley by Burfening, Gardner, and Lattes by bombarding matter with α particles accelerated with the cyclotron.[62]

The properties of the particles contained in large atmospheric showers were discussed by Auger and the problem of the nuclear forces by Rosenfeld from two points of view. In a purely phenomenological approach by means of a nuclear potential or in terms of a nuclear field mediated between any two interacting particles by particles of intermediate mass.

Bhabha treated the general relativistic wave equations and Tonnelat presented the idea of Louis de Broglie of trying to describe a photon as an object composed of two neutrinos.[63]

Heitler spoke on the quantum theory of damping, which is a heuristic attempt to eliminate the infinities of quantum field theory in a relativistic invariant manner, Peierls spoke of the problem of self-energy, and Oppenheimer gave "an account of the developments of the last years in electrodynamics" in which he discussed the problem of the vacuum polarization and charge renormalization with special reference to the recent work of Schwinger and Tomonaga.

A few reports on different topics were also presented to the same conference. Blackett discussed "The Magnetic Field of Massive Rotating

Bodies'' and Teller presented a paper prepared jointly with Maria Goeppert Mayer, on the original formation of the elements.

The conference was closed with some general comments by Niels Bohr ''on the present state of atomic physics'' in which he referred to the satisfactory situation in quantum electrodynamics, but pointed out that the adimensional coupling constant

$$\alpha = \frac{e^2}{\hbar c} \simeq \frac{1}{137}$$

cannot be computed within the framework of the theory itself, indicating the need for a future more comprehensive theory.

The 12th Solvay Conference on ''Quantum Field Theory'' was held in 1961, just 50 years after the first Solvay Conference.[64] This circumstance was celebrated by Niels Bohr, who opened the conference with a speech on ''The Solvay Meetings and the Development of Quantum Physics.''

Once more the development undergone between 1948 and 1961 by the experimental techniques as well as by the theoretical interpretation of subnuclear particles was amazing. In 1952, at the Brookhaven National Laboratory, the first proton synchrotron, the Cosmotron, entered into operation. It produced protons of energies up to 3.2 GeV, and became immediately a controlled source of pions and strange particles of much higher intensity than cosmic rays.

In the same year, Pais[65] suggested that the anomalous property of strange particles could be explained by assuming that they possess an internal degree of freedom, specified by a quantum number, and that various selection rules based on the conservation or nonconservation of this quantum number are operating in the production and decay.

Within 1 year, at the Cosmotron, Fowler et al.[66] discovered a new phenomenon, the associated production in the same collision of two different strange particles: for example, a Λ^0 and a K^0. These observations were in agreement with Pais' approach and prompted its full development: in 1954 Gell-Mann and Pais and, independently, Nishijima[67] were able to define the new quantum number, called strangeness, and, from a detailed analysis of the already rich harvest of new processes, to assign its value to all known particles.

In 1954, a larger accelerator, the Bevatron, producing protons up to 6.5 GeV, went into operation at Berkeley, and marked the end of cosmic rays as a tool for the investigation of subnuclear particles.[68] About 1 year later with this machine, Chaimberlain, Segrè, Wiegand, and Ypsilantis[69] observed the production of antiprotons generated in proton–nucleon collisions. This result provided the long-awaited confirmation of the generality of the 1931 Dirac forecast: his relativistic equation providing an adequate description of the electron and its antiparticle, is valid in general (apart from some correction) for any particle of spin $\frac{1}{2}$.

A further step of fundamental importance was made in 1961 by Gell-Mann and Ne'eman who proposed a classification scheme based on the Sophus Lie group of symmetry SU(3)[70] for the more than 100 particles endowed with strong interactions (hadrons).

About 2 years before the 1961 Solvay Conference two new accelerators producing protons of 28–30 GeV went into operation, one at CERN, near Geneva, and the other at the Brookhaven National Laboratory. Pretty soon, both of them started to enrich even more our knowledge of sub-nuclear phenomenology.

At the 1961 Solvay Conference the situation reached by quantum field theory was reviewed by Heitler, with special regard to the problem of renormalization. The successive speech on the same general theme was given by Feynman who started with a comparison of the predictions of the quantum field theory with experimental results. He discussed the scattering of high-energy electrons by protons, with special emphasis on the proton form factors and the search for deviations from the photon propagator, the comparison between the measured and computed values of the Lamb shift, of the magnetic moment of electron and muon, and of the hyperfine structure of spectral lines.[71] As a general conclusion on the first part of his talk, Feynman stated: "All this may be summarized by saying that no error in the prediction of quantum electrodynamics has yet been found. The contributions expected from various processes envisaged have been found again, and there is very little doubt that in the low-energy region, at least, our methods of calculation seem adequate today." These remarks are valid even today in spite of the fact that the precision reached in the measurements in the low-energy experiment is increased by two or three orders of magnitude[72] and the exploration of high-energy phenomena has been extended from values of the cut-off parameters Λ of 0.6 GeV to more than 100 GeV.[73]

The second part of Feynman's speech dealt with theoretical questions. The first one was the problem of the renormalization of the mass of the electron as well as of particles such as the pion and the kaon which exist in charged ($\pi^{\pm}$, $K^{\pm}$) and neutral (π^0, K^0) states and therefore provide a direct indication of the contribution originating from the electromagnetic field.

A second question was the interaction of photons with other particles which do not allow a sharp separation between quantum electrodynamics and other interactions. "It will not do to say that Q.E.D. is exactly right as it stands, because virtual states of charged baryons must have an influence. Two sufficiently energetic photons colliding will not do just what Q.E.D. in that limited sense supposes; they also produce pions. Or, more subtly, a sufficiently accurate analysis of the energy levels of positronium would fail, for the vacuum polarization from mesons and nucleons would be omitted."

A third and last theoretical item treated by Feynman was "Dispersion Theory," discussed in greater detail by other participants in the Solvay Conference.

"Weak Interactions" were treated by Pais who, starting from Fermi's original theory, discussed the discovery by Lee and Yang,[74] almost 5 years before, of the parity violation by weak interactions, its experimental confirmation,[75] the muon–electron universality,[76] the idea of an intermediate boson as a mediator of weak interaction, and the "two-neutrinos question."[77]

In the following speech, on "Symmetry Properties of Fields," Gell-Mann discussed exact symmetries as well as approximate symmetries: the conservation of the x and y component of the isotopic spin I, broken by electromagnetism and weak interactions, the conservation of I_z, or strangeness, broken by weak interactions and the conservation of C and P separately, also broken by weak interactions. Then, using various arguments, among them the conserved vector current, already recognized in 1955–1958,[78] he stressed the central interest of charge operators and of the equal-time commutation relations among them. "The mathematical character of the algebra which the charge operators generate is a definite property of nature," Gell-Mann said.

He then considered the currents in the Sakata–Okun model for which he derived the expressions for the electromagnetic and weak currents.

At this primitive stage of development of current algebra quarks had not yet appeared on the scene.

Other speeches were by Källen on some aspects of the formalism of field theories, Goldberger on single variable dispersion relation, Mandelstam on "Two-Dimensional Representations of Scattering Amplitudes and Their Application" and finally by Yukawa on "Extensions and Modifications of Quantum Field Theory."

The appearance of the current algebra in the speeches and discussions of this conference really marks the beginning of one of the most fruitful lines of development that has brought us to present views.

The 14th Solvay Conference on "Fundamental Problems in Particle Physics" was held in October 1967.[79] The Conference was presided over by Christian Møller who opened the meeting with a "Homage to Robert Oppenheimer," who had died on 18 February of that year.

Many speeches were of a theoretical nature: Dürr spoke on "Goldstone

Theorem and Possible Applications to Elementary Particle Physics," Haag on "Mathematical Aspects of Quantum Field Theory," Källen on "Different Approaches to Field Theory. Especially Quantum Electrodynamics," and Sudarshan on "Indefinite Metric and Nonlocal Field Theories." Heisenberg gave a "Report on the Present Situation in the Nonlinear Spinor Theory of Elementary Particles."

Chew spoke of the "S-Matrix Theory with Regge Poles," a concept[80] that soon became a stable acquisition in the general picture of subnuclear particles,[81] and Tavkhelidze discussed the "Simplest Dynamic Models of Composite Particles."

Already in 1964 Zweig and Gell-Mann had postulated the existence of quarks as building blocks of hadrons, thus establishing the premises necessary for a dynamical interpretation of the SU(3) already well-established symmetry.[82] At the time of the Solvay conference, however, the quark hypothesis was considered only as a convenient model.

A report by Gell-Mann, also on elementary particles, unfortunately does not appear in the proceedings of the Conference because it was not available at the time of publication. From the contributions to its discussion by W. Heisenberg, C. F. Chew, F. E. Low, S. Mandelstan, R. Brout, R. E. Marshak, S. Weinberg, N. Cabibbo, S. Fubini and others, it appears that the talk by Gell-Mann touched upon the SU_2 and SU_3 groups and their connections with bootstrap mechanism, chiral symmetry, and partial conservation of axial currents (PCAC).[83]

This is the last of the Solvay Conferences on Physics devoted to subnuclear particles in spite of the fact that the progress undergone by this field since 1967 has been amazing. The main reason was that many conferences were held each year on this general subject as well as on various parts of it, so that the Scientific Committee for Physics felt it would be better to turn its attention to other subjects less frequently dealt with in international conferences.

Now it is clearly time to devote yet another Solvay Conference to the last constituents of matter. See note on page 35.

III. THE CONFERENCES ON THE INQUIRE AFTER LAWS IN COMPLEX SYSTEMS

Let me now consider the conferences devoted to solid-state and statistical mechanics.

The first conference of this group is the 4th Solvay Conference on "The Electrical Conductivity of Metals" held in April 1924.[84] The conference was in some way premature. It took place just before the advent of quantum mechanics, in particular 2 years in advance of the first formulation by Fermi of the antisymmetric statistics and the consequent concept of the degenerate electron gas.

The conference was opened with a speech by Lorentz on the theory of electrons he had developed about 20 years before, followed by papers by Joffe on the electrical conductivity of crystals, Kamerlingh Onnes on superconductivity, and Hall on the metallic conduction and the transversal effects of the magnetic field. This last speech was followed by a discussion in which Langevin and Bridgman injected a few interesting remarks.

The 6th Solvay Conference on "Magnetism," held in 1930,[85] was opened by a contribution by Sommerfeld on "Magnetism and Spectroscopy" in which he discussed the angular momenta and magnetic moments of the atoms which had been derived from the investigation of their electronic constitution.

Van Vleck reported on the experimental data of the variation of the magnetic moments within the group of the rare earths and its theoretical interpretation, and Fermi discussed the magnetic moments of the atomic nuclei and their determination from the splitting of hyperfine structure. Pauli treated the "Quantum Theory of Magnetism," with special regard to the paramagnetism of a degenerate Fermi gas of electrons, Weiss dealt with the equation of state of ferromagnets, Dorfman with ferroelectric materials, and Cotton and Kapitza reported on the study of the magnetic properties of various materials in very intensive magnetic fields. The phenomenological theory of ferromagnetism by Weiss is still of interest today, mainly because the microscopic mechanism that gives rise to the dipole–dipole interaction is not yet understood in all its detail.[86]

The 9th Solvay Conference on "Solid State" took place in 1951.[87] The question of "interface between crystals" was discussed by C. S. Smith, grain growth observed by electron optical means by G. W. Ratenau, recrystallization and grain growth by W. G. Burgers, crystal growth and dislocation by F. C. Franck, the generation of vacancies by moving dislocations by Seitz, dislocation models of grain boundaries by Shockley, and diffusion, work-hardening, recovery, and creep by Mott.

It was at that time that Franck and Seitz proposed mechanisms for the multiplication and generation of vacancies by intersection of dislocations[88] explaining the observed softness of crystals and providing models that were subsequently verified by the technique of decoration of dislocations.[89]

The 10th Solvay Conference on "The Electrons in Metals" was held 3 years later, in 1954.[90]

D. Pines examined the collective description of electron interaction in metals; Löwdin, an extension of the Hartree–Fock method to include correlation effects; Mendelson, the experiments on thermal conductivity of metals; Pippard, the methods for determining the Fermi surface; Kittel, resonance experiments and wave functions of electrons in metals; Friedel, primary solid solutions in metals; Fumi, the creation and motion of vacancies in metals; Shull, neutron diffraction from transition elements and their alloys; Néel, antiferromagnetism and metamagnetism; and Frölich, superconductivity with special regard to the electron–electron interaction carried by the field of lattice displacements, which paved the way to the theory that Bardeen, Cooper, and Schrieffer developed between 1955 and 1957.[91] Finally, Mathias dealt with the empirical relation between superconductivity and the number of valence electrons per atom.

The last Solvay Conference that I have rather arbitrarily put in this class is the 17th on "Order and Fluctuations in Equilibrium and Nonequilibrium Statistical Mechanics" held in October 1978.[92]

The conference was divided into four parts to each of which a full day was devoted: the first one treated: "Equilibrium Statistical Mechanics," with special regard to "The Theory of Critical Phenomena"; the second part regarded "Nonequilibrium Statistical Mechanics. Cooperative Phenomena"; the third one, "The Macroscopic Approach to Coherent Behavior in Far Equilibrium Conditions"; and the fourth and last, "Fluctuation Theory and Nonequilibrium Phase Transitions."

Two methods appear to be very powerful for the study of critical phenomena: field theory as a description of many-body systems, and cell methods grouping together sets of neighboring sites and describing them by an effective Hamiltonian. Both methods are based on the old idea that the relevant scale of critical phenomena is much larger than the interatomic distance and this leads to the notion of scale invariance and to the statistical applications of the renormalization group technique.[93]

As pointed out by van Hove in his concluding remarks, the common methodology between high-energy physics and critical phenomena is striking although the conceptual basis is quite different in the two cases. In high-energy physics approximate scale invariance and its calculable breaking are characteristic of the large momentum scale regime (i.e., small space and time intervals), whereas in statistical physics scale invariance and renormalization group methods are applicable to a domain of large space and time intervals.

The approximate methods of renormalization for the investigation of phase transitions in degenerate states[94] were presented to the conference by Kadanoff and by Brezin. The nonequilibrium statistical methods were discussed by Prigogine,[95] followed by Hohenberg who treated critical dynamics. In the third part, Koschmieder discussed the experimental aspects of hydrodynamic instabilities[96]; Arecchi, the experimental aspects of transition phenomena in quantum optics[97]; and Sattinger the bifurcation theory and transition phenomena in physics.[98]

The fourth part included a paper by Graham on the onset of cooperative behavior in nonequilibrium states. Suzuki talked about the theory of instability, with special regard to nonlinear Brownian motion and the formation of macroscopic order, and P. W. Anderson developed a series of interesting considerations of very general nature around the question: "Can broken symmetry occur in driven systems?"

The question in the title can be reformulated by asking how much can be dug out of an analogy between broken symmetry in dissipative structures (such as the ripple marks generated by wind, i.e., an external perturbation, in an otherwise flat surface of sand) and broken symmetry defined as phenomena of condensed matter systems of the kind observed near the critical points. The value of Anderson's discussion is to be seen more in the deepening of the question itself than in the answer that cannot yet be final, and for the moment, according to the author, appears to be more on the negative side.

The contributions presented by Prigogine and by Sattinger to the 17th Solvay Conference on Physics appear as a natural introduction to some of the problems that will be examined at the present Solvay Conference in Chemistry.

Exact symmetries and broken symmetries were the central theme of the 15th Solvay Conference held 8 years before, that is, in 1970 on "Symmetry Properties of Nuclei."[99] This was the second conference on nuclear structure, the previous one being the conference held in 1933 immediately after the discovery of the neutron. In the 37 years that have passed between the first and the second Solvay Conference on nuclear structure the subject has undergone an extraordinary development although the naive hope, generally shared by physicists until 1935, for a full understanding of nuclear dynamics in terms of nucleons interacting with two-body forces, has been completely deluded. A number of models have been developed which, although very different from each other, are not contradictory. Each of them, in some way, enphasizes a particular aspect of some category of nuclei and/or nuclear phenomena, and thus allows an adequate interpretation of a set of their static and/or dynamic properties.

The "Symmetry of Cluster Structures of Nuclei" was discussed by Brink who showed contour plots of nucleon density obtained from Hartree–Fock calculations for simple nuclei such as ^{8}Be, ^{12}C, and ^{20}Ne.

Much attention was devoted to collective models: Mottelson reviewed "Vibrational Motion in Nuclei"; Judd, the use of Lie groups; Lipkin, the

SU(3) symmetry in hypernuclear physics; Radicati, Wigner's supermultiplet theory[100]; Fraunfelder, "Parity and Time Reversal in Nuclear Physics"; Wilkinson, the isobaric analogue symmetry; Aage Bohr, the permutation group in light nuclei; and J. P. Elliot, the shell model symmetry.

The conference was closed by a few concluding remarks by E. P. Wigner, not completely free from critical lines.

At the time of the conference the study of nuclear reactions produced by intermediate energy protons or pions as well as the investigation of collisions between two nuclei of $Z \geq 3$ were still in a rather primitive stage whereas today they constitute a rich field of empirical knowledge and phenomenological interpretation.

IV. THE CONFERENCES ON EXPLORATION OF OUR ENVIRONMENT AT LARGE

I come now to the last group of Solvay Conferences: the three regarding astrophysical problems. The first one, devoted to "The Structure and Evolution of the Universe" was held in June 1958. It was the 11th Solvay Conference in Physics.[101]

The conference took place in a moment of extraordinary expansion of general interest for astrophysical problems due to the first steps made in new observational techniques such as radio signal reception and observations from space vehicles.

This was also the first Solvay Conference in which Einstein's Theory of General Relativity started to be quoted and used as a conceptual structure of fundamental importance for the interpretation of large-scale phenomena.

The theme of the conference was divided into three parts: the first one concerned "General Statements of Cosmological Theory." It was introduced by speeches by Lemaitre, on the "Primaeval Atom Hypothesis and the Problem of Clusters of Galaxies," by Oscar Klein who developed "Some Considerations Regarding the Earlier Development of the System of Galaxies," and by Hoyle on "The Steady-State Theory." This was followed by a talk by Gold, on the "Arrow of Time" and another by Wheeler on "Some Implications of General Relativity for the Structure and Evolution of the Universe."

The second part of the conference was devoted to a "Survey of Experimental Data on the Universe."

In a talk, very important even today, J. H. Oort discussed the "Distribution of Galaxies and the Density of the Universe." Lovell presented "Radio Astronomical Observations Which May Give Information on the Structure of the Universe."

In the third part of the conference, on the "Evolution of Galaxies and Stars," Hoyle presented the then recent and still important work by the Burbidges, Fowler, and himself on the origin of elements in stars.[102]

In the general discussion Bondi asked for tests that could decide between the evolutionary and steady-state universe. The question was premature because the arguments, which later allowed the exclusion of the steady-state theory, are based on the analysis of the blackbody cosmological radiation and of the distribution of the number of radio sources versus flux. The cosmological radiation was discovered only in 1965[103] and the distribution of the radio sources was still highly controversial at the time of the conference. Indeed, in the discussion of this point Lovell stated: "At present the whole of the cosmological interpretation of the radio sources is based on a half a dozen identifications of the Cygnus type." Furthermore, the mechanism leading to radio emission was misunderstood: the prevailing view attributed it to collisions between galaxies.

From the small amount of information I have given it appears clear that in 1958 astrophysics was still in its classical stage. The conference, however, marks a historical step in modern astrophysics because for the first time the physics of neutron stars and collapsed objects was reproposed by Wheeler since 1939 when Oppenheimer and Snyder first presented the existence of black holes.[104]

The 13th Conference on the "Structure and Evolution of Galaxies" took place in October 1964.[105] It was presided over by Robert Oppenheimer, who had succeeded Sir Lawrence Bragg.

The Conference was opened by Ambartsumian who spoke "On the Nuclei of Galaxies and their Activity." He presented new observational data that in his opinion supported the idea, which he had already submitted to the 1958 Conference, that most of the processes connected with the formation of new galaxies and their structure start from the nuclei. This interpretation, however, is shared only by a minority of astrophysicists.

Oort reported on "Some Topics Governing the Structure and Evolution of Galaxies" and described the latest knowledge concerning our Galaxy. Woltjer discussed the "Galactic Magnetic Field."

The structure and evolution of the stars were the subject of the second part. Spitzer discussed the "Physical Processes in Star Formation," a subject that was further developed by Salpeter with special regard to the birthrate function of the stars. W. A. Fowler and Bierman discussed the evolution toward the main sequence and R. Minkowski discussed the data available concerning the supernovae.

Finally, an important survey of the findings about extragalactic radio sources was presented by J. G. Bolton and, in the discussion that followed, Bruno Rossi presented the first observations made by Giacconi et al.[106] in 1962 and by Friedman et al. in 1963 of localized X-ray sources, in particular, Scorpio X-1.[107]

One of the most exciting contributions to the conference was the speech by M. Schmidt who discussed the "Spectroscopic Observations of Extragalactic Radio Sources," with special regard to the interpretation of the red shift of the quasars, discovered about 2 years before.[108] His conclusion, still accepted today by the majority of astrophysicists, was that most likely these red shifts are cosmological.

G. R. Burbidge and E. M. Burbidge, in their report on "Theories and the Origin of Radio Sources," summarized the understanding that had been achieved in the creation of radio waves.

In the final discussion G. R. Burbidge stressed that the problems of energy conversion from a primary energy source to the form of relativistic particles needed for the radio emission are entirely unsolved. In particular, the production of cosmic ray particles requires an acceleration mechanism, of an efficiency at least a few orders of magnitude larger than that of the best man-made accelerators.[109]

In this connection Alfvèn proposed the annihilation of matter and antimatter as a possible source of energy, but also other mechanisms, in particular, some form of release of gravitational energy were examined.

The problem is still open today, but the consensus is indeed that strong gravitational fields should be involved.

The 16th Solvay Conference in Physics, held in September 1973, was entitled "Astrophysics and Gravitation."[110] The progress undergone in many fundamental chapters of astrophysics with respect to the previous conference, held in 1964, was really striking. In particular, X-ray astronomy had won a status comparable with other conventional branches of astronomy.

The Conference was opened with a progress report on pulsars by Pacini, followed by speeches on the observational results on compact galactic X-ray sources by Giacconi, on the optical properties of binary X-ray sources by the Bahcalls, and a review on the physics of binary X-ray sources by Martin Rees.

Among the many communications and invited talks essentially on the same subject, I recall the speech by Pines on "Observing Neutron Stars," by Pandharipande on "Physics of High Density and Nuclear Matter," and by Cameron and Canuto on "The General Review on Neutron Stars Computations."

Black holes were extensively discussed by J. A. Wheeler and the search for their observational evidence by Novikov, followed by communications by Rees and by Ruffini.

Woltjer gave a general talk on "Theories of Quasars," and Martin Schmidt discussed "The Distribution of Quasars in the Universe."

G. R. Burbidge gave a review paper on the "Masses of Galaxies and the Mass-Energy in the Universe" and Hofstadter presented the information available at the time on the recent discovery of bursts of γ-rays.[111]

I am now at the end of my series of flashes on the Solvay Conferences in Physics. I hope that, in spite of its shortness and incompleteness, it may help in stimulating two kinds of considerations. Those of the first kind regard the extraordinary develoment undergone during the last 70 years by our views on the physical world, many parts of which in present days appear to be dominated by a few general concepts, such as those of exact and approximate symmetry, and to be treatable by mathematical procedures such as the application of the renormalization group. The other kind of considerations concerns the role that the Solvay Conferences in Physics have played in the development of physics during the last 70 years, and the unique value they will maintain, even in the future, as sources of information for the historians of science.

References

1. M. Planck, *Verh. Deut. Phys. Ges.*, **2,** 237 (1900).
2. A. Einstein, *Ann. Phys. Ser. 4,* **17,** 132 (1905); **20,** 199 (1906).
3. A. Einstein, *Ann. Phys.*, **22,** 180, 800 (1907); **25,** 679 (1911). Debye developed his model about one year later: *Ann. Phys.*, **39,** 789 (1912).
4. J. Mehra, *The Solvay Conferences on Physics*, D. Reidel, Dordrecht and Boston, 1975.
5. *La Théorie du Rayonnement et les Quanta*, Rapports et Discussions de la Rénunion tenu à Bruxelles, du 30 October au 3 Novembre 1911, Publiés par MM. Langevin et M. de Broglie, Gauthier-Villars, Paris, 1912.
6. Lord Raleigh, *Philos. Mag.*, **49,** 118 (1900); **59,** 539 (1900).
7. J. Larmor, *Proc. R. Soc. London, Ser. A,* **83,** 82 (1909); P. Debye, *Ann. Phys.*, **33,** 1427 (1910)
8. W. Nernst, Göttinger Nachr. 1906 Heft 1, *Zeit. Elektrochemie,* **17,** 265 (1911).
9. The argument goes more or less as follows: high-energy cathode rays, that have a short interaction time with atoms, emit X-rays of an energy (deduced from their penetration) greater than the energy of the X-rays emitted by cathode rays of lower energy. The conclusion is correct, but it is based on the presumption that the penetration of photons is a monotonic increasing function of their energy. This is true only for photons of energy below the threshold for pair production: $2m_ec^2 \simeq 1$ MeV. As I will recall below, pair production was discovered only in 1932, and anyhow the X-ray photons available and considered in 1911 had always energy below $2m_ec^2$.
10. K. Onnes, *Comm. Phys. Lab. Univ. Leyden*, Nos. 119, 120, 122 (1911).
11. *Oeuvres Scientifiques de Jean Perrin*, CNRS 13, Quai Anatole France, Paris (VIIe), 1950.
12. *La structure de la matiere*, Rapports et discussions du Conseil de Physique tenu à Bruxelles du 27 au 31 October 1913, Gauthier-Villars, Paris, 1921.

13. E. Rutherford, *Philos. Mag.*, **21,** 669 (1911).
14. E. Geiger, *Proc. R. Soc. London, Ser. A,* **81,** 174 (1908); E. Marsden, *Proc. R. Soc. London, Ser. A,* **82,** 495 (1909); E. Geiger, *Proc. R. Soc. London, Ser. A,* **83,** 492 (1910).
15. H. Geiger and E. Marsden, *Philos. Mag.*, **25,** 604 (1913).
16. N. Bohr: "On the Constitution of Atoms and Molecules", *Philos. Mag.*, **26,** 1913, p. 1 of July issue; p. 476 of September issue; p. 857 of November issue.
17. A. Debierne, "Sur les transformations radioactives", p. 304 of the volume: *Les idées modernes sur la constitution de la matiere*, Gauthier-Villars, Paris, 1913.
18. E. Amaldi, "Radioactivity, a Pragmatic Pillar of Probabilistic Conceptions," p. 1 of the volume: *Problems in the Foundations of Physics*, edited by G. Toraldo di Francia, North-Holland, Amsterdam, New York, Oxford, 1979.
19. W. Friedrich, P. Knipping, and M. v. Laue, *Ber. Byer. Akad. Wiss.*, 303 (1912).
20. *Atoms et Électrons*, Rapports et Discussions du Conseil de Physique tenu à Bruxelles du 1er au 6 Avril 1921, Gauthier-Villars, Paris, 1923.
21. H. G. J. Moseley, *Philos. Mag*, **26,** 1024 (1913); **27,** 103 (1914).
22. E. Rutherford, *Philos. Mag.*, **37,** 581 (1919).
23. *Électrons et Photons*, Rapports et Discussions du Cinquiènne Conseil de Physique tenu à Bruxelles du 24 au 29 October 1927, Gauthier-Villars, Paris, 1928.
24. These remarks can be found in the few pages of introduction to the album of photographs of the participants in the Solvay Conferences in Physics, published in 1961, on occasion of the celebration of 50 years after the First Solvay Conference. These pages are based on the unpublished paper: Jean Pelseneer, "Historique des Instituts Internationaux de Physique et de Chimie Solvay, depuis leur fondation jusqu' à la deuxiéme guerre modiale," Bruxelles, 1946.
25. A. H. Compton, *Bull. Nat. Res. Council* **XX,** 16 October 1922; *Philos, Mag.*, **46,** 897 (1923).
26. A. H. Compton, *Phys. Rev.*, **22,** 409 (1923); P. Debye, *Phys. Zeit.*, **24,** 161 (1923).
27. L. de Broglie, *Nature (London)*, **112,** 540 (1923); Thése, Paris (1924); *Ann. Phys.* **8**(10), 22 (1925).
28. E. Schrödinger: *Ann. Phys.* **79**(4), 361 (1926); *Naturwissenschatten,* **14,** 664 (1926); *Ann. Phys.* **79**(4), 734 (1926); *Ann.Phys.* **80**(4), 437 (1926); *Ann. Phys.* **81**(4), 109 (1926).
29. W. Heisenberg, *Z. Phys.*, **33,** 879 (1925); M. Born and P. Jordon, *Z. Phys.*, **34,** 858 (1925); P. Dirac., *Proc. R. Soc. London, Ser. A,* **109,** 642 (1925), **110,** 561 (1926); M. Born, W. Heisenberg, and P. Jordan, *Z. Phys.*, **35,** 557 (1926); W. Pauli, *Z. Phys.*, **36,** 336 (1926); W. Heisenberg and P. Jordan, *Z. Phys.*, **37,** 263 (1926); M. Born, *Z. Phys.*, **40,** 167 (1927).
30. C. Davisson and C. H. Kunsmann, *Phys. Rev.*, **22,** 243 (1927); C. Davisson and G. H. Germer, *Nature (London)*, **119,** 558 (1927).
31. G. P. Thomson and A. Reid, *Nature (London)*, **119,** 890 (1927).
32. M. Born, *Z. Phys.*, **38,** 803 (1926).
33. N. Bohr, Atti del Congresso Internazionale di Como, Settembre 1927 (reprinted in *Nature (London)*, **121,** 78, 580 (1928).
34. W. Heisenberg, *Z. Phys.*, **43,** 172 (1927).
35. *Albert Einstein: Philosopher and Scientist*, The Library of Living Philosopher, Inc. Tudor Publishing Company, 1949, 1951, Harper & Row, New York, 1959. See also Ref. 4.

36. See, for example, B. D' Espagnat, *Conceptions de la Physique Contemporaine*, Herman, Paris, 1965, and the following articles in *Rev. Mod. Phys.*, **38**, (1966) by J. S. Bell (p. 447), and D. Bohm and J. Bub (pp. 453 and 470).

37. B. d'Espagnat, *Conceptual Foundations of Quantum Mechanics*, 2nd ed., Benjamin, New York, 1976. See also, by the same author, the article: "The Quantum Theory and Reality," *Sci. Am.*, **241**(5), 128, November (1979), and the critical remarks by V. Weisskopf followed by d'Espagnat answer: *Sci. Am.*, **242**, 8, May (1980).

38. This terminology is used by d'Espagnat who analyses the foundation of this world view in three premises or assumptions: (1) realism, that is, the doctrine that regularities in observed phenomena are caused by some physical reality whose existence is independent of human observation; (2) inductive inference is a valid mode of reasoning and can be applied freely so that legitimate conclusions can be drawn from consistent observations; (3) "Eistein separability," that is, no influence of any kind can propagate faster than the speed of light. Weisskopf, in his critical remarks (Ref. 37), warns about the use of ("misleading") expressions, which suggest that a renunciation to "local realistic views" may be equivalent to a renunciation of "realism" in general. The experiments considered here are in agreement with the predictions of quantum mechanics and therefore provide a clear support to the "realism" inherent to quantum mechanics, which is *different* from classical realism, in particular from "local realistic views."

39. A. Einstein, B. Podolski, and N. Rosen, *Phys. Rev.*, **47**, 777 (1935).

40. J. S. Bell, *Foundations of Quantum Mechanics*, Proceedings of the Enrico Fermi International Summer School, Course 40, Academic Press, New York, 1971.

41. *Structure et Propertés des Noyaux Atomiques*, Rapports et Discussions du Septiénne Conseil de Physique tenu à Bruxelles du 25 au 29 Octobre, Gauthier-Villars, Paris, 1934.

42. J. D. Cockcroft and E. T. S. Walton, *Proc. R. Soc. London, Ser. A*, **137**, 229 (1932).

43. M. L. E. Oliphant and E. Rutherford, *Proc. R. Soc. London, Ser. A*, **141**, 259 (1933).

44. E. O. Lawrence and M. S. Livingston, *Phys. Rev.*, **37**, 1707 (1931); **38**, 834 (1931); **40**, 19 (1932); **42**, 1950 (1932).

45. J. Chadwick, *Nature (London)*, **129**, 312 (1932); *Proc. R. Soc. London, Ser. A*, **136**, 692 (1932).

46. C. D. Anderson, *Phys. Rev.*, **43**, 491 (1933).

47. P. M. S. Blackett and G. P. S. Occhialini, *Proc. R. Soc. London, Ser. A*, **139**, 699 (1933).

48. L. Meitner and K. Philipp: *Naturwissenschaften*, **21**, 468 (1933); J. Curie and F. Joliot: *C. R. Acad. Sci. Paris*, **196**, 1105, 1581 (1933); C. D. Anderson and S. H. Neddermeyer, *Phys. Rev.*, **43**, 1034 (1933); P. M. S. Blackett, J. Chadwick, and G. Occhialini, *Nature (London)*, **131**, 473 (1933); *Proc. R. Soc. London, Ser. A*, **144**, 235 (1934).

49. B. Rossi, *Z. Phys.*, **82**, 151 (1933).

50. W. Heisenberg, *Z. Phys.*, **77**, 1 (1932); **78**, 156 (1932); **80**, 587 (1933).

51. E. Majorana, *Z. Phys.*, **82**, 137 (1933); *Ric. Scient.* **4**(1), 559 (1933).

52. E. Fermi: *Nuovo Cimento*, **11**, 1 (1934); *Z. Phys.*, **88**, 161 (1934).

53. F. Joliot and I. Curie, *C. R. Acad. Sci. Paris*, **198**, 254, 559 (1934); *J. Phys.*, **5**, 153 (1934).

54. E. Fermi, *Ric. Scient.*, **5**(1), 283 (1934); *Nature (London)*, **133**, 757 (1934); E. Fermi, E. Amaldi, O. D'Agostino, F. Rasetti, and E. Segrè, *Proc. R. Soc. London, Ser. A*,

146, 483 (1934); E. Amaldi, O. D'Agostino, E. Fermi, B. Pontecorvo, F. Rasetti, and E. Segrè, *Proc. R. Roy. Soc. London, Ser. A,* **146,** 522 (1935); E. Amaldi and E. Fermi, *Phys. Rev,,* **50,** 899 (1936).

55. *Les Particules Élémentaires*, Rapports et Discussions du huitième Conseil de Physique tenu à l'Université de Bruxelles du 27 Septembre ay 2 Octobre, 1948, R. Stoops, Brussels, 1950.
56. C. D. Anderson and S. H. Neddermeyer, *Phys. Rev.,* **51,** 884 (1937); **54,** 88 (1938).
57. B. Rossi, *Naturwissenschaften* **20,** 65 (1932).
58. H. Yukawa, Proc. Phys.-Math. Soc. Jpn., **17,** 48 (1935).
59. M. Conversi, E. Pancini, and O. Piccioni: *Nuovo Cimento,* **3,** 372 (1945); *Phys. Rev.,* **71,** 209 (1947).
60. C. M. G. Lattes, G. P. Occhialini, and C. F. Powell, *Nature (London),* **159,** 186 (1947).
61. D. Rochester and C. C. Butler, *Nature (London),* **160,** 855 (1947).
62. W. L. Gardner and C. M. G. Lattes, *Science,* **107,** 270 (1948); J. Burfening, E. Gardner, and C. M. G. Lattes, *Phys. Rev.,* **75,** 382 (1949).
63. L. de Broglie, *C. R. Acad. Sci. Paris,* **195,** 862 (1932); **197,** 536 (1932).
64. *La Théorie Quantique des Champs,* Rapports et Discussions du Douzième Conseil de Physique tenu à l'Université Libre de Bruxelles du 9 au 14 Octobre 1961, Wiley-Interscience, New York, and R. Stoops, Brussels, 1962.
65. A. Pais, *Phys. Rev.,* **86,** 513 (1952).
66. W. B. Fowler, R. P. Shutt, A. M. Thorndike, and W. L. Whittermore, *Phys. Rev.,* **93,** 861 (1954).
67. M. Gell-Mann and A. Pais, *Proc. Glasgow Conf. on Nuclear and Meson Phys.,* p. 342, Pergamon Press, London and New York, 1934; K. Nishijima, *Prog. Theor. Phys. (Kyoto),* **12,** 107 (1954).
68. This is true for energies below $\sim 10^3$ GeV, but at extremely high energies ($\geq 10^4$ GeV) cosmic rays will probably not be superseded by accelerators in any forseeable future.
69. O. Chamberlain, E. Segrè, C. Wiegand, and T. Ypsilantis, *Phys. Rev.,* **100,** 947 (1955).
70. M. Gell-Mann, Caltech Report CSTL-20 (1961), *Phys. Rev.,* **125,** 1067 (1962); Y. Ne'eman, *Nucl. Phys.,* **26,** 222 (1961); M. Gell-Mann and Y. Ne'eman, *The Eightfold Way*, Benjamin, New York, 1964.
71. This discussion of the hyperfine splitting of the hydrogen isotopes was still affected by some uncertainty originating in part from the inaccuracy in the value of the electrodynamic coupling constant α.
72. For the experimental tests on quantum electrodynamics at low energy, see, for example, E. Picasso, *Acta Leopoldina*, Suppl. 8 Bd. 44, p. 159 (1976), Deut. Akad. Naturforsh. Leopoldina; F. Combley and E. Picasso: p. 717 of *Proceedings 68th Course on Metrology and Fundamental Constants*, "International Summer School Enrico Fermi, Varenna, July 1976," Societe' Italiana di Fisica, Bologna, 1980.
73. Jade Collaboration: Desy 80/14, March 1980; Pluto Collaboration: Desy 80/01, January 1980.
74. T. D. Lee and C. N. Yang, *Phys. Rev.,* **104,** 254 (1956).
75. C. S. Wu, E. Amber, R. W. Hayward, D. D. Hoppes, and R. P. Hudson, *Phys. Rev.,* **105,** 1413 (1957); R. Garvin, L. Lederman, and M. Weinrich, *Phys. Rev.,* **105,** 1415 (1957); J. I. Friedman and V. L. Telegdi, *Phys. Rev.,* **105,** 1681 (1957); H. Frauenfelder, *Phys. Rev.,* **106,** 386 (1957).

76. B. Pontecorvo, *Phys. Rev.*, **72,** 246 (1947); O. Klein, *Nature (London)*, **161,** 897 (1948); G. Puppi, *Nuovo Cimento,* **5,** 587 (1948); T. D. Lee, M. Rosenbluth, and C. N. Yang, *Phys. Rev.,* **75,** 905 (1949); J. Tiomno and J. A. Wheeler, *Rev. Mod. Phys.*, **21,** 144 (1949).
77. B. Pontecorvo, *J. Exp. Theor. Phys. USSR,* **37,** 1751 (1951); A. Salam, *Proc. Seventh Annual Conf. on High Energy Physics 1957,* Interscience, New York, 1957; J. Schwinger, *Ann. Phys.*, **2,** 407 (1957); N. Cabibbo and R. Gatto, *Phys. Rev. Lett.*, **5,** 114 (1960).
78. S. S. Gerschtein and J. B. Zel'dovich, *JETP (USSR)*, **29,** 698 (1955), translation in *Sov. Phys. JETP,* **2,** 576 (1957); R. P. Feynman and M. Gell-Mann, *Phys. Rev.*, **109,** 193 (1958).
79. *Fundamental Problems in Elementary Particles Physics*, Proceedings of the Fourteenth Conference on Physics at the University of Brussels, October 1967, Interscience, New York, 1968.
80. T. Regge, *Nuovo Cimento,* **14,** 951 (1959).
81. G. F. Chew and S. Frantschi, *Phys. Rev. Lett.*, **7,** 394 (1961).
82. M. Gell-Mann, *Phys. Rev. Lett.*, **8,** 214 (1964); G. Zweig, CERN Preprint Th-492 (1964).
83. Y. Nambu, *Phys. Rev. Lett.*, **4,** 380 (1960); J. Bernstein, S. Fubini, M. Gell-Mann, and W. Thirring, *Nuovo Cimento,* **17,** 757 (1960).
84. *La conductibilité Électrique des Metaux*, Rapports et Discussions du Quatrieénne Conseil de Physique tenu a Bruxelles du 24 au 29 Avril 1924, Gauthier-Villars, Paris, 1927.
85. *Le Magnétisme*, Rapports et Discussions du Sixiénne Conseil de Physique tenu a Bruxelles du 20 au 25 October 1930, Gauthier-Villars, Paris, 1932.
86. J. C. Slater, *The Self-Consistent Field for Molecules and Solids, Quantum Theory of Molecules and Solids*, Vol. 4, Ch. 10, McGraw-Hill, New York, 1974.
87. *L'État Solide*, Rapports et Discussions du neuviénne Conseil de Physique tenu à l'Université Libre de Bruxelles du 25 au 29 September 1951, R. Stoops, Bruxelles, 1952.
88. F. C. Franck and W. T. Read, *Phys. Rev.*, **79,** 722 (1950); F. Seitz, *Imperfections in Nearly Perfect Crystals*, W. Shockley, ed., Ch. 1, Wiley, New York, 1952.
89. J. M. Edges and J. W. Michell, *Philos. Mag.*, **44,** 223 (1953); W. C. Dash, *J. Appl. Phys.*, **27,** 1153 (1956); S. A. Amelinckx, *Phys. Mag.*, **1,** 269 (1956).
90. *Les Éléctrons dans les Métaux*, Rapports et Discussions du dixiénne Conseil de Physique tenu a Bruxelles du 13 au 17 Septembre 1954, R. Stoops, Bruxelles, 1955.
91. J. Bardeen, L. N. Cooper, and J. R. Schrieffer, *Phys. Rev.*, **108,** 1175 (1957).
92. *Order and Fluctuations in Equilibrium and Non-Equilibrium Statistical Mechanics*, G. Nicolis, G. Dewel, and J. W. Turner, eds., Wiley, New York, 1981.
93. K. G. Wilson and M. E. Fischer, *Phys. Rev. Lett.*, **28,** 240 (1972); K. G. Wilson and J. Kogut, *Phys. Rev.*, **12C,** 75 (1974); K. G. Wilson, *Rev. Mod. Phys.*, **47,** 773 (1975).
94. L. P. Kadanoff, *Critical Phenomena*, Proceedings of the International School of Physics "Enrico Fermi," Course 51, M. S. Green, ed., Academic Press, New York, 1971.
95. G. Nicolis and I. Prigogine: *Self-Organization in Non-Equilibrium Systems*, Wiley-Interscience, New York, 1977.
96. E. L. Koschmieder, "Bénard Convection," *Adv. Chem. Phys.*, **26,** 177 (1974).

97. E. Arecchi, in: *Interaction of Radiation with Condensed Matter*, Vol. I, IAEA, Vienna (1977).

98. G. Nicolis and J. F. G. Auchmuty, *Proc. Natl. Acad. Sci. USA*, **71**, 2748 (1974); D. H. Sattinger, *J. Math. Phys.*, **19**, 1720 (1978).

99. *Symmetry Properties of Nuclei*, Proceedings of the Fifteenth Conference on Physics at the University of Brussells, 28 September to 3 October 1970, Gordon and Beach, New York, 1974.

100. E. P. Wigner, *Phys. Rev.*, **51**, 106, 447 (1937); E. P. Wigner and E. Feenberg, *Rep. Prog. Phys., Phys. Soc. London*, **8**, 274 (1941); P. Franzini and L. Radicati, *Phys. Rev. Lett.*, **6**, 322 (1963).

101. "La structure et l'Evolution de l'Univers," Rapports et Discussions de l'onziènne Conseil de Physique, tenu à l'Université de Bruxelles du 9 au 13 Juin 1958, R. Stoops, Bruxelles, 1958.

102. E. M. Burbidge, G. R. Burbidge, W. A. Fowler, and F. Hoyle, *Rev. Mod. Phys.*, **29**, 547 (1957); *Science*, **124**, 611 (1956).

103. A. A. Penzias and R. H. Wilson, *Astr. J.*, **142**, 419 (1965).

104. J. R. Oppenheimer and H. Snyder, *Phys. Rev.*, **55**, 455 (1939).

105. *The Structure and Evolution of Galaxies*, Proceedings of the Thirteenth Conference on Physics at the University of Brusselles, September 1964, Interscience, New York, 1965.

106. R. Giacconi, H. Gursky, F. Paolini, and B. Rossi, *Phys. Rev. Lett.*, **9**, 439 (1962).

107. S. Bayer, E. T. Byram, T. A. Chubb, and H. Friedman, *Science*, **146**, 912 (1964).

108. As announcement of the discovery of quasars one can take No. 4872 of *Nature* (*London*), appeared on March 16, 1963 (Vol. 197), where a set of four articles touch on a few fundamental properties of two radio sources: 3C 273 and 3C 48. The second of the four articles, due to Martin Schmidt, is the core of the whole argumentation. The first article by C. Hazard, M. B. Mackey, and A. J. Shimmins (p. 1037) presents the "Investigation of the Radio Source 3C 273 by the method of Lunar Occultation." In the second paper, by Martin Schmidt (p. 1040) entitled "3C 273: A Star-Like Object with Large Redshift", the author shows that 6 lines (four of the Balmer series, one of Mg II, the other of O III) can be explained with a redshift of 0.158. The third paper by J. B. Oke (p. 1040), is entitled "Absolute energy distribution in the optical spectrum of 3C 273" and the fourth, by J. L. Greenstein and T. A. Matthews (p. 1041), "Redshift of the Unusual Radio Source 3C 48", for which they give a value of 0.3675 as weighted average of six relatively sharp lines.

109. The efficiency of man-made accelerators, defined as the output power in the beam devided by the total power supply is: Berkeley Bevatron, $\sim 3 \times 10^{-4}$; 28 GeV CERN-PS, $\sim 3 \times 10^{-4}$; Frascati 1 GeV electrosyncrotron, $\sim 1 \times 10^{-4}$.

110. *Astrophysics and Gravitation*, Proceedings of the Sixteenth Solvay Conference in Physics at the University of Brussels, 24–28 September 1973, Editions de l'Université de Bruxelles, Brussels, 1974.

111. R. W. Klebesadel, I. B. Strong, and R. A. Olson, *Astrophys. J.*, **182**, L85 (1973).

NOTE: On November 1982 a Solvay Conference devoted to "Higher Energy Physics: What are the possibilities for extending our understanding of elementary particles and their interactions to much greater energies?" was held in the University of Texas at Austin.

Radioactivity, a Pragmatic Pillar of Probabilistic Conceptions.

E. AMALDI

Istituto di Fisica dell'Università - Roma

When, years ago, I read the book by Max JAMMER *The Conceptual Development of Quantum Mechanics* [1], I found very interesting and stimulating the presentation he gives of this fundamental subject. I should say, however, that I was slightly disappointed from Chapter 4, devoted to *The transition to quantum mechanics*, since it appeared to me too short, incomplete and in some way one-sided.

The chapter consists of three sections. The first one, entitled *Applications of quantum conceptions to physical optics*, summarizes very effectively the work and conceptual background of A. FRESNEL, J. J. THOMSON, A. H. COMPTON, G. BARKLA, W. H. BRAGG and W. L. BRAGG, P. DEBYE and a few others. The second section, devoted to *The philosophical background of nonclassical interpretations*, can be, in some way, summarized by the following sentence by JAMMER himself: « certain philosophical ideas of the late ninteenth century not only prepared the intellectual climate for, but contributed decisively to the formation of the new conceptions of the modern quantum theory » [2]. The people quoted are C. RENOUVIER, E. BOUTROUX, F. EXNER, S. KIERKEGAARD, H. HØFFDING, and also H. POINCARÉ, L. DE BROGLIE, N. BOHR, C. G. DARWIN and a few others. The new conceptions are probabilistic conceptions which « differ fundamentally from the traditional notions of probability as used, for example, in classical statistical mechanics. In classical physics probability statements were but an expression of human ignorance of the exact details of the individual event, either because of the insufficient resolving power of our measuring instrument or because of the large number of events involved: the individual physical process, however, was always regarded as strictly obeying the law of cause and effect and the result was always considered as uniquely determined.

The new conception of probability, on the other hand, assumed not only that macroscopic determinism is a statistical effect but also that the individual microscopic and submicroscopic event is purely contingent » [3].

The third and last section of Chapter 4, entitled *Nonclassical interpretation of optical dispersion*, deals with the early work by N. BOHR, J. C. SLATER,

H. A. KRAMERS and by M. BORN, A. LANDÉ, R. LADENBURG and a few others.

With this presentation of the first attempts to solve the fundamental problem of optical dispersion, Chapter 4 is closed and the discussion of the transition to the new probabilistic conceptions is practically finished.

My first impression that this presentation is too hastly and too one-sided did not find further support from successive Jammer works such as the book on the *Philosophy of Quantum Mechanics* [4], since, in this case, an emphasis (or perhaps an over-emphasis) of the philosophical aspects of the subject is justified by the same title of the book.

My original impression was, on the contrary, strengthened by the article of P. FORMAN on *Weimar culture, causality and quantum theory, 1918-1927* [5], where the author states: ... « Jammer did not go very far towards demonstrating his propositions ... » the most important of which is « that extrinsic influences led physicists to ardently hope for, actively search for, and willingly embrace an acausal quantum mechanics ».

The aim of Forman's article (according to him) is not « to fill the gap left by Jammer » between « a variety of late ninteenth century philosophers » and « the development of quantum mechanics by German speaking central European physicists circa 1925 ». His aim is « rather to examine closely the lay of the land on the far side ... » with the result of a « overwhelming evidence that in the years after the end of the First World War, but before the development of an acausal quantum mechanics, under the influence of 'currents of thought' large numbers of German physicists, for reasons only incidentally related to developments in their own discipline, distanced themselves from, or explicitly repudiated, causality in physics ».

One of the main conclusions of Forman is that the « extrinsic influences » suggested by JAMMER are demonstrated in his article « but only for the German cultural sphere ».

Certainly I am not here to deny the interest of Jammer's point of view and the suggestiveness of the above propositions nor the cultural value of Forman's article, which deeply analyses the interrelations between the development of physical sciences in those years and their cultural and philosophical craddle and environment. What I would like to express in the following is my impression that, by looking at the problem of « repudiation of causality in physics » from the most general and far away point of view, one can be brought to over-estimate the « extrinsic influences » outlined above and overlook « intrinsic arguments » inherent to two parallel, almost independent developments. The first one starts from the kinetic theory of gases and passes through statistical mechanics, Planck original definition of quantum, the photons conceived as particles and the relations between emission and absorption of photons by atoms.

The other path, also intrinsic to physics, starts with the accidental discovery of radioactive substances, passes through the experimental recognition of their decay properties and quickly finds its natural settlement in a probabilistic conception which can be accused to be acritical, but has certainly a sound pragmatic ground, uncorrelated or at least extremely loosely correlated to contemporary or pre-existing philosophical lines of thought.

1. – Gradual infiltration of probability's laws into physical sciences.

The first one of these two intrinsic lines of development has been discussed on many occasions, in particular by Stephen G. BRUSH in an article appeared a few months ago, entitled *Irreversibility and indeterminism: Fourier to Heisenberg* [6], which according to the author « might be considered a chapter in the history of the changing meaning of the word *statistical* ». I will not summarize the historical succession of scientific steps forming this path nor will I sketch the line traced by BRUSH. It would take too much time for considerations that are very interesting but outside the main scope of my lecture. I will only mention a few points that appeared to me—already before reading Brush article—as milestones of these developments, omitting the very interesting detailed interconnections.

One of the most important aspects of Brush article, in my opinion, is the emphasis he gives to the gradual, continuous infiltration of probability's laws into physical sciences, to which I will add at the end, as one of my conclusions, the multiplicity of paths that all led to the same final winning-post: quantum mechanics.

Starting from Fourier's theory of heat conduction and a comparison of time-reversible mechanics (in the absence of dissipative forces) and time-oriented heat flow, BRUSH passes in review the first and second laws of thermodynamics with special regard to the work of Sadi CARNOT, Rudolf CLAUSIUS and William THOMSON (Lord KELVIN) and arrives at the kinetic theory of gases, in particular at the work of J. C. MAXWELL and L. BOLTZMANN. The words of these authors are ambiguous so that « in the absence of any explicit statements one might legitimately infer that they tacitly accepted the views of their contemporaries », *i.e.* molecular determinism. « But their equations (points out BRUSH) pushed physical theory very definitively in the direction of indeterminism. As in other transformations of physical sciences ... mathematical calculation led to results that forced the acceptance of qualitatively different concepts. »

From BOLTZMANN, BRUSH arrives to PLANCK, to his work on Boltzmann's statistical interpretation of entropy, and to his successive discovery, in 1900, of the spectral distribution of the black-body radiation.

The two laws of thermodynamics gave rise to well-known debates, which were strongly reinforced in 1865 when CLAUSIUS put them in the simple verbal form: « The energy of the Universe is constant. The entropy of the Universe tends towards a maximum ». These two statements, along with various logical and illogical extensions, and the use of probability methods in kinetic theory were matters of widespread debates, in which various philosophers and even theologians [7] were deeply involved. The fact that the Universe was moving towards a state of maximum probability, a configuration of maximum randomness, a condition of minimum available energy, was deemed to led to a degradation of the energy sources and ultimately to that pessimistic state of affairs that was called « heat-death » of the Universe. Themodynamics accordingly predicted an end of everything as a function of time.

Thus the laws of thermodynamics and their statistical interpretation had, directly or indirectly, an influence on the way of thinking of many philosophers between the end of the last century and the beginning of our century, and these, through a kind of feedback process, sometime had an influence on the general world outlook of physicists of successive generations.

The next milestone, as everybody knows, is the photon theory of light presented by EINSTEIN in sections 8 and 9 of the rather long paper published in 1905 and devoted mainly to the *Entropie der Strahlung* [8].

Max BORN wrote at the beginning of his 1926 paper [9] where he suggests the interpretation of the wave function as probability amplitude: « Dabei knüpfe ich an eine Bemerkung Einsteins über das Verhältnis von Wellenfeld und Lichtquanten an; er sagte etwa, das die Wellen nur dazu da seien, um der korpuskularen Lichtquanten den Weg zu weisen, und er sprach in diesem Sinne von einem *Gespensterfeld.* Dieses bestimmt die Wahrscheinlichkeit dafür, das ein Lichtquant, der Träger von Energie und Impuls, einen bestimmten Weg einschlägt; dem Felde selbst aber gehort keine Energie und kein Impuls zu ». (To this end I pick up from a remark by EINSTEIN on the relationship between wave field and light quanta; he said, more or less, that the waves are present only to show the corpuscular light quanta the way and he spoke in this sense of a *ghost field.* This determines the probability that a light quantum, the bearer of energy and momentum, takes a certain path: no energy nor momentum belongs to the wave field itself.)

In his Nobel lecture delivered in 1954 BORN was even more explicit in affirming a direct connection between his statistical interpretation of the wave function and the interpretation suggested by EINSTEIN of the « square of the optical wave amplitudes as probability density for the occurrence of photons » [10].

The last milestone of this line of development is, of course, the group of papers published by EINSTEIN between 1916 [11] and 1918 [12] on the absorption and emission of radiation by molecules [13]. In the 1918 paper [12] the basic assumptions of his theory are summarized in section 2.

The transition from the higher energy level E_m of a molecule to the lower E_n, during the time interval dt, can take place through two different mechanisms that EINSTEIN calls Ausstrahlung and Einstrahlung. The second one is the stimulated transition. Its probability of occurrence is proportional to the radiation density $\varrho(\nu)$ at the frequency of the emitted photon:

$$\text{(B)} \qquad \mathrm{d}W = B_m^n\,\varrho(\nu)\,\mathrm{d}t\,,$$

where $\nu = (E_m - E_n)/h$ and B_m^n is a constant. In the first process, the Ausstrahlung, the transition is not excited externally. The probability of occurrence in the time interval $\mathrm{d}t$ is given by

$$\text{(A)} \qquad \mathrm{d}W = A_m^n\,\mathrm{d}t\,,$$

where A_m^n is a constant. EINSTEIN comments this assumption as follows: « Das angenommene statistische Gesetz entspricht dem einer radioaktiven Reaktion, der vorausgesetzte Elementarprocess einer derartigen Reaktion, bei welcher nur γ-Strahlen emittiert werden ... ». (The adopted statistical law corresponds to that of radioactive reactions, the elementary process of which is such that only γ-rays are emitted.)

This final remark is of considerable interest from my point of view, since it establishes a connection between the line of development sketched above and the other line of development, that in my opinion is very often forgotten, in some case mentioned [6], but never discussed with sufficient attention and detail

Before passing to this new subject that constitutes the main scope of my seminar, I should recall that the Einstein's papers on the absorption and emission of light by molecules are discussed at length in sect. 3.2 of Jammer book [1], devoted to *The Correspondence Principle*. A few remarks are added by JAMMER that can be summarized as follows: 1) EINSTEIN did not define his conception of probability. Therefore, we are not authorized to consider him, as some authors do, a precursor of the new probabilistic conceptions. 2) Einstein's comparison of his statistical law with that of radioactive disintegration cannot be used as an argument, since at that time radioactive disintegration was generally considered, for example by PLANCK, as a process involving as yet unknown parameters. 3) The absence of a declaration in favour of causality cannot be interpreted as a declaration in favour of the absence of causality. 4) Referring to the related problem of other reactions of an atom or molecule during emission, EINSTEIN stated expressly that « in the present state of the theory the direction of recoil is only statistically determined », alluding thereby to the merely preliminary character of such an approach.

I frankly should say that it seems to me that the remarks 1), 3) and 4) refer only to the personal position of Einstein, but they are not very significant in the historical development of the new probabilistic conceptions. The remark 2) regards the personal position of Planck, who certainly did never work on radioactive decay. Jammer's remarks only underline the personal position of these two great scientists, who were both irreducible determinists, in spite of the fact that a great part of their fundamental contributions was essential for the developments that brought to quantum mechanics.

2. – The discovery of the law of radioactive decay.

A first indication of an activity decreasing with the passing of time was found in Vienna in 1898 by G. C. SCHMIDT, who observed that thorium compounds continuously emit radioactive particles of some kind, which retain their radioactive power for several minutes [14]. Shortly afterwards in 1899, Marie and Pierre CURIE [15, 16] observed a similar effect in the case of radium (and polonium [15]) and called it « radioactivitè induite » (induced radioactivity), while RUTHERFORD [17, 18], in 1900, went on with Schmidt experiments calling « emanation of thorium » « the radioactive particles given out from the mass of thorium compounds in addition to ordinary radiation ».

The expressions « induced activity » and « emanation of thorium » were due to primitive interpretations of essentially the same phenomenon. A few years later these were both replaced by the expression « active deposit », suggested by RUTHERFORD and still used today, for the radioactive bodies (RaA + + RaB + RaC + RaC'+ RaC'' and ThA + ThB + ThC + ThC'+ ThC'') produced in succession by the decay of the emanations of Ra and Th. These are the noble gases belonging to the radium and thorium families, later called radon (Rn) and thoron (Tn) [19].

What the CURIES and RUTHERFORD had observed?

The CURIES had found that all substances placed in the vicinity of a sample of radium acquired a radioactivity which, after the removal of radium, decreases according to an exponential law.

In the experiment of Rutherford a slow current of air passed over a sample of thorium oxide and « carried away the radioactive particles with it and these were gradually conveyed » into an ionization chamber connected to an electrometer arranged to measure the saturation current (fig. 1). Thus RUTHERFORD measured the decay curve (curve A of fig. 2) of a certain amount of emanation brought from its source (thorium oxide) to the measuring instrument and found an exponential curve, for which he wrote the formula

$$n(t) = n_0 \exp[-\lambda t], \tag{1}$$

where $n(t)$ is the number of ions produced per second by the radioactive particles between the plates. This is connected to the measured current $i(t)$ by the relation $i(t) = en(t)$. RUTHERFORD measured also « the rise of current » (curve B of fig. 2), *i.e.* the rise of production of emanation from thorium as a function of time starting from the instant in which all emanation pre-

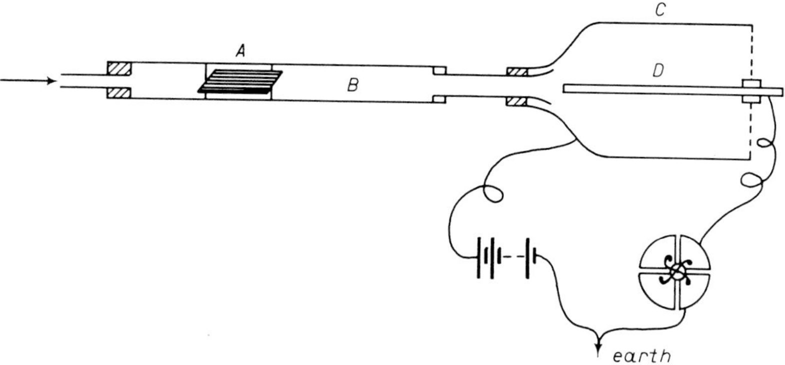

Fig. 1. – Rutherford experimental set-up: a thick layer of thorium oxide was enclosed in a paper vessel A placed inside the long metal tube B. One end of B was connected to a large insulated cylindrical vessel C connected to one terminal of a 100 V battery. Inside C was fixed an insulated electrode D connected to a pair of quadrants of the electrometer. The other pair of quadrants was connected to the other terminal of the battery. A slow current of purified air was passed through the apparatus and carried the emanation of thorium from A to C and then went out through a number of small holes opened in the bottom of C.

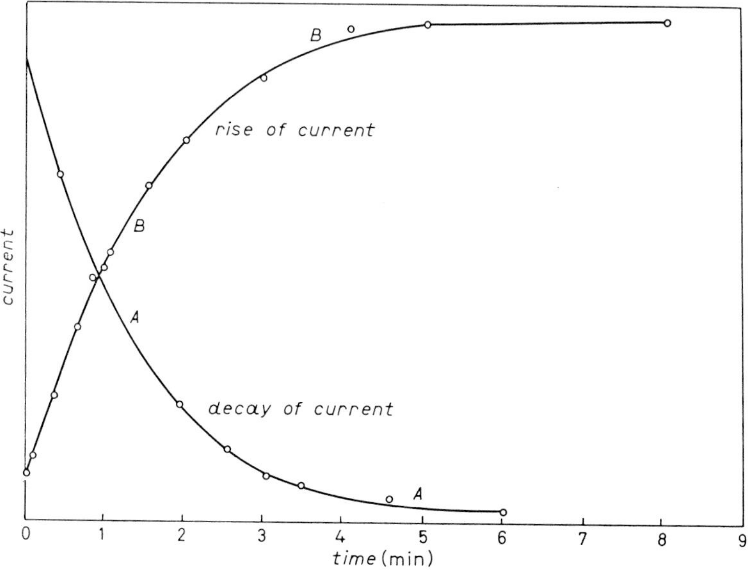

Fig. 2. – Experimental results obtained by RUTHERFORD with the experimental set-up shown in fig. 1.

viously produced was pumped away. For discussing these experimental results RUTHERFORD writes the differential equation

$$\frac{\mathrm{d}n(t)}{\mathrm{d}t} = q - \lambda n(t) \tag{2}$$

(q is the number of ions supplied per second by the emanation diffusing from thorium) and derives the solutions corresponding to the two cases mentioned above:

$$\left|\begin{array}{ll} \text{for } \begin{cases} q = 0 \\ n(0) = n_0 \end{cases} & \text{(decay of current): } n(t) = n_0 \exp[-\lambda t], \\ & \\ \text{for } \begin{cases} q \neq 0 \\ n(0) = 0 \end{cases} & \text{(rise of current): } n(t) = \frac{q}{\lambda}\left(1 - \exp[-\lambda t]\right). \end{array}\right. \tag{3}$$

Thus the decay law was already well established, but the ideas about the nature of the « emanation » were still confused. Two possible interpretations were considered by RUTHERFORD, *viz.*

« 1) That the emanation may be due to fine dust particles of the radioactive substance emitted by the thorium compounds.

2) That the emanation may be a vapor given off from thorium compounds ».

One of the main conclusions of the paper was that the interpretation 1) had to be excluded.

These results of Schmidt, the Curies and Rutherford were the first steps towards the establishment, that came gradually only from systematic experiments, of the general law that any amount of a chemically defined radioactive body, isolated from its source, decreases with the passing of time according to an exponential law.

The experiments described above were made only a few years after the discovery of the radioactivity of uranium made by Henri BECQUEREL [20] in February 1896. The discovery was accidental but made during a systematic investigation of the fluorescence of uranium salts inspired by the discovery made by W. C. RÖNTGEN, in 1895, of X-rays and their remarkable properties.

Becquerel discovery was followed by the discovery of the radioactivity of thorium, made independently in Vienna by SCHMIDT [21] and in Paris by Marie CURIE [22], of polonium [23] and radium [24], both separated chemically from uranium minerals in 1898 by Marie and Pierre CURIE, with the collaboration of G. BÈMONT in the last piece of work.

In none of these cases was observed a time dependence of the activity. Only in the case of polonium such a dependence could have been observed, and it actually was, but only in 1906 [25].

Not only it took a few years to establish the correct interpretation of the active deposit; even the nature of radioactivity remained for years an unsolved puzzle. I may recall, as an example, an interpretation suggested (rather softly) by the CURIES in 1900 [26].

The radiation emitted from radioactive bodies could be a secondary radiation emitted by heavy elements under the action of a primary radiation (similar to X-rays) pervading the whole space and all bodies.

A clear understanding of the nature of radioactivity came by steps. In 1902 the CURIES [27] as well as RUTHERFORD and SODDY [28] express clearly the idea that radioactivity is an atomic phenomenon, more precisely that « radioactivity is a manifestation of sub-atomic chemical change » [28, 29]. RUTHERFORD and SODDY go on and a few lines below say: « The idea of the chemical atom in certain cases spontaneously breaking up with evolution of energy is not of itself contrary to anything that is known of the properties of atoms ... ».

The idea of spontaneous breaking of a chemical atom which thus is transformed into a chemical atom of a different type is further elaborated in various successive papers by RUTHERFORD and SODDY, who start to consider already in 1903 [30] families of radioactive bodies. In this same paper, the authors notice that the exponential law (1) should hold not only for the number of ions produced, but also for the rays emitted by radioactive bodies [31]. Since « the rays emitted must be an accompaniment of the change of the radiating system into the next one produced », the same law should hold for the number $N(t)$ of radioactive atoms of a given type, present at the time t. By differentiating the empirical exponential law, they deduce

$$\frac{\mathrm{d}N}{\mathrm{d}t} = -\lambda N(t) \tag{4}$$

as the general law regulating the behaviour of any radioactive body: « The change of the system at any time is always proportional to the amount remaining unchanged », and « the proportion of the whole which changes in unit time is represented by the constant λ, which possesses for each type of active matter a fixed and characteristic value. λ may therefore be suitably called the *radioactive constant* ».

At the beginning the « activity » measured by the discharge of an electroscope was the quantity decaying exponentially; later it was the current, or the number of ions, still later the number of observed emitted particles and finally, in 1903, the number of decaying radioactive atoms.

In the disintegration theory advanced by RUTHERFORD and SODDY, it is supposed that a definite small portion of the radium atoms (about 1 in every 10^{10} will suffice) breaks up per second. The disintegration of each atom is accompanied by the expulsion of an α-ray or β-particle with great velocity [32]. In the meantime it was shown by RUTHERFORD that the α-rays of radium consist of positively charged bodies of mass about twice that of the hydrogen atom [33].

In the Bakerian lecture, delivered in front of the Royal Society on May 19, 1904, RUTHERFORD reviewed once more the whole subject [34]. The theory of the transformation of a whole succession of radioactive bodies is discussed in general, and is applied to a number of particular cases. In the discussion it is clearly stated that the constant λ « is the fraction of atoms disintegrating per second, and that its inverse is equal to the *average life* of the corresponding radioactive body ».

Furthermore, the value of λ for any substance is independent of all physical and chemical conditions: many authors, already from early times, had attempted, without success, to observe some change of λ as consequence of changes of temperature, concentration, chemical binding, age of the atoms, or due to the presence of intensive electric and magnetic fields [35].

I have dwelt upon certain detail of interest from the point of view of the present discussion, while I have avoided to quote a number of other important results obtained during those years by the CURIES, RUTHERFORD and their collaborators as well as by many other scientists who gave substantial contributions to the study of radioactivity during the first 8 years after Becquerel discovery. A much more complete picture can be found in Rutherford book, published in 1904 [36].

3. – Statistical fluctuations.

The next important step was made in Vienna in 1905 by Egon VON SCHWEIDLER [37], who noticed that, by treating the constant λ multiplied by dt as the probability for one atom to decay during the time interval dt, the exponential decay law can be theoretically deduced without any special assumption about the dynamic processes taking place within the atom, but simply applying the elementary rules of probability calculus. In this paper, presented to the First International Conference for the Study of Radiology and Ionization held in Liège in September 1905, SCHWEIDLER begins by summarizing the situation with the following sentence (fig. 3): « Nach der Vorstellungen, zu denen die Zerfallstheorie der radioaktiven Erscheinungen führt, sind die Atome einer aktiven Substanz instabile Gebilde, denen eine von ihrer Struktur abhangige 'mittlere Lebensdauer' zukommt. Ist $\lambda\,dt$ die Wahrscheinlichkeit, dass ein Atom innerhalb der Zeit dt eine Umwandlung erfährt, so ist die Warscheinlichkeit, dass es eine Zeit t überdauere, gleich $\exp[-\lambda t]$ und $\tau = 1/\lambda$ seine mittlere Lebensdauer ». (According to the picture provided by the disintegration theory of the radioactive phenomena, the atoms of an active substance are unstable systems with a mean life determined by their structure. If $\lambda\,dt$ is the probability that one atom undergoes a transformation in the time interval dt, then the probability that it survives the time interval t is equal to $\exp[-\lambda t]$ and $\tau = 1/\lambda$ is its mean life.)

Uber Schwankungen der radioaktiven Umwandlung

von E. v. SCHWEIDLER

—

Nach den Vorstellungen, zu denen die Zerfallstheorie der radioaktiven Erscheinungen führt, sind die Atome einer aktiven Substanz instabile Gebilde, denen eine von ihrer Struktur abhängige « mittlere Lebensdauer » zukommt. Ist λdt die Wahrscheinlichkeit, dass ein Atom innerhalb der Zeit dt eine Umwandlung erfährt, so ist die Wahrscheinlichkeit, dass es eine Zeit t überdauere, gleich $e^{-\lambda t}$ und $\tau = \frac{1}{\lambda}$ seine mittlere Lebensdauer. Bei einer sehr grossen Anzahl N gleichartiger solcher Atome wird daher, entsprechend dem Gesetz der grossen Zahlen, die Anzahl der nach der Zeit t noch vorhandenen Atome gegeben sein durch $n = Ne^{-\lambda t}$. Es ist selbstverständlich, dass bei einer geringen Anzahl von Atomen der tatsächliche Verlauf ihrer Verminderung von diesem idealen Gesetze abweichen wird, und es soll im Folgenden untersucht werden, ob die durch die Wahrscheinlichkeitsrechnung zu ermittelnde « Streuung » die Grenzen empirischer Nachweisbarkeit erreichen kann.

Es seien N Atome einer Substanz mit der Abklingungskonstante λ gegeben; nach einer gewissen Zeit δ ist für ein bestimmtes einzelnes Atom die Wahrscheinlichkeit noch zu existieren gleich $e^{-\lambda\delta}$, die, inzwischen eine Umwandlung erfahren zu haben, gleich $1 - e^{-\lambda\delta} = \alpha$. Die Wahrscheinlichkeit, dass von den N Atomen die Anzahl x eine Umwandlung erfahren habe, die Anzahl $N-x$ unverwandelt erhalten geblieben sei, ist dann

$$W_x = \alpha^x (1-\alpha)^{N-x} \binom{N}{x}$$

Wie eine einfache Differentiation ergibt, ist

$$W_x = \text{Maximum für } x = \alpha N,$$

also der dem Abklingungsgesetz $n = Ne^{-\lambda t}$ entsprechende Wert ist der wahrscheinlichste. Es lässt sich aber auch die Wahrscheinlichkeit bestimmen, dass x von dem wahrscheinlichsten Werte αN um eine vorgegebene Grösse abweiche.

Fig. 3. – First page of E. von Schweidler theoretical paper on fluctuations of radioactive decay.

Without any knowledge of the causes which determine in single cases the disintegration of a specific atom, we can understand this process as a purely accidental event in the sense of the probability calculus. SCHWEIDLER notices that the exponential law can only be an approximation, because the function $N(t)$ can only take integer values. From the interpretation of this law as a statistical law, it follows that the observable values should fluctuate around the mean value which is provided by the exponential law. SCHWEIDLER in his paper derives the expression of the fluctuations and shows that the variance of the distribution of the actual number of disintegrations approaches $\sqrt{N(t)}$ in the limit of N very large.

Different types of observations can be made on the disintegration process, each corresponding to different treatments of the statistical problem. These, however, are all based on the physical assumption that, for any radioactive atom of an assigned type, the probability that it decays in the time interval dt is given by $\lambda\, dt$, with λ independent of the age of the atom as well as of any external condition.

In the final part of his paper SCHWEIDLER shows that the fluctuations that he has calculated should be easily observable.

The experimental evidence that fluctuations of the order of those computed from the probability calculus do actually take place was provided by K. W. F. KOHLRAUSCH in 1906 [38] (fig. 4) balancing the ionization currents due to two sources of alpha-particles against each other and measuring the fluctuations from the balance by means of an electrometer. Similar observations were made, in 1908, by MEYER and REGENER [39] by balancing the saturation ionization current due to the source of alpha-particles against a Bronson resistance. Experiments were made independently by GEIGER [40] at the same time by balancing one source of particles against another.

I will not try to summarize here all the successive experiments, nor the theoretical considerations mainly of statistical nature that were published for more than twenty years. An excellent presentation can be found in Chapter II of the second edition of the classical book by St. MEYER and E. VON SCHWEIDLER appeared in 1927 [41]. The last of the 29 papers listed in their bibliography was published in 1924.

Some of these papers treat corrections originating from the inertia of the used instruments (electrometers), or from fluctuations of the ionization produced by a single alpha or beta ray, others refer to the fluctuations inherent to the observation of scintillation produced by alpha-particles, others to the fluctuations of the ionization produced by secondary electrons of gamma-rays.

Among the many pubblications I may recall a book appeared in 1913 by L. VON BORTKIEWICH [42] (a.o. Professor an der Universität, Berlin; fig. 5), who treats in some 80 odd pages all possible aspects of the fluctuations and, in particular, *a*) the fluctuations of the number of scintillations observed in preassigned time intervals and *b*) the statistical distribution of the length of time intervals, between two successive decays. Experiments of this second type had been already made by MARSDEN and BARRAT in 1911 [43], and by Mme CURIE, the same year, and in an improved version in 1920 [44].

A different problem was studied experimentally in 1908 and 1910 by RUTHERFORD and GEIGER [45], who measured, by the scintillation method,

Über Schwankungen der radioaktiven Umwandlung

von

K. W. Fritz Kohlrausch.

Aus dem II. physikalischen Institut der Universität in Wien.

(Mit 3 Textfiguren.)

(Vorgelegt in der Sitzung am 21. Juni 1906.)

Die von Rutherford und Soddy begründete Zerfallstheorie geht von der Annahme aus, daß die Atome radioaktiver Substanzen einer Umwandlung unterworfen sind. Der Zerfall erfolgt nach dem Exponentialgesetz $n = Ne^{-\lambda t}$, wo N die Anzahl der Atome zur Zeit $t_0 = 0$, λ eine Konstante und n die Zahl der nach der Zeit t noch vorhandenen Atome bedeutet. Durch Differentiation erhält man $\frac{dn}{dt} = -\lambda n$, das heißt, der Teil der vorhandenen Atome, der in der Zeiteinheit einer Veränderung unterliegt, ist gegeben durch λ; $\frac{1}{\lambda}$ wird als »mittlere Lebensdauer« der betreffenden Atome bezeichnet.

Man gelangt zu diesem Gesetz auch auf anderem Wege. Die Wahrscheinlichkeit, daß ein Atom innerhalb einer gegebenen Zeit eine Umwandlung erfährt, wird desto größer, je größer die Zeit ist. Bezeichnet man die Wahrscheinlichkeit, daß ein Atom in der Zeit Δt, wobei Δt sehr klein sein möge, eine Umwandlung erfährt, mit $\lambda\Delta t$, so wird die Wahrscheinlichkeit w_1, daß dieses Atom in der gleichen Zeit nicht verändert wird: $w_1 = 1 - \lambda\Delta t$; für die Zeiten $2\Delta t, 3\Delta t, \ldots k\Delta t$ erhält man:

Fig. 4. – First page of K. W. F. Kohlrausch experimental paper on fluctuations of the ionization produced by alpha-particles.

the fluctuations of the number of alpha-particles emitted in a given solid angle. The theory for this case had been developed by Bateman (1910-1911) [46].

Other theoretical papers were published by Tatiana Ehrenfest [47] in 1913 and Schrödinger in 1918-1919 [48].

What I said, although incomplete, is sufficient to show that the problem of the statistical nature of the decay process of radioactive atoms was one of

Die radioaktive Strahlung als Gegenstand wahrscheinlichkeits-theoretischer Untersuchungen

Von

L. v. Bortkiewicz
a. o. Professor an der Universität Berlin

Mit 5 Textfiguren

Berlin
Verlag von Julius Springer
1913

Fig. 5. – Frontispiece of L. von Bortkiewicz book.

those at the centre of the attention of many experimental and theoretical physicists, starting from 1905 until 1928 when GAMOW [49] and CONDON and GURNEY [50] applied quantum mechanics to the decay process of nuclei and derived the connection between the mean life and the energy of the emitted alpha-particles which had been established experimentally by GEIGER and NUTTAL [51] seventeen years before.

This whole problem is not mentioned, of course, in the first book by RUTHERFORD appeared in 1904 [36], but is treated in detail in section 75 (pages 186-191) of his 1913 book [52], where he underlines: « It is important to settle whether the emission of alpha-particles follows a simple probability law, *i.e.* whether the alpha-particles are emitted at random in time and space ».

4. – Early models of the nucleus.

The law of radioactive decay and the fluctuations of the number of disintegrations observed in a preassigned interval of time were matters of serious concern for many physicists of the first quarter of our century. As an example of the prevailing attitude I will quote the beginning of an important paper by SODDY [53] appeared in 1909:

« The cause of atomic disintegration remains unknown. It is difficult to construct any model of the disintegrating mechanism, chiefly on account of certain features in connection with the process. In particular may be mentioned the fact that the period of average life of the atoms disintegrating is the same whether newly formed atoms or those which have already survived many times the average period are considered. What may be termed the inevitableness of the process, and its entire independence of all known conditions, suggest that the cause of disintegration is apart from the atom. It is difficult to believe that the cause is resident in space external to the atom. It seems more probable that it exists within the atom and at the same time is uninfluenced by it. The question about to be discussed is whether necessarily only one mode of instability can exist within the atom at the same time ».

The paper goes on treating and solving the theoretical problem of the decay when two or more alternative disintegration processes are allowed.

Most of the physicists working in the field accepted the well-established fact that the clicks of a counter or the little lights observed in the dark on a fluorescent screen are distributed at random and the majority of them did not dare to propose models.

One of the very few exceptions was Andrè DEBIERNE, who had published in part alone in part in collaboration with Mme CURIE, a number of important papers on radioactivity. On various occasions, in particular at the end of a lecture he gave in Paris the 26 January 1912 in front of the Societè Française de Physique [54], DEBIERNE discussed various possible mechanisms and arrived at the following conclusion:

« On est ainsi conduit a faire l'hypothèse qu'il existe réellement un élément de desordre faisant passer les atomes par un grand numbre d'états differents dans un temp très court, mai que cet élément du desordre est distincte de l'agitation thermique ». (Thus one is brought to the assumption that really there is an element of disorder which causes the atoms to pass through a great number of different states, in a very short time interval, but that such an element of disorder is different from thermal agitation.)

The simplest hypothesis, goes on DEBIERNE, consists in assuming that this element of disorder is shut up in the atom. « For example, one could imagine that inside the atom there are infinitely small elements [today we would say « constituents », but why *infinitely* small?] endowed with disordered movements similar to those of the molecules of a gas inside a container. Such an agitation taking place inside the atom could determine, in certain cases, a state of instability followed by the explosion of the atom » [55].

Debierne lecture was one of a series of seminars [54] in the last of which Henri POINCARÉ [56] summarized the fundamental points mentioned by the various speakers in the presentation of their subjects and added a number of stimulating remarks. Some of these are expressed in the form of questions about the many problems still open in 1912, which sound particularly effective in Poincaré's elegant French.

The part of Poincaré lecture concerning Debierne's ideas begins at the end of p. 361 and finishes at about one third of p. 364. POINCARÉ starts by pointing out that Debierne's ideas are the most suitable « in order to measure the complexity of the atom ». He accepts the fact that the law of radioactive decay is a statistical law by saying: « on y reconnait la marque du hazard » (one there recognizes the stamp of chance). A few lines below he underlines that « the chance that presides the radioactive transformations is an internal chance », *i.e.* the atom of radioactive bodies is a world and a world submitted to chance, « mais qu'on y prenne garde, qui dit hazard, dit grand nombres; un monde formé de peu d'éléments obéira à des lois plus ou moins compliquées, mai qui ne seront pas des lois statistiques ...; puis qu'il y a une statistique et par conséquent une thermodynamique interne de l'atome, nous pouvons parler de la température interne de cet atome; eh bien! elle n'a aucune tendence à se mettre en équilibre avec la temperature extérieure, comme si l'atome était enfermé dans une enveloppe parfaitement adiathermique » (but one should be careful, who says chance says large numbers; a world consisting of a few elements will obey more or less complicated laws, which, however, will not be statistical laws ...; since there is a statistics and therefore a thermodynamics internal to the atom, we can talk of a temperature internal to this atom; eh, well! this temperature does not have any tendency to equilibrium with the external temperature as if the atom were shut in a perfectly insulating thermal container).

This is the first time that I found mentioned explicitly the internal temperature of an atom, an idea very interesting, indeed: about 25 years later it became an essential concept of many nuclear models. It was clearly implied by the whole reasoning of Debierne, although I did not find the words « internal temperature » in any of his papers published before 1915 [57].

The ideas of Debierne were reported by Mme CURIE at the second Solvay Conference held in Bruxelles on 27-31 October 1913 [58]. She insisted on the idea of a disorder inside the central part of the atom (the nucleus) where its constituents should move with very high velocity, judging from that of the emitted particles, and on the possibility even to define a kind of internal temperature much higher than the external temperature.

In the discussion that followed the report by Mme CURIE, NERNST, RUBENS, BRILLOUIN, WIEN and LINDEMANN commented in various forms on the possibility of succeeding perhaps in modifying the mean life of radioactive bodies by going to high temperature (perhaps inside the Sun).

RUTHERFORD and LANGEVIN touched the essence of Debierne model. RUTHERFORD said more or less that « the law of the radioactive substances appears to find a possible explanation only as a consequence of the fortuitous troubles taking place according to the laws of probability inside the nucleus. But in the present state of our knowledge it is not possible to formulate clear ideas neither on the constitution of the atomic nucleus nor on the causes of its distintegration ».

Langevin remark runs more or less as follows: « Debierne ideas require a complex structure for each single atom and imply the necessity of a great numbers of parameters for fixing the configuration of the atom. In the case of three bodies, however, the trajectories are already very complicated.

As it has been shown by POINCARÉ—goes on LANGEVIN—, apart from a few exceptional cases of zero probability, any trajectory turns indefinitely approaching asymptotic solutions and then going away from them so that they have the aspects of wires made into balls .... In the case of radioactive atoms one could ask which is the minimum number of degrees of freedom necessary for obtaining the law of probability within the precision of the experimental results. It would be sufficient, for example, to assume that the distribution of the initial phases of the different atoms is sensibly uniform on the surface of constant energy if all atoms have initially the same energy ».

A number of remarks could be made about Debierne ideas and Langevin remarks. These mix two possible types of assumptions. One introduces inside the nucleus a sufficiently large number of degrees of freedom, to produce a random effect. The other transfers the stochastic nature of the decay process to the initial conditions under which the nucleus is produced by its mother. In this second case some machinery should be introduced in the model for assuring the validity of such an assumption for all members of a radioactive family the mean lives of which are spread over an interval of values greater than 20 powers of 10.

A thorough and complete presentation of his ideas was finally given by DEBIERNE in a 22-page long paper published in 1915 [59], shortly after LINDEMANN [60, 61] had developed a quantitative model along similar lines of thought.

In this model the nucleus of a radioactive element contains particles in movement and becomes instable when N independent particles all pass through some unknown critical interval of positions within a short time τ. These particles are taken to be alpha-particles rotating or oscillating with the frequency ν. Their mean energy $E = H\nu$ is equal to the energy of the emitted alpha-particles. Each of them passes through the critical region $\nu = T^{-1}$ times per second so that the probability for one of them to be in the critical region during one period is τ/T. The probability for N of them to be simultaneously in τ in one period is $(\tau/T)^N$, which multiplied by νdt gives the probability that such an event takes place in the time interval dt. If $\mathcal{N}$ atoms are considered, the number of those which become unstable and explode in the time dt is

$$d\mathcal{N} = -\mathcal{N}(\tau\nu)^N \nu\, dt\,,$$

from which it follows that

$$\lambda = \nu(\tau\nu)^N\,. \tag{5}$$

Since the range-energy relation is $R = kE^{\frac{3}{2}}$, LINDEMANN obtains

$$\lambda = \frac{E}{H}\left(\frac{\tau E}{H}\right)^N = \frac{\tau^N}{(Hk^{\frac{2}{3}})^{N+1}} R^{\frac{2}{3}(N+1)}$$

and taking the logarithm

$$\log\lambda = \log\frac{\tau^N}{(Hk^{\frac{2}{3}})^{N+1}} + \frac{2}{3}(N+1)\log R\,,$$

similar to the Geiger-Nuttal relation [51]

$$\log\lambda = a + b\log R\,.$$

Since the experimental value of b is about 53, it follows on comparing the constants that N is about 80.

With the further assumption that the time τ is to be regarded as the time taken by an elastic wave to cross the nucleus, LINDEMANN computed the constant a in terms of known quantities and the unknown radius of the nucleus. By comparison with the experimental value of a he deduced a value of the radius of the radium nucleus of about $4\cdot10^{-13}$ cm, which is too small by about a factor 2.

The theory provided an explanation of the law of radioactive decay as well as of the Geiger-Nuttal law (fig. 6) for reasonable values of the parameters.

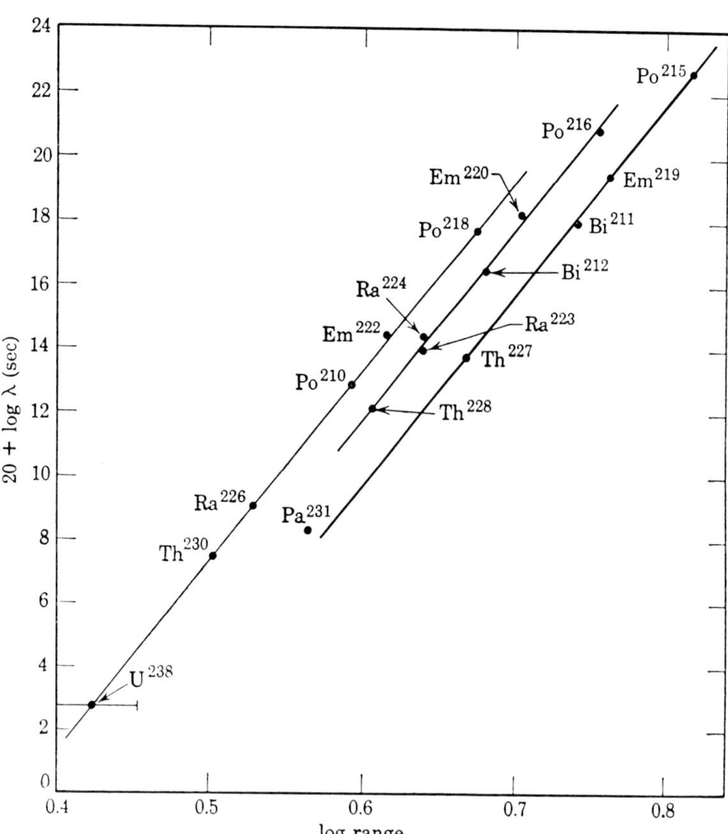

Fig. 6. – Comparison with Geiger-Nuttal rule of the experimental data available to GEIGER in 1921 [62]. This figure is taken from ref. [63].

A few papers were published in the following years [64] attempting to introduce the quantization of the movement of alpha-particles and electrons within the nucleus by applying the same rules used by BOHR for the atomic electrons.

More or less in the same period of time other authors tried to explain the decay law along the lines suggested long before by the CURIE, *i.e.* as due to the action of an external radiation [65]. The most important of these papers is that of Jean PERRIN who devoted about 10 pages of his 103-page paper, trying to extend to the nuclei of radioactive substances his considerations on the chemical processes.

Another type of model was proposed by RUTHERFORD [66] in 1927 for explaining some results obtained by him and CHADWICK [67] and by CHADWICK [61] in 1925, in the study of the elastic scattering at large angles ($\sim 155^{\circ}$) of the alpha-particles of Ra C' (7.68 MeV) from the uranium nucleus

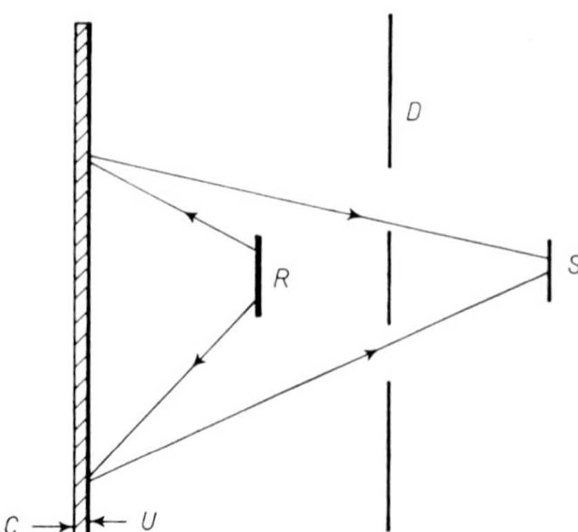

Fig. 7. – Schematic representation of the experimental set-up used in 1925 by CHADWICK to observe the scattering at $\sim 155°$ of the alpha-particles of Ra C' from uranium atom: $R \equiv$ Ra $(B + C)$; $U \equiv$ uranium oxide deposited on $C \equiv$ graphite support; $D \equiv$ diaphragm; $S \equiv$ fluorescent screen used for observing the scintillation by means of a microscope.

(fig. 7). They had found that the observed angular distribution followed very closely that compluted from pure Coulomb potential. The alpha-particles emitted by the decay of the same nucleus have only about one half this energy (4.049 MeV). Thus, it appeared that during the emission process the alpha-particles had to pass through a region of negative kinetic energy (fig. 8). Such a difficulty was avoided in the following model.

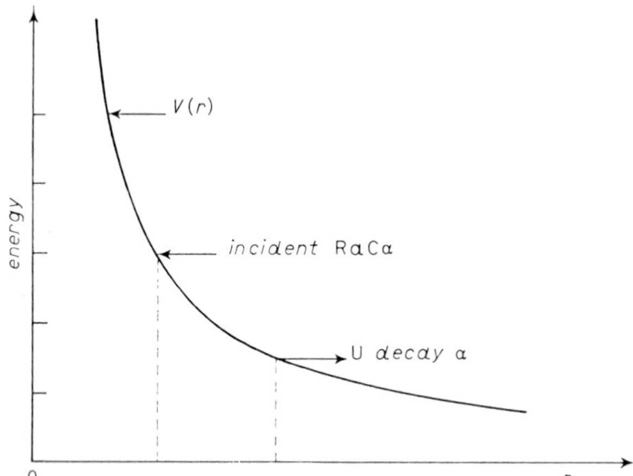

Fig. 8. – The uranium decay alpha-particles are emitted with an energy equal to about one half the energy of the Ra C' alpha-particles used in the scattering experiment of fig. 7. This had shown that the potential $V(r)$ followed the Coulomb law down to distances equal to about one half the distance from which the decay alpha-particles are emitted.

The nucleus of a radioactive atom consists of a very small core of radius less than $1\cdot 10^{-12}$ cm, surrounded at a distance by a number of neutral satellites describing quantum orbits in the field of the central core. At least some of these satellites are neutral alphas in which the two neutralizing electrons are more closely bound than in the free helium atom but not so closely as in the alpha-particle. A neutral alpha is regarded as stable only in very strong electric fields and is held on its orbit by attractive forces due to its polarization in the field of the central core. The quantized orbit of radius between $1.5\cdot 10^{-12}$ and $6\cdot 10^{-12}$ cm is essential to provide the well-defined energy of the emitted alpha.

The disintegration process consists in the *acausal escape* of a neutral alpha satellite from its orbit. When the satellite reaches a distance from the centre of the core where the electric field has a certain critical value, it breaks into two electrons and an alpha-particle: the two electrons are sucked into the core, while the alpha-particle runs away with the observed energy [68].

The model is inconsistent, as it was immediately shown [69], but here it serves to show that the idea of explaining the radioactive decay in term of a spontaneous (or better an acausal) process still in 1927 was alive at the Cavendish Laboratory, in Cambridge, that had been, and still was, the major centre for the study of radioactivity.

Apart from that, the scattering experiments of Rutherford and Chadwick [67] and those (unpublished) by Chadwick alone [61] were made too late for having any influence on the development of quantum mechanics.

They provide, however, one of the most striking examples of a direct observation which imposes quantum mechanics. Apparently they have been forgotten, since they are never quoted in books devoted to the introduction of this fundamental subject.

5. – Final remarks.

A few remarks appear in order at the end of this story, which may appear too long but in reality already constitutes a very shortened version of what happened in the field of radioactivity before the advent of quantum mechanics.

First of all one should keep in mind that the early development of radioactivity took place at the same time as the first steps in the theory of the atom. The existence of the electron was established almost at the same time as the discovery of radioactivity; the quantum of action was introduced at about the same time that the exponential decay law was established, the theoretical prediction and the experimental observation of the fluctuations of the radioactive decay started to be made at the same time of Einstein's theory of photons and of Smolukowski and Einstein theory of Brownian motion.

The people involved in these two developments were always very close to each other, in many cases they worked in the same laboratories. All this went on for more than 15 years before the experiments of Geiger and Marsden established the validity of the Rutherford model, which allowed a separation between the physics of the atom and the physics of the nucleus, and opened the door to the theoretical work of Niels Bohr.

The young BOHR went to Cambridge at the end of September 1911; there he had contacts with LARMOR and J. J. THOMSON and there he met for the first time RUTHERFORD [70]. In the middle of March 1912 BOHR went to Manchester to work in the laboratory of Rutherford, who had been given a professorship there in 1907. RUTHERFORD had collected together in his laboratory a number of the best physicists of those days. Among them H. GEIGER, W. MAKOVER, E. MARSDEN, E. J. EVANS, A. S. RUSSEL, K. FAJANS, H. G. J. MOSELEY, G. HEVESY, J. CHADWICK and C. G. DARWIN.

Before Summer 1912 BOHR started to work on the model of the atom that had been proposed by RUTHERFORD not long before [71], while the second theoretical paper by BATEMAN [46] on the fluctuations and the second paper by GEIGER [72] on the scattering of alpha-particles by nuclei had been received by the *Philosophical Magazine* in January of the same year, and while MARSDEN and BARRAT [43] were just making their experiment on the fluctuations observed in the radioactive decay, as a development of those made by GEIGER in 1908 and by RUTHERFORD and GEIGER in 1910. (Incidentally I may recall that the famous final paper of Geiger and Marsden on the scattering of alpha-particles appeared only in 1913 [73].)

Although the attention of Bohr was mainly concentrated on the model of the atom, it is impossible that the problem of the fluctuations escaped his attention and that the pragmatic attitude of Rutherford and his school did not have a strong influence on him, also on this problem.

The existence of a nucleus at the centre of the atom was just becoming an accepted notion, but its properties, in particular its decay in the case of radioactive elements, were still an « enigma », as Mme CURIE said on many occasions.

A few models for explaining such a behaviour were, of course, proposed. Some of them based on the existence of a hypothetic radiation pervading all space and bodies and provoking the decay of heavy elements, some of a statistical nature such as that proposed by LINDEMANN.

These were perhaps a bit more popular, but not universally accepted. All of them, irrespective of being formally developed or only quantitatively conceived, contained some element of indeterminacy. In Lindemann model this is introduced with the small but nonzero interval of time τ, during which the system becomes unstable. In the case considered by LANGEVIN this element is provided by the initial conditions of the internal motion of the constituents of the nucleus, which are supposed to be distributed at random over a surface of constant energy in the space of phases.

My claim of a historical influence of the study of radioactivity on the general acceptance of probabilistic conceptions in physics clearly does not regard the distinction between statistical and probabilistic points of view, which only years later became a focus of physicists' attention.

It does not concern also the most striking feature of quantum mechanics, *i.e.* the concept of «probability amplitude» and its systematic use as an essential intermediate step for deriving the values of measurable quantities; this concept did not get any inspiration from the investigation of radioactivity. Only the results obtained in 1925 by RUTHERFORD and CHADWICK on the scattering of alphas from uranium were a message of new phenomena requiring this aspect of the theory. The message, however, was not yet construable and in any case it arrived too late for having an influence on the development of quantum mechanics. The only clear experimental indication in favour of the use of «probability amplitude» came from experiments on diffraction of electrons by crystals or molecules.

Max BORN says to have found inspiration from Einstein's theory of photons, but EINSTEIN, in 1916, adopted for the Ausstrahlung the law of radioactive decay.

The value of the development of radioactivity, from the point of view of the acceptance of probabilistic conceptions, lies essentially in the observed phenomena and in the pragmatic attitude that most of the people working in the field had in front of them.

Many other phenomena could have, in principle, played the same role, but radioactivity had two fundamental points of advantage over any other chapter of physics. The mean lives of many radioactive substances were in a range of values very convenient for their observation. Furthermore, the existence of experimental techniques, such as the scintillations, discovered in 1903 by CROOKS and ELSTER and GEITEL [74], and of various types of counters, developed starting from 1908 on by GEIGER and RUTHERFORD and others [75], allowed the observation of the decay of a single nucleus.

I insist on these observational aspects because they provide the basis for the pragmatic attitude of the people working in this field in those days. The situation is not exceptional, as can be seen from many examples, two of which may be mentioned here. The first example is the problem of the value of the velocity of light as it appeared toward the end of the last century. All experiments made during more than half a century [76] were pointing towards the recognition of the invariance of this quantity. Various models were proposed, but were unsatisfactory, so that the invariance of the velocity of light with respect to the reference frame was accepted as an experimental result, still under critical scrutiny. It remained an enigma, repeatedly tested, until EINSTEIN promoted it to the rank of fundamental law.

In more recent years, the θ-τ decay puzzle was based on the equality (within small experimental errors) of the masses and the mean lives of two types of particles: one particle (the θ), decaying into two pions, the other particle (the τ), decaying into three pions, *i.e.* into two final states of opposite parity. Many physicists (especially among the experimentalists working in the field) were ready to accept that the θ and the τ were simply two different decay modes of the same particle, if they had *not* firmly believed in the general validity of the conservation of parity. The puzzle was solved by T. D. LEE and C. N. YANG who proposed a new fundamental law: « parity is violated only by weak interactions », and they indicated the experiments for testing their assumption.

What appears to be the most interesting aspect of the whole story is the convergence of so many lines of thought toward a probabilistic, or, at least, a statistical interpretation as that involved by quantum mechanics.

The line originating from thermodinamics and statistical mechanics, characterized by a sophisticated analysis of the properties of matter and radiation, the line through radioactivity, characterized by an incredible succession of unexpected experimental discoveries, as well as certain lines of philosophical thinking, all appear to converge toward the same target.

These lines, of course, were not independent. Clear reciprocal influences existed between them in all possible combinations. In particular, there were influences of the philosophical thinking on the attitude of the scientists, but there were also strong influences in the opposite direction.

Newton mechanics at the end of 1600, electricity and magnetism in the first part of 1700, thermodynamics and statistical mechanics during the second half of 1800, radioactivity and relativity at the beginning of our century, later quantum mechanics, all became subjects of discussion in much wider circles than those of the people directly involved. They became fashionable matters of conversation, in king's court as well in layman circles.

It is difficult, perhaps impossible, to establish which of the three lines of thought sketched above had more importance for the advent of quantum mechanics. The best is probably to keep in mind all of them.

REFERENCES

[1] M. JAMMER: *The Conceptual Development of Quantum Mechanics* (New York, N. Y., 1966).

[2] M. JAMMER: *The Conceptual Development of Quantum Mechanics* (New York, N. Y., 1966), p. 166-167.

[3] M. JAMMER: *The Conceptual Development of Quantum Mechanics* (New York, N. Y., 1966), p. 170.

[4] M. JAMMER: *The Philosophy of Quantum Mechanics* (New York, N. Y., 1974).

[5] *Historical Studies in the Physical Sciences*, edited by R. MC CORMACH (Philadelphia, Penn., 1971).

[6] S. G. BRUSH: *Journ. Hist. Ideas*, **37**, 603, No. 4 (October-December 1976).

[7] See, for example, E. N. HIEBERT: *The use and abuse of thermodinamics in religion*, *Daedalus*, p. 1046 (Fall, 1966).

[8] A. EINSTEIN: *Über einen die Erzeugung und Verwandlung des Lichtes betreffenden heuristischen Gesichtspunct der Physik, Ann. der Phys.*, **17**, 132 (1905).

[9] M. BORN: *Quantenmechanik der Stossvorgänge, Zeit. für Phys.*, **38**, 803 (1926). An English translation of this article is given at p. 207 of G. LUDWIG: *Wave Mechanics* (London, 1968).

[10] M. BORN: *Die statistische Deutung der Quantenmechanik, Nobelvortrag 1954, Nobel Lectures, Physics, 1942-1962* (Amsterdam and New York, N. Y., 1964), p. 262.

[11] A. EINSTEIN: *Zur Quantentheorie der Strahlung, Mitt. Phys. Ges. Zürich*, **18**, 47 (1916); *Strahlung Emission und Absorption nach der Quantentheorie, Verh. Dtsch. Phys. Ges.* **18**, 318 (1916).

[12] A. EINSTEIN: *Zur Quantentheorie der Strahlung, Phys. Zeits.*, **18**, 121 (1918).

[13] A thorough discussion of this paper and more in general of *Einstein Statistical Theories* has been made by M. BORN at p. 163 of *Albert Einstein: Philosopher-Scientist*, edited by P. A. SCHILPP (New York, N. Y., 1951).

[14] G. C. SCHMIDT: *Wied. Annalen, May 1898*, according to Rutherford quotation (ref. [17]). Most probably it should be identified with ref. [21].

[15] M. CURIE and P. CURIE: *Compt. Rend.*, **129**, 714 (1899), and p. 77, ref. [16]. The effect was observed with radium and polonium. In this second case it was clearly due to incomplete separation from other elements.

[16] *Ouvres de Marie Sklodowska Curie*, recuillies par I. JOLIOT CURIE (Varsovie, 1954).

[17] E. RUTHERFORD: *Phil. Mag.*, **49**, 1 (1900), and p. 220 of ref. [18*a*].

[18] *The Collected Papers of Lord Rutherford of Nelson*, published under the scientific direction of Sir JAMES CHADWICK, F.R.S., Vol. **1** (London, 1962) (*a*); Vol. **2** (London, 1963) (*b*); Vol. **3** (London, 1965) (*c*).

[19] I do not discuss the similar case of actinon (An), the noble gas of the actinium family because discovered later.

[20] H. BECQUEREL: *Compt. Rend.*, **122**, 420, 501, 559, 689, 762, 1086 (1896); **123**, 855 (1896); **124**, 438, 800, 984 (1897).

[21] G. C. SCHMIDT: *Wied. Annalen*, **65**, 141 (1898), paper received on March 24.

[22] M. CURIE: *Compt. Rend.*, **126**, 1101 (1898), paper received on April 12. See also p. 43 of ref. [16].

[23] M. CURIE and P. CURIE: *Compt. Rend.*, **127**, 175 (1898). See also p. 57 of ref. [16].

[24] M. CURIE, P. CURIE and G. BÈMONT: *Compt. Rend.*, **127**, 1215 (1898). See also p. 57 of ref. [16].

[25] M. CURIE: *Compt. Rend.*, **142**, 273 (1906), and p. 314 of ref. [16].

[26] M. CURIE and P. CURIE: *Rapport présentés au Congrès International de Physique*, 1900, Vol. **3**, p. 79; see also p. 106 of ref. [16] (last page).

[27] M. CURIE and P. CURIE: *Compt. Rend.*, **134**, 85 (1902). See also p. 134 of ref. [16].

[28] R. RUTHERFORD and F. SODDY: *Trans. Chem. Soc.*, **81**, 837 (1902), and p. 435 of ref. [18*a*].

[29] This sentence is in Section XII (*General Theoretical Considerations*) of ref. [18*a*], towards the end of p. 455.

[30] E. RUTHERFORD and F. SODDY: *Phil. Mag.*, **5**, 576 (1903), and p. 596 of ref. [18*a*].

[31] These are indicated with a special name (*metabolon*) in order to distinguish them from stable atoms constituting ordinary matter: p. 605 of ref. [18*a*].

[32] E. RUTHERFORD: *Transactions of the Australasian Association for the Advancement of Science* (Dudenin, January 1904), p. 87, and p. 620 of ref. [18*a*].

[33] Only some time later the α-particle was recognized to have a double charge and a mass equal to about twice that of the hydrogen molecule.

[34] E. RUTHERFORD: *Phil. Trans. Roy. Soc.*, **204**, 169 (1904): Bakerian Lecture delivered May 19, 1904; p. 621 of ref. [18*a*].

[35] For an exstensive bibliography of old papers on this problem see at the end of section II.3 (p. 41) of ref. [41].

[36] E. RUTHERFORD: *Radioactivity* (Cambridge, 1904).

[37] E. VON SCHWEIDLER: *Comptes Rendus du Premier Congrès International pour l'Etude de la Radiologie et de l'Jonisation, tenue à Liège du 12 au 14 Septembre 1905, Bruxelles.*

[38] K. W. F. KOHLRAUSCH: *Wien. Ber.*, **115**, 673 (1906).

[39] E. MEYER and F. REGENER: *Verh. Dtsch. Phys. Ges.*, **10**, 1 (1908); *Ann. der Phys.*, **15**, 757 (1908).

[40] H. GEIGER: *Phil. Mag.*, **15**, 539 (1908).

[41] ST. MEYER and E. VON SCHWEIDLER: *Radioactivität* (Leipzig, 1927).

[42] L. VON BORTKIEWICZ: *Die Radioactive Strahlung als Gegenstand Warscheilichkeits-theoretischer Untersuchungen* (Berlin, 1913). Through the kind interest of Prof. W. PAUL of the University of Bonn I had at my disposal, for a few days, a copy of this interesting, and today almost rare, book.

[43] E. MARSDEN and T. BARRATT: *Proc. Phys. Soc.*, **23**, 367 (1911); **24**, 50 (1911).

[44] M. CURIE: *Journ. Phys. Radium*, **8**, 354 (1911), and p. 395 of ref. [16]; **1**, 12 (1920). See also p. 494 of ref. [16].

[45] E. RUTHERFORD and H. GEIGER: *Proc. Roy. Soc.*, **81**, 141 (1908), and p. 109 of ref. [18*b*]; *Phil. Mag.*, **20**, 691 (1910), and p. 196 of ref. [18*b*].

[46] H. BATEMAN: *Phil. Mag.*, **20**, 704 (1910); **21**, 745 (1911). The second paper was made during the two years spent at Bryn Mawr College by Harry BATEMAN (b. Manchester 1882, d. Pasadena 1946), whose contributions to mathematics and theoretical physics refer to solution of potential and wave equations, partial differential equations and integral equations in general, electrodynamics, seismology, relativity, radioactivity, etc.

[47] T. EHRENFEST: *Phys. Zeits.*, **14**, 675 (1913).

[48] E. SCHRÖDINGER: *Wien. Ber.*, **127**, 237 (1918); **128**, 177 (1919).

[49] G. GAMOW: *Zeits. Phys.*, **51**, 204 (1928).

[50] E. U. CONDON and R. W. GURNEY: *Nature*, **122**, 439 (1928).

[51] H. GEIGER and J. M. NUTTAL: *Phil. Mag.*, **22**, 613 (1911); **23**, 439 (1912); **24**, 647 (1912).

[52] E. RUTHERFORD: *Radioactive Substances and Their Radiations* (Cambridge, 1913).

[53] F. SODDY: *Phil. Mag.*, **18**, 739 (1909).

[54] A. DEBIERNE: *Sur les transformations radioactives*, at p. 304 of the volume: *Les idèes modernes sur la constitution de la matière* (Paris, 1913).

[55] Through the kind interest of Prof. J. PRENTKI of CERN and Mme Geneviéve DELVOYE de l'Institut National de Physique Nucléaire et de Physique des Particules, I received the *Notice sur le travaux de M. André Debierne*, prepared by DEBIERNE himself, from which I learned that he already had discussed this same problem in two previous lectures, one in front of La Société Chimique, in 1907, the other at Clark University (USA) in 1909.

[56] H. POINCARÉ: *De la matière et de l'éther*, at p. 357 of the volume quoted in ref. [54].

[57] Perhaps it was mentioned in one of the two lectures of 1907 and 1909 quoted in ref. [55].

[58] M. CURIE: *Sur la loi fondamental des transformations radioactives*, at p. 66 of: *La structure de la matière, Rapports et Discussions du Conseil de Physique tenu a Bruxelles du 27 au 31 Octobre 1913*, Institut International de Physique Solvay (Paris, 1921). See also at p. 507 of ref. [16].

[59] A. DEBIERNE: *Considerations sur le mécanisme des transformations radioactives et la constitution des atomes*, *Ann. de Phys.*, (9) **4**, 323 (1915).

[60] F. A. LINDEMANN: *Verh. Dtsch. Phys. Ges.*, **16**, 281 (1914); *Phil. Mag.*, **30**, 560 (1915); see also p. 324 of ref. [61]. Because of a minor error my expression of λ differs from that of Lindemann by a factor ν.

[61] E. RUTHERFORD, J. CHADWICK and C. D. ELLIS: *Radiation from Radioactive Substances* (Cambridge, 1930).

[62] H. GEIGER: *Zeits. Phys.*, **8**, 45 (1921).

[63] G. C. HANNA: *Alpha radioactivity*, p. 55 of Vol. **3**, *Experimental Nuclear Physics*, edited by E. SEGRÈ (New York, N. Y., 1954). In the quantum-mechanical approach the expression of the decay constant λ can be split into three factors: *a*) the frequency at which the alpha-particle moving inside the nucleus strikes the side of the potential well, *b*) the reflection coefficient of the wall of the well due to the rapid increase of the potential and *c*) the penetrability of the potential barrier. The first term is an exponential and determines the main features of the phenomenon. The factor *a*) corresponds to the factor ν in eq. (5), while the two other factors represent single-particle effects which have nothing to do with the N-fold coincidence factor $(\tau\nu)^N$. Some of the features of the diagram of fig. 6 are not jet fully understood (see p. 98-105 of Hanna's article).

[64] H. TH. WOLFF: *Ann. der Phys.*, (4) **52**, 631 (1917); *Ann. der Phys.*, **60**, 685 (1919); *Phys. Zeits.*, **21**, 175 (1920); A. SMEKAL: *Naturwiss.*, **8**, 206 (1920); *Zeits. Phys.*, **10**, 275 (1922); S. ROSSELAND: *Zeits. Phys.*, **14**, 173 (1923); *Nature*, **11**, 357 (1923); G. KIRSCH: *Naturwiss.*, **8**, 207 (1920); *Phys. Zeits.*, **22**, 20 (1921); W. D. HARKINS: *Journ. Amer. Chem. Soc.*, **42**, 1956 (1920); K. FEHRLE: *Zeits. Phys.*, **16**, 397 (1923).

[65] J. PERRIN: *Ann. de Phys.*, (9) **11**, 5 (1919); E. BRINER: *Compt. Rend.*, **180**, 1586 (1925); A. W. MENZIES and C. A. SLOAT: *Science* (*N.S.*), **63**, 44 (1926).

[66] E. RUTHERFORD: *Phil. Mag.*, **4**, 580 (1927), and p. 181 of ref. [18*c*].

[67] E. RUTHERFORD and J. CHADWICK: *Phil. Mag.*, **50**, 885 (1925); see also p. 143 of ref. [18*c*] and p. 322 of ref. [57].

[68] A similar model was proposed later by D. ENSKOG: *Zeits. Phys.*, **45**, 852 (1927); **52**, 203 (1928).

[69] G. GENTILE jr.: *Rend. Acc. Lincei*, **7**, 346 (1928).

[70] L. ROSENFELD and E. RÜDINGER: *The decisive years*: *1911-1918*, at p. 38 of *Niels Bohr*, edited by S. ROZENTAL (Amsterdam, 1967).

[71] E. RUTHERFORD: *Phil. Mag.*, **21**, 669 (1911), received in April 1911. See also at p. 238 of ref. [18*b*].

[72] H. GEIGER: *Proc. Roy. Soc. A*, **83**, 492 (1910); **86**, 235 (1912), received 25 January 1912.

[73] H. GEIGER and E. MARSDEN: *Phil. Mag.*, **25**, 604 (1913).

Note added in proofs. – N. FEATHER has called my attention to the paper by T. J. TRENN (*ISIS*, **65**, 74 (1974)) where the author shows convincingly that the paper quoted above sent by RUTHERFORD (from Manchester) and published in April 1913 is in fact less of a final version than that sent by GEIGER (from Berlin) to Vienna and published in December 1912 (*Wien. Ber.*, **121**, 2361 (1912)). GEIGER, apparently, took a first draft with him when he left Manchester in September 1912. This was amended in correspondence with RUTHERFORD in October but surprisingly the *Phil. Mag.* paper is essentially based on the first draft. I like also to mention that in the paper read by N. FEATHER at a session on *The work of Frederick Soddy* at the *XV International Congress of the History of Science, Edimburgh, August 1977*, the author uses essentially the same quotations from VON SCHWEIDLER (1905) and SODDY (1909) in a discussion of *Isotopes, isomers and* *the fundamental law of radioactive change* (*Notes and Records, Roy. Soc. London*, **32**, No. 2, 225 (March 1978)).

[74] W. CROOKES and also J. ELSTER and H. GEITEL discovered in 1903 the scintillation method but REGENER was the first in 1908 (*Verh. Dtsch. Phys. Ges.*, **19**, 78, 351 (1908) and p. 54 of ref. [61]) to devise methods of counting scintillations in order to determine the number of alpha-particles incident on the screen.

[75] A kind of wire counter was first developed by RUTHERFORD and GEIGER in 1908 (*Proc. Roy. Soc. A*, **81**, 141 (1908)). These were improved by the same authors and became proportional counters a few years later (*Phil. Mag.*, **24**, 618 (1912)). The point counter was developed by GEIGER the successive year (*Verh. Dtsch. Phys. Ges.*, **15**, 524 (1913); *Phys. Zeits.*, **14**, 1129 (1913)).

[76] See, for example, G. C. WICK: *Introduzione alla teoria della relatività, Lezioni di Fisica Teorica, A.A. 1944-1945* (Roma, 1945).

Reprinted From
Problems in the Foundations of Physics

Section B
Prominent Personalities in 20th Century Physics

On various occasions from the beginning of his career, and with increasing sensitivity and completeness, Amaldi retraced the biographies of those colleagues and friends who were (or had been) particularly close to him. Most of these papers were written for a special occasion, such as a death, commemoration or memorial. Amongst them, a selection of those that refer to Italian physicists has appeared in Part I. This section is dedicated to well-known personalities in European physics. The distinction may not be immediately evident in some cases, such as in the case of Bruno Touschek, which has been included in this section but could just as well be found in the section on Italian post-war physics.

The occasion that prompted the first paper in this section, *A Few Flashes on Hans Bethe Contributions During the Thirties*, was the 91st course of the "Enrico Fermi" International School of Physics held in Varenna in June 1984. Bethe was a speaker and guest of honor at the course devoted to *From Nuclei to Stars*. In the 1930s, he was one of the pioneers in the application of the results of the most advanced research on nuclear physics to questions of astrophysics and, in particular, to problems of energy and stellar evolution. It is not surprising that Amaldi was asked to honor Bethe with a review of his scientific contributions to the "golden era" of nuclear physics. This was due to the fact that Amaldi had known Bethe since the days when he was living in Rome in the early 1930s and was engaged at that time (1984 was the fiftieth anniversary of the discovery of neutron-induced radioactivity by Fermi's group) in a historiographic *tour de force* on the developments of nuclear physics.

This is followed by a brief commemoration of George Placzek, which appeared in *La Ricerca Scientifica* (the journal of the Consiglio Nazionale delle Ricerche), shortly after his sudden death. It is one of Amaldi's first works which is not strictly scientific. Placzek, who was of Czechoslovakian descent and almost the same age as Amaldi and his colleagues in via Panisperna, was a frequent visitor to the Roman institute and had established particularly strong working relationships and friendships with members of the group. Despite its brevity, it is evident from the biography that Amaldi was not only a friend, but was also fascinated by the culture this man represented: hailing from Mitteleuropa and being very cosmopolitan in his outlook, and as a result of having witnessed the events that had overwhelmed central Europe between the 1930s and the 1950s, Placzek possessed all the cultural ingredients which Amaldi, in later years, rediscovered and described with great precision in the biographies of Bruno Touschek and Fritz Houtermans.

In our opinion, the two papers that follow are among the most outstanding examples of Amaldi's historiographic production and the quality of his writing. *The Bruno Touschek Legacy*, published by CERN in 1981 only two years after the Austrian physicist's death, is an intense biography of the fascinating figure of a non-conformist and stateless scientist. In it, Amaldi's usual attention to the strictly scientific aspects does not overshadow a more integral reconstruction, in which the human aspects are sketched more clearly against the backdrop of the great historical events which Touschek lived through. After spending some time in Glasgow, Touschek moved to Italy in the 1950s, where he played a leading role as teacher, theoretician and volcanic source of ideas for new scientific and teaching experiments. In particular, he was responsible for the design and construction of AdA, the first prototype of a storage ring for electrons and positrons, at the National Laboratories in Frascati in 1960. Amaldi began to gather material for this biography in the last three months of Touschek's life, during his frequent visits to his friend in the La Tour hospital in Geneva and from their correspondence after his transfer to a nursing home in Igls, Tyrol. Using a style and method of working which he had begun to use when working on the biography of Enrico Persico, Amaldi continued to accumulate documentation even after Touschek's death: interviews, correspondence with colleagues, and talks with old friends of Touschek. Reading this piece, it is easy to see why Amaldi was, in addition to the strong friendship ties which developed over the years, doubly bound to this man: besides the intrinsic value of his contributions to physics, Amaldi's fascination with the man also derived from the latter's particular mix of Mitteleuropa origins and cosmopolitanism which had characterized Placzek. However, this is brought to the fore much more thoroughly here, especially in the last pages.

The Adventurous Life of Friedrich George Houtermans, Physicist stands apart in this collection of writings for a number of reasons. First, it has never been published before (the only piece in this volume). Second, its length makes it a real biography. Third, the thoroughness of its research, which is seen in the six boxes of documentary material preserved in Amaldi's files, and the length of time which he took to write it: Amaldi began to gather information and write about Houtermans in 1978 while the last, incomplete version of the manuscript published here dates back to October 1987. Thus, this biography is the result of approximately nine years of continual revision and editing. Finally, it is the only piece in this collection which was not directly related to a contingent circumstance or the result of some kind of external pressure, but sprang solely from the author's interest in the subject. Amaldi was fascinated with Houtermans' life and his political, cultural and scientific experiences. Writing about Houtermans, Amaldi wrote about the history and culture and of the Europe of his time, and of the conflicts that he lived through and which marked his generation.

At the suggestion of V. Weisskopf, Amaldi submitted the manuscript to an American publisher towards the end of 1985 but was turned down, the reason being that "it is neither scholarly nor appropriate for a broad trade audience". Although we do not share this opinion, we can understand the reasons why a publisher would hesitate to view this biography as standing on its own merits. There can be no doubt, however, that it fits extremely well into this kind of collection since it reveals the peak of Amaldi's maturity as a researcher and narrator.

Given the fact that the text has never been published before and is partially completed, a few editorial comments are in order. Various successive versions of the work were discovered in Amaldi's files; the one published here, however, is the last and most complete. Only a few minor changes have been made in cases of evident errors. Therefore, it has not been subject to any stylistic editing which the author would probably have carried out prior to its publication. Furthermore, while Amaldi had arrived at a definitive version of the main text before his death, the notes were not in an equally definitive form; Amaldi had obviously intended to return to some of them with further information and biographical data. These have been highlighted in bold. As we have refrained from altering the original manuscript by making any editorial changes, the text and the notes are published in their incomplete state in which Amaldi had left them.

A Few Flashes on Hans Bethe Contributions during the Thirties.

E. AMALDI

Dipartimento di Fisica dell'Università « La Sapienza » - P.le A. Moro 2, 00185 Roma

A few months ago the President of the Italian Physical Society, Renato RICCI, asked me to say a few words on this occasion for expressing to Hans BETHE our gratitude and admiration for the extraordinary work he has produced during many years of scientific activity.

I was very pleased but at the same time embarassed in consideration of the amplitude and variety of Bethe's work. Then it occurred to me that on occasion of a symposium on *Nuclear Physics in Retrospect*, held in Minnesota in 1977 [1], Hans BETHE gave a fascinating contribution that he entitled *The happy thirties.* Since those years are also very dear to me, I have thought to limit the subject of my speech to a few flashes on Hans Bethe contributions during this rather short period. I am aware that by doing so I will not convey to you the correct impression of the dimensions of the complete works carried out by BETHE during almost sixty years of extraordinary scientific work. I will have, however, the possibility of recalling, in the limited time at disposal, the circumstances under which the various papers were produced and the role they played in the development of the physical understanding of many important phenomena.

I like first to recall that Hans BETHE, at the beginning of the thirties, spent two periods at the Institute of Physics of the University of Rome, where Enrico FERMI was professor of Theoretical Physics from 1926 to 1938. During the first period, February-June 1931, BETHE worked on the scattering of relativistic electrons and on the stopping power of charged particles in matter. He studied also the symmetries of crystals, in particular those of crystals in a magnetic field, but did not publish any paper on this subject.

The second visit to Rome took place in the period March-May 1932, at a time when also George PLACZEK and Edward TELLER were there. BETHE devoted most of his time to write his famous review article published in the *Handbuck der Physik*, *On the quantum mechanics of one and two electron problems* [2], but had also the opportunity of publishing a paper in collaboration with FERMI *On the interaction between two electrons* [3], in which the relationships between the expressions derived shortly before by BREIT [4] and by MØLLER [5] and the then recent quantum electrodynamics are clarified.

BETHE had already published, in 1930, in the *Annalen der Physik*, a very extensive paper *On the theory of the passage of fast particles through matter* [6], where he had treated, in the Born approximation, the elastic and inelastic collisions of the incident particle with the atoms of the medium and computed the excitation of optical and X-ray levels of the atoms, the number of primary and secondary ions produced, the velocity distribution of the secondary electrons and the energy loss undergone by the incident particle. Contributions to this very complex problem were given by MØLLER in 1932 [7] and by BLOCH in 1933 [8], but the general frame of the theory remained that developed originally by BETHE, who, on various later occasions during his life, recast his original treatment in more transparent forms which are those we find even in the most recent books.

In Summer 1932 the positive electron was discovered by ANDERSON in cosmic rays [9], a result immediately confirmed by BLACKETT and OCCHIALINI [10], who also observed, in their cloud chamber triggered by Geiger-Müller counters in coincidence, a number of «electron showers». The existence of events of this kind had been already suggested by ROSSI from his experiments with three nonaligned counters [11], but their complexity and variety could be appreciated only by means of cloud chamber's pictures.

The electron theory of Dirac [12] was thus confirmed by these as well as many other experimental papers on the pair production and the annihilation of positrons. The theory also required a detailed application to the new phenomena together with a clarification of the interplay between electrons and photons, especially at high energy. In the period 1932-1935 Hans BETHE published a series of important papers on the bremsstrahlung of relativistic electrons [13], with HEITLER on the stopping of fast particles and the creation of positive electrons [14], on the influence of the screening on the creation and stopping of electrons [15] and on the annihilation radiation of positrons [16].

Apart from their interest *per se*, these papers contain all the cross-sections required for the theory of the electromagnetic showers, which was developed in 1937 by CARLSON and OPPENHEIMER and by BHABHA and HEITLER [17].

In 1914 James CHADWICK [18] had shown that the electrons emitted in beta-decay have a continuous energy spectrum and in the period 1930-1932 Wolfgang PAULI had proposed, as an esplanation of this unespected behaviour, that the emission of the electron is accompanied by the emission of a neutral particle of very small mass that takes away in various cases different fractions of the available energy and momentum. At the end of 1933 Enrico FERMI published a short account of his theory of beta-decay on Pauli assumption of this new neutral particle that he called neutrino. The full presentation of this theory appeared in *Il Nuovo Cimento* and *Zeitschrift für Physik* at the beginning of 1934 [19].

The 15th of January of 1934 Frederic JOLIOT and Irène CURIE presented to the Academy of Sciences in Paris the announcement of the discovery of artificial radioactivity induced by alfa-particles in a few light elements [20]. They referred to this phenomenon as a « new kind of radioactivity », because the artificial radioactive nuclei they had produced emitted positive electrons instead of negative electrons as observed in the case of natural beta emitters.

Shortly after Fermi's paper had appeared in the international scientific press, Hans BETHE and Rudolf PEIERLS sent to the Editor of *Nature* two letters, both with the title *The neutrino.* In the first one, dated February 20, 1934 [21], the authors discuss in qualitative terms the beta-decay of both signs in relation with the mass difference of isobars differing by 1 in atomic number and notice that the possibility of creating neutrinos implies the existence of an annihilation process. They point out that the most interesting among these reactions is the inverse beta-decay

$$\nu + (A, Z) \rightarrow (A + Z \pm 1) + e^{\mp}$$

for which they derive an upper limit of the cross-section of the order of 10^{-44} cm^2 at $E_\nu \sim 2$ MeV. The authors comment this result saying that « it is absolutely impossible to observe processes of this kind with the neutrinos created in nuclear transformations ». A process of this kind was, however, observed in 1956 by COWAN, REINES and coworkers [22], who used the antineutrino from a powerful fission reactor at the Savannah River Plant (USA) and found for the cross-section the value

$$\sigma = (11 \pm 2.6) \cdot 10^{-44}\ \text{cm}^2 .$$

But in 1934 nobody could have foreseen the possibility of constructing nuclear reactors, of using them as neutrino sources and of detecting neutrinos by means of devices of large dimensions.

In the second letter to *Nature*, dated April 1, 1934 [23], BETHE and PEIERLS discuss ways of detecting the neutrino emission. One possible experiment would be to check the energy balance for the artificial beta-decay. Take, for example, the process

$$^{10}\text{B} + \alpha \rightarrow {}^{13}\text{N} + \text{neutron} ,$$

$$^{13}\text{N} \rightarrow {}^{13}\text{C} + e^{+} + \text{neutrino} .$$

When the positron is emitted with the greatest possible energy, the kinetic energy of the neutrino will just be zero. The balance in energy in this case will, therefore, determine the mass of the neutrino. In practice this procedure does not work because of the errors with which are known the masses of the various particles and the energy of the electron.

A second way of deciding the presence of a neutrino would be to observe the recoil of the nucleus in beta-decay. With natural radioactive substances this is in practice impossible because the recoil energy is too small, but the artificial radioactive nuclei are much lighter. The kinetic energy of recoil of a disintegrating ^{13}N nucleus would be of the order of some hundreds of eV if there were no neutrino.

The same idea had been suggested also by PAULI at the 1933 Solvay Conference [24] and was applied, starting from 1936, by a number of experimentalists. All these experiments and their results were reviewed by CRANE in the *Reviews of Modern Physics* of 1948 [25]. A few of them provided evidence in favour of the emission of the neutrino in beta-decay, but their importance for establishing, beyond any doubt, the same existence of the neutrino was shadowed by the work of Cowan and Reines that I have already mentioned above.

In the last part of the second letter to *Nature* BETHE and PEIERLS suggest the existence of a new form of beta instability, *i.e.* by capture by one of the protons inside the nucleus of one of the orbital electrons, a phenomenon envisaged at the same time also by G. C. WICK [26] and observed for the first time in 1938 by ALVAREZ [27].

The particle that during the thirties attracted more attention from both the experimentalists and the theoreticians was the *neutron.* Discovered by James CHADWICK in 1932 [28], it was immediately recognized as one of the constituents of the nuclei: protons and neutrons. Various types of forces acting between these particles had been proposed by HEISENBERG, MAJORANA and WIGNER and their dependence on the distance—and possibly on other variables—was considered by the majority of the physicists as the most important problem.

Artificial radioactivity induced by neutrons was discovered by FERMI in March 1934 [29] and was followed by a long series of experimental results on the processes produced by neutrons [30], on the slowing-down of neutrons (October 1934) [31] and the existence of nuclei with extremely large cross-section for (n, γ) processes induced by slow neutrons (November 1934) [32]. In Summer 1934 CHADWICK and GOLDHABER discovered the photodisintegration of the deuteron [33], an effect that opened the way to precision determinations of the mass of the neutron. According to previous erroneous measurements the mass of this particle was believed to be smaller than that of the proton.

In July 1934 BETHE and PEIERLS sent to the *Proceedings of the Royal Society* an important paper on the *Quantum theory of the diplon* [34]. This was the name they used, following a suggestion by RUTHERFORD, for the heavy isotope of hydrogen, indicated later universally as deuteron.

After a short introduction of general nature on the various kinds of nuclear forces I mentioned above, the authors write the wave equation of the deuteron obtained by introducing into the Schrödinger equation a short-range potential well of Majorana type and deduce the wave function of the ground state of the system.

Then they compute the cross-section for absorption of gamma-rays due to electric-dipole transitions from the ground state of the deuteron to states of the continuum, where the neutron and the proton fly away from each other. They also suggested that the inverse process could provide the mechanism responsible for the capture of a low-energy neutron by a proton with formation of a deuteron. While the rest of the paper is still valid today, the last suggestion is wrong because, at low energies, the process is a S-S transition, which is forbidden by the electric dipole. It can be interpreted as due to a magnetic-dipole transition, as was shown by FERMI some time later [35].

In November of the same year 1934, BETHE and PEIERLS sent to the *Proceedings of the Royal Society* a second paper dealing with the scattering of neutrons by protons [36]. They quote various experiments indicating an isotropic scattering which, however, were «not accurate enough to allow very definite conclusions». Then the authors apply the general method given by MOTT and MASSEY [37] of separating the solution of the Schrödinger equation in the continuum in Legendre polynomials and they arrive to write the differential cross-section for scattering through an angle θ in the form

$$\mathrm{d}\sigma_{\mathrm{el}} = \frac{\pi}{2k^2}\left|\sum_{l=0}^{\infty}(2l+1)P_l(\theta)\left(\exp[2i\delta_l]-1\right)\right|^2 \sin\theta\,\mathrm{d}\theta\,,$$

where δ_l is the phase shift of the wave of angular momentum l due to the potential well representing the neutron-proton interaction.

They notice that, even for incident neutrons of a few MeV, the de Broglie wavelength $\lambda\!\!\!^{-} = k^{-1}$ is much larger than the linear dimension a of the potential well, so that all δ_l are small except for $l = 0$ (S waves) and deduce the simple expressions

$$\mathrm{d}\sigma_{\mathrm{el}} = \frac{2\pi h^2}{M}\,\frac{1}{\frac{1}{2}E+|\varepsilon_{\mathrm{B}}|}\sin\theta\,\mathrm{d}\theta = \frac{4\pi h^2}{M}\,\frac{1}{\frac{1}{2}E+|\varepsilon_{\mathrm{B}}|}\,, \tag{1}$$

where $\varepsilon_{\mathrm{B}} = 2.24$ MeV is the binding energy of the deuteron.

The main approximation adopted in the derivation of these expressions is that for small value of the energy of the incident neutron ($E \leqslant 20$ MeV) the value of the logarithmic derivative at the surface of the potential well of $u_0 = r\psi_0$

$$\left(\frac{1}{u_0}\frac{\mathrm{d}u_0}{\mathrm{d}r}\right)_{r=a}$$

is «practically the same as for the small negative energy ε_{B}», so that one can write

$$\left(\frac{1}{u_0}\frac{\mathrm{d}u_0}{\mathrm{d}r}\right)_{r=a} = -\,\alpha = -\,\frac{\sqrt{M|\varepsilon_{\mathrm{B}}|}}{\hbar^2} \approx -\,2.3\cdot 10^{12}\ \mathrm{cm}^{-1}\,.$$

This approach clearly foreshadowed the so-called *effective-range approximation*, sketched by SCHWINGER in 1947 [38] and fully developed, independently, by BETHE in 1949 [39].

As pointed out by BETHE and PEIERLS, the details of the shape of the potential well were irrelevant, as well as the adoption of Majorana forces. If these are replaced by Heisenberg or Wigner forces, the results remain the same.

The formula for the neutron-proton elastic cross-section, derived by BETHE and PEIERLS, appeared at first to be successful. It provided values in rough agreement with the corresponding experimental results for neutron energies of the order of a few MeV. But when the Columbia University group [40] measured the elastic cross-section of slow neutrons against protons, they found a value more than 14 times larger than the theoretical prediction. The explanation of this serious discrepancy was suggested by WIGNER in a private conversation with BETHE in a noisy subway between Columbia University and Penn Station. WIGNER said [1]: « Well, why do you assume that the singlet state has the same binding energy as the triplet state? Let's assume a different binding energy and all will be well ».

Wigner suggestion of the dependence of the nuclear forces on the relative spin of the two interacting nucleons clearly amounts to writing on the right-hand side of the scattering cross-section (1), in addition to the term corresponding to the scattering in the triplet state, a second term of a similar structure originating from the singlet state. By using the appropriate statistical factors, the new expression for the neutron-proton scattering cross-section takes the form

$$\sigma_{\text{el}} = \frac{4\pi\hbar^2}{M}\left[\frac{3}{4}\,\frac{1}{\frac{1}{2}E+|\varepsilon_{\text{t}}|}+\frac{1}{4}\,\frac{1}{\frac{1}{2}E+|\varepsilon_{\text{s}}|}\right], \tag{2}$$

where ε_{t} (energy of the triplet state) is used for ε_{B} and ε_{s} is the « binding energy » of the neutron-proton system in the singlet state (antiparallel spin). Assuming expression (2) to be correct, a determination of ε_{s} was obtained by comparing the value given by (2) with the experimental results for neutrons in the epithermal region. Such a comparison gave $\varepsilon_{\text{s}} \approx 75$ keV, while the sign of ε_{s} was determined later from experiments with slow neutrons scattered by ortho- and para-hydrogen.

This formula and these numerical values were given in an article by BETHE and BACHER on *Nuclear physics.* A: *Stationary states* published in 1936 [41], which was followed by two other extensive articles, one by BETHE, on *Nuclear physics.* B: *Nuclear dynamics* [42], the other by BETHE and LIVINGSTON, on *Nuclear physics.* C: *Nuclear dynamics, experimental* [43], which covered in the most complete, transparent and, at the same time, brilliant way all aspects of what was known in the field of nuclear physics in the years 1936-37.

These three review articles but in particular the second one, by BETHE alone, on the *Nuclear dynamics*, had an enormous influence on the development of nuclear research in the middle of the thirties. It contains also a number of original contributions that BETHE did not publish in any other scientific journal.

Let me go back for a moment to 1935 for recalling another important paper by BETHE on the theory of disintegration of nuclei by neutrons [44]. Under the assumption that the neutron-nucleus interaction can be described by a potential well, BETHE derives the expressions for the cross-sections for all possible nuclear reactions induced by neutrons, in particular for radiative capture and elastic scattering. Among the many results he finds I will quote the $1/v$ law for the radiative capture cross-section

$$\sigma_c = \frac{k}{v} \tag{3}$$

and the fact that the elastic-scattering cross-section should always represent an appreciable fraction of the capture cross-section. The constant k appearing in (3), in particular cases, can assume very large values, thus providing an explanation of the anomalously large capture cross-section observed in a few elements. This result was deduced, at about the same time, also by FERMI [32], PERRIN and ELSASSER and Guido BECK and HORSLEY [45].

During the same year 1935 two experimental papers by BJERGE and WESTCOTT [46] and by MOON and TILLMAN [47] gave indication that some nuclei had selective absorptions. A more detailed experimental study of this phenomenon [48] brought to the recognition of rather narrow absorption lines in the epithermal region. This result was in clear contradiction with the existence of the universal law (3) for the capture cross-section of all nuclei.

At the beginning of 1936 two important theoretical papers appeared, dealing with the capture and scattering of neutrons by nuclei. The first one was presented by Niels BOHR to the Danish Academy on January 26, 1936, and appeared in *Nature* of February 29 [49]. It contained the idea that by capture of the incident neutron a compound nucleus is formed in an excited state, the mean life of which is long enough for explaining the thin lines observed in the epithermal region.

The other paper, due to BREIT and WIGNER [50], was received by the *Physical Review* on February 15, 1936, and contained the derivation of the well-known one-level Breit and Wigner formula, describing the energy dependence of the capture and scattering cross-sections by a nucleus with a single resonant level.

The two papers were in some way complementary to each other and produced a deep and permanent change in the way of looking at the structure of the nucleus and the mechanism of its transformations.

Hans BETHE immediately appreciated the importance of the new conceptions and published various papers for analysing and completing them.

In a paper with WEEKS and LIVINGSTONE [51], BETHE proposed a method for determining the energy of the neutron resonances by measuring the corresponding absorption coefficient in boron which, with slow neutrons, undergoes a (n, α) process. As was shown by BETHE and coworkers, as well as by FRISCH and PLACZEK [52], by applying Bohr's views to a nucleus as light as that of boron, the resonance width of the (n, α) process is so large that the corresponding cross-section should follow the $1/v$ law over a wide range of energies. This method was employed successfully by a number of authors during 1936-1937 [48, 53]. When, many years later, precision measurements of the boron cross-section as a function of the neutron velocity became possible, it was found that the $1/v$ law was actually respected with great accuracy in full agreement with the old theoretical predictions.

Together with PLACZEK, BETHE published in 1937 an extensive paper on *Resonance effects in nuclear processes* [54] which completes and generalizes the work of Breit and Wigner from two points of view. The authors take into account the spin of the incident neutron and introduce the statistical factor due to its composition with the spin of the target nucleus, which had been neglected in the Breit-Wigner treatment. Furthermore, they generalize the work of these authors to the case of any number of resonant levels. Since Bethe and Placzek paper is based on perturbation theory, their formulae are correct as long as the width Γ of the various resonances is small compared with the distance d between successive levels.

The paper, however, contains many other important results, among which I should recall the study of the influence on the neutron absorption of the Doppler effect due to the thermal agitation of the capturing nuclei and a detailed analysis of all experimental results available in 1936, in particular of the self-absorption experiments.

One of the points stressed by BOHR in the discussion of the compound-nucleus model was that the density of the levels of medium and heavy nuclei had to be appreciably larger than what one had thought until then. This problem had been already tackled by BETHE [55] in the frame of the simplest model, *i.e.* assuming the interaction between the particles to be small with respect to their kinetic energy so that the nucleus could be treated as a gas of nucleons. The same model had been used also by OPPENHEIMER and SERBER [56] for a similar calculation. After the paper by BOHR on the compound nucleus such a model clearly appeared inadequate. The interaction between the constituent particles is large compared to their kinetic energy and the nucleus, in first approximation, corresponds better to a liquid drop.

These various approaches were discussed in depth by BETHE in his article in the *Reviews of Modern Physics* of 1937 [42], where he develops the thermodynamics of the nucleus, deriving a number of general relations such as that between total excitation energy and nuclear temperature or between nuclear entropy and density of nuclear levels. The general derivations are completed by applications to a few important simplified models.

Similar considerations were developed also by BOHR and KALCKAR [57], but the publication of their paper was postponed a few times, so that it appeared a long time after Bethe's article [42], which, in addition, contains a much more complete and, at the same time, more transparent treatment of the problem.

A particularly important contribution was given by BETHE in 1938 with the paper on the production of energy in stars [58].

The first attempt at an explanation of the origin of the energy continuously poured in space by stars was by HELMHOLTZ in 1876 [59] and was based on gravitation. This mechanism, however, is not sufficient by a large factor for supplying the energy irradiated over the ages, for instance, by the Sun. In the 20s EDDINGTON [60] investigated thoroughly the interior constitution of the Sun and arrived at the conclusion that its innermost part is a hot gas mainly of hydrogen and helium, at about $20\cdot10^6$ K and a density about 80 times that of water.

EDDINGTON was very much concerned with the origin of stellar energy that he attributed to « subatomic processes ». He considered, for example, the formation of helium from hydrogen, but did not dispose of the information necessary for judging its feasibility. The same remark holds for its favourite hypothesis, namely the complete annihilation of matter changing nuclei and electrons into radiation. The energy which was to be set free from such a process, if it could occur, is given by Einstein's relation $E = Mc^2$.

This would be enough to supply the Sun's radiation 1500 milliards of years. However, from experiments on the Earth we know that protons and electrons do not annihilate each other over at least 10^{31} y.

In 1928 GAMOW and CONDON and GURNEY [61] developed the quantum theory of α-decay of natural radioactive substances and showed that positively charged particles can penetrate in nuclei even when their energy is not large enough, from the classical point of view, for passing above the Coulomb potential barrier. At about the same time M. VON LAUE and KUDAR [62] suggested the possibility of formation of light elements through the inverse process of α-decay, but found values much too low by orders of magnitude for the probability per unit time to occur even under stellar conditions.

At this point E. D'ATKINSON and F. HOUTERMANS, both experts in Gamow theory for having contributed to its development [63], computed by means of the same theory the probability per unit time of nuclear reactions between nuclei and a proton gas with a density 10^{23} particles/cm^3 and a Maxwellian velocity distribution at the temperature between 1 and 2 in 10^7 K [64]. For heavy elements the probability of processes of this type turned out to be exceedingly small, but for light elements, assuming a collision radius of $4\cdot10^{-13}$ cm, they obtained half-lives ranging from 8 s in ${}^4_2$He to 10^9 y for ${}^{20}_{10}$Ne.

Then they proceeded to estimate, from phenomenological arguments, the probability for a proton of remaining bound in a nucleus through the emission of its excess energy by radiation, and concluded that processes of this kind can provide the mechanism not only for the formation of light elements but also for supplying the energy required for maintaining over the ages the emission of radiation from the stars.

The problem of the nuclear reactions taking place in the stars was taken up again by VON WEIZSÄCKER in 1937 [65], *i.e.* after the discovery of the neutron and the recognition that neutrons and protons are the only constituents of nuclei. At the end of a thorough discussion, WEIZSÄCKER arrives at the following conclusions:

a) at the centre of stars the temperature is large enough for giving rise to nuclear reactions between protons and nuclei of light elements;

b) neutrons can be produced from reactions between the heavy isotopes of hydrogen;

c) heavy elements can be produced by capture of neutrons from light elements.

In the course of his paper WEIZSÄCKER considers as nuclear reactions of interest for the production of energy in stars two kind of processes. The production of D in proton-proton collision taking place through weak interactions

$$^{1}_{1}\mathrm{H}+^{1}_{1}\mathrm{H}\rightarrow ^{2}_{1}\mathrm{D}+\mathrm{e}^{+}+\nu_{\mathrm{e}} \tag{4}$$

computed some time later by BETHE and CRITCHFIELD [66], followed by capture of two successive protons, and the cycle

$$\begin{cases} ^{4}_{3}\mathrm{He}+^{1}_{2}\mathrm{H}\rightarrow ^{5}_{3}\mathrm{Li}\,, \\ ^{5}_{3}\mathrm{Li} \qquad\quad \rightarrow ^{5}_{2}\mathrm{He}+\mathrm{e}^{+}+\nu_{\mathrm{e}}\,, \\ ^{5}_{2}\mathrm{He}+^{1}_{1}\mathrm{H}\rightarrow ^{4}_{2}\mathrm{He}+^{2}_{1}\mathrm{D}\,, \end{cases} \tag{5}$$

which involves the isotope ^{5}He, the properties of which were not yet known.

Shortly later two more papers appeared, one by GAMOW [67] on the importance of selective absorption of thermonuclear reactions for the structure and evolution of stars, the other by GAMOW and TELLER [68] on the rate of energy production in the various reactions involved. The problem was the same already dealt with by D'ATKINSON and HOUTERMANS, but the subsequent development of nuclear physics imposed a revision of the whole treatment.

The formulae derived for the general case were applied by the same authors to the cyclic process (5) considered by WEIZSÄCKER. This, however, lost any interest after experimental evidence was provided of the instability of ^5_2He [69].

In March 1938 GAMOW assembled a small conference in Washington, D.C., under the sponsorship of the Department of Terrestrial Magnetism of the Carnegie Institution. As BETHE wrote years later [70]: « At this conference the astrophysicists told us physicists what they knew about the internal costitution of the stars. This was quite a lot, and all of their results had been derived without knowledge of the specific source of energy. The only assumption they made was that most of the energy was produced near the center of the star ».

A few months later, *i.e.* on September 7, 1938, the *Physical Review* received a fundamental paper by BETHE on the *Energy production in stars* [58]. It contains a thorough discussion not only of the possible nuclear reactions of interest for the problem but also of all its astrophysical aspects. The energy production in stars is entirely due to the combination of four protons and two electrons into an alfa-particle. Such a combination occurs essentially only in two ways. The first mechanism is the one already considered by WEIZSÄCKER, *i.e.* the combination of two protons to form a deuteron with positron emission. The deuteron is then transformed into ^4He by the capture of two additional protons. These captures occur very rapidly compared with the initial process of deuteron formation, which involves weak interactions.

The second mechanism was the today well-known carbon-nitrogen cycle:

$$\begin{aligned}
&{}^{12}_{6}C+{}^{1}_{1}H \rightarrow {}^{13}_{7}N+\gamma\,,\\
&{}^{13}_{7}N \rightarrow {}^{13}_{6}C+e^{+}+\nu_e\,,\\
&{}^{13}_{6}C+{}^{1}_{1}H \rightarrow {}^{14}_{7}N+\gamma\,,\\
&{}^{14}_{7}N+{}^{1}_{1}H \rightarrow {}^{15}_{8}O+\gamma\,,\\
&{}^{15}_{8}O \rightarrow {}^{15}_{7}N+e^{+}+\nu_e\,,\\
&{}^{15}_{7}N+{}^{1}_{1}H \rightarrow {}^{12}_{6}C+{}^{4}_{2}He\,.
\end{aligned}$$

It involves ^{12}C as a catalyst which is reproduced in all cases except about one in 10 000.

Since the C-N cycle involves nuclei of relatively high charge, it has a strong energy dependence. The reaction with ^{14}N is the slowest of the cycle and, therefore, determines the rate of energy production. It goes about as T^{24} near solar temperature. The two cyclic processes $^1H+^1H$ and C-N were found by BETHE to be about equally probable at a temperature of $16\cdot10^6$ K and that at lower temperature the $^1H+^1H$ will predominate, at high temperature the C-N cycle.

In 1938 the central temperature of the Sun was believed to be close to $19\cdot10^6$ K and, therefore, BETHE considered the C-N cycle as the most important source of energy in ordinary stars. Later estimates of the central temperature of the Sun gave, however, a lower value, so that we know to day that the important process is the proton-proton chain of reaction, while the C-N cycle is at work in heavier stars.

This change, however, does not detract value to the paper of Hans Bethe which remains as a fundamental permanent step forward in the development of physics and astrophysics. Its importance was duly recognized with the award of the 1967 Nobel Prize for Physics to Hans BETHE.

As a last flash on Hans Bethe's contributions during the thirthies, I will add a few words about his « meson theory of the nuclear forces », the first part of which appeared in 1939 [71].

This paper represents an application to the case of the deuteron of Hideki Yukawa theory [72], who had proposed, at the end of 1935, a description of the nuclear forces as due to the exchange, between the interacting nucleons, of a particle of intermediate mass, later called the meson. Yukawa's paper, published in a Japanese scientific journal, for some time passed almost unobserved. After the discovery in cosmic rays by NEDDERMEYER and ANDERSON and by STREET and STEVENSON [73] of a particle of mass rather close to Yukawa's prediction, a number of theoreticians in Europe and in the USA started to contribute to the development of the meson theory. In the paper I mentioned above, BETHE presents the results of preliminary computations of the properties of the deuterons carried out for reactor mesons in the two forms of the theory that give « charge independence ». The neutral theory in which the exchanged mesons are only neutral and the symmetrical theory in which charged as well as neutral mesons appear. BETHE computes, for two different cutting off of the potential at short distances, the percentage of D-wave present in the ground state of the neutron-proton system and the quadrupole moment of the deuteron. The problem was tackled by BETHE in greater generality in a successive paper of 1940 [74], where it is shown that the neutral theory gives both the magnitude and sign of the electric quadrupole of the deuteron in good agreement with the experimental results, while the symmetrical theory gives unacceptable results particularly for the cut-off distance which must be chosen considerably larger than the range of the nuclear forces. These results appeared to BETHE very regrettable since only the symmetrical theory gave a natural explanation of the beta-decay and of the extramagnetic moments of neutrons and protons. They were, however, a consequence of the use of vector mesons and were overcome, shortly later, with the recognition of the pseudoscalar nature of the pion.

I will stop here, hoping that what I have reminded you about Bethe's contributions during the thirties is enough for stressing the richness and variety of his extraordinary scientific production.

Dear BETHE we admire you and we are glad of this occasion for expressing you our warmest thanks.

REFERENCES

[1] *Nuclear Physics in Retrospect, Proceedings of a Symposium on the 1930's*, edited by R. H. STEUEWER (University of Minnesota Press, Minneapolis, Minn., 1977).

[2] H. BETHE: *Quantenmechanik der Ein und Zwei Elektronenprobleme*, p. 273-560 of Vol. XXIV, 1. *Quantentheorie* of the *Handbuck der Physik*, edited by A. SMEKAL (Verlag von Julius Springer, Berlin, 1933).

[3] H. BETHE and E. FERMI: *Über die Wechselwirkung von Zwei Elektronen*, *Z. Phys.*, **77**, 296-306 (1932).

[4] G. BREIT: *The effect of retardation on the interaction of two electrons*, *Phys. Rev.*, **34**, 553-573 (1929); *Dirac's equation and the spin-spin interactions of two electrons*, *Phys. Rev.*, **39**, 616-624 (1932).

[5] C. MØLLER: *Über den Stoss zweier Teilchen under Berücksichtigung der Retardation der Kräfte*, *Z. Phys.*, **70**, 786-793 (1931).

[6] H. BETHE: *Zur Theorie des Durchgangs schneller Korpuskularstrahlen durch Materie*, *Ann. Phys.* (*Leipzig*), **5**, 325-400 (1930).

[7] C. MØLLER: *Zur Theorie des Durchgangs schneller Elektronen durch Materie*, *Ann. Phys.* (*Leipzig*), **14**, 531-585 (1932).

[8] F. BLOCH: *Bremsvermögen von Atomen mit mehreren Elektronen*, *Z. Phys.*, **81**, 363-376 (1933); *Zur Bremsung rasch bewegter Teilchen beim Durchgang durch Materie*, *Ann. Phys.* (*Leipzig*), **16**, 285-320 (1933).

[9] C. D. ANDERSON: *The apparent existence of easily deflectable positives*, *Science*, **76**, 238-239 (1932); *Energies of cosmic ray particles*, *Phys. Rev.*, **41**, 405-421 (1932).

[10] P. M. S. BLACKETT and G. P. S. OCCHIALINI: *Some photographs of the tracks of penetrating radiation*, *Proc. R. Soc. London, Ser. A*, **139**, 699-720 (1933).

[11] B. ROSSI: *Über einer Sekundarstrahlung der durchdringenden Korpuskularstrahlung*, *Phys. Z.*, **33**, 304-305 (1932).

[12] P. A. M. DIRAC: *Annihilation of electrons and positrons*, *Proc. Cambridge Philos. Soc.*, **26**, 361-375 (1930).

[13] H. BETHE: *Bremsformel für Elektronen relativisticher Geschwindigkeit*, *Z. Phys.*, **76**, 293-299 (1932).

[14] H. BETHE and W. HEITLER: *On the stopping of fast particles and the creation of positive electrons*, *Proc. R. Soc. London, Ser. A*, **146**, 83-112 (1934).

[15] H. BETHE: *Influence of screening on the creation and stopping of electrons*, *Proc. Cambridge Philos. Soc.*, **30**, 524-539 (1934).

[16] H. BETHE: *On the annihilation radiation of positrons*, *Proc. R. Soc. London, Ser. A*, **150**, 129-141 (1935).

[17] F. CARLSON and J. R. OPPENHEIMER: *On multiplicative showers*, *Phys. Rev.*, **51**, 220-231 (1937); W. HEITLER and H. J. BHABHA: *The passage of fast electrons and the theory of cosmic showers*, *Proc. R. Soc. London, Ser. A*, **159**, 432-458 (1937).

[18] J. CHADWICK: *Distribution in intensity in the magnetic spectrum of the β-rays of radium* ($B+C$), *Verh. Dtsch. Phys. Ges.*, **16**, 383-391 (1914).

[19] E. FERMI: *Tentativo di una teoria dell'emissione dei raggi β*, *Nuovo Cimento*, **11**, 1-19 (1934); *Versuch einer Theorie der β-Strahlen I*, *Z. Phys.*, **88**, 161-171 (1934), received January 16, 1934, and appeared in the March issue.

[20] F. JOLIOT et I. CURIE: *Un noveau type de radioactivité*, *C. R. Acad. Sci.*, **198**, 254-256 (1934); *Artificial production of a new kind of radio elements*, *Nature* (*London*), **133**, 201-202 (1934).

[21] H. Bethe and R. Peierls: *The neutrino, Nature (London)*, **133**, 592 (1934), dated February 20, 1934.

[22] C. L. Cowan, F. Reines, F. B. Harrison, H. W. Kruse and A. D. McGuire: *Detection of free neutrino: a confirmation, Science*, **124**, 103-104 (1956); F. Reines and C. L. Cowan: *Free antineutrino absorption cross section* I. *Measurements of the free antineutrino absorption cross section by protons, Phys. Rev.*, **113**, 273-279 (1959).

[23] H. Bethe and R. Peierls: *The neutrino, Nature (London)*, **133**, 689-690 (1934), dated April 1, 1934.

[24] See W. Pauli intervention at p. 324-325 of *Structure et propriétés des noyaux atomiques*, Reports et Discussions du Septieme Conseil de Physique, tenu a Bruxelles du 25 au 29 October 1933 (Gauthier-Villars, Paris, 1934).

[25] H. R. Crane: *The energy and momentum relations in the beta-decay and the search for the neutrino, Rev. Mod. Phys.*, **20**, 278-295 (1948).

[26] G. C. Wick: *Sugli elementi radioattivi di F. Joliot e I. Curie, Rend. Accad. Lincei*, **19**, 319-324 (1934).

[27] L. W. Alvarez: *The capture of orbital electrons, Phys. Rev.*, **54**, 486-496 (1938).

[28] J. Chadwick: *Possible existence of a neutron, Nature (London)*, **129**, 312 (1932). *The existence of a neutron, Proc. R. Soc. London, Ser. A*, **136**, 692-708 (1932).

[29] E. Fermi: *Radioattività indotta da bombardamento di neutroni* I, *Ric. Sci.*, **5**(1), 283 (1934).

[30] E. Fermi, E. Amaldi, O. D'Agostino, F. Rasetti and E. Segrè: *Artificial radioactivity produced by neutron bombardment, Proc. R. Soc. London, Ser. A*, **146**, 483-500 (1934), received July 25, 1934.

[31] E. Fermi, E. Amaldi, B. Pontecorvo, F. Rasetti e E. Segrè: *Azione di sostanze idrogenate sulla radioattività prodotta da neutroni*, I, *Ric. Sci.*, **5**(2), 282 (1934).

[32] E. Amaldi, O. D'Agostino, E. Fermi, B. Pontecorvo, F. Rasetti and E. Segrè: *Artificial radioactivity produced by neutron bombardment*, II, *Proc. R. Soc. London, Ser. A*, **149**, 522-558 (1935).

[33] J. Chadwick and M. Goldhaber: *A nuclear photoeffect, Nature (London)*, **134**, 237-238 (1934), issue of August 18, 1938; *The nuclear photoelectric effect, Proc. R. Soc. London, Ser. A*, **151**, 479-493 (1935).

(34] H. A. Bethe and R. Peierls: *Quantum theory of the diplon, Proc. R. Soc. London, Ser. A*, **148**, 146-156 (1935), received July 26, 1934.

[35] E. Fermi: *On the recombination of neutrons and protons, Phys. Rev.*, **48**, 570 (1935).

[36] H. A. Bethe and R. Peierls: *The scattering of neutrons and protons, Proc. R. Soc. London, Ser. A*, **149**, 176-183 (1935), received November 19, 1934.

[37] N. F. Mott and H. S. Massey: *The Theory of Atomic Collisions* (Oxford University Press, London, 1933).

[38] J. S. Schwinger: *A variational principle for scattering problems, Phys. Rev.*, **72**, 742 (1947).

[39] H. A. Bethe: *Theory of the effective range in nuclear physics, Phys. Rev.*, **76**, 38-50 (1949).

[40] J. R. Dunning, G. P. Pegram, G. A. Fink and D. P. Mitchell: *Interaction of neutrons with matter, Phys. Rev.*, **48**, 265-280 (1935), received June 8, 1935.

[41] H. A. Bethe and R. F. Bacher: *Nuclear physics.* A: *Stationary states, Rev. Mod. Phys.*, **8**, 82-229 (1936).

[42] H. A. Bethe: *Nuclear physics.* B: *Nuclear dynamics, Rev. Mod. Phys.*, **9**, 69-244 (1937).

[43] H. A. BETHE and M. S. LIVINGSTON: *Nuclear physics*. C: *Nuclear dynamics, experimental*, *Rev. Mod. Phys.*, **9**, 245-390 (1937).

[44] H. A. BETHE: *Theory of disintegration of nuclei by neutrons*, *Phys. Rev.*, **47**, 747-759 (1935), received March 26, 1935.

[45] F. PERRIN et W. M. ELSASSER: *Théorie de la capture sélective des neutrons par certain noyaux*, *C. R. Acad. Sci.*, **200**, 450-452 (1935); *J. Phys. Radium*, **6**, 194-202 (1935); F. PERRIN: *Mecanisme de la capture des neutrons lents par les noyaux légers*, *C. R. Acad. Sci.*, **200**, 1749-1751 (1935); G. BECK and G. H. HORSLEY: *Nonelastic collision cross section for slow neutrons*, *Phys. Rev.*, **47**, 510 (1935).

[46] T. BJERGE and C. H. WESTCOTT: *On the slowing down of neutrons in various substances containing hydrogen*, *Proc. R. Soc. London, Ser. A*, **150**, 709-729 (1935).

[47] J. R. TILLMAN and B. P. MOON: *Selective absorption of slow neutrons*, *Nature* (*London*), **136**, 66-67 (1935).

[48] E. FERMI e E. AMALDI: *Sull'assorbimento dei neutroni lenti* I, *Ric. Sci.*, **6**(2), 334-347 (1935); *Sull'assorbimento dei neutroni lenti* II, *Ric. Sci.*, **6**(2), 443-447 (1935), dated 14 dicembre 1935; E. AMALDI and E. FERMI: *On the absorption and the diffusion of slow neutorns*, *Phys. Rev.*, **50**, 899-928 (1936).

[49] N. BOHR: *Neutron capture and nuclear constitution*, *Nature* (*London*), **137**, 344-348 (1936).

[50] G. BREIT and E. WIGNER: *Capture of slow neutrons*, *Phys. Rev.*, **49**, 519-531 (1936).

[51] D. F. WEEKS, M. S. LIVINGSTONE and H. A. BETHE: *A method for the determination of the selective absorption of slow neutrons*, *Phys. Rev.*, **49**, 471-473 (1936), received March 4, 1936.

[52] G. PLACZEK and O. FRISCH: *Capture of slow neutrons*, *Nature* (*London*), **137**, 357 (1936), issue of February 29, 1936.

[53] H. H. GOLDSMITH and F. RASETTI: *Experiments on residual neutrons*, *Phys. Rev.*, **49**, 891 (1936); *On the resonance capture of slow neutrons*, *Phys. Rev.*, **50**, 328-331 (1936).

[54] H. A. BETHE and G. PLACZEK: *Resonance effects in nuclear processes*, *Phys. Rev.*, **51**, 450-484 (1937).

[55] H. A. BETHE: *An attempt to calculate the number of energy levels of a heavy nucleus*, *Phys. Rev.*, **50**, 332-341 (1936).

[56] J. R. OPPENHEIMER and R. SERBER: *The density of nuclear levels*, *Phys. Rev.*, **50**, 391 (1936).

[57] N. BOHR and F. KALCKAR: *On the transmutation of atomic nuclei by impact of material particles*, *K. Dan. Vidensk. Selsk. Mat.-Fys. Medd.*, **14**, No. 10, 1-40 (1937).

[58] H. A. BETHE: *Energy production in stars*, *Phys. Rev.*, **55**, 434-456 (1939), received September 7, 1938.

[59] H. VON HELMHOLTZ: *Über die Wechselwirkung der Naturkräfte und die darauf bezüglichen neuesten Ermittelung der Physik*, in *Populär wissenschaftiche Vorträge*, Erstes Heft, 2nd edition (Braunschweig, 1876) (English translation: *On the interaction of natural forces*, in *Popular Scientific Lectures* (Dover, New York, N. Y., 1962), p. 59-92.)

[60] A. S. EDDINGTON: *The Internal Constitution of Stars* (Cambridge, 1926).

[61] G. GAMOW: *Zur Quantentheorie des Atomkerns*, *Z. Phys.*, **51**, 204-212 (1928); R. W. GURNEY and E. V. CONDON: *Wave mechanics and radioactive disintegration*, *Phys. Rev.*, **33**, 127-140 (1929).

[62] M. von Laue: *Notiz zur Quantentheorie des Atomkerns*, *Z. Phys.*, **52**, 726-734 (1928); J. Kudar: *Wellenmechanische Begründung der Nernstchen Hypotese von der Wiederenstehung radioaktiver Elemente*, *Z. Phys.*, **53**, 166-167 (1929).

[63] G. Gamow and F. G. Houtermans: *Zur Quantenmechanik des Radioactiven Kerns*, *Z. Phys.*, **52**, 496-509 (1928); R. d'E. Atkinson und F. G. Houtermans: *Zur Quantenmechanik der α-Strahlung*, *Z. Phys.*, **58**, 478-496 (1929).

[64] R. d'E. Atkinson and F. G. Houtermans: *Transmutation of light elements in stars*, *Nature (London)*, **123**, 567-568 (1929); *Zur Frage der Aufbaumöglicheit der Elemente in Sternen*, *Z. Phys.*, **54**, 656-665 (1929); R. d'E. Atkinson: *Atomic synthesis and stellar energy*, *Astrophys. J.*, **73**, 250-347 (1931).

[65] C. F. von Weizsäcker: *Über Element umwandlungen in Innern der Sterne* I, *Phys. Z.*, **38**, 176-191 (1937).

[66] H. A. Bethe and C. L. Critchfield: *The formation of deuterons by proton combination*, *Phys. Rev.*, **54**, 248-254 (1938).

[67] G. Gamow: *Nuclear energy sources and stellar evolution*, *Phys. Rev.*, **53**, 595-604 (1937).

[68] G. Gamow and E. Teller: *The rate of selective thermonuclear reactions*, *Phys. Rev.*, **53**, 608-609 (1937).

[69] J. H. Williams, W. G. Shepherd and R. O. Haxby: *Evidence of the Instability of* ^{5}He, *Phys. Rev.*, **51**, 888-889 (1937).

[70] H. A. Bethe: *Energy production in stars*, p. 135-150 of *Les Prix Nobel en 1967* (Imprimerie Royalle D. A. Nordstedt Söner, Stockholm, 1968).

[71] H. A. Bethe: *The meson theory of nuclear forces*, *Phys. Rev.*, **55**, 1261-1263 (1939).

[72] H. Yukawa: *On the interaction of elementary particles* I, *Proc. Phys.-Math. Soc. Jpn.*, **17**, 48-57 (1935).

[73] S. H. Neddermeyer and C. D. Anderson: *Note on the nature of cosmic ray particle*, *Phys. Rev.*, **51**, 884-886 (1937); J. C. Street and E. C. Stevenson: *Penetrating corpuscular component of the cosmic radiation*, *Phys. Rev.*, **51**, 1005 (1937); *New evidence of the existence of a particle of mass intermediate between the proton and the electron*, *Phys. Rev.*, **52**, 1003-1004 (1937).

[74] H. A. Bethe: *The meson theory of the nuclear forces* II, *Phys. Rev.*, **57**, 390-413 (1940).

Reprinted From
From Nuclei to Stars
© 1985, XCI Corso
Soc. Italiana di Fisica - Bologna - Italy

EDOARDO AMALDI

George Placzek

Estratto da: « LA RICERCA SCIENTIFICA »
Anno 26° - N. 7 - Luglio 1956

CONSIGLIO NAZIONALE DELLE RICERCHE
ROMA

S. p. A. Arti Grafiche Panetto & Petrelli – Spoleto

George Placzek

(1905-1955)

A Zurigo il 9 ottobre 1955 è deceduto George Placzek dopo alcuni mesi della grave malattia che aveva ripreso a tormentarlo poco dopo il suo arrivo in Europa, avvenuto nella tarda primavera dello stesso anno. Era venuto in Europa per passarvi un lungo periodo, ripromettendosi di trascorrere l'anno accademico 1955-'56 presso l'Istituto di Fisica della Università di Roma, grazie ad un assegno di ricerca della Fondazione Guggenheim.

Per quanto tutti gli amici sapessero che il suo stato di salute era malfermo già da varî anni, alla notizia del suo ricovero in una clinica di Zurigo essi avevano sperato in una ripresa, come già altre volte era accaduto nel passato. La notizia della sua scomparsa giunse così inaspettata, destando un profondo dolore in tutti coloro che lo conoscevano, in particolare nei fisici che avevano avuto maggiormente la possibilità di apprezzarne le doti scientifiche ed umane, e, sopratutto, nei fisici italiani a cui egli era legato, da anni, da vincoli profondi.

George Placzek, nato il 26 settembre 1905 a Bruenn, in Cecoslovacchia, aveva compiuto i suoi studi a Praga e a Vienna ove aveva ottenuto il dottorato in fisica nel 1928.

Da quel momento egli aveva iniziato un periodo di peregrinazioni per i grandi centri di studio di fisica in Europa: negli anni 1928-'31 è a Utrecht, ove studia con Kramers e si lega a lui di profonda amicizia. Nel 1931 è a Lipsia nell'Istituto di Deybe e Heisenberg e nel 1931-'32 a Roma, ove stabilisce quei legami scientifici ed umani con il gruppo di giovani che lavorava sotto la guida di Fermi, che lo dovevano riportare in questa università per periodi più o meno lunghi, ogni due o tre anni, per tutto il resto della sua vita.

Nel 1932-'33 e nel 1935-'38 egli lavora presso l'Istituto di fisica teorica di N. Bohr a Copenhagen; nel 1934-'35 è nominato professore di fisica teorica presso l'Università di Gerusalemme e nel 1933-'34 e 1935-'36 professore presso l'Università di Karcov ove lavora con Landau.

Dopo un periodo trascorso, nel 1938, al Collège de France di Parigi, Placzek si trasferisce negli Stati Uniti, ove lo troviamo professore alla Cornell'University di Ithaca (N. Y.) dal 1939 al 1942.

Scoppiata la guerra, va a far parte, con una importante carica direttiva, della Divisione teorica del Gruppo di ricerca di Fisica Nucleare Applicata, organizzato dall'Autorità Britannica a Montreal nel Canadà, ove rimane fino al 1945, anno in cui si trasferisce ai laboratori di Los Alamos negli Stati Uniti. Finita la guerra passa un periodo di circa due anni presso i laboratori della General Electric a Schenectady che lascia poi nel 1948, essendo stato chiamato dall'Institute of Advanced Studies di Princeton, ove doveva rimanere fino alla morte.

I suoi contributi scientifici si riferiscono a molti problemi che vanno da questioni di fisica molecolare a problemi di fisica nucleare applicata e fondamentale.

Un gruppo di lavori, svolti nel periodo che va dal 1929 al 1934, riguarda la teoria quantistica dell'effetto Raman. Egli fu il primo a studiare in maniera sistematica le relazioni che intercorrono fra la diffusione della luce da parte di una molecola e le sue proprietà di simmetria, riuscendo a formulare una elegante teoria generale che costituisce la base di tutti gli sviluppi successivi in questo importante campo di ricerca. Tale teoria è stata da lui raccolta in un articolo fondamentale intitolato « Rayleigh Streuung und Raman Effekt », pubblicato nella seconda edizione dell'Handbuch der Radiologie [15]. L'impiego di questi metodi generali gli permise di risolvere alcuni importanti problemi, come la diffusione della luce da parte di vapori al punto critico, liquidi e cristalli. Inoltre riuscì così ad acquistare, fin dagli anni giovanili, una padronanza del tutto eccezionale dei metodi di indagine della diffusione di radiazioni e corpuscoli da parte di diversi sistemi, padronanza che doveva fare di lui uno dei maggiori esperti della nostra epoca in questo campo.

I suoi contatti frequenti con i gruppi di Roma e di Copenhagen attrassero la sua attenzione, attorno al 1934-'35, verso la fisica nucleare. Durante il suo soggiorno a Copenhagen, nel 1936, pochi giorni dopo la pubblicazione da parte di N. Bohr del suo modello di nucleo, ideato per rendere conto delle risonanze recentemente individuate nell'assorbimento dei neutroni lenti, Placzek pubblica, insieme a Frisch [16], una lettera all'Editore di Nature in cui vengono presentate alcune argomentazioni che portano ad attribuire alla sezione d'urto di cattura dei neutroni lenti da parte del boro e del litio il ben noto andamento di proporzionalità inversa con la velocità. Tale andamento, noto come legge 1/v, costituì a quell'epoca un termine di paragone di grande importanza nella soluzione di numerosi problemi della allora nascente « spettroscopia dei neutroni lenti » e fu, nel giro di alcuni anni, verificato sperimentalmente con grande precisione grazie all'impiego di raffinati spettrometri.

L'anno successivo, insieme a Bethe, pubblica un lavoro fondamentale sugli effetti di risonanza nei processi nucleari [17] nel quale le idee di Bohr e il formalismo impiegato da Breit e Wigner per dedurre la loro « formula a un sol livello », vengono fusi ed elaborati in una teoria generale che rappresenta ancor oggi il primo grande passo verso una trattazione teorica generale dei processi nucleari.

Nel 1939, insieme a Bohr e a Peierls, pubblica su Nature [18] una breve lettera all'Editore sul cosidetto « teorema ottico » che regola l'assorbimento e la diffusione dei neutroni veloci da parte dei nuclei, problema su cui gli stessi autori stesero una più adeguata trattazione che fu fatta circolare privatamente in una ristretta cerchia di amici, ma mai pubblicata a causa delle circostanze belliche che avevano separato gli autori.

Un vasto e importante gruppo di ricerche riguarda il rallentamento dei neutroni che egli studiò sia prima della guerra, per quanto riguarda i neutroni presenti nella radiazione cosmica [19], che durante e dopo la guerra in relazione con la sua attività rivolta al calcolo del moderatore [20, 21, 22, 23, 24, 26] e della moltiplicazione del neutroni [25] nei reattori nucleari. Questi lavori, svolti in parte in collaborazione con giovani allievi, hanno spesso la caratteristica di messa a punto di questioni difficili e delicate su cui altri autori avevano precedentemente sorvolato.

Parte almeno dei risultati in essi contenuti sono raccolti nel volume pubblicato in collaborazione con Case e de Hoffmann sotto il titolo « Introduction to the theory of neutron diffusion. Vol. I » [31] il quale è basato sui corsi di lezioni tenuti da Placzek nell'estate 1949 a Santa Monica e Los Angeles in California.

Dopo la guerra egli raffina ulteriormente l'analisi dei processi di diffusione elastica ed anelastica da parte di liquidi e di cristalli e questo sia in connessione con il problema di stabilire l'esistenza ed entità della interazione neutrone-elettrone, in base ai risultati di esperienze di diffusione dei neutroni [29], sia allo scopo di sviluppare metodi adeguati per l'analisi della struttura dei liquidi e dei cristalli medesimi [27, 32, 33, 34].

In questo importante ultimo gruppo di lavori ebbe la collaborazione di Van Hove il quale può essere considerato come un suo allievo in questo campo e certo come l'ultimo suo grande amico.

Da questo breve esame dei lavori pubblicati da Placzek risulta chiaramente l'importanza del suo contributo allo sviluppo di varî, fondamentali capitoli della fisica contemporanea. I lavori pubblicati però non sono sufficienti a dare una piena visione dello scienziato. Mentre era sempre pronto a studiare a fondo i problemi cercando di sviscerarne completamente la natura, Placzek era pigro a scrivere i risultati ottenuti. La redazione di un lavoro era per lui uno sforzo enorme, tanto che spesso nella sua vita finì col non pubblicare risultati anche importanti da lui chiaramente stabiliti in forma definitiva. Così per esempio egli pubblicò solo una parte [14] del lavoro svolto insieme a Landau nel periodo trascorso a Karcov.

Questo aspetto del suo carattere era dovuto in buona parte al suo desiderio di approfondire sempre più l'analisi dei fenomeni, desiderio che si rifletteva spesso in un senso di insoddisfazione di fronte ai risultati già conseguiti o alla forma in cui egli stesso o i suoi collaboratori riuscivano a formularli. Dotato di uno spirito critico vivacissimo era pronto nel comprendere i problemi e nel metterne in evidenza tutti gli aspetti, doti queste che, accoppiate alla sua disinteressata generosità, facevano di lui un consigliere prezioso, anzi insuperabile.

Queste caratteristiche di Placzek scienziato erano però un semplice riflesso delle sue qualità di uomo. La sua cultura generale, al di fuori del campo della fisica, era anormalmente vasta e profonda e ancorata a solide basi classiche. Non soltanto parlava e scriveva correntemente molte lingue, dal russo allo spagnolo, dal danese all'italiano, dall'olandese al francese, ma conosceva la storia dei rispettivi paesi di cui aveva letto, in testo originale, la letteratura classica e moderna. Una volta trasferitosi negli Stati Uniti, aveva, per esempio, approfondito in maniera non comune la storia del suo nuovo paese d'adozione, tanto da conoscerne particolari noti usualmente solo a professionisti.

Ma la sua curiosità di capire e conoscere sempre fatti nuovi, grandi e piccoli della vita umana facevano di lui un individuo tipicamente europeo. Sembrava quasi che avesse assorbito, durante la sua prima giovinezza, doti apparentemente diverse e contradditorie possedute dai varî popoli che allora costituivano l'impero Austroungarico.

La tragica distruzione della sua famiglia, avvenuta ad opera dei Nazisti durante la guerra, e le vicende politiche che il suo Paese aveva in seguito subito, lo avevano in qualche modo allontanato dal suo ambiente di origine, di cui parlava con gli amici intimi solo se interrogato, ma non lo avevano allontanato dall'Europa per cui negli ultimi anni auspicava una ripresa economica e culturale. E quando parlava dell'Europa si riferiva sopratutto ai piccoli Paesi come la Danimarca o l'Olanda ove era stato da giovane e a cui era legato anche attraverso la moglie Els olandese. Ma sopratutto si riferiva all'Italia a cui era così affezionato da non limitarsi ad ammirarne le virtù ma da comprenderne benevolmente i difetti. Conosceva la nostra letteratura e la nostra storia come ben pochi italiani, canticchiava le canzoni popolari delle varie regioni e raccontava le storielle studentesche con gusto fine ed acuto. Dopo la guerra si era grandemente adoperato in tutti i modi per agevolare la ripresa della fisica in Italia, appoggiando i giovani che desideravano recarsi negli Stati Uniti con borse di studio o cercando egli stesso di insegnare quanto sapeva in modo così profondo. Il suo penultimo viaggio in Europa era stato nella primavera ed estate 1953 e lo aveva dedicato a svolgere brevi corsi di lezioni sul rallentamento e la diffusione dei neutroni presso le Università di Roma e Milano. Aveva poi trascorso insieme ad amici un breve periodo di vacanze a Riccione e a Venezia, periodo che sia lui che la moglie ricordavano spesso con gioia.

Ma desiderava stare in Italia più a lungo e così si era interessato per avere un assegno Guggenheim che permettesse ad Els ed a lui di vivere e lavorare nel nostro Paese per molti mesi, l'anno accademico 1955-'56. Roma li attraeva moltissimo non solo per i vecchi amici che vi ritrovavano ma per il suo valore storico ed artistico che entrambi sapevano apprezzare giustamente, sia pure da diversi punti di vista: George conosceva le gallerie d'arte e gli angoli pittoreschi della città, la vita dei quartieri moderni e i libri antichi e rari che sapeva scoprire nelle biblioteche della città o in quella Vaticana.

Era da anni membro della Società Italiana di Fisica alla cui vita e sviluppo si interessava; si rallegrava profondamente quando qualche amico diceva di considerarlo un fisico italiano.

È anche per questo e sotto questo aspetto che oggi vogliamo ricordarne il nome e la figura con profondo affetto.

EDOARDO AMALDI

BIBLIOGRAFIA *

[1] *Ponderomotorische Wirkungen des Lichtes auf Gelandene Submikroskopische Körper im Elektrischen Felde*, « Zs. f. Phys. », **49**, 601, 1928.

[2] *Zur Theorie des Ramaneffektes*, « Zs. f. Phys. », **58**, 585, 1929.

[3] *Zur Dichten-und Gestaltbestimmung Submikroskopischer Probekörper*, « Zs, f. Phys », **55**, 81, 1929.

[4] *Ueber die Lichtzerstreuung beim kritischen Punkt*, « Phys. Zeits », **31**, 1052, 1930.

[5] *Ueber den Ramaneffekt beim kritischen Punkt.*, Proc. Amsterdam, **33**, 832, 1930.

[6] *Polarisations messungen am Ramaneffekt von Flussigkeiten*, (with W. R. van Wyk), « Zs. f. Phys. », **67**, 582, 1931.

[7] *Intensitat und Polarisation der Ramanschen Streustrahlung mehratomiger Molekule*, « Zs, f. Phys », **70**, 84, 1931.

[8] *Contribution to Leipziger Vortrage*, 1931, page 71.

[9] *Ueber das kontinuierliche Ramanspektrum und sein verhalten beim Kritischen punkt.* (with W. R. van Wyk), « Zs. f. Phys », **70**, 287-292, 1931.

[10] *Evidence for the spin of the photon from light scattering*, « Nature », **128**, 410, 1931.

[11] *Ramaneffekt des gasförmigen ammoniaks*, (with E. Amaldi), « Naturwissenschaften », **20**, 521, 1932.

[12] *Ueber das Ramaspektrum des Gasförmigen Ammoniaks*, (with E. Amaldi) « Zs. f. Phys. », **81**, 259, 1933.

[13] *Die Rotationsstruktur der Ramanbanden mehratomiger Moleküle*, (with E. Tell), « Zs. f. Phys. », **81**, 209, 1933.

[14] *Struktur der unverschobenen streulinie*, (with L. Landau), « Phys. Zeits der Sowjetunion », **5**, 172-173, 1934.

[15] *Rayleigh Streeung und Raman Effekt*, article in « Hdb. der Radiologie » 1934, Vol, VI, 2, p. 209.

[16] *Capture of Slow Nautrons* (with O. Frisch) « Nature », **137**, 357, 1936.

[17] *Resonance effects in nuclear processe* (with H. Bethe) « Phys. Rev. », **51**, 450, 1937.

[18] *Nuclear Reactions in the Continuous Energy Region*, (with N. Bohr and R. Peierls), « Nature », **144**, 200, 1939.

[19] *Interpretation of neutron measurements in cosmic radiation* (with H. Bethe and S. A. Korff», « Phys. Rev », **57**, 573, 1940.

[20] *On the Theory of Sloving down of Neutrons in Heavy Substances*, « Phys. Rev. », **69**, 423, 1946.

[21] *The spatial distribution of neutrons slowed down by elastic collisions*, « U. S. Atomic Energy Commission », MCDD-2 66 pag., 1946.

[22] *The concept of albedo in elementary diffusion theory* (MIT nuclear Science and Engineering Seminar 39) « U. S. Atomic Energy Commission », NP-160, 16 pages, 1947.

[23] *Milne's Problem in Transport Theory* (with W. Seidel), « Phys. Rev. », **72**, 550, 1947.

[24] *Angular Distribution of Neutrons Emerging from a Flane Surface.* « Phys. Rev. », **72**, 556, 1947.

[25] *A theorem on neutron multiplication* (with G. Volkoff), « Canad. J. Res. », A, **25**, 276, 1947.

[26] *Theory of Slow Neutron Scattering*, « Phys. Rev. », **75**, 1295, 1949.

[27] *Effect of Short Wavelenght Interference on Neutron Scattering by Dense Systems of Heavy Nuclei* (with B. Nijboer and L. Van Hove), « Phys. Rev. », **82**, 392, 1951.

[28] *Correlation of Position for the Ideal Quantum Gas. Proceedings of the Second Berkeley Symposium on Mathematical Statistics and Probability*, 1950: 581-588 (1951).

[29] *The scattering of neutrons by Systems of Heavy Nuclei*, « Phys. Rev. », **86**, 377-388 (1952).

[30] *Scattering of X-Rays by Atoms*, « Phys. Rev. », (2) **86**, 588, 1952.

[31] *Introduction to the Theory of Neutron Diffusion*, vol I, (by) K. M. Case, F. de Hoffmann (and) G. Placzek), « Los Alamos, Los Alamos Scientific Laboratory », 1953, vii, 174, p.

[32] *Crystal Dynamics and Inelastic Scattering of Neutrons* (with L. C. P. Van Hove), « Phys. Rev. », (2), 93: 1207-1214, 1954.

[33] *Incoherent Neutron Scattering by Polycrystals*, « Phys. Rev. », (2) 93, 895-896, 1954.

[34] *Interference Effects in the Total Neutron Scattering Cross-Section of Crystals* (with L. C. P. Van Hove), « Il Nuovo Cimento », (1), 233-256, 1955.

* Questa bibliografia non pretende di essere completa ma certamente contiene i lavori più importanti di George Placzek.

George Placzek
(1905-1955)

George Placzek died in Zurich on October 9, 1955, after several months in which the serious illness that had plagued him since his return to Europe in late spring of that year intensified. He had intended to stay in Europe for some time, with a research grant from the Guggenheim Foundation for a year (1955-56) at the Physics Institute of the University of Rome.

Since his health had already been poor for some years, his friends hoped, on hearing that he had been admitted to a clinic in Zurich, that he would recover as he had other times. Thus, news of his death was unexpected and caused the deepest sorrow in all those who knew him, particularly the physicists who had had the opportunity to value his scientific and human qualities and, above all, the Italian physicists with whom he had very close ties.

George Placzek was born in Bruenn, Czechoslovakia on September 26, 1905. He studied in Paris and Vienna, where he received his degree in physics in 1928.

That was when he started his wanderings from one important European center of physics to another: from 1928 to 1931, he was in Utrecht, where he studied under Kramers and became a good friend. In 1931, he moved to the institute of Debye and Heisenberg in Leipzig, and in 1931-32 he moved on to Rome. There he established the scientific contacts and friendships with the group of young physicists working under Fermi that were to bring him back to that university every two or three years for the rest of his life.

In 1932-33 and from 1935 to 1938, he worked at N. Bohr's Institute of Theoretical Physics in Copenhagen; in 1934-1935, he was named professor of theoretical physics at the University of Jerusalem and in 1933-34 and 1935-36, he was professor at the University of Kharkov, where he worked with Landau.

After a time at the Collège de France in Paris in 1938, Placzek moved to the United States, where he received a position as professor at Cornell University at Ithaca (NY) from 1939 to 1942.

When the war broke out, he took on an important executive job with the Theoretical Division of the applied nuclear physics research group organized by the British Authorities in Montreal, Canada. He remained there until 1945, the year in which he left for Los Alamos in the US. After the war, he spent almost two years at the General Electric laboratories in Schenectady, New York, and in 1948, was called to the Institute of Advanced Studies at Princeton, where he remained until his death.

His scientific contributions deal with a multitude of problems ranging from molecular physics to applied and fundamental nuclear physics.

One set of works written between 1929 and 1934 refers to the quantum theory of the Raman effect. He was the first to carry out a systematic study of the relations between the light diffusion of a molecule and its symmetrical properties, and formulated an elegant general theory which formed the basis for all further developments in this important field of research. The theory was set down in a fundamental article entitled "Rayleigh Streuung und Raman Effekt", published in the second edition of the *Handbuch der Radiologie* [15]. Use of these general methods allowed him to solve some important problems such as the diffusion of light by vapours at the critical point, by liquids and by crystals. Furthermore, it gave him an early mastery of the methods of investigation of the scattering of radiation and corpuscles by different systems which was to make him one of the greatest experts of his time in that field.

Around 1934-35, his frequent contacts with the groups in Rome and Copenhagen drew his attention to nuclear physics. During his stay in Copenhagen in 1936, only a few days after Bohr's publication of his model of the nucleus, developed to explain the resonances recently detected in the absorption of slow neutrons, Placzek published a letter to the editor of Nature together with Frisch [16]arguing that the well known trend of inverse proportionality to velocity could be attributed to the capture cross-section of slow neutrons of boron and lithium. That trend, known as the 1/v law, was at that time an important point of reference in solving numerous problems linked with the then nascent "spectroscopy of slow neutrons" and was experimentally verified in only a few years, thanks to the development of sophisticated spectrometers.

The following year, Placzek published a fundamental paper on the effects of resonance in nuclear processes [17] that combined and elaborated on Bohr's ideas and the formalism used by Breit and Wigner in deducing their "one level formula" to produce a general theory which still represents the first major step towards a general theoretical treatise of nuclear processes.

In 1939, Placzek wrote a brief letter to the editor of Nature together with Bohr and Peierls [18] on the so-called "optical theorem" regulating the absorption and the scattering of fast neutrons by nuclei. The three later wrote a more complete treatment of the subject which was circulated privately in a close circle of friends but never published because of the war and the authors' separation.

A vast and important group of studies on neutron slowdown were carried out both before the war (neutrons present in cosmic radiation)[19] and during and after the war (calculation of the moderator,[20, 21, 22, 23, 24, 26] and the multiplication of neutrons in nuclear reactors)[25]. These studies were frequently carried out with young students and often involved providing answers to delicate and difficult points that other authors had overlooked.

A part of the results are contained in the book published in collaboration with Case and de Hoffmann entitled *Introduction to the Theory of Neutron Diffusion, Vol. I*,[31] based on a course of lectures given by Placzek in summer 1949 at Santa Monica and Los Angeles, California.

After the war, Placzek further refined his analysis of processes of elastic and inelastic scattering in liquids and crystals both in connection with the problem of establishing the existence and the magnitude of neutron-electron interactions on the basis of experimental results of neutron scattering[29] and in order to develop adequate methods for analysis of the structure of those liquids and crystals.[27, 32, 33, 34] He was assisted in his work by Van Hove, who may be considered his pupil in this field and certainly his best friend in his later life.

This brief overview of Placzek's published works reveals the importance of his contribution to the development of various fundamental areas of contemporary physics. His published works are not sufficient, however, to provide a complete picture of the scientist. While always ready to delve into the details of a problem to uncover its most profound nature, he neglected writing up results. Drafting papers was such a tedious job that he often failed to publish important results that he had established in a definitive manner. For example, he published only a part[14] of the work done with Landau during his stay in Kharkov.

This side of his character was largely the result of his desire to explore phenomena more deeply, a desire which often manifested itself in a sense of dissatisfaction with results already obtained and the form in which he or his collaborators managed to formulate them. His sharp critical sense allowed him to understand problems readily and to highlight their many aspects, both talents which, combined with his disinterested generosity, made him a precious, I might say unequalled advisor.

These qualities of Placzek the scientist were, however, mere reflections of his qualities as a man. His culture outside of the field of physics was vast and profound and based solidly on classic foundations. He not only spoke and wrote a number of languages fluently, from Russian to Spanish, Danish, Italian, Dutch and French, he also knew the history of each of those countries, and had read their modern and classic literature in the original. After moving to the United States, he studied the history of his newly adopted country so thoroughly that he knew details that were common knowledge only among professionals in the field. But his thirst for knowledge and understanding of the facts--both major and minor--of human life were typically European. It almost seemed as though he had absorbed the apparently different and contradictory talents of the various peoples making up the Austro-Hungarian Empire during his childhood.

The tragic destruction of his family by the Nazis during the war and the subsequent political vicissitudes in his country somehow detached him from his homeland, of which he spoke only with his most intimate friends when directly questioned. But it did not detach him from Europe, which he hoped would witness an economic and cultural recovery. When he spoke of Europe, he referred mainly to the small countries like Denmark and Holland where he had lived as a child and to which he had ties through his Dutch wife Els. But above all he referred to Italy, of which he was so fond that he not only admired its virtues, but was able to have a benevolent understanding of its defects. He knew Italian history and literature as few Italians do, he sang folk songs of the various regions and recounted anecdotes with great wit. After the war, he made every effort to facilitate the recovery of physics in Italy, by helping students who wished to study in the United States with scholarships or offering them personal contact with his vast knowledge. He dedicated his last trip to Europe in the spring and summer of 1953 to a brief course of lectures on neutron deceleration and scattering at the Universities of Rome and Milan and then took a short vacation with friends in Riccione and Venice, a period that he and his wife later often recalled with pleasure.

Since he wished to stay in Italy longer, he had looked into the possibility of a Guggenheim grant allowing him and Els to live and work in our country for a few months during the 1955-56 academic year. But friends were not the only attraction that drew the two to Rome. They both loved it for its historic and artistic treasures which they appreciated from various points of view. George knew all the art galleries and picturesque corners of the city, but he was also acquainted with life in the modern neighbourhoods and the ancient and rare books that he discovered in the libraries of the city and the Vatican.

George Placzek was an active member of the Società Italiana di Fisica and was extremely pleased when friends called him an Italian physicist. It is also for these reasons and in this light that we would like to remember him with extreme affection today.

Edoardo Amaldi

CERN 81-19
23 décembre 1981

ORGANISATION EUROPÉENNE POUR LA RECHERCHE NUCLÉAIRE
CERN EUROPEAN ORGANIZATION FOR NUCLEAR RESEARCH

THE BRUNO TOUSCHEK LEGACY

(Vienna 1921 – Innsbruck 1978)

Edoardo Amaldi
University of Rome, Italy

GENEVA
1981

ABSTRACT

A biographical portrait of Bruno Touschek, an Austrian physicist who, before and during the Second World War, went through the dramatic adventures of a non-pure arian young person. Later he worked in Germany, Great Britain and Italy.

Touschek was the first to propose chiral symmetry. He was the initiator and main driving force in the early developments of e^+e^- colliding machines. His wide scientific culture, his ingenuity and enthusiasm for any new challenging problem and the search of its solution were essential elements in determining the extraordinary influence Bruno had on the work of his younger colleagues. "He led an intense and vigorous life and by his example and friendliness helped many colleagues and friends to achieve greater happiness and awareness in their own lives".

PU-TP/jmr-pe-msv-hm

CONTENTS

1. THE DEPARTURE OF BRUNO TOUSCHEK

On 25 May 1978, Bruno Touschek died in the Medical Ward of the University Hospital, Innsbruck, as a result of the last of a series of hepatic comas. He had been suffering from this illness for several years. He had had it in a serious form since February 1977, when he was taken to the Medical Ward II of the Policlinico of the University of Rome. This "dramatic collapse" as he wrote to me on 29 March 1977, a few days after he had returned home, "is somewhat providential as it has convinced me more than any preaching to put an end to this childish alcoholism, which has led me to my climacteric, and has made me realize that my Bursche[1] days are over". The doctors had already explained, however, that not only his liver but also his kidneys were in a bad condition, so that it was not possible to carry out any major clinical or surgical treatment.

After a fresh collapse which had caused an ever greater irritability towards his family—an irritability characteristic of his illness—he was taken at the beginning of July 1977 to the Medical Ward I of the Policlinico, where he remained for practically the whole summer. In the meanwhile, he had been appointed "Senior Visiting Scientist" at CERN for a year from the autumn of 1977. As soon as he was fit to face the journey, he moved to Geneva at the beginning of October.

Another attack in the middle of November forced him to enter the Cantonal Hospital of Geneva. A little later he was transferred to the Hospital of La Tour, near the CERN Laboratories in Meyrin. During his whole period in Geneva, Touschek's various friends and colleagues took an interest in his health, in particular Giorgio Salvini, who had gone to Geneva at the end of August 1977 on a year's sabbatical leave from the University of Rome. CERN's Director-General for Research, Leon Van Hove, himself took an interest in Touschek's problems and asked the CERN doctors to check on his state of health.

The nearness of the Meyrin laboratories made it easier for him to receive visits from many physicists, his friends, of different nationalities—above all Italians, Germans and Austrians—so that every day he saw different people with whom he discussed his state of health, his family problems and—above all—physics.

He always put forward, in an original and unexpected way, a point of view of substantial value on matters not yet sufficiently clarified. As he had always done, he read a great deal of literature and, in particular, history, and he had a great interest in figurative arts.

Nevertheless, Bruno was not happy in the La Tour Hospital, chiefly because the staff spoke French, and he felt that he did not know it well enough. It is true that he did not speak French so well as English and Italian, which he spoke fluently, using precise expressions—even if they were sometimes unusual or betrayed in their origin a Viennese mentality. But he understood everything and could express any thought in this language too. He probably succeeded in doing this by a much greater effort of concentration than he needed in order to express the same needs or thoughts in German.

One day, during his stay at the La Tour Hospital, he picked up the phone next to his bed, with which he communicated regularly with his wife and younger son in Rome or the elder one in London, and managed to book a room in the Sport Hotel in Igls, 5–6 km from Innsbruck.

When I went to see him in the La Tour Hospital on 27 and 28 February, he spoke to me with a certain amount of enthusiasm about this plan of his and of his success in arranging to be transferred on 8 March 1978 from Geneva to Innsbruck by a car, put at his disposal by CERN.

In a letter to me dated 2 May 1978 he spoke of the "very comfortable journey" and said that the hotel was without doubt the best that one could find, "with the staff always smiling, excellent food, and a 20-metre indoor swimming pool at 28°, with also a doctor in attendance." In the same letter he also wrote of "the Alps in Springtime" and of having "enjoyed the balcony, from which one can see the Patscherkofel."[2]

In a note written the day before, he gave me news of his "state of health", saying that between 25 February and 1 May he had been in the Nursing Home five times (the first time with a stomach haemorrhage) and had had six attacks of hepatic coma. He discussed in a detached way—I would say almost humorous, had his situation not been so tragic— the possible causes of "hepatic coma" and its "immediate effects". He had been transferred from the Sport Hotel to the Psychiatric Hospital in Innsbruck, and then to the Medical Ward of the University more than once until the final crisis in which he died. Because of a railway strike in Italy his wife Elspeth did not arrive from Rome until the day after his death.

In his letter of 1 May, Bruno wrote to me: "I can write—still badly—and can read; I am still a little weak even for short walks in the village.... I do everything at a snail's pace. So far, the only unwise action has been to hold a seminar in Innsbruck... Cap and Rothleitner, (dean)[3] took care of me with Ernst[4], Valentino[5], etc., while I was in hospital. Everything went well, except that after 30 minutes I had to sit down".

As Rothleitner wrote to me, when together with Cap he visited Bruno for the first time in the clinic in Innsbruck:

"...he was very weak but still had a strong will to live. He told us: 'I have been in coma and I have forgotten everything. I should start again from the beginning: I should again learn to speak, I should learn everything again'. He kept on his table a heft, in which he noted the important thoughts as soon as he found them again. On top of the first page I read: 'Cogito ergo sum'.

"In spite of his weakness, he expressed the desire to give a seminar and offered a number of themes...."

The news of his death was a very severe blow to all of his friends, especially the physicists at the University of Rome and the Frascati Laboratories (Laboratori Nazionali di Frascati del CNEN). The announcement was made by his friends on 31 May 1978 in Information Bulletin No. 4, which, in just over a page, attempts to portray the personality and achievements of Bruno Touschek[6]:

"He was one of the few physicists able to speak authoritatively in an extremely wide range of physics fields, from elementary particles to statistical mechanics and accelerators. His intelligence was that of a genius, and was inexhaustible....

"His reflections were to cross the destiny of our laboratories in 1960, when Touschek started to give thought to the possibility of building storage rings for electrons and positrons....

"Bruno Touschek was not only extremely intelligent, but was also endowed with enormous drive and enthusiasm. It so happened that the young scientific environment at the (Frascati) Laboratories was wisely receptive to this proposal, and in the space of a year AdA was constructed. The prime contributor to the designing and construction of this project was our much regretted colleague Giorgio Ghigo.

. . . .

"It is precisely for this reason that his memory will remain alive throughout the future existence of the Laboratories."

Various articles which appeared during the following days in the daily press stressed his gift as a designer[7] and his middle-European culture, which incorporated a certain Anglo-Saxon empiricism[8].

At 3.30 p.m. on 7 July 1978 an official tribute was paid to Bruno Touschek in the main hall of the National Laboratories of the CNEN at Frascati, by Carlo Bernardini and Giorgio Salvini, who gave an address after a short introduction by Renato Scrimaglio, Director of the Laboratories.

The ceremony, which took place in a hall packed with physicists of all ages from many Italian universities, was concluded with the unveiling of a memorial erected in the centre of the area where ADONE is situated: the AdA magnet, bearing the inscription "Bruno Touschek 1921–1978".

An account of this ceremony, published in the *Corriere della Sera* of 22 July 1978 and entitled "Who was the man of No. 137" recalled Bruno Touschek's great interest in the scientific investigation of fundamental problems. One such example is the value of the electric charge of electrons and protons, the square of which, expressed in adimensional form, is the reciprocal of 137. "Because the real problem is the number of this room", Bruno suddenly remarked to a young friend during a visit he received at the La Tour Hospital, where he occupied room No. 137. "This is the problem around which I have hovered throughout my life, without success."[9]

2. HIS YOUTH[10)]

Bruno Touschek was born in Vienna on 3 February 1921, and was the son of Franz Xaver Touschek, a Staff Officer in the Austrian Army, who had fought on the Italian front in the First World War, and of Camilla Weltmann.

Owing to a very serious form of Spanish flu, contracted during the epidemic that struck all Europe in 1918, his mother remained in very poor health, so that Bruno always saw her in bed or, at best, lying on a sofa.

Bruno's father had left the Army and entered the reserve at the age of 31, with the rank of Major, when in 1932 the power in Austria was taken over by Dollfuss' Christian-Socialists, in reality Clerico-Fascists, who tried to gain support from Mussolini's Italy and Horthy's Hungary against the threat of Germany's annexation of Austria. Franz Touschek found a job in an employment agency and in this way established contacts with many building and industrial firms.

When Hitler took over control of Germany on 30 January 1933[11)], the German pressure on Austria was strongly increased. On 19 June 1933, Dollfuss had succeeded in declaring the Austrian National-Socialist Party illegal, but, on 25 July 1934, a group of 154 members of this party, wearing uniforms, burst into the Federal Chancellor's Office in Vienna, and murdered Dollfuss.

The international reaction, in particular that of Mussolini's Government, which immediately sent four Divisions to the Brenner Pass, prevented the immediate annexation of Austria by Germany. Dollfuss' successor was his party-companion Kurt Schuschnigg, who tried to save Austria's independence by following a policy of detente with Hitler's Germany. But the Austro-German agreement, signed on 11 July 1936, contained concessions that spelt disaster for Schuschnigg and his country. The Austrian National-Socialist Party was reconstituted, with a strong renewal of antisemitism, the roots of which in Austria dated back to the years 1880–1890[12)].

Bruno had attended school in Vienna and, at the beginning of the summer of 1937, he had completed the 8th class of the Piaristen Gymnasium, that is a year before the Abitur (state examination), when he was told that he could no longer attend school because he was of mixed blood, as his mother was Jewish.

He stayed away from school but he had many friends that he met in cafés, and this kept him in touch with what was happening. With the beginning of war in sight, the Austrian High Command was already making preparations. In particular, the Abitur examination of 1938 was brought forward to February, so that a large number of young men would be immediately available as junior officers. A friend who attended another school suggested that he sat the exam at a different school as an external student without making any mention of his real position. Bruno took his advice and sent in his application to the Director of Education for the Schottengymnasium. He was allowed to take the exam and passed it very well in all subjects except Greek, in which he was declared, at first, "nicht genügend" (insufficient), but the decision was later changed to "genügend", as an order had arrived to pass large numbers of young men who would soon be needed as reserve officers.

Thus he passed his state examination and in February 1938 he went to Rome for the "school-leaving holiday" according to the tradition of the bourgeoisie of that period.

Around the end of that same month, Vienna entered a period when the Schuschnigg government was engaged in a death-struggle, which ended on 13 March 1938 with the proclamation of the "Anschluss" of Austria by Hitler's Germany. This occurred without Great Britain and France taking any measure, and with the consent of Mussolini who, first with the Abyssinian war (1935–36) and later through participation in the Spanish civil war (1936–39), had once and for all espoused Hitler's cause.

Bruno had thought of studying engineering in Rome and so he began to attend the first two-year course in engineering in the spring of 1938. He attended, in particular, Francesco Severi's course on "Mathematical analysis". In the meanwhile, however, he had applied for a visa to enter Great Britain in order to study Chemistry in Manchester. He was told that this could be obtained through an organization established in Vienna and run by the Quakers, who were very active in that period, as in other dramatic circumstances, trying to save people persecuted for political or racial reasons. Towards the summer of that year he returned to Vienna and not long after went on holiday with his family, namely his father and his second wife, Rosa Reichel; Bruno's mother had died in 1931.

Following the Hitler-Stalin Pact, at the beginning of September 1939 the Second World War broke out, when the Russian and German armies entered Poland from opposing fronts.

Franz Xaver Touschek was invited to re-enter active service, in order to collaborate with the German authorities in maintaining good relations with the groups he had worked with during the previous few years. He declined the invitation, however, and Bruno, even years later, was clearly pleased at his father's refusal. I realized, however, from conversations I had with him many years later, that at the height of the racial persecutions, Bruno suffered at the thought that he was a burden on his father, and that he was unintentionally damaging his career as well as his private life, as a living testimony of his first marriage to a Jewish girl.

With the war raging in Europe, every possibility of going to study in Great Britain had vanished. As a result Bruno remained in Vienna and started to attend the University courses in physics and mathematics, trying to avoid attracting attention. By the third term he was clearly the best in his course, so much so that he gave a talk at the preseminar (or seminar for students) on the Markov double sum series, prepared in verse. But in June 1940 he received a notice that he could no longer attend the University for racial reasons.

Luckily, some time earlier he had, with the help of Paul Urban[13], studied the first volume of the famous treatise *Atombau und Spektrallinien* written by Arnold Sommerfeld[14], then Professor at Munich University. Touschek had spotted a few minor errors and, encouraged by Edmund Hlawka[15], wrote to Sommerfeld. Sommerfeld replied asking Touschek to read also the second volume of the same treatise which, at that time, was one of the best of its kind for both clarity and mathematical rigour. In the preface of the second edition of this second volume, Sommerfeld thanked Bruno Touschek for his critical review of the text.

When Bruno was expelled from the University of Vienna, Urban endeavoured, as I will tell later, to obtain the support of Sommerfeld. Sommerfeld wrote a letter of introduction to Paul Harteck[16], who was teaching in Hamburg, and Bruno moved to that town, where nobody knew of the "racial imperfection" of the young Austrian. That of course, was not true for Sommerfeld nor for Harteck or a few other professors who were perfectly well aware of it.

Harteck was a chemical physicist well known for his work on the production of heavy water (1934), on the chemistry of deuterium compounds (1937–38), and on artificial radioactivity and neutron physics (since 1938). He welcomed Touschek and advised him on how to behave when approaching the other professors, particularly one of them, Professor P.P. Koch[17], who had developed a high-sensitivity method for the photometric analysis of X-ray plates. Following Harteck's advice, Bruno studied this method in great detail. When, therefore, Koch questioned him during his admission examination, Touschek was in a position to answer in a very competent way, to the great satisfaction of his examiner.

In order to keep himself Bruno was forced to work; in fact, he had to do several jobs simultaneously. There were periods when he had to do four or five jobs at the same time. In addition, he did not have a fixed residence, but frequently moved so that he could not be easily found. In Hamburg he worked for a long time for the Studiengesellschaft für Elektronengeräte, an industry affiliated to the Dutch firm Philips, where "drift tubes", forerunners of the klystron (tubes in which the transit time is the same for all electrons), were being developed. This was a very important problem at that time for high-frequency communications.

At the University, Bruno attended various courses, without being registered, in particular the courses on theoretical physics, at the invitation of W. Lenz[18], who gave a course on relativity, and H.J.D. Jensen[19] who, about twenty years later, in 1963, won the Nobel Prize for Physics with Maria Goeppert-Mayer for the nuclear shell model.

During this period he was also in frequent contact with H. Suess[20], who is well known for his study of the abundance of chemical elements in the Universe. His uncle Eduard Suess[21] had been a famous geologist, who had collaborated with the Curies and with Rutherford, and owned the largest collection of meteorites in the world. This had stimulated Suess to study and closely examine the problem of abundance.

For long periods "Touschek lived in the flat of Professor Lenz in Hamburg ... and he had considerable difficulty bringing the old and often sick man to the cellar when the bombers came."[22]

Once in a train in Berlin he met a girl, M. Hatschek—she too was half Jewish—who worked in a factory that had changed its name from *Lowenradio*, typically Jewish, to that of *Opta*. Miss Hatschek introduced Touschek to the management, who employed him in a section directed by Dr. Egerer, working on the

4

development of Brown's small tubes (i.e. cathodic oscillographs) for television. At that time Egerer was also Chief Editor of the scientific magazine *Archiv für Elektrotechnik*. Bruno worked at Opta for a long time, even after Dr. Egerer had left it to work only for the *Archiv für Elektrotechnik*. Egerer had Bruno's help in this work too, and it was thus that, at the beginning of 1943, Touschek heard of a proposal presented by Rolf Wideröe[23)] to construct a 15 MeV betatron. The proposal was kept secret because of its possible applications. Such secrecy, to tell the truth, appears today and certainly would have seemed to me (and to many others) rather curious even at that time[24)]. The *Physical Review* of 1940–1 contained the papers by D.W. Kerst[25)] in which he described, with an abundance of detail, the 2.3 MeV betatron that they had conceived, designed, constructed, and put into operation at the University of Illinois, together with the theory, practically complete, of the orbits of the electrons, which had been developed by Kerst and R. Serber[26)]. Furthermore, it was already clear that the betatron could be employed only as a source of X-rays used mainly for medical purposes. Reading Wideröe's proposals Touschek had the impression that the relativistic treatment of the stability of the orbits contained some mistakes. He wrote to Wideröe, who replied and invited him to go and work with him when, towards the end of 1943, he was ordered to build a machine of this type. So Touschek began working with Wideröe, R. Kollath[27)], and G. Schumann[164)], to develop a betatron. His principal contribution at that period was the use of the Hamiltonian formalism to study the orbits of circular machines. As Wideröe wrote to me: "He was of great help to us in understanding and explaining the complications of electron kinetics, especially the problems associated with the injection of the electrons from the outside to the stable orbit where they are being accelerated. Touschek showed that the process could be described by a Painlevé differential equation."[22)]

The machine was constructed in the Röntgenröhrenwerk (X-ray tubes workshop) of C.H.F. Müller, at Pfuhlsbüttel near Hamburg. In the autumn of 1944, the betatron began to function at 15 MeV.

At the same time Touschek devoted a great deal of time to the development of fluorescent screens for radar tubes in which, in addition to the instantaneous blue fluorescence, a delayed reddish flash was produced. This work was done for the Strategic Command of Berlin in a building of a control tower which—being situated in the Tiergarten (Zoological Gardens) of Berlin—was practically unattackable. At this period he was also following a course in theoretical physics on superconductivity held by M. von Laue at the University of Berlin.

Once "in the evening Berlin had been attacked by bombers and heavily hit. In his neighbourhood many of the houses were burning and confusion was great. When he came out in the street with his heavy bookpacks he had no way of getting to the railway station. But he was lucky. In the confusion someone had left a small electric goods stacker (Elektrostapler) in the street. He placed his books on the machine and drove off. He had never ridden on such a device before but everything went well; he got to the station and then caught a train to Hamburg."[22)]

More or less at the same time he had got into the habit of going to the Chamber of Commerce in Hamburg, where there was a room in which one could read all the foreign newspapers. These repeated visits of his caused people to notice him, with the result that at the beginning of 1945 he was arrested by the Gestapo on racial grounds. At first, Wideröe went to see him in prison and brought him "some food, his dear books and, even more important, cigarettes."[22)] During these visits, Rolf and Bruno continued to talk of the betatron.

It was in prison that Touschek conceived the idea and developed the theory of "radiation damping" for electrons circulating in a betatron, which he wrote in invisible ink in the pages of Heitler's book *The quantum theory of radiation*[28)].

Around the end of February, or the beginning of March 1945, an order arrived to transfer the prisoners from Hamburg prison to a concentration camp in Kiel. Touschek had a very high temperature but was nevertheless ordered to leave the prison. He carried with him a heavy package of books and while he was marching, escorted by the SS, in the outskirts of Hamburg, he felt ill and collapsed into the gutter at the side of the street[29)]. An SS officer took out his pistol and, pointing at his head, shot at him, wounding him behind the left ear. It was not a serious wound but he lost a lot of blood. As they thought he was dead, the column with the SS guards went on. A short time after, a group of civilians gathered on the edge of the road, discussing whether the prisoner abandoned in the gutter was dead or not. Really Touschek was still conscious and could

Photograph of the 15 MeV betatron constructed near Hamburg in 1943–1944 by Wideröe, Kollath, Schumann and Touschek (from the book by Kollath, see Ref. 27)

hear their conversation, but, as they went on for a long time, at a certain point he got up and to the general surprise asked where the nearest telephone was. They pointed to a building not far away and he went there. It proved to be a hospital, and he was treated there, but the Greek director told the police, who arrested him again and transferred him to the prison of Altona.

As Bruno said, this was a "prison of bats"[30)] where everything was extremely old. In particular, the guards and staff were all very old and kind to the prisoners. On Sundays a number of Czech prisoners of war were brought in and they did various odd jobs such as cutting wood for the stoves.

In the meanwhile the betatron group, in particular Kollath and the machine itself, were transferred to Wrist (in Holstein, near the Danish border), where some time later (probably in June 1945) the English arrived. Touschek was freed and went to Wrist, where the English asked him if he was willing to go with the troops as an interpreter. Having thought over the proposal, Touschek refused, also because at that moment both the occupying troops and the Germans—in particular the peasants in the country—were all extremely violent and killed each other practically without reason or purpose.

At the beginning of 1946, Bruno succeeded in going to Göttingen, where he had been attracted by the presence of a large number of physicists and the existence of a 6 MeV betatron. This machine had been constructed by K. Gund[31)] in the Siemens Reiniger Laboratories at Erlangen, and had been transferred to the Institute of Physics of the University of Göttingen as a place that, presumably, would be left more or less alone by the Allied Authorities in Germany. W. Paul[32)], who at the time was "Privat Dozent", had succeeded in extracting the electron beam from the machine, and started to use it for a few experiments, in particular for the study of the disintegration of deuterium[33)]

$$e + D \rightarrow e' + p + n .$$

Bruno arrived in Göttingen just at that time and came into contact with R. Becker[34)], O. Haxel[35)], H.C. Kopfermann[36)], W. Heisenberg[37)], F.G. Houtermans[38)], L. Prandtl[39)] and C.F. von Weizsäcker[40)].

During the summer, Bruno obtained the title of Diplomphysiker with a thesis on the theory of the betatron, made under the supervision of Becker and Kopfermann. A short time later, he was appointed a "wissenschaftliche Hilfskraft" (research worker) at the Max Planck Institute of Göttingen[41)], where he began to work under the direction of Heisenberg. During this period he did two pieces of work, one on the double beta decay [6] and the other on the branching points of the solutions of Schrödinger equations [7].

Before passing on to the period spent by Touschek in Glasgow, it seems in order to add two historical notes, one concerning the manner in which Paul Urban managed to help Bruno to go to Hamburg, the other on the origin and development of betatrons, which are circular machines in which electrons are accelerated by the electric field generated by the time variation of the magnetic flux linked with the electron orbit.

For the first point, I reproduce a few parts of a long letter sent to me by Paul Urban in June 1980[42)]. From 1931 to 1939 Urban had been employed in the Technical Division of the Austrian Railroads and had been at the same time in charge of a course at the Institute of Physics of the University of Vienna. Urban writes:

"This had been before the chair held by Professor Hans Thirring which had remained vacant, when the latter had been removed from his teaching because of his political ideas[43)]. The chair had been entrusted to a temporary substitute, Professor Ludwig Flamm (son-in-law of Boltzmann) of the Technische Hochschule. Later E. Fues was called to Vienna from Wroclaw as one of the three taken into consideration, Weizsäcker, Sauter and Fues. Fues occupied this post until the final collapse of [Nazism]....

"I took Bruno with me to give him the possibility, together with a few other similar cases, such as Koch, Fränkl, etc., to work with me and use the library undisturbed. He immediately appeared to me to be a talented person but rather difficult because he was self-opiniated.

"Having been dismissed from my State employment (State Railroads) on 13 March 1938 because of my 'hostile' behaviour towards the NSDAP (National-Sozialist Deutsche-Arbeiter Partei) under the existing regime (§ 4 of the Law on the New Regulation concerning State Employment) also my position at the University had become very uncertain. My political convictions were so well known that I could not become a 'Dozent' but I had to content myself with the title of 'Dr. habil.' in order to prevent me from coming into contact with the students through teaching. At that time I was working on the theory of the experimental work of my friend R. Haefer: Experimental research aiming at verifying the quantum mechanical theory of electron-emission (tunnel effect) published in the *Zeit.f. Phys.* **116**, 604 (1940). I wanted to give a lecture on this subject at the seminar led by Sommerfeld. As is well known, Sommerfeld had been sent away from his institute and had been replaced by Professor J. Müller (hydrodynamics). Professor Clusius allowed the old Sommerfeld to hold a small seminar at the Institute of Chemical Physics at the University of Munich at Amalienstrasse, where his friends and admirers could meet and discuss with him. Taking advantage of my seminar, fixed for 24 November 1942, I wanted to introduce Touschek to those important people and provide a job for him in Germany, since his presence in Vienna had already become difficult. My principal, Professor Theodor Sexl, was always called on the telephone from the Second Institute for Experimental Physics (Professor Dr. Stetter) and requested to take measures concerning me, should I still be gathering with students who were not of pure race. At the end, I was forced to lend the books privately to my protégés and had to meet them frequently at my home at 28, St. Veitg (Vienna 13), where my mother also provided us with food.

"Thus we went to Munich together and I gave my lecture in the presence of Sommerfeld and other famous physicists (Touschek took care of the projection of the slides).

"In this way, I obtained a job for Touschek at Hamburg, so that he could leave Vienna. I do not know how he fared after this, since I had lost all contact with him. When we separated I gave him, however, a research subject that I recommended to him very warmly: the double beta decay. Later, he was able to obtain his diploma on the basis of a paper on this theme.

"I should not wish to keep you in the dark about another interesting fact which we lived. All the papers of Einstein, Laue, etc., were cut away with a razor blade from our scientific journals and books in order to prepare an auto-da-fè and thus cancel the 'Judaic spirit'. The evening before this event, Sexl and I saved some of these publications, so that we could complete our library after the final collapse [of Nazism]. A person who shared my opinion, Hofrat (Councillor) R. Chorherr (at that time retired) helped us with extraordinary ability in these sad circumstances."

I pass now to the origin and development of the betatrons. The first suggestion of constructing accelerators based on this principle was proposed by J. Slepian[44] in order to produce X-rays. In the short text of this patent, however, the electrons moved in a region where the magnetic field remained constant with the passing of time. Consequently, the electrons could undergo only a very small acceleration. Shortly after (autumn 1922) and independently from Slepian, Rolf Wideröe proposed a scheme not essentially different from that of the present betatrons and started, at Aachen, some experimental research[45], whose negative result was due in part to the lack of a theory of the stability of the orbits, in part to the absence of a thin conducting layer on the internal surface of the acceleration chamber. Wideröe noticed that the electric charges deposited on the glass wall of the chamber destroyed the equilibrium orbit of the electrons.

A series of experimental as well as theoretical papers followed these first steps[46], which led in 1939 to the construction by Kerst et al., of the University of Illinois, of a small betatron of 2.3 MeV[25,26]. At that time Kerst did not know about the work done previously by Slepian, Wideröe and a few others.

The search carried out during the war by Wideröe and others near Hamburg, and by K. Gund and others[47], first at Erlangen and later at Göttingen, certainly has an important place in the story of the development of accelerators[48].

In a short note entitled "Das erste europäische Betatron für 15 MeV" (The first 15 MeV European Betatron) prepared by Wideröe for the Röntgen Museum in Lennep, the construction and end of the machine constructed near Hamburg is described as follows:

"...The construction of the Betatron was made possible in 1943 by the Air Force (Colonel Geist). Wideröe prepared the design and computations; C.H.F. Müller constructed the machine in the Röntgenröhrenwerk, at Pfuhlsbüttel near Hamburg; tests of the betatron were made by R. Kollath, G. Schumann and R. Wideröe, with a remarkable support by R. Seifert. B. Touschek developed the various theoretical investigations. Towards the end of the war, the machine was transferred to Wrist, in Mittelholstein, and at the end of 1945 was brought to England as booty of war and deposited in the Woolwich arsenal (near London), where it was used for the non-destructive examination by means of X-rays of thick steel plates. Here the betatron disappeared."

3. THE GLASGOW PERIOD

In February 1947 Touschek moved to Glasgow on being awarded a fellowship of the Department of Scientific and Industrial Research (DSIR) and started to be interested in the construction of the 350 MeV Synchrotron, initiated more or less at that time under the direction of P.I. Dee[49)], who has kindly sent me the following recollection:

"My association with Bruno Touschek began in April 1947 when he was brought to my office (under guard!) for an interview. This had been arranged by Dr. Ronald Fraser[50)] (a friend of mine), who had met Bruno when serving on a post-war Allied commission which was visiting laboratories in Germany and elsewhere. Touschek had expressed a wish to work in a British laboratory, and Fraser knew that I had recently come to Glasgow to try to construct a nuclear physics centre in the university here.

"I was quickly impressed by Touschek's obvious ability, his extensive knowledge of Physics and his enthusiasm, and I arranged forthwith for him to have a research appointment in the department, which at that time had only one staff member on the theoretical side.

"In the following five years Touschek took an active part in the expansion of the department and worked closely with Professor Gunn[51)] who, after his appointment in 1949, formed a strong theoretical team in parallel with the expansion of experimental work. Over this period Bruno became a close collaborator and personal friend. He was a person with immense vitality and enthusiasm. He was very clever and very original. He was also untiringly energetic and extrovert. Bruno led his life to the full extent in all situations and at all times. His enthusiasms were many and, although often brief, were exploited in a manner which most people would have found utterly exhausting.

"Naturally over this early period he gave me many problems! The first was his housing. A room in a small lodging house seemed satisfactory for a while but after a short 'holiday', which he spent potato picking in the north of Scotland, under spartan conditions, but fortified by the prospect of an early return to his comfortable room in Glasgow, this arrangement came to an abrupt end. On his return he found that the landlady had changed his curtains without prior consultation and, enraged by this destruction of his anticipated homecoming, he immediately returned the curtains to the manageress with a demand for instant restoration of the original ones. After a few further abortive attempts to find agreeable lodgings we seemed finally to reach a solution by installing him as a paying guest with a local resident. This however came to an end on a Sunday morning when, during my lunch, I answered the door to find Bruno on the doorstep, very dishevelled and agitated and exhibiting a severely bruised eye. It transpired that during lunch his host had spoken very rudely to his wife and Bruno's attempts to teach him marital civility had ended in a violent physical encounter. After this event my wife and I decided that the only solution was to give Bruno the top room in our house in the university and for him to eat with us, this hopefully not only to put an end to my searches for accommodation, but also perhaps to provide a present restraint on my own behaviour. Our house in the university was an old one on five floors, with rather steep communicating stairways. During the year or two which followed I never met Bruno on these stairs. His transit times from top to bottom and in reverse were always so short that there was negligible probability of an encounter. Touschek's varied and intense enthusiasms added much to the lives of the members of my family. On a neighbouring court my wife taught him to play tennis but never succeeded in providing sufficiently long periods of play to satisfy him, despite the fact that she has an abnormally low heart rate and is normally never tired under any other known circumstances. Frequently, near midnight, he would raid my study for a quick game of chess or some similar activity. At chess he was very quick and enterprising but often too original, or too obsessed with some new plan, to be regularly successful. At five minute games however, he was invincible, often perhaps because he would keep one finger firmly pressed upon his timing button so that my clock would be registering almost throughout and his rarely, if at all, until my time limit had elapsed.

"Bruno's passion for novelty and independence knew no bounds. When he decided to have a desk made by a local carpenter he produced detailed drawings which the carpenter was forced to follow despite his strong reluctance. The end product was a desk having drawers with no backs and sides (to avoid dusty corners and edges) and which were to serve as withdrawable trays. This scheme might have been successful for a person of less ebullient character, but with Bruno's rapidity of movement the result, in use, was a progressive and systematic transfer of the contents of all the drawers to the bottom level.

Bruno Touschek in Glasgow (1949)

"On many occasions Bruno joined us in climbs on Scottish mountains. On these he would gradually discard and hide items of clothing, the final stage being completed in almost complete nudity. This also had the advantage that he could immerse himself in any small pools or burns which we came across on the way. He was an excellent swimmer with a dolphin like action, but here again he was rash and adventurous, once swimming to an island on Loch Lomond which even he felt to be too remote to risk the return swim. Fortunately he managed to hail a passing boat which returned him, blue with cold, to our picnic site.

"Bruno's impatience with the slowness of behaviour of normal people often had very amusing consequences. Once having been allocated a new room in the laboratory and being unable to wait for a proposed redecoration, he embarked personally and without warning on this activity, during a weekend when the department was otherwise unoccupied. Apparently he soon found that proceeding systematically was very dull and boring, so he covered various areas at random as the spirit moved him. By Monday morning the room had a nightmarish patchwork appearance, whilst the fine teak block floor (left uncovered during the operation) was now coated with thousands of spots and streaks, so numerous as to give the impression that perhaps this had been intended. The situation was seemingly irrevocable and further exacerbated by a local contractor working in the department at the time, who took his friends and visitors to see the workmanship of what he claimed to be the university's own works department. This led to official protests and the placing of an embargo on any work on that room by university staff. My only course was to lock the room up for a period to allow passions to subside.

"I hope I have not given the impression that Bruno did not normally produce tidy and systematic work. In fact his written work was always very tidy and ordered, set out in stylish calligraphy and overall with a presentation which had quite an artistic flavour. Indeed, he had quite marked artistic talent and could make sketches showing a fine sensitivity and economy of line. My daugthers still have some beautiful bookmarkers which he put into their birthday presents. I once even tried to persuade him to develop and exploit this ability but his reaction was that to do so would spoil the pleasure he derived from such an occasional pastime.

"I have perhaps not written enough about Bruno's work because I think others can do better than I can, but I must refer to the delight he always experienced in the solution of a problem or the presentation of a piece of scientific argument. Requiring the proof of a theorem he would only rarely bother to consult the literature. In this there was a degree of arrogance. He seemed to assume that his own proof would probably be shorter and neater. Despite his excitement when reaching the desired result there was no lasting pride or boast. For a brief instant he would beam with satisfaction, perhaps self satisfaction, but moments later all would be forgotten and he would be away with something new.

"In Bruno's personal relationships he was normally very polite, friendly and loyal. He could certainly be hot-tempered, angry and emotional but in the end he was basically kind and reasonable. When I met him in Rome in 1962 he was much more stable and much quieter in demeanour but still very considerate and kind. He took great pains to ensure that my wife and I had a happy and interesting stay in that beautiful city. I think he knew that we were rather raw and inexperienced travellers and did much to make us feel welcome and at home there.

"I am deeply sad to realise that he has gone from us all. He was a person who led an intense and vigorous life and one who, by his example and friendliness, helped me and I expect many others to achieve greater happiness and awareness in their own lives."

P.I. Dee
25.5.1979

4. HIS SCIENTIFIC WORK AT GLASGOW (1947–1952)

Touschek's friendship and collaboration with Dee enabled him to study in depth the problems related to the working of the synchrotron, and he published an article on its characteristics some years later [10]. That same year, he was awarded his Ph.D. with a thesis on nuclear excitation and the production of mesons by electrons, of which Gunn was the internal rapporteur and Rudolf Peierls[52)] the external one. Immediately after this he was appointed "Official Lecturer in Natural Philosophy" at the University of Glasgow, a position he held until he left for Rome.

The subject of his thesis already reveals the nature of his scientific interests at the time. In a series of papers [some of which were written with I.N. Sneddon[53)]] he dealt with the problem of nuclear excitation by electrons [1, 5], nuclear models [3], the density of the energy levels of the nuclei [4], the evaluation of the position of the lowest level excited by electric dipole transitions, and the quadrupole moments of the nuclei in the shell model [11, 12]. Other papers concerned the problem of meson production: by electrons (with Sneddon [8, 9] and in proton-proton collisions (with J.C. Gunn and E.A. Power [13, 14]). Other papers study the divergence in quantum field theory [2] and the perturbative treatment of the bound (closed) states in quantum field theory [16].

In September-October 1950, Walter Thirring[54)] came up to Glasgow as Nuffield Fellow, and met Touschek. In those months they worked together [15] on the covariant formulation of the Bloch-Nordsieck method to solve the general electrodynamic problem in the presence of an external current. They saw that in this case perturbation theory cannot be used, since the perturbed state is orthogonal to the unperturbed state.

In a paper written with Roy Chisholm [19] Touschek discussed the spin orbit coupling in nuclei as essentially due to the exchange of pions in an S state between the nucleon and the rest of the nucleus. The spin orbit interaction thus calculated, however, has the wrong sign. After Touschek had left Glasgow, Ernest Laing, a student of Chisholm[55)] showed that the sign of the spin orbit interaction becomes the correct one because of an enhancing factor originating from a nucleon self-energy term due to the simultaneous emission and reabsorption of a pion in a P state[56)]. However, an extremely high density of nucleons in the nucleus was necessary to obtain the right order of magnitude of the interaction. Many years later, in his paper with A. Bietti ("Scalar mesons and the nuclear spin-orbit coupling" [55]) Touschek showed that introducing an interaction between the two pions in an S state so as to create a resonance (with $mc^2 \simeq$ 380 MeV, as suggested by some experiments) one obtains a scalar spin orbit potential, which is basically correct for reasonable densities of the rest of the nucleus.

In other papers written with W.K. Burton, Touschek examined the commutation relationship [18] and Schrödinger's dynamic principle [20].

As Chisholm relates, in Glasgow Bruno had bought a motor cycle "which had a special feature, independent suspension of the front forks. During the first fortnight with this vehicle, he fell off it twice. He was convinced that he was not to blame for these accidents, and the evening after the second he settled down to make a complete study of the dynamics of the motor cycle. This took him about nine or ten hours, ending in the early hours of the next morning. His study showed that the new degree of freedom which had been introduced by the makers was unstable. He sent his full analysis to the manufacturers together with a letter beginning 'Dear Assassins, ...' The model was withdrawn from the market shortly afterwards. Some of those who suffered as a result of this accident were members of the first year Physics class, who had a lecture on Mechanics from Bruno the next day. Since he had prepared no other material they were treated to a lecture on Mechanics of the motor cycle and the bicycle; you can imagine that they would find this fairly difficult."

5. BRUNO ARRIVES AND SETTLES IN ROME (1952)

In December 1952 Bruno Touschek moved to Rome, to which he had always been attracted owing to cultural and family ties. It was in Rome that his aunt Ada resided, having come there many years before. She was his mother's sister, and had married an Italian, Gaetano Vannini. Aunt Ada was the owner and joint manager with her husband of an agency in Rome representing an Austrian firm (Garvens s.r.l.), which specialized in the manufacture of water pumps and irrigation systems. His aunt and uncle also owned a house in the neighbourhood of Albano and, as they had no children, were always happy when their nephew visited them, although Bruno sometimes could not conceal a certain intolerance for the care and advice which his aunt Ada lavished on him. Their relationships, which could be glimpsed from the incidental remarks made by Bruno, were often reminiscent of those which existed between Berty Wooster and Aunt Agatha in the books by P.G. Woodhouse, devoted to the butler Jeeves, although the background, typically Viennese, was quite different from that which prevailed in Britain.

His mother's mother, Joseffa, had come to Rome to live in 1938, but at the beginning of the war her daughter Ada persuaded her to return to Vienna, thinking that it would be safer there. During 1941, however, on a date which I have not been able to determine, she was arrested by the Nazis and sent to the Theresienstadt extermination camp, where she died shortly after.

Rather more than for family reasons, however, Bruno Touschek was attracted to Rome as a result of his acquaintance with Bruno Ferretti, owing to their papers which appeared in the scientific press.

In 1948, Ferretti[57)] was invited to be Professor of Theoretical Physics at the University of Rome, a post previously held by G.C. Wick, who had accepted an offer from the University of Notre Dame at South Bend (Indiana).

Ferretti, in fact, had been in Rome until he was appointed Professor of Theoretical Physics at the University of Milan in 1947, and on his return had succeeded in strengthening the group of young theoretical physicists from Rome and instilling considerable life into it. I should mention that among the members of the groups were the following, in order of age and training: E. Corinaldesi, M. Verde, B. Zumino, G. Morpurgo, R. Gatto, F. Fabri and C. Bernardini.

In September 1952, Bruno Touschek went to visit Ferretti at the Guglielmo Marconi Physics Institute. A few hours after their first meeting, spent discussing mutual scientific questions, they established such a marked professional respect and personal attachment for each other that Touschek decided to remain permanently in Rome. This became possible because he was appointed to the post of researcher (grade R2, equivalent to that of an Extraordinary Professor) in the Rome Section of the Istituto Nazionale di Fisica Nucleare, of which I was Director at that time[58)].

Bruno Touschek and Bruno Ferretti never published a joint paper, perhaps because both were too individualistic in their manner of thinking, and because they had complementary qualities. This was so much the case that they were always ready to engage in a detailed discussion but had difficulty in following a systematic approach in solving a specific problem. Their daily discussions of the very diverse and most difficult problems of theoretical physics provided, however, for many years, an extremely strong incentive to a deep understanding of these very problems, not only by themselves, the two main protagonists, but also by other young theoreticians, who had studied or were studying in those years at the Institute of Physics.

This form of discussion came to an end in 1954, when Bruno Ferretti moved to Bologna and was replaced by Marcello Cini. But Touschek's influence on the group of theorists continued to have an effect for many years, and began to diminish only from 1960 onward, when his interests shifted towards the possibility of constructing accelerator machines in order to study the processes produced by electron-positron collisions, of which we shall speak later.

When Touschek moved to Rome, the main interest which he shared with Ferretti was the construction of a quantum field theory which would also include bound states, i.e. a theory which would go beyond perturbation methods.

Ferretti pointed out to me that Touschek was among the first to maintain that it was possible to construct a unified theory of electromagnetic and weak interactions, the first example of which was actually constructed by S. Weinberg and A. Salam shortly after[59)]. A passing reference to this idea had already been given in a work by Touschek on the neutrino theory [43] but was dealt with in a more detailed manner in a subsequent paper prepared with the collaboration of I.M. Barbour and A. Bietti [51], and we shall come back to this later (see Section 13). Ferretti had some correspondence on this subject with Touschek.

In those early years of his life in Rome, Bruno Touschek owned a motor cycle which he called Josephine; he claimed that when he had been out late drinking in a pub or at a friend's Josephine knew how to bring him home safely.

In 1955, Bruno returned to Glasgow to marry Elspeth Yonge, the daughter of a well-known professor of Zoology at the University of Edinburgh, Sir (Charles) Maurice Yonge[60]. Elspeth gave Bruno two sons, Francis in 1958 and Stefan in 1961. The Touscheks first lived in a flat in Via Mancinelli (Piazza Vescovio), then in Via Saliceto, in a flat which had a terrace overlooking a most beautiful garden with trees. They then moved to Viale Regina Margherita, at the corner with Via Morgagni, and finally to No. 23 Via Pola.

Each time he looked for accommodation and found a vacant flat in a district which he liked he went to the doorkeeper and asked "Do any officers live here?" The reason for this question, he explained to those that accompanied him or to his friends at the Institute next day, was that "it is impossible to cohabit in the same building as officers". This profound antimilitarism also emerged on other occasions. For example, as Careri recalls, when in 1957 "Sputnik II" was launched carrying a living being for the first time (the she-dog Laika), Bruno went around the Institute and, rubbing his hands out of satisfaction, said "it is rather a costly method but you will see that gradually we will get rid of all the military by sending them out into space."

On another occasion one night on his motor cycle he ran into the rear of a large car at a cross-roads, flew right over the top of it and injured his skull during the fall. He was immediately taken to the Neurological Clinic, which at that time was directed by Professor Ugo Cerletti.

The injured person was in a state of great agitation and even if what he said was not understandable, he appeared to be of German tongue. The next morning, the doctor of the Psychiatric Section asked Dr. Valentino Braitenberg, of the Bolzano province, to go from the laboratory where he was working to fill in the hospital sheet for the newly admitted person. As Braitenberg relates "... a first superficial examination (that would not have lasted more than a few minutes) allowed my colleagues to determine that he was not Italian and to suspect psychosis. The injured man had declared he was a theoretical physicist, Vice-Director of the Scuola di Perfezionamento di Fisica Nucleare and a specialist in time reversal (see Section 6). I found Bruno sitting on his bed in the ward, still rather angry but already occupied in observing with interest the spectrum of mental alienations displayed by the surrounding patients. He wore a turban applied by the first-aid doctors when they treated the wound that he had received during the fall. The slight concussion which he suffered in combination with the high alcohol content in his blood had caused a state of agitation, as frequently happens, as a result of which the police decided to apply the rules of psychiatry rather than those of the Penal Code. We immediately became friends. His story was convincing and not at all psychotic, his German was delicious, rich and precise, his humour was uncontrollable even in such embarrassing circumstances. I wrote his clinical story with his help. The only slightly abnormal detail was the daily quantity of wine, but Bruno gave good reasons: tennis, the scarcity of water in aqueducts, etc. I thought it appropriate to give him the "routine" sermon. Bruno answered that his liver was his and that if I wanted to associate myself with his habits, he would be very glad to bring me to Nemi the following Sunday. I accepted and after that for many months we spent almost all week-ends together, Bruno, my future wife, whom I had just met at that time, and myself. We started to talk of cybernetics, there was a seminar on the computer machines in which I participated as a guest, and finally the avalanche of cybernetic activities that carried me with it and transformed my life".

On arriving in Rome Touschek spoke a peculiar form of Italian which was a combination of Latin and English.

Bruno Ferretti's wife, Maria, who at that time worked at the Institute as a "Nostromo" (a combination of a secretary, purchasing employee and store-keeper of materials and instruments), recalls the following expressions which were typical of Bruno: "... il giusto tecnico del tornare", instead of "la tecnica giusta per voltare (l'automobile)"; "... c'era tant'acqua che si nagiava e la padrona diceva che ero io che avevo pizzicato le pipe..." (pinched the pipes); "... mia zia è tanto affezionata ai suoi cannelli lunghi lunghi (bassotti)..." and so on. But as years went by, his Italian became excellent, although he continued to use his own characteristic turns of phrase.

At home, he had two cats, a black one, which he called Planck, and a striped cat which he called Pauli. After making Pauli undergo a series of intelligence tests, Bruno decided that he was extremely intelligent. He made him undergo another test but Pauli failed to pass this one: after very careful preparation, Bruno gave him a tin of sardines and a key to open it, but Pauli did not open the tin and so did not eat the sardines.

Shortly after his arrival, we acquired the habit of playing tennis together two or three times a week. Sometimes we were joined by Francesco Calogero. We always played very seriously and enjoyed ourselves very much. On one occasion, a few years later, we were both taken with an irresistible desire to wear out our opponent. We played for a long time, putting everything into it, each trying to make the other run as much as possible. We played much longer than usual and finally we gave up, exhausted but satisfied. During the night I felt ill; by all appearances, I had taxed my heart unduly. I was seen by a doctor whose harsh words were combined with those which I inevitably received from my wife who, on this as on every other occasion, pointed out how foolish I was, although I had almost reached my fifties.

For many years I kept quiet about this incident, but one day, much later, Bruno by chance told me that on that particular night he had not felt well either.

6. THE FIRST PERIOD OF HIS SCIENTIFIC WORK IN ROME (1952–1960)

The first work carried out in Rome, at the beginning of 1953, shortly after his arrival, was in collaboration with M. Sands[61] of the California Institute of Technology. Sands was in Rome on a "fellowship" awarded by the Fulbright Foundation. The work [17] followed a few months after the discovery of strong focusing[62] and concerned the errors of magnets' alignment in synchrotrons based on this principle. The authors showed that certain ideal orbits (i.e. calculated without alignment errors), which were periodic in a single revolution, are transformed by alignment errors into open orbits which are secularly unstable. This was found to be the case in the vicinity of certain values of the betatron frequency. But far from such "resonances" there are however stable orbits. They also produced estimates for the misalignment tolerances necessary to give rise to a reasonable amplitude of perturbed motion. At the same time, similar calculations were made also by other authors[63], who are quoted, together with that made by Sands and Touschek, in all treatments of the stability of strong focusing accelerating machines[64].

The short paper written with E. Fabri on "The mean lifetime of the τ meson" [22] concerns a problem on which work had been started by an experimental group at the Institute[65,66] shortly after the discovery of this new particle[67]. The problem of the possible existence of correlations between the energies of the three pions emitted during the decay $\tau^+ \to \pi^+ + \pi^- + \pi^+$ had been tackled but not solved in a general manner in 1953[66], a time when the experimental data available were not statistically significant to provide an answer.

Touschek's work contains a detailed discussion of the relation between the mean lifetime and angular momentum J of the τ meson. In particular, the authors notice that this particle decays into two identical charged pions which necessarily should be into a state of parity $P = +1$ and angular momentum ℓ even, and a third pion of opposite charge and of angular momentum λ relative to the centre of mass of the two others. They concluded from this observation that the parity of the τ meson should be equal to $-(-1)^{\lambda}$. On the basis of certain estimates, the authors also noted that with statistically significant data it should be possible to distinguish at least the case $J = 0$ from that of $J = 1$.

This work slightly preceded that of Fabri[68], in which the problem was re-examined in detail and clarified with the introduction of the same graph submitted a little earlier, without Fabri's knowledge, by R. Dalitz[69] and which today is known in the literature as the "Dalitz-Fabri plot".

The "Report of the Committee on τ Mesons" [26] is the final report made by the Committee of "experts" appointed at the "International Congress on Heavy Unstable Particles and on High-Energy Events in Cosmic Rays", held in Padua during 12–15 April 1954[70]. This report, together with similar reports prepared by participants at the Congress, re-examines the situation of each of the new particles, paying special attention to the experimental values of the electric charge, mean lifetime, rest energy, and decay modes.

The paper [25] is a discussion, prepared by Touschek in conjunction with Fabri, on the final states produced during the capture of K particles, of which at that time only seven events[71] had been observed, which according to Touschek could not be attributed to a single mechanism.

The paper with G. Stoppini, entitled "Phenomenological description of photo-meson production" [31] presents certain phenomenological expressions for the differential and integral collision cross-section, for the photoproduction of pions on a nucleon, valid in the region between the threshold and the first resonance ($J = 3/2$, $I = 3/2$).

These expressions were obtained by introducing suitable corrections to the perturbation expression, in a similar manner to that used in the Born approximation with distorted waves, which is currently used in nuclear physics. The fundamental idea is to include phenomenologically the two main causes which make the perturbation method unsuitable: i.e. the correction to the final state due to the pion-nucleon interaction (in state 3/2, 3/2) and the interaction due to the anomalous magnetic moment of the nucleon.

The expressions which the authors derived in this manner are in very good agreement with experimental results. The general trend of the paper follows the ideas by G. Chew[72], although it differs from these in certain important details.

In addition to these papers, some of them of an engineering and others of a phenomenological nature, Touschek published during these years a whole series of papers, some of which tackle problems relating to calculation methods, whereas others deal with fundamental questions or questions of principles.

In Rome (1955)

To the first category belong papers [21], [23] and [35]. G. Morpurgo, who graduated in Rome in 1948 and went to work in Chicago in 1952, returned in August 1953 and found Touschek in Rome. He had had the idea of studying, in the case of a solvable model, such as that of G. Wentzel[73)], what difference there was between the exact solution and those calculated with the Tamm-Dancoff method or in the perturbation approximation. He spoke to Touschek about it; they started to discuss the problem jointly and rapidly came to the conclusion which was submitted to the Congress of the Società Italiana di Fisica, which took place in Cagliari on 23–27 September 1953, and also sent to Nuovo Cimento [21].

In this paper the authors clarify the relation between the eigenvalues of a system closed inside a box of finite volume and the phase shifts of the various waves that appear in collision theory (i.e. in the limits of infinite volume). In particular they deduce an elegant relation between the bound states and the phase shifts using the Wentzel model in S-wave.

This paper to a certain extent contributed to reducing the interest in the Tamm-Dancoff method[74)].

This work also led to a discussion with M. Cini during one of his trips to Rome from Turin where he was the assistant of G. Wataghin. Cini had already devoted himself to the Tamm-Dancoff method[75)]. On that occasion Cini, Morpurgo and Touschek applied an approximate but not perturbative method, previously developed by Cini and S. Fubini[76)], to the solvable case of Wentzel's model and showed that, in that case, it was superior to the Tamm-Dancoff method [23].

Morpurgo and Touschek had found, when discussing between them, that there were many more fundamental difficulties in fully understanding the "time reversal" operation than in understanding the "parity" operation.

During a trip to Naples they spoke about this to L. Radicati, who had recently been appointed Professor of Theoretical Physics at Naples University. This gave rise to a collaboration between the three of them, facilitated by the fact that Radicati came to Rome each week and stayed with Touschek, who lived in a flat on the fourth floor of the mansion at No. 3 Via Saliceto.

Radicati recalls certain evenings he spent in Touschek's flat, when he was visited by various friends. The conversation, which followed an open course and was imperceptibly guided by Bruno, ranged over a wide variety of subjects, from literature to politics, painting to physics, the latter being rarely touched on, and then only lightly. During May and June, the characteristic atmosphere of the city was clearly perceptible from the terrace, where their conversation was punctuated by witticisms and ironical remarks, most of which were made by Bruno.

In the paper "On time reversal" [24], the authors made a detailed analysis of this operation within the framework of classical and quantum mechanics. This paper was amply reviewed by F.J. Dyson[77)], in *Math. Rev.*, who introduces his comments with the following remarks: "In the extensive literature devoted to the problem of time-reversal in quantum mechanics, this is one of a few papers which add substantially to the original discussion by E.P. Wigner ...."

There then follows a short paper by Morpurgo and Touschek [27], in which the authors provide a rigorous proof of the irreversibility of a system composed of a neutral scalar meson coupled with a nucleonic field, which had already been discussed in paper [24].

In a subsequent paper, entitled "Time reversal in quantized theories" [28] Morpurgo, Radicati and Touschek extend the considerations presented in the first paper on this subject [24] to the case of field theory.

Papers [29] and [30] by Morpurgo and Touschek are extensions of the same types of definition, techniques, and procedure to the parity and charge conjugation operations.

A final paper, also written with Morpurgo a short time after [37], entitled "Parity conservation in strong interaction", was never published, because when they received the galley profs, Morpurgo had the impression that the paper was rather trivial. They did, however, make it known by circulating it in the form of an internal report of the Institute[78)] and making a presentation of its essential points at a Congress held in Padua in 1957, as a comment to a paper by another author. In the paper it is shown that if one assumes invariance of the pion-nucleon system with respect to time reversal and rotation in isospin space, the invariance with respect to parity and charge conjugation also follows automatically. The same result was found by G. Feinberg as well as others some time later[79)].

In the various papers on time reversal, Bruno had also concerned himself with the problem of parity which was specifically dealt with in the papers "Parity conservation and the mass of the neutrino" [32] and "The mass of the neutrino and the non-conservation of parity" [33], both of which are very important, particularly the former. In the first of these two papers, Touschek is the first to introduce what was much later referred to as *chiral symmetry*. Abdus Salam[80)] had proposed the operation of *discrete symmetry*

$$\psi \to \gamma_5 \psi, \tag{1}$$

where Touschek introduced the operation of *continuous symmetry*

$$\psi \to e^{i\alpha\gamma_5}\psi . \tag{2}$$

He had proposed this operation in order to define the conservation of the leptonic number in the presence of parity violation, which is also the definition adopted today, for which the leptonic number of the neutrino is equal to its helicity. Later it was recognized that this continuous symmetry introduced by Touschek intervenes in a very general manner, for example in the algebra of currents, as was pointed out by Gell-Mann[81)].

In this paper [32], received by Nuovo Cimento on 26 January 1957, Touschek quotes the famous paper by T.D. Lee and C.N. Yang, in which it is suggested that parity is not conserved by weak interactions[82)]; he also quotes the paper by Mrs C.S. Wu and collaborators[83)], announcing the experimental confirmation of the predictions made by Lee and Yang with regard to ^{60}Co decay. In fact, Touschek quotes an issue of *Time Magazine* of 28 January 1957, which appeared a few days later, in which the news was published for the first time. This suggests that the quotation was added when corrections were made to the proofs. Touschek's paper was, however, fully written, or almost so, before the experimental confirmation of parity non-conservation in weak interactions.

It should also be said that both Touschek and Salam proposed the neutrino two-component theory, and that the uncertainty between the two possible interactions (S + P + T) and (V − A) had already been solved by Touschek in favour of the latter[84)] in the second of the two papers discussed here, on the basis of the asymmetry observed in the decay of the muon in the paper by R. Garwin, L. Lederman and M. Weinrich on the $\pi \to \mu \to$ e process[85)].

At that time everything was in agreement with the assumption that the weak interaction was of the (V − A) type, with the exception of the experimental results relating to ^{6}He, which were, however, erroneous as was found shortly after. In his second paper, Bruno concludes in favour of (V − A), since only this interaction was compatible with the sign of the asymmetry observed in the $\pi \to \mu \to$ e process.

Another paper—the last in collaboration with Radicati—is entitled: "On the equivalence theorem for the massless neutrino" [34]. It concerns the magnetic moment of the neutrino. The authors show that, at the limit of the mass of the neutrino equal to zero, the magnetic moment of this particle tends to zero as a consequence of the invariance of the wave function under the operation of chiral symmetry (2).

This paper was written after Radicati had moved from the University of Naples to that of Pisa, where Bruno regularly went in order to lecture on "field theory" at the Scuola di Perfezionamento in Fisica at this University (see Section 7).

The paper written with Cini on "The relativistic limit of the theory of spin 1/2 particles" [35] links up with the two papers on the neutrino properties [32, 33], of which we have already spoken. When Touschek spoke of them to Cini, stressing the interest of the invariance of the wave function of the neutrino under the operation (2), the problem arose as to whether it was not possible to deal with the case of a particle having a non-zero mass, starting, as zero approximation, from the case of a zero-mass particle. The problem appeared, to a certain extent, as the reverse of that dealt with by L.L. Foldy and S.A. Wouthuysen[86)], who had developed a systematic theory, i.e. valid at all orders, for the case of Dirac particles having a kinetic energy much lower than the rest energy.

The problem solved by these two authors is to find a canonical transformation which eliminates from the Hamiltonian any uneven operator, i.e. any operator which (like the α of Dirac) connects the first two components with the second two of Dirac's spinor. Foldy and Wouthuysen had found the general expression for this transformation and shown that, when it was cast in the customary form of a perturbative series, the most important term was that of the mass.

The view taken by Cini and Touschek is the opposite: they start from the solution of Dirac's equation, valid for a momentum p vastly greater than mc (i.e. rigorously valid for zero mass), in which the spinor has only two components of opposite helicity, and try to find a canonical transformation which takes into account the finite value of the mass. The result is that in this case too there is a canonical transformation which separates the major components of the spinor from the small components. This transformation, expressed in perturbative form, involves operators which are both even and odd, the most important of which is $\exp\{ia\gamma_5\}$.

This formalism was developed by Cini and Touschek in view of its use for the electrons emitted in weak decays where, generally, they have a momentum p much larger than mc.

Papers [36], [38] and [40] concern essentially the same problems, and represent an extension of the concepts introduced in papers [32] and [33]. The most important of these three papers is [36] entitled "The symmetry properties of Fermi Dirac fields". In this paper a presentation and discussion are given of the first example of non-Abelian chiral symmetry, although it is expressed in a different form from that which is now more customary. This is probably the reason why this important result obtained by Touschek in practice has never been duly attributed to him.

The reason for this kind of presentation of the subject is that Touschek dealt with this problem taking his inspiration from Heisenberg's non-linear theory[87)], in which an attempt is made to construct all the possible fields starting from the Majorana-type Fermionic fields, whereas, following the discovery of parity non-conservation[82)], use is made of the Dirac-type fields and a separation is made from the outset between the fields of different chirality, i.e. left-handed and right-handed.

In paper [33], which has already been discussed above, Touschek had introduced the symmetry operation (2) and shown that the requirement for "free particle" fields to be invariant under such an operation is sufficient to guarantee that the mass of the corresponding particle is zero. This result is taken up and generalized in paper [38] entitled "A note on the Pauli transformation", where Touschek shows that this is applicable also in the presence of interactions. In this rather formal paper, Touschek seeks, among other things, to expound Heisenberg's theory in an axiomatic form.

7. BRUNO TOUSCHEK'S CONTRIBUTION TO TEACHING AND HIS UNIVERSITY CAREER

From the academic year 1953–54 until 1961–62, Touschek made a substantial contribution to teaching with a course he gave at the Scuola di Perfezionamento in Fisica at the University of Rome, of which he became also Vice-Director (see Section 5). This school had been set up in 1952 with the idea of creating a system in Italy which would be very similar to an American postgraduate school. For this purpose I had had lengthy discussions on this problem with various colleagues, in particular Bruno Ferretti, and had consulted Enrico Fermi and Bruno Rossi to obtain first-hand information on what was being done at the University of Chicago and the MIT[58)].

In the capacity of Vice-Director of the Scuola di Perfezionamento Bruno Touschek made for years a remarkable contribution to the preparation of the syllabus and the choice of the teachers, but the course he gave was even more important. The course changed its title as the years went by: "Cosmic rays and subatomic particles" from 1953–54 to 1955–56, "General particle theory" from 1956–57 to 1959–60, and then "Field theory" in 1960–61. Although different all these titles corresponded to the presentation of the same subject, i.e. elementary particle theory, which was updated every few years. In 1960–61 he also gave a six-months' course on "Particle accelerators", dealing with certain theoretical aspects of these machines, in which Bruno Touschek had at that time developed a renewed interest (see Section 9). The following year, 1961–62, he also gave a few lectures of a course in which other teachers (N. Càbibbo, M. Cini and G. Jona) participated: "Complements to theoretical physics". In the following years, however, he preferred to break off his teaching activity to have more time to devote himself fully to the theoretical and experimental research which was being carried out at Frascati and Orsay on and with AdA (see Section 9). He again took up teaching at the Scuola di Perfezionamento of the University of Rome in 1966–67 with the course entitled "Complements to theoretical physics, 2nd part" which he continued to give until 1970–71.

The courses which Touschek gave had a marked influence on many pupils as each of them graduated. Among the many who submitted their theses whith Bruno I should mention: for the "laurea" in physics, N. Cabibbo, F. Calogero, P. Guidoni (who graduated in 1958), A. Putzolu (in 1961), Giovanni Gallavotti (in 1963), Paolo Di Vecchia (in 1965), Aurelio Grilli (in 1968), and for the "laurea" in mathematics Etim Etim (in 1965) a young Nigerian who came to Italy with a scholarship from "AGIP Mineraria".

Bruno Touschek did a considerable amount of teaching at Pisa as well. He was responsible for the course on "Field theory" at the Scuola di Perfezionamento of that University in 1956–59 [66], not only to increase his income, but especially because his trips to Pisa enabled him to maintain the contacts with Radicati, which resulted in their collaboration on weak interactions [34].

Radicati recalls Touschek's brilliant and extremely clear lectures, which to a large extent followed the pattern of the course he gave on the same subject at the Scuola di Perfezionamento of the University of Rome during the years 1953–54 to 1966–67.

It was by means of these lectures that Touschek had a considerable influence on various young physicists at Pisa such as A. Di Giacomo, P. Menotti and L.E. Picasso.

Bruno usually caught a train which left Rome at 12 a.m., went to the restaurant car, where he would eat and drink without too much restraint, arrived in Pisa somewhat tipsy and then went to sleep; then, shortly after 6 p.m. he gave his lecture in full form and in a brilliant manner.

In addition to this, he was also one of the lecturers on various courses at the Scuola Internazionale di Fisica of the SIF (Società Italiana di Fisica) which were given in Varenna. There was the Sixth Course, which took place from 21 July to 9 August 1958, on the "Mathematical problems of the quantum theory of particles and fields" directed by A. Borsellino, and the Eleventh Course which was held from 29 June to 11 July 1959 on "Weak interactions", directed by Radicati. In addition, the SIF gave Touschek the responsibility for organizing and directing the Ninth Course, which was held from 18 to 30 August 1958, on "The physics of pions".

The lecture which was given at the Sixth Course [39] is also signed by W. Pauli. It is entitled: "Report and comment on F. Gürsey's 'Group structure of elementary particles'". It is a paper which is original only in part, and sets forth some considerations which Pauli had communicated to him in a letter (see Sections 1 and 2) combined with what Touschek had in essence already stated in paper [36]. Pauli had put forward the idea of trying to find very general symmetries and had applied it to the case of $n = 2$ Majorana-type Fermionic fields. In this lecture Touschek takes up the same idea and applies it to the cases with $n = 1, 2$, and 4 Majorana-type Fermionic fields, but without adding anything basically new compared with paper [36].

His inaugural lecture at the Ninth Course of the Scuola Internazionale di Varenna on pion physics, organized and directed by Touschek himself, is presented in publication [40], whereas publications [41] and [42] are lectures which he gave during this same course. Paper [41] concerns the 'Fixed source meson theory", and contains an elegant exposition of Chew's ideas[72]. Paper [42] concerns "Elementary considerations in photo meson production" and, in addition to providing a general statement of the problem, contains the same phenomenological considerations he had put forward with Stoppini in paper [31].

The lectures on the "Theory of the neutrino", given by Touschek at the Eleventh course of the Scuola di Varenna on "Weak interactions" is contained in publication [43] and that given to the Scuola Primaverile, organized by E.R. Caianello in Naples in 1960 [47], summarizes the discussion of the chiral transformation (2), which had already been published in papers [32] and [33].

Touschek's contribution to these courses, and especially to the Ninth Course, went, however, well beyond his own always very original and brilliant lectures. His own participation in the majority of the discussions and the observations he made quite frequently during the lectures given by other teachers were characterized by a vivacity and enthusiasm which remained imprinted in the minds of all those who attended them.

In recognition of all these contributions, the Società Italiana di Fisica awarded the "SIF prize" for teaching merit to Bruno Touschek, on the occasion of its 44th National Congress held in Palermo from 6 to 11 November 1958[88].

From the academic year 1960–61 to the end of 1968–69, the Faculty of Mathematical, Physical and Natural Sciences put Bruno Touschek in charge of the course of "Statistical mechanics", which had never been given before at the University of Rome. The course, which was intended for 4th year physics and mathematics students, had, like the others, a substantial influence on various students, such as Giovanni Gallavotti and Gian Carlo Rossi, the second of whom collected the lectures given by Touschek (1965–66), producing them later in the form of a book entitled *Statistical mechanics*, published by Boringhieri in 1970 [67] (Section 13).

Touschek gave up the course in statistical mechanics when, in 1969, he was appointed "Professore Aggregato" at the Faculty of Science, where he was asked to give a newly-introduced course on "The mathematical methods of physics", which he continued to give for the rest of his life. I will come back to this point later.

During the academic year 1970–71 he was asked to give a course on 'Statistical mechanics" at the Scuola Normale in Pisa, the contents of which were basically the same as those of the book which he wrote in collaboration with G.C. Rossi [67].

Finally he organized and directed the Summer School on "Physics with Intersecting Storage Rings" that took place from 16 to 26 June 1969 at Villa Monastero on Lake Como[89].

Among the teaching staff I can recall R. Gatto, J. Haissinski, A.N. Lebedev, G.K. O'Neill, G. Pellegrini, M. Sands, K.G. Steffen and R. Wilson. Bruno did not give any lecture, but with his continuous presence and his extremely lively participation in all discussions instilled life into the course, which is still remembered today by all the participants as a particularly agreeable and interesting intellectual experience.

As I have already said, from the beginning of his stay in Italy Bruno Touschek held an R2 post at the Istituto Nazionale di Fisica Nucleare (INFN), which was equivalent in pay (and in prestige, but only in the restricted field of the INFN and physicists in general) to the post of Extraordinary Professor in an Italian university. The problem of getting Bruno into a State post remained unsolved for many years, since the reform of Italian schooling, introduced by G. Gentile when he was Minister of Education in 1928, had short-sightedly specified that a certificate of Italian nationality was required for candidates wishing to sit university competitions.

After the war, the physicists' community, and in particular the SIF, had on many occasions drawn the attention of the Minister of Education to the urgent need for a change in the law, in order to allow foreigners also to participate in our university competitions, as had been authorized by the Casati law of 1859[90]. The reason for this was that it would certainly have a beneficial effect in general for the Italian universities, and because of Bruno Touschek's own specific case. Despite the fact that more than one Minister, and in particular Giuseppe Medici, was firmly convinced that this was a reasonable proposal, it had fallen by the wayside following rejection by the Consiglio Superiore della Pubblica Istruzione.

This shortcoming in Bruno Touschek's career considerably upset him—so much so that in 1959 he seriously considered applying for Italian citizenship. After all, he had already lived in our country for many years and so far had felt very happy there. He did, however, very strongly resent having to collect all the documents which are required by law for a formal application of this type. Nevertheless, I had hopes of convincing him to ask for the essential documents to be sent from Vienna. As it seemed, furthermore, that a description of his previous scientific and teaching activities might speed up the application he had to make personally, and seeing that he refused to take any action in this direction, I wrote to P.I. Dee asking him to send me an official statement concerning Bruno's activity at Glasgow University. Dee, who was not in the habit of such formalities, was rather taken by surprise, but nevertheless sent me a document signed by the registrar of Glasgow University, decorated with a few very attractive stamps! But on a return from a trip to Vienna in the summer holidays, Bruno told me that his father had got wind of the matter and asked to see his passport (which was Austrian, since he had still taken no effective action), and announced that following a discussion with his father he had decided not to apply for naturalization. I told him that I fully understood his reasons, including those of his father, and that I realized that if it was not possible to help him embark on a university career in Italy it was purely the fault of the short-sightedness of Italian law.

In order to provide a partial remedy for this situation, which had become even more intolerable because some of his Italian pupils were embarking, quite rightly, on their university careers, while Bruno was still grade R2, the INFN undertook formal measures to promote him from grade R2 to grade R1, for which the pay was the same as that of an Ordinary Professor, but which was not the same from the viewpoint of security, stability and pension, nor did it have the same social prestige.

The promotion took effect on 1 July 1963, on the unanimous approval of a commission appointed by the Consiglio Direttivo of the INFN, composed of E. Persico, B. Ferretti and G. Puppi, to examine the situations of Bruno Touschek and Ernesto Corinaldesi, who were unable to follow a normal university career because they were of foreign nationality[91)].

A much overdue solution to the problem of his university career was found in 1969 when the post of "Professore Aggregato" was introduced in the Italian Universities by a law where no mention is made of certificate of Italian nationality. He was thus appointed, in November 1969, Professore Aggregato to the Facoltà di Scienze M.F.N. of the University of Rome and was asked to give a course on "The mathematical methods of physics".

When, in 1973, the law came into force concerning the "Urgent Measures for the University" (law No. 580 of 1 October 1973) containing—*inter alia*—details of the transfer procedure to enable Aggregate Professors to become Extraordinary Professors, Touschek was appointed on 24 October 1973 Extraordinary Professor of "Mathematical methods of physics", on a unanimous proposal by the Consiglio di Facoltà di Scienze M.F.N. of the University of Rome.

In Italy, one normally becomes an Ordinary Professor after three years of teaching as an Extraordinary Professor. The decision is taken on the basis of the views expressed by the Faculty and a specific commission appointed by the Minister, concerning the applicant's scientific and teaching activities.

Bruno, however, refused to submit the application and the documents required by the Ministry, since he considered these bureaucratic formalities as an unbearable obligation, lacking in all consideration. All of these formal procedures, including filling in the application were finally performed by colleagues at the Istituto Guglielmo Marconi. The result was that it was not until early 1978 that the Ministerial commission could be convened and Touschek appointed Ordinary Professor.

In the meantime, however, the Accademia Nazionale dei Lincei appointed him Foreign Associate on 26 September 1972, and Bruno was highly pleased at this act of recognition, which to some extent offset the bitterness he felt towards the Italian universities.

During the sixties I got the impression on a few occasions that the very aspiration of Bruno was for a university career in Austria. Paul Urban has recently informed me that when Professor Theodor Sexl died, in

1967, Walter Thirring, who was already professor of Theoretical Physics in Vienna, proposed Bruno Touschek as Sexl's successor. At the beginning Bruno was very much in favour of such an appointment, but did not take an immediate decision and kept the Faculty in suspense for about two years. To the renewed insistences of Urban, Bruno finally answered personally: "I am from Vienna, my wife is from Scotland and therefore it is better for our children to grow up in a third country".

8. OTHER ASPECTS OF HIS NATURE AND ENJOYABLE RENEWED CONTACTS WITH HIS OWN COUNTRYMEN

Bruno Touschek was not only very keen on tennis, but also on swimming and underwater diving. Whether it was a little alpine lake, a Scottish "loch" or the lakes of Bracciano, Albano, or Nemi, Bruno was always ready do dive into the water even when the weather was far from warm. His passion for this became much greater at a later date, when his children were five or six years old and could enjoy swimming with him. He went with them in person to take swimming lessons and took an enormous delight in their progress. When he arrived at the Institute in those years he would speak to all of his friends about his children's success in sport or would give a vivid description of his own latest underwater adventure and the long, patient tactics he employed in order to catch a huge perch, which had been watching him for a long time from a creek among the rocks.

In order to keep himself and his young children in trim without having to go to public pools, he had set up in the courtyard of his flat, on the ground floor of No. 23 Via Pola, a 3 × 3 metre plastic swimming pool. Shortly after, however, he was forced to replace it by a smaller pool, because the ceiling of the garage under the courtyard was unable to carry the weight much longer.

During his conversations with his friends he referred from time to time to a few short stories of "Graf Bobby" and his friends "Baron Mucki" and "Herr Poldi", three typical figures in Austrian humour at the end of last century[92)]. They are rather decadent people who always meet the questions and circumstances of life in an inept manner or make remarks which are grammatically correct but quite illogical and inspired by detailed criticisms of all possible situations. Quite often, Bruno would re-evoke on the spot a short story featuring Graf Bobby, which was appropriate for the situation he and his friends were in at that particular moment, and would comment that he felt he was himself somewhat like Graf Bobby.

He would have a profuse supply of these remarks when he was feeling a little heady from a glass of wine that he drank adding the remark that "The superego is soluble in wine".

As Braitenberg wrote to me[5)] "in an attempt of self-analysis many years ago Bruno, a strong Freud follower (... perhaps for Viennese solidarity reasons), had noticed he had completely and irremediably lost the image of his mother. He knew he had seen her many times even when in his early 'teens'; he knew that he could count on much precise information about her existence and about the daily relations between mother and son but did not succeed in evoking her image. He believed, according to the canons of psychoanalytical orthodoxy, that from the removal of this emotion-charged complex there sprang out for the rest of his life the obscure forces which he did not feel he mastered. He refused, however, the suggestion of submitting himself to psychoanalysis. A first encounter with a Freudian psychoanalyst was not followed up. On that occasion Bruno told me he was afraid of losing the source of energy from which research stemmed, a price he was not prepared to pay for his serenity."

An event which was important for Bruno and certainly contributed to bringing him back closer to his country of origin was an invitation he received in 1975 to participate as an Austrian scientist at the "Staatsfeiertag" (State Holiday). After all, this was the first official recognition he had received from his country.

At the end of the Second World War, Austria's keen desire to reacquire independence, which was always strongly felt by at least part of the population, had been met by the formation of a national government, headed by the socialist Karl Renner and supported by a coalition of popular and socialist parties. Austria had, however, remained under Allied control, divided into four areas occupied by American, British, French and Russian troops[12)].

After long negotiations and Chancellor Raab's trip to Moscow, on 15 May 1955, the "Staatsvertrag" (State Treaty), which guaranteed Austria's independence, was signed at Belvedere Castle in Vienna by the foreign ministers of the four occupying forces and of Austria.

A few weeks later, the Republican National Council unanimously voted a constitutional law which established Austria's permanent neutrality. With this began the withdrawal of the occupational troops, ending on the 15 October 1955, with the departure of the Russian contingent.

This date is marked as the "Staatsfeiertag", which is celebrated every year and for which invitations are sent to Austrian citizens residing abroad, each time representing a different category. In 1975 it was the scientists' turn. One of the persons who helped prepare the invitations was Victor Weisskopf[93)], who had been a close friend of socialist Prime Minister Bruno Kreisky ever since high-school, where they had been members of the "Bund Sozialistischer Mittelschuler" (Socialist Association of Intermediate Schools), at the beginning of the 20's.

9. THE PERIOD OF THE e^+e^- RINGS

On 7 March 1960, Bruno Touschek held a seminar at the Laboratori Nazionali di Frascati, where he demonstrated for the first time the importance of a systematic and thorough study of electron-positron collisions (e^-e^+) and how this could be achieved, at least in principle, by constructing a single magnetic ring in which bunches of electrons and positrons circulate at the same energy E, but in opposite directions[94)]. By installing suitable particle detectors near to the parts of the ring where the bunches of opposite sign intersect, it is possible to study the particles emitted in all the reactions produced at the centre-of-mass energy $2E$, since the centre-of-mass of the two colliding particles is stationary in the laboratory reference frame (Fig. 1a).

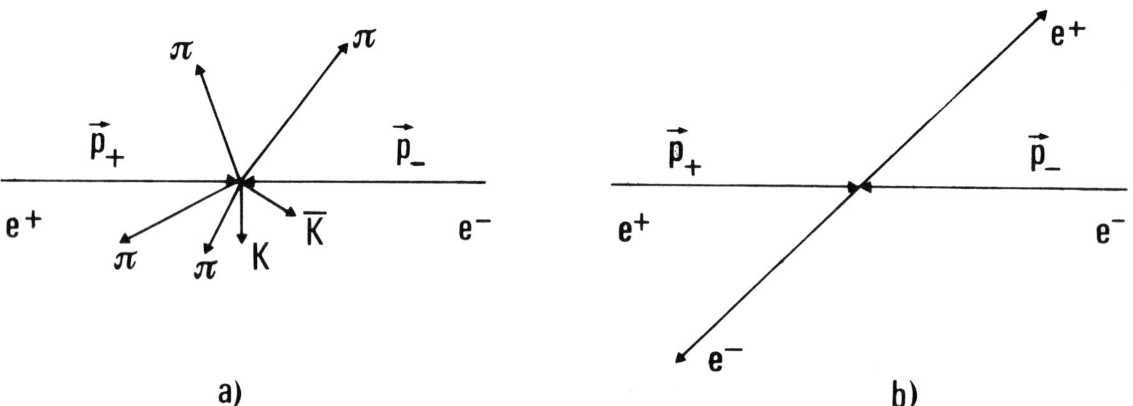

Fig. 1 Various processes are observed in e^+e^- collisions:
(a) production of $n \geq 2$ hadrons (π, p, K, ...);
(b) elastic scattering e^+e^-.

As I will clarify below, the talk of Bruno was more elaborate. It contained, however, already in this first part, two of the various arguments which made his proposal interesting:

i) It made it possible to obtain a considerable *kinematic advantage* for any process between particles. One of the most important parameters of such a study is always the relativistic invariant $W = \sqrt{s}$, i.e. the energy in the centre-of-mass of the two colliding particles. In the case of the usual accelerators in which a particle of mass M_1, energy and momentum E_1 and $\vec{p}_1$, collides with a particle of mass M_2 at rest, one has

$$W = \sqrt{(E_1 + M_2c^2)^2 - c^2\vec{p}_1^{\,2}} \tag{3a}$$

which, for $E_1 \gg M_1, M_2$ becomes

$$W = \sqrt{2M_2c^2E_1} \quad , \tag{3b}$$

which shows that W grows in proportion to the square root of the energy E_1 given by the accelerator to the incident particles.

In the case of two particles that collide with equal and opposite momenta ($\vec{p}_1 = -\vec{p}_2$) one has

$$W = \sqrt{(E_1 + E_2)^2 - c^2(\vec{p}_1 + \vec{p}_2)^2} = E_1 + E_2 \tag{4a}$$

If the two particles have the same mass so that $E_1 = E_2 = E$,one has

$$W = 2E \, , \tag{4b}$$

i.e. W goes up in proportion to the energy of each of them.

This aspect of the problem had been already pointed out by Kerst et al.[95] and O'Neill[96] in 1956. These authors, however, had considered only the case of e^-e^- collision obtained with the bunches of electrons circulating in opposite directions in two magnetic rings tangent to each other at a point where the collisions take place[97].

None of the articles of the two American groups mention the work done by Wideröe and of which I shall talk below.

Following O'Neill's proposal[96,97], a group at Stanford University had even started to design and construct a machine of this type[98], which was the first one producing very interesting scientific results on the e^-e^- collision[99].

ii) If the circulating particles have equal and opposite electrical charges, a considerable *constructional advantage* is obtained because a *single* magnetic ring can be used in which the particle bunches circulate in opposite directions. Also this point had been made by Wideröe who, years before, had discussed it with Touschek.

In his talk Touschek emphasized, however, two other important aspects of his proposal.

iii) The electron-positron system (i.e. e^-e^+) has the same quantum numbers as a neutral boson, so that at high energies it should become an electromagnetic particle source which is especially useful for studying strong interactions and electrodynamics. It also offered a number of various possible "two-body reactions", i.e. reactions in which, starting from the initial state e^+e^-, a final state is reached in which there is only one electron and one positron (elastic collision), or only two other particles, since the initial electron and positron disappear simultaneously owing to annihilation.

iv) Touschek also pointed out that, in any process which begins with the annihilation of a particle and its antiparticle (of initial equal and opposite momenta: $\vec{p}_+ = -\vec{p}_-$), the relativistic invariant q^2, known as four-momentum transfer, is always time-like, i.e.

$$q^2 = -(E_+ + E_-)^2 = -4E^2 < 0 \quad , \tag{5}$$

i.e. it enters a region of values which can be reached only through a few other processes in which e^+e^- pairs are produced by a (real or virtual) photon[100] or in hadron-hadron collisions[101].

The interest of processes of the first of these types was indicated in 1961 also by S. Drell and F. Zachariasen[102].

Two other arguments in favour of this line of research that were not indicated by Touschek but recognized years later, are the following:

v) The detailed study of storage rings has shown that these machines have an extremely high energy resolving power

$$\frac{\Delta W}{W} = 10^{-3}.$$

It is just this extraordinary property that has made possible a detailed study of extremely narrow resonances such as the J/ψ, ψ', etc.

vi) As it was clearly understood only many years later[103], the e^+e^-collision opens up the possibility of studying two-photon processes (Fig. 2)

$$e^+ + e^- \to e^+ + e^- + X \tag{6}$$

in which the particle X produced has the value $c = +1$ of the charge conjugation quantum number. This is a line of investigation of great interest for the 1980's[104].

All of the arguments discussed by Touschek, and their brilliant exposition, made a considerable impression on everyone present, including the then Director of the Laboratori Nazionali di Frascati, Giorgio Salvini, and Carlo Bernardini, Gianfranco Corazza and Giorgio Ghigo.

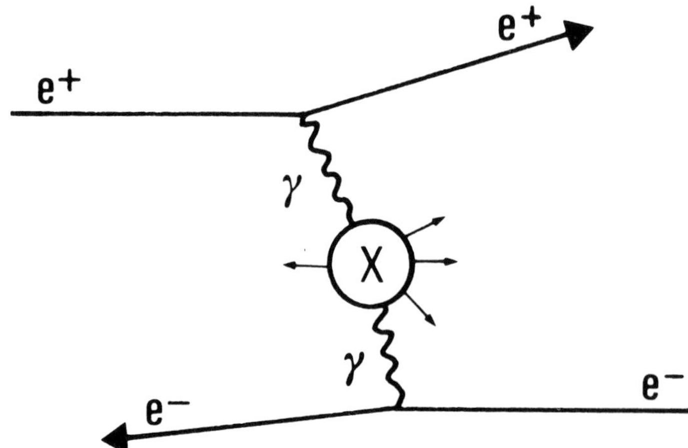

Fig. 2 Two-photon mechanism in e^+e^- interactions.

During the same day, the three last-mentioned persons began to work with Touschek on a project for the first e^-e^+storage ring, essentially designed as a protype for checking the feasibility of accelerators based on the ideas set forth by Touschek during the seminar.

This first machine for the study of collisions between a particle and an antiparticle was known as AdA (Anello di Accumulazione e^+e^-). The story of its design, construction and use has already been written about in detail by people who worked with Touschek during those years[105,106], and consequently only the main headings will be described here.

Touschek had, therefore, immediately found his first collaborators, but had also quickly found the financial resources. Giorgio Salvini, who had immediately realized the importance of the proposal, succeeded, shortly after this, in obtaining from the CNEN (of which Felice Ippolito was Secretary-General) an extraordinary grant of 20 million lire for the construction of the AdA prototype. This sum was almost entirely spent on the construction of the magnet, which was designed in a few days in a very original manner by Giorgio Ghigo[107] and produced with the assistance of G. Sacerdoti. A description of the project is set out by Ghigo in an internal note of the Laboratori Nazionali di Frascati, dated 8 December 1960[108].

It should first be said that the idea of constructing machines based on the collision of two particle beams, instead of one particle beam which strikes a fixed target, was devised for the first time by Rolf Wideröe in the late summer of 1943, during his holiday at Tuddal, near Telemark in Norway[22]. As Wideröe wrote himself:

"On a nice summer day I was lying on the grass seeing the clouds drifting by and then I started speculating what happens when two cars collide. If we have a car moving with the velocity v colliding with a resting car of equal mass, the dissipated energy will be $1/4\ mv^2$ (inelastic collision), whereas two cars with the velocity v having a head-on collision would dissipate four times as much energy (mv^2) is spite of having only twice the energy before the collision. This clearly demonstrated that head-on collisions have to be avoided for cars, but might be very useful for protons[109].

"When I discussed this idea with Touschek later, he did not appear very impressed at the moment. All he said was 'It is something obvious and trivial and cannot be patented'."[22] But Wideröe nevertheless managed to obtain a patent in May 1953[110], when he had already been working for many years in the Brown Boveri research laboratories in Baden, Switzerland. This was some three years before Kerst et al.[95] suggested, independently, that use should be made of the collision of two equal particle beams, such as a proton-proton or electron-electron beam, circulating in two different magnet rings.

In his patent, Wideröe discussed collisions between equal particles (proton-proton), and different particles (proton-deuteron), or particles of opposite charge (electron-nucleus, in particular electron-proton), suggesting various possible systems for the magnetic rings illustrated by four different figures. The collision always occurs in a ring (Reaktionsröhre): in the case of particles bearing an opposite charge, these are kept on their orbit by the magnetic field itself. In the case of particles of equal charge, Wideröe suggests the utilization of electric fields, but does not give any details about their design.

A person who seems to have come to appreciate, at about the same time as Touschek, the importance of point i) (kinematic advantage) and, for the e^+e^-case, point ii) (constructional advantage) is G.I. Budker (1918–1977)[111)] of Novosibirsk, as he himself stated on various occasions subsequently[112)], and as is confirmed by the rapid construction of the VEPP 2 machine[113)]. The other attempts by Budker and the Stanford-Princeton group always concern machines composed of two tangential rings for e^-e^- collisions[114)].

As Wideröe wrote[22)]: "It was only after Touschek had 'broken the ice' with his small AdA and later on with ADONE that people got really interested in this principle. Today it is one of the leading ways to study elementary particles".

The AdA project is described in the paper by Bernardini et al. [46], from which I have taken Table 1, and the few further details given below.

Table 1

Some characteristic parameters of AdA [46]

Magnet weight	8.5 t
Outside diameter	160 cm
Number of straight sections	4
Useful magnet gap	5 cm
Cavity frequency	147.2 MHz
Energy of each beam	200 MeV

With AdA it was possible to attain a maximum centre-of-mass energy W of 400 MeV (or slightly less), which could be obtained only with a beam of positrons having an energy $E_1 = 160$ GeV in collision with the electrons of a fixed target. This example shows, amongst other things, the kinematical advantage [point i)] of intersecting-beam machines over fixed-target machines.

AdA's energy was more than sufficient to give rise to three important two-body processes:

$$e^+ + e^- \rightarrow \begin{cases} e^+ + e^- \\ \mu^+ + \mu^- \\ \pi^+ + \pi^- \end{cases}$$

But it was not, in fact, possible to study them owing to the extremely low number of electrons and positrons circulating in AdA.

The electrons were injected with a beam of γ-rays produced with the Frascati electron synchrotron. The beam was directed at a metal target inside AdA's high-vacuum chamber, and produced e^+e^- pairs. The electrons of one sign were captured in a closed orbit and were then bunched and accelerated by the RF system. Those of the opposite sign were deflected by the magnetic field towards the outside of the machine and were lost. Once the electrons of one sign had been injected, the whole machine, together with its base, which was equipped with wheels (Fig. 3), was moved along a track at right angles to the beam, until it had covered a distance sligthly larger than the orbit diameter. In this new position, the injection of electrons of opposite sign was repeated, using a second metal target located symmetrically with respect to the centre of the magnet. With this injection the electrons approached the equilibrium orbit, as a result of radiation losses starting from the metal targets which lay slightly outside the equilibrium orbit.

Touschek discussed the theory of this injection procedure in an internal note of the Laboratori Nazionali di Frascati [45].

As the work progressed, the group was expanded with the addition of three new collaborators (U. Bizzarri, G. Di Giugno, and R. Querzoli) and the support of experts in the field of RF cavities (A. Massarotti and M. Puglisi) and magnet construction (G. Sacerdoti).

Fig. 3 Photograph of AdA mounted on its movable support.

AdA began operation on 27 February 1961 [48], a date which had a special meaning for Bruno, since it was on that day a few years earlier that his aunt Ada had passed away.

This initial result had required the solving of certain complicated technological problems, especially that of constructing a chamber with a vacuum of not more that 10^{-9} Torr. With his pioneering work on the production and measurement of vacuums up to 10^{-11} Torr, Corazza succeeded in fulfilling this requirement which was essential for the circulation of electrons for a time long enough to allow the construction of this type of machine. The operating level of the machine was described very precisely by Touschek himself during a lecture on storage rings which he gave to the Centro Linceo Interdisciplinare (see Section 14):

"We had stored 80 electrons (or positrons) (one never knew which, since the discussions on signs were unending and conflicting). The work of measuring the decay curve was left to Peppino Di Giugno alone, at that time the youngest member of our '*équipe*'. At 7 a.m. I received a phone call: 'There are still eighteen left, can I kill them?' My reply can be easily guessed. It was the experimental proof that it was possible to obtain mean lifetimes of many hours, in this particular case, five hours, which was essential in order to attain significant intensities for sufficiently long periods to allow the measurement of the collision cross-section of the various processes."

The presence of even a small number of electrons circulating in the machine was determined by using a photomultiplier to observe, through a plastic window, the synchrotron light emitted tangentially to the orbit in one of the "straight sections" of the machine, where, in fact, the average radius of the equilibrium orbit was not infinite but simply twice that in the quadrants. Figure 4 shows part of the recording of the output of the

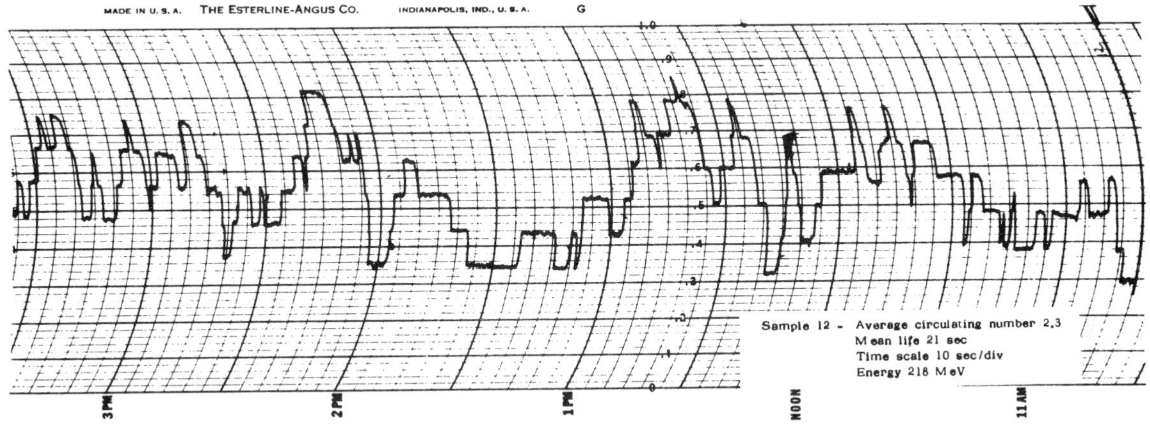

Fig. 4 Recording of the current of the photomultiplier used for observing the light emitted by single electrons circulating in AdA.

photomultiplier, where one can clearly see the sharp changes due to the loss (down steps) or capture in the orbit (up steps). The synchrotron light emitted by a *single electron in orbit* could in fact be seen through a telescope, as could, incidentally, easily be calculated from the electron energy (E = 200 MeV), the curvature radius of the orbit and the fact that the period taken by the electrons to make one turn [at a velocity which differed from c only by $(m_e c^2/E)^{1/2}$, i.e. one part in 800] was about 14 nanoseconds. A similar recording to that shown in Fig. 4 was sent to me by Touschek on the morning following the second night of AdA's operation, and during the next night I myself went to see with my own eyes the synchrotron light which was emitted by a single electron. Bruno took an immense pleasure in showing this phenomenon which, to a certain extent, was commonplace, but at first sight appeared incredible. His enthusiasm was extreme when P.I. Dee and his wife made a trip to Rome at that particular time.

A significant experiment on e^+e^- collisions could not, however, be made with such an inefficient injection system as that offered by the facilities available at the Laboratori Nazionali di Frascati. Appropriate arrangements were then made with Orsay Laboratory near Paris, where there was a linear accelerator producing electrons up to 1 GeV, which was obviously much more suitable as an injector for AdA. In the summer of 1962, AdA was "moved to France over the Alps on a large 'truck'; the vacuum chamber was blanked off and the titanium battery-fed pump kept in operation at a pressure which did not exceed 10^{-8} Torr during the trip. The only problems encountered were with the Customs Officers (who were understandably surprised)." [106)]

At Orsay the Frascati group was joined by Pierre Marin and François Lacoste. The latter collaborated for a short while but then left the laboratory to devote himself to other research work, and was replaced by Jacques Haissinski, who, with his Docteur ès sciences thesis, remains the best biographer of AdA[105)].

At Orsay, AdA was installed in an intermediate station (the 500 MeV "salle de cible") of the linear accelerator (LINAC), which, when operated in the correct manner, allowed the injection of 4300 electrons (positrons) per second, at a medium LINAC intensity of 0.5 μA.

The first experiment to be carried out at the machine's new location was to make extremely careful measurements of the mean lifetime τ of the bunches circulating in the machine. The new measurements, however, carried out at the beginning of 1963 at 195 MeV per beam, showed a new phenomenon which very soon became known as the "Touschek effect" |52|.

The Italo-French group found that the results of the measurements of the mean lifetime of each beam could be represented with good accuracy by the expression

$$\frac{1}{\tau} = \frac{1}{\tau_0} + aN \quad ,$$

where τ_0 is the mean lifetime measured when the bunches of electrons have a very low density, and N is the number of particles contained in the beam. They also found that the coefficient a depended very much on the energy

$$a \sim E^{-9/2}$$

(for E in the 100–200 MeV range). As C. Bernardini[106)] writes: "Touschek found the mechanism of the phenomenon by spending one night working on it (during which he went away from the room in which we lived, distraught by the din of the RF system's fans and the chattering of the internal telephone with the LINAC operators who were a very long way away). He observed that: i) the problem had to concern a mechanism within the circulating bunches, ii) it should be a collision effect, since the loss was proportional to the number of pairs of electrons in a bunch. He then observed that, by describing the stability of the beams in terms of a three-dimension potential well, the longitudinal walls of the well were much lower than the transverse walls and consequently the transfer, owing to collision, of the transverse energy into longitudinal energy necessarily resulted in electron losses. An approximate calculation immediately gave the correct expression for τ, in the 100–200 MeV range; we then completed the theory of the effect by extending it to each

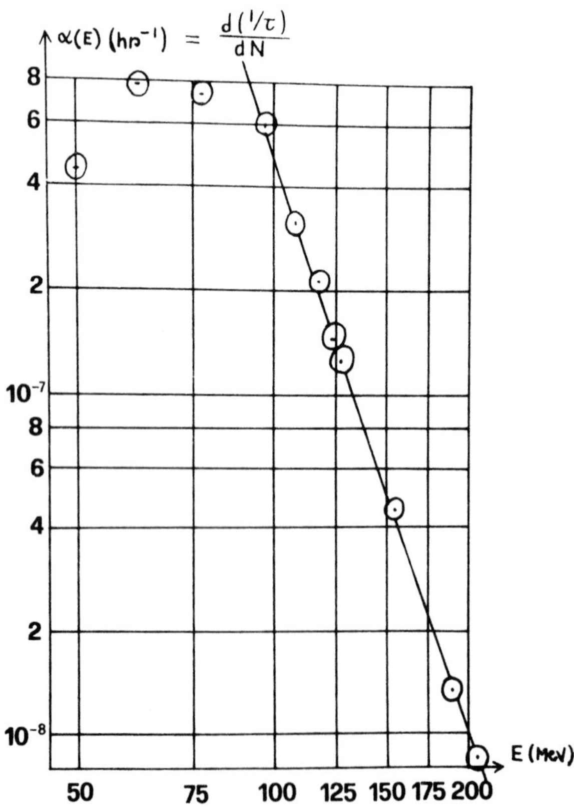

Fig. 5 Comparison of the theory of the "Touschek effect" with experimental data. (Touschek's original drawing.)

energy (Fig. 5). The immediate conclusion was, however, that since the loss was comparable with the injection velocity for $N = 4 \times 10^7$ incident particles, we were unable to go beyond this level (at 200 MeV) and in fact even at $2-3 \times 10^7$ particles saturation had been virtually attained."

Since the effect concerned was due to collisions between particles belonging to the same bunch, one way of reducing it was to decrease their density without reducing their number. In other words, it was necessary to increase the volume of the bunch. Since the vertical dimension was much smaller than the horizontal dimension, it was possible, by coupling the betatron oscillations between the two modes, to "boost" the beam during the whole injection period, i.e. when the effect was most harmful. This result was obtained by means of a quadrupole coil and orienting its axes in a suitable manner[106)]. In this way, the Italo-French group succeeded in increasing the mean lifetime of the bunches by a factor of three.

The initial programme, which was to demonstrate experimentally that two beams could be made to intersect successfully and observe the e^+e^- collision, was based on the revelation, in coincidence, of the two γ-rays emitted in the process.

$$e^+ + e^- \rightarrow \gamma + \gamma.$$

The cross-section of this reaction is, however, too small to be observed even with the intensity of the beams produced at Orsay. Touschek's group consequently resorted to observing single bremsstrahlung

$$e^+ + e^- \rightarrow e^+ + e^- + \gamma, \tag{7}$$

the cross-section of which is 1000 times greater than that of the former process.

In order to distinguish the γ-rays emitted in reaction (7) from those diffused by the residual gas, Touschek and his collaborators took advantage of the fact that the γ-rays diffused by molecules of gas only move forward with respect to the direction of the incident electrons. It follows that the number of photons C revealed per unit time is linked to the numbers N_1 and N_2 of particles present in the two beams by the relation [53]

$$C = aN_1 + bN_1N_2, \tag{8}$$

valid when the gamma-rays are revealed in a forward direction with respect to beam 1. The first of the two terms on the right of Eq. (8) represents the contribution from the residual gas, and the second the effect of the e^+e^- collision. It follows from Eq. (8) that

$$\frac{C}{N_1} = a + bN_2 \quad , \tag{9}$$

i.e. the ratio C/N_1 must be a linear function of N_2, namely of the intensity of the other beam. From measurements of this type, Touschek's Italo-French group succeeded in determining the parameter b (Fig. 6), thus demonstrating that the electrons belonging to the two beams did effectively interact [53].

In the period devoted to the experimentation on and with AdA, Bruno participated with extraordinary intelligence and efficiency in all the various phases of the work. In particular he prepared with great care the machine runs, writing a detailed programme of the measurements to be made, and at the end of the run, he always summarized the work done in order to have a clear and ordered view of the results. Figure 7 shows, as an example, the first page of one of these "Summaries of Results" after a few runs made at Orsay.

During the summer of 1963, the Brookhaven National Laboratory held the "1963 Summer Study on Storage Rings, Acceleration and Experimentation at Super-High Energies", at which Touschek presented a report [50] entitled "The Italian storage rings". At about the same time, C. Bernardini submitted, on behalf of the Frascati group, a report to a Dubna symposium entitled "Lifetime and beam size in electron storage rings" [54], which was referred to in a brilliant article on the symposium entitled "The accelerators of the future", which appeared in *Pravda* on 25 August 1963.

The interest in the e^+e^- storage rings was now extremely keen not only in Frascati but also in many other laboratories.

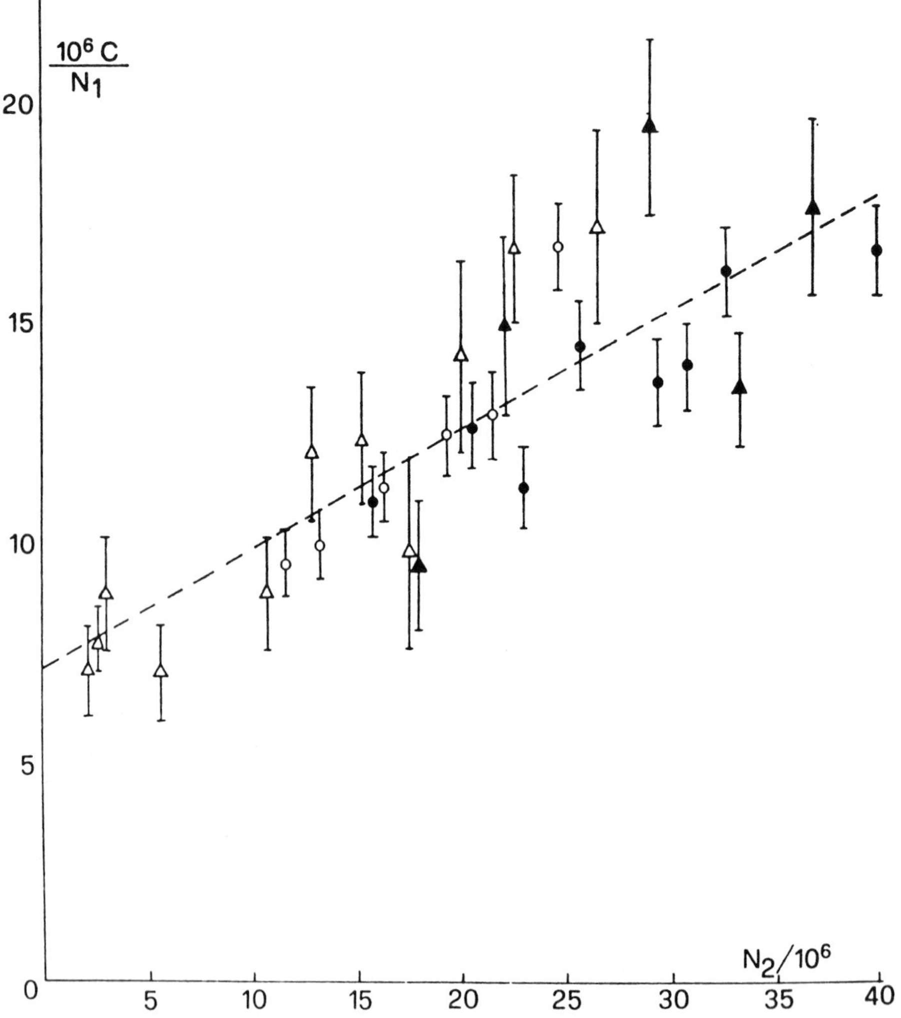

Fig. 6 Counting rate C per particle (C/N_1) in a bunch as a function of the number of particles in the opposing bunch, N_2. (Touschek's original drawing.)

①

(1) Observation of single electrons with the electron counter.
HV = 1800, dark current at the beginning of the measurement 40 nA.
We observe the decay of 20 electrons:
from 222 nA to 176 nA
∴ 46 nA /20 electrons. 2.3 nA/electron
1 μA = 435 electrons.

(2) 1st attenuation. We accumulate a current of 1.46 μA. The dark current is 0.054 μA the effective current therefore 1.406 μA corresponding to 1.406 × 435 = 611 electrons.
The voltage is changed to 1480 volts.
We read 0.146 μA. Dark current 0.024 nA & therefore 0.122 μA effective. This gives 611 : 0.122 = 5.01×10^3 electrons/μA.

(3) 2nd attenuation. A current of 2.215 μA is accumulated. This corresponds to 2.215 − 0.024 = = 2.191 μA = 1.096×10^4 electrons.
Voltage changed to 1220 volts. Current 0.221 μA; dark current 0.018. Effective current 0.203 μA. This gives 1.096×10^4 : 0.203 = 5.4×10^4 electrons/μA.

(4) 3rd attenuation. Current of 1.15 μA accumulated ≈ 1.13 μA = 6.11×10^4 electrons.
At 1000 volts 0.115 μA; dark current 0.015 ∴ 0.100 μA effective. Efficiency 6.11×10^5 electrons/μA

Fig. 7 Part of the summary of the conclusions reached in a few machine runs made at Orsay, hand-written by Touschek.

Already in the course of the same year, 1960, at the Laboratori Nazionali di Frascati a start was made, under the direction of Fernando Amman[115)], on the design of ADONE (great AdA), which was planned for extending the study of the processes generated in e^+e^- collisions up to centre-of-mass energies $W = 2E = 3.0$ GeV. In Section 11 I shall come back to the contribution made by Touschek to this programme.

In 1964 in Orsay a decision was taken to construct ACO, an e^+e^- strong-focusing ring for a centre-of-mass energy of $W = 2E = 1.1$ GeV [116)]. Its rapid construction and operation were possible also because of the existence in that laboratory of the already-functioning linear accelerator, which was used for the injection of the electrons.

In 1965 a modification of the Cambridge Electron Synchrotron (CEA) of Harvard-MIT was also decided in order to rapidly dispose of two e^+e^- colliding beams[117)] at $W= 2E= 5.0$ GeV (Table 2).

In Section 12 I shall come back to the later development of the line of research based on e^+e^-storage rings, considered from a general point of view.

10. BEAM STABILITY AND RADIATIVE CORRECTIONS

Touschek's "experimental period" had thus come to an end, although this was not true of his interest and participation in the development of the e^+e^- rings. For a few years more his attention was concentrated on two fundamental problems also in connection with the design and construction of ADONE. The first was the problem of the stability of the electron and positron ultrarelativistic beams, and the second was that of radiative corrections.

The first work on beam stability was made in collaboration with C. Bernardini [44] and dates back to 1960, in other words the period prior to the construction of AdA. It concerns the losses incurred by the electron bunches which circulate within a synchrotron at a velocity close to that of light. As has already been said, the fact that the electrons in a bunch are not rapidly lost during the motion is represented by a potential well which moves at the velocity of the centre of mass of the bunch and in which the electrons remain trapped. Both the problem of the longitudinal (synchrotron) oscillations and that of the transverse (betatron) oscillations are dealt with in this manner, except that the potential well related to the longitudinal oscillations is much shallower and shows a sharp edge (stability limit) which depends on the radiofrequency voltage amplitude. Naturally, the containment of the beam is never complete, and there are always electron losses which it is very important to calculate.

The problem was tackled by R.F. Christy[118)], who had calculated the losses due to synchrotron oscillations. Matthew Sands, who was also at Caltech, had carried out certain measurements on the electron synchrotron of that laboratory, and had communicated his experimental results[119)], which disagreed with Christy's predictions, to Fernando Amman. The latter started to take a few measurements with the Frascati electron synchrotron and found that, in agreement with Sands, the losses were considerably higher than those computed by Christy.

Informed by Amman, Bernardini began to work on the theory of this problem; pretty soon, Bruno Touschek joined him and the problem was quickly solved by their joint work.

For low excitation the quantum levels of the electrons inside the potential well are essentially those of a harmonic oscillator. But as the excitation increases, an ever-increasing anharmonicity appears since the top of the potential well becomes closer. The levels, however, are generally so close to each other as to form a continuum.

The transition from one state to the other are due to the recoil caused by the emission of the photons of the synchrotron light. Bernardini and Touschek tackled the problem as a diffusion process, described by a Fokker-Planck equation, and reached the conclusion that the loss of electrons was mainly determined by processes involving a large number of quantum jumps, whereas those calculated by Christy, which involve one or a few transitions, provide a much lesser contribution. In their work, Bernardini and Touschek derive the expression for the lifetime of a bunch of electrons in terms of the attenuation of the synchrotron oscillations and a parameter, between zero and one, related to the amplitude of the potential provided by the synchrotron's resonant cavities.

The paper written with E. Ferlenghi and C. Pellegrini [56] concerns the instability of a circulating beam of electrons caused by signals transmitted by the actual bunches to the walls of the chamber (and to various components of the high-frequency cavity, all conductors). These signals generate currents, which in turn produce fields that have a delayed phase effect on the beam bunches.

These instabilities, due to the wall's resistivity, had been observed at MURA[120)] and later in a few proton synchrotrons. The theory had been elaborated by L.J. Laslett, V.K. Neil and A.M. Sessler[121)] for the ISR (Intersecting Storage Rings) of CERN. Touschek, Ferlenghi and Pellegrini extended the theory to the more complex case of bunched beams.

The other problem, closely related to experiments on the reactions produced in e^+e^- rings, was that of the radiative corrections, where Touschek was assisted by various young collaborators. The first approach to this complex problem is set out in the thesis of Gian Carlo Rossi, who graduated with Touschek in 1966. The case concerned two beams moving in directions which formed between them an angle slightly less ($\sim 15°$) than 180°.

Touschek's general idea was to find a reasonable compromise between a strictly defined geometry of the final state, in which the radiative corrections could be calculated more easily but are, in terms of percentage, large, and an open geometry in which the corrections are smaller but difficult to calculate since they involve many perturbation orders.

The problem was tackled more thoroughly in collaboration with Etim Etim [57], and subsequently extended, and to some extent concluded, in the subsequent work with Etim Etim and G. Panchieri [58].

The case under consideration is the calculation of the cross-section observed experimentally in the processes involving the production of a particle A and its antiparticle $\bar{A}$, i.e. processes of the type

$$e^+ + e^- \rightarrow A + \bar{A},$$

where A can be a muon, a K meson, or some other particle.

In the case, for example, of a K meson, if the initial energy $W = E_+ + E_-$ is equal to or greater than the rest energy of the ϕ meson ($m_\phi c^2 = 1020$ MeV) the production of $K^\pm$pairs takes place essentially by passing through the intermediate stage "ϕ", i.e. in accordance with the scheme

$$\begin{gathered} e^+ + e^- \rightarrow \phi + n\gamma \ (n = 1, 2, ...) \\ \phi \rightarrow K^+ + K^-, \end{gathered}$$

and the collision cross-section observed experimentally is given by the product of two factors

$$\sigma_{\exp} = \sigma_0(\phi;\omega) \cdot p(\omega), \tag{10}$$

where the first factor is the collision cross-section for production of the resonance ϕ, and the second represents the radiative correction, i.e. the probability of emission of any number of photons whose total energy is $\omega = W - m_\phi c^2$. As Bruno said, "If the radiative corrections are not properly administered, the collision cross-section $\sigma_{\exp}$ diverges".

The approach developed by Bruno and his collaborators in these papers is based on the use of the Bloch-Nordsieck[122)] method, and consists of a procedure which makes it possible to summate the contributions due to the emission of $n \geq 2$ photons in the limit for very low values of both the energy of the actual photons and of the resolving power of the instruments used to detect them.

By very simple reasoning, it is found that in the first order the following formula is valid

$$\sigma_{\exp} \simeq \sigma_0 \left(1 - \beta \ln \frac{W}{\omega}\right) \quad , \tag{11}$$

which was used by various authors. In Eq. (11) ω is the resolving power expressed as total energy of the emitted photons. This logarithmic dependence of $p(\omega)$ on ω is an obvious consequence of the fact that the bremsstrahlung spectrum is of the type $d\omega/\omega$. However, as we shall see shortly, Eq. (11) is incorrect by a large factor.

In papers [57] and [58], Touschek and his collaborators showed that by summing over all the possible orders (i.e. over all the processes in which any number of photons is emitted of global energy ω) Eq. (11) is replaced by the exact expression

$$\sigma_{\exp} = \sigma_0 \cdot \omega^\beta . \tag{12}$$

$\beta = \beta\,(W,\omega,\theta)$ is a function calculable in the individual cases, which, for $v \rightarrow c$, assumes values close to $\beta = 0.07$.

Introducing into Eq. (12)

$$\omega^\beta = e^{\beta \ln \omega} = 1 + \beta \ln \omega + ..., \tag{13}$$

we immediately see that the radiative correction calculated to the first order is very much over-valued. The expression (13) contains the factor $\beta \simeq 0.07$, which Bruno Touschek, with his own particular jargon, referred to as the "Bond factor" [123)].

11. THE CONTRIBUTION OF TOUSCHEK TO ADONE

From an examination of the publications and other printed documents, the person interested in the historical development of the e^+e^- rings might be led to think that Bruno Touschek had only a marginal role in the design and construction of ADONE. And this would seem strange, having in mind how profoundly and directly he had been involved in the construction and experimentation on and with AdA. As Fernando Amman wrote to me[124)]: "One should say first of all that at the beginning (1960) the liaison between the AdA group and the embryo of what later became the ADONE group had been rather tight, and this through the action of Bruno, Carlo Bernardini and Gianfranco Corazza; the first theoretical competences on the ring, as well as the technological competences (in particular for the vacuum) evolved in a closely interknit manner. The contacts were very tight during the ADONE design phase (1963–64) which coincided with the measurements on AdA. In this period, the key figure in the liaison work was Gianfranco Corazza, whose role in both projects has perhaps not yet been adequately underlined.

"I should here recall the profound link between Bruno and Gianfranco. The latter was the person whom Bruno trusted most for his capacity to overcome any technological obstacle. To this one should add that the calm and reassuring nature of Gianfranco was the best complement to Bruno's creative anxiety.

"Bruno lived intensely the various phases of ADONE: the design as well as the construction period (1965–67) and especially the difficult year 1968 during which various beam instabilities were studied and cured. We did not meet frequently, but with regularity. Bruno was afraid of causing me to waste precious time, as he would say to C. Bernardini, and therefore followed almost daily the activity of the group, keeping in contact with Gianfranco Corazza. He was afraid of the dimension of ADONE: he considered it an enterprise of industrial type and with this he justified his reverential awe; but at the same time he felt that this was really the materialization of his original idea. Whenever a new problem came up on which he was certain to be able to contribute, Bruno was present, without the need to look for him; an example was the case of the transverse instabilities (1965).

. . . .

"Much time and effort was devoted by Bruno to the Committee for Experimentation with ADONE which started to operate in 1966; in this body he was a strong believer in the importance of a preparation of the experimental equipment adequate to the machine, but at the same time he was against all-embracing and monopolizing approaches. I think that altogether this has been an experience considered negative by Bruno in the sense that it led to foreseeing those difficulties in the use of the machine that actually occurred later."

Between the tendency to assign all, or almost all, the available resources to a single group that thus could have disposed of high-performance equipment and the opposite tendency of dividing the same funds between various groups, each by necessity endowed with an apparatus of limited performance, it was certainly not easy to find the right compromise! The solution finally adopted involved an excessive fragmentation of the financial means, with consequences not completely favourable from the scientific stand-point, and a certain disappointment to Bruno Touschek and Fernando Amman.

"What I would like to underline" continues Amman, "and I hope with this not to distort the perspective of the merits of single persons, is the substantial unity of the AdA-ADONE enterprise well beyond the official documents; this unity was achieved through the contributions of Bruno, Carlo Bernardini, Gianfranco Corazza, Ruggero Querzoli (and myself). As regards the part of which I had the direct responsibility, ADONE, I am absolutely convinced how important this unity was in proceeding in a difficult enterprise in far from ideal conditions (I recall that after Medici there was the difficult period of Lami-Starnuti as President of CNEN, and then that of Tanassi!)[125)]. This support and collaboration inside the Laboratories was assured by Bruno, Carlo and Querzoli; I feel also indebted for all that I learnt during the first years of AdA, which was of great utility to me in the subsequent period.

"How does Bruno fit into all that: he was the initiator, but also the element of continuity during the ten golden years of the Laboratories, the person that had a great idea and allowed it to be materialized by others; his scientific and human qualities, I believe, were decisive in maintaining the connections which have been essential in achieving success; if success there has been, or in the limits in which there was success. For these reasons I believe that one should accord to Bruno a primary role in the adventure of the electron and positron storage rings, an adventure that saw the emergence on the international level of an Italian laboratory more than that of individual Italian physicists....

"The epilogue of this adventure was not consistent with its beginning for a series of reasons among which the contestation in Frascati was the most apparent, but certainly not the only one; the seeds were already perceptible in the meeting of the Commission for the Experiments with ADONE that took place before 1969, in which started to appear the difficulty of organizing a serious and coherent experimental effort of Italian groups in Frascati due in part to financial limitations and to the competition from CERN.

. . . .

"This first operation of ADONE with a single beam started on 8 December 1967, and during the course of 1968 the instabilities were studied and cured. In April/May 1969 the first measurements of the luminosity of the two beams were made. The machine was then opened for the installation of the straight sections for the experiments and the protest movement started while we were closing the vacuum system (this work was actually completed in the subsequent days when the Laboratory's activities were stopped) and precisely Friday 30 May 1969. The more or less regular operation was resumed on 19 September 1969."[126].

12. IMPACT ON THE DEVELOPMENT OF SUBNUCLEAR PARTICLE PHYSICS OF THE STUDY OF e^+e^- HIGH-ENERGY COLLISIONS

If one looks back over the history of subnuclear particle physics during the last twenty years it clearly emerges that the study of the e^+e^- high-energy collision constituted and still constitutes the main means of obtaining more detailed knowledge in this field. Furthermore, as is obvious from the three preceding sections, the contribution made by Bruno Touschek at the outset and during the development of this approach is absolutely exceptional.

Not only was Touschek the first, at the beginning of 1960, to sense the fundamental importance of this line of research, but he himself contributed to a whole series of subsequent developments such as: the designing of the prototype AdA, in which he was involved directly, the design and performance of the first experiments with this machine which led him to measure the lifetime of the particle bunches and finally to work out the complete theory of the phenomenon, which is correctly referred to in the literature as the Touschek effect. Subsequently Bruno studied one of the instabilities of the beams, caused by the fact that these interact through the (conducting) wall of the vacuum chamber. He provided a substantial clarification of the problem of radiative corrections, which constituted an essential ingredient for the derivation from the experimental counting rates not only of the absolute values but also of the relative values of the collision cross-sections of the various processes. Finally, he made a considerable contribution by discussions, calculations, and advice, to the design of ADONE, which is the first machine of this type, whose design and construction reflect the most advanced optimization criteria at the end of the sixties.

Touschek's direct influence on those who were close to him was outstanding.

Already in a seminar held in 1960 at the Istituto Guglielmo Marconi a few weeks after the one at Frascati, Touschek spoke on the same subjects but in a broader and more complete manner. He also listed at least 16 two-body reactions and the need to take radiative corrections into account.

One of the important consequences of this stimulating action was that N. Cabibbo and R. Gatto[127)] as well as L.M. Brown and F. Calogero[127)] worked out expressions for the collision cross-sections of all the processes foreseeable in the sixties, as final channels of e^+e^- high-energy collision. These results were obtained within the framework of quantum electrodynamics, completed with the introduction of phenomenological form factors to take into account the effect of strong interactions. Together with certain classical results [as far as electrodynamics is concerned[103)]], these constituted the fundamental term of comparison for the analysis of the data from all the experiments made at ACO, ADONE and, to a large extent, the machines which immediately followed.

These experiments, in particular those performed with ADONE[128)], can be divided into two categories, depending on whether they concern purely electrodynamic processes or processes involving strong interactions such as all of those in which a production of hadrons is observed.

The experimental study of the electrodynamic processes performed at ADONE allowed further improvements of the limits of the validity of quantum electrodynamics[129)], and to demonstrate for the first time the photon-photon interaction processes[130)].

The experimental study of the processes with the emission of hadrons showed that an abundant multi-hadronic production begins to occur just at the energies of ADONE[131)]. More precisely, by determining experimentally the ratio

$$R = \frac{\sigma_h}{\sigma_{\mu^+\mu^-}} = \frac{\sigma(e^+e^- \to \text{hadrons})}{\sigma(e^+e^- \to \mu^+\mu^-)} \tag{14}$$

between the collision cross-section for production in e^+e^- collisions, of at least two hadrons and $\mu^+\mu^-$ pairs, one began to find, already at the energies of ADONE ($W \leq 3$ GeV), a greater value than that calculated by Brown and Calogero[127)], on the basis of purely electrodynamic considerations.

This phenomenon constituted, among others, one of the first proofs of the "parton" structure of hadrons[132)] and the need to identify these partons with quarks [i.e. particles having charge 1/3 and 2/3, of three different types (colours)].

There can be no doubt about the fact that it was Touschek who opened up this major approach, which, right from the beginning in 1960, was in competition, with regard to the study of electrodynamic processes and the possible production of new particles, with the observation of the wide-angle emission of e^+e^- pairs in high-energy proton-proton collisions[133].

Both of these approaches led, approximately at the same date[134], to the discovery of the J/ψ, but the wealth of information which could be gathered from a study of all the final channels of the processes initiated in the e^+e^- channel is incomparably greater. In particular, the determination of the ratio (14) is obviously possible only when starting from an e^+e^- collision, which, when studied experimentally at increasingly high energies and with higher resolutions, led to confirmation that multi-hadronic production, which could be recognized but was certainly not conspicuous at ADONE's energy, becomes rapidly a dominating aspect of the phenomenological framework. It was by following this course that the discovery was subsequently made of the ψ' and later of the D and F mesons[135], whose existence called for the intervention of a new quantum number, charm, as theoretically predicted by S. Glashow, J. Iliopoulos and L. Maiani[136]. All of these discoveries made the greatest contribution to the substantial change in the theoretical framework of the subatomic particles as from 1970.

Although the process

$$e^- + e^+ \leftrightarrow p + \bar{p}$$

was indicated for the first time by Drell and Zachariasen[102] for its particular interest as a means of entering the "time-like" region, its experimental study was tackled[137] only following a theoretical treatment[138] developed as a natural extension of that relating to one of the many processes which we can today refer to as the "Touschek reactions".

Table 2, taken from the article by Robert R. Wilson on the next particle accelerators[139], shows in chronological order, and at the same time in increasing energy, the complete family of "storage rings". The list begins with AdA and the first machines mentioned at the end of Section 9 and ends with the accelerators of this type, which are still being constructed or are in the course of design. The list includes a single ring for the only mode e^-e^-, two rings of the pp type, and five of the $p\bar{p}$ type. All of the others are of the e^-e^+ ($\equiv e\bar{e}$) type, i.e. machines following the line of research recognized and initiated by Bruno Touschek. It should be added that also the machines of the type $p\bar{p}$ represent a natural extension and variant to the type $e\bar{e}$.

For his physical insight and design work relating e^+e^- rings and his contributions to the development of this type of machine, the Accademia Nazionale dei XL awarded Bruno Touschek the 1975 Matteucci Medal[141].

Table 2[a)]

Storage rings

	Year of entry into service	Type	Particles accelerated	Energy per beam (GeV)	Centre-of-mass energy (GeV)	Luminosity[b)] ($cm^{-2} s^{-1}$)
AdA, Lab. Naz. Frascati, Italy (dismantled)	1961	Single ring	e^+e^-	0.25	0.50	*10^{24}*
Princeton-Stanford, USA (rings dismantled)	1962	Tangential rings	e^+e^-	0.56	1.1	*10^{28}*
VEPP 2, Novosibirsk, USSR	1964	Tangential rings	e^+e^-	0.7	1.4	*10^{28}*
ACO, Orsay, Paris, France	1965	Single ring	e^+e^-	0.5	1.0	*5×10^{28}*
ADONE, Lab. Naz. Frascati, Italy	1969	Single ring	e^+e^-	1.5	3.0	*3×10^{29}*
ISR, CERN, Geneva, Switzerland	1971	Single ring	pp	31.0	62.0	*4×10^{31}*
CEA-Bypass, Cambridge, Mass., USA	1971	Single ring	e^+e^-	2.5	5.0	*2×10^{28}*
SPEAR, Stanford, USA	1972	Single ring	e^+e^-	4.2	8.4	$5 \times 10^{31} - 10^{32}$
DORIS, Hamburg, Germany	1974	Single ring	e^+e^-	4.5	9.0	$5 \times 10^{29} - 10^{32}$
VEPP-2M, Novosibirsk, USSR	1975	Single ring	e^+e^-	1.3	2.6	
DCI, Orsay, Paris, France	1975	Intersecting rings	e^+e^-	3.7	7.4	10^{32}
VEPP-3, Novosibirsk, USSR	1977	Single ring	e^+e^-	3.0	6.0	10^{30}
VEPP-4, Novosibirsk, USSR	1978	Single ring	e^+e^-	7.0	14.0	10^{31}
PETRA, Hamburg, Germany	1978	Single ring	e^+e^-	19.0	38.0	10^{32}
CESR, Cornell Univ., USA	1979	Single ring	e^+e^-	8.0	16.0	10^{32}
ISR $p\bar{p}$, CERN, Geneva, Switzerland	1981	Intersecting rings	$p\bar{p}$	31.0	62.0	10^{29}
PEP, Stanford, USA	1980	Single ring	e^+e^-	18.0	36.0	10^{32}
SPS $p\bar{p}$, CERN, Geneva, Switzerland	1981	Single ring	$p\bar{p}$	270.0	540.0	10^{30}
Fermilab $p\bar{p}$, Batavia, USA	1982	Single ring	$p\bar{p}$	1000	2000	10^{30}
VEPP, Novosibirsk, USSR	?	Single ring	$p\bar{p}$	23.0	46.0	?
ISABELLE, Brookhaven, USA	1986	Intersecting rings	pp	400	800	10^{33}
LEP, CERN, Geneva, Switzerland	Late 1980's	Single ring	e^+e^-	86	172	10^{32}
UNK, Serpukhov, USSR	Late 1980's	Intersecting rings	$p\bar{p}$	3000	6000	?

a) The whole of this table has been taken from an article by R.R. Wilson dated January 1980[139], except for certain data *(in italics)* taken from an article by C. Bernardini in 1976[140]. The values given by this latter author always relate to the initial luminosity, whereas Wilson's values relate to subsequent stages in the development of single machines or are simply the "design values" which seem reasonable to attain.

b) The luminosity is the factor to apply to the collision cross-section of a process so as to obtain the number of events which take place per second.

13. OTHER RESEARCH ACTIVITIES DURING HIS FINAL PERIOD IN ROME

Despite Bruno's marked commitment in the development of the e^+e^- rings, he managed to produce, from 1963 onwards, various interesting theoretical papers on a wide range of subjects. The first, in chronological order, was that which he wrote with I.M. Barbour and A. Bietti entitled "A remark on the neutrino theory of light" [51], in which he takes up the old idea put forward by L. de Broglie[142)] [46] and P. Jordan[143)], i.e. to describe photons as an appropriate state of a system composed of two neutrinos. This description was criticized by M.H.C. Pryce[144)], who had shown that it was impossible to construct a circularly polarized photon in this manner. When, much later, S.A. Bludman[145)] suggested that the existence of two different neutrinos could reinstil life into this idea, Touschek, Barbour and Bietti developed a generalization of the original Jordan proposal by introducing a "weighting" factor to distribute the momentum of the photon between the two component neutrinos.

They found, however, that in this more general scheme the resulting photon was *longitudinal,* i.e. not physical, in accordance with what Pryce had found in a much more particular case. This result could, *inter alia,* also be deduced by using the invariance (2) discovered by Touschek.

Later, B. Ferretti and G. Venturi[146)] observed that this negative result was due to the use of only neutrino S states, but that if one also took into account neutrino waves with ℓ different from zero, it was possible to construct photons with the correct states of polarization.

They showed that the difficulties encountered by previous authors were not related to symmetries [i.e. not with Pryce's theorem[144)]] but with causality: starting from a positronium state with any ℓ, it is possible to obtain, with a suitable procedure for passing to the limit for $e = 0$ and $m_e = 0$, a set of ℓ photons (having a zero mass and any ℓ) with spins which are all parallel. Later, these ideas were discussed privately by Ferretti also with Heisenberg, who agreed with the idea that the difficulties were not due to Pryce's theorem, but rather to the use of corpuscular models instead of the corresponding fields.

Another subject to which Bruno devoted his attention during those years, in connection with his university course, was that of statistical mechanics, on which he wrote a fine book with G.C. Rossi (see Section 7) and to which he made two contributions: one [59] is a critical examination of the temperature definition in the relativistic case, the other [60] a discussion of the connection between microscopic reversibility and macroscopic irreversibility.

As Rossi relates about the writing of this book on statistical mechanics [67], Touschek prepared a rather schematic text for each chapter, in English, which he handed to Rossi who rewrote it in Italian so that, as Bruno had requested, "the students could understand it". Touschek was not, however, easily satisfied and the text was, in practice, rewritten three or four times. The work of reading and discussing the subsequent drafts was carried out jointly with a bottle of fine red wine next to Bruno, who repeatedly helped himself and, in a friendly manner, invited his collaborator to do the same.

The book sets out the main part of the work done on "Covariant statistical mechanics" [59] and on "Statistical reversibility" [60]. The first concerns the problem which for a number of years had been dealt with in the literature concerning the law of transformation of the absolute temperature T under Lorentz transformations. In reality, T is completely defined only if it is measured within the centre-of-mass reference frame. But if the thermometer is in motion with respect to this, the answer is ambiguous, since it depends on the physical law adopted for its definition. If the law adopted is that of energy equipartition, i.e. it is stated that $kT/2$ is the kinematic energy per degree of freedom of a thermodynamic system in equilibrium, or the starting point is the relativistic invariant

$$dS = \frac{\delta Q}{T} \quad ,$$

T is clearly an energy, apart from a constant, and thus is transformed as the *time*variable. If, however, the basis taken is the law of the ideal gas

$$pV = RT,$$

and p is defined as the trace (invariant) of the stress tensor, then T is transformed as a *length* parallel to the motion.

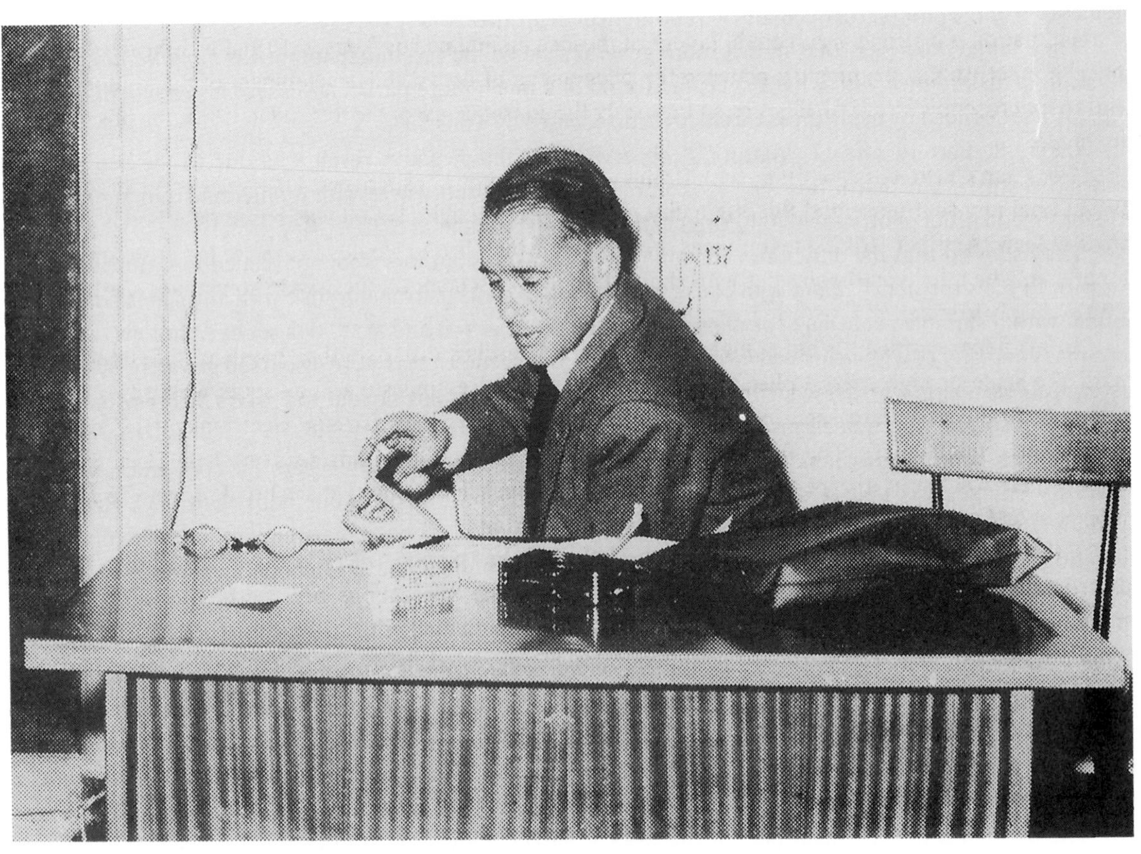

Touschek in Catania (1964)

In paper [60], which is much less important and original, Touschek discusses another problem of thermodynamics, namely the link between the reversible microscopic world and the irreversible macroscopic world. His observation is that the macroscopic world appears to us irreversible because we always start from conditions which are extremely ordered and consequently it is necessary to wait a very long time before any fluctuation occurs which is large enough to allow the observer to speak of a macroscopic process in which the entropy diminishes. As a check of this viewpoint, he observes that, if it is correct, the law of rise in time of a fluctuation must be the same, except for time reversal, as that by which a fluctuation is normally observed to disappear. This equality was in fact found to be true by Rossi and Touschek by simulating on a computer the development and decay of the fluctuations of the time average, in an interval of time T, of a dynamic macroscopic variable (Chapter 8, paragraph 49, Figure 49.1 of his book [67]).

The fact that there is an exponential law with the same time constant during the development and decay of a fluctuation is a strong argument in favour of the idea maintained by Touschek, and by others before him, that the macroscopic asymmetry between the phenomena of decay and development of fluctuations which lead to macroscopic irreversibility are, in fact, only the consequence of the method in which the observations are made.

For Gian Carlo Rossi, working with Bruno was a "grandiose" experience. One was always struck by his exceptional physical sense and the originality of his line of thought, which was based on a purely personal point of view. Another striking feature was the speed at which Bruno would always see the grotesque side of situations, which he would point out with short but effective sentences; these were sometimes scathing, and always deeply ironical[147)].

In his paper entitled "What is high energy" [62], Touschek discusses the "possible" milestones which mark the scale of high-energy phenomena. He describes these "milestones" as spots where two or three different "branches", developed independently in low-energy physics (strong, electromagnetic, and weak interactions), might meet. The first point of reference is placed, according to Touschek, at $m_G c^2$ where, despite their weakness, the perturbation theory of the weak interactions must break down. This should happen at $mc^2 = G^{-1/2} \simeq 300$ GeV, where G is Fermi's constant.

Other milestones are placed by Touschek at the points where weak interactions "reach" electromagnetic interactions and strong and electromagnetic interactions become of the same order of magnitude[148)]. The curious thing, pointed out by Touschek, is that all of them appear to be of the same order of magnitude, as if they were part of some superior design, suggested also by the fact that the relationship

$$m_n = \left(\frac{m_\mu}{m_e}\right)^{(n-1)/2} m_e = (206)^{(n-1)/2} m_e$$

seems to be grossly fulfilled; for $n = 1$ one has the electron, for $n = 2$ nothing (?), for $n = 3$ the muon, for $n = 4$ the proton, for $n = 5$ the intermediate boson W, and for $n = 6$ the particle of mass m_G defined above.

Apart from these numerological considerations, I like to recall that Bruno always stressed, as an essential point to be kept in mind in the decisions about the construction of a new accelerator, that its energy should be above a "foreseeable threshold" so that it could open up the possibility of throwing at least a glance over a new phenomenological panorama.

Bruno's influence on his young students and on those who had recently graduated was very marked, even when their interests were not directly those which concerned Bruno at that moment. This was particularly true in the case of Luciano Pietronero, who came into contact with Bruno immediately after graduating in 1971.

Bruno suggested that he should re-examine a classic problem which had been set by Hans Thirring in 1918[149)], but had not yet been fully resolved. As a result of this research, an internal report of the Istituto Guglielmo Marconi had been written with the title: "On rotating reference systems in Einstein's theory of gravitation"[150)]. The manuscript of this report was revised by Touschek when Pietronero paid a visit to Sperlonga, where Bruno was spending several days' holiday in July 1971. Pietronero still remembers the scene with Bruno, his son Francis, and himself sitting at a small table facing the sea, with a bottle of wine, while a record of Viennese cabaret music was being played on the gramophone. On this occasion, as on others, Touschek would amaze his listeners with his brilliant conversation, the speed at which he would grasp the meaning of human situations, like scientific problems, "as if he had an extra gear", to use one of Pietronero's expressions. He would go right to the heart of the problem with extreme clarity, as is also clear from the "Lectures on mathematical methods for students of physics"[66].

Pietronero was convinced that this first paper should also be signed by Bruno, who had helped him more or less directly, but, to his astonishment and embarrassment, Bruno would hear nothing of this. He continued to be interested in Pietronero's work on this subject or related problems, for some time with a substantial direct influence on him[151)], but very soon after allowing him to find his way by himself[152)].

14. THE CRISIS IN THE UNIVERSITY AND RESEARCH LABORATORIES. SOME CULTURAL CONTRIBUTIONS BY TOUSCHEK OUTSIDE THE UNIVERSITY

The protest movement started at the end of January 1968 at the University of Rome and the Istituto di Fisica Guglielmo Marconi. It involved students and even more young people who had graduated in recent years and wanted a change in the university, since the chances of entry were very few. Touschek was deeply moved by the overwhelming majority of arguments adopted by the demonstrators. He was even more disturbed by the views enforced by a small group who assumed the sole authority to understand the situation and the consequent right of taking even violent action against the others.

There were certainly many reasons for the unease in the Italian Universities, but these were always relegated to the second or third place, or were simply overlooked in view of extremely abstract statements of principle, which were almost always borrowed, without any adaptation, from historical or political situations in other countries (USA or France).

As almost always happens in such circumstances, some of the young dissidents, especially during the initial period, were incited by noble and disinterested proposals, even though they were naïve and did not relate to the true situation; but they were immediately followed by all of those who wanted to take advantage of any upheaval in order to have the chance of embarking on a career which was not selective at the outset, and was guaranteed for the rest of their lives. The quality of the persons who, from time to time, assumed the role of leaders of the agitation, and their proposals, worsened over the years, and Bruno, who was extremely sensitive to these changes, became increasingly alarmed.

When, in 1973 or 1974 he was strongly criticized personally by the students, who started to accuse him of being a Nazi, because he tried to maintain a serious tone (even if on occasions this was rather difficult) during his lectures and examinations, he almost completely stopped attending the Institute. He worked at home, where he was joined by his collaborators. It was thus that the paper which he produced in 1974 entitled "What is high energy?" [62], contains the rather polemic reference to the location of the work as: "Garvens S.p.A., Rome, Piazza Indipendenza, Italy".

It was about this time that he finally changed his attitude with regard to Italy. When he settled in Rome, and over all the subsequent years—especially during the e^+e^- ring period—Bruno felt fully at ease in this country. From time to time he was irritated by the red tape necessary when completing some official formality (his application to renew his Italian residence permit, to move his phone from one flat to another and so on), but he felt, and often openly said, that in the physics environment the work was carried out in a receptive framework, which was flexible enough to adapt to rapid changes in ideas, and simplified the implementation of the corresponding programmes.

The first major shock to this view which he had of Italy occurred in 1963–64 as a result of the Ippolito trial[153)], a well-known case which was sparked off by a series of rather superficial articles by Giuseppe Saragat[154)], ending with a verdict which was not only unjust, but also resulted in paralysing the entire Italian bureaucracy, especially in the research organizations. The latter was subsequently further affected by the Marotta and Giacomello cases[155)], which occurred slightly later and for rather similar reasons. There was, in fact, a whole series of extremely serious legal cases, which involved people who had spent many years of their life in promoting the scientific and technological development of our country, to ensure—often with much success—that Italy did not lag behind the other European countries.

The ensuing overall paralysis of the bureaucratic system had also had a serious effect on the new vacancies, and in this way had been one of the elements added to the many other important and justified reasons for the protest.

All of this had dismayed Bruno Touschek, who told me of his disappointment, sometimes expressing his relief that he had rejected the idea, admittedly of short duration, of changing his nationality.

The protest movement in 1971 had also affected the Laboratori Nazionali di Frascati, as was already said at the end of Section 11. As soon as ADONE was ready to operate, a strike prevented its utilization from 30 May to 19 September, and even during the successive months the work was resumed but without the commitment and enthusiasm which the type and performance of the machine deserved.

It was in this climate of a university in the throes of a crisis, and research laboratories which were only half operating, that Bruno Touschek began to take an interest in teaching problems at the high schools.

Another reason was that his children Francis and Stefan, who had attended since they were small the Swiss school in Rome (where the teaching was given in German), had now started to attend high school.

In November 1972, Bruno Touschek attended the second Incontro di Serapo, organized by Giulio Cortini for a group of high school teachers, which, under the guidance of Cortini, had begun to concern itself with the problem of "restricted relativity". After a full day of discussions, at a time when everyone else was going to bed, Bruno sat down at a table in the hotel room, with a bottle of cognac and a glass in front of him, and started to write the "skeleton" of his lectures on relativity. Next morning, he was still writing there, faced by an almost empty bottle and a 25-page manuscript beautifully written, without any corrections.

On his return to Rome, he gave, at the invitation of Professor Lina Mancini Proia a set of four lectures at the Liceo Virgilio on this subject. A few months later (March 1973) he prepared an internal report of the Istituto Guglielmo Marconi on the "Course on restricted relativity" [67], telephoned Professor Piera Salvetti and, after obtaining the necessary agreement, gave seven lectures at the Liceo Mamiani during March and April. In these lectures, he adopted the fairly usual approach of deducing the Lorentz transformation from the invariance of the velocity of light with respect to the reference system, and the essentially linear nature of the relations sought. In Bruno's opinion, this method of deduction had the merit of showing students how a theory is constructed.

The lectures, which were always brilliant, proved highly successful among the students, who found them somewhat difficult but were fascinated by Bruno's personality and liveliness.

Bruno's appointment as a foreign member of Accademia Nazionale dei Lincei, in 1972 (see Section 7), opened up a possibility for him to use his extraordinary energy in another type of teaching activity.

At its meeting on 9 March 1975, the Accademia Nazionale dei Lincei examined and discussed the report submitted by the Commission appointed for this purpose, concerning the possibility of setting up, within the Academy, and as part of the Centro Interdisciplinare di Matematica e Fisica, a "Science Centre", the main purpose of which was to communicate the latest results of scientific work to the widest possible public. The Commission was composed of A. Carrelli, G. Salvini and B. Touschek.

During the discussion, Bruno stated that he was immediately available on a "full-time" basis to set under way the programme proposed by the Commission, arranging a series of meetings with the people who were interested in this idea, so that arrangements could be made as soon as possible for a series of lectures of an "episodic" type. The main feature of these lectures was to demonstrate the "dynamics of science", whilst ensuring a "humanistic appearance". These expressions were used in a circular letter which Bruno Touschek sent on 13 March 1975 and which received enthusiastic response from many people.

In this way, a series of lectures took place, recorded on video tape, under the general title "Study of living science". From that moment, Touschek devoted himself with outstanding energy to producing these lectures, in which he was helped by his son Francis. As can be seen from the names of the speakers and the titles of the lectures[156)], a very remarkable range of subjects was dealt with. In fact, the overall collection represents a "living and permanent document" of the development of many important chapters in physics, which remains in the safekeeping of the Accademia dei Lincei. The lectures proved extremely successful, as was shown from the number of persons who attended, mainly young people, many of whom usually found it impossible to find a seat.

The lecture concerning the teaching of quantum theory [69], which was given by Bruno himself, although very valuable, does not form part of this scientifically and historically interesting set of lectures.

When Bruno Touschek moved to Geneva, at the beginning of October 1977, he was now in poor health, and it was not clear whether he would return to Italy to work and live permanently.

In this post of Visiting Scientist to which Touschek had been appointed, he was to collaborate in the development of the method of stochastic cooling of electron (and proton) beams, as proposed and developed by S. van der Meer at CERN[157)], in preparation for its use in the SPS of CERN (see Table 2).

In reality, Bruno hoped to find a post of professor or research worker in the UK, about which he had often said rather unfavourable things, but, in view of the upheaval in Italy, appeared to be an island of salvation.

After arriving in Geneva, Bruno Touschek had already completed on 16 January 1978 a manuscript on stochastic cooling [63], but its contents were not in keeping with the approach developed at CERN by

Bruno Touschek in Geneva (Hospital of La Tour, 1978)

S. van der Meer and others. Three days earlier, he had called van der Meer from La Tour Hospital and tried to explain the essential point of his paper, but as can easily be understood, this proved insufficient.

This was, however, the last scientific paper produced by Bruno whose health was now extremely poor. As I was told by R. Hagedorn, who frequently went to visit him at the La Tour Hospital, just before he left for Innsbruck, one could not fail to be taken aback by his physical and mental condition.

15. BRUNO'S ARTISTIC BENT AND HIS FINAL CULTURAL INTERESTS

As a researcher, Bruno Touschek struck everyone by the originality of his thinking, the Cartesian clarity of his approach, and his enthusiasm in what he himself or others were doing.

Since the earliest days after his arrival in Rome, Bruno had acquired the habit of knocking on the door of my study at least three or four times a week, when he arrived at the Institute, rather early in the morning on his way to his own study. He would come in and tell me about his latest thoughts, usually those of the night before, concerning the problem which he was concentrating on, or about an interesting result achieved by one of his young pupils and collaborators. His enthusiasm and incisive remarks were extremely stimulating and pleasant to hear. Even when he was talking of scientific problems he would very often introduce a subtle degree of humour, which would emerge from his texts and especially his drawings.

As stated by P.I. Dee (see Section 3), he possessed an unusual skill in caricaturing his surroundings and local customs, which he would draw with a pen on the first piece of paper which came to hand, during the degree examination or Faculty sessions, or during the various meetings of the commissions or working groups dealing with the activities of the Institute or of the Laboratori Nazionali di Frascati.

This skill was a very marked characteristic on his mother's side of the family. She was rather good at drawing and this was even more true of her brother, Oscar Weltmann, a well-known doctor and dilettante painter.

The 24 sketches reproduced at the end of this biography have been taken from a large collection, which friends and relatives gathered from the wastepaper basket or the table where Bruno had left them at the end of a committee meeting or examination session.

One of the "leitmotivs" which is frequently represented is the "self-injury" from which the whole of our society and each one of us, individually, suffers (Nos. 1–3). This disease, which according to Bruno was particularly serious in the Italian university environment, is represented in various ways, such as a man who is planting a nail in his knee or eye or who is firing a gun overhead into the top of his spine.

His caricatural approach, which is often very amusing and sometimes grotesque, is always present, and in certain cases (Nos. 4–10) is the only real purpose of the drawing.

Drawing No. 8 is a caricature of myself, which Bruno drew at the time when, as a result of the Ippolito Case, suspicion was cast on all those who, like myself, had been or were members of the Directing Committee of the CNEN, and had defended the organization and the broad lines which it had followed for many years. Drawing No. 9 illustrates the "superego of the motorcyclist" and therefore also has a self-critical content.

In other drawings (Nos. 11–13) it was the Degree Examination Commission which received the attention of the caricaturist, or it might be one of its members who is playing, rather shyly behind his seat, with a yo-yo, or another member involved in the "hearing of theses". In this instance the improper use of the Italian enhances the inherently comic nature of his drawing.

Other drawings represent typical laboratory scenes: a woman looking through a microscope, embarrassed by the presence of a small fly which is flying around the panel located on the wall (No. 14), and a discussion concerning the direction of the magnetic field on the basis of the "three-finger rule" (No. 15) which Bruno drew on a page of the record of measurements made with AdA at Orsay (see Section 9), during the Symposium on Storage Rings held in that laboratory from 26 to 30 September 1966. This drawing was printed in the proceedings of the symposium as the initial page of the session on "Magnetic Detectors — Radiative Corrections".

Further drawings (Nos. 16–19) concern the life at the Faculty of Mathematical, Physical and Natural Sciences: the contempt of the Council of the Faculty for an article of a recent law or a ministerial circular (No. 16), the decision taken by the chemistry professors who, failing to reach agreement on which of them should be proposed as Director for all of the chemical activities performed at their department, had decided to keep the single-professorship institutes (No. 17), and the disagreements which arose at a certain moment among mathematical colleagues (No. 18). Drawing No. 19, concerning the introductory nature of the courses attains the heights of efficiency in its schematic symbolism.

Still other drawings (Nos. 20–22) concern the period of the protest movement (1968–1976). No. 20 represents an "assembly" in the large hall of the Istituto di Fisica Guglielmo Marconi. No. 21 represents a group of unidentified persons who wanted to enter the Institute which was occupied by the "students", at the time when Giorgio Careri was Director of the Institute (1968–70), and was therefore accused, by all of the factions, of extremely serious and absurd failings and problems of management. No. 22 shows the discussion for the choice of the Director of the Istituto di Fisica Guglielmo Marconi, which Bruno Touschek had renamed "Istituto Maria Montessori", to stress the attitude taken by a part of the teaching staff, whose only thought was to allow the students to do whatever they wished. The symbol CB stands for "Carlo Bernardini".

Drawing No. 23 is an example of drawing which contained a fundamental contradiction, and No. 24 is an example of those based on the merging of two different concepts or objects. In this case the combination was between a lynx, which is the emblem of the Accademia Nazionale dei Lincei, and the six-legged dog, symbolizing the Ente Nazionale Idrocarburi.

Some of these drawings recall those of Egon Schiele (1890–1918) who, together with Gustav Klimt (1862–1918), an admirer of and sometimes inspired by oriental art, was among his preferred painters. These artists of the Viennese Secession had always attracted him by their culture and sensitivity, as well as by their almost sickly refinement, but this attraction grew immensely in the final months of his life. When he was in Geneva, in autumn 1977, he had purchased from somewhere or other a batik[158)] by a Java painter. It bore geometric designs, with strong and at the same time dark colours, giving a decorative pattern, which is somehow reminiscent of certain decorative paintings by Klimt, rather more because of the differences than certain remote resemblances.

Talking with Valentino Braitenberg, who had visited him in Innsbruck a few days before his death, Bruno expressed the desire to read a biography of Ludwig Wittgenstein[159)]. He had noticed, he said, in himself a desire of identification with the philosopher, his fellow-citizen, perhaps due also to the already remote readings of logic he had made at the gymnasium in Vienna. He had the impression of having neglected the more general aspects of knowledge. Valentino, however, did not succeed in providing him in time with the Wittgenstein biography.

It was in the final phase of his life that it became easier for his Italian friends to grasp the profound reasons for his enthusiasm for *Das Glasperlenspiel, Versuch einer Lebensbeschreibung des Magister Ludi Josef Kneckt* by Hermann Hesse (1877–1962)—his enthusiasm for this Utopia in which the various figures fluctuate between a real and symbolic existence, in an imaginary future country, where, in about two centuries, mankind has succeeded in overcoming the present world and society, characterized by frequent wars, by wild individualism and by a culture reduced to "feuilletons", that is to the "third page" ("potted culture" page) of the newspapers. The new society is guided, morally and culturally, by an intellectual aristocracy, which, through the study and meditation, always deeper, of the form and contents of music and mathematics in all their aspects, has succeeded in producing a new order, which, at the end, reveals itself to be without a way out, based on a game, extremely refined but sterile, like so many others.

At the hospital, first in Geneva and later in Innsbruck, Bruno read with great interest books on history and literature, mainly those regarding the Viennese life of the beginning of our century, and specially concerning the books of Karl Kraus (1874–1936)[160)], and on Karl Kraus[161)] and Gustav Klimt[162)].

Karl Kraus had been always his favourite author: Kraus had founded, in 1899, the review *Die Fackel* which he wrote virtually unaided for some 37 years, hinting at the pending collapse of the Habsburg Empire, satirizing the monstrous day-to-day events, and putting to shame the police chiefs who had committed murder, as well as the criminal financiers.

Kraus' famous aphorisms[163)] were probably the source of Bruno Touschek's paradoxical expressions or remarks. In the 1950's and 1960's he would often refer to his country of origin in scathingly critical terms. However, on reaching the end of his life, he seemed to find rest and contentment only by re-immersing himself in the culture of that Viennese and partly Jewish atmosphere of the beginning of the century, which had been his background and had so profoundly affected his youth.

Acknowledgements

An initial draft of these biographical notes was sent at the end of January 1980, with a request for remarks and comments, to the following persons who knew Bruno Touschek: U. Amaldi, F. Amman, C. Bernardini, A. Bietti, W.K. Burton, V. Braitenberg, N. Cabibbo, F. Calogero, R. Chisholm, M. Cini, M. Conversi, G.F. Corazza, P.I. Dee, Etim Etim, B. Ferretti, J.C. Gunn, J. Haissinski, P. Marin, G. Morpurgo, R. Peierls, W. Paul, R. Querzoli, L. Radicati di Brozolo, G. Salvini, M. Sands, I.N. Sneddon, W. Thirring, P. Urban, R. Wideröe. I am grateful for the many comments I received, most of which have been used in the final text.

LIST OF PUBLICATIONS OF BRUNO TOUSCHEK

[1] B. Touschek, "Excitation of nuclei by electrons", *Nature* **160**, 500 (1947).

[2] B. Touschek, ""Note on Peng's treatment of the divergency difficulties in quantized field theories", *Proc. Cambr. Phil. Soc.* **44**, 301 (1948).

[3] I.N. Sneddon and B. Touschek, "Nuclear models", *Nature* **161**, 61 (1948).

[4] I.N. Sneddon and B. Touschek, "A note on the calculation of the spacing of energy levels in a heavy nucleus, *Proc. Cambr. Phil. Soc.* **44**, 391 (1948).

[5] I.N. Sneddon and B. Touschek, "The excitation of nuclei by electrons", *Proc. Roy. Soc.* **193**, 344 (1948).

[6] B. Touschek, "Zur Theorie des doppelten Beta Zerfalls", *Z. Phys.* **125**, 108 (1948).

[7] B. Touschek, "Zum analytischen Verhalten Schrödinger'scher Wellenfunktionen", *Z. Phys.* **125**, 293 (1949).

[8] I.N. Sneddon and B. Touschek, "The production of mesons by electrons", *Nature* **163**, 524 (1949).

[9] I.N. Sneddon and B. Touschek, "The production of mesons by electrons", *Proc. Roy. Soc.* **199**, 352 (1949).

[10] B. Touschek, "Das Synchrotron", *Acta Phys. Austriaca* **3**, 146 (1949).

[11] B. Touschek, "An estimate for the position of the lowest dipole level of a nucleus", *Philos. Mag.* **41**, 849 (1950).

[12] B. Touschek, "Quadrupole moments and the nuclear shell model", *Philos. Mag.* **42**, 279 (1951).

[13] J.C. Gunn, E.A. Power and B. Touschek, "The production of mesons in proton-proton collisions", *Phys. Rev.* **81**, 277 (1951).

[14] J.C. Gunn, E.A. Power and B. Touschek, The production of mesons in proton-proton collisions", *Philos. Mag.* **42**, 523 (1951).

[15] W. Thirring and B. Touschek, "A covariant formulation of the Block-Nordsieck method", *Philos. Mag.* **42**, 244 (1951).

[16] B. Touschek, "A perturbation treatment of closed states in quantized field theories", *Philos. Mag.* **42**, 1178 (1951).

[17] M. Sands and B. Touschek, "Alignment errors in the strong focusing synchrotron, *Nuovo Cimento* **10** 604 (1953).

[18] W.K. Burton and B. Touschek, "Commutation relations in Lagrangian quantum mechanics," *Philos. Mag.* **44**, 161 (1953).

[19] R. Chisholm and B. Touschek, "Spin orbit coupling and the mesonic Lamb shift", *Phys. Rev.* **90**, 763 (1953).

[20] W.K. Burton and B. Touschek, "Schwinger's dynamical principle", *Philos. Mag.* **44**, 1180 (1953).

[21] G. Morpurgo and B. Touschek, "Remarks on the validity of the Tamm-Dancoff Method", *Nuovo Cimento* **10**, 1681 (1953).

[22] E. Fabri and B. Touschek, "La vita media del mesone τ", *Nuovo Cimento* **11**, 96 (1954).

[23] M. Cini, G. Morpurgo and B. Touschek, "A non-perturbation treatment of scattering and the Wentzel example", *Nuovo Cimento* **11**, 316 (1954).

[24] G. Morpurgo, L. Radicati and B. Touschek, "On time reversal", *Nuovo Cimento* **12**, 667 (1954).

[25] B. Touschek, "A speculation on the capture mechanism of K-mesons", *Suppl. Nuovo Cimento* **12**, 281 (1954).

[26] E. Amaldi, E. Fabri, T.F. Hoang, W.O. Lock, L. Scarsi, B. Touschek and B. Vitale, "Report of the Committee on τ Mesons", *Suppl. Nuovo Cimento* **2**, 419 (1954).

[27] G. Morpurgo and B. Touschek, "Remarks on time reversal", *Nuovo Cimento* **1**, 201 (1955).

[28] G. Morpurgo, L. Radicati and B. Touschek, "Time reversal in quantized theories", *in Proc. of the Glasgow Conf. on Nuclear and Meson Physics, 1954* (Pergamon Press, London, 1955).

[29] G. Morpurgo and B. Touschek, "Space and time reflection of observable and non-observable quantities in field theory", *Nuovo Cimento* **1**, 1159 (1955).

[30] G. Morpurgo and B. Touschek, "Space and time reversal in quantized field theory", *Suppl. Nuovo Cimento* **4**, 691 (1956).

[31] G. Stoppini and B. Touschek, "Phenomenological description of photo meson production", *Suppl. Nuovo Cimento* **4**, 778 (1956).
[32] B. Touschek, "Parity conservation and the mass of the neutrino", *Nuovo Cimento* **5**, 754 (1957).
[33] B. Touschek, "The mass of the neutrino and the non-conservation of parity", *Nuovo Cimento* **5**, 1281 (1957).
[34] L. Radicati and B. Touschek, "On the equivalence theorem for the massless neutrino", *Nuovo Cimento* **5**, 1693 (1957).
[35] M. Cini and B. Touschek, "The relativistic limit of the theory of spin $\frac{1}{2}$ particles", *Nuovo Cimento* **7**, 422 (1958).
[36] B. Touschek, "The symmetry properties of Fermi Dirac fields", *Nuovo Cimento* **8**, 181 (1958).
[37] G. Morpurgo and B. Touschek, "Parity conservation in strong interactions", not published, but circulated as an internal report (1958). (See Ref. 78.)
[38] B. Touschek, "A note on the Pauli transformation", *Nuovo Cimento* **13**, 394 (1959).
[39] W. Pauli and B. Touschek, "Report and comment on F. Gürsey's 'Group structure of elementary particles'", *Supp. Nuovo Cimento* **14**, 205 (1959).
[40] B. Touschek, "La fisica dei pioni", *IX Corso della Scuola Intern. di Fisica della SIF sulla "Fisica dei pioni"*, tenuto a Varenna dal 18 al 30 agosto 1958, organizzato e diretto da B. Touschek, *Suppl. Nuovo Cimento* **14**, 218 (1959).
[41] B. Touschek, "Fixed source meson theory", *IX Corso della Scuola Intern. di Fisica della SIF sulla "Fisica dei pioni"*, tenuto a Varenna dal 18 al 30 agosto 1958, organizzato e diretto da B. Touschek, *Suppl. Nuovo Cimento* **14**, 242 (1959).
[42] B. Touschek, "Elementary considerations on photo meson production", *IX Corso della Scuola Intern. di Fisica della SIF sulla "Fisica dei pioni"*, tenuto a Varenna dal 18 al 30 agosto 1958, organizzato e diretto da B. Touschek, *Suppl. Nuovo Cimento* **14**, 278 (1959).
[43] B. Touschek, "The theory of the neutrino", *Rendiconti del XI Corso della Scuola Intern. di Fisica della SIF sulle "Interazioni deboli"*, tenuto a Varenna dal 29 giugno al 11 luglio 1959, organizzato e diretto da L. Radicati (Nicola Zanichelli, Bologna, 1960).
[44] C. Bernardini and B. Touschek, "On the quantum losses in an electron synchrotron", *Nota Interna LNF* 60/34, 27 aprile 1960.
[45] B. Touschek, "A Study of the mechanism of injection into a storage ring", *Nota Interna LNF* 60/50 (1960).
[46] C. Bernardini, G.F. Corazza, G. Ghigo and B. Touschek, "The Frascati Storage Ring", *Nuovo Cimento* **18**, 1293 (1960).
[47] B. Touschek, "A note on the transformation $\psi' = \exp(i\gamma_5 a)\psi$", *in Lectures on field theory and the many body problem* (ed. E.R. Caianello) (Academic Press, New York, 1961), pp. 173–183.
[48] C. Bernardini, U. Bizzarri, G.F. Corazza, G. Ghigo, R. Querzoli and B. Touschek, "Progress Report on AdA (Frascati Storage Ring)", *Nuovo Cimento* **23**, 202 (1962).
[49] C. Bernardini, U. Bizzarri, G.F. Corazza, G. Ghigo, R. Querzoli and B. Touschek, "L'anello di accumulazione AdA per elettroni e positroni da 250 MeV", *Ric. Sci.* **32**, 1, 137 (1962).
[50] B. Touschek, "The Italian Storage Rings", Proc. 1963 Summer Study on Storage Rings, Accelerators and Experimentation at Super-High Energies, BNL, Brookhaven, USA, 1963.
[51] I.M. Barbour, A. Bietti and B. Touschek, "A remark on the neutrino theory of light", *Nuovo Cimento* **28**, 452 (1963).
[52] C. Bernardini, G.F. Corazza, G. Di Giugno, G. Ghigo, J. Haissinski, P. Marin, R. Querzoli and B. Touschek, "Lifetime and beam size in a storage ring", *Phys. Rev. Lett.* **10**, 407 (1963).
[53] C. Bernardini, G.F. Corazza, G. Di Giugno, J. Haissinski, P. Marin, R. Querzoli and B. Touschek, "Measurements of the rate of interaction between stored electrons and positrons", *Nota Interna LNF* 64/33 (1964), *Nuovo Cimento* **34**, 1473 (1964).
[54] C. Bernardini, G. Corazza, G. Di Giugno, G. Ghigo, J. Haissinski, P. Marin, R. Querzoli and B. Touschek, "Lifetime and beam size in electron storage rings", *Proc. Int. Conf. on High Energy Accelerators*, Dubna, 1963 (Atomizdat, Moscow, 1964), p. 332.

[55] A. Bietti and B. Touschek, "Scalar mesons and the nuclear spin orbit coupling", *Nuovo Cimento* **35**, 582 (1965).

[56] E. Ferlenghi, C. Pellegrini and B. Touschek, "The transverse resistive wall instability of extremely relativistic beams of electrons and positrons", *Nota Interna LNF* 65/27 (1965), *Nuovo Cimento* **44B**, 253 (1966).

[57] E. Etim and B. Touschek, "A note on the administration of radiative corrections", *Nota Interna LNF* 66/10 (1966).

[58] E. Etim, G. Panchieri and B. Touschek, "The infra-red radiative corrections for colliding beam (electrons and positrons) experiments", *Nuovo Cimento* **51**, 276 (1967).

[59] B. Touschek, "Covariant statistical mechanics", *Nuovo Cimento* **58**, 295 (1968).

[60] B. Touschek, "Statistical reversibility", *Atti del Convegno Internazionale sul tema "Metodi valutativi nella fisica mathematica",* Roma 15–19 dicembre 1972, Accademia Nazionale dei Lincei N. 217 (1972).

[61] B. Touschek, "On the measurement of temperature in covariant statistical mechanics", *Symposia Mathematica* **12**, 35 (1973).

[62] B. Touschek, "What is high energy?", *Phys. Lett.* **B51**, 184 (1974).

[63] B. Touschek, "An analysis of stochastic cooling", *Nota Interna LNF* 79/6(R), 16 gennaio 1979, posthumous publication.

* * *

[64] B. Touschek, *Lezioni sulla teoria dei campi,* tenute all'Istituto di Fisica dell'Università di Pisa, A.A. 1957–58.

[65] B. Touschek, *Lezioni di elettrodinamica quantistica,* Appunti tratti dalle lezioni da Marko Marinkovic, Scuola di Perfezionamento in Fisica dell'Università di Roma, A.A. 1965–66.

[66] B. Touschek, *Lezioni di metodi matematici,* Istituto di Fisica Guglielmo Marconi, Università di Roma, A.A. 1969–70.

[67] B. Touschek and G.C. Rossi, *Meccanica statistica* (Boringhieri, Torino, 1970).

[68] B. Touschek, *Corso di relatività ristretta,* Istituto di Fisica, Università di Roma (1973).

[69] B. Touschek, *Sull'insegnamento della teoria dei quanti,* Accademia Naz. Lincei, Contributi del Centro Linceo Interdisciplinare di Scienze Matematiche e Loro Applicazioni, N. 3 (1975).

$PC = P_0C_0 e^{i\gamma \cdot T_3 + i\sigma S}$
$T = T_0 e^{-i\gamma T_3 \mp i\sigma S}$
aggiuntiva Geometria I.
docenti su-indicati.
fisica generale I.
L'anno scorso
corsi domenicali.

1

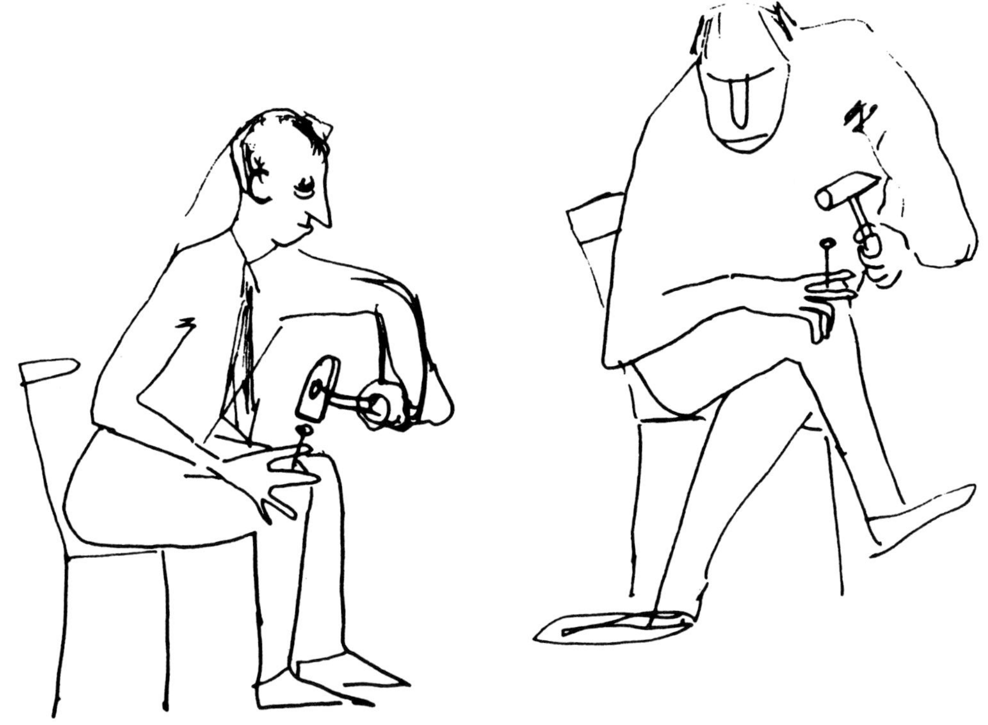

2

95.

KINSEY-REPORT

3.5
0.5
0.5
4.5
1.2
5.7
81.3
88.0

95

5

Ich habe mit die Touscheflaschen Feder geschreiben
er ie

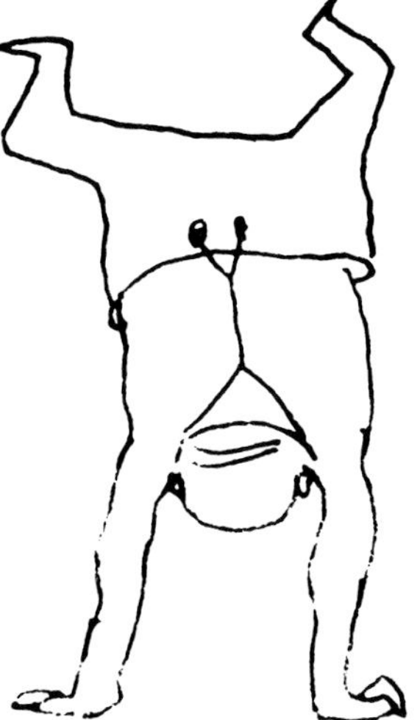

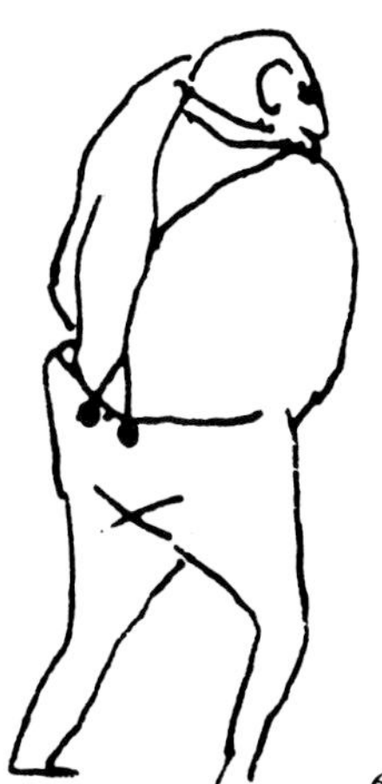

6

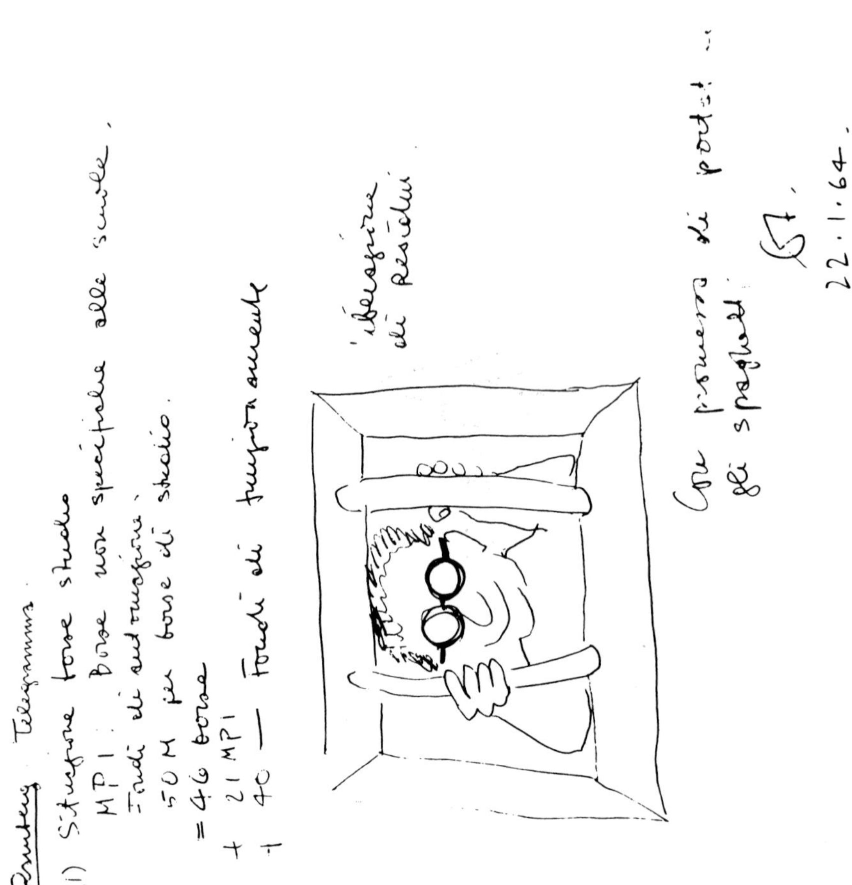

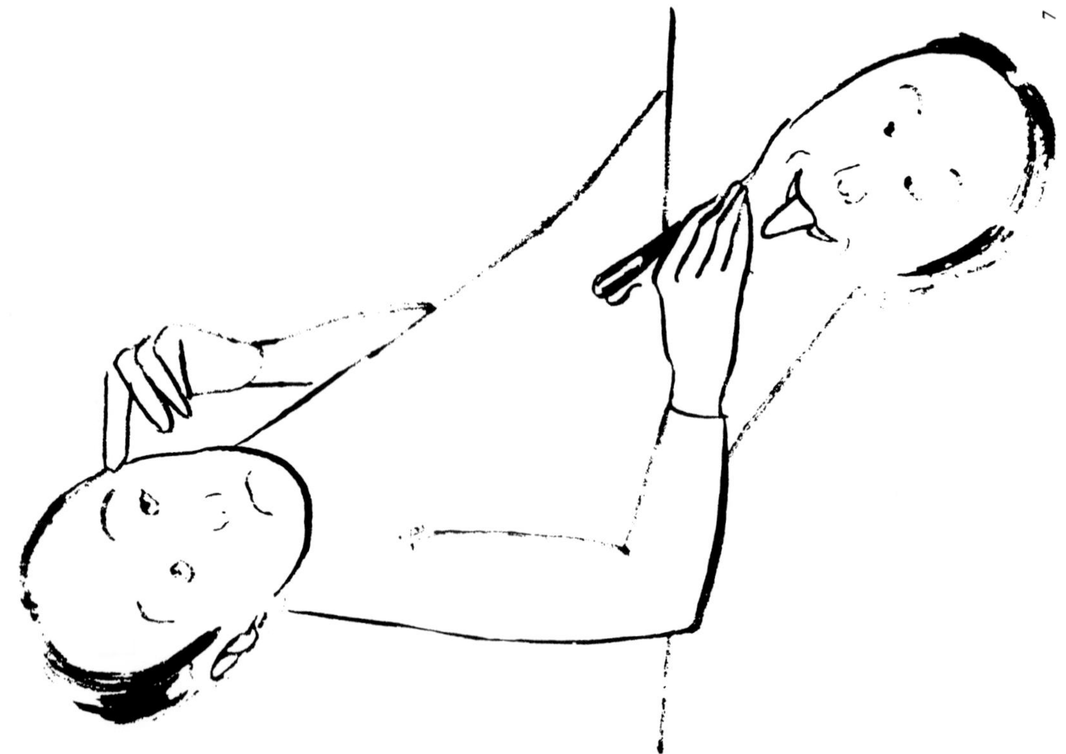

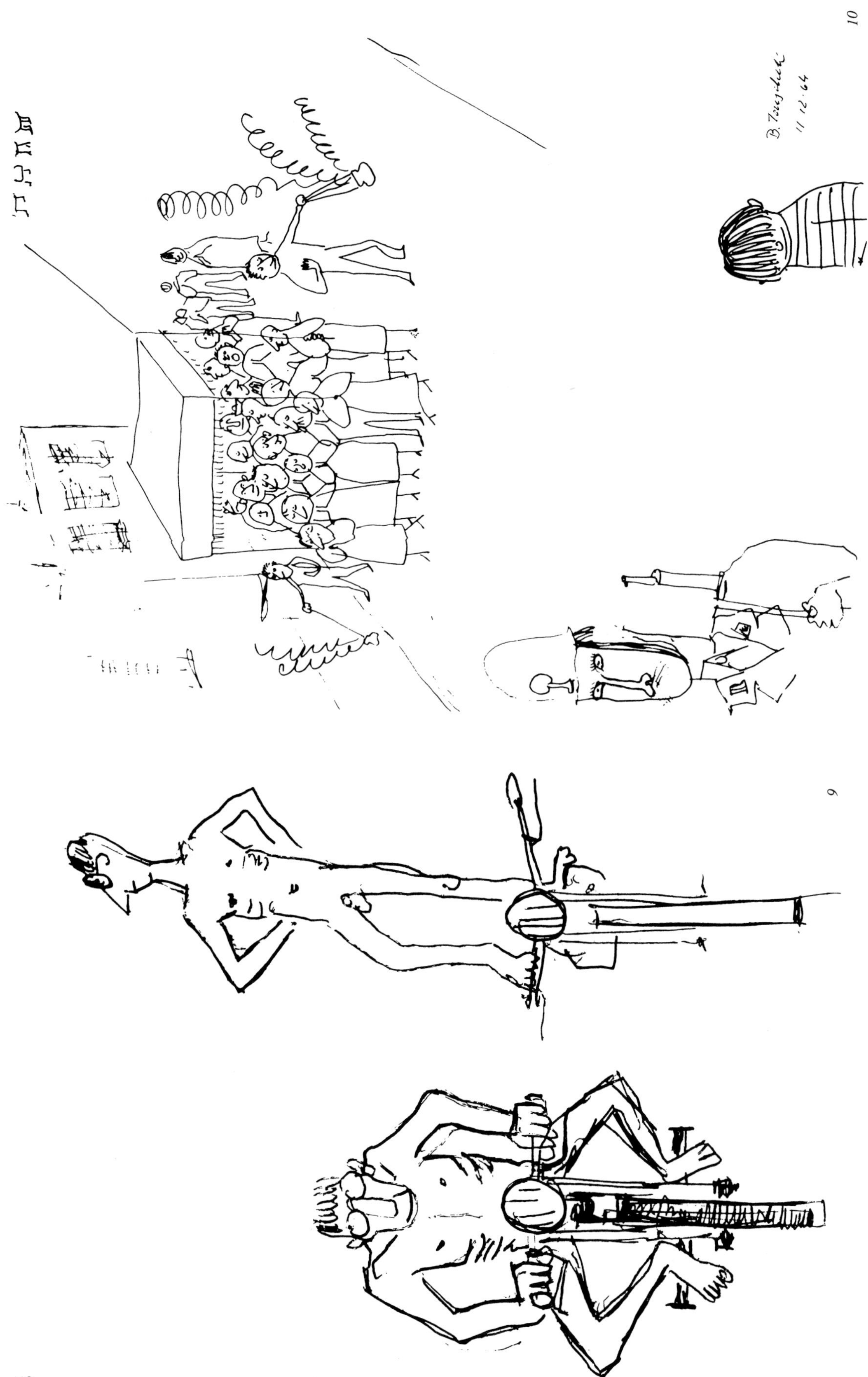

62

11

ESAMI DI LAUREA
L'AUSCULTAZIONE DELLE TESINE
$E = \frac{m_0 c^2}{\sqrt{1-\beta^2}}$
22-2-58.

12

13

14

MAGNETIC DISCUSSION

16

65

SIAMO TUTTI DIRETTORI
CAN-CAN CHIMICO

17

§ x dx
E ANCHE LA MATEMATICA
CODICE
CODICE
IN RICERCA DEL MODUS VIVENDI

18

19

20

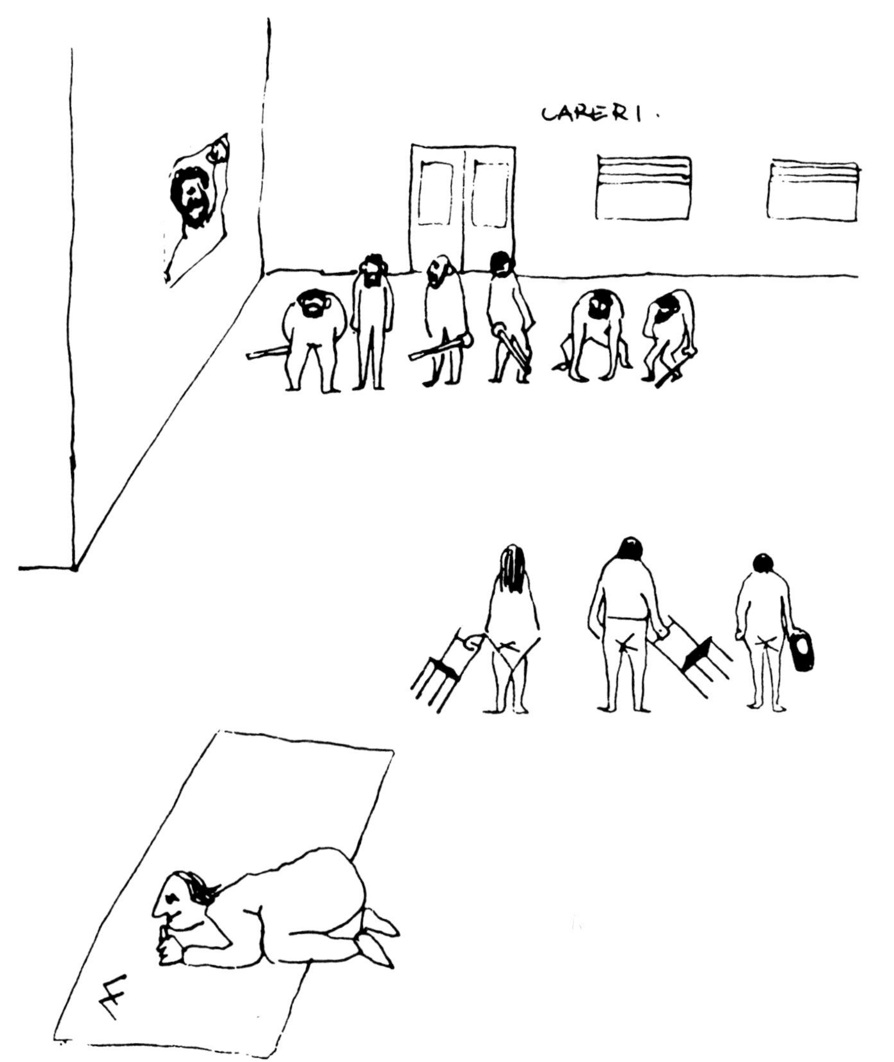

21

Direzione dell' Istituto di Fisica Maria Montessori.
la Scelta del Direttore.

22

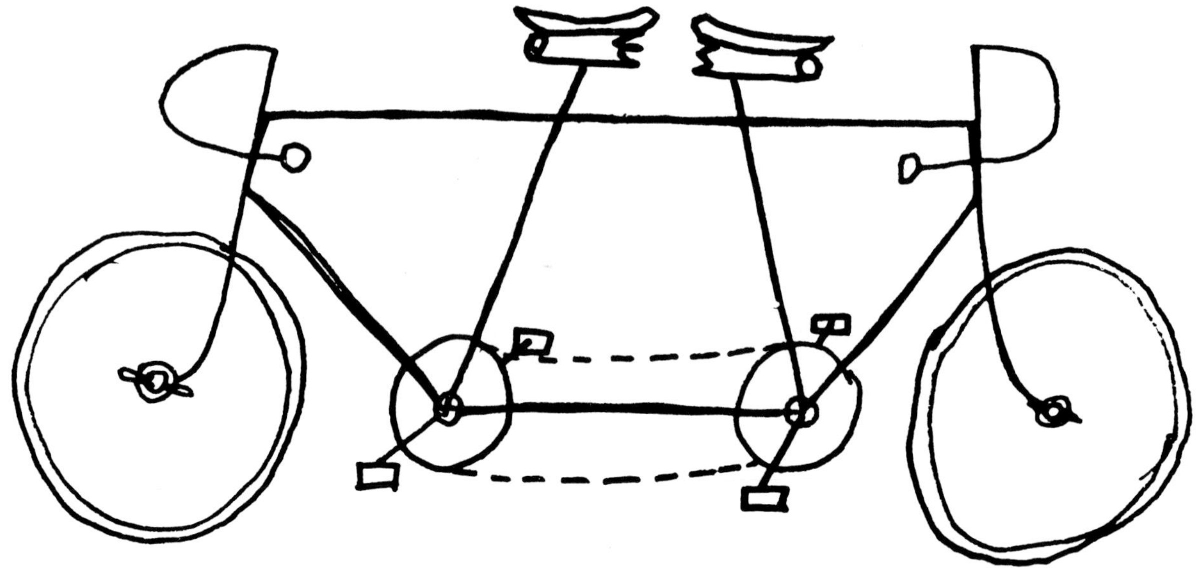

PROBARE ET REPROBARE ! 23

24

REFERENCES AND NOTES

1) Student in the sense of "goliard" (wandering student).

2) In reality he was referring to the Nordkette, as Valentino Braitenberg pointed out to me.

3) Ferdinand Cap (b. 1924), Professor of Theoretical Physics at the University of Innsbruck, has worked in elementary particle physics and on general relativity. He had met Touschek for the first time during the fifties on the occasion of a few international conferences. Joseph Rothleitner, Professor of Theoretical Physics at the University of Innsbruck, met Touschek for the first time at the University of Heidelberg about 15 years ago. At my request he has sent me a letter (dated 8 October 1980) about the last weeks of the life of Bruno in Innsbruck.

4) Ernst Gartner (b. Vienna), designer and painter, teaches arts and drawing at the Gymnasium in Reutte, Tyrol. He had been a schoolmate of Bruno from the primary schools in Vienna and remained his closest friend during all his life.

5) Valentino Braitenberg (b. Bolzano, 1926) studied medicine and specialized in neurology and psychiatry at the University of Rome. After a few years devoted to research on the brain in Germany and the USA, he entered, under the influence of Bruno Touschek and Eduardo Caianello, into a research group of the Institute of Theoretical Physics of the University of Naples, which later became the Naples Section of the National Group for Cybernetics of the Consiglio Nazionale delle Ricerche (CNR). From 1961 to 1968 Braitenberg was Associate Professor of Cybernetics at the University of Naples. Called to a chair of Biology and Applied Science at CalTech, he however accepted the almost contemporary offer of the codirection of the Max Planck Institut für Biologische Kybernetic, in Tübingen. On my request Braitenberg sent me a long letter, dated 27 July 1980, parts of which have been used here.

6) Laboratori Nazionali di Frascati: *Bollettino di Informazione* n. 4, 30 aprile 1978. The announcement of B. Touschek's death was dated 31 May and given on the first page of the *Bollettino* n. 4, dated 30 April but which appeared a few days after 1 June.

7) C. Bernardini: *Paese Sera*, 4 June 1978: "In Italy, Touschek became a designer", with the subheading "Loss of a major scientific personality".

8) A. Bietti, *l'Unità*, 13 June 1978: "From the Great Era in Physics", with the subheading: "Death of Bruno Touschek".

9) Ugo Amaldi, *Corriere della Sera*, 22 July 1978: "Who was the man of No. 137", with the subheading "Commemoration in Frascati of the physicist Bruno Touschek".

10) When drafting this paragraph I used the notes which I wrote on 28 February 1978, when I went to visit Bruno Touschek at the La Tour Hospital. I asked him if he was prepared to give me a chronological account of certain occasional remarks made in the past, and on receiving an affirmative reply, I took a few notes with his permission. On returning to Rome I arranged my notes, had them typed and sent them off to Igls asking him to fill in the gaps, and correct any errors. Bruno did not receive my manuscript until 29 April, and replied on 2 May with a long series of small corrections to my text, which is reproduced here in full. The only additions are a few historical references and more detail concerning the work performed at Hamburg, which I received in the letters from Wideröe (Ref. 22), and on the work at Göttingen, which I obtained from Paul (Ref. 32).

11) W.L. Shirer, *The Rise and Fall of the Third Reich* (Secker and Warburg, London, 1960).

12) Hellmut Andics, *50 Jahre unseres Lebens, Österreichs Schicksal seit 1918* (Verlag Fritz Molden, Vienna, 1968).
Erich Zollner, *Geschichte Österreichs* (Verlag für Geschichte und Politik, Vienna, 1970), 4. Auflage. French translation: *Histoire de l'Autriche, des origines à nos jours* (Horvath, Vienna–Munich, 1965).
See also: Silvio Furlani and Adam Wandruszka: *Austria e Italia, Storia a due voci* (Jugend und Volk, Vienna–Munich, 1973; Cappelli Editore, Bologna, 1974).

13) Paul Urban (b. Purkersdorf, near Vienna, 1905) obtained an engineering diploma (electrotechnics and machine construction) in 1928 at the Technische Hochschule in Vienna, and a Ph.D. (in physics and mathematics) at the University of Vienna in 1935. He has worked in industries (1928–30), in the Technical Section of the Austrian State Railroads (1931–39), and as Assistant (to Professor Hans Thirring) at the Institut für Theoretische Physik of the Universities of Vienna (1940–45) and Innsbruck (1945–46). Finally he became Professor of Theoretical Physics at the University of Graz (1947–1975), of which, since 1975, he has been "Professor Emeritus". He is the author of more than one hundred papers dealing with quantum mechanics, atomic and nuclear physics and elementary particle theory. He is also the author of a book of considerable interest: *Topics in applied quantum electrodynamics* (Springer Verlag, Vienna–New York, 1970), and edited, in collaboration with his pupil, Walter Thirring: "The Schrödinger equation", Lectures presented at the International Symposium 50 years Schrödinger Equation, *Acta Phys. Austr. Suppl.* **17** (1977).

14) Arnold Sommerfeld (1868–1951) studied mathematics at the University of Könisberg and in 1893 went to Göttingen where he made his "habilitation" under the supervision of Felix Klein, whom Sommerfeld considered always his "master". In 1900 Sommerfeld was appointed Ordinary Professor of Mathematics at the Technische Hochschule of Aachen, where for six years he collaborated with a few high-level engineers in the solution of a number of technical problems (resonance phenomena in bridges, construction of locomotives, construction of ships, etc.). In 1906 he was called to the chair of theoretical physics of the University of Munich that he kept until his retirement. Sommerfeld was one of the first supporters of the theory of special relativity of Einstein, which constituted one of the many subjects he used to deal with in his many courses of lectures. Starting from 1920 he made many important contributions to quantum theory, on the use of which he published, in 1919, a famous treatise *Atombau und Spektrallinien* (Vieweg, Braunschweig, 1919), of which a few editions appeared in the successive years. During the twenties Sommerfeld made various important contributions to the quantum theory developed by Heisenberg, Born, Schrödinger, Bohr, Dirac, and others, and to which he devoted a further volume: *Atombau und Spektrallinien: Wellenmechanischer Ergänzungsband* (Vieweg, Braunschweig, 1929). Sommerfeld is well known not only for his many important papers and this fundamental book, but also for his six volumes of lectures in theoretical physics (Leipzig, 1942–62). Among his many pupils it is enough to recall: E. Fuess, H. Hoül, W. Kassel, W. Lenz, W. Pauli, W. Heisenberg, H.A. Bethe. An extensive biography, in which also his vicissitudes during the nazi regime are recalled, has been published by Ulrich Benz: *Arnold Sommerfeld*, Vol. 38 of the collection "Grosse Naturforscher" published under the direction of Dr. Heinz Degen (Wissenschaftliche Verlagsgesellschaft m.b.H., Stuttgart, 1975).

15) Edmund Hlawka (b. 1916), Professor of Mathematics at the University of Vienna has made fundamental contributions, especially to the theory of numbers.

16) Paul Harteck (b. 1902), Professor of Chemical Physics, and subsequently Rector (1948–50) of the University of Hamburg and (since 1951) Distinguished Research Professor of Physical Chemistry at the Rensseleer Polytechnic Institute, Troy, N.Y. He is the author of more than 150 papers on experiments on para- and ortho-hydrogen, deuteron plus deuteron nuclear reactions, separation of hydrogen isotopes, artificial radioactivity, diffusion of slow neutrons, isotope separation by diffusion, and the chemistry of the Earth's atmosphere.

17) P.P. Koch (1879–1945). His most well known scientific contribution goes back to the years before 1912, when he developed high sensitivity and great accuracy method [Ann. Phys. (Germany) **30**, 841 (1909); **34**, 377 (1911); **39** 705 (1912); **40**, 797 (1913); **41**, 115 (1913)] for the photometer analysis of the X-ray photographic plates obtained by B. Walter and R. Pohl [Ann. Phys. (Germany) **29**, 331 (1909); **38**, 507 (1912)] with a wedge-shaped slit, the width of which, of a few μm in the upper part becomes of a few $\mu\mu$m at its lower end. A beam of X-rays impinging on this slit produces a diffraction figure, whose maximum becomes wider moving from the upper to the lower part of the slit. By this method Koch gave evidence for the wave nature of X-rays before the famous work by W. Friedrich, P. Knipping and M. von Laue [Ber. Bayer. Akad. Wiss., 303 (1912); Ann. Phys. (Germany) **41**, 971 (1913)]. Koch's method and results are described in some detail in Chapter 3 on "Die Röntgenspektrum" of the 1919 edition of the book by Arnold Sommerfeld, *Atombau und Spektrallinien*.

18) W. Lenz [Frankfurt (Main), 1888–1952)] studied in Göttingen (1906–1908) and Munich (1908–1912) and later worked on various developments of quantum mechanics with G. Wentzel, W. Pauli, P. Jordan, A. Unsöld and J.H.D. Jensen. See P. Jordan, "The life of W. Lenz", *Phys. Bl.* **13**, 269 (1957).

19) H.J.D. Jensen (1907–1973), Professor of Theoretical Physics at the Universities of Hamburg and Heidelberg, author of numerous works on nuclear physics; shared with Maria Goeppert-Mayer the 1963 Nobel Prize for Physics, for their discoveries concerning nuclear shell structure. No biography of Jensen has been published, according to his wishes, *Phys. Bl.* **29**, 233 (1973).

20) H. Suess (b. 1909), subsequently Professor of Theoretical Physics at the University of California (La Jolla), author of numerous papers on the abundance of the elements, and later on the radioactivity of the atmosphere and hydrosphere. Of particular importance is the paper, which he wrote in collaboration with O. Haxel and H.J.D. Jensen, on nuclear shell structure.

21) Eduard Suess (1831–1914), born in London of Viennese parents, was the most famous geologist of the second half of the nineteenth century. Starting from 1857 he was Professor of Geology at the University of Vienna, and, from 1888, President of the Austrian Academy of Sciences. His name is bound up with the work *Antlitz der Erde* (Vienna, 1883, 1909), *La face de la Terre* (Paris, 1905, 1909), a grand comparative synthesis of the knowledge at the end of the nineteenth century of the structure of the terrestrial globe.

22) At my request, R. Wideröe sent me on 10 November 1979, a long letter concerning his collaboration with Bruno Touschek. Other letters followed with further information and details.

23) Rolf Wideröe (b. Oslo, Norway, 1902), gained a Degree in Engineering at Karlsruhe, conceived the betatron in 1922, and submitted a thesis (Aachen, 1927) in which he set out the bases for the multiple acceleration of charged particles. He constructed the first European betatron (Hamburg 1943–44), and later a number of other machines

of this same type for therapeutic use, working in the laboratories of Brown Boveri (Baden 1946–49). Wideröe has made a few other inventions in the field of accelerators; in particular, in 1943, while working in Hamburg, he proposed for the first time the use of storage rings for high-energy particles, in order to study nuclear reactions produced in the collision of particles moving with the same energy but opposite velocity [German Patent No. 876279 (1943)].

24) The betatron, as a source of particles for the production of nuclear reactions, certainly was much inferior to the cyclotron and to the voltage multiplier, both already very much used during the thirties.

25) D.W. Kerst, *Phys. Rev.* **58,** 841 (1940); **59,** 110 (1941); **60,** 47 (1941).

26) D.W. Kerst and R. Serber, *Phys. Rev.* **60,** 53 (1941).

27) R. Kollath (1900–1978), Professor of Physics at the University of Mainz, author of papers on collisions between slow electrons and ions against the molecules of gases. "At that time Kollath belonged to the AEG Research Laboratory (founded and directed by C. Ramsauer). He also had problems because his wife was Jewish" (Ref. 22). He is the author of the book *Teilchenbeschleuniger* (Friedr. Vieweg and Sons, Braunschweig, 1955).

28) W. Heitler, *The quantum theory of radiation* (Clarendon Press, Oxford, 1936).

29) In his letter (Ref. 22), Wideröe relates the same episode, but with two unimportant variants: the first is that the SS official aimed at Bruno's head because he had stopped to tie his shoe-lace; the second was that he was found by the doctor in the roadside gutter. I have adhered to the account which Bruno Touschek gave me.

30) Bruno, talking with me, used this expression in Italian, but Sir Rudolf Peierls has suggested that probably he referred to the operetta "Die Fledermaus" of Johann Strauss, where a very permissive prison is described.

31) K. Gund (1907–1953) born in Vienna, studied in his native town and, starting from 1931, worked, in Vienna, in the laboratories of Siemens und Halske. In 1936 he transferred to Siemens Reiniger Werke, Erlangen (Germany) where, in 1941, he started to develop a 6 MeV betatron for medical applications.

32) W. Paul (b. 1913), Professor of Experimental Physics at the University of Bonn, author of numerous papers on atomic physics. He introduced high-energy experimental physics with accelerators in Germany.

33) W. Paul, *Naturwissenschaften* **36,** 31 (1949).

34) Richard Becker (1887–1955), Professor of Theoretical Physics at the Technische Hochschule in Berlin, and at the University of Göttingen, author of numerous papers and excellent books on atomic physics, ferromagnetism, and plasticity.

35) O. Haxel (b. 1909), Professor of Physics at the University of Heidelberg, author of numerous papers on nuclear physics. Particularly important is the paper in collaboration with H.J. D. Jensen and H. Suess on the shell model of nuclei.

36) H.C. Kopfermann (1895–1963), Professor of Physics at the Universities of Göttingen and Heidelberg, author of numerous important papers on spectroscopy and nuclear physics, and of the book *Kernmomente* (Akademische Verlagsgesellschaft, Leipzig, 1940). See the biography by V. Weisskopf in *Nucl. Phys.* **52,** 177 (1964) where the list of his papers is also given.

37) Werner Heisenberg (1901–1974), Professor of Theoretical Physics at Leipzig, Berlin, Göttingen and Director of the Max Planck Institut für Physik und Astrophysik in Munich. One of the founders of quantum mechanics, and author of many fundamental papers and books on atomic and molecular physics, ferromagnetism, cosmic radiation, and elementary particles. For an extensive biography see: Armin Hermann, *Heisenberg*, Rowohlt Monographien (Rowohlt Taschenbuch Verlag GmbH, Reinbek bei Hamburg, 1976 and 1977).

38) F.G. Houtermans (1903–1966), Professor of Experimental Physics at Kharkov, Göttingen and Berne, author of numerous papers on spectroscopy, nuclear physics, and on the determination of the age of rocks based on the measurement of their content of uranium and potassium and their decay products. Of particular importance are certain papers dated 1929, written in collaboration with G. Gamow on the alpha decay of nuclei, and others, written in 1930–31, in collaboration with R. d'E. Atkinson on the production of energy by means of nuclear reactions inside stars.

39) Ludwig Prandtl (1875–1953) was the founder of the boundary layer theory and the originator of the German school of aerodynamics. Among his many students in Göttingen the most notable was Theodore von Karman. By their competitive efforts the problem of describing turbulent flow was clarified in the mid-1920's.

40) C.F. von Weizsäcker (b. 1912), a pupil of W. Heisenberg, has made various important contributions to physics and astrophysics. He developed the liquid-drop model of the atomic nucleus which led him to the derivation of the so-called Weizsäcker mass formula. He was also one of the first people to recognize that the energy irradiated by stars is provided by certain nuclear reactions taking place at their centre. Weizsäcker has also devoted a considerable part of his activities to philosophy and politics.

41) The Kaiser-Wilhelm-Gesellschaft was founded in Berlin in January 1911 and dissolved in Berlin, in June 1960, because after the Second World War it was not possible to continue it without some important modification and under the same name. On 26 February 1948, there was founded in Göttingen the Max-Planck-Gesellschaft, which assumed that part of the patrimony of the previous Kaiser-Wilhelm-Gesellschaft which had been saved from the war. For more details see *50-Jahre Kaiser-Wilhelm-Gesellschaft und Max-Planck-Gesellschaft zur Förderung der Wissenschaften, 1911–1961,* Beiträge in Dokumenten, Göttingen, 1961.

42) Paul Urban has sent me three letters, dated 30 June, 4 July and 16 September 1980. The most important information is contained in the first one.

43) In the article "In Memory of Hans Thirring" written by Paul Urban, immediately after his death, one can read that the antimilitarism of Hans Thirring had begun already during the First World War, and his interest in the problems of peace had been strongly reinforced in the middle of the thirties when international tension increased owing to the advance of the Japanese troops in the Far East, and of the Italian Troops in Abyssinia.

44) J. Slepian, X-ray tube, USA Patent No. 1645305 asked for 1 March 1922 and granted 10 October 1928.

45) R. Wideröe, *Arch. Elektrotech.* **21**, 387 (1928).

46) For a detailed account of the development of this idea, see for example: R. Wideröe, "Das Betatron", *Z. angew. Phys.* **5**, 187 (1953); "Die ersten zehn Jahre der Merfach-Beschleunigung, Einige historische Notizien" *Wiss. Z. der F. Schiller Univ., Jena* **13**, 491 (1964).

47) K. Gund, *Naturwissenschaften* **34**, 343 (1947); K. Gund and H. Reich, *Z. Phys.* **126**, 383 (1949); K. Gund and W. Paul, *Nucleonics* **7**, 36 (1950).

48) See also: W. Paul, "Early days in the development of accelerators", *Aesthetics and Science*, Proc. Int. Symposium in honour of R.R. Wilson, 27 April 1979 (Fermi Nat. Accel. Lab., Batavia, 1979), p. 25.

49) Philip Ivor Dee (b. 1904), Professor of Natural Philosophy at the University of Glasgow, is well known in nuclear physics as an author of numerous papers, among which I should mention his collaboration with James Chadwick in the discovery of the neutron. He is the author of the third of the three papers which appeared together in *Proc. R. Soc., London, Ser. A,* **136** (1932):
J. Chadwick, "The existence of the neutron";
N. Feather, "The collisions of neutrons with nitrogen nuclei";
P.I. Dee, "Attempt to detect the interaction of neutrons with electrons".
From 1939 to 1945, Dee was a superintendent of the Telecommunications Research Establishment, where he was in charge of the group responsible for the development of airborne centimetric radar devices.

50) R. Fraser: research physical chemist at Cambridge, where he worked for a few years after having been a lecturer at Aberdeen University.

51) John Currie Gunn (b. 1916), Professor of Natural Philosophy and Head of that department since 1973, at the University of Glasgow, is author of a long list of papers on theoretical physics.

52) Sir Rudolf Peierls (b. Berlin, 1907) studied theoretical physics in Berlin, Munich, Leipzig, and at the ETH of Zurich, under Sommerfeld, Heisenberg, and Pauli. He has been Professor of Applied Mathematics at the University of Birmingham (1937–63) and of Theoretical Physics at the University of Oxford (1963–74). In addition to many important papers on the theory of solids, nuclei, and fields, Peierls has published a few books: *Quantum theory of solids* (1955); *The laws of nature* (1955); *Surprises in theoretical physics* (1979). He is the editor of *A perspective of physics*, Vol. 1 (1977), Vol. 2 (1978).

53) Ian Naismith Sneddon (b. 1919), Simson Professor of Mathematics at the University of Glasgow, is author of almost one hundred papers on Fourier transforms and mixed boundary value problems in potential theory.

54) Walter Thirring (b. 1927), Professor of Theoretical Physics at the University of Vienna, author of many major papers on field theory and relativity.

55) Roy Chisholm (b. 1926) is now Professor of Applied Mathematics at the University of Kent. He has also worked at Glasgow, Cardiff, CERN, Texas, and Dublin. He has done research in quantum field theory, especially on

computational methods and on gauge theory, and in non-linear approximation theory, in particular on multivariate and other generalizations of Padé approximation.

56) E.W. Laing, *Philos. Mag.* **46**, 106 (1955).

57) B. Ferretti (b. Bologna, 1913), Assistant and later deputy of E. Fermi at the University of Rome. Subsequently Professor of Theoretical Physics at the University of Milan (1947), Rome (1948–1956) and Bologna (from 1956 onwards). Author of many important publications concerning quantum electrodynamics, field theory, and elementary particles.

58) E. Amaldi, "The years of reconstruction", *Giornale di Fisica* **20**, 186 (1979); *Scientia* **114**, 29 (1979).

59) S. Weinberg, *Phys. Rev. Lett.* **19**, 1264 (1967); **27**, 1688 (1971).
A. Salam and J. Strathdee, IC/71/145, Int. Center Theor. Phys. Trieste (ICTP).

60) Sir (Charles) Maurice Yonge, C.B.E., F.R.S. (b. 1899), Professor of Zoology at the University of Glasgow, has devoted a great part of his activities to research on marine biology and has occupied administrative national and international positions. Among his publications I recall: *The seas* (1928); *A year on the great barrier reef* (1930); *British marine life* (1944); *The sea shore* (Collins, London–New York, 1949); *Oysters* (Collins, London–New York, 1960); *Physiology of mollusca* (ed. with K.M. Wilbur) (Academic Press, New York, 1964–1967).

61) M. Sands (b. 1919), at that time Professor of Physics at the California Institute of Technology. The friendship between Touschek and Sands went on for the successive years and gave rise to a renewed tight collaboration in 1968–69 on problems concerning the electron storage rings. In 1952–55 Sands participated in the design and construction of the first electron accelerator of energy greater than 1 GeV (1.5 GeV Synchrotron of CalTech), in 1959 he was the first to propose a design of a 300 GeV proton-synchrotron, together with W.K.H. Panofsky, in 1963–69 he was in charge of the SLAC project during the building and first operation of the 3 kilometre linear electron accelerator. He is the author of various books, among which I recall (in collaboration with W.C. Elmore): *Electronics: experimental techniques* (McGraw-Hill, New York, 1949); and (in collaboration with R.P. Feynman and R.B. Leighton): *The Feynman lectures in physics* (3 volumes) (Addison-Wesley, Reading, Mass., 1963–64). In recent years Sands became Faculty Dean at the University of California, Santa Cruz.

62) E.D. Courant, M.S. Livingston and H.S. Snyder, *Phys. Rev.* **88**, 1190 (1952).

63) E.D. Courant and H. Snyder, Brookhaven Internal Report 1953 (unpublished) and *Ann. Phys. (USA)* **3**, 1 (1958).
J.B. Adams, M.G. Hine and J.D. Lawson, *Nature* **173**, 926 (1953).

64) E. Persico, E. Ferrari and S.E. Segrè, *Principles of particle accelerators* (Benjamin, New York, 1968).
A.A. Kolomensky and A.N. Lebedev, *Theory of cyclic accelerators* (North Holland, Amsterdam, 1962).

65) Reported at the Third Annual Rochester Conference on High Energy Physics, Rochester, 1952 [see *Proceedings* (Univ. Rochester, NY, 1953), p. 50] and Royal Society Conference on Heavy Mesons, London, 1953.

66) E. Amaldi, G. Baroni, C. Castagnoli, G. Cortini and A. Manfredini, *Nuovo Cimento* **10**, 937 (1953).
E. Amaldi, G. Baroni, G. Cortini, C. Franzinetti and A. Manfredini, *Suppl. Nuovo Cimento* **12**, 181 (1954).
E. Amaldi, *Suppl. Nuovo Cimento* **4**, 179 (1956).

67) R.M. Brown, U. Camerini, P.H. Fowler, H. Muirhead, C.F. Powell and D.M. Ritson, *Nature* **163**, 47 (1948).

68) E. Fabri, *Nuovo Cimento* **11**, 479 (1954).

69) R. Dalitz, *Philos. Mag.* **44**, 1068 (1953). Dalitz presented his results at the Congrès international sur le Rayonnement cosmique, organized by the University of Toulouse, under the patronage of IUPAP, with the support of UNESCO, at Bagnères-de-Bigorre (6–11 July 1953).

70) Report of the International Congress on Heavy Unstable Elementary Particles and on High-Energy Events in Cosmic Rays, Padua, 12–15 April 1954, *Suppl. Nuovo Cimento* **12**, 163 (1954).

71) C. Dilworth, A. Manfredini, G.D. Rochester, J. Waddington and G.T. Zorn, "Report of the Committee on K-Particles", submitted to the Conference of Ref. 70, *Suppl. Nuovo Cimento* **12**, 433 (1954).

72) G.F. Chew, *Phys. Rev.* **89**, 591 (1953);
G.F. Chew and F. Low, *Phys. Rev.* **101**, 1570 (1956);
G.F. Chew, "Theory of pion scattering and photoproduction" (University of Illinois, Urbana, Illinois, 1956), p. 1–140;

M. Cini and S. Fubini, "General properties of the fixed source meson theory", *CERN Symposium on High Energy Accelerators and Pion Physics* (CERN, Geneva, 1956), Vol. 2, p. 171–172.

73) G. Wentzel, *Helv. Phys. Acta* **25**, 569 (1952). See also J.M. Blatt, *Phys. Rev.* **72**, 466 (1947).

74) See, for example: S.S. Schweber, H.A. Bethe and F. de Hoffman, *Mesons and fields* (Row, Peterson and Co., Evanston, Ill., 1956).

75) M. Cini, *Nuovo Cimento* **10**, 526, 614 (1953).

76) M. Cini and S. Fubini, *Nuovo Cimento* **11**, 142 (1954).

77) F.J. Dyson, *Math. Rev.* **17**, 438 (1956).

78) This report is quoted in: D. Amati and B. Vitale, *Fortschr. Phys.* **7**, 375 (1959) and S. Fubini and J.D. Walecka, *Phys. Rev.* **116**, 194 (1959).

79) G. Feinberg, *Phys. Rev.* **108**, 898 (1975);
S.N. Gupta, *Can. J. Phys.* **35**, 1309 (1957);
V.G. Soloviev, *Nucl. Phys.* **6**, 618 (E7, 791): The problem is discussed on p. 65 of G. Morpurgo, "Strong interactions and reactions of hyperons and heavy mesons", *Ann. Rev. Nucl. Sci.* **11**, 41 (1961).

80) Abdus Salam, *Nuovo Cimento* **5**, 299 (1957).

81) M. Gell-Mann and Y. Ne'eman, "The weak current of the hadrons", Chapter 8 of *The eightfold way* (W.A. Benjamin Inc., New York, 1964), pp. 171–206.

82) T.D. Lee and C.N. Yang, *Phys. Rev.* **104**, 254 (1956).

83) C.S. Wu, E. Ambler, R.W. Hayward, D.D. Hoppes and R.P. Hudson, *Phys. Rev.* **105**, 1413 (1957).

84) The same choice was also made at the same time by other authors: R.P. Feynman and M. Gell-Mann, *Phys. Rev.* **109**, 193 (1958);
R.E. Marshak and E.C.G. Sudarshan, *in Proc. Int. Conf. on Mesons and Recently Discovered Particles* Padua-Venice, 1957 (Padua, 1957) and *Phys. Rev.* **109**, 1860 (1958).

85) R. Garwin, L. Lederman and M. Weinrich, *Phys. Rev.* **105**, 1415 (1957).

86) L.L. Foldy and S.A. Wouthuysen, *Phys. Rev.* **78**, 29 (1950).

87) W. Heisenberg, *Z. Naturforsch. a* **12**, 177 (1957).

88) The following comments are made on page 3 of the Bulletin of the Società Italiana di Fisica No. 12, of 4 March 1959:
The "Prize for Teaching Merit" amounting to 1,000,000 Lire, offered by the University of Palermo, by the Sicilian Regional Committee for Nuclear Research and by the University of Palermo's Institute of Physics, has been awarded to Professor BRUNO TOUSCHEK, of Austrian nationality, for the following reasons: "Bruno Touschek, born in Vienna, has lived in Italy since the end of 1952 being a Professor at the Scuola di Perfezionamento di Fisica Nucleare at the University of Rome. Over these years, he has published the results of his outstanding theoretical research, most of the work being performed in collaboration with colleagues and pupils. The research relates to the stability conditions of orbits in strong-focusing accelerators, the decay of the τ meson, the Tamm-Dancoff method, time reversal, the capture of negative heavy mesons with hyperon production, the conservation of leptonic number and the properties of symmetry in spinor fields. Some of the results contained in these papers have been recognized internationally: examples of these are the study of heavy meson capture and time reversal. This research activity is difficult to separate from his teaching activities which have always gone well beyond his official obligations. Up to 1954, the latter consisted of giving a course at the Scuola di Perfezionamento in Fisica Nucleare on 'Cosmic rays and elementary particles', and from 1954 onwards a course on Field theory. After 1956 these same courses were given by Touschek at the University of Pisa, in a series of short periodic visits. The often very lively and sometimes almost frenzied discussion of the subjects dealt with in his courses gave rise not only to a substantial part of the papers which carry his name, but also to various others from students who attended these courses. His personality and his human qualities have brought him close to his colleagues and collaborators, with whom he has shared his outstanding enthusiasm for the problems to which his attention is devoted at the particular moment. This happy combination of a researcher and master puts Touschek right in the forefront for the award of the Prize for Teaching Merit of the Società Italiana di Fisica".

89) *Proc. Int. School of Physics "Enrico Fermi", XLVI Course*, directed by B. Touschek (Academic Press, New York, and Periodici Scientifici, Milan, 1971).

90) The Casati Reform goes back to November 1859. Count Gabrio Casati (1798–1873) was Minister for Public Education in the Lamarmora Cabinet for only 6 months, during which he drew up the "Casati Law", which remained in force with slight modifications up to the Gentile Reform of 1928. The need to suppress the certificate of Italian nationality for applications to participate in university competitions had been discussed in 1960 by G. Bernardini and myself with the Minister for Public Education, Giuseppe Medici, especially on 4 and 5 February, when the Minister came to Geneva to inaugurate the 29 GeV CERN Proton Synchrotron. This proposal was discussed and accepted also by all of the Italian physics professors, as is recalled also by A. Rostagni, during the meetings which were held at the Academia dei Lincei for the university syllabus reform, between the end of 1956 and the beginning of 1957. For the moment, however, we have been unable to find the corresponding documents or the papers in which the Consiglio Superiore della Pubblica Istruzione gave its negative reply. This probably happened in the period 1958–62.

91) After a long stay abroad, E. Corinaldesi acquired United States citizenship.

92) *Graf Bobby, Baron Mucki and Poldi, 123 mal in Wort und Bild*(Fischer Taschenbuch Verlag, Frankfurt am Main, September 1976).

93) Victor F. Weisskopf (b. Vienna, 1908), studied theoretical physics at Göttingen (Ph.D. in physics in 1931) and then became assistant of Heisenberg (Leipzig, 1931) and Schrödinger (Berlin, 1932), with periods spent in Copenhagen (with Niels Bohr) and Kharkov, USSR (with L. Landau). In 1933–35 Weisskopf was assistant of Pauli at the Zurich Federal Polytechnic. Later he again worked in Copenhagen, where he remained until 1939, when he moved to Cornell University (Ithaca, NY, USA). In 1943 he was offered a chair at the University of Rochester (N.Y.), but the same year went to Los Alamos, New Mexico, to work on the applications of nuclear energy. From 1945 to 1960 he was a Professor at the Massachusetts Institute of Technology, that he left to become Director General of CERN (1961–1965). After 1965 Weisskopf went back to his MIT chair. He is the author of a wide spectrum of important papers dealing with quantum mechanics, field theory, nuclear physics, and particle physics. He is also the author of a few remarkable books.

94) Fernando Amman has pointed out to me that a first mention of e^+e^- rings was made by Bruno at a meeting held in Frascati on 17 February 1960, in the frame of a series of discussions organized by Giorgio Salvini and devoted to the future of the Laboratori Nazionali di Frascati. There still exists a copy of the minutes of the meeting, prepared by Dr. Icilio Agostini, then Administrative Secretary of the Laboratories.

95) D.W. Kerst, F.T. Cole, H.R. Crane, L.W. Jones, L.J. Laslett, T. Ohkawa, A.M. Sessler, K.R. Symore, K.M. Terwilliger and Niels Vogt Nilsen, *Phys. Rev.* **102**, 590 (1956) received 26 January 1956. At the beginning of their paper, these authors write: "... The possibility of producing interactions in stationary coordinates by directing beams against each other, has often been considered, but the intensity of beams so far available have made the idea unpractical. Fixed field alternating gradient accelerators offer the possibility of obtaining sufficiently intense beams so that it may now be reasonable to consider directing two beams of approximately the same energy at each other... ."

96) G.K. O'Neill, *Phys. Rev.* **102**, 1418 (1956), received 13 April 1956. In a footnote O'Neill writes that between the mailing and the publication of his Letter to the *Physical Review* he had become aware that similar suggestions had been made also by W.M. Brobeck of the Berkeley Accelerator Group and by D. Lichtenberg, R. Newton and M. Ross of the MURA Group.

97) G.K. O'Neill, "The storage ring synchrotron", *Proc. CERN Symposium on High-Energy Accelerators and Pion Physics*, Geneva, 1956 (ed.: E. Regenstreif) (CERN, Geneva, 1956), Vol. 1, pp. 64–65.

98) W.C. Barber, B. Gittelman, G.K. O'Neill, W.K.H. Panofsky and B. Richter, Stanford University Report HEPL 170 (June 1959). This report contains a description of the e^-e^- tangent storage rings and the proposal for its construction in view of an electron-electron scattering experiment as a test of the limits of QED.

99) W.C. Barber, B. Gittelman, G.K. O'Neill and B. Richter: "Test of quantum electrodynamics by electron-electron scattering", *Phys. Rev. Lett.* **16**, 1127 (1966).

100) Typical examples are:

$$\gamma + N \rightarrow N + e^+ + e^-$$
$$\pi^- + p \rightarrow N + e^+ + e^-$$

101) Typical examples are:

$$p + N \rightarrow e^+ + e^- + X$$
$$p + \bar{p} \rightarrow e^+ + e^- + X$$

102) S. Drell and F. Zachariasen, *Electromagnetic structure of nucleons* (Oxford University Press, London, 1961), pp. 18, 19.

103) L.D. Landau and E.M. Lifschitz, *Sov. Phys.* **6**, 244 (1934), had considered the processes

$$e^+ + e^- \to e^+ + e^- + \text{or} \begin{cases} e^+ + e^- \\ \mu^+ + \mu^- \end{cases}$$

in the frame of pure electrodynamics. Processes of this type were later re-examined by F.E. Low [*Phys. Rev.* **120** 582 (1960)], F. Calogero and C. Zemach [*Phys. Rev.* **120**, 1860 (1960)] and later by others: A. Jaccarini, N. Arteaga-Romero, G. Parisi and P. Kessler [*Compt. Rend.* **269B**, 153, 1129 (1969); *Nuovo Cimento* **4**, 933 (1970)]; V.E. Balakin, V.M. Budnev and I.F. Ginzburg [*Zh. Eksp. Teor. Fiz. Pis'ma* **11**, 559 (1970); *JETP Lett.* **11**, 388 (1970)]; S. Brodsky, T. Kinoshita and H. Terazawa [*Phys. Rev. Lett.* **25**, 972 (1970); *Phys. Rev.* **D4** 1532 (1971)]; H. Terazawa [*Rev. Mod. Phys.* **45**, 615 (1973)].

104) See, for example: *Proc. Int. Workshop on γ-γ Collisions*, Amiens, 8–12 April 1980 (ed.: G. Cochard) (Springer, Berlin, 1980).

105) J. Haissinski, *Expériences sur l'anneau de collisions AdA*, Thèse, Orsay Série A, No. d'ordre 81, soutenue le 5 février 1965.

106) C. Bernardini, "La storia di AdA", *Scientia* **113**, 27 (1978).

107) Giorgio Ghigo (b. Turin, 1929–d. Rome, 1968); in 1948 he started to study engineering at the Turin Polytechnic, but some two years later was influenced by Gleb Wataghin to change over to physics, taking his degree in 1953. After a year spent studying cosmic rays in the Turin Section of the INFN, Ghigo joined in September 1954 the Magnet Group of the Accelerator Section of the same Institute. As a highly skilled instrument engineer, he contributed substantially to the design and construction of the magnet for the Italian electron synchrotron, developing various high-quality instruments (flow-meters, magnetometers, quantum-meters, etc.). In 1959 he was appointed Machine Director of the Laboratori Nazionali di Frascati. He then participated in the AdA project and later in ADONE. After moving to Naples for two years he constructed, in 1962, in connection with the research by Edoardo Caianello and Valentino Braitenberg, a completely transistorized machine capable of simulating 100 neurons, the purpose of which was to clarify certain aspects of the functioning of the brain. During the same period he dealt with various other problems of electronic simulation of the structure and/or functioning of the brain. On returning to Frascati he resumed his activities as a skilled instrument engineer, but was suddenly taken ill with an incurable disease and died on 15 March 1968, thereby leaving a serious gap among the researchers of the Laboratori di Frascati, and was sadly missed by all his friends.

108) G. Ghigo, "Preliminary discussions on AdA", Internal Memorandum No. 62, 8 December 1960, Laboratori Nazionali di Frascati del CNEN.

109) *Particle Accelerators*, Vol. 3, No. 2, p. 127 (April 1972). The piece of news is reported by John Blewett.

110) Deutsches Patentamt, Patentschrift Nr 876279 Klass 21g Gruppe 36, Ausgegeben am 11. Mai 1953: Dr. Ing. Rolf Wideröe, Oslo, ist als Erfinder genannt worden: Aktiengesellschaft Brown, Boveri & Cie, Baden (Schweiz). Anordnung zur Herbeiführung von Kernreaktionen.

111) A.P. Aleksandrov, L.M. Barkov, S.T. Belayev, Ya.B. Zel'dovich, B.B. Kadomtsev, A.A. Logunov, M.A. Markov, D.D. Fyutov, V.A. Sidorov, A.N. Skrinski and B.V. Chirikov, "Academician Gersh Itskovic Budker (obituary)", *Usp. Fiz. Nauk* **124**, 731 (1978).

112) E.M. Abramyan et al., "Studies of colliding electron-electron, electron-positron and proton-proton beams at the Institute for Nuclear Physics, the Siberian branch of the USSR Academy of Sciences" (in Russian), *in Proc. Int. Conf. on High-Energy Accelerators,*Dubna, 1963 (Atomizdat, Moscow, 1964), p. 274. [English transl. of Proceedings (USAEC, Oak Ridge, Tenn., 1965), p. 334].

113) G.I. Budker, *Sov. Phys. Usp.* **9**, 534 (1966).

114) It is in order to notice that at the International Conference on High-Energy Accelerators, held at Brookhaven during the summer of 1961, the Soviet physicists did not appear at the last moment, but their contributions were published and did not contain any mention of possible e^+e^- storage rings, while a report on AdA and a preliminary study on ADONE were presented to the conference, in addition to a paper on the limits to space-charge due to beam-beam effects. When, in 1963, Amman and Bernardini visited Novosibirsk, the ring e^-e^- (VEPP 1) was in operation and VEPP 2 was in the construction stage. Therefore it appears reasonable to conclude that the activity on e^+e^- rings was started after 1961 and no document is known which proves the contrary.

115) F. Amman, C. Bernardini, R. Gatto, G. Ghigo and B. Touschek, "Storage ring for electrons and positrons (ADONE)", Internal Report No. 68 of the Laboratori Nazionali di Frascati, 27 January 1961;

F. Amman, M. Bassetti, M. Bernardini, G.F. Corazza, L. Mango, A. Massarotti, C. Pellegrini, M. Placidi, M. Puglisi and A. Tazzioli, *Ric. Scient.* **32**, Parte 1, 197 (1962).

116) A. Blanc-Lapierre, R. Beck, R. Belbeoch, B. Boutouvrie, H. Bruck, L. Burnod, X. Buffet, G. Gendreau, J. Haissinski, R. Jolivot, G. Leleux, P. Marin, B. Milman and H. Zyngier, "Projet d'un anneau de stockage à Orsay pour électrons et positrons d'une énergie maximale de 450 MeV", Rapport CEA No 2363 (LAL 1081), avril 1964.
Also consult the various articles published in: *Symposium International sur les Anneaux de Collisions à Electrons et Positrons*, held at Saclay (Paris) from 26 to 30 September 1966 (eds.: H. Zyngier and E. Cremieu-Alcan) (Presses universitaires de France, Paris, 1966).

117) This modification of the CEA, indicated in the literature as a "bypass", presupposes the invention of the so-called "low-β", made by K. Robinson and G.A. Voss [Cambridge Electron Accelerator Report TM 149 (1965)]. The "bypass" started operation in 1971, but the luminosity never exceeded the value of 2×10^{28} $cm^{-2}s^{-1}$. It was dismantled in 1973. For further overall information, see for example:
G. Giacomelli, "High-energy laboratories in the United States", *Giornale di Fisica* **12**, 93 (1971).

118) R.F. Christy, "Synchrotron beam loss due to quantum fluctuations in the radiation", California Institute of Technology (1957), unpublished.

119) M. Sands, "Observation of quantum effects in an electron-synchrotron", California Institute of Technology (1956), presented at the West Coast Meeting of the APS in December 1956.

120) C.P. Curtis, A. Galonsky, R.H. Hilden, F.E. Mills, R.A. Otte, G. Parzen, C.H. Pruett, E.M. Rowe, M.F. Shea, D.A. Swenson, W.A. Wallenmeyer and D.E. Young (MURA: Midwestern Universities Research Association), *Proc. Int. Conf. on High-Energy Accelerators*, Dubna, 1963 (Atomizdat, Moscow, 1964), p. 20.

121) L.J. Laslett, V.K. Neil and A.M. Sessler, Lawrence Radiation Laboratory Report UCRL 11090 (1963) and *Rev. Sci. Instrum.* **36**, 436 (1965). See also: C. Pellegrini and A.M. Sessler, *Sanford Linear Accelerator Center Storage Ring Summer Study*, 1965 (SLAC, Stanford, Calif., 1965), Report SLAC-49, p. 61;
E.D. Courant and A.M. Sessler, *ibid.*, p.36.

122) F. Bloch and A. Nordsieck, *Phys. Rev.* **52**, 54 (1937).

123) James Bond, secret agent 007, in a long series of books by Ian Fleming.

124) Letter of 21 May 1980.

125) According to the law establishing the foundation of the Comitato Nazionale per l'Energia Nucleare (CNEN) valid at the time, the Minister of Industry and Trade was also President of the Consiglio Direttivo of CNEN. This function was fulfilled, during the period of AdA and ADONE, by: the Honourable Emilio Colombo (1960–63), the Honourable G. Togni (1963), Senator Giuseppe Medici (1963–1964), Senator Edgardo Lami Starnuti (1965–1966), the Honourable G. Andreotti (1966–68) and the Honourable Mario Tanassi (1968–69). The Honourable G. Togni was not favourable to CNEN, Senator Lami Starnuti and the Honourable M. Tanassi, differently from the others, showed little interest in the direction of CNEN and in the life and activities of the Laboratori Nazionali di Frascati.

126) F. Amman, R. Andreani, M. Bassetti, M. Bernardini, A. Cattoni, V. Chimenti, G.F. Corazza, D. Fabiani, F. Ferlenghi, A. Massarotti, C. Pellegrini, M. Placidi, M. Puglisi, F. Sosò, S. Tazzari, F. Tazzioli and G. Vignola, "Two-beam operation of the 1.5 GeV electron-positron storage ring ADONE", *Lett. Nuovo Cimento* **1**, 729 (1969).

127) L.M. Brown and F. Calogero, *Phys. Rev. Lett.* **4**, 315 (1960);
N. Cabibbo and R. Gatto, *Phys. Rev. Lett.* **4**, 313 (1960); *Nuovo Cimento* **20**, 185 (1961); *Phys. Rev.* **124**, 1577 (1961);
R. Gatto, *Proc. Aix-en-Provence Int. Conf. on Elementary Particles*, 1961 (CEN, Saclay, 1962), Vol. 1, p. 487.
R. Gatto, *Ric. Scient.* **32**, Part 1, 161 (1962).

128) An overall picture and the way it changed subsequently can be gathered from the following articles from reviews, lectures at summer schools, and papers submitted at the invitation of International Congresses:
M. Grilli, Invited report on "Preliminary results with ADONE", *Proc. Daresbury Study Weekend No. 1 on Vector Meson Production and Omega–Rho Interference*, 12–14 June 1970 (eds.: A. Donnachie and E. Gabathuler) (Daresbury, Nuclear Physics Laboratory, Nr. Warrington, Lancs., 1970) DNPL-R 7, p. 215.
Richard Wilson, Invited report on "Lepton-hadron interactions and quantum electrodynamics", *Proc. 15th Int. Conf. on High-Energy Phys.*, Kiev, August 1970 (Naukova Dumka, Kiev, 1972), p. 219.
G. Salvini, Lectures to the Scuola di Erice on "Electromagnetic production of hadronic resonances", *Elementary*

processes at high energy (ed.: A. Zichichi) (Academic Press, New York, 1971), Part A, pp. 322–383.
M. Conversi, Invited report on "Experiments on electron-positron colliding beams", *Proc. Daresbury Study Weekend No. 4 on Lepton and Photon Physics in Europe,* 1-3 October 1971 (ed.: A. Donnachie) (Daresbury, Nuclear Physics Laboratory, Nr. Warrington, Lancs., 1971), DNPL-R 19, pp. 87–128.
C. Bernardini, Invited report on "Results of e^+e^- reactions at ADONE, *Proc. Int. Symposium on Electron and Photon Interactions at High Energy*, Ithaca, 1971 (Cornell Univ. Lab. Nuclear Studies, Ithaca, 1972), p. 38.
V. Silvestrini, Invited report on "Electron–positron interactions", *Proc. 16th Int. Conf. on High Energy Physics,* Chicago-Batavia, 1972 (National Accelerator Laboratory, Batavia, 1972), pp. 1–38.
G. Salvini, Invited report on "Researches in Frascati on the reactions $e^+e^- \to e^+ + e^- + X$: The results of the $\gamma\gamma$ group", at the Int. Colloquium on Photon-Photon Collisions in e^+e^- Rings, 3–4 September 1973, *J. Phys.* C2-C2-8;
M. Conversi, Invited report on "Experiments with e^+e^- colliding beams", *Proc. Seminar on ep and ee Storage Rings,* Hamburg, 8–12 October 1973 (Hamburg, 1974), DESY 73-66, pp. 121 – 172.
L. Paoluzi, "e^+e^- colliding beams physics", *Acta Phys. Polon.* **B5**, 839 (1974);
G. Salvini, Seminar on "e^+e^- physics in Frascati", Ettore Majorana Center, Erice, July 1974, not published, and address given to the LX Congress of the SIF, Bologna, 1974, not published;
A. Zichichi, "Why (e^+e^-) physics is fascinating", *Riv. Nuovo Cimento* **4**, 498–532 (1974);
P. Monacelli and F. Sebastiani, "Recent experimental results in e^+e^-physics and the new particles, *Riv. Nuovo Cimento* **6**, 449 (1976);
A. Zichichi, "Old and new problems in subnuclear physics", *Riv. Nuovo Cimento* **6**, 529–584 (1976);
M. Conversi, "e^+e^- physics", *Proc. Symposium on Frontier Problems in High Energy Physics*, Pisa, 4–5 June 1976 (Scuola Normale Superiore, Pisa, 1976), pp. 65–88.

129) The first verifications of quantum electrodynamics with e^+e^- rings were carried out at Novosibirsk with VEPP 2 [V.A. Sidorov, *Proc. 4th Int. Symposium on Electron and Photon Interactions at High Energy*, Liverpool, 1969 (Daresbury Nuclear Physics Laboratory, Nr. Warrington, Lancs., 1969), p. 22] and at Orsay with ACO [J.E. Augustin et al., *Phys. Lett.* **31B**, 673 (1970)]. Shortly after this there were the initial results with ADONE [B. Bartoli, B. Coluzzi, F. Felicetti, V. Silvestrini, G. Goggi, D. Scannicchio, G. Marini, F. Massa and F. Vandi, *Nuovo Cimento* **70A**, 603 (1970); B. Borgia, F. Ceradini, M. Conversi, L. Paoluzi, W. Scandale, G. Barbiellini, M. Grilli, P. Spillantini, R. Visentin and A. Malachiè, *Phys. Lett.* **35B**, 340 (1971); C. Bacci, G. Penso, G. Salvini, R. Baldini-Celio, G. Capon, C. Mencuccini, G.P. Murtas, A. Reale and M. Spinetti, *Lett. Nuovo Cimento*2, 73 (1971) on the $e^+ + e^- \to \gamma\gamma$ processes; C. Bacci, G. Penso, G. Salvini, R. Baldini-Celio, G. Capon, C. Mencuccini, G.P. Murtas, A. Reale, M. Spinetti and B. Stella, *Lett. Nuovo Cimento* **3**, 709 (1972) on the process $e^+ + e^- \to e^+ + e^- + e^+ + e^-$; C. Bacci, G. Parisi, G. Penso, G. Salvini, B. Stella, R. Baldini-Celio, G. Capon, C. Mencuccini, G.P. Murtas, M. Spinetti and A. Zallo, *Phys. Lett.* **44B**, 530 (1973) on the process $e^+ + e^- \to e^+ + e^- + \gamma$; M. Bernardini, D. Bollini, P.L. Brunieri, E. Fiorentino, T. Massam, L. Monari, F. Palmonari, F. Rimondi and A. Zichichi, *Phys. Lett.* **45B**, 510 (1973); V. Alles-Borelli, M. Bernardini, D. Bollini, P. Giusti, T. Massam, L. Monari, F. Palmonari, G. Valenti andA. Zichichi, *Phys. Lett.* **59B**, 201 (1975)].

130) F. Ceradini, M. Conversi, S. D'Angelo, M.L. Ferrer, L. Paoluzi, R. Santonico, G. Barbiellini, S. Orito, T. Tsuru and R. Visentin, *J. Phys. (France)*, Colloque C2, supplément au No 3, Tome **35**, p. C2-9 (mars 1974).
G. Barbiellini, S. Orito, T. Tsuru, R. Visentin, F. Ceradini, M. Conversi, S. D'Angelo, M.L. Ferrer, L. Paoluzi and R. Santonico, *Phys. Rev. Lett.* **32**, 385 (1974).

131) The first results obtained with ADONE by the three groups named "$\gamma\gamma$", "$\mu\pi$" and the "boson" were submitted by Mario Grilli at the Darebury Study Weekend in June 1970, and by Richard Wilson to the Kiev Conference in August 1970 (Ref. 131). The first original work on multiple hadron production is by the "boson" group: B. Bartoli, B. Coluzzi, F. Felicetti, V. Silvestrini, G. Goggi, D. Scannicchio, F. Massa, G. Marini and F. Vandi ("boson" group), *Nuovo Cimento* **70A**, 615 (1970). More complete results were submitted to the First EPS Conference on Meson Resonances and Related Electromagnetic Phenomena, Bologna, 1971 (Proc.: edited by A. Zichichi, published by Editrice Compositori, Bologna, 1972), by Conversi for the "$\mu\pi$" group (p. 471), by Salvini for the "$\gamma\gamma$" group (p. 481) and Zichichi for the "BCF" group (p. 489). Other original papers, still on this same subject are:
G. Barbarino, F. Ceradini, M. Conversi, M. Grilli, E. Iarocci, M. Nigro, L. Paoluzi, R. Santonico, P. Spillantini, L. Trasatti, V. Valente, R. Visentin and G.T. Zorn (gruppo "$\mu\pi$"), *Lett. Nuovo Cimento* **3**, 689 (1972);
C. Bacci, R. Baldini-Celio, G. Capon, C. Mencuccini, G.P. Murtas, G. Penso, A. Reale, G. Salvini, M. Spinelli and B. Stella (gruppo "$\gamma\gamma$"), *Phys. Lett.* **38B**, 551 (1972);
M. Bernardini, D. Bollini, P.L. Brunini, T. Massam, L. Monari, F. Palmonari, E. Rimondi and A. Zichichi (gruppo "BCF"), *Phys. Lett.* **44B**, 393 (1973); **46B**, 261 (1973).
In the review article by Monacelli and Sebastiani (Ref. 128) an account is given of the results of an overall analysis of the Frascati data, made by G. Salvini: a) in a seminar of e^+e^- physics in Frascati, presented to the International School of Subnuclear Physics, Ettore Majorana Center, Erice, July 1974) (unpublished); b) in a talk to the LX Congress of the SIF, Bologna, 1974 (unpublished). Another analysis of the results of the BCF group is given by Zichichi in his article dated 1974 (Ref. 128).

132) For a detailed examination of these problems, see, for example: R.P. Feynman, "Photon-hadron interactions", *in Frontiers in physics* (W.A. Benjamin Inc., Reading, Mass., 1972).

133) For the WAEP (wide angle electron pairs) see:
R. Richter, *Phys. Rev. Lett.* **1**, 114 (1958); R.B. Blumenthal, D.C. Ehn, W.L. Faissler, P.M. Joseph, L.J. Lanzerotti, F.M. Pipkin and D.G. Stairs, *Phys. Rev.* **144**, 1199 (1966); E. Eislander, J. Feigenbaum, N. Mistry, P. Mostek, D. Rust, A. Silverman, C. Sinclair and R. Talman, *Phys. Rev. Lett.* **18**, 425 (1967); J.C. Asbury, W.K. Bertram, U. Becker, P. Joos, M. Rhode, A.J.S. Smith, S. Friedlander, C.L. Jordan and S.C.C. Ting, *Phys. Rev.* **161**, 1344 (1967).
For the WAMP (wide angle muon pairs) see:
A. Alberigi Quaranta, M. De Pretis, G. Marini, A. Odian, G. Stoppini and L. Tau, *Phys. Rev. Lett.* **9**, 226 (1962); J.K. De Pagter, A. Boyarski, G. Glass, J.I. Friedman, H.W. Kendall, M. Gettner, J.I. Larrabee and R. Welstein, *Phys. Rev. Lett.* **12**, 739 (1964); J.K. De Pagter, J.I. Friedman, G. Glass, R.C. Chase, M. Gettner, E. von Goeler, R. Weinstein and A.M. Boyarski, *Phys. Rev. Lett.* **17**, 767 (1966); D.J. Quinn and D.M. Ritson, *Phys. Rev. Lett.* **20**, 890 (1968).

134) The discovery of the J/ψ particle was announced in the papers published on 2 December 1974 by *Phys. Rev. Lett.*, by the MIT Group, led by C.C. Ting, who had observed at the AGS of the Brookhaven National Laboratory a very strict resonance at 3.1 GeV in the $p + p \to J \to e^+ + e^-$ process, and by the SLAC Group, headed by B. Richter, who had observed at SPEAR the same resonance in the process $e^+ + e^- \to \psi \to e^+ + e^-$. In the same issue of *Phys. Rev. Lett.* a confirmation of this discovery is given by a super-group of the Laboratori Nazionali di Frascati, which had learnt, by a telephone conversation with Ting, of the existence of this resonance, and which had modified slightly the operation mode of ADONE in order to observe it, since the latter was slightly above the energy range for which the Frascati machine had been designed. The index of the issue of *Phys. Rev. Lett.* of 2 December 1974 (Vol. **33**, No. 23) is as follows:
"Experimental observation of a heavy particle J": J.J. Aubert, U. Becker, J.P. Biggs, M. Chen, G. Everhart, P. Goldhagen, J. Leong, T. McCorriston, T.C. Rhoades, M. Rhode, Samuel C.C. Ting, Sau Lan Wu and Y.Y. Lee: p. 1404.
"Discovery of a narrow resonance in e^+e^-annihilation": J.-E. Augustin, A.M. Boyarski, M. Breidenbach, F. Bulos, J.T. Dakin, G.J. Feldman, G.E. Fischer, D. Fryberger, G. Hanson, B. Jean-Marie, R.R. Larsen, V. Lüth, H.L. Lynch, D. Lyon, C.C. Morehouse, J.M. Paterson, M.L. Perl, B. Richter, P. Rapidis, R.F. Schwitters, W.M. Tanenbaum, F. Vannucci, G.S. Abrams, D. Briggs, W. Chinowsky, C.E. Friedberg, G. Goldhaber, R.J. Hollebeek, J.A. Kadyk, B. Lulu, F. Pierre, C.H. Trilling, J.S. Whitaker, J. Wiss and J.E. Zipsce: p. 1406.
"Preliminary result of Frascati (ADONE) on the nature of a new 3.1 GeV particle produced in e^+e^-annihilation": C. Bacci, R. Baldini-Celio, M. Bernardini, G. Capon, R. Del Fabbro, M. Grilli, E. Iarocci, L. Jones, M. Locci, C. Mencuccini, G.P. Murtas, G. Penso, G. Salvini, M. Spano, M. Spinetti, B. Stella, V. Valente, B. Bartoli, D. Bisello, B. Esposito, F. Felicetti, P. Monacelli, M. Nigro, L. Paoluzi, I. Peruzzi, G. Piano Mortari, M. Piccolo, F. Ronga, F. Sebastiani, L. Trasatti, F. Vanoli, G. Barbarino, G. Barbiellini, C. Bemporad, R. Biancastelli, M. Calvetti, M. Castellano, F. Cevenini, F. Costantini, P. Lariccia, S. Patricelli, P. Parascandalo, E. Sassi, C. Spencer, L. Tortora, U. Troya and S. Vitale: p. 1408 (1649E).

135) For an overall picture of these new particles, see the various invited reports published in *Proc. 1977 Int. Symposium on Lepton and Photon Interactions at High Enery*, Hamburg, 25–31 August 1977 (DESY, Hamburg, 1977), and G. Feldman, invited report on "e^+e^-annihilation", *Proc. 19th Int. Conf. on High Energy Physics*, Tokyo, 23–30 August 1978 (Physical Society of Japan, Tokyo, 1979), pp. 777–789.

136) S. Glashow, J. Iliopoulos and L. Maiani, *Phys. Rev.* **D2**, 1285 (1970).

137) M. Conversi, T. Massam and A. Zichichi, *Nuovo Cimento* **40A**, 690 (1965).

138) A. Zichichi, S.M. Berman, N. Cabibbo and R. Gatto, *Nuovo Cimento* **24**, 170 (1962).

139) R.R. Wilson, "The next generation of particle accelerators", *Sci. Am.* **242**, 26 (1980).

140) C. Bernardini, "Colliding beams in the future", *J. Phys. (France)*, Colloque C2, Supplément au No 2, Tome **37**, p. C2-67 (février 1976).

141) The Matteucci Medal has been awarded on some forty occasions, starting in 1870. Among the award winners were: A. Righi (1882), K. Röntgen (1896), H. Poincaré (1905), P. Zeeman (1912), A. Einstein (1921), N. Bohr (1923), E. Fermi (1926), W. Heisenberg (1929), F. Rasetti (1931), I. and F. Joliot (1932), W. Pauli (1956).

142) L. de Broglie, *Compt. Rend.* **195**, 862 (1932); **197**, 536 (1932).

143) P. Jordan, *Z. Phys.* **93**, 464 (1935).

144) M.H.C. Pryce, *Proc. Roy. Soc. Ser. A* **165**, 247 (1938).

145) S.A. Bludman, *Nuovo Cimento* **27**, 751 (1963).

146) B. Ferretti, *Nuovo Cimento* **33**, 264 (1964); see also: B. Ferretti and G. Venturi, *Nuovo Cimento* **35**, 644 (1965).

147) For example, when speaking of himself, in those years, he would often say he was the "shit-pump mantenuto" referring to the Garvens pumps (see Section 5) which were used some time for cesspools.

148) This last convergence was based, however, on experimental data, completely outdated now, but that at the time seemed to indicate a linear increase of R with s. Such an increase is clear in the data presented by B. Richter ["Plenary Report on $e^+e^- \to$ hadrons", *Proc. 17th Int. Conf. on High Energy Physics,* London, July 1974 (Rutherford Laboratory, Chilton, Didcot, 1974), p. IV-37]. It was due, as it was understood later but was already suspected then (see, for example, B.H. Wiik, "Parallel session on e^+e^- interactions", p. IV-1 of the same volume), to the (unresolved) J/ψ and to its radiative tail and, in minor part, to the threshold of the "charm".

149) H. Thirring, *Z. Phys.* **19**, 33 (1918); **22**, 29 (1921).

150) L. Pietronero, "On rotating reference systems in Einstein's theory of gravitation", Internal Memorandum No. 337, 29 September 1971, Istituto di Fisica Guglielmo Marconi, University of Rome.

151) L. Pietronero, "The mechanics of particles inside a rotating cylindrical mass shell", *Ann. Phys. (USA)* **79**, 250 (1973).

152) L. Pietronero, "Gravitational interpretation of the centrifugal and Coriolis force", *Istituto di Alta Matematica, Symposia Mathematica* **12**, 57 (1973); "Mach's principle for rotation", *Nuovo Cimento* **20B**, 144 (1973).

153) Felice Ippolito (b. Naples, 1915), Professor of Applied Geology at the University of Naples, and Secretary General of the CNEN since its foundation, was accused in 1963 of administrative irregularities and prosecuted. The heavy sentence which he received was mainly on account of breaches due to "oversights". In reality, they concerned administrative measures decided by the Commissione Direttiva of the CNEN, presided by the Honourable Emilio Colombo, Minister for Industry and Commerce, and of which I myself was a member. These measures were essential either in order to comply with the undertaking which the Italian Government had made with respect to EURATOM, or in order to complete the CNEN programmes within the prescribed time scale. The repercussions of the Ippolito Case are easy to grasp: in 1963 Italy had begun to become a fairly advanced country in the use of nuclear energy. Today, in 1980, it is certainly one of the lowest on the list.

154) Between the 10 and 17 of August 1963, Giuseppe Saragat published, via a press agency, four critical articles concerning Italy's nuclear policy and the management of the CNEN, which led to the Ippolito Case and to the partial paralysis of the Italian Nuclear Organization. The visit paid by Saragat to the CERN Laboratories in Geneva on 27 August 1963 was intended to show that his position in regard to the development of the energy applications of nuclear physics had nothing to do with his interest in fundamental research. The four articles by Saragat and his statements issued at CERN were published by his followers and admirers in a booklet entitled *Putting nuclear policy into order* (Editoriale Opere Nuove, 1963), which is still very interesting to read.

155) Domenico Marotta (1886–1974), a chemist, who conceived and founded the Istituto Superiore di Sanità (ISS), which he directed with extreme broadmindedness and understanding of scientific, health, and organizational problems between 1935 and 1961. In 1965 he was prosecuted for some very minor administrative irregularities, which solely concerned certain formal aspects of his management of the ISS. This legal action embittered the final years of his life, even though the majority of his colleagues did not fail to show their unswerving respect and admiration [see D. Bovet, "Domenico Marotta", Celebrazioni Lincee 91 (1975); G.B. Marini Bettolo, "Domenico Marotta", *Commentarii Pontificia Academia Scientiarium,* Vol. 3, No. 12 (1975)].
Giordano Giacomello (1910–1968), Professor of Pharmaceutical Chemistry at the University of Rome from 1948 onwards, and author of a considerable number of scientific works, replaced Marotta in 1961 as Director of the ISS. Even more unjustly and unjustifiably he too was involved in the same legal action and, griefstricken, died shortly after the sentence [see G.B. Marini Bettolo, "Giordano Giacomello", Celebrazioni Lincee No. 43 (1971)].

156) List of lectures recorded on video tape, from the programme "Living Science"
(1)—B. Touschek, *The birth of the quantum* (2/4/1975)
(2)—E. Amaldi, *The neutron* (9/4/1975)
(3)—P.A.M. Dirac, *The story of the positron* (15/4/1975)
(4)—G. Careri, *From the atomic hypothesis to Bohr's atom* (30/4/1975)
(5)—E. Segrè, *The chemical elements* (7/5/1975)
(6)—R. Wideröe, *History and principles of high-energy accelerators* (28/5/1975)
(7)—B. Touschek, *Comments on R. Wideröe's lecture* (31/7/1975)
(8)—E. Bombieri, *Classification of surfaces* (22/5/1975)
(9)—M. Caputo, *Applications of elasticity theory to the study of earthquakes* (4/6/1975)
(10)—G. Colombo, *Review of current investigations into the solar system* (11/6/1975)
(11)—T. Regge, *Superfluidity* (18/6/1975)

(12)—G. Sansone, *Certain developments in the theory of ordinary differential equations over the last fifty years* (18/12/1975)
(13)—G. Bernardini, *Matter and antimatter: conceivable symmetries and objective reality* (17/12/1975)
(14)—E. Picasso, *Measurements of the anomalous moment of the electron and muon* (16/1/1976)
(15)—B. Segre, *An approach to the four-colour problem* (12/2/1976)
(16)—M. Conversi, *The mu meson* (11/2/1976)
(17)— A. Carrelli, *The symmetries in classical physics* (13/2/1976)
(18)—B. Touschek, *The harmonic oscillator* (25/2/1976)
(19)—V. Telegdi, *The P, C, T symmetries in atomic and particle physics* (24/4/1976)
(20)—T. Regge, *The method of dimers in statistical mechanics* (5/5/1976)
(21)—G. Chiarotti, *The Volta effect and the battery* (7/2/1977)
(22)—B. Touschek, *On restricted relativity*
1—*The pre-history of the theory* (17/5/1977)
2—*Relativistic kinematics* (19/5/1977)
3—*Relativistic dynamics* (24/5/1977)
(23)—L. Pietronero, *On the gravitational constant* (24/5/1977)

157) S. van der Meer, CERN ISR-Po/72-31, "Influence of bad mixing on stochastic acceleration", CERN SPS/DI/PP/Int. Note 778; "Design study of a proton-antiproton colliding beam facility", CERN/PS/AA 78-3; "Stochastic cooling theory and devices", *Proc. Workshop on Producing High Luminosity High Energy Proton-Antiproton Collisions",* Berkeley, 27–31 March 1978 (LBL, Berkeley, Calif., 1978), LBL 7574: Conf. 780345, p. 73.

158) "Batik" is a local form of art of the inhabitants of Java, in which cotton materials are coloured with various designs. The drawing (Javanese: *batik* = point or drawing) is made with liquid wax on a white background. The material is then immersed in a bath of dye, and when the layer of wax is removed with boiling water the pattern stands out against the coloured background.

159) Ludwig Wittgenstein (b. Vienna, 1889 - d. 1951). After having frequented the Technische Hochschule in Berlin in order to study engineering, he dedicated himself to mathematical logic, fundamental problems of arithmetic, and philosophy of science. Initially he was strongly influenced by the works of Frege and Russell, from whom he later completely separated on philosophical views. His most famous works are: *Tractatus Logico-Philosophicus* [edition in German and English, with an introduction by Bertrand Russell (Kegan Paul, Trench, Trubner and Co., London, 1922)], which he wrote in his early years, and his posthumous work *Philosophische Untersuchungen* [English transl. by C.E.M. Anscomde: *Philosophical investigation* (Basil Blackwell, Oxford, 1953)]. He was primary school teacher in Austria and University Professor in England, musician, and architect. Although he was not a member of it, he was one of the main inspirers of the "Vienna Circle".

160) For example, Karl Kraus, *Briefe an Sidonie Nadherny von Borutin (1913–1936)* (Deutscher Taschenbuch Verlag, Munich, 1977), Vols. 1 and 2.

161) Hans Weigel, *Karl Kraus oder Die Macht der Ohnmacht* (Deutscher Taschenbuch Verlag, Munich, 1972).
Werner J. Schweiger, *Das Grosse Peter Altenberg Buch* (Paul Zsolnay Verlag, Vienna-Hamburg, 1977).

162) C.M. Nebe Hay, *Gustav Klimt: sein Leben nach zeitgenössischen Berichten und Quellen* (Deutscher Taschenbuch Verlag, Munich, 1976).

163) Karl Kraus, "Sprüche und Widersprüche (1909)" *in Beim Wort genommen* (Munich, 1955).

164) **Note added in proof:** Gerhard Schumann (b. Dresden 1911) studied in Halle and Leipzig, where he worked with Smekal on the mechanical resistance of glasses under traction. In 1950 he moved to Heidelberg where he worked under O. Haxel. Later he studied the "fall out" by means of the filter method and became an expert in the exchange phenomena in the atmosphere.

The adventurous life of Friedrich Georg Houtermans, physicist (1903-1966)

- P R E A M B L E -

It was during spring 1978, I believe, that the Italian Television transmitted, in three one hour instalments, the film "The Confession" directed by Costa Gravas, inspired by the book by Arthur and Lise London "L'aveu"(1). The book is an autobiographic account of the adventure passed by Arthur London in Prague in 1957.

He was an authoritative member of the Czecho-Slovak Communist Party, put under trial, initially because suspected of deviationism from the prevailing party-line, but later accused of connection with state enemies.

The film had been circulated at least two years before and I knew, more or less, its content from newspaper critics and friendly conversations. But when I saw it on the television screen, I was struck by the similarity of certain general aspects as well as many details of the trial with those that had been told me in the early fifties by a physicist who had gone through a similar adventure twenty years before in USSR: Friedrich Georg Houtermans.

I knew his name in the thirties because of some important paper he had published in those years in scientific journals.

When in 1937 my friend Georg Placzek arrived in Rome from USSR, he had mentioned Houtermans as one of the young physicists gone to Kharkov to participate in the construction of a socialist society and recently in serious political troubles.

I received a letter from him, from Berlin in 1942, after, as I learned later, he had succeeded in getting out of a German prison to which he had been transferred from the Lubianka in Moscow.

I started to meet him rather frequently around 1950, on occasion of various scientific meetings taking place in France, Italy and Switzerland, where he had been appointed professor of physics at the University of Bern.

Stimulated by the adventures of Arthur London, interpreted in the film with impressive realism by Yves Montand, I asked myself whether the adventures of Fritz Houtermans had been written or recorded by some of his colleagues and friends.

I knew, of course, a very nice but rather short biography published in Helvetica Physica Acta, shortly after his death, by his collaborator and friend Professor J. Geiss(2). But only his more important scientific papers are mentioned there without any reference to or mention of his adventurous public as well as private life. These are very shortly mentioned in the "Obituary" published by Physics Today(3). I found that another of his friends, Professor Martin Teucher, the 26th October 1966, had given a "Gedenkrede" in front of the Deutsche Physikalische Gesellschaft (4) and prepared a few years later, a short biography for the Neue Deutsche Biographie(5). But also these notes are short and deal mainly with Houtermans' scientific work, except for a few very expressive lines on his imprisonement in Teucher's speech.

On the occasion of some of my trips to Geneva, I mentioned the problem to Charles Peyrou, working at CERN since 1957, who had been Professor of Physics at the University of Bern from 1954 to 1960 and a colleague and close friend of Houtermans.

In conversation with Peyrou and Geiss, whom I also met a few times in those years, the idea of an extended biography of Houtermans gradually grew in my mind. It was strengthened by a few exchanges of views with W.Gentner, W.Jentschke, W.Paul and V.Weisskopf.

On the occasion of a trip to the United States I visited Houtermans' first wife, Charlotte Riefenstahl, in her house in Bronxville, N.Y. the 3rd of March 1979. Not only was her conversation of great help, but she even put at my disposal copies of a number of documents of extraordinary importance for the reconstruction of Houtermans' family background, youth and university period and for the years he passed in Russian and German prisons. The nature and content of these documents is explained in References(6) to (8).

Further information and more documents (9) were given me by Charlotte Riefenstahl and Paul Boschan (10) on occasion of a second visit I paid to Charlotte the 25-29 April 1980 in Bronxville.

Information on later periods of Houtermans' life have been given me on occasion of a trip to Bonn (5 to 9 May, 1981) by Professors Wolfgang Paul and W.Walcher, Mrs Dr. Ilse Bartz Haxel and Professor Otto Haxel and of two trips to Bern (25-26 September 1984 and 20-21 March 1985) by Professor Johannes Geiss and Mrs. Loore Houtermans, the fourth and last wife of Fritz Houtermans. I had also repeated exchanges of views and letters with Professors Giuseppe Occhialini and Constance Dilworth Occhialini.

Towards the end of my work I came in contact with Professor David Holloway of the Center for International Security and Arms Control of Stanford University, who has written a book on the early Soviet nuclear program, from the development of nuclear physics in the 1930s to the testing of thermonuclear weapons in the mid-1950s(71). He called my attention on a few sources of indirect information in USSR in the 1940s(161) or 1950s(162).

At a still later stage of my work I was contacted by Dr. A. Kramish, author of a book on a Berlin friend of Houtermans, Dr. Paul Rosbaud, who appears only tangentially in Fritz's biography but apparently was the "British Master Spy in Nazi Germany"(89). I have used some of the information given on Fritz Houtermans by both these authors, Professor Holloway and Dr. Kramish.

I should express special thanks to Professor Holloway in consideration of the fact that his work has not yet been published.

My manuscript was read by a number of old friends or acquaintances of Fritz Houtermans listed below, many of whom were kind enough to communicate me their comments (c).

I am particularly indebted to Doctor Charlotte Riefenstahl, Professor Wolfgang Paul, Professor W. Walker, Professor Johannes Geiss.

The pictures of Houtermans were given me by Charlotte Riefenstahl and Johannes Geiss.

Rome, October 1987 Edoardo Amaldi

The persons that have read the manuscript are: I.Bartz, C. Dilworth Occhialini (c), J. Geiss (c), O. Haxel (c), G. Occhialini (c), W. Paul (c), E. Picciotto (c), C. Riefenstahl (c), E. Tongiorgi (c), W. Walcher (c), V. Weisskopf (c), E. Weisskopf (c), C.F. v. Weizsacker.

1. - Fritz's birth and family background -

Friedrich Georg Houtermans, or Fritz to his parents, Otto Houtermans and Elsa Waniek, was born in Danzig the 22 January, 1903.

The first name of Fritz's father was Oscar, but he did not like it and used, during all his life, the name Otto. He had inherited from his father Joseph Cornelius Houtermans (1848-1921), a large estate with a beautiful house in Zoppot as well as other properties.

Charlotte, the first wife of Fritz, wrote for her children[6a] the following information on Fritz's grandfather. She had collected them from Fritz's mother Elsa, who, « probably, had embroidered it slightly. She must have met Joseph Houtermans in Wien when he and his wife stopped there on their way to Italy where they spent months at a time in their villa in Capri » ...

« Joseph was of Dutch origin, being born in Voerendaal of a catholic middle class family. He became an architect and must have been quite successful. He went to Germany and obtained a contract to build the military harbour in Bremerhaven. This work brought him not only fame and money but also honours bestowed on him by the Kaiser ». Fritz's stepbrother Peter keeps a silver trowel on which are engraved his various contributions and the emperor's signature.

« At that time he was still a young man, unmarried, anxious to see the world. Once he went to Monte Carlo with a lot of money, gambled in the Casino and lost all his properties».
This happened around 1869-1870 when in Italy the fourth and last independence war was raging. The kingdom of Piedmont, in three previous wars (1848, 1859 and 1866) had succeeded in unifying almost completely the territory of the Italian peninsula. Only the "Pontificial State", Trentino and Venezia Giulia were not yet a part of the Italian kingdom.

In 1870 king Vittorio Emanuele II had declared war on the Vatican State and his troop had started to enter its territory. Joseph Houtermans, enrolled in the Pope's Zuaven Battalion, went to Rome and took part in the defence of the pontifical State until September 20th, 1870, when the Italians entered the Holy City. Fritz's stepbrother Peter keeps a diary of their grandfather Joseph, which covers the period from 20th July, 1869, when he left Voerendaal, to October 2nd, 1870, when he was back at home[11]. He returned to work again, met Lotte Strathman, who lived near Bielfeld in Spenge, and married her. He constructed the Fort Thorn in Posen, became a businessman and earned a fortune dealing in real estates. His residence finally was Danzig where his only son Otto was born. He died in Thorn, West-Preussen, in 1921.

Otto Houtermans (1878-1936) had studied law but he did not practice it until rather late in his life when he became a bank director. He had inherited enough money for living in an opulent way without really been obliged to do any regular work.

He married Elsa Waniek (1878-1942) who was born and brought up in Wien. She was half Jewish, since her mother belonged to the Karplus family, well known mainly because it owned the liberal Viennese daily journal "Das Wiener Tageblatt".

She was used to a rather sophisticated atmosphere and a high level intellectual life, not offered in

Zoppot. Charlotte «gathered from her conversation many years later that she detested the luxuries in Zoppot, which might have been conditioned by the residence of the German Crown Prince, Friedrich Wilhelm von Hohenzollern who was wild in the eyes of German citizens, spent a great deal of money and seemed to follow any whim [12]. Elsa did not care about eating food out season, listening to what she considered flat jokes. She missed the music and the theatre and the beautiful museums in Wien. After three years of marriage she left Zoppot, took her young son back to her home town and was divorced from Otto Houtermans.

Otto married again and with his second wife "Mimi" (Suzanne Helmhold) went on an adventurous trip in German South West Africa and became mayor of Windhuk. Their first son Peter was born during the year they spent there (1912). Otto visited also Buenos Aires but they must not have liked life in Argentina either because they were soon back in Zoppot. Otto became director of a bank and they had two more children: Rosemary (b. 1916) and Hans (1919-1980) ».

Fritz only met his father when he was fifteen years old. Charlotte and Fritz accepted him because he had a sense of humour and also easily forgave the radical jokes or strange behaviour on Fritz's part. Otto understood nothing of Fritz's interests or his professional ambitions and Fritz disliked Otto's kind of life. In spite of the distance between them, Fritz had a sentimental attachment to his father's family, especially to Hänsi, whom he met again in 1945 in Göttingen where Hänsi studied physics.

2. - His youth in Wien-

Back to Wien in 1905 Elsa Waniek rented «a large apartment which was next to that of her sister Lili and her husband Rudolf Gmeyner, a well established Viennese lawyer. They had three daughters: Anni (Anna) about two years older than Fritz, Tully (Alice) about the same age and Kitty (Elisabeth), the baby, born shortly after the arrival of Elsa and Fritz in Wien. Both flats comprised the whole second floor of the house.

Elsa was a very intelligent self-possessed intellectual person that in today's jargon would be indicated as a "liberated woman".

She settled down, organized her life and went back to school, finished the gymnasium and entered the University where she studied chemistry and biology, ending with a doctorate. The topic of the dissertation was: "Is pure water poisonous?" She had tremendous energy and enjoyed a very intense social life dominating the lives of her gentle sister, her three nieces and son. The preponderant female environment in which Fritz grew up was only relieved by the presence of Lilli's husband. His position in the family was unique. He was adored by all, as the best, the wisest, to whom everybody had to look up to, if not to serve»[6a]. Charlotte adds in her notes: «I have never understood how Elsa, who thought herself intellectually equal to any man, could believe that men were so utterly superior, that women definitely were meant to serve them and play the part of, at best, beloved slaves» ...

«During the summer months the whole family went to their country house in Baden, near Wien. There was a large garden with fruit trees. The children loved to climb them, eat cherries, spiting the stones into the grass. Lili discovered that soon enough, forbade it strictly and went back into the house. The children remained in the tree, went on eating cherries, but now swallowed the stones also. The effect was terrible tommy ache, and, probably, some punishment too.

In the house there were frequent musical evenings. The enormous living room made the grand piano look small. Fritz and his cousins must have listened very often to chamber music, trios and string quartets.

Tully studied the piano and acquired a Ph.D. in music. Fritz was not considered gifted enough, though he loved classical music all his life. Anni was the poet of the family and the baby sister was too young to be considered.

The emphasis in their education was literature and the classics. Fritz's mother had studied science. She had taught at the University during the first World War, when there was a scarcity of teachers, but later on devoted her life to the study of history of arts and philosophy. She studied Nietzsche and involved herself in Hindu philosophy. With her phenomenal memory she seemed to me (Charlotte) a living encyclopaedia. The small family estate was not large enough for her enterprising energy. She began to lecture for private groups on history of arts, combining this with tours through the museum. Her main thoughts, her main interests were Nietzsche and Goethe.

In Wien there was a group of young women who sat at her feet and adored her. She herself was in turn fascinated and influenced by another dominating and very intellectual woman, Laura, who led a salon in Berlin, where she ruled supreme.

Elsa was under her spell for several years, but was very likely, too intelligent to be dictated to. What is still puzzling me as I write this story, is the extraordinary philosophy on life which pervaded Elsa and her circle. They combined the loftiest ideals as outlined by the philosophers, with the humblest notions of serenity they needed in daily life».

Charlotte remembers one tale where Laura and her disciples were in the mountains, found an empty shepherd's hut and began a thorough spring cleaning, washing the dirty shirts etc. The shepherd must have been tremendously annoyed when he returned, also he might have appreciated it.

Another "liberated woman", that had also an influence on the formation of Fritz and his cousins, was Genia Schwarzwald, usually called "Tante Genia" or "Fraudoktor"[10]. She had created a circle of many people interested in music, literature, politics and economics and organized summer camps for the young people of the Viennese "elite schools" in different vacation resorts, such as Grundlsee, Ischl and so on. Karl Popper, Rudolf Serkin, the quadrefolium (Paul Lazarsfeld, Joseph Gluecksman, Ludwig Wagner and Alexander Weissberg-Czybulski) were used to participate in these summer colonies. Fritz Houtermans and his cousin Tully went to some of them.

«When Fritz was in his teens, life became difficult, in general, because of the first World War (1914-18), and for him, in particular, because of tensions at home between the group of the four children and the three grown up people, who had developed rather complicated mutual relations among them. These were due in part to the presence of a single grown up man with two women, in part to the dominating nature of Elsa, who, at one time or another, made to feel all the others not to be up to the intellectual standards she had set for herself.

Fritz started to show such a difficult behaviour that his mother, through her friend Anna Freud, arranged for Fritz to be taken care of by Anna's father, Sigmund Freud. The sessions, however, did not last long because when Fritz realized that he had to relate his dreams to him, he and his imaginative cousin Anni, began to invent dreams, which Freud soon discovered and stopped the psychoanalytic treatment[6c]».

Already at the age of 8 Fritz was fascinated by stars, minerals and butterflies and a few years later, by mathematics and also by history. The literary and musical atmosphere at home left its imprint on him but he never played an instrument or wrote.

«In a way he was somewhat of an outsider since nobody took his interests seriously. His mother realized that he was very intelligent and somehow furthered his inclinations, but he was rather rebellious and tempted her authority but never succeeded to upset her Olympian calmness.

Fritz went to the Akademische Gymnasium well known for counting among its graduates many young people that later became famous in one way or another: for example the poet Hugo von Hoffmannsthal (1874-1929), the mathematician Richard v.Mises (1883-1953) and the physicist Erwin Schrödinger (1887-1961). Fritz became a close friend of his classmate Paul Boschan[10], and also of Rudolf Serkin and Alexander Weissberg-Czybulski.

Years later Serkin became one of the greatest pianists of his time, and Weissberg a recognized physico-chemist, whose life-line crossed that of Fritz Houtermans again in later periods as we shall see in the following.

In those years Houtermans, Boschan and a few other of their schoolmates, deepened the reading of "Plato Dialogues" and "Presocratics" and discussed thoroughly "Das ideal der inneren Freiheit bei den griechischen Philosophen" by Heinrich Gomperz[13], whose father, Theodor Gomperz was the well known author of the "Greek Thinkers"[14].

A copy of the book "Aufruf zum Socialismus" by Gustav Landauer[15], given by Fritz to Boschan, was inscribed by him with quotes from the "Novum Testamentum, Graece et Latine"[16] rendering of the sermon of the mountain and a quote from Friedrich Adler's "Vor dem Ausnahmegericht". Fritz used to cite Landauer's statement "Den Marxismus prophezeit aus dem Kaffee Satz" (The Marxism makes predictions from the tea-leaves)».

Around 1917 Fritz contributed, together with a few other of his friends, to the journal "Der sozialistische Mittelschüler". At the time of the assassination by Friedrich Adler of the Austrian prime minister, Count Stürck, both Fritz and Adler belonged to the "Verein für Sozialwissenschaften" and Fritz wrote an article on "Der sozialistische Mittelschüler" supporting violence as a necessary means to obtain political changes in connection with discussing the defense of Friedrich Adler before the extra-ordinary court[17]. This gave rise to a long and rather heated discussion with his friend Paul Boschan who insisted on the need for finding more gradual means for reaching radical changes, but the «stifling circumstances which had led to the emigration of many active intellects from Austria, were so horrible to make him (Fritz) impatient»[10].

The difficult life conditions, after the first World War, but even more the family stresses, had started to make Fritz's life in Wien not easy at all. Even forty years later on various occasions Houtermans said: "Danke Gott wenn Du noch eine Mutter habt, aber ich hatte zwei" (Thanks God if you have a mother, but I had two). The final blow was his dismissal from the Akademische Gymnasium after he read inside the school, on occasion of the 1st May, the "Communist Manifesto" in front of a crowd of schoolmates.

Thus Fritz went for the last year of school to the Landeserziehungsheim in Wickersdorf in Germany (Thüringen) which in those years was one of the best progressive schools in Central Europe[10]. Following the general scheme established by the director of this school Gustav Wyneken[18], pupils of different ages and sexes lived there in small groups under the care of a teacher who succeeded to create something of a family life for all of them.

The Landeserziehungsheim was one of the educational enterprises created in the course of the "deutsche Jugendbewegung" which appeared here and there in Germany and Austria in those years. Wyneken was involved in a dispute with the less progressive parts of the Youth Movement. "Fraudoktor's" saloon and colonies, the intellectual students of the Akademische Gymnasium and the educational radicalism of the Wickersdorf school characterized the general background relevant to Fritz's life story.

At Wickersdorf, Fritz made lifelong friends, one being Irmine von Holten, the other one Heinrich Kurella. Irmine, later, married R. d'E.Atkinson, an English physicist and astrophysicist who worked for his doctorate under James Franck at the same time as Fritz, and collaborated with Fritz in the late 20s. Heinrich Kurella later became Sitzredacteur (permanent editor) of the "Rote Fahne", the journal of a rather small group of communists which included Ernst Fischer and Manes Sperber [19]. In Austria, at that time, there was a strong socialist party supported by the working class. The communists were few and in great part "crazy intellectuals", at least in the opinion of the other leftist Austrians belonging to more or less the same circles at that time. Weissberg most probably became a member of the communist party, at least later, but at that early time he appeared to have a political connotation in some way in between those of the socialist party and the communist group.

Fritz Houtermans probably never became a member of the communist party, but at that time was extremely attracted by the "crazy intellectuals" who, under many respects, conformed to his anticonformist views.

In 1933 the Nazi party, was becoming more and more threatening and Kurella had to escape from Berlin[20]. He went to Zurich and became the chief editor of the illegal newspaper Inprecor (International Press Correspondence). But about one year later, the Swiss Intelligence identified his activity and expelled him from the country.

Kurella went to USSR but, shortly later, was arrested put under trial, condemned and finally executed as an anti-communist agent. He was one of the many victims of the massive and devastating political repressions that, at the end of the second "five-years plan" shook the USSR, which according to Stalin's official declarations, was entering into the socialist era[(21)].

3. - Student in Göttingen -

These were the years of the Weimar Republic. Its constitution, approved by the National Assembly of Weimar, on August 11, 1919, established that the German Reich was a republic, whose sovereignty derived from the people.

It came to an end the January 28, 1933 when the 86 years old President of the Republic, Fieldmarschall von Hindenburg suddenly dismissed the chancellor, General Kurt von Schleicher and, two days later, nominated Adolf Hitler new chancellor of the German Reich[(23)].

Fritz Houtermans studied physics at the University of Göttingen from 1922 to 1927 under James Franck who put him to work on the excitation of atoms by means of the mercury ultraviolet line of wavelength λ = 2537 Å /1/. In 1926 at the preseminar of Franck, Fritz met a girl from Bielefeld, who was also studying physics: Charlotte Riefenstahl. She was the daughter of G.Riefenstahl, journalist and magazine editor.

«During those years Fritz spent many vacations with Kurella exploring Switzerland and Nothern Italy by bicycle. On one of these trips they run out of money, Kurella went back home but Fritz befriended a test-driver of the FIAT works, who persuaded him to come along to Naples. The car had no body but in spite of its many inconveniences, the two young men had a great fun. The Neapolitan family of the driver took him in, asked what was his profession or trade and when he explained in his broken Italian that he was a physicist, they translated it to mean he was a plumber or electrician. Fritz stayed quite a while selling buttons to his only customer who was impressed that he knew who "Cavour" was, but nevertheless never paid him. His last job was at the harbour loading cargos. The other long-shore men liked him, and when they noticed that he could barely lift those heavy sacs, they let him keep track of counting the sacs and did the heavy work for him. Shortly before he was hired as a deck-hand to sail to New Zealand, Fritz's mother, frantic about his disappearance, succeeded in spotting his address in Naples and sent money and a laconic telegram asking him to return immediately to Germany. His return was urgent, since it would have been impossible for James Franck to reserve his research room any longer»[(6a)].

In those years the University of Göttingen was one of the three outstanding centres of German mathematical and physical tradition. Berlin and Munich were the two others.

Among the Göttingen mathematicians the outstanding figure had been David Hilbert (1862-1943), who however was ill during much of Weimar period. His interdisciplinary spirit was continued by his younger colleague Richard Courant (1888-1972) whose activity was oriented towards a tight collaboration with the physicists. The treatise, first appeared in 1924, on "Methoden der mathematischen Physik" by Courant and Hilbert was a clear manifestation of this attitude. It became in those years an important tool for the use of advanced mathematical methods by theoretical as well as experimental physicists. Even today it is one of the best books of this type.

The Institutes of Mathematics and of Physics were on the Bunsenstrasse. Among the mathematicians in Göttingen in the 20s one should recall, beside Hilbert and Courant: Edmund Landau (1877-1938) prominent in number theory, Emmy Noether (1882-1935)[(25)] in algebra and Herman Weyl (1885-1955) in the theory of relativity, group theory and mathematical foundations of quantum theories.

At the Institute of Physics Max Born (1882-1970) had the chair of Theoretical Physics. He had been one of the founders of quantum mechanics. Among his many papers on this subject, the most outstanding one was his interpretation of the square of the modulus of a wave function at a certain instant and point of space as the probability for a particle (for example an electron) to be at that time at that point. He shared with W.Bothe the 1954 Nobel Prize for Physics. To Max Born the Prize was awarded "for his fundamental research in quantum mechanics, especially for his statistical interpretation of the wave function".

A number of younger theoreticians spent months or years working in his institute at that time. Let me recall among the German young people: Werner Heisenberg, Wolfgang Pauli, Friedrich Hund, Max Delbrück, W.M.Elsasser, Maria Goeppert Mayer, and among the foreigners: Enrico Fermi and G.C.Wick from Italy, E.U.Condon, Robert Oppenheimer and H.P.Robertson from U.S.A., P.A.M.Dirac from Great Britain, E.A.Hylleraas from Norway, A.Fock from USSR, Victor Weisskopf from Austria and L.Rosenfeld from Belgium[(26)].

In the same building, but at different floors, there were two Institutes for Experimental Physics: the first Physical Institute, directed by Robert Pohl (1884-1976), the second Physical Institute directed by James Franck (1882-1964).

Pohl was a distinguished experimenter, especially in optics. He was the author of a well known university textbook with a beautiful series of experimental demonstrations.

The interests of Franck were completely oriented towards the new atomic physics. His insight into fundamental problems complemented Born's tendency toward formalism[(27)]. He had shared the 1925 Nobel Prize in Physics with Gustav Hertz (Chapter 4) "for the discovery of the laws governing the impact of an electron upon an atom".

The young people of the Physics Institutes of the University of Göttingen had taken the habit of discussing physics not only at the University but also at the "Cafe Cron & Lanz" where they were used to go all together. The marble tables were used for writing formulas or sketching experimental set-ups during the almost daily heated discussions.The resonant voice of Fritz, very often dominating all the others, could be heard all over the Café.

All the students of the group were rather poor mostly because of the inflation. The only exceptions were Oppenheimer and Houtermans who received allowances from their parents.

Almost always Fritz paid for everybody. They had also taken the habit of spending hours in the bookshop Peppmüller.

Fritz was a very witty man. He also possessed an almost inexhaustible store of Viennese Jewish stories and jokes that he was used to tell in connection with the most different situations. Many years after, these stories were collected and published by a few of Fritz's pupils in Bern, who later became professors of physics in various German speaking Universities(28).

Once Fritz, with the permission of Franck announced at the Colloquium the presence in Göttingen of a Russian professor and then introduced into the lecture room two dancing bears, whose owner he had met in the street just before.

In Göttingen the Preseminar, directed jointly by Born and Franck, was one of the stimulating gathering of each week, where graduate students presented papers and where new ideas were discussed. Once, during 1926, Werner Heisenberg gave a lecture on the development of quantum mechanics which was one of the most awe inspiring outlines of the state of this new field. At the end the audience applauded him with great enthusiasm. Pascual Jordan (1902-1980) who was present in the lecture room took part in the following discussion.

In spring 1927 the Annual Meeting of the Physical Society of Germany (Gauverein) was held in Hamburg, and many of the Göttingen young physicists, including Fritz and Charlotte, went over there. At the meeting they had the opportunity of seeing and listening P.A.M.Dirac from U.K. and W.Pauli from Austria.

During spring and summer 1927 Robert Oppenheimer (1904-1967), already graduated at Harvard University, Fritz Houtermans, Walter Elsasser and Charlotte Riefenstahl passed Ph.D. examinations in short succession.

Their Committee was composed by Max Born, James Franck, Richard Courant and Gustav Tammann (1861-1938). Charlotte's thesis, under the physical chemist Tammann, was on the crystallization of gold and silver. Elsasser's and Oppenheimer's theses, both under Max Born, dealt with the theory of collisions, and the quantum theory of the continuous spectrum. Finally Houtermans' thesis, under Franck, was "On the band fluorescence and the photoelectric ionization of mercury vapour". Divided into two parts it appeared in Zeitschrift für Physik in 1927 /2/.

During that academic year Fritz and Charlotte, or Fissel and Schnax, as they were called by their friends, had become great friends and had taken the habit of walking in the streets at night. The town, however, was so small that they always reached the free country pretty soon.

The evening before Oppenheimer's departure from Göttingen they went to his house, where they met for the first time Georg Uhlenbeck (b. 1900)(29) just arrived from Rome. Chatting about his Rome stay as a preceptor of the son of the Dutch Ambassador to Italy and where he had established friendly relations with Enrico Fermi and his colleagues, Uhlenbeck expressed his great admiration for the Divina Commedia of Dante Alighieri, in particular for the 5th Canto of the "Inferno" which had enchanted him. In wish they all joined, enthusiastically, to read it to each other. Later on - not having enough copies - they borrowed a copy from the library, a priceless first edition.

A few interesting remarks about Fritz and Charlotte can be found here and there in Elsasser's book [(26b)].

In the meantime George Gamow (1904-1968) arrived in Göttingen from Berlin, where shortly before he had published his first paper on the theory of emission of alpha-particles from radioactive nuclei[(30)]. The discussions between Gamow and Houtermans brought to a new formulation of the same ideas from a more phenomenological point of view, which allowed a better fit of the experimental data /3/.

At about the same time Fritz Houtermans got a job at the Technische Hochschule in Berlin-Charlottenburg, and Charlotte went to U.S.A. as an instructor in the Physics Department of Vassar College in Poughkeepsie, N.Y. Before her departure she received a bouquet of a hundred roses from Fritz.

4. - Assistant in Berlin -

In Berlin there was an impressive collection of scientific talents belonging to four or five different research institutions. Max Planck (1858-1947), the first to quantize the energy of the electromagnetic oscillator (1900), was professor of theoretical physics at the University of Berlin. He had been awarded the 1918 Nobel Prize for Physics "in recognition of the services he rendered to the advancement of Physics by his discovery of energy quanta". His pupil, Max von Laue (1879-1960) had been awarded the 1914 Nobel Prize for Physics "for the discovery of the diffraction of X-rays by crystals". He was an outstanding figure not only as a scientist but also as a man, as we shall see later.

When Max Planck retired, in 1927, his successor was the Austrian physicist Erwin Schrödinger (1887-1961) founder of wave mechanics. He shared with P.A.M.Dirac the 1933 Nobel Prize for Physics "for the discovery of new productive forms of atomic theory".

Albert Einstein (1879-1954) was in Berlin from 1913 to 1933, first (1913) with a chair of theoretical physics at the Prussian Academy of Science, and later (1914-1933) as director of the Kaiser Wilhelm Institute for Physics. His works on general relativity and on exchange of energy between molecules and radiation were produced in that period.

At the Technische Hochschule there was another outstanding group of physicists: Gustav Hertz (1887-1975), professor of experimental physics, who had been awarded the Nobel Prize for his work with James Franck, and Richard Becker (1887-1955) professor of theoretical physics and author of many papers and books on atomic physics, ferromagnetism and plasticity. For some time Becker's assistant was the Hungarian Eugene Wigner (b. 1902), who a few years later emigrated to U.S.A. and, in 1944, was awarded the Nobel Prize for Physics "for his contributions to the theory of the atomic nucleus and elementary particles".

The Kaiser Wilhelm Institute for Physical Chemistry and Electrochemistry was headed by Fritz

Haber (1868-1934), winner of the 1918 Nobel Prize for Chemistry "for the synthesis of ammonia from its elements", and the Kaiser Wilhelm Institute for Chemistry was guided by Otto Hahn (1879-1968) and Lise Meitner (1878-1968). In 1963 Hahn received the Nobel Prize for Chemistry for the discovery of fission of heavy nuclei.

A point of encounter of all physicists working in Berlin was the seminar led by Max von Laue at the Institute of Physics of the University.

For various reasons Berlin was also the organizational centre of German physics, a leadership accepted by the majority of the other physicists but not by Wilhelm Wien (1864-1928), professor of experimental physics at the University of Munich, by Philipp Lenard (1862-1947) and Johannes Stark (1874-1957) who were also hostile to modern physics and anti-semitic. All of them were very recognized experimenters and were awarded the Nobel Prize for Physics: Lenard, in 1905, "for his work on cathode rays"; Wien, in 1911, "for his discoveries regarding the laws governing the radiation heat", and Stark, in 1919, "for the discovery of the Doppler effect in canal rays and the splitting of spectral lines in electric fields".

Both Lenard and Stark voiced their views about the deteriorating influence of Jewish scientists (in particular Albert Einstein) and in general of the Jewish way of thinking, in various meetings and in writing, gradually more explicitly, starting from the early 20s and reached, already in 1923, forms of expressions very similar to those used by Hitler in the first volume of Mein Kampf, appeared about one year later.

The outstanding figure at the University of Munich was Arnold Sommerfeld (1868-1951), professor of theoretical physics, who, differently from Wien, always closely collaborated with his colleagues in Berlin and had a number of Jewish students. He was the author of a number of fundamental papers and beautiful textbooks which have been studied by many generations of young physicists in Germany as well as abroad.

Werner Heisenberg (1901-1976) and Hans Bethe (b. 1906) had been two of his many talented students. Bethe was also his assistant for a few years.

At the Technische Hochschule in Berlin Fritz Houtermans was first assistant of Gustav Hertz but his main duty was to teach in the "Praktikum", managed by Professor W.H.Westphal. Robert d'E. Atkinson (1898-1982) had come with Fritz from Göttingen and was one of the many assistants conducting the advanced laboratory. Walter Elsasser and Alexander Weissberg joined this group for a while.

In 1929 Atkinson and Houtermans published a more complete and systematic treatment of the theory of the alpha decay of heavy nuclei /4/ /7/ and used their formulas for computing from the experimental values of the energy of the emitted alpha-particle and of the decay constant the radius and height of the potential barrier for each single nucleus of the uranium-radium family. The smooth dependence of their results on the atomic mass represented a considerable progress with respect to all previous publications.

At about the same time they tackled a completely new problem: that of the formation of elements in stars with production of energy /5/ /6/. Already M.von Laue and Kudar independently(31) had suggested the possibility of formation of light elements through the inverse process of alpha-decay, but found values much to low by orders of magnitude for the probability per unit time to occur even under stellar conditions. Atkinson and Houtermans, however, computed the probability per unit time of nuclear capture in light nuclei not of alpha particles but of protons under the conditions expected at the centre of a star.

As typical conditions they used a density of at least 10^{23} protons/cm^3 and a temperature between 10 and 20 million degree Kelvin. For heavy nuclei the probability of proton capture turned out to be exceedingly small, but for light elements, assuming a collision radius of $4x10^{-13}$ cm, they obtained halflives ranging from 8 sec in $_2He^4$ to 10^9 years for $_{10}Ne^{20}$. Then they proceeded to estimate from phenomenological arguments the probability for a proton of remaining bound in the nucleus through the emission of its excess energy by radiation.

From a detailed analysis of the capture of protons by light nuclei they concluded that processes of this kind can provide the mechanism not only for the formation of light elements, but also for supplying the energy required for maintaining over the ages the emission of radiation from the stars.

The authors discussed their paper with Gamow before publication during the Christmas holidays spent to do some skiing in Zürs am Arlberg. The results are very close to those given by modern calculations, as Gamow pointed out years later(32).

Gamow also noticed that the derivation by Atkinson and Houtermans involves two mistakes, which compensate each other and which, in 1929, were generally accepted by all physicists. For the proton-nucleus collision cross section the authors used the geometric cross section and not the square of the de Broglie wave length of thermal protons. Furthermore they attributed the emission of photons to dipole transitions while we learned during the 30s that nuclei radiate quanta of higher angular momentum.

As a conclusion of this group of important papers Houtermans published in 1930 a comprehensive review article in the Ergebnisse der exakten Naturwissenschaften /8/ entitled "Recent works on the quantum theory of the atomic nucleus".

Shortly later Houtermans started to work with Max Knoll and W.Schulze on the electron microscope with special regard to the construction of the magnetic lenses /9/ /11/. Houtermans and Knoll took a patent on the design of magnetic lenses and the first model of their microscope was kept for a rather long time in the living-room of the apartment of Fritz and Charlotte, who in the meantime had returned from USA as we shall see below. With his work on the electron microscope Houtermans became "privatdozent" with Hertz. In a paper on the prehistory and early history of the electron microscope, presented by Ruska at an international conference held in Australia in 1974(33), Houtermans and Knoll are quoted as the first to have introduced the de Broglie wave length of the electron as the important physical quantity that should be considered in connection with the resolving power of the electron microscope.

During the Berlin period Houtermans published two more papers, one on a light source particularly designed for the investigation of the fluorescence light of metal vapours /10/, the other on the experimental study of the absorption and other optical properties of fluorescent media by means of modulated light /12/.

From the beginning of this paper Houtermans stresses the interest of this method for measuring the absorption of spectral lines which start from excited states of atoms or molecules. The modulated light was produced by a periodic source emitting a continuous spectrum which was detected by means of a photocell connected to an alternated current amplifier tuned on the same frequency. The absorbing material, typically a tube filled with a gas crossed by a continuous electric discharge, and, possibly, the instruments necessary for the spectral analysis of the transmitted spectrum, were inserted between the source and the photocell which recorded the modulated light but was insensitive to the radiation emitted continuously by the absorber. The method was tested by Houtermans by measuring the absorption lines emitted by the caesium vapour present in an appropriate electric arc.

The problem of the absorption of light from excited states of atoms or molecules is very closely connected with those taking place in the laser, which, however, was discovered about 50 years after. According to what Houtermans was used to say years later, in the 30s he had even thought about the "licht lawine" (light avalanche), but I was not able to find these words in his writings although it is quite possible that he arrived rather close to this idea, which is essential for the discovery of the laser.

In Berlin Houtermans devoted part of his time also to help G.Hertz who, in those years, was the first to develop the diffusion method for separating isotopes and to apply it successfully to neon(34) and hydrogen(35). Houtermans, making use of the remarkable experience as an atomic spectroscopist he had acquired at the Franck school, determined the relative abundance of the isotopes by photographing and measuring the spectra of the enriched samples. He was actually the first to measure the hyperfine structure of artificially separated isotopes(36).

In autumn 1931 appeared in Berlin a new brilliant young physicist from Wien: Victor Weisskopf (b.1908) who had been asked by Erwin Schrödinger to replace for one semester his assistant, F.London (1900-1954), who was spending a few months in United States.

Weisskopf had studied in Wien during the academic years 1926-27 and 27-28, mainly under the guidance of Hans Thirring (1888-1976) and had completed his formation in Göttingen with Max Born and James Franck, passing his Ph.D. examination at the end of spring 1931. His thesis on the theory of resonant fluorescence(37) was made under the supervision of Born, but the choice of the problem clearly belonged to the area of interest of Franck, who exerted a considerable influence on the young Austrian. He was also influenced by Eugene Wigner who came often to Göttingen from Berlin, and with whom Victor Weisskopf published two important papers on the width of the spectral lines and the meanlife of the atomic states(38).

From Göttingen Weisskopf went to Leipzig where he stayed until March 1932, at the Institute of

Werner Heisenberg, another of the great centres at that time. Besides Heisenberg there was Friedrich Hund (b. 1896), professor of spectroscopy, Felix Bloch (1908-1984), assistant of Heisenberg, C.F. v.Weizsacker (b.1912), an exceptional student, and a number of other remarkable young physicists coming from everywhere.

5. - Fritz's first marriage -

Charlotte came back from the United States in 1929 and went to live in Berlin. Actually her contact with Fritz had been maintained all the time by letters. In Berlin they lived in Halensee and saw each other constantly.

Late in 1929 or early in 1930 Fritz and Charlotte paid a visit to Fritz's father. Charlotte describes her first encounter with Otto as follows: «We went by train to Danzig and from there to Zoppot, where Otto Houtermans lived with his second wife and their three children Peter, Hänsi and Rosemary, who, of course, were quite grown up, Peter studying in Munich or Bonn and only Hänsi, I believe, while in the gymnasium.

Not used to any opulence and more adjusted to the utilitarian way of life in the United States, I was impressed by the vastness of the house, the layers of oriental rugs, the lovely edged wine glasses with hunting scenes, etched especially for Otto in Bohemia. It was a pleasant visit, but strangely enough, the family impressed me less than the ancient city of Danzig. It may be that I was under Fritz's influence, who made fun of everything which was not strictly academic. Otto Houtermans was big, jovial and easy going. His main interests were photography and hunting. The house was filled with trophies reminders of some of his many trips to the various hunting grounds in Europe»[(6a)].

On occasion of a physics conference held in Odessa (USSR) during August 1931 Fritz and Charlotte went over there by train. During the one week-long trip some tension arose, from time to time, among them for completely futile reasons. Typical the following little story.

They had just passed the USSR border, when they thought to try to practice in Russian but her two slim Russian grammars had been placed in the big suitcase. Fritz said: "All right, let's start with the first declination", and Charlotte answered too quickly with her suggestion: "Let's decline the 'table'; we always started with table in all languages. Do you remember mensa, mensae, ...?" "I will not" Fritz answered in an unreasonably violent mood. "Why shall I always, from now on all my life, in all languages, always start with: the table, of the table, to the table ... It's absurd, utterly bourgeois and reactionary" [(6b)]. After a while he added "I shall decline the elephant". They looked at each other, their faces were blank. They did not know how to stop it ... and they both looked for a way back, ... to regain their lost happiness, ... but they only fumbled, trying to make him give up his absurd demands, to convince her of the despicable conservatism of her view-point. It continued for many miles ... At last the train stopped: nothing outside but very bare fields. They were in the next to the last coach and had no way of seeing the station. Further ahead doors banged, feet tromped in, voices, yelling, calling. After a while Charlotte asked "What are they shouting?" "I told you so" said Fritz, "they shout elephants". Then Charlotte was really furious and of course did not believe it. The train restarted and just then they saw where all the people were running to. Just outside the station there was another train with unusually shaped cars, huge doors, some wide open and out of theese looked tranquilly not less than four elephants. It was a circus train![(6b)]

Fritz felt this encounter as a proof of the soundness of his views about learning languages and Charlotte, in this as well as in other circumstances of their life, remained with the impression that Fritz had a kind of divination capacity. But, if I can comment very softly, apparently this virtue popped out only in unimportant circumstances.

In Odessa Fritz and Charlotte went to live in a hotel in the vicinity of the steps well known from the Russian film "Revolt of the Potemkin", directed by Sergei M. Eisenstein. At the end of the conference the participants were invited to a trip by boat on the Black Sea which lasted a few days and brought to Sebastopol, Yalta and Batum in Caucasus.

Fritz and Charlotte married in Batum, where the documents were issued in Grusinian language. Their marriage witnesses were D.Iwanenko and W.Pauli who had recently divorced from his first wife. Also Rudolf Peierls and his Russian wife, Eugeniia Kanegiesser, were present[(39)]. She was a Leningrad physicist and they had married less than one year before, after they had met in 1929 on occasion of a trip of Rudolf to the USSR.

After their marriage Fritz and Charlotte went via Tiflis and Kharkov to Moscow. They stayed near the Kremlin in the Grand Hotel and met there, by good luck, a very interesting group, which accompanied the Indian poet Rabindranath Tagore. They knew his nephew Shomendranath, who acted as his uncle guide and interpreter, and they were thus included in the magnificent tour prepared for the poet. Part of the group was also the daughter of Albert Einstein, Margaret, who was there with her husband Marianov, a Russian theater actor. Before going back to Berlin the Houtermans also spent a few days in Leningrad (Hotel Astor).

Not long after their marriage Fritz and Charlotte were visited by Otto Houtermans, who had taken the habit of stopping in Berlin to see his son and his wife on his trips to Munich.

It was on occasion of the first of these visits that Otto cut Fritz's allowance "because two can live cheaper than one", which astonished Charlotte and also made their life more complicated. An agreable aspect of these visits was an invitation to both of them to have dinner at the fashionable and expensive Kempinsky's Restaurant, on the famous Kurfürstendamm.

According to Charlotte's description «these dinners were sumptuous. We usually - if it was the right month - ate oysters and we consumed as much as possible because we were always hungry. I remember that we occasionally went to Kempinsky by ourselves and usually ordered "kleine Zwischengerichte", cheese and coffee, which was the cheapest of all possible dinners and gave one the excellent feeling of having dined well. After dinner we often went to our apartment ... Otto Houtermans trusted me to make good coffee, but he did not trust my cooking, and I believe he was quite right. He really was very amusing though his stories were on the heavy side, while our enthusiasm for exciting problems in modern physics did not produce any reaction in him». On the contrary he was proud both of Fritz and his first wife Elsa, who had earned her doctorate degree in the meantime, and after Fritz had passed his Ph.D. examination, he financed for both of them a trip to Spain which was a wonderful experience. Charlotte continued: «They could not speak Spanish, but were very successful speaking Latin with what they thought was a Spanish accent, hoping to be understood. And apparently it worked.

While his father's visits were very nice in a way, they also depressed us because we had so little money which did not bother us in the ordinary way, except when we saw the bill at Kempinsky's».

6. - The life and work in Berlin -

After their marriage the Houtermans moved into an apartment across the road from the laboratory. It was one end of the long barracks built after World War I as living quarters for students. According to Charlotte[(6d)]: «Life in Berlin between 1929 and 1933 was unforgottable. The city vibrated with new ideas in practically all fields. ... The Kurfürstendamm, Berlin's entertainment center, was alive with new plays, films, concerts. The numerous cafés were the meeting places for the intelligencia. All social life was of extraordinary vigor, and the levels of conversations were higher than I ever experienced before and since.

We had many many friends - Fritz seemed to attract people. He was always full of ideas, he told stories, witty jokes, he was interested in a great number of different things, running the gamut from physics to music, to economy, to politics. Wolfang Pauli came to visit one Christmas, George Gamow and Lev Landau were frequently in Berlin ... Then there was Michael Polanyi with his interest in economics and politics, his niece Eva Striker, a gifted ceramic designer, who later married Alex Weissberg, Manes Sperber, a writer who had been a student of Adler's. Alex Weissberg-Czybulski was a physicist and engineer, and one of the most intelligent persons with an astonishing command and knowledge of history, politics, Marxism, to which one should add his gift of quoting and reciting Rilke by the hour.

The small house and the tiny garden were always bursting with guests. It was not unusual to have 35 people dropping in for tea».

One evening almost every week the Houtermans invited their colleagues and friends to what Fissel called "Eine kleine Nachtphysik" paraphrasing Mozart's "Eine kleine Nachtmusik". Their first child Giovanna was born in Berlin in 1932 and, as wrote Charlotte, "brought us great happiness".

«Parallel to all this high pitched excitement though ran the ever growing awareness of the slowly increasing power and influence of the National Socialist Party. People became more and more afraid to speak openly, to be linked rightly or wrongly with some political parties. Our apartment was raided by Nazi students, who in those early years though fanatics were poorly initiated in Nazi theories. When raiding bookcases, they did not know what was - in their own belief - controversial and what was not. I remember a raid on an apartment building, where many actors and writers lived. Some were imprisoned on trumped up charges, but were allowed visitors who brought food and books. The Nazi guard inspecting it all, seized one of the books: Karl Marx 'Das Kapital'. Very pleased he exclaimed: You dumb communist, here is a good book for you to study.

After the raid on our apartment the Technische Hochschule forced the Nazi students to apologize, thus protecting us for a little time at least. Any liberal attitude was interpreted as anti-nazi or communistic. Since the police itself was under attack, they became gradually unable to help an ordinary citizen, whether jewish or not. Thus fear and uncertainty and anticipation of terrors to come were increasing every day.

After the raid we inspected our bookcases for incriminating materials. Fritz had bought books, pamphlets, periodicals, representing the whole gamut of the political spectrum. Some books like Marx seemed to us to be classics, some liberal writings legitimate criticisms, others maybe doubtful or even dangerous. We began a private purge destroying compromising papers by tearing them apart and flushing them down the toilet, since we had no facilities to burn them. We had to stop this soon, because the Nazi students next door might have become suspicious. Fritz then remembered that he had some relatives on his mother's side, an elderly jewish couple, who - we hoped - might have a stove, and asked them whether he could burn the rest of the material in their kitchen. He was asked to come late in the evening, so they could be sure that their non-jewish maid was asleep in her upstairs room. Fritz left our house with his large package, very apprehensive whether he would manage to cross the dark Savigny Place unmolested. When he arrived at his uncle's flat, he felt very nervous and became suddenly conscious of the danger for these nice people. But his uncle did not allow him to apologize. Never mind, he said, it's nothing new. Father did it also in '48.

As time went on, the Nazis got stronger and bolder. Then came the horrible day, when jewish homes were attacked and vandalized, pianos thrown out of the windows ... people were arrested, the police was helpless. When they could not protect anybody any longer, when one day the jewish faculty members were not allowed to enter the university, it was time to think of leaving. People began to emigrate, smuggling money and valuables if they were suspected to have no legitimate reason for going. It was easier for those who had relatives abroad or were invited to new jobs. Everybody who could help to make connections to foreign universities, suggesting jobs. Fritz was involved in these projects, but the great pillar who helped so many people was Max von Laue. Fritz was only 1/4 jewish on his mother's side and in no immediate danger. But since he did not look like the Nazi prototype, more like an Italian than a blond Nazi, and since he never would have said "Heil Hitler" his safety might not have been of long duration. When Vicki Weisskopf came through Berlin from Wien on his way to Copenhagen, I asked him to do something for Fritz abroad because he was not taking any steps on his own behalf. Vicki made the connection with the Electrical and

Musical Industries, His Masters Voice, in Hayes, England. The head of their research department was Isaac Shoenberg (1880-1963) who was born in Pinsk, in Russia, had moved with his family to Great Britain in 1914, where he made a brilliant career as an applied scientist and television pioneer[(40)]».

7. - Emigration to U.K. and USSR -

The dissolution and end of the Republic of Weimar took place mainly in the period from 1931 to 1933 while the period 1933 to 1934 is indicated by historians like W.L. Shirer[(24)] as that of the nazification of Germany. The complexity of this amazing tragedy, that occurred in Germany and involved pretty soon the whole of Europe and later the entire world, certainly can not be reduced to a few names, dates and figures. But as a hint at some of the wings of the huge stage, in a little corner of which takes place our story, I will recall very briefly indeed, a few emblematic events of those years.

In September 1930 on occasion of the national elections in Germany, the National-Socialist Party, led by Adolf Hitler, obtained six and one half millions of votes to be compared with slightly more than eight hundred thousand of two years before.

The 10 October 1931 Hitler was for the first time received by the President of the German Republic, the Fieldmarshall Paul von Hindenburg.

At the beginning of 1932 Hitler put his candidature at the elections for the presidency of the Republic and started to campaign against the about 85 years old Hindenburg, who was at the end of his seven years mandate. The result of the elections, that took place the 13 March 1932 are

Hindenburg	18651497 votes	49.6%
Hitler	11339446 votes	30.1%
Thälmann	4983341 votes	13.2%
Düstelberg	2557729 votes	6.8% .

At the end of a period of great confusion and instability, the last chancellor of the Weimar Republic, Schleicher, 57 days after his appointment, the 28 January 1933, resigned his mandate; two days later, the 30 January 1933, President Hindenburg nominated Hitler chancellor. In his cabinet the National-Socialists represented a minority, while the most important departments fell to "conservatives" and those remaining to "independent experts". But that day the Republic of Weimar came to an end after fourteen years of unsuccessful attempts to create in Germany a working democracy.

The 27 February the Reichstag (i.e. the Parliament House) was set on fire by an insane communist incendiary who had followed an idea born in the minds of Göbbels, Göring and other

Nazi leaders. Immediately the Nazis accused the Communist Party of this crime against the new government and taking advantage of the consternation of the public opinion, next day, 28 February, Hitler imposed to Hindenburg the signature of a degree "for the protection of the People and the State" by which the seven articles of the constitution regarding the personal and civil rights were abolished. Furthermore it authorized the Reich's Government to assume full powers in the Federal States, if necessary, and to apply the capital punishment for a certain number of crimes, including that of "serious perturbation of peace" by armed people.

The Communist Party was banished, about four thousand communist State employees and a great number of socialdemocratic and liberal leaders were arrested, including members of the Reichstag who were supposed to enjoy parliamentary immunity.

The 5 March 1933 democratic elections took place for the last time during Hitler's time. The Nazis collected about 17.3 million votes, with an increase of 5.5 millions and thus reaching 44 per cent of the votes, their allied Nationalists guided by von Papen and Hugenberg, obtained only 3.1 million votes (8%), the Party of the Center 4.4 million votes; together with its allied, the Bavarian Catholic Party they realized 5.5 million votes (14%). The Social democrates essentially maintained their position with 7.2 million votes (19%) and the Communist still collected 4.8 milions votes (12%) with a loss of one million votes with respect to previous elections.

The solemn ceremony of the opening of the first Reichstag of the Third Reich took place on the 21 March, and two days later Hitler presented to the Parliament the "Gesetz zur Behebung der Not von Volk und Reich" (Law for the elimination of the state of need of People and State) which gave full power to his cabinet. In five short articles this law took from the Parliament the legislative power, the control on the Reich budget, the approval of treaties with foreign States and the initiative of amendments to the constitution, all of which were transferred for four years to the Reich's Cabinet.

The 2nd of August 1934 Hindenburg died and three hours later it was announced that the offices of Chancellor and President of the Reich were unified in the person of Adolf Hitler, who was also the supreme commander of the armed forces.

That's enough for sketching the general frame in which our story took place.

Thus during the winter 1932-33 both Fritz and Charlotte started to realize that they also had to leave the country as soon as possible. Before doing that Fritz went once more to Zoppot, via the Danzig corridor [41], to say goodbye to his father. As Charlotte wrote[6c]:

«This visit to Zoppot was a short one and ironical in a way that his father's bourgeois tastes in living actually saved him from a possible imprisonment at that time. Fissel was a chain smoker and, what I used to call, a chain newspaper reader. Hence, when he left Zoppot, he was loaded with a stock of the latest newspapers, which ranged from the very right to the very left, including some economical statistical surveys, that were probably issued by either the social- democrat or communist party. Passing the Danzig corridor, during which time the train was sealed, a Nazi police detachment raided and searched the train and found Fissel's compromising newspapers and accused him to be a communist. Fritz defence consisted in pointing out that well informed persons ought to be familiar with all points of view, and gave the right wing papers as evidence.

These arguments made no impression, but something else saved him. In his hurried and rather disorganized way he had picked up, together with his books, his father cellar-book leather bound with his list of wines, when bought, when laid down, price etc. The police took this impressive document as his personal property, and concluded that the owner of such a cellar could not possibly be a communist.

Fissel arrived in Berlin free, but rather shaken. He never saw his father again and we learned of his death in 1936 when we were in Kharkov. It was said that he died of a heart attack, while on a hunt, aiming to shoot a deer».

In spring 1933 Fritz went first to Copenhagen and from there to England. Charlotte followed him in June, packing and shipping their belongings and saying goodbye. The last person Charlotte saw was Professor von Laue who had messages to take abroad. As Charlotte wrote(6d):

«I went by train to Rotterdam, after having been seen off at "Bahnhof Zoo" by a large group of friends, we might never see again. Crossing the border into Holland was nerve wrecking. I had to face the Nazi patrols and their inspections and questions. On the boat we were safe finally. It was peaceful in spite of the crying baby who was thoroughly upset and would not sleep».

The Houtermans stayed in Cambridge for a few weeks, where they met again the Blacketts(42) they had first seen years before in Göttingen and became friends with Giuseppe Occhialini(43) who stayed in the same boarding house. As soon as they found a place to stay, they moved to Hayes, Middlesex, because Fritz wanted to be near his place of work.

«Settling down in this small town was rather an anticlimax after the high pitched atmosphere of the last weeks in Berlin and the intellectual companionship with the friends in Cambridge. After the constant tension in Berlin with all its fears and pressures, Hayes presented us with quitness and peace. It was unfortunately the calmness of a suburb, a very tiny place with ugly little houses, semi- detached, with identical gardens in streets with pretentious names. Fritz had to contend with a well organized working day and its fixed hours. He was used if necessary to work all night if a research project required it. To stop exactly at 5 p.m. as well as to begin in the morning at a given time was a great burden for him.

We were not quite as isolated as we had feared during the first weeks. Our concerns for the refugees and the friends who were still in danger inside Germany kept us busy and brought all kinds of people from London to discuss ways of helping. Fritz got a rather princely salary from "His Masters Voice" and was able to spend great part of it on these financial projects. The Quakers in London had organized the Coordinating Committee and collected names and curricula vitae of scientific personnel. They tried to connect them with jobs abroad: in Turkey, India, Scotland, England ... Leo Szilard(44) worked with them and devoted all his time and energy to this task. Fritz made contacts with professors in Oxford and Cambridge, collecting signatures for letters pleading for the release of political prisoners.

Our house was constantly full of guests, many of them refugees who stayed for days or weeks or even months. Hayes was quite far from London, but the stream of visitors kept on coming. Fritz seemed to attract people. He was very hospitale, generous, always full of ideas of how to help, how to organize whatever was needed. Fritz and Lange[(45)] equipped the upstairs kitchen as a darkroom, where they developed micro-photography and finally succeded to reproduce whole pages of the London Times in the size of a postage stamp. The idea was to send news and information into Germany to people who had no other means of learning what was going on abroad. These tiny films, often undeveloped, were pasted under the postage stamps or into cigarette packages.

But in spite of all these useful and important activities, in spite of the visits of so many friends, the isolation in Hayes became more and more oppressive. We went of course to London to parties, at the Blacketts where we met the French scientists: Auger[(46)], Perrin[(47)] and the Joliot-Curies[(48)]. Somehow this contributed only to Fritz's misery instead of alleviating it. He missed the scientific inspiration, the exchange of ideas, which is the essence of university life.

One day Sasha Leipunski (1903-1972) arrived from Kharkov. He had a two year stipendium to work in Cambridge in the Rutherford laboratory. He spent all his weekends in Hayes. Sasha was loved by everybody. He was "simpatico", gentle, entertaining, pleasant and very intelligent. It was he who told Fritz about the scientific work in the Physics Institute in Kharkov. His interests were the same as Fritz's. Sasha painted the scientific possibilities in such rosy colours that Fritz began to anticipate a possible renaissance there for his scientific ambitions. Hayes did not look like a permanent solution for him, the USA was closed, and now Sasha was opening a new door. Sasha offered him a contract with half a salary paid in foreign valuta, guaranteed vacations abroad, a house or an apartment. It all sounded so good, especially against the background of Hayes and became more tempting as the months went by. In order to accomodate more refugees, the salary fund from which Fritz was paid, was now to be divided between more people, a curtailment which would have made the financial assistances he had promised to others impossible. So in spite of all the warnings of friends, of all the negative aspects of such a step, Fritz finally accepted.

When Pauli visited us and warned over and over again not to go, he did not listen any more. Our political instincts were dormant. What was alive was the Hate of the Nazi regime, the danger and the terror of the Third Reich, the possible dangers and risks awaiting us in Russia were minimized by comparison. It was the time of the Kirov murder[(49)], and if nothing else this alone should have warned us and prevented us from leaving England»[(6d)].

8. - Professor in Kharkov -

The director of the Ukrainian Physico-Technical Institute in Kharkov, Professor Sinielnikov, had obtained for the Houtermans a rather large apartment with rooms on two different floors where they lived with Charlotte Schlesinger, with the nickname "Bimbus". She was a distinguished pianist emigrated not long before from Germany. Shortly after their arrival she got a very serious form of dissentery that kept the Houtermans in great anxiety.

Apart from this episode their life at the beginning was rather pleasant. They had a maid, Marussya, who helped in keeping the house, and when the 4 November 1935 their second child Jan was born, they also got a nurse to feed him. They were, however, surprised to see that the campus of the Ukrainian Institute was under control of guards with guns with mounted bayonets.

At a census taking they were asked to fill some form containing a number of questions, such as date and place of birth, names of parents, previous addresses and jobs and so on. To one of the questions about religion, Fritz answered by writing in the appropriate space "Jewish".

Charlotte was rather astonished at such an answer since Fritz had only a quarter of "Jewish blood" and anyhow both of them did not practice any religion. An argument followed between the two, that was closed by Fritz with the sentence: "Ich wollte mich herausmendeln!", where the verb is derived from the name of Gregor Mendel, the man who around 1860 discovered the fundamental laws of biological heredity.

Shortly after their arrival at Kharkov, the Houtermans received the luggage they had shipped from London through Arcangel. It consisted of a single large box, about 3 meters long 1.5 meters wide and 1.5 high, which contained everything: matrasses, pottery, books etc. An inspector sent by the local police searched very carefully all their belongings, in particular their books. The man was rather puzzled by the seven Bibles, of different types and in different languages and by the Rilke's "Geschichten von lieben Gott". It took them some time to persuade the inspector that these books were not brought in by them for doing religious propaganda.

Fritz also got an assistant, V.Fomin with whom he started to work on slow neutrons, as it appears from the names of the authors of a few papers published between 1936 and 1937 /13-18/.

The first of these papers /13/ reports on the observation of a new weak long lived activity observed in tantalium bombarded with slow neutrons, while the others /14/ - /18/ deal with the influence of the temperature of the moderator on the absorption of thermal neutrons (or better the so called group C neutrons) detected by means either of the activity induced in silver with 2.2 minutes halftime, or by a boron lined ionization chamber coupled to a linear amplifier. From their experimental results the authours conclude that: (a) in the thermal region the capture cross section for neutrons follows the 1/v-law in boron and silver while cadmium should have a resonance; (b) in a hydrogeneous moderator, like paraffin or water, the neutron spectrum should be (roughly) Maxwellian at room and higher temperatures, while at liquid hydrogen it does not reach thermal equilibrium.

Besides Houtermans there were many other young physicists who under the pressure of Hitler in Germany had started to visit the USSR looking for better living and working conditions.

As I said before Rudolf Peierls had visited the USSR in 1929, and even found his wife over there.

Alexander Weissberg-Cybulski had moved to Kharkov in 1932 with the idea of remaining there for good. He was a man of very wide culture, a marxist of very strong convictions and apparently the man with the better political formation and stronger dialectic capacity than all the others of the group. He had sent invitations to move to Kharkov to many other German speaking young physicists, now in troubles because of the Nazification of Germany and Austria. He was "managing director" of the Physics- Technical Institute.

Among the people invited by Weissberg there was Vicki Weisskopf who went over there for the first time in 1932 and stayed about eight months. He had got a professorship offer from Kiev and therefore went back to USSR in 1936 and revisited Kharkov for a few weeks after there had been a considerable accumulation of other physicists who had left fascist Germany and were eager to find a place to work.

Among these was George Placzek(50) who had already passed the academic year 1933-34 in Kharkov and had started to work with L.D.Landau on the scattering of light(51). He returned to Kharkov in 1935-36 after having spent the academic year 1934-35 as professor of physics at the new University of Jerusalem. He had in mind to write a second paper with Landau which would have concluded their work, but this plan was never completed because of his sudden departure from Kharkov as I will tell below. George Placzek has been one of my dearest friends. I had met him in Leipzig, where I spent ten months in 1931 at the Physikalisches Institut der Universität, directed by Peter Debye (1884-1966), working on X-rays diffraction by liquids and where I had the opportunity of becoming acquainted with many of the young people appearing in this story. In particular I became immediately a great friend of Placzek's who spent the successive academic year 1931-32 in Rome working with me on the Raman effect of the ammonia molecule. He was born in Brno, the main city of Moravia, then a part of the Austrian Empire. After the first World War he had become a citizen of the new state Czechoslovakia. He had studied physics in Prague and Wien, where he had passed his Ph.D. examination in 1928. He spent the following years 1928-31 in Utrecht. George spoke fluently about ten languages and had a fairly extensive knowledge of the corresponding literatures. For example, he had learned Italian, reading the "Decameron", and was used to quote Dante, Petrarca or Ariosto or discuss current politics in the very language of Boccaccio. After his first visit to the Istituto di Fisica di Via Panisperna(52) in 1931-32, he had taken the habit to stop in Rome for a few weeks, or even months, on occasion of any of his frequent displacements from North to South or East to West or viceversa. On each of these visits he told us facts about what was going on in the various places which could not be learned from Italian newspapers because of the fascist censorship.

In 1936 one of the subjects of passionate discussion everywhere in USSR, was abortion. The revolution had granted free abortion and the Great Soviet had encouraged people to discuss the problem deeply, in particular how to limit the number of abortions per year. But in 1937 the Soviets passed an abortion law with rather strong limitations which had become a matter of unsatisfaction especially for the intelligentia. Weisskopf tells me that the following joke was going around in Kharkov: "In USSR abortion is always free but there is such a red tape that it takes at least 9 months before it can take place".

Life conditions, however, were not easy in USSR in those years. In particular food supplies were very scarce. In Kharkov wives had to go around almost every day trying to buy from the black market, i.e. directly from farmers, eggs, chicken or something else, to add to the small rations. The bachelors, among the foreign physicists, lived all together in a single large apartment house. One of them in turn, almost everyday, did not work, but went around looking for extra food for all of them. In practice only one satisfactory meal per day was possible for almost all grown up people.

This unsatisfactory food supply was mainly due to the failure of the "integral collectivization" of agriculture. While the First (1928-32) and Second (1933-37) Five Years Plans for the development of heavy industry (oil, carbon, electricity, etc.) were successful, the "collectivization" was a failure, and, "after the revolution, the most terrible and stormy internal process" [(22a)]. The production of cereals between '26 and '32 decreased from 77 to 70 million tons, the number of heads of cattle decreased between '28 and '33 from 277 to 177 millions and so on. Furthermore the Kulaks, i.e. the farmers rich enough to have other people at their dependence for working their estate, were destroyed as a social class and the effort of organizing the farmers' families almost completely in Kolchozes, i.e., collective country concerns, was rather unsuccessful. For example, between '33 and '35 the individual working private farms were reduced by 4.8 millions but only 2.1 millions entered in kolchozes. The others disappeared from the villages. Before the collectivization there were about 25 millions of <u>dvorez</u>, i.e. family units in the country. At the beginning of '33 there were still 24 millions but these were reduced to 21 millions in '35 and to slightly less than 20 at the end of '37 [(22b)].

From conversations I had with G.Placzek in those years I heard the following story, which was confirmed and enriched by some further details by Vicki Weisskopf many years later (in 1976).

Once Martin Ruhemann, a German physicist working in low temperature, who like Houtermans had emigrated from Germany to USSR, gave a reception for all physicists present in Kharkov, Russian as well as foreigners.

Not long before Placzek had received the offer of a permanent chair and a few of the people present asked him what was his decision.

Placzek, in his typical joking mood, answered that he was ready to accept provided five conditions were fulfilled. But what are these conditions? His friends asked.

The first one was that his salary should be not inferior to a certain satisfactory value, the second one that about one third of the salary had to be paid in dollars or pounds because he wanted to spend every year two or three months abroad in order not to loose contact with the international scientific community. The third condition was that at least two young people who could work with him had to be paid in some form. The fourth condition was that the economic treatment of these young people had also to be decent. Finally the fifth condition, was that the "Chasain must go" -

"Chasain" in Russian means "boss" and at that time, was the name used to indicate Stalin. Everybody laughed and commented humorously.

The wife of Ruhemann, who was a very rigid party member, reported this little story to the Communist Party in Kharkov with the result that shortly later the local newspaper started to attack the foreign physicists with gradually heavier and more explicit accusations of being German spies.

Weisskopf, Placzek and most of the others understood immediately the general trend taken by the situation and rapidly left the USSR.

Weisskopf went to Zurich where he stayed three years, married Ellen Tvede in 1934, and after a short return to Copenhagen went to USA in 1937.

George Placzek suddenly reappeared in Rome and told all of us of the Istituto di Fisica that in the USSR a period of collective insaneness had began. But in Italy also the atmosphere was deteriorating rapidly. The precursory signs of the racial laws were already noticeable. Placzek left Italy, went to the great safe harbour of Copenhagen and from there, in 1938, to the Collége de France in Paris. About one year later he moved to Cornell University (Ithaca, N.Y.) where he could enjoy once more the collaboration with H.A.Bethe(53). They had published together, already in 1936, an important paper on the neutron resonances.

During the Second World War George Placzek was entrusted with an important directive position in the Theoretical Division of the Anglo-Canadian group, organized by the British Authorities in Montreal (Canada) for the development of Applied Nuclear Physics. In 1945 he was transferred to Los Alamos. After the war he worked, for about two years, at the Research Laboratories of the General Electric Company in Schenectady and from 1948 onwards at the Institute for Advanced Studies in Princeton, N.J.

But let me go back to Kharkov and to what happened of Fritz and Charlotte Houtermans and their two children.

9. - The beginning of the great trials -

The story I have reported above should be seen as the trigger that determined the events that during 1937, in Kharkov, involved the physicists of the Ukrainian Physiko- Technical Institute. What happened, there and at that moment, however, was only a part of a much wider stream of events that passed through the USSR, alarmed everybody and caused great anxiety in the small group of the Ukrainian Institute.

A few among Houtermans' friends were arrested already at the beginning of 1937. The first was Eva Striker, wife of Alexander Weissberg.

She was the nice of the physicist Polanyi and was a ceramic pattern designer. Alex and Eva had met and married in Berlin, where Alexander most probably was mainly working for the German communist party. A few months after they had moved to Kharkov, they were divorced, and she had gone to live in Moscow, where the Secret Police (NKVD) get hold of her at the beginning of 1937. A few days later Konrad Weisselberg, who had married an Ukrainian girl, was arrested in Kharkov. He was sentenced to 10 years and some time later was found dead in his prison cell.

The third to be arrested was Weissberg on March 1, 1937. He had tried to get Eva out of prison. As we shall see later for a few years his adventures were not too dissimilar from those of Fissel.

Those tragic events alarmed Fritz and Charlotte who were extremely disturbed by the rapidity with which the situation in USSR was deteriorating. A way had to be found, as quickly as possible, to get out of the country. With this secret programme in mind at the beginning of summer 1937, Charlotte with Giovanna went by boat from Leningrad via the Kiel channel to London where they learned of the bombing of Guernica in the Spain civil war. Officially the trip was for a vacation, but its real goal was to ask their British friends to invite Fritz to go to England to give a course or at least a few lectures.

The friends of the Houtermans, however, did not believe what she was saying, or at least, did not take it seriously enough. In the meantime Fritz in Kharkov noticed signs of impatience by local authorities about Charlotte's trip abroad and sent a telegram asking her to come back as soon as possible, and she returned immediately.

Shortly later two policemen came to the Ukrainian Physico-Technical Institute in Kharkov looking for the assistant of Fritz Houtermans, V.Fomin. They informed him that his brother, a ski instructor in Caucasus, had been arrested and asked him to follow them to the quarter of the "Secret Police" for a few questions. Fomin obtained the permission to go upstairs for a few minutes in order to collect a few personal things and books, drank a good deal of sulphuric acid from a bottle they had in the laboratory and then jumped from the window and died.

Fritz Houtermans was terribly disturbed. He talked without interruption all the successive night in a state of complete madness. Charlotte did not know what to do to calm him and she also was in a state of great desperation. Finally Charlotte called Leipunski, who was then Institute director. During the beginning of the purges he had become rather distant, because he was probably afraid for his own safety. But now he came to assure Fritz that nothing would happen to him, that Fomin's death would not involve him.

A few days later three policemen came to Houtermans' house during the night. Fritz was convinced they had come to arrest him, but it was not so. The policemen asked only the address of Fomin's apartment. Fritz must have been so frightened, that waking up the next morning he imagined to be in prison and could hardly believe to be still at home.

Many people in the group had the impression that in Moscow the political pressure could be less strong, especially if they were allowed to continue their work in one of the research Institutes of the Academy of Sciences. Lev D.Landau had already moved over there.

Thus Fritz Houtermans left Kharkov for Moscow, while Charlotte remained in Kharkov with the children for a few weeks more. Then she went to Moscow and went to live in Landau's house, while he had moved to the house of a Russian friend.

Fritz Houtermans had already asked for permission to leave the USSR and go abroad with the whole family and their personal goods and books. They were told that this was possible but they had to present a detailed list of their belongings. In order to overcome the red tape Fritz went every day to the Custom House in Moscow hoping for some progress in clearing their papers.

It was the 1st of December 1937 when he was arrested while at the Custom House waiting for some news about his request. He was brought to the Lubianka, the famous Moscow prison.

The same day Anja, the wife of Peter Kapitza, went to visit Charlotte and had to give her a sedative. The Russian (jew) mathematician Rubin asked Charlotte immediately to leave the apartment of Landau because her presence was a danger to Landau himself. She left the children to Anja Kapitza and kept the key to Landau's house for a few days while trying to locate Fritz. She had no passport because Fritz had it when arrested. Helped by Peter Kapitza, she was given it in a office of the Lubianka. After that Kapitza found her a hotel room, where she now easily could stay with her children.

She decided to leave for Riga as soon as possible. The evening before her departure Charlotte, with Bimbus, who at that time was also in Moscow, went to visit the sister of Grisha Weller, a musician who was in charge of the Kharkov Radio Station.He had been sent to Siberia some time before.

The trip was all right but when they arrived not far from the Latvian border, Charlotte was asked to get out of the train and was housed with her two children in a building normally used for housing the railroad employees. They were kept there for about ten days without receiving any explanation about the interruption of their trip, but were asked continuously about the whereabouts of Fritz and the reason for Charlotte travelling alone. Had she admitted that she knew Fritz had been arrested, they would immediately have arrested her too. It was the presence of the small children, who according to a Russian law could not be arrested, that saved them all.

On the 16th of December the authorization arrived from Moscow, she was permitted to get into the train again and arrived in Riga the same day with the two children in good health and one hundred Swiss Francs plus one hundred Belgian Francs in her pocket, sewed in the linings of their fur caps.

The nightmare was not yet finished but she felt very much relieved, in spite of the anxiety for Fritz in a Moscow prison. Immediately she sent a telegram to the Bohr Institute in Copenhagen asking for an invitation to go over there. She got a positive answer and some money, obtained a permit to enter Danmark through the help of Niels Bohr and took the boat for the new destination.

On Christmas Eve Charlotte arrived in Copenhagen where Christian Moller [(54)] met her at the boat. She was invited by him to a dinner in a restaurant with its Danish elegance, was overhelmed by it and the friendly and warm reception since her arrival. It was almost unbelievable after the drabness of their lives in the USSR. The next morning at the Institute for Theoretical Physics, Charlotte met with Niels Bohr[(55)] and a great many other physicists; Gamow[(56)], Rosenfeld[(57)], and many others were there and eagerly asked about the fate of Fritz, Landau, Weissberg and others. She remained in Copenhagen for some time more and then went to London, after having planned with Bohr how to help the people still in troubles in the USSR.

In April 1939 Charlotte with the two children went to the United States on an immigrant visa to meet Fritz's mother, who was immigrated in 1936 on invitation by one of her pupils. Through friends, Edna Carter and Monica Heaba, at Vassar College, Charlotte obtained a small research stipendium.

Though continuously writing to Fritz in various prisons, Charlotte and Elsa never had any answer and did not know whether he was alive. At that time Charlotte got in contact with Eleanor Roosevelt, who via the ambassador Steinberg found out that Fritz was alive. Constant letters and questions were the only things to prevent a prisoner might be forgotten.

In the U.S.A. Charlotte succeeded in making a career as a physics teacher in various women colleges: from Vassar College (1939-42) she went to Radcliffe College (1944-45) in Cambridge Mass., she worked at the research laboratory of the Polaroid Company in Cambridge, Mass., in 1944-45, but went back pretty soon to teach at Wells College (1945-46). She finally succeeded, in 1946, to Maria Goeppert Mayer(58) at the Sarah Lawrence College, in Bronxville, N.Y., when Maria moved to the University of Chicago (1946). Charlotte taught physics in Bronxville for 22 years and even after retirement, in 1968, she gave for four years more (1968-72) courses at the Manhattan Ville College, combined with some teaching at the Summer School Course at Sarah Lawrence College (1972-76).

Attempts to get out of prison Houtermans and Weissberg were made also from Europe. Iréne Curie, Frédéric Joliot-Curie(48) and Jean Perrin (1879-1942)(59) sent in June 1938 a telegram and a letter to Stalin and the State Attorney General of the USSR Vishinsky, the texts of which are given below(60).

Telegram: State Attorney General of USSR, Moscow. Same text to : Stalin, Kremlin, Moscow.

Please send information on the fate of well-known physicists Alexander Weissberg, arrested in Kharkov the March, 1, 1937, and Friedrich Houtermans arrested in Moscow the December, 1, 1937. Stop. Their detention threat to provoke a political campaign from the enemies of USSR and at the same time is incomprehensible to the friends of USSR since they are convinced that Weissberg and Houtermans are unable of actions hostile to the Socialist Construction and that their arrest is a serious mistake of subordinated organs. Stop. Please pay attention to this case, we underline its political meaning, and urge a prompt answer.
Signed by:
Irène Joliot-Curie, previous undersecretary of State, Nobel Prize;
Jean Perrin: previous undersecretary of State, Nobel Prize;
Frédéric Joliot-Curie, Professor at the Collège de France, Nobel Prize.

Letter to the State Attorney General of USSR (Vishinsky), Moscow
Paris, June 15, 1938

Very estimable Mr. Attorney General! The undersigned, friends of the Soviet Union, believe it to be their duty to bring the following facts to your attention. The imprisonment of two well-known foreign physicists, Dr. Friedrich Houtermans, who was arrested on December 1, 1937, in Moscow, and Alexander Weissberg, who was arrested on March 1 of the same year in Kharkov, has shocked scientific circles in Europe and the United States. The names of Houtermans and Weissberg are so well-known in these circles that it is to be feared that their long imprisonment may provoke a new political campaign of the sort which has recently done such damage to the prestige of the country of socialism and to the collaboration of the USSR with the great Western democracies. The situation has been made more serious by the fact that these scientific men, friends of the USSR who have always defended it against the attacks of its enemies, have not been able to obtain any news from Soviet authorities on the cases of Houtermans and Weissberg in spite of the time which has gone by since their arrest, and thus find themselves unable to explain the step that has been taken.

Mrs. Houtermans and Weissberg have many friends among the most important scientists such as Professor Einstein of Pasadena, Professor Blackett of Manchester, Professor Niels Bohr of Copenhagen, who are interested in their fate and will not abandon this interest. Mr.Weissberg, one of the founders and editor of the journal "Zeitschrift für Physik" in USSR, has been invited by Professor Einstein to the University of Pasadena; because of his arrest he could not accept the invitation. Similarly, Dr. Houtermans has been invited for scientific work by a London Institute and at the moment of his arrest at the Custom House of the Moscow Station, was just on the point of leaving.

The only official information available on the reasons that have brought to the arrest of Mr. Weissberg, is a communication from the Soviet authorities of March 1937 to the Austrian Embassy in Moscow, according to which Weissberg is accused of having made espionage in favour of Germany, and to have participated in the preparation of an armed insurrection in Ukraina. No official information is given on Houtermans' case.

All those who know personally Weissberg and Houtermans are sincerely convinced that they are faithful friends of USSR, unable of any hostile action. They are deeply convinced that the accusations against Mr. Weissberg are absurd and should be attributed to a serious misunderstanding which requires an immediate clarification from both the political and human point of view.

Responsible personalities of the USSR have recently indicated, in official declarations, that in the course of the purge campaign, necessary in that country so heavily threatened from the inside as well as from the outside, some errors - inevitable in such critical times - have been committed by subordinated organs. The same personalities have underlined the urgent necessity of eliminating the errors and occasional abuses.

The undersigners and all friends of both these accused people are convinced that one is in front of a misunderstanding of this kind.

Therefore they address to the State Attorney General, calling his attention on Mrs Weissberg and Houtermans and ask him, for the reputation of USSR in the foreign scientific circles, to undertake all steps necessary for the immediate release of these two gentlemen. The political meaning of this question authorizes us to transmit to Mr. Stalin a copy of this letter by means of the Soviet Embassy in Paris.

In consideration of the urgency of the matter, we ask you the courtesy of providing us with a prompt answer. With the expressions of our more sincere consideration, honorable State Attorney General, we remain

Irène Joliot-Curie: previous undersecretary of State, Nobel Prize
Jean Perrin: previous undersecretary of State, Nobel Prize
Frédéric Joliot-Curie, Professor at the Collége de France, Nobel Prize.

10. - The years of prison -

All the information we have about the years of prison of Fritz Houtermans are the two letters that he wrote shortly after liberation to his mother the 1st of August and 28th of September 1940 (7), the "Chronological Report of My Life in Russian Prisons" (8), that he wrote in spring 1945, and a book published in London in 1950 under the title: "Russian Purge and The Extraction of Confession"(61) by F.Beck and W.Godin. This book was translated into other languages. It was recommended to me by Fritz himself in the late fifties, when I asked about his experience in Russian and German prisons. The false names Beck and Godin stand for Friedrich G.Houtermans and Konstantin Feodossovitch Shteppa (1896-1958) who had been for one and a half year in the same cell of the Kiev prison.

In their book the two authors try to give a general view of the events that took place in those years in Russian prisons but they avoid giving any precise date or name so that the whole story becomes in some way detached from their personal adventures and inevitably loses vividness and human touch.

To make up for this to some extent, the authors devote Chapter 7 to "Three Cases" which are described in somewhat greater detail. The first of the three cases is that of Konstantin Shteppa himself and I will come back to it later on (Chapter 12 and 15).

No detailed information about "Houtermans' case" can be found in the book and, therefore, after many doubt, I decided to reproduce here the "Chronological Report" in spite of the fact that in its "Post Scriptum" Fritz Houtermans recommends not to use it for publication. Many years, however, have passed after Houtermans' death and all, or almost all, the other people mentioned in this document died. With the vividness of Houtermans' case the "Chronological Report" adds substantially to the picture we already have from accounts published by other people that went through similar adventures in USSR.

"Chronological Report of My Life in Russian Prisons"(8).

«On December 1st 1937 I was arrested at the Custom House in Moscow, where I was preparing my property to get locked through for my departure from Russia. I was immediately brought to the Lubianka prison where I was shown the order of arrest, dated from Kharkov from November 27th, on account of Paragraph 28 (political reasons). After a quarter of an hour I was brought to the big Butyrka prison into a cell for 24 men. Gradually this cell was filled until it held 140 men, sleeping on and under wooden boards, about 2-3 men per m^2. While still in Moscow, 11 days after my arrest I was called by an officer of the NKWD, to give a full confession of my alleged counterrevolutionary activities on behalf of the German fascist government, but no concrete charge was brought up against me; only the names of a number of my - Russian and foreign - colleagues from the Kharkov Physical Technical Institute were mentioned as being members of a counterrevolutionary organization, as Shubnikov, Landau, Ruhemann, Weissberg, Fomin, etc. I was told that if I gave a full confession I would be immediately be sent abroad. Of course I did not make a false confession and denied any activity against the USSR.

On January 4, 1938 I was brought up in a prisoner car by railway to Kharkov and put into the prison Kholodnaja Gora in Kharkov, in a cell which was still more overcrowded than that in Moscow, but without any sleeping accomodation so that we all had to lie on the floor. I remained there till January 10th when I was brought into the central Kharkov prison of the NKWD, into a cell perfectly clean and not too overcrowded. Here many fellow prisoners tried to persuade me to give a false confession of things of my own invention as they had done themselves since sooner or later I had to do anyhow in order to save a lot of trouble. The same day I was asked to give confession again by a questioning official, named Drescher, who threatened to beat me and to get anything out of me.

In the evening of January 11th began an uninterrupted questioning of 11 days with the exception of 5 hours interruption the first day and about 2 hours on the second day. No concrete charge was brought up against me as in nearly all cases of people I have seen in Russian prisons and I was told to give all "facts" myself. The only two questions that were asked were: "Who induced you to join the counterrevolutionary organization" and "whom did you induce yourself?"

Three officials questioned me in turns of about 8 hours each, the first two days I was allowed to sit on a chair, later only on the edge of a chair and from the 4th day I was forced to stand nearly all day, I was always kept awake, and when I fell from lack of sleep I was brought up by means of cold water that was poured out on my face. The chief official who led the questioning was named Pogrebnoi.

The night to January 22nd, shortly after midnight, Pogrebnoi showed me an order of arrest of my wife and another order to bring my children into a home "besprisornis" under a false name so that I would not be able to find them ever again. I was of the opinion that they were all still in Moscow. I have learned since that they had left shortly after my arrest so all I was told was bluffing but in the state of weakness after nearly 10 days without any sleep I fell for it.

In this state I fell unconscious nearly every 20-30 minutes but I was awaked every time and my feet were so swollen that my shoes had to be cut off. I was beaten little, only occasionally, and not with instruments as many other prisoners I have seen and I was told by them that the treatment I had to undergo myself was very mild indeed if compared with what they had to endure in beating. At the end I declared that I was ready to sign any statement they wanted under the conditions that my family were to be sent abroad immediately and I would be shown a letter from abroad by my wife telling me her whereabouts, after 3 months. In case I would not get such a letter I would revoke any statement I made. I signed a short statement as they asked me, that I was sent to the USSR by the German Gestapo, for espionage. Then I was able to eat luxuriously and got tea and was sent to sleep to my cell where I slept for about 36 hours. Then I was asked upstairs again and there I wrote a long confession of about 20 pages in German and I was very careful to give only names of people whom I knew to be abroad, or whose evidence against me - of course forced by 3rd degree methods also - was shown to me. I had to write about espionage, sabotage and counterrevolutionary agitation and I was absolutely free to invent anything I liked, no corroboration by facts or by other evidences as to material facts being needed. I made nuclear physics the theme of my espionage, though at that time no technical applications of nuclear physics were known, since fission was not yet discovered but I wrote a lot of phrases that the nuclear energy is there and that it needed only the right way to start a chain reaction in the way of popular novels on this matter. Another instrument I wrote I had spied upon was an instrument for measuring absolute velocities of airplanes by the number of magnetic lines of force which went through a coil, a device contradicting the law of conservation of energy, and being obviously a perpetuum mobile. I intentionally made my confession as stupid as possible in order to be able to test that it is nonsense in case of a trial and I put in a short statement in English in ciphered form that I was under third degree torture and that all I wrote was pure invention.

During the last year of my being in Kharkov many people I knew and I knew perfectly well as being innocent had been arrested already and it was stated that they all had given evidence to be guilty. I did not know then about the methods how these statements were forced from the people but I had told my wife in case a signature would ever be forced from me, I would leave out the full stop after my signature, and in case my signature were given by my freewill I would always put a full stop after my name. I had opportunity to do so and I left out the full stops in the written confession. My written confession was translated into Russian and I was left alone and was not troubled anymore till August 1938, living till March in a clean prison cell not too overcrowded in the central Kharkov prison. On March 17th I was called again and a letter from my wife dated from Copenhagen was given to me. The same day I was transferred to the Kholodnaya Gora prison in Kharkov, to a small cell, rather dirty and very overcrowded where I remained till August 2nd.

Food was very little and we suffered rather from hunger. The rations consisted of 600 gr black bread containing more water than ordinary bread (equivalent to about 500-550 gr of ordinary bread), about 15-20 gr sugar, a mug of soup containing little nourishing value and 1-2 spoons of porridge of some kind a day, from fair estimates made by physicians I met and by myself about 500-1100 cal per day. Food was always given regularly and I don't know of any cases that prisoners were not given their rations. Treatment by prison officials was hard but not sadistic, but there existed cells were conditions were much worse for people who had not given the confession of evidence wanted. I remained there till August 2nd when I was sent to Kiev in "Stolypin car", a special sort of railway car for prisoners. I remained in Kiev till October 31st 1938 where I was asked to give more evidence especially against a friend of mine, Professor Leipunski, member of the party and an absolutely sincere man. From prisoners in my cell I learned that he was arrested in another cell in Kiev and a man in my cell tried to persuade me to give evidence against him and told me what I should say. No specially hard pressure was used against me then and therefore I did not give any evidence against him or against Professor Obreimov, another member of our institute, that I was asked to give. Prison conditions in Kiev were much better than in Kharkov, the rooms being very clean and food a little bit better. It was hard though because it was not allowed to sleep in the day time.

On October 1st 1938 I was sent back to Kharkov and put into a clean cell in the central prison. Prisons were not so overcrowded any more at that time, but still there were 1-2 prisoners per m^2 of room. I was not questioned till January 1939 again when I was asked to sign an application for Soviet citizenship. For that case they promised me the leadership of a big institute for my research work to be built by the NKWD itself but I did not consider that offer to be sincere, having met foreigners in prison cells who had agreed to such an offer without having been released and therefore I said I could talk about this matter only after release and after having communication with my family. This time was the only one when I got some of the things that were sent to me by my wife and from Mrs. Cohn-Vossen, a friend of mine in Moscow.

I got a blanket and a few pieces of underwear. I did not get any letter nor any money that was sent to me from abroad, as I have learned since. This was rather bad because all the time there was the possibility to buy some additional food supply and smoking material for about 20 roubles a fortnight and this helped a great deal but since I had less than 100 roubles on me when I was arrested I nearly never could make use of this possibility and therefore I had lost about 18 kg in weight and became more and more feeble. I could not think of a revocation of my confession the year before and when I was asked to give more evidence against persons as Obreimov whom I knew to be in USSR, I declined but I confirmed my former confession not wanting to have all the trouble over again. On the new evidence they were not pressing very hard.

In February 1939 I was sent again to Kiev where I was put again in the central prison but in an underground cell without any daylight (artificial light was in all cells during the night) which was very humid. I was asked again by a new official to give evidence against Obreimov and Leipunski and I was threatened to be beaten in case I refused and shown written evidence of both of them against me in their own handwriting. I was very weak then, I could hardly walk about and so I decided to confirm their statements on counterrevolutionary activity about myself and that I knew about theirs. I put in some slight discrepancies concerning dates etc. with their evidence and my evidence was accepted. Again I was told I would be sent abroad.

In May 1939 I was asked by the People Commissar of the Interior of the Ukraina himself to give evidence against Professor Fritz Lange a good physicist and friend of mine who was working in Ukrainian Physical-Technical Institute and also against Professor Landau, Professor Ioffe(62) and Professor Kapitza all of them being prominent physicist of the USSR. He told me he knew well that all of them were active spies and members of a counterrevolutionary organization and they only wanted me to confirm this. I said I knew nothing about it but did not try to revoke my own statements given earlier. This confirms the fact I had often heard in prison cells about, especially by men who once had been officials of the NKWD themselves, that it is quite usual to collect evidence about conterrevolutionary activity of prominent people who are not arrested at all in case their arrest should be effectuated later on. Neither Lange, nor Joffe or Kapitza ever have been arrested as far as I learned since.

No paper or books were allowed in prison cells and therefore it was nearly impossible to do any work. Yet from the very beginning of my prison time I decided to work under all conditions and since it was the only field I could do it in I started already at the end of 1937 to think about problems of the theory of numbers. All I knew was Euclid's proof about the existence of an infinite number of primes(63) and I started thinking on the problem whether there exists an infinite number of the type 6x + 1 and 4x + 1 also, while for the 6x - 1 and 4x - 1 I could find Euclid's proof to hold with a slight alteration off hand.

I had no writing materials but I tried to write some numbers with matches on a piece of soap or on places of the wall where it could not be seen, but I had to extinguish it all every day before leaving the cell for the toilet.

I thought about that problem for more than a year and finally in Kiev in the first days of March I found that any form $x^2 + xy + y^2$ with x, y being relative prime cannot contain any other factor than primes of 6x + 1 type or 3, and the sum of the squares of relative primes contains only primes 4x + 1 or 2.

After solving this problem I found Fermat's theorem (I only learned its name after I left the prison as with all theorems I found) and quite a number of theorems in elementary theory of numbers.

When I found on August 6th an elementary proof for Fermat's famous problem for n = 3, which as I have learned since is essentially the same as Euler's, by descente infinite, I got very excited about it because I did not know Euler's elementary proof to exist, so that I made application to the People Commissar of the Ukraina to get paper and pencil. (I said I wanted to work out an idea of mine on a method in radioactivity which might be of economic importance). When my petition was not granted I went on hunger strike (only declining food, not water). I was alone in a cell then and succeeded in getting paper and pencil after 8 days of hunger strike by which I was very much weakened since I was in a bad state when I started. I wrote a number of theorems, I had found the so called indices of theory of numbers, a theorem of Lucas and a new proof of a theorem of Sylvester which is in course of publication at the Jahresbericht upon the advice of Professor van der Waerden(64) whom I told about my prison studies in the theory of numbers.

I even could keep writing materials when Professor Melamet (a philosophy Professor from Odessa) was put into my cell and I remained there, so that I could make steady progress in the theory of numbers.

In August all my evidence I had given 1 years 1/2 earlier was rewritten and I was asked together with Professor Obreimov for a so called "double questioning" in which he - of course it was all pure invention - stated before my eyes that I had induced him while still in Berlin to do espionage work for the Nazis - though at the time of his visit to Berlin the Nazis were not in power and a quite small party. I affirmed all his statements because I did not want and in the state of health I was in, could not afford to go another time over all the tortures again by which I was threatened.

Suddenly on September 30th 1939 I was called out and brought to the station in a closed car and was brought to Moscow. I did not know about the war till January 1940. The isolation of prisoners is extreme in Russia, the only source of information being what is told by newly arrested prisoners and I had not seen such people for a considerable time.

In the train I saw that the official who had questioned me before, travelled with me in the same train and in Moscow I was brought immediately into the central prison of the NKWD on the Lubianka. While I was still in the showerbath, everybody arriving there has to go through, I was already called for being questioned. I was brought into a luxuriously furnished room in which a man in the uniform of a general of the NKWD sat and beside him in civilian clothes a very intelligent looking man who presided and who asked me politely to sit down. Then he asked me what I felt guilty of. I asked: "Do you want to hear what confession I signed or do you want facts?" "Of course facts" he replied. "This is the first time I am asked this question within these walls" I said. "But since you want to have them, the only thing I feel guilty of is that I stole a pair of underwear in the Kharkov prison a year ago estinguishing the prison stamp on them by calciumcloride in the toilet. That's all". "And what about your confession?" He asked. "That's all pure invention!" Then he asked who had forced me to give a confession and by what means I was forced. I gave all the names as far as I knew them and all details. "We are going to get it all clear", he said shortly and I was brought back to my cell, a good cell, where I was alone.

I liked that better, by the way, since I could work. Everything was extremely clean and I got books, very good ones, too, special food in quite sufficient quantity and a package of cigarettes every day.

Though my Kiev manuscript had been taken away from me when I entered the Lubianka, I got writing materials again without any effort and I went on to occupy myself with what I have learned since to be "Pells problem" and other things in theory of numbers. In this cell I remained without being called a single time until the beginning of December 1939, being all the time alone. After all I had passed it was a treat.

In the first part of December I was called up again by another official who asked me absolutely correctly about everything and I answered all questions correctly. When I asked to write or to cable to my family (I supposed them to be in England from where I had last heard from them in August 1938) he said I would soon be sent out. I then asked especially not to be sent to Germany and he made a note of it.

About a week later I got new clothes and was sent to Butyrka prison into one of the big cells where I had been 2 years previously, but it was not overcrowded then. All people in the room were Germans, not all of them foreigners, some had taken the Soviet citizenship. Among them I was glad to meet another Professor, Professor Fritz Noether(25), former Professor at the Breslau University for applied mathematics and later refugee living in Tomsk. He had been arrested as German, - though being a Jew - and was forced to invent an espionage story, also. But in contradiction to my case a sentence of 25 years of imprisonment had been passed on him. Shortly after my arrival he was removed from the cell and I have never heard about him since. In this cell also we all got special food in sufficient quantity and cigarettes and we had the impression that we were kept there because most of us were in a very bad state of health and they did not want to send us abroad like that. Most of people were German workers, skilled workers, or engineers, specialists and many of them former communists. Among them was Hugo Eberlein(65), friend of Lenin and Liebknecht and former member of the executive control committee of the Komintern, president of the communist fraction of the Prussian Landtag for many years. He had been beaten severely also like nearly all of them. Some were called out and presumably sent abroad, some arrived directly from camps in Siberia and the far North.

In March I was called out alone and asked to sign a paper that I was not to tell about what I have seen in Russian prisons and that I would agree to do secret work for the USSR abroad. This I signed because I had learned from many people that most of them were asked to sign such a paper, otherwise one would be kept indefinitively.

I again asked as a condition not to be sent into Germany and this was promised to me by the official who made me sign the paper.

On April 17th 1940 some of us were gathered into another cell in the same building and on April 30th we all were called out, a sentence was read to each of us, that we were condemned to be exiled from the USSR by a special court of the NKWD, and we were transported in a prison car to Brest-Litowsk where we all were taken over by the officials from the Gestapo.

We were not set free, but taken to a German prison in Siala Podlaska, a small town near the frontier line and after some day we were all transferred to the citadel of Lublin.

Isolation was not as strong as in Russian prisons, the regime was more military and food and accomodation conditions much worse that the last time in Moscow. Every day we heard the songs and the noises from drunken Gestapo officers below our windows, while we learned that every day about a hundred Poles and Jews were executed in the prison court.

We had passed the frontier on May 2nd 1940 and were transported to Berlin on May 25th. Some of us were brought into "Nazi-Rückwandererheim" where they were set free after a few days, but some of us, among them also I, were brought to the police prison on Alexanderplatz. By the way the only prison in my experience, where I have met lice. Here I met people from concentration camps who told me about German camps and a well experienced communist who advised me how to behave before the Gestapo.

A week later I was brought to a small prison at the Gestapo headquarters in the Prinz-Albrecht-Strasse where I was asked about my Russian experiences, why I had left Germany and gone to Russia, and about some communist friends of mine in Germany before 1933. I told them I had known those people but I did not know about any illegal activity of theirs confirming my information nevertheless on such people I knew to be abroad. I was asked to give an account of my Russian experiences, which I did, also mentioning by precaution the paper I had been made to sign but not the fact that I had asked not to be sent to Germany.

On July 16th finally I was set free. A few days later I met Professor von Laue from whom I learned the whereabouts of my family. As soon as he had heard that I was in Germany in a Gestapo prison he went there himself, brought me some money and did all he could to accelerate my liberation.

May, 19th 1945 F.G.Houtermans

P.S. I have been asked several times by many people to publish something on my Russian experiences since the war between Germany and Russia began. I always declined strictly because I do not want any propagandistic conclusions to be drawn from my experiences. I also do not want this information to be used for publication. I want to emphasize again that apart from the treatment by which false confessions were forced from people in Russian prisons by the questioning officials under special order from the government nearly no facts have been brought to my knowledge indicating sadistic or even uncorrect treatment in prisons by prison officials in the execution of their duties. It is my opinion that most of the atrocity stories being told by prisoners about the executions are incorrect. I don't know any case that a prisoner has seen the execution of another one though I have frequently met people who had been sentenced to death and had sat in a so called "death cell" after revision of their sentences».

11. - An overview of the situation in Central Europe -

Following the adventures of Fritz Houtermans we have lost the overall view of the general political scene in Europe. Which had been the most important changes that took place during the six years between the departure of Fritz from Berlin just fallen under Hitler's power, and Fritz's liberation from the Gestapo prison in Berlin? Only a few of the most dramatic events of that period can be mentioned here.

The 25 July 1934 the Federal Chancellor of Austria, Dollfuss, leader of the Christian-Socialists, in reality clerico-fascists, was murdered in his office in Wien by a group of 154 members of the dissolved Austrian National- Socialist Party.

The immediate annexation of Austria by Germany was prevented by the international reaction, in particular that of Mussolini's government, which promptly sent four divisions to the Brenner Pass.

Dollfuss was succeeded by his party companion Kurt von Schuschnigg, who tried to save Austria's independence by following a policy of détente with Hitler's Germany. But the Austro-German agreement, signed on 11 July, 1936, contained concessions that spelt disaster for Schuschnigg and his country. The Austrian Nazional-Socialist Party was reconstituted, with a strong renewal of antisemitism, the roots of which in Austria dated back to the year 1880-1890(66).

Around the end of February 1938, Wien entered a period when Schuschnigg's government was engaged in a death struggle, which ended on March, 13, 1938 with the proclamation of the "Anschluss" of Austria by Hitler's Germany.

This occurred without Great Britain and France taking any measure, and with the consent of Mussolini, who first with the Abyssinian war (1935-36), and later, through participation in the Spanish civil war (1936-39), had once and for all espoused Hitler's cause.

The September 29-30, 1938 at an encounter in Munich, Great Britain and France accepted the dismembrement of Czechhoslovakia, with the transfer of a few of its border provinces to Germany, Poland and Hungary.

Under Hitler's pressure the 15 March, 1939 the Czechoslovakia ceased to exist, without the slightest initiative from France and Great Britain which had solemny promised at the Munich meeting to garantee the mutilated Czechoslovakia from any aggression.

On August 23, 1939 a "Non aggression Pact" was signed in Moscow by Hitler's Foreign Affairs Minister von Ribbentrop and his colleague from USSR, Molotov. This Hitler-Stalin, or Ribbentrop-Molotov Pact was in reality an agreement on the partition of Poland, while Hitler gave to Stalin free hand on the East-Baltic countries, a situation rather familiar at the times of German kings and Russian tzars.

At the dawn of September 1, 1939, a date fixed by Hitler since the 3 of April of the same year(!)(24), the German troops crossed the Polish border at many points in the direction of Varshaw. Stalin's troops entered Poland from the eastern border. In a few days Great Britain and France declared war to Germany in spite of their inadequate military preparation. It was the beginning of the Second World War and at the same time the climax of tensions in the left oriented part of the public opinion of the "free world".

The natural and most general reaction was that a pact of agreement between Hitler and Stalin was incredible, almost impossible. But common sense pretty soon won: opposite extremes, quite often, are much closer than suggested by reason.

A variety of judgements were expressed here and there on Stalin's behaviour under these circumstances. Many people thought that the signature by Stalin of a Pact with Hitler was an untolerably unmoral deed. A smaller fraction tried to defend Stalin's position as an unpleasant but necessary procedure for giving the USSR some time for a military preparation to resist a most probable attack of Hitler in the future.

Still others felt that the Hitler-Stalin pact was fully justified because, after all, the real enemy number one of the new socialist societies were the capitalistic countries and therefore any action which increased forces against them was a good thing.

That's enough as a general overview. I would only add a small reflection on the transfer from the USSR to the German prisons of many German people like Houtermans.

Most of them were communists or very close to the communist party and all of them asked the USSR authorities to send them to any other place but to Hitler's Germany. But this understandable desire was not taken into account: following the von Ribbentrop-Molotov Pact they were all handed over to the German Gestapo. The only exception was made for those among them that were jewish: they were all sent to concentration or annihilation camps.

12. - More about Fritz and Konstantin -

The name of Konstantin Shteppa does not appear in the "Chronological Report of My Life in Russian Prisons", because each of the authors, for years, even when it was no more necessary, kept the habit they had taken during the purges in USSR, of avoiding to give any information about their friends.

We have, however, plenty of information about Shteppa's origin and life (67) (68). Son of an Orthodox priest who had married the descendent from an Ukrainian family of noblemen, Konstantin had interrupted his studies at the University of Petrograd for military service, first in the tsar's armies, during the war (1914-16), and then in the civil war with the "Whites"(69). After the war he completed his studies and in 1927 was awarded a Doctorate of Philosophy in History. In 1930 he had been appointed professor of ancient and medieval history at Kiev University. At the same time Shteppa was senior Research Associate of the Academy of Science of the Ukrainian SSR, where he became chairman of the Committee for Byzantine Studies.

Shortly after 1930, having concentrated on demonology and ancient legenda, Shteppa shifted his attention to social revolts in ancient Rome and especially in Roman Africa. This subject was to remain his special interest for the rest of his life. At the same time he reviewed the whole analysis of antiquity. Politics inescapably crept into his professional and personal life.

Shteppa was arrested in February, 1938, at the height of the purges. His troubles with the authorities had started in 1937 with an expression of view about Joan of Arc. As Shteppa wrote years later(68):

«In a lecture on the Hundred Years' War I had described the famous French heroine as nervous and highly strung. Dimitrov, secretary general of the Komintern(70), had announced, at the last congress that the French Communists had denied the right of the Fascists to claim Joan of Arc, the French national heroine, as a champion of their fascist philosophy. In other words the French Communists wished to appear to be good patriots, for Joan of Arc had shared in the people's struggle for liberation from a foreign oppressor.

That was the line of Dimitrov's argument. But I, by calling Joan of Arc highly strung in my lecture, had detracted from her significance as champion and representative of the struggling people, and had thereby also failed to respect the statement of a party leader. Thus I had deviated from the general party line in the matter of national movements and had exposed myself as a bourgeois scholar. That is what I was accused of in an article in the University wall-newspaper.

Joan of Arc was followed by Midas. To illustrate a point in one of my lectures I had mentioned the legend of Midas, I think in connection with the invention of money. The legend had no particular importance in the context, and I had merely mentioned it in passing, so it may be that I did not sufficiently emphasize some minor aspect of it. Probably, however, I told an unfamiliar version of it.

Meanwhile Stalin, in one of his speeches on the cleavage between officials and the party masses, had mentioned the legend of Antaeus. It was now suggested to me by my critics, among them my own assistants, the "representatives of the younger generation", that only a bourgeois professor - and my name was mentioned - would neglect these myths and distort their context, thus setting his students a bad example, while Stalin, the wisest and most brilliant leader of the party and of the workers of the whole world, showed the greatest respect for ancient mythology, and even quoted it in confirmation of his conclusions. The political significance of my mistake, in the opinion of my critics, lays in the fact that I had insufficient respect for the party leader's authority and did not accept his statements as my guiding principle. It sounds like a joke, but it was a joke with serious consequences.

The Midas incident was followed by more and more accusations, though the attacks on me were still isolated and individual. I felt more and more clearly that these were merely the preparations for the big offensive that was still to come, and sure enough this began in the later autumn of 1937, when the whole country was in the grip of an unprecedented purge. Eventually my turn came. For many days and nights I was the object of a "checking-up process" at meetings and sessions in the university attended by my colleagues, assistants and students. At the same time essays and articles violently criticizing my work appeared in the most varied periodicals and newspapers»(61).

Arrested in February 1938, Shteppa was held as a political prisoner until after the outbreak of World War II, in September 1939.

His first encounter with Fritz in the Kiev prison was described by Konstantin Shteppa and written down by his daughter Aglaya(9b).

«I entered the cell of the Soviet prison which contained a single piece of furniture - a wooden bunk-bed. Immediately I was schocked ... On the top bed lays a corpse. The man's face was grey and the skin was so thin that one could see every bone under it. I was terrified ...

"Is it possible that 'They' got to be so cruel? That the degree of their mockery went to the point of putting the dead and alive together?" ... It was my first thought after I looked at his face.

In a while he opened his eyes. He stared at me with a look of expectation that I would bring him all kinds of news which he needed desperately.

"Are you new? I can tell by the way you walk, you move and you look". He said in his broken Russian. He lifted himself up and offered his thin hand for a handshake. "My name is Fritz Houtermans, a German ... a physicist ... a former member of the socialist party ... former emigrant from fascist Germany ...former director of Institute of Science in Kharkov ... former human being ... and who are you"?

Those were my first weeks in prison, so didn't forget how to smile. I shook his hand and introduced myself properly. "Professor Konstantin Feodossovitch Shteppa, professor of history of the University of Kiev".

"I am pleased to meet you" - Fritz Houtermans replied - "I think we will have much in common". "Do you smoke?" "No, sorry, I don't". "This is a pity. If you did I would hope to occasionally smoke your stubs. You know that I don't get any money from the outside. My family is abroad. I tried several times a hunger strike but they don't care, so I gave up, and you can see the only results", as he pointed out his skinny to distress limbs. "I also used to exchange rations for cigarettes" - He added with sorrow in his voice . - I answered - "My wife is allowed to send money every month, I don't know for how long, but as long as I will have money we shall share".

"You are a lunatic" - he said frankly - "nobody does this here. Everyone cares only for one thing, to survive. And one can not survive on what they give you ... Forget about this; just tell me all the news ... when did 'They' arrest you? What are the news from the West?»

Aglaya Konstantinova Shteppa goes on as follows(9b):

«This was the atmosphere under which my father and ... Fritz Ottonowich met. They stayed in one cell for several months. They became great friends. They hated one another on some occasions and loved on the others. They told to one another everything about themselves, about their families, their experiences, their knowledge. They just had to talk. They remembered the books they had read and the letters they received. They talked about good and bad, beautiful and ugly, about wonders and terrors. For that period of time they became one being, one soul. They had the same fears, the same agonies, the same distress and hopelessness. They both were in the hands of fate or free-will of the NKWD. Either one was completely sure that the other will never come out alive.

My father kept his word and shared his small allowance with Fritz Houtermans. This made him feel sore, but there was no other solution, so he accepted.

When one was called out for interrogation, the other one prayed. They prayed together. One Catholic, the other Ortodox, but the cell made them compromise. They shared their sorrows, they swore that if someone would come out alive, he would take care of the remaining families. Both had a wife and two children left behind. And there was very little hope for either one.

The fate brought them together and fate separated them. When they said good-bye it was meant forever. But will of God is not ours to know. They both came out of prison at different times and under different circumnstances. They met again in a free world and stayed friends. They wrote a book which they started in their thoughts while still over there, in a cell of NKWD building on Korolenko 33 in Kiev».

13. - A few other physicists' political troubles -

It would be very interesting indeed to find out what happened of all scientists gathered in 1933-37 at the Ukrainian Physical-Technical Institute in Kharkov. The information available, however, is fragmentary and therefore it is difficult to reconstruct a picture as complete as it would be desirable.

Besides a few cases already mentioned above (Chapter 10 and 12) I have found that I.V.Obreimov, one of Ioffe's earliest students, a specialist in crystals physics and first director of the Ukrainian Physico-Technical Institute, served ten years in prison, while his successor, Sinielnikov, in charge during Houtermans' times, was condemned to death and executed in 1937 or 1938.

Almost all other leading scientists of the Physico-Technical Institute in Kharkov also were in troubles in that period: Shubnikov, head ot the low temperature laboratory, Leipunski, head of the nuclear disintegrations laboratory, Gorksy, head of the X-rays department, Weissberg, head of the experimental low temperatures station, and Landau, head of the department of theoretical physics, were all arrested, while Ruhemann, head of the second low temperature laboratory, was deported.

Within a year about half of them were released except Shubnikov and Gorsky, who died in imprisonment[71]. Leipunski was soon back at work in the Institute while Alex Weissberg, who had been arrested months before Fissel (Chapter 10) was also handed over to the Nazis by the Russians after the Ribbentrop-Molotov treaty. Being Jewish, he was put into a concentration camp in Poland, from which he escaped and joined the Polish underground. He survived the war and lived in Paris later on. He published a book on his adventures: "Hexensabbat" [72] which also has been translated into other languages.

The adventures of Lev Davydovich Landau, one of the greatest Soviet theoretical physicists of our century, appear to be of the greatest interest. It is not proved that his arrest, on 28 April, 1938, was directly related to the purge of the scientists moved to Kharkov during the early thirties. What is sure, however, is that this short but dramatic interlude, belongs to the same great stream of events.

Many biographies of Landau have been published in Russian. A beautiful one, in English, is due to P.L.Kapitza and E.M.Lifshitz[(73)], who give us, respectively a general outline of his life and work and a short but substantial survey of his scientific production. Additional information about his way of working, relationships with other scientists and little amusing stories, which throw light on Landau's nature, can be found in the lecture given at CERN by F.Janouch in June 1978[(74)].

Landau was born in Baku in 1908, and being an "enfant prodige", entered the University of Baku in 1922. In 1924 he moved to the Department of Physics of the University of Leningrad, where he completed his course the day before his 19th birthday in 1927.

He became a post-graduate student at the Leningrad Physical-Technical Institute of the Academy of Science of the USSR, founded immediately after the revolution by Abram F.Ioffe (1880-1960)[(62)] who had been a student of Röntgen (1845-1923) and became well known for his work on solid state physics. In Leningrad at that time there was no senior scientist of note to set up a theoretical school. Paul S. Ehrenfest (1880-1933)[(75)] had lived and worked in St.Petersburg but later had been invited to the University of Leiden to take the chair which had become vacant on H.A.Lorentz's retirement. He left behind in Russia several talented students and some of them led the teaching of theoretical physics while Landau was at the University of Leningrad.

When 18 years old, Landau published his first paper on quantum mechanics which was just emerging. In 1928 he went abroad for the first time and passed one and a half year in various centers such as Göttingen (Max Born), Leipzig (W.Heisenberg), Berlin (M.v.Laue etc.), Cambridge (P.A.M. Dirac) and Copenhagen (N.Bohr). Everywhere he impressed everybody for his originality and talent and established lasting friendships.

In 1931 he returned to the Physical-Technical Institute in Leningrad which was the chief physical institute in the USSR and had grown considerably. Not long before it had branched out into other scientific centers all over the Soviet Union. The more important of these were three, in Tomsk, Sverdlovsk and Kharkov, the capital of Ukraina. The Ukrainian Physical-Technical Institute in Kharkov, directed by I.V.Obreimov, was mainly devoted to solid state physics and low temperature.

Landau went to Kharkov in 1932. He had already a fame as theoretical physicist and represented the main scientific attraction for the young German speaking physicists looking for a place where to live and work outside Hitler's political sphere.

Some of Landau's main works belong to his time in Kharkov: on the theory of second-order phase transition, the kinetic equations for particles with a Coulomb interaction, and the theory of the intermediate state in superconductivity.

In 1937 he left Kharkov to go to the Institute of Physical Problems created in Moscow in 1935 to allow P.L. Kapitza to continue research he had began in Cambridge.

A year later Landau was joined by E.M.Lifshitz (1 - 19), his closest student and friend, and his co-author of the many volumes "Course of Theoretical Physics"[(76)].

According to Janouch[(74)] the main reason for Landau to leave Kharkov for Moscow was a difference with the rector of the Kharkov University about Landau's method for testing physics students. According to some of the people present in Kharkov in those years, the heavy political atmosphere prevailing in the Ukrainian capital also had a weight on his decision.

The day after Landau's arrest, Pjotr Kapitza wrote a letter to Stalin asking for the reason of the imprisonment of a scientist, member of the Institute for Physical Problems of the Academy of Science, placed under his directorship. Such an arrest, Kapitza added in his letter, had the effect of increasing the already existing "gap" between the scientists and the Country.

Since this letter remained without answer, six months later Pjotr Kapitza wrote again, but this time the letter was addressed to Prime Minister Molotov. He wrote that the reasons of Landau's arrest were still unknown and that in the meantime he, Kapitza, had discovered helium superfluidity and the only man that could explain the new phenomenon was Lev Landau still kept in prison.

A few days later Molotov let Kapitza know that the Minister of Interior, Beria, had entrusted his deputy with receiving him. At the Ministry Kapitza was received by a group of three people: Beria's deputy and two generals who had on the table in front of them a very thick dossier. They showed it to Kapitza, told him that it contained the documents proving all crimes committed by Landau, and invited him to have a look at them in order to become aware of the kind of person he was taking into such consideration. They asked him also what were his reasons for trying to protect such a criminal.

Kapitza thought that in order to avoid to be beaten, Landau, as most other people in the same circumstances, had signed any possible declaration of guilt presented to him. Therefore he declined to look at the dossier saying that it was not his but their business to judge Landau's behaviour. He added that he wished to discuss with them which could be the reasons that had induced a scientist like Landau to commit all those crimes. His speach, interrupted here and there by discussions with the examining judges, went on for about three hours, during which he also declared to be ready to assume on himself the responsibility of Landau's behaviour if he were released.

Some time later Kapitza was called again and required to sign a declaration to take on himself the responsibility of Landau's behaviour. As a result Landau was released at the end of April 1939 just one year after his arrest.

Since then Landau used to say to his close friends that Kapitza had saved his life and that what he had done for him, required more courage than entering a tiger's cage.

Landau's achievements in science are very well known and cover an exceptionally wide range of problems. The acknowledgement of his scientific merits was shown by a number of academic awards in the USSR and abroad. He was elected Member of the Academy of Science of the USSR in 1946; awarded State Prizes on three occasions (1946, 1949 and 1953), the Lenin Prize in 1962; given the title of Hero of the Socialist Effort in 1953 for his scientific activity and for fulfilling government projects, twice awarded the Order of Lenin and several other medals. In 1962 he was awarded the Nobel Prize for his pioneer research in the theory of condensed state, especially liquid helium.

This interlude in Landau's life was not the only circumstance in which Kapitza gave proof of extraordinary civil courage.

When, in 1948, Stalin placed Beria at the head of the USSR Atomic Energy Organization, Pjotr Leonidovich Kapitza prepared a letter to Stalin containing a number of considerations, all indicating that such an appointment was a wrong decision. In the country, pointed out Kapitza, there were a number of scientists with all scientific, technical and human qualities desirable for such an important position. He added, as an example, that such an appointment was a mistake similar to that of nominating director of the Bolshoi Theater a man that did not know the musical notes.

Before sending the letter he called to his home three of his closest friends(76) and read it to them. All of them said that to send such a letter was suicidal and tried to convince him to desist from his intention. Pjotr Leonidovich, however, answered that it was his citizen's duty to intervene at the highest level. He, however, accepted a few minor changes and sent his letter to Stalin.

About one week later he received an answer from Stalin who thanked him for the criticisms and asked eventually for further information. Kapitza was very happy, showed Stalin's letter to his friends pointing out that this was a clear sign that times were changing and criticisms could be made when necessary.

Stalin, however, had passed Kapitza's letter to Beria who started a well planned action of attack on what Kapitza had made as chairman of the National Committee in charge of applied research on production and use of compressed gases. Some time later Kapitza was openly accused of having led the Committee to take wrong decisions which had brought considerable economic damages to the USSR.

He was removed not only from the Committee but also from the direction of the Institute of Physics Problems of the Academy and compelled to live in a private house (dacha) he had in the periphery of Moscow.

When, seven years later, in 1955, Beria was executed, Pjotr Kapitza went back to his activity and previous work.

Pjotr Leonidovich Kapitza (1894-1984) was born in Kronstadt, near St.Petersburg (later Leningrad). He had started his scientific career in Ioffe's section of the Electromagnetics Department of the Petrograd Polytechnical Institute, completing his studies in 1918. Here, jointly with N.N. Semenov, he prepared a method for determining the magnetic momentum of an atom interacting with an inhomogeneous magnetic field, later used in the Stern- Gerlach experiments that brought the discovery of the spin space quantization.

At the suggestion of Ioffe, in 1921, Kapitza went to the Cavendish Laboratory to work with Rutherford. In 1923 he discovered the linear dependence of resistivity on magnetic field for various metals placed in very strong magnetic fields. The importance of these results was fully appreciated in Cambridge in general and in particular by Lord Rutherford.

Kapitza was nominated Clerk Maxwell Student of Cambridge University in 1923-26 and one year later Assistant Director of Magnetic Research at the Cavendish Laboratory (1924-1932).

By the use of a special alternator, Kapitza was able to produce fields up to 300,000 gauss, and to carry out experiments showing the existence of new phenomena in conduction and magnetostriction. Since most of these phenomena are more pronounced at low temperatures, a hydrogen liquefaction plant was added in 1929 and in 1930 the Royal Society made a special donation of £ 1500 to enable a new laboratory to be built to house the original apparatus, together with a helium liquefaction plant. Kapitza was nominated at the same time Messel Research Professor of the Royal Society (1930-34) and Director of the new laboratory, named the Royal Society Mond Laboratory (1930-1934), which later was partially financed also by the Department of Scientific and Industrial Research.

It was characteristic of Kapitza that in 1930 he was not satisfied to take over existing designs of helium liquefiers but began immediately to work on the construction of a new type of liquefier which required no liquid hydrogen. This liquefier is an illustration of Kapitza's special technical gift, for it incorporates a piston type engine, which works down to the temperature of liquid helium. In summer 1934 Kapitza was able to start to carry preliminary experiments using strong magnetic fields combined with liquid helium temperatures.

It was just at that time that I had the great pleasure to meet him for the first time. At the beginning of July of that year 1934, Emilio Segrè and I went to the Cavendish Laboratory and remained there for almost two months. During our visit, mainly devoted to talking to people working at the Cavendish on nuclear reactions and in particular to those working on neutron physics, we visited the Mond Laboratory where we were received and shown around by Pjotr Kapitza. We still remember the great impression we had of his preparation, brightness and human kindness. Since then only many years later did I have the pleasure to see him again on occasion of the Pugwash Conference, held in Nice, in 1968[(77)].

In September 1934, i.e. less than one month after our visit to Cavendish and Mond Laboratories Kapitza left Cambridge for USSR to attend the Mendele'eff Congress.

He had visited his country almost every summer since he had started his research activity in Cambridge. During these visits he gave lectures and advises on the construction of new institutes.

It came, therefore, as a shock to his colleagues to learn, in October 1934, that Kapitza's return passport had been refused and that he had been ordered to begin the construction of a new laboratory in Russia.

Such news made an enormous impression not only in Great Britain but everywhere in Europe and United States.

The reasons underlying this action were given in the News Chronicle by the Soviet Embassy in London: "Pjotr Kapitza is a citizen of the USSR educated and trained at the expenses of his country. He was sent to England to continue his studies ... Now the time has arrived when the Soviet urgently needs all her scientists. So when Professor Kapitza came last summer, he was appointed as director of an important new research station which is now built in Moscow"[78].

Lord Rutherford in a letter to The Times of 29 April, 1935, expressed his concern about the whole story but shortly later he contributed to prepare an agreement concerning the sale to the Government of the USSR of the large generator for the production of strong magnetic fields, together with the associated apparatus and a duplicate of the helium liquefier to allow the continuation of Kapitza's work in Moscow. With the sum received new equipment was bought for the continuation of the work in Mond Laboratory[79].

When, in 1978, Leonidovich Kapitza was awarded the Nobel Prize for his discovery of superfluidity, all physicists throughout the world were very happy to see the recognition of his outstanding piece of work which emerges from a full life devoted, with great success, to the progress of science[80]. But everybody that had the great luck of meeting him or at least of knowing some detail about his life, was also satisfied or even moved, that such a recognition was given to a person endowed with such extraordinary human values.

14. - Finally out of prison! -

Let me go back, very rapidly, to when Fritz Houtermans was transferred to the Gestapo prison at Alexanderplatz in Berlin. After a few days one of the fellow-prisoners was informed that he would be released pretty soon. The news produced a considerable excitement among the prisoners and many of them gave the fortunate fellow messages for relatives and friends to be communicated as soon as he was out of prison. Fritz gave him the name of an old friend and colleague at the Technische Hochschule, with the recommendation simply to say on the telephone "Fissel is in Berlin".

After a first astonishment Dr. Robert Rompe[81], who received the message understood its real meaning and concluded "If Fissel is in Berlin, he sure is in jail!" He went immediately to see Max von Laue and communicated him the message he had received and its obvious interpretation. Laue went to look for Houtermans' name in the list of prisoners kept in appropriate offices, found where he was and went to visit him carrying some food and money. Immediately afterwards he took all necessary steps to get him out of prison, as he had made on many other similar occasions, and always with success.

Friedrich Georg Houtermans was finally out of prison! One of his first acts was to use a considerable part of the money he had received from von Laue for buying an expensive gold pen. Immediately he started to exchange letters with his wife Charlotte and his mother.

The attachment of Fritz to his family can be appreciated from the beginning of the letter he sent from Berlin to his mother, the 28 September 1940. The other letters are all of similar tone.

9th Letter, Berlin 28/9/1940

«My Mammy,

what a good luck: yesterday a postcard from Schnax, early this morning one from you and both your letters; I also live only waiting for the mail and often I come suddenly home in order to see if there is some letter. I have not written to you for such a long time, but I believe that you read my letters to Schnax and she those to you. I have no secret for you. I am so proud of Jan, that I tell to everybody he has received 185 at the intelligence test, but I am a little afraid that he may grow up in the conviction of his importance, and this could damage him, but he is welcome to have a little more than I.

You can not imagine how terrible it is that he grows without me, obviously only for what concerns me, because without any doubt he has all what he needs and this comforts me. But I loose his best age and that of Bamsi (Giovanna), if, however, the fate is that one loses his children to the world and to opinions and perspectives that one believed overcome since long time, to problems that do not appear any more in an acute form, they might, eventually, return to you only after many years, if they return completely. Perhaps one has to give them a lot, avoiding that they notice it, one should not explain to them everything, because, if they have some value, sometimes they will open the eyes, if they will need it; and if you are needed you only have to wait.

Anyhow, nature is very complicated. All intelligence tests do not make a man of somebody and, without talent one can perhaps obtain more than with talent, once one has understood that one should work with a high degree of efficiency, because everybody has good ideas, the problem is that of carrying them out.

For what concerns me, my mathematics, that I have found, has been such a decisive event that I nearly do not care of finding now everything in Euler, Gauss and others, so that there is almost nothing new in what I have discovered. Still I regret it. Anyhow my manuscript remains to my son and will always remain as a prove that the spirit can not be destroyed.

I have now a little book by Fermat and I can describe more or less how far I have arrived ...».

The remaining three quarters of the letter are devoted to a summary of the results he had found during the years of prison in the "theory of numbers" /28/ and information about aunt Lilly and the shock of his cousin's Tully suicide. She was a gifted pianist, became an anthroposoph(82) and emigrated to London.

There are only a few words about his future work in v.Ardenne Laboratory and some sentence about his reading again of books of Burkhard and of "my old Laotse".

At the end, referring to history in general, Fritz writes:

«The history of the Church and of China should be read. The Chinese empire and the Church are the wisest institutions that did ever exist and now I can permit me to say that one should not become too excited about a pair of hundred years. Everything goes in order ...».

Already in July, Fritz also sent a short scientific paper to the "Naturwissenschaften", which appeared in the issue of August, 1940, i.e. about one month after his liberation from the prison /19/.

In ten lines, under the title "Half-time of radio-tantalum", Houtermans reports the results of measurements completed in October 1937, the publication of which, as he says in a footnote, "has been postponed until now because of external reasons". He expresses his thanks to Herr Kurschatov(89) for help in the irradiation of the samples and to Fräulein Poluschkina for help in the measurements with counters. As address of the author he gives: Berlin-Charlottenburg 2, Uhlandstrasse 189, i.e. his private address.

The real meaning and aim of his short communication was to inform all his friends and scientific acquaintances that he was back in Germany and to his work.

Now clearly, there were new problems to be solved. First of all what to do with him? How to provide him with the means for survival?

The solution was found once more by von Laue, who recommended him to Manfred Baron von Ardenne for a research position in the "Laboratorium für Elektronenphysik" he had founded, in 1928, in Berlin-Lichterfelde Ost. Von Ardenne (b. Hamburg, 1907), a remarkable inventor and applied physicist, financed his private laboratory by means of contracts with industries, orders from ministerial departments and royalties from the many (600 in all his life) patents regarding radiocommunications, electronmicroscopes, mass spectrometers, etc.

Fritz entered the "Laboratorium Manfred von Ardenne" the 1st of November 1940(84) and immediately started to work very hard. Already at the beginning of May of the same year, appeared in the Archiv für Elektrotechnik /20/ a paper of Fritz, in collaboration with Karl-Heinrich Riewe, "On the action of space-charge on a beam of charged particles (shaped) by a rectangular window" and in July of the same year a second one by Fritz alone in the Annalen der Physik "On the energy required for the separation of isotopes" /21/. In this paper Houtermans extends and completes the previous work by Walcher (86) by discussing for various separation procedures, the enrichment factor, the transport per unit time of the desired isotope and the amount of energy required for its separation.

The subjects of these two papers clearly belonged to the typical research lines cultivated in von Ardenne's Laboratory.

A third paper, dated August 1941, was given by Houtermans to von Ardenne as an "internal report" of the Laboratory, but did never appear in print because of the importance of its military and political implications. In this report /22/, about 30 typewritten pages long, Houtermans discusses the possibility of producing energy by means of nuclear chain-reactions based on the fission of heavy elements. It is divided in seven sections concerning: (1) the general point of view; (2) the processes in competition with fission which can produce an undesired reduction of the neutron density; (3) the chain reaction based on fission produced by fast neutrons; (4) the chain reactions based on the use of thermal neutrons; (5) the possibility of realizing a nuclear chain reaction with thermal neutrons; (6) the chain reaction in the case of a finite volume of the system; and finally (7) the meaning of a chain reaction at low temperature as a source of neutrons and a device for producing nuclear transmutations.

The author quotes from the beginning the theory of nuclear fission published in 1939 by Bohr and Wheeler(87) as well as the experimental papers by the Paris and Columbia University groups concerning the emission of secondary neutrons in fission(88).

The most interesting points of Houtermans' paper are: (a) the conjecture derived from the theory of nuclear fission(87) that the nuclide of mass number 239 and atomic number Z = 94 (called by its discoverers plutonium: $_{94}Pu^{239}$, Chapter 16) should undergo fission even with slow neutrons; (b) the impossibility of a chain reaction with fast neutrons using ordinary uranium (U^{238} 99.3%; U^{235} 0.7%); (c) the advantage as fissionable material of the new nuclide (Pu^{239}) with respect to U^{235} because the use of the latter requires the separation of this rather rare isotope from the about one hundred more abundant U^{238}; (d) the possibility of constructing a bomb based on nuclear chain reaction, and finally (e) the possibility of constructing what today is called a fast breeder reactor as the most rational procedure for the exploitation of nuclear energy and the only one which allows the utilization of the whole energy content of natural uranium. In a conversation he had years later with Giuseppe Occhialini and Connie Dilworth (Chapter 18) Houtermans said clearly that all his considerations were "highly hypothetical" since he did not know the various cross sections and the fact that plutonium would show thermal fission was only derived from Bohr's and Wheeler's considerations. This lack of essential data explains the wrong view contained in his paper that a nuclear reactor would have to be run at low temperature (about - 100° C).

The paper was sent by von Ardenne to the Post Minister Ohnesorge who was well informed about the research activities carried on by the Baron and his collaborators. In 1934 von Ardenne had started to develop for the Post Ministry communication devices based on decimeter radiowaves, and in 1940 had received a substantial financial support for the construction at the Lichtenfelde Laboratory of a one million volts van der Graaf that was first used as a generator of fast electrons and, later, converted into a deuteron accelerator used for providing an intensive source of neutrons.

The same paper arrived also to the physicist Otto Haxel (b. 1909) who during the war was serving as a navy officer in charge of the supervision of the scientists working in the Uranverein (Chapter 16). Following the agreement of Haxel's superior, Admiral, and the assent of the leading scientists of the Uranverein, Houtermans' paper was not forwarded to the upper political authorities and the strategic command where it would have triggered a greater interest in the researches carried on by the physicists of the Uranverein. Apparently also the information channel passing through Ohnesorge did never bring this information on the "higher levels".

Other papers by Houtermans in that period concerned the experimental determination of the cross section of a few elements for thermal neutrons /23/ and the absolute determination of the number of neutrons emitted by a source /24/ by means of a variant of a method used by Fermi and myself in 1936[(89)], the essence of which was communicated by Fissel to me by letter in spring 1942.

15. - Shaken by world wide storms. Fritz's second family -

On June 22, 1941 Hitler attacked the USSR. The German troops started to enter the USSR territory and, in three weeks, advanced 450 miles on the central front, from Bialystok to Smolensk, in the North passed through the Baltic States and advanced towards Leningrad, and in the South moved towards Dniepr and Kiev, the capital of Ukraina. Kiev fell in German hands on September 16 while German units were already 150 miles farther. The battle inside and around the city came to an end only on 26 with the surrounding and surrender of a large contingent of Russian troops (665,000 according to the German command).

Shortly later in Berlin Houtermans was asked to take part in a mission sent by the German authority to Ukraina to collect scientific instruments from the Universities and other research Institutes. Fritz accepted and went over there. The trip lasted only a few days but was considered by most of his friends in Germany as well as abroad, as an act of collaborationism. Many years later I asked him why he had accepted to participate in such a mission, and he answered that he had thought to succeed in giving some help to his old Russian friends.

This is what appears also from what the daughter of Konstantin Shteppa wrote years later[(9b)]. Aglaya describes in her broken English, translated from Russian, Fritz's visit to Kiev as follows:

«I met F.O. (Fritz Ottonovich Houtermans) for the first time in 1941 in Kiev. It was a time of beginning of a war, and a very hard period of life for everyone.

F.O. came to Kiev, asked about my father, found out that he was there, and meeting was arranged. German occupation forces at that time proved already to be cruel, unjustical and based on force and terror. It seemed strange to meet a man, who was a German, but was not "Nazi" at all. It felt like a miracle. F.O. came to our apartment...We remembered so well father's story about friendship in prison. F.O. entered our house as a friend and was welcomed as a relative... The warm feelings towards him remained the whole future life in our hearts... There was no other person, who would prove to be a more sincere and true friend as F.O. It gave us a confort to know, that person, one like him exists. The whole world seemed to be not as bad. If there is one like him, there is a hope...

I shall never forget, how he managed to send a little food parcels to us, which contained a small amount of different food products, probably his own and his willing friends' rations. My mother used to shed tears over those packages...»

But which were the past adventures and present situation of Konstantin Shteppa and his family?

The «experience of that year and a half he had passed in prison, forms, as it were, a caesura in Shteppa's life. If until then he was adjusted to Soviet life -whatever his mental reservations and grievances - primarily absorbed in the pursuit of his scholarly work, henceforth he was forcibly "politicized", concerned with a search for answers to the bedeviling dilemmas of intellectual life under totalitarism»[(67)].

Shortly after Hitler's attack on the USSR it became clear that in a short while the German troops would occupy Kiev and Shteppa decided not to leave the city.

«In the first days after the departure of the Russians, he hoped for a better future. The remaining colleagues elected him Rector of the Kiev University. But the tragedy was soon to unfold. The Germans closed the University. Arrests, abuse, shooting, and hanging rapidly assumed proportions unprecedented even in the worst days of Soviet terror. The next two years were to cause Shteppa the greatest amount of anxieties and soul-searching, involving him in choices which most of his friends found impossible to understand and leading to espouse positions which were to net him public recrimination in later years. Having committed himself initially to the imperative of collaboration with any system that was willing and able to topple the Bolsheviks, he felt constrained to stick to his commitment. Having broken intellectually and emotionally with what Bolshevism stood for, and having by his wartime cut himself off from the Soviet cause, he left for Germany before the Red Army returned to Kiev in 1942» [(67)].

About one year before the departure of the whole family, Eric, the second child of Kostantin Shteppa, followed the dream of his life to see the Western World, and at the first occasion volunteered for labour in Germany.

He belonged to a group of young Russians that was brought to work for the German railroads. After months of life in a camp under very hard conditions, a few of them - including Eric - tried to escape, but were captured, severely beaten and put in a "straf-lager" (punishment camp).

Here Eric contracted a very serious lung illness but was forced to go to work in spite of his health conditions. He fainted, was declared dead and his death was communicated to his parents.

An unknown truck driver, who found him on the side of a highway, picked him up and delivered him to an hospital. Some time later, Eric succeeded in communicating with Fritz, who settled "his case" with the German authorities, brought Eric to live with him in Berlin for a few weeks and arranged his trip back home in Kiev.

In Berlin Houtermans met Doctor Paul Rosbaud, born in Austria and, at the time, the editor of the German scientific periodical Naturwissenschaften, published by the firm Springer-Verlag. They became close friends also because of their common scientific and cultural roots and cosmopolitan interests. Only forty years later the "secret life" of Rosbaud as a spy of the British Intelligence Service in Hitler's Germany, was revealed to the wide public[(89)].

Through Paul Rosbaud in Berlin Fritz met, in 1942, Charles Peyrou (b.1918), a young French physicist, who had studied in Paris at the École Polytechnique, under Louis Leprince-Ringuet[(90)]. He had been called in the French army at the beginning of the Second World War, had fallen prisoner on the Maginot Line in October 1941 and sent first to a war-prisoner camp, later to work in Berlin-Charlottenburg. Years later Houtermans and Peyrou became colleagues at the University of Bern (Chapter 21).

In the period 1942-43 Houtermans published three more papers, concerning the production of RaE through slow neutron capture in Bismuth /25,26/ and the photonuclear reaction in Beryllium /27/. These works were carried on in collaboration with Ilse Barts, a chemical engineer working in the von Ardenne laboratory. She was a beautiful slender girl with black hair and grey-green incredible eyes.

Fritz fell in love with her, obtained the divorce from Charlotte Riefenstahl, who was in the United States, an enemy country, and married Ilse in 1944. They had three children: Pieter, born in Gera (near Ronneburg, see below) in 1944, who became a mathematician at the University of Hannover, Elsa born in Göttingen in 1946, an artist living in Hamburg, and Cornelia, also born in Göttingen in 1947 and also living in Hamburg as an architect.

16 - An outline of the early development of applied nuclear energy in Germany -

In order to see in the right perspective some of the adventures and scientific activities of Fritz Houtermans in those years I feel it necessary to summarize very briefly the effort made in Germany during the war for developing the utilization of nuclear energy.

The information about the actual work made by German scientists is found in a four pages article by Werner Heisenberg "On the work in Germany towards the technical utilization of atomic energy"(91) and in the second of two volumes describing in some details the researches made in Germany during the war in the fields of biophysics, nuclear physics and cosmic rays(92). This volume consists of three chapters numbered from 5 to 7, since the first 4 form volume one. A part of chapter 5 (i.e. 5.2: Measurement methods for neutrons) is written by Houtermans /36/. The sections of interest for the present discussion are: "7.1 General Researches for the Preparation of the Construction of a Uranium Reactor" by W.Heisenberg and K. Wirtz, and, to a minor degree, "7.2. The Contribution of Fast Neutrons to the Neutron Multiplication in Uranium" by O.Haxel.

Detailed information of historical interest can be found in a number of books, a fairly long list of which is given in the bibliography (93-97). Each of them has been conceived and written from a different point of view, and reaches, of course, corresponding conclusions. One of the most objective presentation appears to be that given by David Irving(97), as one a priori would have expected because of the rather long time elapsed from the end of the war to its publication. But all the other books also have their clear, historically justified value and meaning. For instance, one voices the attitudes and views of all those that had suffered for years for the Nazi increasing domination in Germany and Europe, had fought for years against it, and finally saw their military supremacy smatched and their scientific-technical effort practically irrelevant in a field of paramount importance. Another book is a reaction to the generalization of the conclusions of the first one of those that had worked seriously, sometimes in very difficult conditions and always at a manpower scale too modest for the problems to be solved. Another one is the report of a high-rank officer who has the pleasure of showing how complex and unusual had been the duty, he had succeeded in bringing to full success.

Here we are not trying to express judgements or comparisons of this kind; our desire is to try to understand individuals, normal individuals, and even some, who, like Fritz Houtermans, did not completely conform flat normality.

Nuclear physics, as any other part of science, always had been the product of international effort, resulting from contributions put in common by any person or group of persons irrespective of their nationality, religious belief or political creed.

This was the situation until the time of the discovery of nuclear fission by Hahn and Strassmann at the Kaiser Wilhelm Institut für Chemie in Berlin-Dahlem at the end of 1938(98).

A number of important papers which were essential also for the applications of nuclear energy were published in the few months period that preceded the break of the international scientific unity.

In Sweden, Lise Meitner and her nephew Otto Frisch, already in January 1939, understood the physical nature of fission and deduced the very large value of the energy released in this process(99a). Similar conclusions independently were reached, in Berlin-Dahlem by Siegfrid Flügge and von Droste(99b), two young physicists of Otto Hahn's group.

Many experimentalists among which first of all Frisch(100) and Joliot(101), carried out experiments which allowed the observation of the recoiling nuclei produced in uranium fission and the measurement of their large energy(102). Bohr and Wheeler in Copenhagen developed the general theory of nuclear fission(86). Part of their results was independently obtained by many other authors(102).

Joliot and coworkers in Paris(103) and independently two groups of Columbia University, one led by Fermi, the other by Szilard(104), were the first to present experimental evidence for the emission, on the average, of a number of fast neutrons larger than 2 in each uranium fission process. This specific aspect of uranium fission was essential for opening the door towards the release of nuclear chain reactions. These were discussed in a paper by S.Flügge appeared in June 1939 in Naturwissenschaften under the title "Is it possible to exploit the energy contained in atomic nuclei for technical purposes?"(105) Flügge also developed something like a theory for the production of nuclear energy and gave hints as to how one should proceed in practice for building an "uranium machine". Similar considerations were developed also by other researchers at about the same time, but by then the authors refrained from publishing such results(106).

Two types of applications of nuclear energy were envisaged from the beginning: 1) an energy release under controlled conditions for various peaceful applications, and, perhaps, 2) the construction of a new type of explosive devices.

A practical solution of each of these two central problems still required, however, a number of ideas about many particular aspects, the majority of which also involved relatively accurate values of many nuclear constants which, by necessity, had to be found from appropriate experiments.

The most promising solutions of problem 1) clearly were based on the use of slow neutrons and therefore a neutron moderator was an essential ingredient of any machine: pile or nuclear reactor in the Anglosaxon terminology, Uran Brenner (Uranium burner) in German.

The solution of problem 2) requires the use of fast neutrons. It involves a detailed knowledge of the properties of uranium 235 which in great part can be found only after a successful solution of problem 1) was tested and a detailed investigation of the behaviour of a nuclear reactor was carried out.

The solution of problem 1) requires: (a) an accurate experimental determination of all the cross sections of the elements present in the reactor for any process produced by neutrons and in many cases their dependence on neutron energy; (b) large amounts of uranium, possibly enriched in the isotope 235, and anyhow of very high purity; (c) a moderator sufficiently effective, of high purity and in sufficient amount; (d) a design of the whole reactor which would reduce to a minimum the losses of neutrons taking place through their absorption in various parasitic processes inside the reactor or their escape through its surface.

Already in spring 1939 a group of foreign-born physicists centered on the Hungarian Leo Szilard (1898- 1964) started efforts in U.S.A. both at restricting publication and at getting government support in view of the possible military use of nuclear energy.

The first contact with the government was made by G.B. Pegram of Columbia University in March 1939[(107)]. In July L.Szilard and E.Wigner[(108)] conferred with Einstein, who wrote a letter to President Roosevelt explaining the desirability of encouraging work in this field because of the danger that German scientists could arrive first to constructing weapons of unprecedented power.

The President received the letter in the fall 1939 and appointed a Committee, known as the "Advisory Committee on Uranium" which represents, in an embryonal form, the U.S. Government interest in the field of applied nuclear energy.

First the "Metallurgical Laboratory" at the University of Chicago and later the Manhattan Project with the Los Alamos Laboratories at its centre, represented financial and organizational efforts of unprecedented dimensions[(109)].

On April, 1940, at the meeting of the Division of Physical Sciences of the National Research Council, the formation of a censorship committee was proposed for controlling publication in all American scientific journals.

The "Reference Committee", set up a little later that spring, was organized to control publication policy in all fields of possible military interest, with special regards to papers on uranium fission.

Already a few months before, however, Leo Szilard had taken the initiative of convincing scientists in United States, Great Britain and France to take action for preventing the publication in scientific periodicals of any result concerning possible military application of nuclear energy.

Szilard's advice was followed by the great majority of the physicists working in U.S.A. and Great Britain, while in France Frederic Joliot did never answer to the letter he received from

Szilard on early February 1939 and continued the publication of his results(110).

The reason of this attitude has been analyzed, but, perhaps, not deeply enough. A circumstance that should not be forgotten is that at the beginning of 1939, in the French communist press as well as in the German (and Italian) fascist information mass media, news had started to appear which, in some way, were paving the way to the acceptance of the Hitler-Stalin Non-Aggression Pact, signed in Moscow a few months later.

Sometime a joint letter sent by Paul Harteck, professor of physical chemistry at the Hamburg University(111) and his assistant Dr. Wilhelm Groth, to the Ministry of War, on April 24, 1939, two days after the publication of the Paris physicists' paper in Nature(103), is quoted as the first solicitation to the German Government to start an action on the field of nuclear energy in view of possible military applications. The authors outlined, in simple terms, that, due to the discovery of Hahn and Strassman, it had now, in principle, become possible to produce a new type of powerful explosives. After stressing the importance of Joliot's results(103), they continued that while in USA and Britain great emphasis was placed on research in nuclear physics, the subject had been neglected in Germany. One thing above all was important: «That country which first makes use of it has an unsurpassable advantage over the others»(112).

Information about this new possibility, however, reached at about the same time the Government level through a completely different channel. Immediately after the publication of the results on the emission of secondary neutrons in fission(103), Professor Wilhelm Hanle presented a short paper to the Physics Colloquium in Göttingen on the employment of uranium fission in an energy producing reactor. After the Colloquium, Haule's chief, Professor Georg Joos told him that this was a development which they could not keep for themselves and at once wrote a letter to the Reich Ministry of Education. The ministry acted promptly. They deputed Professor Abraham Esau(114), former professor of physics at the University of Jena and at the time President of the Physikalisch-Technische Reichsanstalt (Reich Bureau of Standard) and head of the Ministry's Reich Research Council, Physics Section. He drew up a short list of scientists to attend the first conference, headed of course by Professor Otto Hahn. Hahn was happy to withdraw: he had a previous lecture engagement in Sweden, and deputized Professor Josef Mattauch(115), recently arrived in Dahlem from Wien to take the place of Lise Meitner.

The meeting took place in all secrecy on April 29, 1939, at the Ministry's building at Unter den Linden in Berlin(116). Dr. Dames, head of the Ministry Research Department, voiced his disquiet at the way in which Hahn had been able to publish his vital discovery to the world. Mattauch reacted with vehemence to this critical remark to the behaviour of his new chief and the reproaches were not repeated. After Joos and Hanle had outlined the stage reached by nuclear research abroad and in Germany, the practicability of building a "uranium bureau" (or uranium reactor) was examined. Also the possibility of constructing uranium bombs for war goals was explicitly mentioned.

A few considerations were also formulated by Esau as a result and conclusion of the discussion: the first was to secure at once all available uranium stocks in Germany. A second one was that the most important nuclear physicists in the country should be co-opted to a joint research group under the overall administration of Esau, and so on. The last recommendation was that the participants were obliged to keep the meeting secret.

According to a typewritten memo signed by S.Flügge(117), the same evening Mattauch told Flügge even the details of the discussion and Flügge that had not participated in the meeting and did not feel bound to secrecy, decided immediately of intervening and preventing by means of a publication that "this possibility were withdrawn from the eyes of publicity". Thus he wrote "the review article" that appeared in Naturwissenschaften of June 9, 1939(105) and used this opportunity, as a representative of the Deutsche Allgemeine Zeitung, for publishing an extensive article on the same subject on this daily newspaper on 15 August 1939.

This was the beginning. To summarize the main lines of what happened later, I will follow Heisenberg's report(91) which is an accurate summary of the project as its participants viewed it in retrospect after the war:

«Around the outbreak of the war news arrived in Germany from North America, that American military organs had provided financial means for developing researches on the problem of atomic energy. In view of the possibility that from the Anglo-Saxon side atomic weapons were developed, a Forschungstelle (Research Establishment) was created within the Heereswaffenamt (Army Ordinance Department) and put under the leadership of Colonel Eric Schumann, who was in charge of proving the possibility of the technical utilization of atomic energy.

Already at the end of September 1939, many nuclear physicists and experts in various connected fields were ordered to collaborate to this research program and Kurt Diebner was committed with its administration.

This Uranunternehmung (Uranium Enterprise) or Uranverein (Uranium Association or Club), as it was commonly indicated, consisted of four main research groups.

I. A group led by Werner Heisenberg at the Kaiser Wilhelm Institut für Physik in Berlin-Dahlem with ramifications at the Institut für Physik of the Leipzig University; this group included K. Döpel, C.F. von Weizsäcker(118) and others.

The Kaiser Wilhelm Institut für Physik was located nearby the Kaiser Wilhelm Institut für Chemie, where Hahn and Strassmann had made their discovery in 1938 and where Hahn and collaborators continued during all the war their work on the chemical identification of fission products, which was regularly published in scientific journals.

II. A group led by Paul Harteck, at the University of Hamburg, and which included P.Jensen, W.Groth and K.Beyerle.

III. A group led by Walther Bothe(119) at the Kaiser Wilhelm Institut in Heidelberg, which included Bothe's pupils: W.Gentner and H.Maier-Leibnitz(120).

IV. A group led by General Karl Becker, chief of the technical Services of the Wehrmacht and professor of "Explosive Materials" at the Technische Hochschule in Berlin. This group was located at Kummersdorf (near Berlin, a military place for experimental ballistics) and was the only one with a military structure and attitude.

Connections of these four groups also existed with smaller groups or individuals working in other laboratories or universities: for example with G. Stetter and K. Lintner at the University of Wien and Houtermans working in Berlin, first at the von Ardenne Laboratory, later at the Physikalisches Institut of the Physikalisch-Technischen Reichsanstalt.

For decision of Schumann the Kaiser Wilhelm Institut für Physik in Berlin-Dahlem became the scientific centre of the whole Uranverein and therefore was transferred from the Kaiser Wilhelm Gesellschaft, a private foundation established in 1911, to the Heereswaffenamt. Its director, Peter Debye (1884-1966), of Dutch nationality, had either to change nationality or go away. Debye offered his resignation, and went to U.S.A. (1940). He was replaced by Werner Heisenberg, who until then had been professor of Theoretical Physics at the University of Leipzig».

It is impossible to summarize here the many results obtained by such a large number of competent people through years of work, from Autumn 1939 to the end of April 1945 when the U.S.A. troops occupied Heigerloch where the "Burner No8", i.e. the more advanced subcritical reactor constructed by the German scientists could have been brought to operate at zero power by the addition of about 50% more uranium and heavy water[(91)].

The structure and the direction of the Uranverein went through two changes. The first one took place the 26 February 1942 when the results obtained until then were reported to Bernhard Rust, a Obergruppenführer of the S.A.[(121)], for years Hannover Gauleiter[(122)] and Hitler's friend since 1920, and at the time Reich Minister for Education. Also a few leaders of war oriented research were present.

A number of essential cross sections had been measured (W.Bothe, W.Gentner, H.Maier-Leibnitz, 1940), the necessity of using U^{235} pure or almost pure for the construction of bombs had been recognized, the possibility of using heavy water (D_2O) or pure graphite as moderators had been theoretically clarified (W.Heisenberg, K.Döpel 1939), the minimum dimensions of a reactor had been explored, the neutron absorption by the U^{238} resonances had been studied theoretically (S.Flügge) and the necessity of separating the uranium from the moderator by shaping the fissionable material in layers or blocks had been recognized already in 1939 (P. Harteck).

The production of very pure uranium oxide (U_3O_8) had been studied and entrusted to the Auer-Gesellschaft and the fusion of the produced metallic uranium powder to the Fa. Degussa, Frankfurt/M. These developments went on very successfully.

The enrichment of the important isotope U^{235} by using Clusius-Dickel separation tubes with uranium hexafluoride (UF_6) turned out to be impossible (1941) while such an enrichment by means of a centrifuge gave promising results about one year later (Harteck, Groth, Beyerle, 1942).No attempt, however, was made of beginning such an isotope enrichment at a large scale.

In summer 1940 Weizsacker mentioned the fact that the nuclei of mass 239 and atomic number 93 (Neptunium) and 94 (Plutonium) produced by neutron capture by the U^{238} present in a reactor, should show a behaviour with respect to fission similar to that of U^{235}. This is one of the ideas contained in the Internal Report of Houtermans from von Ardenne Laboratory /22/ that we have already discussed in Chapter 14. In Heisenberg's paper(91), however, only Weizsäcker is mentioned in this connection.

The neutron absorption of technically pure graphite was determined by the Heidelberg group (W.Bothe, P.Jensen) who arrived at the conclusion that even extremely pure carbon had not the properties required for its use as a moderator in a nuclear reactor. This was a mistake of unclear origin: it could be due to impurities (such as H_2 or N_2) present in the graphite and not taken in due consideration, or to imperfections of the theory used for interpreting the experimental results.

All the German effort was concentrated on the use of heavy water (D_2O) as moderator, the production of which took place at the Rjukan plant of the Norsk Hydro in Norway where the production was increased by a factor between 10 and 20 by improvements introduced by P.Harteck, H.E.Süss(123), J.H.Jensen(124) and K.Wirtz.

Also a number of subcritical structures of increasing dimensions and improved design, all composed of heavy water and uranium (U_3O_8 uranium at the beginning, later metallic uranium) were constructed and tested starting from the first one (K.Döpel) in 1940.

At the meeting of February 1942, the Reichsminister Rust decided the transfer of the Uranverein from the Heereswaffenamt to the Reichsforschungrat (Reich Research Council) and appointed as new director Esau, then President of the Physikalisch-Technische Reichsanstalt.

Shortly later, the 6 June 1942, the results obtained by the Uranverein were reported to Albert Speer, Reichsminister of Armaments and War Production. The situation was as follows:

«There was a clear proof that the technical utilization of atomic energy in a uranium reactor was possible. In addition one could expect that in such a reactor one could produce an explosive for atomic bombs».

No researches, however, had been pursued on the technical aspects of the atomic bomb, for example on its minimal dimensions. More emphasis was given to establishing whether the energy liberated in a nuclear reactor could be used for the operation of machines, «since it appeared that this goal could be reached more easily and with limited (financial) means.....»(91).

After this meeting Speer decided that the project, carried on until then at a small scale, had to be strengthened. The only feasible goal could be the construction of a uranium reactor for the production of energy to be used for the operation of machines. From this moment on experts of the German Navy took part in the discussions in view of the possible use of nuclear energy for the propulsion of war ships.

The Kaiser Wilhelm Institut für Physik was given back to the Kaiser Wilhelm Gesellschaft and the leadership of the whole Uranverein passed from Esau to Gerlach(125), who was in charge of the "Sparte" (Section) Physik in the Reichsforschungrat.

Important progresses were made in various directions but the difficulties of the German industries had started to be considerable because of the raids of the allied air forces. The production of uranium and of blocks of fused metallic uranium proceeded with considerable difficulty.

In spring 1943 the electrolytic plants of the Norsk Hydro in Norway was destroyed by a parachuted command. In October of the same year the destruction was completed. At that time the Uranverein had at its disposal 2 tons of heavy water and about 2 tons of metallic uranium.

On February 15, 1944 the Kaiser Wilhelm Institut für Chemie in Berlin-Dahlem was destroyed and the group working at the nearby Institut für Physik was in great part displaced from Berlin to Hechingen (now Baden-Württenberg). On order of Gerlach a cellar excavated in the rocks, was rented in the village of Haigerloch and the more advanced subcritical structure was mounted there. It consisted of 1.5 tons of D_2O and 1.5 tons of uranium, surrounded by a blanket of 10 tons of graphite and equipped with cadmium rods for its control (K.Wirtz, E.Fischer, F.Bopp, P.Jensen, O. Ritter). This structure gave a multiplication by 7 of the neutrons emitted by a source placed at its center. «The material at disposal at Haigerloch was not sufficient for reaching the instability point. Probably the addition of a relatively small quantity of uranium was enough but this could not be done because the transports from Berlin to Haigerloch could no more arrive. On the 22 of April [1945] Haigerloch was occupied and the material was seized by the American troops»(91).

The group of physicists of the Uranverein (Bagge, Bothe, Heisenberg, Weizsäcker, Wirtz and others) including Hahn who had not participated in these activities, was arrested by the Allied Forces and kept prisoners in Britain until the spring 1946(126). They were extremely surprised when they learned on August 6, 1945, that the first atomic bomb had been dropped on Hiroshima. They could not imagine that somebody else could have been so much ahead of them!

Asked later, even from English or American sides, about the reason one had not attempted also in Germany the construction of atomic bombs, Heisenberg said(91): «The simplest answer one can give to this question sounds as follows: because this enterprise could not have succeeded before the end of the war».

In U.S.A. the group led by Fermi used from the beginning graphite as moderator and the chemical industry succeeded in providing increasing amounts of this material of gradually higher purity. As a result the 2nd of December of 1942, at the University of Chicago the first nuclear reactor was operated successfully.

One should also recognize that the design of the many subcritical systems that the Chicago group constructed and tested before autumn 1942 appear simpler and more suited for a clear interpretation of the experimental results than the few corresponding heavy water-uranium devices developed by the Uranverein.Perhaps the experiment carried out with the Burner No.8 could have been successful if its form had been a sphere and not a cylinder.

I have mentioned above that during summer 1940 both Houtermans and Weizsäcker suggested that the nucleus that today we call plutonium 239 (Pu^{239}) should show the same properties of U^{235} i.e. it should undergo fission under the action of slow neutrons.

Similar conclusions were reached by Bretscher in Great Britain and Joliot in France(98) and by Fermi and Segrè in United States, who around Christmas 1940 discussed the possibility of producing Pu^{239} with the Berkeley cyclotron in a quantity large enough for establishing its nuclear properties, in particular its fission cross section for slow neutrons(127). During the first months of 1941 Kennedy, Seaborg, Segrè and Wahl (128) succeeded in preparing, by means of the Berkeley cyclotron, about one microgram of this new isotope and to show that, as expected, it undergoes fission with slow neutrons.

Such an experiment could not have been carried out in Germany where the Heidelberg cyclotron (W.Gentner) was ready for a first test only in 1944.

From Einstein's letter to Roosevelt as well as from the beginning of Heisenberg's article, mentioned above, the main argument for a secret development of nuclear energy in view of its peaceful as well as military applications was, for both belligerents of the Second World War, the fear that the "enemy could arrive first".

In this psychological struggle, unavoidable in this as well as in any other fought or cold war, Fritz Houtermans found the way to inject his contributions.

Robert Jungk, in his book on the history of atomic bomb(129) mentions Houtermans at many points and in particular at page 114 where he wrote:

«In 1941 the chemical expert Professor Reiche, who had escaped from Germany a few weeks before, arrived at Princeton. He brought a message from Houtermans to the effect that the German physicists had hitherto not been working at the production of the bomb and would continue to try, for as long as possible, to divert the minds of the German military authorities from such a possibility. This news was passed on from Princeton to Washington by another scientist who had emigrated to America, the physicist Rudolph Ladenburg. But it does not seem ever to have reached those actually engaged on the atomic project ... ».

In 1943 the anxiety of the scientists working in the Metallurgical Laboratory, the Chicago phase of the Manhattan Project, was increased by another message they received from Fissel: "Hurry up - we're on the track" was the warning to his colleagues(130). I tried to get more information by correspondence with the author of the book containing this news, who suggested to contact E.Wigner. But also Professor Wigner did not add much, except that the cable was sent from Switzerland.

Apparently this message had an influence on the engagement of the scientists working in the Manhattan Project. It was clearly inspired by the deep antifascist attitude of Houtermans, but seen today, in retrospective, it appears very naive and unjustified or at least inspired by unjustified presumptions.

17 - From Berlin to Ronneburg -

In 1944 Houtermans went from von Ardenne Laboratory to the Physikalische Technische Reichsanstalt (PTR) in Berlin, which from 1933 to 1943, was under the presidency of Johannes Stark (1874-1957).

When von Paschen retired from the presidency of the PTR, Wilhelm Frick, minister of the interior, named Stark in his place, in spite of the fact that von Laue had informed him that the great majority of the scientists rejected Stark's candidacy(131).

At the PTR Houtermans was expected to take up again research work on neutron physics, but because of the frequent bombardment of Berlin by the Allied Air Forces, during the same year 1944, the Department of Physics of the PTR was transferred to Ronneburg (in Thüringen, about halfway between Leipzig and Weimar).

Fissel who also had moved to Ronneburg with Ilse, started to do some research work. The paper "On a phenomenological relation between the strength of a source of neutron and the maximum density of slow neutrons in a hydrogenous medium" /29/, received by the Physikalische Zeitschrift on June 18, 1944 from the PTR, was probably sent from Berlin shortly before Fritz moved to Ronneburg.

The considerations presented in this note are in some way connected with certain aspects of the problem he had treated in the previous paper on the absolute determination of the intensity of a neutron source /24/. The semiempirical relationship found by Houtermans is interesting but remained outside the main lines of development of neutron physics because it was not very useful.

In Ronnenburg Fritz felt rather unhappy because the tobacco war ration was not sufficient for the chain smoker he was. He then wrote, on PTR paper, a letter to the "Land Zigaretten Fabrik" in Dresden saying that he was carrying on research on the absorption of light by "fog and smoke" for which about one kilogram of tobacco powder was needed. In consideration of the national interest of these researches, please send it to PTR Ronneburg, care of Dr. F.G.Houtermans.

The answer was prompt and positive! Pretty soon Fritz received a large package containing a mixture of tobacco and tobacco powder that had been collected from the waste of cigarette machines. Fritz had to work very carefully for hours and hours to separate, by means of meshes, the tobacco from the powder. Finally he succeeded in recovering almost a kilogram of tobacco that he started to smoke with the greatest physical and moral pleasure.

In a few months this first shipment came to an end and Fissel thought to repeat the game. He wrote that the researches on "fog and smoke" had undergone a satisfactory progress but their termination required further experimentation and, therefore, a second tobacco shipment from Land Zigaretten Fabrik.

He forgot, however, to note on this second order, that the bill was to be sent to PTR-Ronneburg. So, finally, it came into the hands of the president who did not like Fissel and now saw a welcome opportunity to fire him.

Once again the problem was that of finding the way to allow the survival of Fritz, Ilse and their first child. Fritz went to see Heisenberg and Weizsacker and asked their help. They immediately arranged a meeting with Walter Gerlach who was "Der Beauftragte (authorized representative) des Reichsmarschalls für Kernphysik".

In the meeting the decision was taken to send as soon as possible Houtermans to work in

Göttingen with Hans Kopfermann (1895-1963), a pupil of James Franck, who had been assistant of Haber in Berlin, had worked in Copenhagen, and was well known for his spectroscopic work on hyperfine structures of optical spectra and for his clear convictions against national socialism.

Gerlach provided a letter of authorization for the railroad ticket and thus Houtermans with his second wife Ilse and the baby moved to Göttingen. It was spring 1945.

Once more in Göttingen there was a great concentration of physicists, since, for various reasons its famous University was considered a safer place than any institution in Berlin or Leipzig. Beside Kopfermann there were Richard Becker, L.Prandtl and W.Walcher.

A new problem for Fritz Houtermans arose shortly after his arrival in Göttingen. He was a 42 years old tall man and could be mobilized at any moment. To avoid this it was necessary to prove that he was involved in a research program of importance for the defence of the Reich and with a high priority. Kopfermann wrote the necessary letters and his pupil Wolfang Paul[(132)], by bicycle, took them to a village in the Harz Mountains, about 50 kilometers from Göttingen. Here was the administrative section of the "Osenberg Organization" which was in charge of the problem of the utilization of scientists in appropriate work, bringing them back from the front, whenever necessary (and possible). As a result of this action Houtermans became formally a collaborator of Richard Becker, who was in charge of a research program concerning the demagnetization of ships, a problem of considerable interest in connection with the magnetic mines used in the war on the seas. But the war came to the end in Europe on May 7, 1945 with the unconditional surrender of Hitler's Germany, and Houtermans' scientific activity could soar freely on various subjects.

One day, before the end of the war, walking in the street in Göttingen, Fritz met, with the greatest surprise, Konstantin Shteppa. After he had left Kiev with his family and settled in Plauen, the advance of the USSR troops had forced him to escape from what became the Russian zone.

He had just arrived in Göttingen, had left his family at the railroad station and was looking around for his dear friend. Fritz was delighted to help the Shteppas to settle in West Germany. The authorities did not allow them to stay in one place more than three days, outside from the camps for displaced foreigners, but these camps offered only one solution: repatriation.

The Shteppas were wandering from town to town, absolutely hopelessly, to find a place to stay. But Houtermans obtained the permit for the family to remain in Göttingen, found rooms where they could live and obtained a job for Aglaya in a factory. Aglaya was about twenty years old, she had married about two years before, but her husband had been killed during the Kiev battle.

Konstantin Shteppa had obtained the German citizenship and since Eric was still under age, he had also become a citizen of Germany. He was drafted into the German army, shortly later was captured by the Russian, declared a collaborationist of Germany, and sent in a re-education camp. Only after the death of her father (1958) did Aglaya learn through an office of the Red Cross in Turkey, that he was living in Siberia working as a stone cutter. Some time later he started to correspond with his mother and after a few years succeeded to go to live in Rostov.

In West Germany Aglaya met an American soldier, Bill Corman, married him and moved to the USA where Bill became an architect.

Shteppa and his wife also moved to the USA in 1952, and for the next three years he was associated with the "Research Program on the USSR", a subsidiary of the East European Fund, established to help refugee scholars return to scholarly work under American standards and conditions. At the time of his death in 1958, the manuscript of a book "Russian Historians and the Soviet State" had been completed. It was published shortly later with the assistance of a grant from the Research program on the USSR(133). According to a well recognized reviewer "in term of factual content and documentation this is the most substantial study which has yet been made of the relationship between historical scholarship and politics in the Soviet Union."

18. - In Göttingen again -

During the seven years spent in Göttingen, from 1945 to 1952, Houtermans worked at the Physikalisches Institut der Universität and the Max Planck Institut, directed by Werner Heisenberg.

This was one, actually the next to last, of his most productive periods. Beside his old theme of research, neutron physics, Houtermans tackled a few subjects completely new to him, and started to develop a deep interest in the methods currently used for determining the age of the rocks.

The papers concerning neutron physics /31/ /36/ /38/ /45/ are a natural continuation of his previous work at Kharkov and Berlin. In paper /31/ Fritz presents an experimental determination of the Beryllium cross section for the (n, 2n) process produced by the neutrons of a Polonium plus Beryllium source. The report /36/ devoted to "Measurement Methods for Neutrons" is Houtermans' contribution to the presentation made by the German physicists of their work during the war(92b). It has been already mentioned in Chapter 16. The first of the two papers in collaboration with his pupil Martin Teucher /38/ is a development of a research line already tackled by Fritz alone in a previous publication /24/, while the second one /45/ concerns the anelastic scattering undergone in lead by the neutrons of a Polonium plus Beryllium source. All other papers, however, concern research problems completely new to Houtermans.

A paper of 1947 is in collaboration with Jensen, who in 1963 shared with Maria Goeppert-Mayer the Nobel Prize for Physics for their discoveries concerning the nuclear shell structure. Under the title "On the Thermal Dissociation of Vacuum" /34/ Houtermans and Jensen discuss some of the consequences of Dirac's hole theory. They notice that the density of electron-positron pairs present in a "free from particles" volume in thermal equilibrium is small at low temperature ($kT \ll m_ec^2$), but increases rapidly with increasing temperature, reaching, for $kT \cong 137\ m_ec^2$ ($T \cong 10^{11}K$), values larger than one pair per elementary volume ($v_o = 4\pi/3(e^2/m_ec^2)^3$), thus giving rise to a "close packed structure of elementary particles" in space. They point out that this effect had been neglected in all discussions about many cosmological problems on the origin and abundance of the elements in the Universe, where temperatures of this order of magnitude had just started to be considered.

They add that at these temperatures also neutrinos and mesons [muons] should be present in considerable amounts and raise (but do not discuss in detail) the question of thermal equilibrium under these conditions.

The idea of an abundant presence of electron pairs in thermal equilibrium remained, in its essence, in present treatments of the evolution of the Universe during the first seconds after the initial big-bang(134), but any reference to the classical radius of the electron has disappeared. We know that this length has nothing to do with the structure of the electron, which is a "point-like particle" or at least an object of linear dimensions hundred times smaller than e^2/m_ec^2.

In the same paper the authors also consider the polarization of the real electron-positron pairs around any electric charge and its temperature dependence, a problem clearly related to the polarization of vacuum, which usually is computed today at low temperature, where the virtual positron-electron pairs give by far the dominating contribution.

In a paper of 1946 /30/, in collaboration with Pascual Jordan, one of the founders of Quantum Mechanics, the authors tackled the possibility of observing experimentally a very slow variation in time of the beta-decay constant. The possibility of phenomena of this kind was first pointed out by Dirac in 1937(135) who had noticed a remarkable numerical coincidence among fundamental quantities in physics: the time t elapsed since the beginning of the Universe, expressed in terms of a unit fixed by the constants of atomic theory, say the unit e^2/m_pc^3, turns out to be of the order 10^{40}, i.e. of the <u>same</u> order of magnitude as the ratio γ of the electric to the gravitational force between two protons.

Dirac suggested such a coincidence to be due to some deep connection in Nature between cosmology and atomic theory: if such a law exists and holds not only at the present epoch but for all times, then, for example, in the distant future, when the epoch has become 10^{50}, also γ should be of the same order of magnitude.Since it is rather reasonable to assume that the atomic constants c, e^2, h and m_p remain constant, such a variation of γ with time requires a decrease of the gravitational constant G proportional to t^{-1}.

A further study of cosmology leads to the appearance of other very large dimensionless numbers. These all turn out to be of the order of 10^{40} or, sometimes, 10^{80}. By a natural extension of the forgoing ideas, Dirac suggested all those numbers of the order of 10^{40} to increase proportionally to time t, and all those of the order 10^{80} to increase proportionally to the square of t.

These ideas of Dirac were the point of departure of a few papers by P.Jordan(136) who developed his cosmological considerations by incorporating Dirac's suggestion within the frame of the theory of General Relativity.From the start Jordan points out that while there are observational data in favour of the "cosmological constancy" of dimensionless quantities such as m_p/m_e, as well as of the time interval e^2/m_pc^3 and the coupling constant of nuclear forces, nothing can be stated about the constancy of the beta-decay coupling constant. He also noticed that an experimental test of a possible time variation of the last coupling constant requires measurements of a precision at the limit of the technology of the time.

This problem is discussed in detail by Houtermans and Jordan, who devote the first section of their paper to the derivation of the law of beta-decay and outer-electron capture on the basis of Dirac-Jordan's cosmological theory and to the discussion of the possibility of an experimental test of the time variation of the beta-decay constant. In the second section they deal with the hypothesis proposed by Weizsacker that the A^{40} present in the atmosphere is produced by outer-electron capture in K^{40}. They estimate the half time of this process on the assumption that the K^{40} present in the Earth crust is the only source of the A^{40} observed in the atmosphere. As a result they deduce, from the K^{40} half-life ($1.42x10^9$ years), the ratio k of outer- electron capture to total decay which is obtained as a function of the age of the atmosphere. In the third section the authors discuss the consequences of the decay law with respect to age determination by the Rb-Sr method. Finally they examine the constancy of the beta-decay constant using the end point of the Th-series in specimens of different geological ages.

Two other papers refer to surface phenomena taking place in the electric discharge /47/ and the electrolytic separation /56/ and two others /50/ /52/ report on the experimental determination of the isotopic shift of the spectral lines of the lead isotopes 206, 208, 210 (RaD) which indicate, in agreement with the nuclear shell model, that, for a number of the constituent neutrons N = 126, there is a rather large variation of the nuclear volume, as it had previously been observed for N = 82.

The remaining papers of this period regard the mean life and use of various radioactive substances for the determination of geological ages and the pertinent experimental techniques: the half-life of Uranium and the Pb/U method /32/ /33/ /35/ /46/ /49/ /53/, the half life of RaE /51/ /55/ /57/, which was conveniently used by Houtermans in a modified Pb/U method, the mean life of Potassium 40 /42/ and the decay of Rubidium /37/ /39/ /40/ dealt with in papers in collaboration with Otto Haxel.

The discrepancy found between the values of the half- life of Rb^{87} obtained from geological data and by direct measurement of the decay electrons appeared to indicate a substantial difference between the time scale of alpha- and beta-decay in the sense postulated by the cosmological theory of Jordan(136).

Houtermans and Haxel observed /40/, by means of two counters in coincidence, that the Rb^{87} nucleus emits simultaneously two electrons per disintegration: one of nuclear origin, the other due to the almost hundred per cent conversion of a gamma ray. By taking into account only the electrons of nuclear origin the discrepancy mentioned above is eliminated and the time scales of alpha- and beta-decay of U^{238} and Rb^{87} turn out to be in agreement.

I will not try to discuss here a few interesting developments of the Pb/U method introduced by Houtermans in the papers of the Göttingen period mentioned above, because he continued also at later times to pursue this line of research and we will come back to it in chapter 21.

In this section I like to mention his interest for the use of nuclear emulsion technique in problems of geology and mineralogy /44/ /48/ and in particular his paper with Buttlar "On the Determination by the Photographic Method of the Activity Contain of Mangan Module of Deep-Sea" /41/.

In order to become acquainted with this technique Houtermans went to Bruxelles where Occhialini had created one of the most advanced centres of research in this field. The Belgian physicists, still very sensitive on the subject of "war time collaborators", had heard various rumours about Houtermans: that he had been a Nazi spy in Russia and had collaborated in the German war effort, and in particular in atomic bomb research. There was even the rumour that he had deposited at the Post Office in Berlin-Charlottenburg the application for a patent of a fissionable device of military interest.

In order to understand what he had actually done, at Fritz's arrival in Brussels, «Occhialini started asking questions, but lost patience, went off to a cinema and left me (Constance Dilworth) to carry out a "third degree"». Dilworth took notes during the conversation, which were summarized in one and half typewritten pages already mentioned in Chapter 14[(137)].

It came out that for priority reasons, Fritz wished to deposit a document about some of the ideas he had developed in the internal report /22/ but did not want to make them available to the Nazi. Therefore he thought to "bury" his ideas in a place (the Postal Office) where nobody would have paid any attention to it.

Turning back to Fritz's family life in Göttingen, some difficulties started to arise between him and Ilse already in 1950. Fritz's attention for any pretty woman he had the occasion to meet here and there, was much less important than the irritability he had developed in front of any small difficulty of daily life.

One evening of 1951, W.Paul and his wife were sitting at home when the door bell rung. Fritz and Ilse were there with a bottle of wine, and asked to enter since they wished to celebrate an important decision they had taken. With a glass of wine in their hands they announced the decision to divorce.

In the meantime Houtermans had received the offer of the chair of experimental physics from the University of Bern. He accepted and took up the new job starting from September 1952.

19. - Houtermans' third family -

When Fritz decided, in 1944, to marry Ilse Bartz, he did not inform his first wife Charlotte of his desire of divorcing from her but took advantage of a law which exempted from such a "formality" any person living separated from his wife or husband, for more than five years. The law, promulgated by Hitler's government, had been used for interrupting many marriages of "pure arians" to Jewish people.

As an announcement to Charlotte of his second marriage, Fritz sent her a reprint of one of the papers he had published in collaboration with Ilse.

Charlotte was already aware of the existence of these papers, which had been pointed out to her by Otto Oldenburg, an old friend from Göttingen and now professor of physics at Harvard University. But only a few months after the arrival of the "reprint" Charlotte learned of the divorce of Fritz from her and of his marriage with Ilse.

In 1951 Charlotte went to Europe in order to visit her mother in Bielefeld. Coming back, with Casimir(138), from a conference in Copenhagen, Fritz went to see her. After a first rather stormy encounter, Fritz told Charlotte of difficulties he had with Ilse, expressed to her his affection and insisted that she should remain in Europe and go immediately to live with him. But such an unprepared radical change was not possible for Charlotte. She had commitments at Sara Lawrence College, and Giovanna had to finish school in the United States.

Charlotte came back to Europe with Giovanna and Jan a first time during summer 1952. At their landing in Rotterdam, they were received by Fritz who had organized a vacation in Uberlingen am Bodensee. All the children were there with Fritz and Charlotte: Giovanna, Jan, Pieter, Elsa and Coja.

In spring 1953 Charlotte and her two children were awarded fellowships from three different organizations, which facilitated their second trip to Europe. After their landing, she drove with a friend through France and met Fritz again in Bagnères de Bigorre, in the French Pyrénées, where, from July 6 to 12, 1953, took place "Le Congrés International sur le Rayonnement Cosmique". That week was just in the middle of the Second Expedition to Sardinia for the study of cosmic rays at high altitude by means of balloon flights(139), in which also the Bern physicists Houtermans and Teucher were involved (see below).

Among the many participants in the Bagnères de Bigorre Conference I should recall P.S.M.Blackett, B.Rossi, G.Bernardini, C. Powell, L.Leprince-Ringuet, G.Occhialini, C.Dilworth, etc., many of which were old friends or more recent friends of Houtermans. I also was there and thus had the opportunity of taking up some of the conversations I had started with Fritz not long before when he passed through Rome on his trip to Sardinia.

During the conference we read in the international press that recently Beria had been condemned and executed. I will always recall the excitement of Fritz when we talked about this event.

After the conference, Fritz, Charlotte with Giovanna and Jan went to Marseille, where Fritz and Charlotte left the children with a friend. They themselves visited James Franck in Kreuznach Spa, near Bern, who was there for a few weeks' cure for heart troubles. Asked for advice by Fritz about his planning of marrying again Charlotte, Franck expressed a clear negative view, which, however, was not followed by Fritz in spite of his almost filial devotion.

From Kreuznach they went to Hamburg for participating in a meeting of people that had been communist or very close to them, in the thirties, had moved to USSR with the idea of helping in the construction of the socialist society and that, once over there, had passed very awkward adventures and had the great luck to be still alive(140).

Towards the end of August 1953, Charlotte arrived in Bern. At their marriage the August 28, the witnesses were Giovanna, Jan and W.Pauli.

The "third family" of Fritz that settled in Bern included two children, Pieter and Elsa, but not Coja, who was only 5 years old, and still deserved the devotion of her mother. Giovanna was in Tubingen as a Fullbright student and Jan at the University of Rochester (USA).

Houtermans and Charlotte went to the annual conference of the Italian Physical Society that in 1953 was held in Cagliari from 23 to 27 September. Ilse Barts was also there.

In October of the same year a meeting was organized at the Department of Physics of the University of Bern, by Houtermans and Teucher for the distribution of the packages of nuclear emulsions exposed to cosmic rays at high altitudes among the participants in the Second Expedition to Sardinia(139). I was also there, and one evening when the work was ended, Fritz and Charlotte invited all participants in the meeting to their house where all enjoied their hospitality and the friendly atmosphere created in great part by the cordial and witty interventions and jokes of Fritz.

During the winter 1953-54 Fritz started to drink abundantly and the relationship with Charlotte deteriorated rapidly. On March 1, 1954 she left Bern and went to Paris where she remained for a few weeks and had the opportunity to meet once more old friends like Manes Sperber and Weissberg-Cybulski. From Paris she went to Bristol where, from April to September, she worked in Cecil Powell's Laboratory at the microscope in the investigation of events produced by cosmic rays in nuclear emulsions.

At the end of August Charlotte went back to the United States and took up again her teaching work at Sarah Lawrence College. The divorce from Fritz was accorded her in 1954. It was a shock for both of Charlotte's children, in particular for Jan, who by now was 18 years old and had started to study physics at the University of Rochester.

20 - Fritz's fourth family - His departure -

In 1955 Fritz married Lore Muller, the sister of the wife of his stepbrother Hansi, that he had met a few years before in Göttingen. For their marriage Wolfgang Pauli sent Fritz the following telegram: "The usual congratulations".

Lore had already a few years old daughter, Sabine (b.1951) that some years later was adopted by Fritz.

In 1956 Fritz and Lore got a child, Hendrich, who unfortunately a few years after the death of his father, died, at the age of seventeen, in a car accident.

In 1957 Houtermans took a sabbatical year and went to the United States where he spent months at the California Institute of Technology (Pasadena, California) and the Scripps Institution of Oceanography (San Diego, La Jolla, California). In both places he was in close contact with well known figures in the Earth's Sciences: Harrison Scott Brown (b.1917), Samuel Epstein (b.1919) and Gerald J.Wasserburg (b. 1927) in Pasadena, and Roger Randall Dougan Revelle (b.1909), Director, and Hans Eduard Suess (b.1909) in La Jolla.

As we shall see in more detail in chapter 21, nuclear geology had become perhaps the most important part of the scientific program set up by Houtermans in Bern. Another subject was the investigation of particle physics by means of nuclear emulsions, and Lore started to take part as a microscopist, in the work of the group composed of eight or ten people that was active in this field under the direction of Teucher.

Lore remembers these years with great pleasure not only for the trip to the United States but also for a number of other trips to many places in Europe, in particular to Italy where they visited many towns that combined artistical and natural interest either with new friendships, like Pisa and Rome, or with Fritz's youth recollections, like Naples.

The year 1961 marked a breach in Houtermans' life. He was supposed to go to a far east country and therefore had to do a certain number of preventive injections. He had a rather strong reaction with high fever, and when at a very early hour of the day he tried to get up for greeting Lore's daughter that was going out for participating in an excursion for seeing the sunrise, he fell and hit his head against the banister of the stairs. The concussion he had was not too serious, and after six months the neurologist declared him normal. The accident, however, had triggered some damage that the doctors never really identified. The hard experience of his life, his demanding work and life style, in particular his rather heavy drinking habit during the last few years, caught up with him. Thus, from about 1962 on he was never again able to work with the enthusiasm and the success which had characterized the first eight or ten years of his life in Bern.

At the beginning of autumn 1965 a doctor made him a diagnosis of a lung cancer: a dark stain was clearly visible in the radiographies. After about three months of irradiation the stain disappeared and Fritz started to regain part of the weight he had lost. He felt much better and enjoyed again sitting in his house, at his working table, and writing the paper on the "History of the K/Ar-Method of Geochronology" for the book dedicated to Wolfang Gentner's 60th birthday /135/. Lore still remembers when Fritz went out from their house to mail his manuscript.

Shortly after, the March 1, 1966 he had a stroke: a lung's artery broke, his heart stopped and he suddenly died.

21 - Houtermans' scientific work and influence in Bern -

Heinrich Greinacher (1880-1974) had been Professor of Physics and Director of the Institute of Physics at the University of Bern for many decades. He had invented the voltage multiplier(141), that became well known after Cockcroft and Walton, at Cambridge (U.K.) in 1932, re-invented it and applied it for constructing the accelerator they used in the discovery of the first nuclear reactions produced by accelerated particles(142). Greinacher had also constructed the first ionization chamber connected to a linear amplifier capable of detecting the ionization produced in a gas by a single alpha particle or fast recoiling proton(143). But the full power of this new detection technique was exploited by others, in particular by Chadwick in the discovery of the neutron(144).

In spite of these remarkable contributions the Bernese Institute of Physics, under Greinacher's direction, remained small and in some way provincial.

When he reached the age of retirement, the Bernese Faculty looked for a successor. At that time all the chairs of experimental physics in Swiss universities were occupied by students of Paul Scherrer (b.1890-1969), the Director of the reputed Physics Institute at the Federal Institute of Technology in Zurich (ETH). The Dean of the Faculty, at that time the astronomy professor Max Schürer (b.1910), André Mercier (b.1913), Professor of Theoretical Physics, and the Director of the Federal Bureau of Standards, Hans König (b.1904), asked for advice from Paul Scherrer in Zurich for the right successor. They liked, however, to have a wide appreciation of all possibilities, and went to the yearly meeting of the German Physical Society in Heidelberg and consulted Hans Kopfermann of the University of Göttingen (chapter 17). Kopfermann mentioned to the Bernese delegation Fritz Houtermans who - as we have described - occupied quite an unsatisfactory position in Göttingen in Kopfermann's Institute. In contrast to Houtermans, Kopfermann was a very balanced personality but he shared with Houtermans the sense of wit and humour that was characteristic of physicists of their generation.

However, when Kopfermann mentioned this nomination to Fritz Houtermans, he was already having second thoughts which he expressed to Houtermans in terms "I wonder whether you could get along with the staid Bernese bourgeoisie" whereupon Houtermans tried to reassure him "I would not worry so much, I am sure they don't replicate them by spores".

A few weeks later, Houtermans was invited to Bern for a talk that gave the other faculty members the opportunity to look him over. There were, of course, competitors, and in the end, the Bernese faculty and government had to choose between the left leaning Houtermans and a German that had some reputation for leaning toward the other side of the political spectrum. In the end, faculty and government decided themselves for Houtermans, because he could introduce new research fields in Switzerland . This was the right decision as it turned out pretty soon.

On his part, Houtermans had the choice between several offers of professorships: Graz in Austria, one or two places in Germany and Bern. He decided himself for Bern. Perhaps his remark that his decision was biased towards Bern because this was the town where Albert Einstein did his famous work at the beginning of this century, was meant only half jokingly.

The members of the Institute in Göttingen remember the day when Houtermans was leaving the town where he had lived and worked for seven years. After a proper farewell party everybody accompanied him to the train. With his salary in Göttingen, he could not afford a sleeping car, but he reassured them that brandy was the best sleeping car. There is no record of the circumstances of his arrival at Bern, but with a decisiveness and initiative which surprised his friends, he began to build up a small but active institute doing excellent research. With his magnetic personality, he succeeded in getting Charles Peyrou from the École Polytechnique in Paris to assume the newly created position of Extraordinary Professor in Houtermans' "Institute" and for a while also Walter Thirring, now Professor of Theoretical Physics in Wien, joined him. He brought from Göttingen with him Martin Teucher (1921-1978) who had done his PhD with Houtermans in the field of nuclear physics and Friedrich Begemann (b. 1927) as a graduate student and a year later, after finishing his thesis on lead isotopes with W.Paul, Johannes Geiss (b. 1926) joined the Institute.

At Bern he found three young Swiss who were eager to start their PhD work with the new professor: Hans Oeschger (b.1927), Christoph Burckhardt (b.1927) and Walter Winkler (b.1927). These young men filled all the scientific positions available. Houtermans put M. Teucher in charge of building up a high energy physics group using the emulsion method. Geiss was to introduce mass spectrometry and Hans Oeschger got as a thesis theme a pet project of Houtermans, low level counting, with the aim to apply it to the carbon 14 dating method that a short while ago had been invented by Willard F. Libby(145).

The influence of Houtermans not only on the young men in his Institute but also on many colleagues in the faculty was enormous. He brought to Bern an internationalism which meant at least in the physical sciences a new spirit. He was fortunate to have arrived at the right moment at Bern because Alexander von Muralt (b.1903), Professor of Physiology at the University of Bern and a very influential man in Switzerland, had just single handedly created the National Science Foundation. Up to that point it was very difficult in Switzerland to pursue modern experimental physics outside the ETH in Zurich (Scherrer's Institute) because the Universities in Switzerland are cantonal and the cantonal governments did not see themselves in a position to support expensive research.

Houtermans had instilled in the young men working with him a sense of urgency and devotion to science that very soon produced interesting and scientifically significant publications: the anomalously high radioactivity in the lead of fumaroles of Vesuvio, the first lead age of a rock published from a European laboratory, several significant papers with Geiss and Eberhardt on the large variations of the isotopes of lead in galenas and their interpretation in terms of geologic history and the age of the Earth. Under Houtermans' leadership, Hans Oeschger constructed a new type of anticoincidence counter which was for many years unsurpassed in the important aspect of background rejection.

In the field of nuclear geology Houtermans established close cooperation with the Institute of Ezio Tongiorgi(146) in Pisa and of Edgar Picciotto(147) in Brussels. Some of his young Bernese collaborators spent many weeks in Pisa and Brussels helping to build up mass spectrometry and low level counting equipments there, while Bern profited from Picciotto's talent as a nuclear chemist and from Tongiorgi's wide ranging interests and knowledge in the Earth sciences and in archaeology.

However, success is a mixed blessing for an Institute because the young people working with Houtermans got offers to other places and were eager to widen their horizon. First Begemann went to the University of Chicago to work with William Libby and later became director of the Max Planck Institute for Chemistry at Mainz. A year later Geiss followed to work with Harold C.Urey(148) and Martin Teucher also went to the United States and later became one of the directors of Desy laboratory in Hamburg. But in the meantime, the younger students, such as Peter Eberhardt (b.1931) and Peter Signer (b.1929) had grown, and Hans Oeschger stayed at Bern after his thesis was completed and they were able to carry on with the work very well.

When, in 1952, Houtermans accepted the offer of the chair of experimental physics in Bern, he was promised by the authorities of the University the creation of a new building for the Institute of Physics. For various financial and bureaucratic reasons this was ready only nine years later.

The new institute was called by Houtermans "Institut für Exakten Wissenschaften" (Institute for Exact Sciences) for stressing that the research programme concerned many parts of applied mathematics, theoretical physics, experimental physics (such as cosmic rays and particles physics), astronomy and nuclear geology /125/(149).

Beginning in about 1959, Houtermans with his collaborators had began to cooperate with the mineralogist Professor Ernst Niggli (b.1917) to date rocks. After some years a separate "Laboratory for Isotope Geology" in Niggli's Institute grew out of this collaboration.

When, starting from 1962, the scientific leadership of Houtermans was strongly reduced because of his health conditions (chapter 20), his "boys" as he called his younger collaborators, were suddenly left to themselves. Geiss had returned from the Oceanographic Institute in Miami to succeed Peyrou, gone to CERN, as an Extraordinary Professor, and he, Oeschger, Eberhardt and Debrunner, all very young men, faced the challenge of continuing the research and the functioning of the Institute without Houtermans' leadership, and they managed to meet this challenge with success. There were difficulties with the National Science Foundation to continue the support of the research work and they were helped by senior friends of Houtermans, Hans König (b.1904), Klaus Peter Meyer (b.1911) who became the Director of the newly created Institute of Applied Physics, and Max Schürer. Also A. von Muralt helped in avoiding that the very promising work started by Houtermans did not remain an episode in the history of the University of Bern. When Houtermans died in 1966, Geiss became his successor as Director of the Institute. He was able to convince Beat Hahn (b.1921) to assume the responsibility for the High Energy Physics. At that time, Max Keller (b.1919), a high ranking civil servant, was responsible for university matters in the cantonal government. Among civil servants this was outstanding for his insight, very good in judging people and he was determined to use the rather good financial situation of his government to create a modern University. Geiss and his colleagues established good relations with Max Keller and the result is that a powerful institute was gradually built up which is generally now considered one of the best research Institutes in Switzerland.

Geiss ventured into the new field of space research and together with Eberhardt and Signer did a solar wind experiment with the Apollo program and this experiment became very popular in Switzerland, helping the Institute to obtain support. After some urging by Giuseppe Occhialini, Geiss became also active with the newly formed European space research organization and together with Eberhard and some younger people he built up a well known research group in space physics and planetary physics, working successfully with the European Space Agency and with NASA. Hans Oeschger continued to work in nuclear geology applying low level counting methods to Earth Science problems. His group is in Europe now on the forefront of this field that is particularly significant for environmental problems. H.Debrunner continued with his thesis theme given to him by Houtermans, cosmic rays. He has not only instrumentation on the Jungfraujoch Scientific Station, but he succeeded A. von Muralt as the Director and President of the Jungfraujoch Scientific Station. Thus, the seeds planted by Houtermans, his internationalist approach to science, his refusal of even considering secondary scientific questions and his taste for the interdisciplinary approach had a lasting and even today visible effect on the University of Bern.

Concerning the "staid Bernese bourgeoisie", Houtermans and they got along very well. They are quite willing to accept an unconventional character like Houtermans - if he is a "foreigner" - and to this day the elder generation in Bern remembers many stories of Houtermans whom they consider an original who has instilled some spirit of adventure and unconventionality into their University.

Some more details about Houtermans' scientific production during the Bern period can be of interest to many people.

His contributions to the investigation of cosmic rays concern the systematic recording of the nucleonic component at the Jungfraujoch station (3450 m.s.l.) /94/ /116/ /124/, the daily periodic variation of cosmic rays /113/, the correlation between Forbush decreases of cosmic rays and satellite drag /112/, and the sudden increase of the nucleonic component observed in cosmic rays on May 4, 1960 /118/.

Two papers concern particles physics /68/ /107/. The first one, in collaboration with W.Thirring /68/ contains an estimate of the solar neutrinos produced in the carbon-cycle and a discussion of their mean free path for absorption inside the Sun. The attempt was premature; the paper was written in 1954, just before the publication of the experimental results by Cowan, Reines et al.[(151)] on the inverse beta-decay induced by a beam of neutrinos generated inside a large nuclear reactor. The authors derive an estimate of the neutrino-electron cross section from Fermi's theory of beta-decay[(150)] by taking into account only the processes corresponding to the decay of the nucleon and the muon, and obtain a much to small value. Only in 1958 by expressing the weak interaction Lagrangian as a current- current interaction, Feynman and Gell-Mann showed[(152)] that it contains also lepton-lepton terms which give an appreciably larger direct neutrino-electron interaction.

The second paper on particle physics /107/ is the report on the experimental work carried out by a rather large group of the Bern Institute with the nuclear emulsion technique on a sample of 1600 K^- mesons of 130 MeV mean kinetic energy, produced by the Bevatron in Berkeley. It contains evidence for the $K^- \to 2\pi$ and $K^- \to 2\mu$ decay in flight, a determination of the differential scattering cross section and of the relative reaction rates for the production of $\Sigma^{\pm}$, Σ° and Λ°, in nuclear emulsions.

Two other papers of 1960 by Houtermans alone /117/ /119/ deal with the "Maser condition" in the spectra emitted by dissociating molecules, a subject closely connected with some ideas he had glimpsed many years before (Chapter 4).

Almost all other papers of the Bern period regard nuclear geology, studied by means of the more appropriate techniques for measuring the relative abundances of stable isotopes and radioactive isotopes and for stimulating thermoluminescence in meteorites or rock samples.

In particular the paper /77/, in collaboration with Bot, Geiss, Niggli and Schürmann, contains the first determination of the age of a rock performed in Europe.

Continuing the work carried out in Göttingen in 1947, and pursued also by Holmes at about the same time /153/, Houtermans went on improving the determination of the age of the Earth from the isotopic composition of lead. The method, already employed in the papers /35/ and /46/, starts from the results obtained by Nier(154).

Lead found in terrestrial rocks or meteorites has stable isotopes of masses 204, 206, 207 and 208. The isotope 204 is not the product of radioactive disintegrations while the isotopes 206, 207 and 208 are the final products of the radioactive families beginning with ^{238}U (= UI), U^{235} (= AcU) and ^{232}Th and are frequently called RaG, AcD and ThD. A sample of a rock or meteorite is characterized by the measured values of the following ratios

$$\alpha = \frac{(^{206}Pb)}{(^{204}Pb)}, \quad \beta = \frac{(^{207}Pb)}{(^{204}Pb)} \quad \mu = \frac{(^{238}U)}{(^{204}Pb)}$$

where ^{A}X means the number of atoms of the isotope A of the element X found in the sample at present. The values of α, β and γ we observe today are clearly determined by the past history of the sample. In the simplest possible model one can consider the time w elapsed from the formation of the lithosphere until today (t = 0) and the present age p of the lead mineral (for example a galena, PbS), i.e. the time elapsed from when the sample separated from the rock or magma, containing, in addition to lead, also a certain amount of uranium and thorium. In this model one has

$$(^{206}Pb)_p = (^{206}Pb)_w + (^{238}U)_w \left\{ 1 - \exp -[\lambda(w - p)] \right\}$$

where

$$\frac{(^{238}U)_o}{(^{238}U)_w} = \exp[-\lambda w]$$

Dividing this equation by (^{204}Pb) one obtains

$$\alpha - \alpha_w = \mu \left[\exp(\lambda_w) - \text{ex}[\,(\lambda_p)\right]$$

$$\beta - \beta_w = \frac{\mu}{139}\left[\exp(\lambda'_w) - \exp(\lambda'_p)\right.$$

where λ' is the decay constant of 235U and $139 = ({}^{238}U/{}^{235}U)_0$. From these ratios it follows

$$\frac{\beta - \beta_w}{\alpha - \alpha_w} = \frac{1}{139}\,\frac{\exp(\lambda'_w) - \exp(\lambda'_p)}{\exp(\lambda_w) - \exp(\lambda_p)}$$

In a plane with α in abscissa and β in ordinate this equation represents a family of straight lines, each corresponding to a different value of the parameter p, but all passing through the same point (α_w, β_w). These straight lines are called <u>isochrones</u>.

Through the same point (α_w, β_w) pass also all the <u>development curves</u> representing the dependence of β on α for any fixed set of values of α_w, β_w and μ. In principle each measured sample (α, β, μ) allows the determination on the "Bern graph" /67/ of the corresponding isochrone and development curve which pass through it and cross again at the point (α_w, β_w).

By applying this kind of analysis to the first measurements carried on by Patterson et al of the University of Chicago(155) for the troilite phase of the Canon Diablo meteorite, and assuming that this material has the same age as the lithosphere, Houtermans arrived in 1953 /61/ to derive a fairly good estimate of the age of the Earth

$$w = (4.5 \pm 0.3) \times 10^9 \text{ years.}$$

A similar analysis was carried out also by the Chicago group, which obtained the same result(156).

As I mentioned in Chapter 18, Houtermans introduced also a very convenient variant of the Pb/U method consisting in deriving the value of ${}^{206}Pb/{}^{298}U$ from the ratio RaD/Pb where RaD(= ${}^{210}Pb$) decays into RaE which, because of its short half-life, reaches in a few weeks the radioactive equilibrium with its mother. Because of their higher energy the beta rays of RaE are measured with greater accuracy than those of RaD. This method was applied by Begemann, Butlar, Houtermans, Isaac and Picciotto /62/ for determining the age of Uranium minerals.

A few other papers, in collaboration with Begemann and Geiss /63/ /66/ /74/ /130/, concern the radioactivity of the lavas of Vesuvium, others /79/ /84/ /85/ /86/ /87/ /100/ /101/ the determination of the age of the galenas of Madagascar and of the yttrocrasite of Katanga. In paper /109/, in collaboration with Herr, Gfeller and Oeschger, Houtermans studied the presence of 36Cl in meteorites. This work represents one of the first successful attempts at using the radioactivity induced by cosmic rays in meteorites. Still other papers concern the isotopic analysis of Osmium in iron meteorites and earth sample /110/ /126/ leading to the determination of the corresponding ages.

Houtermans gave also interesting contributions to the development of the instrumentation required by his geological work. He and Oeschger constructed proportional counters for the determination of very weak beta activities /78/ /99/ /105/ and in collaboration with a Milan-Pisa group detectors of alfa activity in very low concentrations /98/.

In another series of papers Houtermans used the thermoluminescence as a method for investigating the thermal and irradiation history of various minerals and stones /73/ /89/ /93/ /95/ /104/ /120/ /132/. The more important of this set of papers is /104/ in collaboration with Grgler and Stauffer. In paper /96/ Houtermans and Stauffer proposed the use of thermoluminescence for dosimetry, on a variant of which the authors took also a patent.

In paper /115/, in collaboration with Grögler and Stauffer, Houtermans was one of the first to use thermoluminescence for dating also ceramics and tiles.

In another paper of 1966, in collaboration with Grögler, Geiss and Grnenfelder /133/ he applied the isotopic analysis of lead for establishing the origin of Roman lead pipes and lead bars.

Finally the paper on the production of Kr^{81} (T1/2 ~200,000 years) by cosmic rays /134/ has opened a line of research that later became significant, mainly through the work of Houtermans' pupils, for dating water masses.

22 - Why we remember him -

The most important contribution given to science by Fritz George Houtermans consists in the work, carried out in Göttingen in 1928-29 in collaboration with R.d'E.Atkinson, on the formation of light elements and the consequent energy production in the centre of stars /5/ /6/ (Chapter 4). These papers, appeared years before the discovery of the neutron (1932), are quoted in all review articles reporting on the history of the interpretation of stellar energy in terms of nuclear processes. They provided the motivation for the International Astronomical Union (IAU) for assigning, on occasion of its meeting in Sidney in August 1973, the name of Houtermans to one of the lunar craters located near the eastern hedge of the visible disk of the Earth satellite (lunar long. + 87.0, lunar lat. - 9.3)[(158)].

The rest of his scientific production reaches from atomic spectroscopy to electron microscope technique, from neutron physics to cosmic rays and from theory and experiments on subnuclear particles to nuclear geology, keeping always a very good level and frequently showing the clear signs of a very brilliant and versatile mind.

The Bern graphs, introduced by Houtermans for determining the age of a rock from the isotopic composition of lead and uranium (Chapter 21), are sometime called, even today, "Houtermans plots". In the field of nuclear geology, he has opened a few other research lines and has created in Bern a lasting school of research which is the only one in Switzerland and one of the few well established in the world.

Apart from these scientific achievement the case of Houtermans is almost unique for the adventurous life imposed on him by the political events of our century. Only in the 20th century a life like that of Fritz could take place.

On some occasion the situations that Fritz had to face were extremely difficult and delicate and this explains why, sometime, his behaviour was not as clear and transparent as his relatives and friends would have desired.

All the people who worked with him have kept a very high consideration of him, not only as scientist but also as human being. For instance, in a letter of 23 October 1985, that E. Picciotto sent me after having read these pages, he summarized his views as follows:

« ... Houtermans était un personage de roman, avec ses grandeurs, ses faiblesses, ses conflits interieurs, le tout règi par son humanitè, sa générosité et son humour tellement personnel, qui était evidemment le défaut qu'aucun règime totalitaire ne peut admettre. C'est pourquoi je crois que seul un romancier de grand talent aurait pu rendre le personage, aux dèpenses evidemment et inèvitablement de la fidelité historique que vous avez si scrupuleusement observée».

Talking and talking over again about Houtermans with Giuseppe Occhialini, he reminded the verses of Oscar Wilde:

"For he who lives more than one life
More than one death must die"

For Beppo, Fissel during his life had a number of little deaths.

The first one was his flight from Germany and from scientific work for going to Middlesex. The second one was his Russian disillusion and his serving in Russian prisons under the accusation of being a Nazi spy , culminated in an exchange "together with Nazi spies". The third one was his return to Kharkov for trying to help his old friends, for finding out what was happened of Leipunski and many others overthere. Still years later, Fissel anxiously asked to his interviewing friends in London (Occhialini and George [(159)] in 1950): "Did they tell you I went to Kharkov in a Nazi uniform? ... that is not true! Did Rosbaud tell you I had a military cap? ... that is not true!"

The fourth little death of Fissel was his divorce from Charlotte, without asking her consensus, for marrying Ilse. About four years had passed from his arrest in Moscow, and when finally free, after more than two and half years of prison, life in Berlin was extremely hard. From the end of August 1940, after a few heavy bombings by the Germans of London and other British cities, the Royal Air Forces had started to do frequent raids on Berlin. It was in the atmosphere of nerves' tension created among the people in an underground shelter by a heavy bombing of the city, that between Ilse and Fritz happened the unavoidable, which brought Fritz to marry the girl, of whom he was in love since months.

The fifth and last little death was the abandonment by Ilse followed by the return of Charlotte and the second separation from her, everything happening in an incredible short span of time.

Of all these little deaths or crises for Fritz the most serious was the third one. Already at the time of the trip from Kharkov to London of Charlotte and Giovanna in summer 1937 (Chapter 9),

Houtermans strongly desired to survive for convincing his friends in the West that USSR was quite different from what all of them had thought. But nobody took seriously their appeals! He had also friends overthere, in Kharkov, and with the German occupation they were certainly in an extremely difficult situation.

The thought of how, eventually, he could try to help them, was the one that dominated his mind, and cancelled any other consideration about completely different possible interpretations of his behaviour.

Around 1950 Fritz Houtermans fully convinced of his fair intentions a few open minded people like the Occhialinis, but not most of his more ideologically rigid old friends in France, Great Britain and Belgium. He was condemned for his trip to Kharkov also by some of his Russian colleagues like Kurchatov, Iwanenko(160), Leipunsky and Silienikov. Other Russian physicists, however, spoke well of him for having saved the electrostatic generators of the Physico-Technical Ukrainian Institute by convincing the Nazi authorities that there was no point in dismantling this equipment and moving it to Germany because it was obsolete.

There is a short book published in Kharkov in 1944(161) which describes how the Kharkov physicists set to work to restore the electrostatic generators once Kharkov had been liberated, thus proving that this equipment was still there.

In a book (in Russian) dealing with Soviet nuclear physics from 1932 to 1945, written by a Soviet science fiction writer on the basis of interviews with physicists(162), Houtermans is portrayed as a very good physicist and as someone very glad to have found a new homeland. The author also hints that he was <u>not firm</u> in his left-wing political views and that he was something of an anti-semite.

This last accusation is so impudently false to throw strong doubts on the sincerity of the others.

For those who met him personally, it is impossible to forget his enthusiasm for science, his devotion to research, his friendliness for anybody and his readiness to participate in any conversation, discussion or action aiming to a deeper understanding of a new observation or idea. In spite of the hard experience he had in his life, he maintained a joyful look at life contributing, in many cases, in helping friends, pupils and colleagues to achieve greater happiness and appreciation of their own work and life.

According to Giuseppe Occhialini we frequently cross figures of this kind in the past and present literature and we like them. But usually we are not ready to accept them when we meet them in real life.

Dimitri Karamazov is an innocent, accused of infamies, even of his father's murder; the reader likes him mainly for his intelligence, but then discovers that he is "a screwball".From time to time Fissel was a romantic hero with the cynicism of Heine von Kleist, the officer in the book by Guy de Maupassant. But on most occasions he appeared more similar to Puck or Robin Goodfellow, the goblin of the British heath, who, as an agent of the king of the Fairies is ultimately beneficent but has an independent love for mischief.

According to Connie Dilworth we remember Fritz Houtermans not only(163) "... as individual but also as a representative of a culture that is lost. The latter is what he had in common with Touschek(164). Both were essentially "Mittel Europa", Germanic in the best sense, of the anarchist breed; in revolt against the Prussian element, and against bureaucracy. Individualistic, romantic, but with a crystalline cynicism with regard to any form of fanaticism: enthusiastic and in love with life, but life of the city.

Professionally, both were much better than their work, their intelligence was at least equal to that of the most successful of their contemporaries. The lack of success was in part due to being in the wrong place at the wrong time.

Houtermans was an innocent, of the tribe of Peter Pan. His refusal to grow up, to become serious, was the basis of his charm. On the other hand - as all Peter Pans - he never understood women. His four marriages, three wives and many children do not denote the profligate but rather a child seeking the road to the Never Never Land.

Part of this was his warmth and generosity. He may have lied, betrayed and twisted truth to feed his vanity, but he was not mean. He had no ideal, but was idealistic in the sense that he did not live for today or tomorrow. He lived in the stream of history, egoistic but not egocentric, that his culture forbade. Above all, life was a great joke, a continuous laugh against the lumpen-bourgeoisie.

They don't make them like that anymore. That's why we remember him".

- LIST OF PUBLICATIONS BY F.G.HOUTERMANS -
(compiled by his collaborators of the University of Bern)

/1/ 1926 "Über die Bandenfluoreszenz des Quecksilberdampfes"; Vortrag, 18. Juli 1926; Verhandl.d.Deutschen Physikal. Ges. 3., 7.Jg.37.

/2/ 1927 "Über die Bandenfluoreszenz und die lichtelektrische Ionisierung des Quecksilberdampfes"; Inaugural-Dissertation, Göttingen, erschienen in: "Über die Bandenfluoreszenz des Quecksilberdampfes", Zeit. f.Physik 41, 140-154. "Zur Frage der lightelektrischen Ionizierung des Quecksilberdampfes", Zeit.f.Phys. 41, 619-635.

/3/ 1928 "Zur Quantenmechanik des radioaktiven Kerns"; in collaboration with G.Gamow; Zeit.f.Physik 52, 496-509.

/4/ 1929 "Zur Wellenmechanick des radioaktiven Kerns", in collaboration with R.d'E.Atkinson; Verhandl. d.Deutschen Physikal.Ges. 3., 10.Jg. 1,10.

/5/ 1929 "Zur Frage der Aufbaumglichkeit der Elemente in Sternen"; in collaboration with R.d'E.Atkinson; Zeit.f.Physik 54, 656-665.

/6/ 1929 "Transmutation of the lighter elemens in stars"; in collaboration with R.d'E.Atkinson; Nature 123, 567-568.

/7/ 1929 "Zur Quantenmechanik der α-Strahlung"; in collaboration with R.d'E. Atkinson; Zeit.f.Physik 58, 478-496.

/8/ 1930 "Neuere Arbeiten ber Quantentheorie des Atomkerns"; Ergebnisse der exakten Naturwissenschaften 9, 123-221.

/9/ 1932 "Über geometrisch-optische Abbildung von Glhkathoden durch Elektronenstrahlen mit Hilfe von Magnetfeldern (Elektronenmikroskop)"; in collaboration with M.Knoll and W.Schulze; Verhandl.d.Deutschen Physikal. Ges. 13(2), 23-24.

/10/ 1932 "Über eine neue Form von Lichtquellen zur Anregung der Resonanzfluoreszenz von Metalldmpfen, insbeson dere des Quecksilbers"; Zeit.f.Physik 76, 474-480.

/11/ 1932 "Untersuchung der Emissionsverteilung an Glkathoden mit dem magnetischen Elektronenmikroskop"; in collaboration with M.Knoll and W.Schulze; Zeit.f. Physik 78, 340-362.

/12/ 1933 "Über Absorptionsmessungen und andere optische Untersuchungen leuchtenden Stoffen mit Hilfe der Wechsellichtmethode"; Zeit.f.Physik 83, 19-27.

/13/ 1936 "Radioaktivität in Tantal durch Neutronenbestrahlung in collaboration with V.Fomin; Phys. Zeitschr.d.Sowjetunion 9, 273-274.

/14/ 1936 "Slowing down of neutrons in liquid hydrogen"; in collaboration with V.Fomin, A.I.Leipunsky and L.W.Schubnikow; Phys. Zeitschr. d. Sowjetunion 9, 696-698.

/15/ 1936 "Absorption of thermal neutrons in silver at low temperatues"; in collaboration with V.Fomin, I.W.Kurtshatov, A.I.Leipunsky, L.Schubnikow and G.Shtshepkin; Nature 138, 326-327.

/16/ 1936 "Neutron absorption of boron and cadmium at lowtemperatures"; in collaboration with V.Fomin, A.I.Leipunsky, L.B.Rusinov and L.W.Schubnikow; Nature 138, 505.

/17/ 1936 "Über die Absorption thermischer Neutronen in Silber bei niedrigen Temperaturen"; in collaboration with V.Fomin, I.W.Kurtschatow, A.I.Leipunsky, L.W.Schubnikow and G.Shtshepkin; Phys.Zeitschr.d. Sowjet union 10, 103-105.

/18/ 1937 "The absorption of group C-neutrons in silver, cadmium and boron at different temperatures"; in collaboration with A.I.Leipunsky and L.Rusinov; Phys. Zeitschr. d.Sowjetunion 12, 491-492.

/19/ 1940 "Halbwertszeit des Radiotantal"; Natuwiss. 28, 578.

/20/ 1941 "Über die Raumladungswirkung an einen Strahlgeladener Teilchen von recheckigem Querschnitt der Blende"; in collaboration with K.H.Riewe; Archiv f. Elektrotechnik 35, 686-691.

/21/ 1941 "Über den Energieverbrauch bei der Isotopentrennung", Ann.Physik 40, 493-508.

/22/ 1941 "Zur Frage der Auslsung von Kern-Kettenreaktionen"; Communication from the Laboratorium Manfred von Ardenne; unpublished report, 115-145.

/23/ 1941 "Über Wirkunsquerschnitte einiger Elemente für thermische Neutronen"; Zeit.f. Physik 118, 424-425.

/24/ 1942 "Ein Neutronenintegrator, eine Anordnung zur Messung der Ergiebigkeit von Neutronenquellen nach der Methode von Fermi und Amaldi"; Physik.Zeitschr. 43, 496-503.

/25/ 1942 "Über die Entstehung von RaE aus Wismut durch den (n,)-Prozess mit langsamen Neutronen"; in collaboration with I.Bartz; Naturwiss. 30, 758-759.

/26/ 1942 "Zur Frage der Einfangung thermischer Neutronen in Wismut"; in collaboration with I.Bartz; Verhandl.d.Deutschen Physikal.Ges. 3., 23.Jg.(2), 67.

/27/ 1943 "Über den Kernphotoeffekt im Beryllium"; in collaboration with I.Bartz; Physik. Zeitschrift 44, 167-176.

/28/ 1943 "Über einen elementaren Beweis für die Existenz unendlich vieler Primzahlen der Form (2 px + 1) und eine Verallgemeinerung des Euklidischen Beweises für die Existenz unendlich vieler Primzahlen; Manuscript Christmas 1943, 5 pages.

/29/ 1944 "Über eine halbempirische Beziehung zwischen der Ergiebigkeit einer Neutronenquelle und der maximal erreichbaren Dichte langsamer Neutronen in einen wasserstoffhaltigen Medium"; Physik.Zeitschrift 45, 258-264.

/30/ 1946 "Über die Annahme der zeitlichen Vernderlichkeit des -Zerfalls und die Mglichkeiten ihrer experimentellen Prfung"; in collaboration with P.Jordan; Zeit. Naturforschg. 1, 125-130.

/31/ 1946 "Über den (n,2n)-Prozess am Beryllium mit Neutronen einer (Po + Be)-Quelle"; Nachr.d.Akad.d.Wissensch. Göttingen, Math.-physik.Klasse 52-54.

/32/ 1946 "Die Isotopenhufigkeiten im natrlichen Blei und das Alter des Urans"; Naturwiss. 33, 185-187.

/33/ 1946 "Nachtrag zu der Mitteilung: Die Isotopenhufigkeiten im natürlichen Blei und das Alter des Urans"; Naturwiss. 33, 219.

/34/ 1947 "Über die thermische Dissoziation des Vakuums"; in collaboration with J.H.D.Jensen; Zeit.Naturforschg. 2a, 146-148.

/35/ 1947 "Das Alter des Urans"; Zeit.Naturforschg. 2a, 322-328.

/36/ 1948 "Messverfahren für Neutronen; Field Information Agencies Technical (Fiat)- Review 14, 13-47.

/37/ 1947 "Über Koinzidenzen beim -Zerfall und Zerfalls konstante des Rubidiums"; in collaboration with O.Haxel; Vortrag. Abstract, Physikal.B.letter Jg. 1947, 9, 319.

/38/ 1948 "Die Zahl der von einer (Ra + Be)-Quelle emittierten 'schnellen' Neutronen"; in collaboration with M.Teuscher; Zeit.f.Physik 124, 700-704.

/39/ 1948 "Gleichzeitige Emission von zwei Elektronen beim radioaktiven Zerfall des Rubidium 87"; in collaboration with O.Haxel; Zeit.f.Physik 124, 705-713.

/40/ 1948 "On the half-life of Rb87"; in collaboration with O.Haxel and M.Kemmerich; Phys. Rev. 74, 1886-1887.

/41/ 1950 "Photographische Bestimmung der Aktivittsverteilung in einer Manganknolle der Tiefsee"; in collaboration with H.v.Buttlar; Naturwiss. 37, 400-401.

/42/ 1950 "Die Halbwertszeit des K40"; in collaboration with O.Haxel and J.Heintze; Zeit.f.Physik 128, 657-667.

/43/ 1950 "Über Absorptionsmessungen an -Strahlen"; in collaboration with D.Vincent; lecture given by F.G.Houtermans on occasion of the Meeting of the Deutschen Physikalischen Gesellschaften in Bad Nauheim. Abstract, Verh.d.Deutschen Physikal.Ges 1950, p.71.

/44/ 1951 "Die Kernemulsionsplatte als Hilfsmittel der Mineralogie und Geologie"; Naturwiss. 38, 132-137.

/45/ 1951 "Das Primrspektrum der schnellen Neutronen einer (Ra + Be)-Quelle und deren unelastische St osse in Pb. II."; in collaboration with M.Teuscher; Zeit.f.Physik 129, 365-368.

/46/ 1951 "Über ein neues Verfahren zur Durchfhrung chemischer Alterbestimmungen nach der Blei-Methode"; Sitzung sberichte d.Heidelberger Akad.d.Wiss., Math.-naturw. Klasse, 2. Abh., Sitzung vom 24.2.1951, 123-136.

/47/ 1951 "Die Elektronenemission von Metalloberflchen als Nachwirkung einer mechanischen Bearbeitung oder Glimmentladung"; in collaboration with O.Haxel and K.Seeger; Zeit.f.Physik 130, 109-123.

/48/ 1951 "Photographische Messung des U- und Th-Gehaltes nach der Auflagemethode"; in collaboration with H.v. Buttlar; Geochim. Cosmochim.Acta 2, 43-61.

/49/ 1951 "Eine neue Ausfhrungsform der Uran-Blei-Methode zur Altersbestimmung uranhaltiger Mineralien"; Abstract, Discussions-Conference on Problems of Nuclear Physics and Cosmic Rays held in Heidelberg, 1-3 July, 1951.

/50/ 1951 "Die Isotopenverschiebung zwischen RaD und den stabilen Bleiisotopen"; in collaboration with P.Brix, H.v.Buttlar and H.Kopfermann; Nach.d.Akad.d. Wissensch. Göttingen, Math.-Phys.-Chem.Abtl. 7.12.1951.

/51/ 1952 Die Halbwertszeit des RaE"; in collaboration with F.Begemann; Zeit.Naturforschg. 7a, 143-144.

/52/ 1952 "Die Isotopieverschiebung zwischen RaD und den stabile Blei-isotopen"; in collaboration with P.Brix, H.v.Buttlar and H.Kopfermann; Zeit.f.Physik 133, 192-200.

/53/ 1952 "Les résultats préliminaires des mesures d'âge de la pechblende de Shinkolobwe par la méthode du RaD"; in collaboration with F.Bergemann, H.v.Buttlar, N.Isaac and E.Picciotto; Bull.de la Soc.belge de Gol., de Palontol. et d'Hydrol. Tme LXI, Sance du 17.6.1952, 223-226.

/54/ 1952 "Absoluteichungen energiereicher -Strahler mit dem 4 -Zhlrohr"; in collaboration with L.Meyer-Schtz meister and D.H.Vincent; Zeit.f.Physik 134, 1-8.

/55/ 1952 Nachtrag zu "Die Halbwertszeit des RaE"; in collaboration with F.Begemann; Zeit.Naturforschg. 7a, 763.

/56/ 1952 "Über den Einbau von Schwefel bei der elektrolytischen Abscheidung von Kupfer aus Kupfersulfatlosung und die Messung des Ionenradius bei der Bildung monomolekularer Oberflchenschichten"; in collaboration with D.Vincent und G.Wagner; Zeit.f.Elektrochemie -Berichte der Bunsengeselschaft f.physik.Chemie, 56, 944-946.

/57/ 1952 "Herstellung einer Radium-D-E-F-Standard-Lsung"; in collaboration with F.Begemann; Sitzungsberichte der sterr.Akad.d.Wiss. Wien, Math.-naturw. Klasse, Abtl. IIa, 161, Mitt.d.Instituts f.Radiumforschung, Nr.492, 245-249.

/58/ 1952 " -Spektren"; in collaboration with J.Geiss and H.Mller; Landolt-Bernstein, Zahlenwerte und Funktionen, I.Bd. Atom- und Molekularphysik, 5 Teil Atomkerne und Elementarteilchen, 414-470.

/59/ 1953 "Relative Eichmessungen radioaktiver Isotope"; in collaboration with L.Meyer-Schtzmeister, E.Schmid and D.H.Vincent; Strahlentheraphie 91, 135-148.

/60/ 1953 "Empirisches ber Eigenschaften der -Mesonen"; Komische Strahlung, 2.Aufl., herausgegeben von Werner Heisenberg, Springer Verlag, 11-131.

/61/ 1953 "Determination of the age of the earth from the isotopic composition of meteoritic lead"; Il Nuovo Cimento 10, 1623-1633.

/62/ 1953 "Application de la mthode du RaD la mesure de l'ge 'chimique' d'un mineral d'Uranium"; in collaboration with F.Begemann, H.v.Buttlar, N.Isaac and E.Picciotto; Geochim.Cosmochim. Acta 4, 21-35.

/63/ 1954 "Isotopenzusammesetzung und Radioaktivität von rezentem Vesuvblei"; in collaboration with F.Begemann, J.Geiss and W.Buser; Il Nuovo Cimento 11, 663-673.

/64/ 1954 "Über die Lokalisierung von Wismut in Gesteinen mit Kernemulsionsplatten"; in collaboration with M.Debeauvais, E. Jger and W.Buser; Tschermaks Min. Petr. Mitt. 3. Folge 5(1-2), 129-136.

/65/ 1954 "Problems of nuclear geophysics"; I. Cosmic radiation in past, II. The distribution of - radiactivity in rocks and minerals, III. Determination of the age of the earth from the isotopic composition of meteroritic lead; Supplemento al Nuovo Cimento N.2 , 11, 390-405.

/66/ 1954 "Isotopenzusammensetzung und Radioaktivität von rezentem Vesuvblei"; in collaboration with F.Begemann, J.Geiss and W.Buser; Helv.Phys. Acta 27, 175.

/67/ 1954 "Berne graphs"; in collaboration with P.Eberhardt, J.Geiss and P.Signer; distributed

1954.

/68/ 1954 "Zur freien Weglng von Neutrinos"; in collaboration with W.Thirring; Helv.Phys. Acta 27, 81-88.

/69/ 1954 "L'età della Terra e dell'Universo"; Supplemento al Nuovo Cimento, 12, 17-25.

/70/ 1954 "Bestimmung extremer Th/U-Verhlnisse durch Aktivittsvergleich radioaktiver Bleiisotopen"; in collaboration with F.Begemann, W.Buser and H.R. von Gunten; Chimia 8, 259-260.

/71/ 1955 "Brennstoff -Reserven für die Gewinnung von Kernenergie"; Bull.schweiz. elektrotechn.Ver. 46, 62.

/72/ 1955 "Isotopenverhlnisse von 'gewhnlichem' Blei und ihre Deutung"; in collaboration with P.Eberhardt and J.Geiss; Zeit. Physik 141, 91-102.

/73/ 1955 "Le applicazioni delle misure di deboli termoluminescenze alla geologia e alla mineralogia"; Atti del I Convegno di Geologia Nucleare, Roma, 4.4.1955, 9-12.

/74/ 1955 "Il piombo vulcanico del Vesuvio e di Vulcano"; in collaboration with P.Eberhardt, J.Geiss, W.Buser and H.R. von Gunten; Atti del I Convegno di Geologia Nucleare, Roma, 4.4.1955, 50-56.

/75/ 1955 "Radioaktive und stabile Isotope als Hilfsmittel der angewandten Forschung in Chemie und Physik"; Chimia 9, 256-257.

/76/ 1955 "Über Blei- und Schwefelisotopenverhltnisse in Bleiglanzen"; in collaboration with P.Eberhardt and J.Geiss; Helv.Phys. Acta 28, 339-341.

/77/ 1955 "Preliminary note on age determination of magmatic rocks by means of radioactivity"; in collaboration with H.M.E.Schrmann, A.C.W.C.Bot, E.Niggli and J.Geiss; Geologie en Mijnbouw, Nieuwe Serie, 17e Jaargang, Nr.9, 217-223.

/78/ 1955 "Proportionalzhlrohr zur Messung schwacher Aktivitten wiecher -Strahlung", in collaboration with H.Oeschger; Helv.Phys. Acta 28, 464-466.

/79/ 1956 "Mesures d'ge de quelques galnes de Madagascar"; in collaboration with H.Besairie, P.Eberhard and P.Signer; Compt.Rend des sances de l'Academie des Sciences, 242, 317-319.

/80/ 1956 "Klassische und moderne Physik"; Lecture; Abstract in "Der Bund" vom 25.2.1956, p.7.

/81/ 1956 "Les lments radioactifs en tant qu'horloge gologique"; La Suisse Horlogere, 71e Anne, No. 1, 3-6.

/82/ 1956 "Etude de la radioactivité de météorites métalliques par la méthode photographique"; in collaboration with S.Deutsch and E.E.Picciotto; Geochim.Cosmochim. Acta 10, 166-184.

/83/ 1956 "Radioactivity of iron meteoriotes by the photographic method"; in collaboration with S.Deutsch and E.E.Picciotto; Nature 177, 885-886.

/84/ 1956 "Les 'ges conventionnels' des galnes de certains gisements de plomb du Maroc"; in collaboration with G.Choubert, P.Eberahrdt, J.Geiss and P.Signer; Compt. Rend. des sances de l'Acad. des Sciences, tme 243, 286-288.

/85/ 1956 "Deuximme série de mesures d'âge de galenes de Madagascar"; in collaboration with H.Besairie, P.Eberhard and P.Signer; Compt.Rend. des sances de l'Acad. des Sciences, tme 243, 544-545.

/86/ 1956 "Mesure de l'âge de l'yttrocrasite de Mitwaba (Katanga) par la methode au plomb. II. Mesures isotopiques"; in collaboration with P.Eberhard, J.Geiss, H.R. von Gunten and P.Signer; Bull. de la Soc.Belge de Gol., de Palont. et d'Hydrologie, Bruxelles, tme 65, 251-256.

/87/ 1956 "Second preliminary note on age determinations of magmatic rocks by means of radioactivity"; in collaboration with H.M.E.Schrmann, A.C.W.C.Bot, J.J.S. Steensma, R.Sevinga, P.Eberhardt, J.Geiss, H.R. von Gunten and P.Signer; Geologie en Mijnbouw (Nw.Ser.) 18e Jaargang, 312-330.

/88/ 1956 "Kernenergie-Reserven"; Schweiz. Bauzeitung, 74 Jg. (49), 761-766.

/89/ 1957 "Messung der Thermolumineszenz als Mittel zur Untersuchung der thermischen und der Strahlung sgeschichte von natrlichen Mineralien und Gesteinen"; in collaboration with E.Jger, M.Schn and H.Stauffer; Ann.d.Physik 20, 283-292.

/90/ 1957 "Les projects de recherches physiques pour l'Anne Gophysique Internationale"; La Suisee Horlogre 72, 533-537.

/91/ 1957 "Radioaktivität und Lebensbedingungen"; Atomenergie Mitteillungsblatt d. Delegierten f.Fragen der Atom energie, No. 1, 25-29.

/92/ 1957 "Radioaktivität und Alter der Erde"; Die Naturwissen., 44, 157-163.

/93/ 1957 "Thermolumineszenz als Mittel zur Untersuchung der Temperatur und Strahlungsgeschichte von Mineralien und Gesteinen; in collaboration with H.Stauffer, Helv.Phys. Acta 30, 274-277.

/94/ 1957 "Die Bedeutung der Forschungsstation auf dem Jungfraujoch für die Erforschung der Kosmischen Strahlung"; in collaboration with M.Teuscher; Experientia, Suppl. VI, 33-35.

/95/ 1957 "Natural thermoluminescence of silicate meteorites; Cosmological and geological impications of isotope ratio variation"; Proc. of an Informal Conference, June 13-15, 1957, Nuclear Science Series, Report No. 23, 31-36 (Publication 572, National Academy of Sciences - National Research Council, Washington, D.C. 1958).

/96/ 1958 "Thermoluminenszenz als Mittel zur Dosimetrie"; in collaboration with H.Stauffer; Wissenschaftliche Fragen des zivilen Bevlkerungsschutzes, Garmisch, 135-137.

/97/ 1958 "Materie und Antimaterie"; Technische Rundschau 50, 4, 1-2.

/98/ 1958 "Study on -radioactivity in low concentration"; in collaboration with R.Dugnani Lonati, U.Facchini, I.Iori and E.Tongiorgi; Il Nuovo Cimento, Serie X, 7, 133-141.

/99/ 1958 "Proportionalzhlrohr zur Messung schwacher Aktivitten weicher -Strahlung"; in collaboration with H.Oeschger; Helv.Phys. Acta 31, 117-126.

/100/ 1958 "On a correlation between the common lead model age and the trace-element content of galenas"; in collaboration with L.Cahen, P.Eberhardt, J.Geiss, J.Jedwab und P.Signer; Geochim. Cosmochim. Acta 14, 134-149.

/101/ 1958 "Correlation between the isotopic composition of the lead of galenas and their silver content"; (A) in collaboration with L.Cahen, P.Eberhardt, J.Geiss and P.Signer;

Geochim. Cosmochim. Acta 14, 152-153.

/102/ 1958 "Alpha radioactivity of iron meteorites"; (second letter) in collaboration with S.Deutsch and E.Picciotto; Geochim. Cosmochim. Acta 14, 173-174.

/103/ 1958 "Radiation damage as a research tool for geology and prehistory"; in collaboration with N.Grgler and H.Stauffer; Supplemento agli Atti del Congresso Scientifico, Sezione Nucleare, 5a Rassegna Internazionale Elettronica e Nucleare, 16-20 giugno 1958, Roma, 275-285.

/104/ 1958 "The use of thermoluminescence for dosimetry and in research on the radiation and thermal history of solids"; in collaboration with N.Grgler and H.Stauffer; Proc. Sec.United Nations Internat. Conf. on the Peaceful Uses of Atomic Energy, Geneva 1. -13.9.1958, P/235 Switzerland, Vol. 21, 226-229.

/105/ 1958 "Special low-level counters"; in collaboration with J.Geiss, C.Gfeller and H.Oeschger; Proc. Sec. United Nations Internat. Conf. on Peaceful Uses of Atomic Energy, Geneva 1. - 13.9.1958, P/236 Switzerland, Vol. 21, 147-149.

/106/ 1958 "Determination of extreme Th/U ratios in minerals: A radiochemical method for determination of Thorium"; in collaboration with H.R. von Gunten and W.Buser; Proc. Sec. United Nations Internat. Conf. on the Peaceful Uses of Atomic Energy, Geneva, 1. - 13.9.1958, P/250 Switzerland, Vol. 2, 239-241

/107/ 1958 "Physics of K-particles"; in collaboration with Y.Eysenberg, W.Koch, E.Lohrmann, M.Nikolic, M.Schneeberger, P.Waloscheck and H.Winzeler; Proc.Sec. United Nations Internat. Conf. on the Peaceful Uses of Atomic Energy, Geneva 1. - 13.9.1958, P/256 Switzerland, Vol. 30, 220-222.

/108/ 1958 "Monitoring of fall-out radioactivity of rain and snow waters by means of liquid counters"; in collaboration with C.Mhlemann; Proc. Sec. United Nations Internat. Conf. on the Peaceful Uses of Atomic Energy, Geneva 1. - 13.9.1958, P/1959 Switzerland, Vol. 18, 559-562.

/109/ 1959 "Cl36 in Meteoriten"; in collaboration with Ch.Gfeller, W.Herr und H.Oeschger; Helv.Phys. Acta 32, 277-279.

/110/ 1959 "Isotopenanalysen des Osmiums aus Eisenmeteoriten und irdischen Proben"; in collaboration with W.Herr, E.Merz, J.Geiss and B.Hirt; Helv.Phys. Acta 32, 282-283.

/111/ 1959 "Über die Gefahren und ntigen Vorsichtsmassnahmen bei der Herstellung und Verarbeitung von thorium haltigen Glsern in der optischen Industrie"; Glas- Email-Keramo-Technik 10, 11, 429-433.

/112/ 1960 "Correlation between Forbush decreases of cosmic radiation and satellite drag"; in collaboration with H.Debrunner; Space Research, Proc. First Int.Space Sci. Symp. Nice, Januar 11. - 16.1960, North-Holland Publ. Comp., Amsterdam, 37-45 (Ed. N.K.Kallmann Bijl.).

/113/ 1960 "La préiodicité solaire diurne du rayonnement cosmique"; in collaboration with H.Debrunner and W.Lindt; Archives des Sciences (Genve) 13, 141-149.

/114/ 1960 "Die Blei-Methoden der geologischen Altersbestimmung" Geolog. Rundschau 49, 168-196.

/115/ 1960 "Über die Datierung von Keramik und Ziegel durch Thermolumineszenz", in collaboration with N.Grgler and H.Stauffer; Helv. Phys. Acta 33, 595-596.

/116/ 1960 "Registrierung der Nukleonenkomponente der kosmischen Strahlung am Jungfraujoch"; in collaboration with H.Debrunner and W.Lindt; Helv. Phys. Acta 33, 596-597.

/117/ 1960 "Über die Erfllung der 'Maser' -Bedingung bei Zer falls-Spektren von Moleklen"; Helv.Phys. Acta 33, 597-598.

/118/ 1960 "Der pltzliche, kurzzeitige Anstieg der kosmischen Strahlung vom 4.Mai 1960 nach der Registrierung der Nukleonenkomponente am Jungfraujoch"; in collaboration with H.Debrunner and W.Lindt; Helv.Phys. Acta 33, 706-708.

/119/ 1960 "Über Maser-Wirkung im optischen Spektralgebeit und die Mglichkeit absolut negativer Absorption für einige Flle von Moleklspektren (Licht-Lawine)"; Helv.Phys. Acta 33, 933-940.

/120/ 1961 "Thermoluminescence glow curves as a research tool on the thermal and radiation history in geologic settings"; Summer Course on Nuclear Geology, Varenna 1960, Pisa, Laboratorio di Geologia Nucleare, 233-253.

/121/ 1961 "Common lead and common occurrences of such elements which have a radiogenic isotope"; in collaboration with A. Eberhardt; Summer Course on Nuclear Geology, Varenna 1960; Pisa, Laboratori di Geologia Nucleare, 371-401.

/122/ 1961 " -Koinzidenzmessung zur zerstrungsfreien Messung des Gehaltes von Meteoriten an Positronenstrahlen und -aktiven Isotopen"; in collaboration with Ch. Gfeller, H.Oeschger and U.Schwarz; Helv.Phys. Acta 34, 466-469.

/123/ 1961 "Versuch zur Datierung von Eisenmeteoriten nach der Rhenium- Osmium- Methode"; in collaboration with W.Herr, W.Hoffmeister, B.Hirt and J.Geiss; Zeit. Naturforschg. 16a, 1053-1058.

/124/ 1962 "Die Tagesschwankungen der Nukleonenkomponente der kosmischen Strahlung am Jungfraujoch"; in collaboration with H.Debrunner; Helv.Phys. Acta 35, 137-146.

/125/ 1962 "Warum 'Institut für exakte Wissenschaften'"; Schweiz. Hochschulzeitung, 35. Jahrg. Heft 2, 75-81.

/126/ 1962 "Os187-isotope abundances in terrestrial and meteoritic osmium and an attempt to determine Re-/Os- ages of iron meteorites"; in collaboration with W.Herr, W.Hoffmeister, J.Langhoff, J.Geiss and B.Hirt; Radioisotopes in the Physical Sciences and Industry, International Atomic Energy Agency, Vienna, 29-36.

/127/ 1962 "Age determinations on lead ores"; in collaboration with P.Eberhardt, J.Geiss and P.Signer; Geol. Rundschau 52, 836-852.

/128/ 1962 "Radiation effects in space and thermal effects of atmospheric entry by thermoluminescence measurements on meteorites"; Technical Report AF EOAR, grant 61- 51, 29 pp.

/129/ 1963 "Radiation effects in space and thermal effects of atmosferic entry"; Technical Report AF EOAR, grant 62-116, 19 pp.

/130/ 1964 "Lead of volcanic origin"; in collaboration with A.Eberhardt and G.Ferrara; Isotopic and Cosmic Chemistry, North-Holland Publ. Comp., Amsterdam (Ed. H.Graig, S.L.Miller and G.J.Wasserburg), 233-243.

/131/ 1965 "The physical principles of geochronology"; Colloque International de Gochronologie absolue, Nancy, 3-8 Mai 1965. Sciences de la Terre, tme X, Nos. 3-4, 231-244.

/132/ 1966 "Thermoluminescence of meteorites"; in collaboration with A. Liener; J.Geophys. Res. 71, 3387-3396.

/133/ 1966 "Isotopenuntersuchngen zur Bestimmung der Herkunft rmischer Bleirohre und Bleibarren"; in collaboration with N.Grgler, J.Geiss and M.Grnenfelder; Zeit. Naturforschg 21a, 1167-1172.

/134/ 1966 "A search for cosmic-ray-produced 81Kr"; in collaboration with S.Aegerter, H.Oeschger and Rama; Earth Planet. Sci.Lett. 1, 256-258.

/135/ 1966 "History of the K/Ar-method of geochronology"; in "Potassium Argon Dating" (Eds. O.A.Schaeffer and J.Zhringer) Springer Verlag, Berlin, 1-6.

- B I B L I O G R A P H Y -

(1) Arthur and Lise London: "L'aveu".

(2) J.Geiss: Professor Dr. Friedrich Georg Houtermans, Helv. Phys. Acta, 39 (1966) 169-171: Bericht über die Tagung der Schweizerischen Physikalischen Gesellschaft, in Bern vom 29 und 30 April 1966.

(3) Physics Today, April 1966, p.126.

(4) The scientific meeting, held on Tuesday 20 October 1966 of the Deutsche Physikalische Gesellschaft, was opened with the "Gedenkrede" of Friedrich Georg Houtermans, pronounced by Professor Dr. Martin Teucher.

(5) Neue Deutsche Biographie, Band 9, 1972, edited by the Historische Kommission der Bayerischen Akademie der Wissenschaften, Duncker & Humblot, Berlin.

(6) Charlotte Riefenstahl (Fritz Houtermans' first wife) gave me copies of the following manuscripts she had written in various periods of her life for either her children (a, b, c) or close friends (d):
(a) A twentysix pages manuscript she wrote for her children about the family background of Fritz Houtermans;
(b) "And he was always right", ten typewritten pages, dated September 1948;
(c) A twentyone typewritten pages written after Fritz's death, during a vacation she took in July 1966 in Brione, high on the Lugano Lake.
(d) A seven typewritten pages circulated among a few friends in 1981 by Charlotte Riefenstahl

(7) Two letters by Friedrich Houtermans from Berlin to his mother, the first dated 1st August, 1940, the second 28th September, 1940.

(8) "A Chronological Report of My Life in Russian Prisons", consisting of 6 typed pages by F.G.Houtermans and dated May 19th, 1945.

(9) Copies of the following manuscripts by Aglaya Shteppa Corman have been given me by Charlotte Riefenstahl:
(a) A 5 pages letter by Aglaya Shteppa Corman to Charlotte, dated February 27, 1962, concerning the bad health of Fritz and his relations with Eric Kostantinovich Shteppa.
(b) "In Memory of Professor Dr. F.Houtermans", 7 typewritten pages by Aglaya Shteppa Corman.

(10) Paul Boschan a classmate of Fritz Houtermans and later a mathematician, who emigrated to USA where he worked for the Institute of Applied Econometrics. He gave me a number of useful information the 26th April 1980, when I visited him with Charlotte Riefenstahl in his house at 181 New Hamstead Road, New City, N.Y. 10956 and in later correspondence.

(11) Die Belagerung (1869-1870) Tagebuck des Joseph Houtermans, geb. am 10.2.1848 in Voerendaal, gest. am 12.8.1921 in Thorn: Zum 100 Jahrestag der Belagerung Roms herausgegeben von Albert Houtermans, Munchengladbach, und aus dem Niederländischen bertragen von seinen Patenonkel Christian Henskens. Eggelshoven, Niederlande. Das Original stellte freundlicherweise zur Verfügung: Peter Houtermans, Düsselsdorf. Fussnoten vonArno Houtermans.

(12) Kronprinz Friedrich Wilhelm of Hohenzollern ( )

(13) About H.Gomperz see in the "Michigan Quarterly Review", winter 1979, the article by Karl Popper "Three Worlds", footnote on p.13.

(14) Theodor Gomperz: "The Greek Thinkers".

(15) Gustav Landauer: "Ausruf zum Socialismus"

(16) Novum Testamentum, Graece et Latine, Eberhard Nestle, United Bible Society, London.

(17) Friedrich Wolfang Adler (Vienna 1879, Zürich 1960), Austrian politician, socialdemocratic, antimilitaristic, who, for protesting against the first World War, killed the Austrian Prime Minister Stürck. Condamned to death and amnistiated after the fall of the monarchy, he became a member, after the revolution, of the Austrian National Council. Exiled in 19 , he became secretary of the International Union of the Socialist Workers unti 1940. From 1940 to 1946 was in New York as President of the Austrian Labour Committe.

(18) **Gustav Wyneken**

(19) Manes Sperber from Yugoslavia is the author of the book "Der verbrannte Dornbusch".

(20) In the book "Out of the Night" (Alliance Book, New York 1941) Jan Valtin presents the history of the communist party in Germany before and during Hitler's time.

(21) See, for example, Vol 1, Chapter IV of Ref (22b).

(22) See, for example:
(a) Roy A.Medvedev: "Let History Judge, the Origins and Consequences of Stalinism", Vintage Books, New York, 1973.
(b) G.Boffa: "Storia dell'Unione Sovietica", Arnoldo Mondadori Editore, Milano (1976).

(23) See, for example, Chapter 6 of Ref (24).

(24) See, for example: W.L.Shirer: "The Rise and Fall of the Third Reich".

(25) The brother of Emmy Noether, Fritz Noether, professor of applied mathematics at the Breslau University, left Germany and went to the USSR with his wife and children, time before Fritz Houtermans and met him in the Butyrka prison as we shall see in Chapter 10.

(26) For more detail about the situation in Göttingen in those years see:
(a) M.Born: "My Life, Recollections of a Nobel Laureate", Taylor & Francis Ltd (London, 1978);
(b) W.M.Elsasser: "Memoirs of a Physicist in the Atomic Age", Science History Publications (New York, 1978).

(27) Remarks in this sense were made independently by E.Fermi and V.Weisskopf.

(28) "Leonium und andere Anekdoten um den Physikprofessor Dr. F.G.Houtermans" Herausgeber: Haro v.Buttlar, Bochum, 1982
Mitautoren: F.Begemann (Mainz), K.Hintermann (Hansen b.Brugg), H.Oeschger (Bern), I.Wandt (Hannover), W.Winkler (Würenlingen).

(29) George Eugene Uhlenbeck, born 1900 in Batavia (Indonesia), professor of physics at Utrecht (1935-1939), Ann Arbor (Mich., 1939-1961), Rockefeller Centre (New York, 1961-). Well known for the discovery, in 1925, in collaboration with S.Goudsmit, of the spin of the electron and for a number of important contributions to statistical mechanics.

(30) G.Gamow: "Zur Quantentheorie des Atomkerns", Zeit.f.Phys. 51 (1928) 204-212. The same ideas were published independently at the same time by: R.W.Gurney, E.V.Condon: "Wave Mechanics and Radioactive Disintegration", Nature (London) 122 (1928) 439; Phys.Rev. 33 (1929) 127-140.

(31) M. von Laue: "Notiz zur Quantentheorie des Atomkerns", Zeit.f.Phys. 5 (1928)726-734; J.Kudar: "Wellenmechanische Begründung der Nerststchen Hypotese von der Wiederentstehung radioaktiver Elemente", Zeit.f.Phys. 53 (1929) 166-167.

(32) G.Gamow: "My Early Memories of Fritz Houtermans", preface to the volume "Earth Science and Meteorites" dedicated to F.G.Houtermans on his sixtieth birthday, compiled by J.Geiss and F.D.Goldberg, North-Holland Publ.Co. Amsterdam, 1963.

(33) E.Ruska: "Zur Vor- und Frü-geschichte des Elektronenmikroskope", Eigth International Congress on Electron Microscopy, Camberra 1974, Vol 1, 1-5.

(34) G.Hertz: "Ein Verfahren zur Trennung von gasförmigen Isotopengemischen und seine Anwendung auf die Isotope des Neons", Zeit.f.Phys. 79 (1932) 108-121.

(35) G.Hertz: "Reindarstellung des schwerer Wasserstofflsotops durch Diffusion", Naturwiss 21 (1933) 884.

(36) P.Brix: "50 Jahre Kernvolumeneffekt in den Atomspektren", Phys.Bl. (1981) 181-183.

(37) V.Weisskopf: "Zur Theorie der Resonanzfluoreszenz", Ann.d.Phys. 9 (1931) 23-66.

(38) V.Weisskopf, E.Wigner: "Berechnung der natürlichen Linienbreite auf Grund der Diracschen Lichttheorie", Zeit.f.Phys. 63 (1930) 54-73; "Über die natürliche Linienbreite in der Strahlung des harmonischen Oszillator", Zeit.f.Phys. 65 (1930) 18-29.

(39) Sir Rudolf Peierls (b.Berlin, 1907) studied theoretical physics in Berlin, Münich, Leipzig, and at the ETH of Zurich, under Sommerfeld, Heisenberg, and Pauli. He has been Professor of Applied Mathematics at the University of Birmingham (1937-63) and of Theoretical Physics at the University of Oxford (1963)-74). In addition to many important papers on the theory of solids, nuclei, and fields, Peierls has published a few books: "Quantum theory of solids", (1955); The laws of nature", (1955); "Surprises in theoretical physics", (1979). He is the editor of "A perspective of physics", Vol.1 (1977), Vol.2 (1978). His autobiographic book ("Bird of Passage", Princeton University Press, 1985) is particularly interesting.

(40) For more information see: "Shoenberg, Sir Isaac", in Dictionary of National Biography.

(41) After the First World War Poland was reconstitued as an independent state with a way out to the Baltic Sea obtained by transferring to it the German Länder of Poznan and Pomerelia. This strip of Polish territory, about 50 kilometer wide, was called Polish Corridor or Danzig Corridor. It separated the Republic of Danzig and the German Land of East Prussia from the main part of Germany. During Hitler's period the trains connecting Germany to Danzig and East Prussia were closed and kept under strict watch by the Nazi police while travelling along the Danzig Corridor.

(42) P.M.S.Blackett (1897-1974) was awarded the 1948 Nobel Prize for physics for his development of the Wilson cloud chamber method and his discoveries therewith in the field of nuclear physics and cosmic rays. For a detailed biography, see: Bernard Lowell, Biographical Memoires of Fellows of the Royal Society, Vol.1 (1975) pp.1-115.

(43) Giuseppe P.S.Occhialini (b. Fossombrone (Pesaro) 1907) studied physics at the University of Florence passing his laurea examination in 1929. In 1930 he became assistant at the University

of Florence and the successive year with a fellowship of the Italian Consiglio Nazionale delle Ricerche (CNR) went to Cambridge where he started to work with P.M.S.Blackett, with whom he developed the technique of triggering a cloud chamber by means of two or more counters in coincidence. In 1937 he became associate professor at the University of S.Paulo in Brasil. When in 1942 Brasil entered the Second World War, he was obliged to give up his academic position until 1945 when Italy became cobelligerent. At the end of that war, Occhialini went to England where, after abortive attempts to serve the Allied cause, he moved to Bristol as DSIR fellow of the laboratory directed by C.F.Powell. He worked at this University until 1948, when he was offered a research position at the Centre de Physique Nucleaire in Bruxelles. In 1949 he was appointed professor at the University of Genoa and in 1952 he moved to the same chair at the University of Milan. His remarkable experimental work was devoted to particle physics, cosmic rays and, in later time, space research. His most outstanding contributions are the works with Blackett on the positron and electromagnetic showers and those with C.F.Powell which led, among others, to the discovery of pions.

(44) L.Szilard (1898-1964) very brilliant physicist, and later biophysicist, of Hungarian origin, who emigrated from Germany to Great Britain and later to USA. For a detailed biography see: E.P.Wigner: Biographical Memoirs, National Academy of Sciences 40 (1969) pp.337-347.

(45) Fritz Lange (-), physicist and engineer worked at the AEG (Berlin) on high voltage production. Under Hitler's pressure he emigrated to Great Britain and later to USSR.

(46) **P.Auger**

(47)

(48) Frédéric Joliot and Irène Joliot-Curie were awarded the 1935 Nobel Prize for chemistry in recognition of their synthesis of new radioactive elements. For their biographies see: P.M.S.Blackett: Jean Frédéric Joliot (1900-1958) Biographical Memoirs of Fellows of the Royal Society Vol.6 (1960) pp.87-105; J.Teillac: Irène Joliot Curie (1897-1956) Nucl.Phys. Vol.4 (1957) pp.497-502.

(49) Kirov, pseudonym of Sergej Mironovic Kostrikov, was member of the Politburo, Secretary of the Central Committee and the first Secretary of the Leningrad oblom (regional committee). He gave great contributions in solving complex problems of industrial development and collectivization of agricolture. He represented the tendency towards an internal "détente" after the great successes obtained in the industrial development. At the beginning of the XVIIth Party Congress, early in 1934, a group of Party officials had a talk with Kirov, touching on the need to replace Stalin. But Kirov would not agree either to get rid of Stalin or to be elected Secretary General. On December 1, 1934, a shot in the back killed Kirov in front of the door of his office. The shot was fired by a young Party member, Leonid Nikolaev, but immediately the rumor went around, in USSR and abroad, that the murder had been prepared by Stalin. This hypothesis seems to be backed by a few rather transparent hints given by Nikita Chruscev. The assassination of Kirov marks the beginning of the great political purges in USSR.

(50) E.Amaldi: "G.Placzek (1905-1955)", Ric.Scient. 26 (1956) 2037-2041;
L. Van Hove: "George Placzek (1905-1955)", Nuclear Phys. 1 (1956) 623-626.

(51) L.D.Landau, G.Placzek: "Struktur der Unverschobenen Streulinie", Phys.Zeit. Soviet Union 5 (1934) 172-173.

(52) The Institute of Physics of the University of Rome was located in the building, constructed between 1877-1880, in Via Panisperna 89A, and remained there until autumn 1936, when it was transferred to a new building at the Città Universitaria. All or almost all the work of Fermi and collaborators and pupils was carried on in the building of Via Panisperna.

(53) H.A.Bethe, born in 1906 in Strasbourg, has been a pupil and assistant to A.Sommerfeld in Munich. He emigrated to Great Britain in 1933 (Manchester and Bristol) and to USA in 1935 (Cornell University). He received the 1967 Nobel Prize for Physics for his contribution to the theory of nuclear reactions, especially his discoveries concerning the energy production in stars.

(54) Christian Möller (1904-1980), a pupil of Niels Bohr, studied the interaction between two relativistic electrons before the presently accepted methods of quantum electrodynamics, deriving the formula which bears his name. Subsequently he has played an important role in the development of both nuclear theory and quantum electrodynamics. He gave also important contributions to the theory of General Relativity.

(55) Niels Bohr (1885-1962), one of the greatest physicists and scientists of our century, not only was the first to quantize the energy states of the atom, but also inspired, for years, the work of many other first class physicists who contributed to develop the quantum mechanical description of atoms and nuclei. Among the many books devoted to Niels Bohr as man and scientist, see, for example: "Niels Bohr, a centenary volume" edited by A.P.French and P.J.Kennedy.

(56) George Gamow (Odessa, 1904 - Boulder, Colorado, 1968) immediately after his Ph.D. at the University of Leningrad (summer 1928) travelled to Göttingen, where he made his first major contribution to physics: the theory of nuclear alpha decay. During all his life he produced many other first class contributions to nuclear physics, astrophysics, cosmology and biology. He is also the author of many very brilliant popular books on science. His autobiography ("My world line", Viking Press, New York, 1952) is fascinating.

(57) Leon Rosenfeld (1904-1974), born in Belgium, studied at many of the leading theoretical centers in Europe. He has been professor in Liege, Utrecht (1940), Manchester (1947-57) and Copenhagen (since 1958) at the newly established Nordic Institute for Theoretical Physics (Nordita). Rosenfeld has been active in many fields of physics, including nuclear physics, field theory, statistical mechanics and history of science. He has been one of the closest collaborators of Niels Bohr.

(58) **Maria Goeppert Mayer** (1906-1972).

(59) **Jean Perrin**

(60) These texts can be found in Ref (71).

(61) F.Beck, W.Godin: "Russian Purge and the Extraction of Confession", Hurst and Blackett Ltd (London, 1951). This book was written during the second Göttingen period (Chapter 18) with the help of Höxter, an American journalist of German origin.

(62) **A.F.Ioffe** (1880-1960).

(63) A "prime number" is a natural number, other than 1, having no divisors except itself and the integer 1.

(64) **Van der Waerde**

(65) **Hugo Eberlein**

(66) Hellmut Andics: "50 Jahre unseres Lebens-Österreichs Schicksal seit 1918", Verlag Fritz Molden (Vienna) 1968.

(67) See the Foreword by Alexander Dallin to Konstantin Shteppa's book (Ref (133)).

(68) See chapter 7 of Ref (61): the first of the three cases discussed there is that of Konstantin Shteppa.

(69) "Whites" and "White army" were expressions used in contrapposition to "Reds" and "Red army" for indicating the Russian troops that during 1918-19 fought on the Russian territory a desperate and unsuccessful war against the communist government.

(70) In 1919 the Bolshevik Party, that one year before had adopted the name Communist Party, promoted the constitution of Komintern, or Third International, i.e. a unitary organization of all communist parties which were represented in proportion to the number of their members and strictly bound to follow the decisions taken by its organs.

(71) **David Holloway**

(72) A.Weissberg-Cybulski: "Hexensabbat", Pocket-book Suhrkamp no 369, 1977.

(73) P.L.Kapitza, E.M.Lifshitz: "Lev Davydovitch Landau (1908-1968) Biographical Memoirs of Fellows of the Royal Society 15 (1969) 141-158.

(74) F.Janouch: "Lev D.Landau: His Life and Work", CERN 79-03, 28 March 1979 (Geneva, 1979).

(75) **Paul S.Ehrenfest**

(76) They were Landau and Alikanian, Alikamow and E.M.Lifshitz.

(77) **Pugwash**

(78) "Prof. P.Kapitza and the USSR", Nature (London) 135 (1935) 755-756.

(79) "Dr. P.Kapitza Apparatus and the USSR", Nature (London) 136 (1955) 825.

(80) **(biografia Kapitza)**

(81) Robert Rompe was a distinguished physicist who worked mainly in industry, in particular for the Studiengesellschaft für Elektrische Beleuchtung (Osram). After the war he remained in the Deutsche Democratik Republik and reached important positions such Staatssekretar of the Kultur Ministerium.

(82) "Antroposophy" was a movement founded by the Austrian philosopher Rudolf Steiner (1861-1925), who derived it from Theosophy by assigning greater importance to the nature and destiny of man.

(83) **Kurschatov**

(84) See the Section "Houtermans als Mitarbeiter und Forscher" at pp.135-136 of Ref (63).

(85) Manfred von Ardenne: "Ein Glückliches Leben für Technik und Forschung, (Autobiographie)", Verlag der Nation (Berlin, 1973).

(86) W.Walcher: "Isotopentrennung", Ergebnisse d. exact. Naturwiss 18 (1939) 155-228.

(87) N.Bohr, J.A.Wheeler: "The Mechanism of Nuclear Fission" Phys.Rev. 56 (1939) 426-450, dated June 28, 1939.

(88) H.v.Halban, F.Joliot, L.Kowarski: "Liberation of Neutrons in the Nuclear Explosion of Uranium", Nature (London) 143 (1939) 470-471; "Number of Neutrons Liberated in the Nuclear Fission of Uranium", ibidem 680;
H.v.Halban, L.Kowarski, P.Savitch: "Sur la capture simple des neutrons thermiques et des neutrons de resonance par l'uranium", C.R. 208 (1939) 1396-1398;
H.Anderson, E.Fermi, L.Szilard: "Neutron Production and Absorption in Uranium", Phys.Rev. 56 (1939) 284-286;
L.Szilard, W.H.Zinn: "Emission of Neutrons by Uranium", Phys.Rev. 56 (1939) 619-624.

(89) E.Amaldi, E.Fermi: "On the absorption and slowing down of neutrons", Phys.Rev. 50 (1936) 899-928.
A.Kramish: "The Griffin, The Story of Paul Rosband, Britain's Master Spy in Nazi Germany", Houghton Mifflin Company, 52 Vanderbilt Avenue, New York, N.Y.10017 (1986). Fragmentary information about Rosbaud's work for the British Intelligence Service is given in Ref (97).

(90) **Louis Leprince-Ringuet**

(91) W.Heisenberg: "Über die Arbeiten zur technischen Ausnutzung der Atomenergie in Deutschland", Naturwiss. 33 (1946) 325-329. It is published in an abridged translation in Nature 160 (1947) 211.

(92) "Naturforschung und Medizin in Deutschland, 1936-1946", Field Information Agencies Technical (FIAT) - Review.
(a) Band 13: "Biophysik, Teil I", herausgegeben von B.Bajewsky and M.Schön;
(b) Band 14: "Kernphysik und Kosmische Strahlen, Teil II", herausgegeben von W.Bothe and S.Flügge, Verlag Chemie - GmbH - Weinheim, Bergstrasse, 1950.

(93) Samuel A.Goudsmit: "Alsos", P.Schuman, New York (1947); reprinted by Tomash Publisher, Los Angeles (1983).

(94) Erich Bagge, Kurt Diebner, Kenneth Jay: "Von der Uranspaltung by Calder Hall", Rowohlt Hamburg (1957). The contribution by Bagge and Diebner first appeared in the Rowohlt Deutsche Enzyklopedie, the contribution by Jay appears with the agreement of the U.K. Atomic Energy Authority (London).

(95) General Leslie R.Groves: "Now It Can Be Told", Harper & Brothers (1962).

(96) R.Gerwin: "Atoms in Germany, a report on the state and development of nuclear research and nuclear technology" Econ-Verlag, GmbH, Dusseldorf-Wien (1964).

(97) David Irving: "The Virus House", William Kimber and Co.Ltd., 6 Queen Anne's Gate, London (1967); or "The German Atomic Bomb", Simon and Schuster, New York (1967).

(98) O.Hahn, F.Strassmann: "Über den Nachweiss und das Verhalten der bei Bestrahlung des Urans mittels Neutronen entstehenden Erdalkalimetallen", Naturwiss.27 (1939) 11-15, received December 22, 1938.

(99) (a) L.Meitner, O.R.Frisch: "Disintegration of Uranium by Neutrons: A New Type of Nuclear Reaction", Nature (London) 143 (1939) 239-240, dated January 16, 1939; "Products of Uranium Nucleus", ibidem 471-472.
(b) S.Flügge, v.Droste

(100) O.R.Frisch: "Physical Evidence of the Division of Heavy Nuclei with Neutron Bombardment", Nature (London) 143 (1939) 276, dated January 16, 1939.

(101) F.Joliot: "Preuve experimental de la rupture explosive des noyaux d'uranium et de thorium sous l'action des neutrons", C.R.Acad.Sci. Paris 208 (1939) 341-346, presented at the meeting of January 30, 1939.

(102) For an excellent review of the work carried on before December 6, 1939 see: L.A.Turner: "Nuclear Fission", Rev.Mod.Phys. 12 (1940) 1-29.

(103) H.von Halban, F.Joliot, L.Kowarski: "Liberation of Neutrons in the Nuclear Explosion of Uranium", Nature (London) 143 (1939) 470-471; "Number of Neutrons Liberated in the Nuclear Fission of Uranium", ibidem 680; "Energy of Neutrons Liberated in the Nuclear Fission of Uranium Induced by Thermal Neutrons", ibidem 939;
M.Dodé, H.von Halban, F.Joliot, L.Kowarski: "Sur l'énergie libérés lors de la partition nucleaire de l'Uranium", C.R.Acad.Sci. Paris 208 (1939) 995-997.

(104) H.L.Anderson, E.Fermi, H.B.Hanstein: "Production of Neutrons in Uranium Bombarded by Neutrons", Phys.Rev. 55 (1939) 797-798.
L.Szilard, W.H.Zinn: "Instantaneous Emission of Fast Neutrons in the Interaction of Slow Neutrons by Uranium", ibidem 56 (1939) 619-624.

(105) S.Flügge: "Kann der Energie Inhalt der Atomkerne technisch nutzbar gemacht werden?", Naturwiss. 27 (1939) 402-410.

(106) This was done in particular by O.Frisch and R.Peierls in Great Britain.

(107) **G.B.Pegram**

(108) **E.Wigner**

(109) H.D.Smyth: "Atomic Energy for Military Purposes", Rev.Mod.Phys. 17 (1945) 351-471.

(110) S.R.Weart: "Scientists in Power", Harvard University Press, Cambridge, Mass. (1979).

(111) **Paul Harteck**

(112) The full text of this letter is given in the book by General Groves, Ref(95).

(113) **Georg Joos**

(114) Abraham Esau (-) was a leading authority on high-frequency electronics. He was politically active and had followed the rising star of nationalism in Germany.

(115) **Joseph Mattauch**

(116) According to D.Irving, who reports this conference at the beginning of chapter 2 of his book (Ref.(97)). the participants were: Esau (chairman), Joos, Hanle, Geiger, Mattauch, Bothe and Hoffmann, and the Ministry's representative, Dr.Dames.

(117) I use here a memo by S.Flügge, slightly longer than one typewritten page and dated: Göttingen, 3 October 1945. He had sent a copy of it to Charlotte Riefenstahl, first wife of F.G.Houtermans, who years later passed it to E.Amaldi on occasion of one of his visits to her (see the Preface).
Flügge's publication had also an epilogue: probably in autumn 1941. At that time Diebner, Wirtz and Stetter tried to obtain from Flügge a declaration that he had already worked in connection with the Heereswaffenamnt before his publication. They were trying to attribute to him a "priority" with respect to the French group. To this remark that such a declaration was not true, it was answered that, this was his "patriotic duty". Flügge's argument that the French group had published before him initially was not believed. Only later, when he presented the published papers, they renounced to get the declaration from Flügge.

(118) **Weizsäcker**

(119) **Walther Bothe**

(120) **Wolfang Gentner**

Hans Meier Leibniz

(121) S.A. means "Sturm Abteilung". It was the name of the paramilitary corp of Hitler's party. "Öbergruppen Führer" was a high rank officer of S.A.

(122) Gauleiter was the highest authority of Hitler's party in a province.

(123) **H.E.Süss**

(124) **J.H.Jensen**

(125) **W.Gerlach**

(126) The ten German scientists detained in Great Britain (at Farm Hall) were: Dr. Eric Bagge, Dr. Kurt Diebner, Prof. Walter Gerlach, Prof. Otto Haber, Prof. Paul Harteck, Prof. Werner Heisenberg, Dr. Horst Korseling, Prof. Max von Laue, Prof. Karl-Friedrich von Weizsäcker, and Dr. Karl Wirtz.

(127) E.Segrè: "From X-rays to Quarks. Modern Physicists and their Discoveries", Freeman, S.Francisco (1978).

(128) J.W.Kennedy, G.T.Seaborg, E.Segrè, A.C.Wahl: "Properties of 94 (239)", Phys.Rev. 70 (1946) 555-556, dated May 29, 1941, voluntary withheld from publication until the end of the war.

(129) R.Jungk: " Brighter than a Thousand Suns", Harcourt, Brace and Co., New York (1958).

(130) J.P.Lash: "Eleanor and Franklin, based on Eleanor Roosevelt's private papers", W.W.Norton & Co. Inc. New York. For this episode see p.704-705. Houtermans' message is discussed also by:
F.Herneck: "Eine alarmierende Botschaft", Spektrum 7 (1976) 32-34.

(131) See pp. 85-102 and 108-11 of: A.D.Beyerchen: "Scientists under Hitler", Yale University, New Haven and London (1977).

(132) **Wolfgang Paul**

(133) Konstantin F.Shteppa: "Russian Historians and the Soviet State", pp.IV-437, Rutgers University Press, New Brunswick, New Jersey, 1962. Posthumous Publication.

(134) See, for example: S.Weinberg: "Gravitation and Cosmology". "Pinciples and Applications of the General\ Theory of Relativity", John Wiley & Sons, New York (19).

(135) P.A.M.Dirac: "A new basis for cosmology", Proc.Roy.Soc. A165 (1938) 199-208.

(136) P.Jordan: "Bermerkungen zur Kosmologie", Ann.d.Phys. 36 (1939) 64-70; "Über die Entstehung der Sterne", Phys. Zeit. 45 (1944) 183-190.

(137) I express my warmest thanks to G.P.Occhialini and C.Dilworth Occhialini for these informations.

(138) H.B.G.Casimir, born in the Hague in 1909, has been for many years extraordinary professor at the University of Leiden and Director of the Physics Laboratory of Philips at Eindhoven and later President of the Royal Academy of Sciences of Amsterdam. He worked in close contact, especially during his formation period, with Niels Bohr in Copenhagen, and W.Pauli in Zurich. He has worked as a theoretical physicist on the applications of the group theory to quantum mechanics [Casimir operator] on the thermodynamics of superconductors and on the Van der Waals forces [Casimir effect]. In his memory's book "Haphazard Reality, Half a Century of Science" (Harper and Row Publ., New York, 1983) he mentions Houtermans, his life's adventures and jokes at various points (pp.133 and 220 to 223.

(139) Three international expeditions, promoted by C.F. Powell of the University of Bristol and based on the support of the Milan, Padua and Rome Universities, were made in the Mediterranean area in the years 1952, 1953 and 1954. The Physical Laboratories of the following Universities took part in the second one (June-July 1953): Bern, Bristol, Brussels, (Université Libre), Catania, Copenhagen, Dublin, Genoa, Göttingen (Max Planck Institute),

London (Imperial College), Lund, Milan, Oslo, Padua, Paris (Ecole Politechnique), Rome, Sydney, Turin, Trondheim, Uppsala.

(140) "Encounter, literature, arts, current affairs", December 1955.

(141) H.Greinacher: "Über eine Methode Wechselstrom mittels elektrischer Ventile und Kondensatoren in Hochspannten Gleichstrom unzuwanden", Zeit.f.Phys. 4 (1921) 195.

(142) J.D.Cockcrof, E.T.S.Walton: Proc.Roy.Soc. A136 (1932) 619-630; ibidem 137 (1932) 229-242.

(143) H.Greinacher: "Eine Neue Methode zür Messung der Elementarstrahlen", Zeit.f.Phys. 36 (1926) 364; "Über die Rekistirerung von α-und H-strahlen nach der neuen elektrischen Methode", Zeit.f.Phys. 44 (1927) 319.

(144) J.Chadwick: Nature (London) 129 (1932) 512; Proc.Roy.Soc. A136 (1932) 692-708.

(145) Willard Frank Libby (1908-1980) at the end of his career was director of the state wide University of California Institute of Geophysics and Planetary Physics. He discovered the natural alpha-particle radioactivity of samarium and invented the method for determining the age of archeological artifacts by means of the radioactivity of ^{14}C. For this work he was awarded the 1960 Nobel Prize for Chemistry.

(146) Ezio Tongiorgi (b. 1913 in Milan), after the accomplishment of his studies at the University of Pisa, has devoted most of his research activity to paleobotanic, geology and paleontology of the Quaternary. From 1953 he has started to apply nuclear physics methods to geological problems, creating in Pisa a well known laboratory of nuclear geology.

(147) Erza Edgard Picciotto (Italian citizen b. in Istanbul, Turkey, 1921) has studied in Brussels, where he became "Docteur en Sciences" of the Université Libre in 1952. He made most of his carrier at the same University, except for a number of study periods spent at foreign institutions. His research activity refers to the geochemistry of stable and radioactive isotopes in rocks, ocean and atmosphere. In the period 1957 to 1966 Picciotto concentrated his work on the chemistry of antartic ices, taking part in a number of Belgian and American expeditions to the Antartic. He likes to remind that his orientation and formation as a chemist was influenced by Irène Curie and mostly by Giuseppe Occhialini and Friedrich Houtermans.

(148) Harold Clayton Urey (1893-1981) is best known for his discovery in 1932, in collaboration with F.G.Brickwedde and G.M.Murphy, of the deuterium, for which he was awarded the 1934 Noble Prize for Chemistry. In those years he became the leading authority in isotope chemistry and later one of the major experts in the problems of the history of the solar system and the origin of life.

(149) See, for example, the article: Kurt Meyer: "Kernphysik und Erdenspinschaften", Neue Zurcher Zeitung, March 7, 1966.

(150) C.L.Cowan, F.Reines, F.B.Harrison, H.W.Kruse, A.D. McGuire: "Detection of Free Neutrino: a Confirmation", Science 124 (1956) 103;
F.Reines, C.L.Cowan: "Free Antineutrino Absorption Cross Section I, Measurements of the Free Antineutrino Absorption Cross Section by Protons", Phys. Rev. 113 (1959) 273.

(151) E.Fermi: "Tentativo di una teoria dei raggi beta", Ric.Scient. 4(2) (1933) 491; "Versuch einer Theorie der β-Strahlen I", Zeit.f.Phys. 88 (1934) 161.

(152) R.P.Feynman, M.Gell-Mann: Phys.Rev. 109 (1958) 193.

(153) The first substantial paper by A.Holmes is: "An Estimate of the Age of the Earth", Nature (London) 157 (1946) 680-684.

(154) A.O.Nier: "The Isotopic Constitution of Radiogenetic Leads and the Mesaurements of Geological Time II", Phys.Rev. 55 (1939) 153;
A.O.Nier, R.W.Thompson, B.F.Murphy: III, ibidem 60 (1941) 112.

(155) C.Patterson, G.Tilton, M.Ingram, H.Brown: "Concentration of uranium and lead and the isotopic composition of lead in meteoritic material", Phys.Rev. 92 (1935) 1234-1235.

(156) C.Patterson, G.Tilton, M.Ingram: Bull.Geol.Soc.Amer. 64 (1953) 1461.

(157) The first good measurement of atmospheric K^{81} was done by H.H.Looshi, H.Oescher: ^{37}Ar ^{37} and Kr^{81} in the Atmosphere, Earth Planet Sci.Lett. 7 (1969) 67-71. They obtained 0.10± 10% disintegrations per minute per liter. According to more recent results one has 0.08 d/min x liter.

(158) "New Names on the Moon" in "Sky and Telescope", March 1974, p.170-171.

(159) **Georg**

(160) Iwanenko expressed his views about this point to Occhialini, while the three others told their opinion on this matter to Weisskopf.

(161) **Kharkov (1944)**

(162) **Snegov: "The Creators"**

(163) I express my thanks to Constance Dilworth Occhialini for her letter of 11 November 1985.

(164) **Bruno Touschek**

Section C

European Physics and CERN

CERN has been called "the greatest monument to the work of Edoardo Amaldi". Involvement with the founding and growth of the large European laboratory for high energy physics indeed constitutes the core of Amaldi's scientific activity: from the pioneering days when the idea was first conceived through to the planning stage, and from the interim period to the final creation of the permanent organisation and the subsequent developments which established CERN as a leader in the field. Amaldi was one of the key figures involved from the very beginning when the project, the result of a joint European effort to establish a large laboratory, began to materialize in the late 1940s. He served as the Secretary-General of the "provisional CERN" during the two crucial years from 1952 to 1954, which led to the final ratification by member states of the convention, hence giving life to a permanent organisation. He also supported the first Director-General, Felix Bloch, by acting as the Deputy Director-General from September 1954 to February 1955 when Bloch was replaced by Cornelis Bakker. Amaldi became the President of the Scientific Policy Committee of CERN from 1958 to 1960, a member of the CERN Council from 1960 to 1973, the Vice-President of the Council from 1960 to 1962, and its President in 1970 and 1971.

The first two texts in this section refer to the early years of CERN, from the late 1940s to approximately the mid-1950s. Although there is some overlap between them, as they both deal with the same events over the same period of time, they nonetheless look at the story from perspectives that are different enough to deserve separate consideration. The first, *Niels Bohr and the Early History of CERN*, was one of Amaldi's two contributions to the conference held in Rome in February 1985 on the centennial anniversary of Niels Bohr's birth (the other can be found in Section B in Part I of this volume). It gives an account of how the idea of the joint laboratory was conceived and the preliminary stages of construction, and pays particular attention to the role played by Bohr, one of the most authoritative speakers in European physics, in the process. It is interesting to note that in preparing his talk, Amaldi could rely not only on his personal recollections and the documentation he had collected throughout his years at CERN, but also on the extensive research conducted by the team of historians who were just about to complete the first volume of the history of CERN (which was published in 1987). In fact, their results had already appeared in provisional form as a series of CERN reports in 1983. As a member of the Advisory Committee of the CERN History Project, Amaldi was in touch with them and had closely followed their progress.

The Beginning of Particle Physics: From Cosmic Rays to CERN Accelerators is an after-dinner address delivered by Amaldi at the international conference held in Bristol in July 1987 to celebrate the fortieth anniversary of the discovery of the pi meson. The original programme of the conference had listed Victor Weisskopf as the after-dinner speaker. However, as illness prevented Weisskopf from attending the meeting, Amaldi received an invitation in June to speak on the same topic in his place, with "many thanks for agreeing to talk at such short notice".

The third and last paper of this section covers later ground in CERN's history. John Adams appeared on the CERN scene in the early 1950s as a member of the group working on the design and construction of the PS, the proton synchrotron. It had been selected as the new laboratory's first goal and eventually came into operation at an energy of 25 GeV (later to become 28) under Adams' direction at the end of 1959. Later, his mastery in accelerator engineering and ability as a manager led to the successful completion of the programme of the SPS, the large proton synchrotron, which together with the intersecting storage rings (ISR) opened the second phase of CERN's experimental facilities in the mid-1970s. Under Adams' direction, the SPS was actually completed and reached its planned energy of 300 GeV in 1976, the same year in which Adams was appointed the Director-General of CERN together with Leon Van Hove. Their mandate expired at the end of 1980. *John Adams and His Times* is the text of a lecture delivered by Amaldi at CERN on 2 December 1985, and is the first John Adams Memorial Lecture following Adams' death in 1984.

Niels Bohr and the Early History of CERN

Edoardo Amaldi*

My contribution is based only in part on my memory and personal diary of those years. I have also used the remarkable work carried out by the group of young historians.[1] led by Armin Hermann, professor of History of Science and Technology at the University of Stuttgart. Until now, this work has been circulated in the form of "preliminary reports" as a preparation of a detailed history of CERN – from the beginning to 1965 – in two volumes of about 600 pages each, that will be published by North Holland. The first volume will appear in 1986, the other about one year later. From this extensive presentation, John Krige, one of the members of Hermann's group, keeping in contact with his colleagues, will extract a more concise history, contained in a single volume of 300 pages and addressed to a wider public.

I had also the priviledge of consulting the record on magnetic tape of the speech that Leon van Hove, who has been Director General of CERN from January 1976 to December 1980, pronounced on May 6, 1985, at CERN on occasion of the celebration of Niels Bohr centenary [2] .

* E. Amaldi, Dipartimento di Fisica, Università di Roma "La Sapienza".

[1] In addition to Armin Hermann, from the Federal Republic of Germany, the group is composed of: John Krige from South Africa, Dominique Pestre from France, Ulrike Mersits from Austria, all full time, and Lanfranco Belloni from Italy, part time.

[2] In the afternoon of May 6, 1985, CERN celebrated the centenary of Niels Bohr with two speeches: Abraham Pais: "A tribute to Niels Bohr", CERN 85-17, 13 November, 1985; Leon van Hove: "Niels Bohr and the creation of CERN".

For time reasons I can not summarize here the succession of events that, starting from the late 1940s brought to the creation of CERN on September 29, 1954. At that date the provisional Organization, established on February 1952, came to an end, passing to the permanent Organization the task of pursuing and developing the work of planning, designing and constructing the European laboratory and its two accelerators.

I should only recall that already at the end of the 1940s in many European and overseas places, scientists were becoming aware of the continuously increasing gap between the means available in Europe in the field of nuclear physics and elementary particles and the means available in the United States, where a few high-energy accelerators had started to produce results, while others had already reached advanced stages of construction or design. It was becoming more and more evident to many physicists, in Europe and Overseas, that such a situation could be changed only by a considerable effort made in common by many European countries. These views of the scientists were also in harmony with those of many politicians.

In those years in many European countries, in particular in France, Italy, West Germany and Belgium, the idea of moving towards some form of economic and/or political unification of at least a considerable part of the old continent, was considered of primary importance by many authoritative politicians, who adopted it as a guiding principle of their immediate and long range program of action.

I remember that in the years 1948-1950 the various problems arising in connection with the construction of a large accelerator by an international organization were examined in Rome in frequent discussions between Bruno Ferretti, professor of theoretical physics, and myself, and in letters exchanged with Gilberto Bernardini, who, in those years, was at Columbia University, where he had been invited by I.I. Rabi. I remember that I became aware that similar problems were discussed in other European countries, in particular in France, when I heard of the European Cultural Conference held in Lausanne in December 1949. At the meeting a message from Louis de Broglie was read by Raoul

Dautry, Administrator of the French Commissariat à l'Energie Atomique. In the message the proposal was made to create in Europe an international research institution without mentioning, however, nuclear physics or fundamental particles. To the message of de Broglie Dautry added the proposal that the Conference study the ways of strengthening collaboration in two fields, in astronomy and astrophysics by building powerful telescopes and all necessary auxiliary material, and in the field of atomic energy, by setting up a centre with all the required modern apparatus.

As we see from this example, at that time the opinions were still rather vage about the type of research to be tackled by the new organization and even more about the nature of the collaboration to be established.

In June 1950 the General Assembly of UNESCO was held in Florence and Isidor I. Rabi, who was a member of the delegation from the United States, made a very important speech about "the urgency of creating regional centres and laboratories in order to increase and make more fruitful the international collaboration of scientists in fields where the effort of anyone country in the region was insufficient for the task". In the official statement, approved unanimously by the General Assembly along the same lines, neither Europe nor high-energy physics were mentioned. But this specific case was clearly intended by many people, in particular by Rabi himself and by Pierre Auger, who was Director of the Department of Natural Sciences of UNESCO.

I knew Rabi very well since 1936, and on occasion of his trip to Florence, I had with him a thorough conversation about the future European laboratory.

A further endorsement of this idea came from the International Union of Pure and Applied Physics (IUPAP), which was at that time under the presidency of H.A. Kramers from the Netherlands. I was one of the vice-presidents and asked Kramers, at the beginning of the summer 1950, to include the discussion of Rabi's proposal, with specific reference to Europe and high-energy physics, in the agenda of the meeting of the Executive Committee of IUPAP that was to take place at the beginning of September of the same year at MIT in Cambridge, Mass. As a conclusion of a rather long discussion I was asked by the Executive Committee of IUPAP to get in contact with Rabi and with physicists from various European countries in order to clarify the aims and structure of the new organization and to help in the coordination of the different efforts. My first step was to write to Auger who in the meantime has presented the problem to the Conference on nuclear physics held in Oxford during the month of September where, according to reports of Auger and Ferretti (also present in Oxford), in the discussion that followed Auger speech, Niels Bohr proposed "to begin by building a big apparatus to accelerate particles" where by "big accelerator" was meant a machine providing about 1 GeV.

After the UNESCO Conference in Florence Auger had the authority to act but there was no money appropriated on the scale required for a detailed expert study of such a project.

Exchanges of views between Auger and Denis de Rougmont, director of the European Cultural Centre (ECC), founded at the already mentioned Lausanne meeting of 1949, brought Denis de Rougmont to convene in Geneva on December 12, 1950 a committee for scientific co-operation for discussing Rabi proposal at the UNESCO conference in Florence. The meeting was announced very late and many people could not participate because of other commitments they had taken for the same days. For example, I had already accepted to participate in a conference on elementary particles, taking place in India and therefore asked to be replaced by my colleague and friend B. Ferretti. The participants in the December 1950 meeting in Geneva were (in addition to a few people of ECC): P. Auger, P. Capron (B), B. Ferretti (I), H.A. Kramers (NL), P. Preiswerk (CH), G. Randers (N), M. Rollier (I), and J.L. Verhaege (B). The Commission concluded its works with a series of recommendations, the most important of which was "the creation of an international laboratory centred on the construction of an accelerator capable of producing particles of an energy superior to that foreseen for any other accelerator already under construction", i.e. the cosmotron of Brookhaven (3 GeV) and the Bevatron of Berkeley (6 GeV). The Commission also discussed and endorsed the estimates brought by Ferretti of the cost of such a machine obtained by comparison with the cost of the two American machines mentioned above.

Immediately after this meeting G. Colonnetti, President of the Consiglio Nazionale delle Ricerche of Italy, R. Dautry, Administrator of the French Commissariat à l'Energie Atomique and J. Willelms, Director of the Fond National de la Recherche Scientifique in Belgium made available to Auger some funds which, all together, amounted to about $ 10.000. This sum, although very modest, was sufficient for Auger to initiate the first steps for arriving at the planning and construction of a large particle accelerator.

At the beginning of 1951 Auger established a small office at UNESCO and invited me to Paris, at the end of April, to discuss the constitution and composition of a Working Group of European physicists interested in the problem. The first meeting of this "Board of Consultants" was held at UNESCO, in Paris, from 23 to 25 of May 1951. The members of the Board were: E. Amaldi (I), P. Capron (B), O. Dahl (N), F. Goward (UK), F.A. Heyn (NL), L. Kowarski and F. Perrin (F), P. Preiswerk (CH), and Alfvèn (SW) was also present in place of I. Waller.

Two goals were considered: a long-range, very ambitious, project of an accelerator second to none in the world, and in addition the construction of a less powerful and more standard machine which could allow at an earlier date experimentation in high-energy physics by European teams. For various reasons in the paper summarizing the conclusions reached by the consultants only the large machine was esplicitly mentioned and underlined.

These conclusions were made immediately known, but were received by various people in completely different ways.[3] This became very clear on occasion of the colloquium, sponsored by IUPAP, on "Problems of Quantum Physics" that was organized by Niels Bohr and Stephan Rozental and took place in Copenhagen from 6 to 10 July, 1951. This colloquium was followed by the 7th General Assembly of IUPAP, that also was held in Copenhagen,

[3] Studies in CERN History: D. PESTRE: CHS-2, June, 1983; CHS-3, March, 1984; CHS-9, May, 1984.

from 11 to 13 of the same month. During these two meetings a number of discussions took place in the lecture rooms as well as in the corridors and during the meals or the entertainments, about the programme worked out by the group of UNESCO consultants.

Bohr and Kramers that, as I said before, was president of IUPAP, started to be afraid of the consequences, especially of the financial consequences, of this programme that appeared too ambitious to them. They doubted the wisdom of starting immediately the construction of a large machine. It was better, they thought, to procede by steps, for example by constructing first a small machine, an idea that had been stressed by Perrin and Verhaege at the meeting of the Board of Consultants and had been proposed also by other people (for example, Wideroe, in Switzerland).

In discussions parallel to the IUPAP meeting the agreement was reached of recommending to proceed in three phases: the first phase should be devoted to the study of two accelerators, one big, the other rather small, and also to the creation of an institute for advanced studies. The second phase should be centered on the construction of the small machine, while the construction of the large machine should be tackled only in the third phase, starting, perhaps around 1955 or later.

This agreement was clearly a compromise and did not satisfy neither the people that favoured the programme proposed by the UNESCO Board of Consultants, nor those that shared the views of Bohr and Kramers. During summer 1951 James Chadwick from UK who was an old friend of Bohr, by correspondence with him and on occasion of a long visit he paid to Copenhagen, became a supporter of the idea of starting with an international coordination around the Copenhagen Institute and an international use of the large facilities that could become available: in Liverpool a 400 MeV Synchrocyclotron and in Uppsala a 200 MeV Synchrocyclotron.

The possible construction of a large machine could be examined in a later stage.

The people of UNESCO and of the Board of Consultants proceeded in their work and made attempts for taking into account the line of thought of Bohr, Kramers and Chadwick. There was no difference in principle as it is shown, for example, in a letter of October 1951 of Bohr to Auger and where Bohr refers to "the great European effort with which, in principle, everybody so deeply sympatizes". The question was what to do and in which order.

The Board of Consultants held two other meetings one in October, the other in November, but did not succeed in appeasing the worries of the other side. These worries became particularly clear when the 28 and 29 November, 1951, all the leading scientists from the Nederlands and Norway gathered in Kjeller, Norway, for the inauguration of the research reactor that had been constructed together by the Dutch and the Norwegian.

During this same period Auger was preparing an official "meeting of government representatives", in order to take the first important intergovernamental step towards CERN. He was very worried to see that the difficulties had remained and called another meeting of the consultants on December, 14, 1951.

Three days later, on December 17, 1951 started, in Paris, the "Conference on the Organization of studies relating to the establishment of a European Nuclear Physics Laboratory" that had been prepared by Auger and his staff but, formally, was called by the Director General of UNESCO. Shortly after the opening of the meeting the Dutch delegation introduced a compromise solution trying to combine the two approaches. This compromise resolution was the central topic of discussions but the delegates were not able to reach an agreement. Everybody felt, however, that a remarkable progress had been made towards a solution agreable to all participants, and that a re-consideration of the whole matter was desirable. Under these circumstances the meeting decided, on December 20, to interrupt its work and to resume the discussions in a second session to be held in February in Geneva.

The second session of the UNESCO meeting took place in Geneva on February 12 to 15, 1952 (Fig. 1).

An agreement was easily reached. This was embodied by a formal intergovernamental document which was signed the February 15, by the Representatives of eleven European States, but not by the Representative of United Kingdom, that, at that stage, desired

Fig. 1 - Niels Bohr greets Paul Scherrer, Chairman of the UNESCO meeting, at the opening of its second session in Geneva, on February 12, 1952. The third gentleman appearing in the photograph is Messieur Picot, Chairman of the Department of Education of the Cantoon of Geneva, who represented Switzerland and the Republic of Geneva in offering to CERN the Meyrin site.

to keep the position of "Observer". The title of this important document was: *Agreement constituting a Council of Representatives of European States for planning an International Laboratory and organizing other forms of co-operation in Nuclear Research.*

This title shows the spirit of compromise pervading the document. This feature is further stressed by Article 1, which reads:

A Council of Representatives of European States is hereby constituted for planning an International Laboratory and organizing other forms of co-operation in Nuclear Research.

At the second session in Geneva the proposal was introduced by the Danish delegates to have a large scientific conference in June 1952, in Copenhagen, in order to discuss in depth the scientific aims and the basic scientific equipment of the new laboratory.

The word "Council" used in the Agreement of February 1952, still appears today in the acronim CERN: Conseil Européenne pour la Recherche Nucléaire.

The Agreement had to be ratified by the parliaments of the Member States, since the financial contributions were fixed in its Annex. Enough ratifications were obtained rather rapidly so that on May 2, 1952 the Agreement came into force and the Council could start its work.

The first decision of the provisional Council was the creation of four groups and the nomination of the officers in charge of them.

Two groups had the task of designing and starting the construction of the two machines, a third group that of taking care of the laboratory and the subsidiary equipment.[4] The fourth Working Group, that could start immediately to operate, was the Group for Theoretical Studies, "to be based in Copenhagen" and Niels Bohr accepted of becoming its leader, what was not so light for a man of his stature, already at the age of about 67. He also became a regular participant of the Council Meetings of the provisional Organization. In the same meeting I was nominated Secretary General, with the task of coordinating all the activities of the new Organization, in particular the work of the four groups.

The Executive Group, composed of the leaders of the four groups and the Secretary General, who chaired it, had regular meetings where the most important decisions were taken about all problems arising in a rapidly growing organization. Niels Bohr took part if not in all, certainly in many of these meetings and with

[4] The leaders of these three groups were: C.J. Bakker (NL) for the SC, O. Dahl (N) with F. Goward (UK) as deputy for the PS, and L. Kowarski (F) with P. Preiswerk (CH) as deputy for the Lab. group.

his wide vision of all cultural and general problems, his kindness and wisdom, gave a great help in points of substance as well as of stile.

The large Copenhagen scientific Conference, proposed in February by the Danish Delegation, took place, as foreseen, in June 1952. On purely scientific reasons the conclusion was reached by a overwelming majority of the physicists participating in the Conference, that "Europe should try to construct a proton-synchrotron for energies between 10 and 15 GeV".

Immediately after the Copenhagen Conference, the June 20-21, the provisional Council held its 2nd session in Copenhagen. The conclusions reached by the scientific Conference were presented to the Council by Werner Heisenberg. The Council accepted them and approved also another Danish proposal that fitted into the notion of other forms of co-operation in nuclear research. Namely another international co-operation in the field of cosmic rays. The result was that one year later, in 1953, a very large expedition of baloon flights was organized in Sardinia with the participation of many European universities. It gave data for studying the properties of the new particles produced in nuclear emulsion by high altitude cosmic rays.

The 3rd session of the provisional Council was held in Amsterdam on October 1952 and took three important decisions that I should mention here:

1) They decided where the European Laboratory should be. They decided unanimously for the Geneva site.

2) They decided that the big machine should be a proton-synchrotron of about 30 GeV. The change of energy was due to the rediscovery, in August 1952, by Courant, Livingston and Snyder of the "strong focusing" principle[5] which allowed a remarkable reduction of the amount of iron and copper necessary for constructing a

[5] E.D. COURANT, M.S. LIVINGSTON, H.S. SNYDER: "The Strong Focusing Synchrotron. A New High Energy Accelerator", *Phys.Rev.*, *88*, 1952, 1190-1196. I have used the word "rediscovered" because the same principle had been found by the Greek engineer Christophilos in 1950. In the same year he applied for a USA patent, but his application was not noticed by the experts in the field and therefore remained ignored.

large magnet for a fixed value of the magnetic field and, therefore, permitted to reach a higher energy of the accelerated particles for the same total cost of the machine.
3) The Council created a committee on other forms of co-operation [6] and Niels Bohr accepted to chair it.

I should add that after the adoption of the strong focusing principle in the construction of the proton-synchrotron, Niels Bohr became much more favourable to this machine which, by involving the application of a new principle, was loosing the unpleasant feature of a scaled up version of already existing accelerators.

After the decision of the provisional Council to place the European laboratory in the vicinity of Geneva, many Scandinavian physicists felt necessary to take steps concerning the future, in the long run, of the Copenhagen Institute, where Niels Bohr for about three decades had created and lead the most extraordinary institute of theoretical physics even seen. The February 17, 1953 a meeting took place in Göteborg in Sweden of Danish, Norwegian and Swedish physicists, where they elaborated the plan to establish, with the other Scandinavian countries Finland and Iceland, a Nordic Institute for theoretical atomic physics to be based in Copenhagen. This is the institute that became Nordita. They also realized that Geneva will become the centre of high energy experimental physics and inavoidably will attract also the corresponding theory, but that, in spite of that, there was room for something else for maintaining the long tradition of international collaboration that had prevailed over decades in Copenhagen under the leadership of Niels Bohr. The Nordita became one of the leading institutes in theoretical low energy nuclear research, it achieved two Nobel Prizes in 1975, Aage Bohr and Ben Mottelson, and pioneered the collaboration with the East, with the Soviet Union and the countries of the Soviet block and, more recently, with China.

Returning to the early history of CERN I should recall that in

[6] The other members of this committee were: E. Amaldi, Secretary General, M.S. Dedijer (Y), W. Heisenberg (FRG) and F. Perrin (F). The Committee's task was of making proposals concerning other forms of co-operation and particularly the selection of candidates wishing to work with the existing facilities put at disposal of the Council.

parallel with the activities mentioned above the lawyers prepared the "Convention for the Establishment of the European Organization for Nuclear Research". This Convention was signed in Paris on July 1, 1953 and entered into force when a prescribed point in the ratification procedure was reached on September 29, 1954.

The Council of the permanent Organization held its first session in Geneva on October 7-8, 1954 and among many other important decisions appointed Felix Bloch Director General of the (permanent) Organization and created the Scientific Policy Committee (SPC). The first chairman of the SPC was Werner Heisenberg, and Niels Bohr accepted to be a member.

Bohr continued to be the leader of the theoretical group in Copenhagen until the September 1, 1954, when he was replaced by his pupil and collaborator Christian Møller. This group continued its activity until the end of 1956. At its 6th Session the Council decided the December 14, 1956, the termination of the CERN Theoretical Group in Copenhagen and the end of the activity of C. Møller as its leader. About one month later the Council, at its 7th session, created the Theory Division in Geneva, with Bruno Ferretti as leader.

Niels Bohr did not participate in all meetings of the SPC but whenever he went to Geneva for such a meeting he took an active part. For example in one of these meetings, in November 1955, Bohr intervened in the discussion of the future programme of research to be carried out with the synchrocyclotron, not too far from completion, pointing out that not all countries would be able to send a complete research team, and that one had to foresee the formation of teams composed of individuals from various countries.

In another SPC meeting of November 1956, in a discussion about bubble chamber pictures, Bohr stressed the importance of automatic scanning and measuring methods and suggested that the pictures after the first analysis at CERN, could be rescanned in other laboratories, perhaps, for other types of events. At the same meeting the SPC accepted the proposal of terminating the existence of the theoretical group in Copenhagen on October 1, 1957. Bohr gave an account of the activities carried out in Copen-

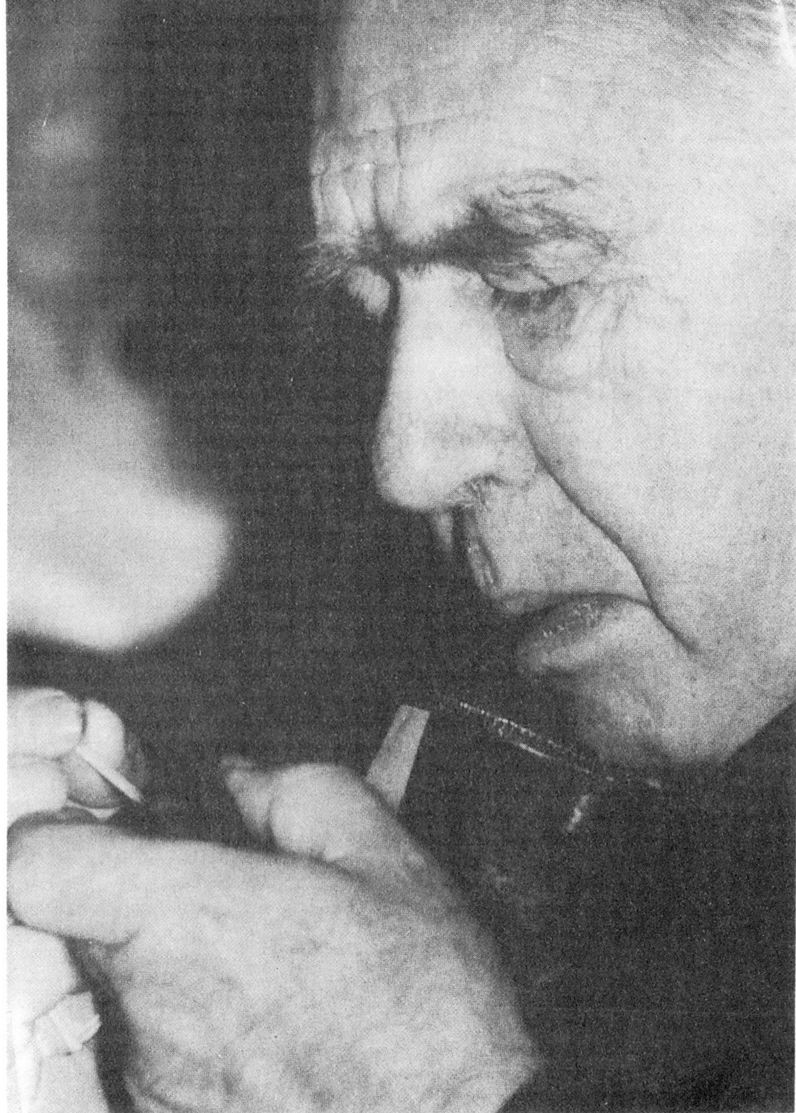

Fig. 2 - A photograph of Niels Bohr, taken at CERN, in April 1953, while he is trying to light his pipe.

hagen and pointed out that the Theory Division at CERN had now two tasks:
1) To carry out theoretical studies and provide guidance in direct connection with the experimental work made with the accelerators;
2) To provide the advanced education of the young physicists, which until then had been taken care of by the Copenhagen group.

In another meeting of the SPC, held in April 1959, an important discussion took place about the experimental research programme of the PS which was expected to be put in operation in a few months. Actually a full energy beam circulated in the PS on November 24 of the same year. Bohr intervened in the discussion for supporting the concept of "mixed teams" proposed by Massey of London, i.e. teams including, in addition to outside scientists, also a few CERN staff which would facilitate the utilization of the technical services of the Meyrin Laboratory by the outside scientists and technicians.

The last meeting of the SPC attended by Niels Bohr was the 22nd meeting, held in Geneva on November 25, 1961 (I recall that N.Bohr died on November 18, 1962). In November 1961 UK had financial difficulties and had requested a reduction of CERN budget. Some people had proposed the closing down of the synchrocyclotron but Niels Bohr said to be so strongly against such a decision "that the possibility of doing so should not even be mentioned". He stressed that this machine is "an integral part of CERN which produced remarkably good results".

As I said before the proton-synchrotron entered into operation on November 24, 1959. A cerimony of inaguration of this machine took place at CERN on February 5, 1960. This occasion was a great feast for all the people that had conceived or contributed to plan and construct CERN and its machines. That day, in front of a large audience composed of authorities of Member States, staff of CERN and scientists from many laboratories and Universities in Europe, Niels Bohr pressed the botton putting the PS into operation (Fig. 3).

I like to stress today, about 25 years later, that Niels Bohr was the right person to do that for many reasons, a few of which

Fig .3 - Photograph of Niels Bohr pressing the button to put the protonsynchrotron into operation on occasion of the inauguration of this machine on February 5, 1960.

should be mentioned here as a conclusion of my speech.

The first point is that the Institute of Theoretical Physics of the University of Copenhagen, created and led by Niels Bohr for a few decades starting from the early 20s, has been always a quite exceptional centre of investigation and thinking about the more fundamental laws of nature, based on the collaboration between scientists coming from any part of the world. When after the Second World War a number of people started to think, here and there, about the possibility of creating an European Centre of research of completely new dimensions, all of them had in mind, consciously or unconsciously, this unique example, unique not only for the scientific results achieved there, but also for the close and effective collaboration between people coming from different countries that Niels Bohr had promoted and developed.

The fact that Niels Bohr adhered from the beginning to the general idea of an European venture in the field of high energy physics, has been also of paramount importance. His authority, not only as a scientist, but also as an upholder of all basic human values, was so high to give an extraordinary strength to the endeavour to create CERN.

His participation in the life of CERN during its first years of existence, as the leader of the Theoretical Group in Copenhagen, and as a Member of the SPC, has provided an unvaluable guarantee in front of the Governments of the Member States and a reassuring sign of the continuity between the great atomic and nuclear physics tradition of the 20s and 30s, with the set of problems that the physicists and engineers of younger generations entering in this new adventure, were facing.

All these people, and I belong to them, recall Niels Bohr with profound admiration, lasting gratitude and great affection.

Paper presented at 40 Years of Particle Physics, Bristol, 22–24 July 1987 109

The beginning of particle physics: from cosmic rays to CERN accelerators

Edoardo Amaldi
Department of Physics, University of Rome "La Sapienza", Rome, Italy.

ABSTRACT: The author summarizes the most relevant events that brought to the creation of CERN as a permanent Organization the October 7, 1954. The first initiatives started to appear after the middle of the 1940's and lead, the February 15, 1952, to the creation of a provisional Organization which accomplished its task in the successive two and a half years. At the end the author reminds the glorious tradition of Great Britain in particle physics and concludes with an appeal for avoiding that in the near future it may be voluntarily interrupted.

1. INTRODUCTION

The Second World War was ended, also in the Pacific, before the end of August 1945 and already on 24 October of the same year the United Nations Organization (UN) was born during a conference held in San Francisco. The decision to create the United Nations Educational, Scientific and Cultural Organization (UNESCO) was taken in London in November 1945, by the delegates of forty-five nations. The constitution of this worl-wide Organization took place in Paris in 1946. Its main inspirers and organizers originated from European countries and Europe was the geographic region where these ideas, shortly afterwards, found the most fertile ground for the generation and development of regional organizations not closed in itself but open towards the rest of the world.

In this climate a number of mission oriented organizations were set up to tackle problems belonging to specific scientific or technological areas by an effort made in common by many European countries.

The European Organization for Nuclear Research (CERN) was the first European organization of this type to be created. The first official step took place at the General Assembly of UNESCO in Florence in 1950, a provisional Organization came into being in 1952, and the final permanent Organization - still in activity and evolution today - started to operate in 1954.

The history of CERN is of considerable interest not only from the point of view of high-energy physics, but also in general as the first example of a research laboratory of intergovernamental nature, created in Europe, that has been operated successfully for slightly more than thirty years, thus providing a remarkable model for the creation of other international organizations.
North-Holland has recently published the first volume of the "History of CERN" prepared by Armin Hermann, professor of History of Science at the University of Stuttgard, helped by a few younger historians. This 600 pages volume covers the years from 1949 to 1954. A second volume is scheduled for publication at the end of 1988. It will cover the period 1955-1965.

From the extensive presentation contained in the already published volume, John Krige, one of the members of Hermann's group, keeping in contact with his colleagues, will extract a more concise CERN history, contained in a single volume of 300 pages and addressed to a wider public.

My speech, of course, will be of a completely different nature. I did not consult the archives of the Foreign Ministries or of the Research Councils of the Members States of CERN or of other intergovernamental organizations. What I say is based on a few well known documents, on my personal diary of that period and on a few reports and lectures, prepared years ago by Kowarsky or by me.

Everybody agrees that the early history of CERN can be divided into three periods. The first one includes the first initiatives and extends from the middle of the 1940s to 15 February, 1952. On this date the representatives of eleven European governments signed in Geneva the so-called Agreement, establishing a provisional organization with the aim of planning an International Laboratory and organizing other forms of co-operation in nuclear research.

The second period, the so-called planning stage, extends from February 1952 to 1 July, 1953 when the Convention establishing the final organization was signed by the representatives of twelve European Governments.

Finally, the third period, usually called interim stage, runs from July 1953 to 29 September, 1954, when a prescribed point of the ratification procedure by the Parliaments of the Member States was reached and the Convention entered into force.

I limit my presentation to these three early phases because of their peculiar nature and also because I was directly involved in them.

Niels Bohr (left) and G.P.Thomson in Geneva, 12-15 February 1952.

Pierre Auger in Geneva, 12-15 February 1952.

2. THE FIRST INITIATIVES

It is not easy to fix a date for the beginning of the first stage, the stage of the first initiatives.

I may start by mentioning that already in October 1946 a proposal was submitted by the French Delegation to the "UN Economic and Social Council" of creating "United Nations Research Laboratories" and that the "UN Atomic Energy Commission" created also in 1946 with the aim of bringing atomic weapons under some form of international control, was one typical circle were the problem of international collaboration, especially in nuclear physics, was raised and discussed.

These, as well as another initiative appeared in 1949 and involving the Netherlands, Belgium and the Scandinavian countries, refer mainly to low energy physics and its applications and, therefore, should be considered, in my opinion, more as precursors of the idea of Euratom than of CERN.

In those years in many European countries, in particular in France, Italy, West Germany and Belgium, the idea of moving towards some form of economic and/or political unification of at least a considerable part of the old continent, was considered of primary importance by many authoritative politicians who adopted it as a guiding principle of their immediate and long range program of action.

Another favourable element was the scientific, technical and administrative experience gained in a few countries, during and immediately after the war, of wide and complex organizations operating in the field of the nuclear sciences and their applications. This experience had brought about the creation in United States of a few big research laboratories such as the Argonne National Laboratory and the Brookhaven National Laboratory. In particular the Brookhaven National Laboratory had been created and run - very successfully, - by the Associated Universities Inc. Furthermore the dimensions of the geographic region involved and of the laboratory were very similar to those of a possible future European research establishment. For Europe, however, a program of research had to be chosen completely free from any limitation or restriction originating from military, political or even industrial secrecy.

At the same time in many European and overseas places, scientists were becoming aware of the continuously increasing gap between the means available in Europe for research in the field of nuclear physics and elementary particles and the means available in the United States, where a few high-energy accelerators had started to produce results, while others had already reached advanced stages of construction or design. It was becoming more and more evident that such a situation would be changed only by a considerable effort made in common by many European nations.

To these one should add a further important element: immediately after the Second World War cosmic ray research had a very high level in Western Europe and was in part carried out through successful international collaborations.

The discovery, in 1946, by Conversi, Pancini and Piccioni that the cosmic ray mesotron is a weak interacting particle and not a Yukawa particle, the discoveries in 1947 by Lattes, Muirhead, Occhialini and Powell of the π-meson and of its decay into the muon, and of the first two strange particle by Rochester and Butler (a Λ° and a θ°) are the most brilliant results of an extensive and rich production obtained in Western Europe.

As I have pointed out in a lecture I gave at the Varenna School in summer 1972, the mountain laboratories in Switzerland (Jungfraujoch), France (Pic du Mid) and Italy (Testa Grigia) and even more the nuclear emulsions laboratories of the Universities of Bristol and Bruxelles, led respectively by Powell and by Occhialini, had become in those years points of encounter of young physicists originating from many different countries.

The life in common in mountain huts and the co-ordination of the experiments planned by different groups paved the way to the idea of wider and more ambitious collaborations which were popping out in various places in Europe.

Cosmic ray physicists, of course, were interested in a European laboratory only if devoted to high-energy physics. I remember that in the years 1948-1950 the various aspects of the problem, including energy and cost of machines, were examined in Rome in frequent discussions between Bruno Ferretti and myself and in letters exchanged with Gilberto Bernardini, who in those years was at Columbia University where he had been invited by I.I.Rabi.

I remember that I became aware that similar problems were discussed in other European countries, in particular in France, when I heard of the European Cultural Conference held in Lausanne in December 1949. At the meeting a message from Louis de Broglie was read by Raoul Dautry, Administrator of the French Commissariat à l'Energie Atomique. In the message the proposal was made to create in Europe an international research institution without mentioning, however, nuclear physics or fundamental particles. To the message of de Broglie Dautry added the proposal that the Conference study the ways of strengthening collaboration in two fields, in astronomy and astrophysics by building powerful telescopes and all necessary auxiliary equipment, and in the field of atomic energy, by setting up a center with all the required modern apparatus.

As we see from these examples, at that time the opinions were still rather confused about the type of research to be tackled by the new organization and even more about the nature of the collaboration to be established.

In June 1950 the General Assembly of UNESCO was held in Florence and Isidor I. Rabi, who was a member of the delegation from the United States, made a very important speech about "the urgency of creating regional centres and laboratories in order to increase and make more fruitful the international collaboration of scientists in fields where the effort of anyone country in the region was insufficient for the task". In the official statement approved unanimously by the General Assembly along the same lines, neither Europe nor high-energy physics were mentioned. But this specific case was clearly intended by many people, in particular by Rabi himself and by Pierre Auger, who was Director of the Department of Natural Sciences of UNESCO.

I knew Rabi very well since 1936, when I had visited Columbia University for the first time. Since then I had revisited him in his laboratory a few times, before and after the Second World War. In June 1950, on occasion of his trip to Italy, I had with him a thorough conversation about the future European laboratory.

A further endorsement of this idea came from IUPAP which was at that time under the presidency of Kramers. I was one of the vice-presidents and I had asked Kramers, at the beginning of the summer 1950, to include the discussion of Rabi's proposal, with specific reference to Europe and high-energy physics, in the agenda of the meeting of the Executive Committee of IUPAP that was to take place at the beginning of September of the same year in Cambridge, Mass. Although Kramers could not preside at the meeting because of bad health, the problem was discussed at length under the chairmanship of Sir Darwin assisted by the Secretary General Fleury. As a conclusion I was asked to get in contact with Rabi and with physicists from various European countries in order to clarify the aims and structure of the new organization and to help in the co-ordination of the different efforts. My first step was to write to Auger who in the meantime had presented the problem to the conference on nuclear physics held in Oxford during the month of September.

After Rabi's proposal Auger had the authority to act but there was no money appropriated on the scale required for a detailed expert study of such a project.

In December 1950, Denis de Rougemont, director of the European Cultural Centre (which was founded at the already mentioned Lausanne meeting of 1949) in agreement with Auger called at Geneva a commission for scientific co-operation for discussing Rabi proposal. The participants, in addition to a few people of ECC, were: P.Auger, P.Capron (B), B.Ferretti (I), H.A.Kramers (NL), P.Preiswerk (CH), G.Randers (N), M.Rollier (I), and Verhaege (B).

The Commission concluded its works with a series of recommendations, the most important of which was "the creation of an international laboratory centred on the construction of an accelerator capable of producing particles of an energy superior to that forseen for any other accelerator already under construction", i.e. the cosmotron of Brookhaven (3 GeV) and the Bevatron of Berkeley (6 GeV).

The Commission also discussed and endorsed the estimates brought by Ferretti of the cost of such a machine obtained by comparison with the cost of the two American machines mentioned above.

Immediately after this meeting G.Colonnetti, President of the Consiglio Nazionale delle Ricerche of Italy, R.Dautry, Administrator of the French Commissariat à l'Energie Atomique and J.Willelms, Director of the Fond National de la Recherche Scientifique in Belgium made available to Auger some funds which, all together, amounted to about $ 10 000. This sum, although very modest, was sufficient for Auger to initiate the first steps for arriving at the planning and construction of a large particle accelerator.

W.Heisenberg (left) and A.Hocker, delegates from the Federal Republic of Germany. Geneva, 12-15 February 1952.

Alfen from Sweden in Geneva, 12-15 February 1952.

At the beginning of 1951 Auger established a small office at UNESCO and invited me to Paris at the end of April to discuss the constitution and composition of a working group of European physicists interested in the problem.

The first meeting of this "Board of Consultants" was held at UNESCO, in Paris, at the end of May 1951. The members of the "Board were: E.Amaldi (I), P.Capron (B), O.Dahl (N), F.Goward (UK), F.A.Heyn (NL), L.Kowarski and F.Perrin (F), P.Preiswerk (CH), and Alfven (S) in place of I.Waller.

Two goals were immediately established: a long-range, very ambitious, project of an accelerator second to none in the world, and in addition the construction of a less powerful and more standard machine which could allow at an earlier date experimentation in high-energy physics by European teams.

In this and a few successive meetings of the Board of Consultants a few other features of the new organization were also examined, as preparatory work for a conference of delegates of governments that was convened by UNESCO in Paris, in December 1951. This conference took place under the chairmanship of Franois De Rose, a French diplomat who had been involved in discussions with French scientists (P.Auger, F.Perrin, L.Kowarski) about the future European laboratory from the time of the UN Commissions mentioned above, and was elected, a few years later, President of the Council of CERN. It led to the signing of the Agreement, which took place in Geneva in February 1952.

3. THE PLANNING STAGE

The first problem tackled by the Council of the provisional organization created by the Agreement was the nomination of the officers responsible for the appointment of the remainder of the staff and for planning the laboratory.

Bakker from the Netherlands was nominated Director of the Synchro-Cyclotron Group, Dahl from Norway (with Goward from the U.K. as deputy) Director of the Proton-Synchrotron Group, Kowarski from France (with Preiswerk from Switzerland as deputy) Director of the Laboratory Group, which had to take care of site, buildings, workshops, administrative forms, financial rules, etc., Niels Bohr Director of the Theoretical Group and finally E.Amaldi as Secretary General with the task of maintaining cohesion between the four groups of which the provisional organization was composed.

Almost all these people had worked on the Board of Consultants, nominated by UNESCO, during the first stage and contributed later to the creation of the new organization and to the development of its activities.

In July of the same year, 1952, an international nuclear physics conference was held in Copenhagen; on that occasion the type of accelerator to be built as the main goal of the new European organization was amply discussed. A report of the conclusions reached by the participants was presented by Heisenberg to the Council which held its Second Meeting in Copenhagen immediately after the Conference. Thus the decision was taken that the Proton-Synchrotron Group should explore the possibility of constructing a 10 GeV proton-synchrotron which, at that time, represented the biggest machine in the world.

During the month of August Dahl and Goward went to Brookhaven in order to study in detail the Cosmotron (with a maximum energy of about 3 GeV) that was very close to completion. During their two-week visit, and in some way in connection with the discussions going on in relation to the European project, Courant, Livingstone and Snyder came out with the "strong-focusing principle".

This important discovery came soon enough to allow a change of the plans of the provisional organization: with the approval of the Council, the PS Group embarked on the study of a strong-focusing PS of 20-30 GeV instead of thc weak-focusing 10 GeV machine considered until then.

During the summer of 1952 four sites were offered for the construction of the new laboratory: one near Copenhagen, one near Paris, one in Arnhem in the Netherlands and one in Geneva.

After long and lively discussions the site in Meyrin near Geneva was unanimously selected.

Another point I would like to touch about those times concerns the participation of the European nations.

All the European members of UNESCO had been invited to the Conference opened in Paris in December 1951 and closed in Geneva on 15th February 1952, but no response came from the countries of Eastern Europe. Furthermore while the Agreement was signed at that date by the representatives of eleven countries, the Convention establishing the permanent organization was signed (the July 1st 1953) by twelve countries.

The difference was due to the U.K. which, at the beginning, was rather cautious in committing itself to take part in the new organization. The U.K. government preferred to remain in the formal position of an observer during the first two stages, while, by signing the Convention, it became a full-right member of the permanent organization.

I remember that in the autumn of 1952 it was decided that Bakker, Dahl and myself should go to Brookhaven to take part in the dedication of the cosmotrom that was foreseen for December 15.

Our trip was already arranged when I was called on the telephone from London by Sir John Cockcroft who had been from the start very much in favour of the participation of the U.K. in the new venture.

As a consequence of our conversation on the telephone I decided to leave earlier and to pass through London on my way to Brookhaven.

In London I went to the D.S.I.R. (Department of Scientific and Industrial Research) where I met its chairman Sir Ben Lockspeiser, who a few years later was elected president of the permanent CERN.

After rather long discussions about various organizational and financial aspects of the project as a whole, Sir John and Sir Ben brought me to Lord Cherwell who, at that time, was a member of the Churchill government.

Lord Cherwell appeared to be very clearly against the participation of the U.K. in the new organization. As soon as I was introduced in his office he said that the European laboratory was to be one more of the many international bodies consuming money and producing a lot of papers of no practical use. I was annoyed and answered rather sharply that it was a great pity that the U.K. was not ready to join such a venture which, without any doubt, was destined to full success, and I went on by explaining the reasons for my convictions. Lord Cherwell concluded the meeting by saying that the problem had to be reconsidered by His Majesty's Government.

When we left the Ministry of Defense, where the meeting had taken place, I was rather unhappy about my lack of self- control, but Sir John and Sir Ben were rather satisfied and tried to cheer me up.

A few weeks later Sir Ben Lockspeiser wrote me an official letter asking that the status of observer be given to the U.K. in the provisional organization, and the D.S.I.R. started to regularly pay "gifts", as they were called, corresponding exactly to the U.K. share calculated according to the scale adopted by the other eleven countries.

In spite of its very particular legal position, the U.K. gave in practice fundamental support to the provisional organization and was the first among the European countries to ratify the Convention.

4. THE FIRST SCIENTIFIC RESEARCHES UNDER THE AEGIS OF CERN IN ITS EARLY STAGES

The exploration of high altitude by means of balloons to study cosmic radiation had been tried extensively with success in the U.K., particularly by the Bristol Goup, around 1950. This group thus arrived at the discovery of the charged τ – and K-meson.

In 1951 it was suggested that a study of high energy events produced by cosmic radiation could be more conveniently carried out at lower latitudes. At a latitude of 40°, for example, only that part of the primary radiation which has an energy above 7 GeV per nucleon is allowed by the magnetic field of the Earth to enter the atmosphere. Thus, at these latitudes the magnetic field of the Earth removes a large number of primaries which are not very efficient in producing the new secondary particles because of their relatively low energy. The nearer one goes to the equator the stronger this effect is.

Thus a first international expedition was planned in 1952 under the sponsorship of CERN that recently had entered its "Planning Stage" (Sect.3). Naples and Cagliari were chosen as the most convenient of the available bases in the Mediterranean. Thirteen universities took part in the expedition which met some success. In particular was the first successful attempt to recover balloons at sea. Furthermore, the expedition made a survey of the winds at high altitudes which was of great importance for the expedition of the subsequent year. The results of this first expedition were briefly reported to the Third Session of the Council of CERN held on 4-7 October 1952 in Amsterdam.

A second expedition took place, also to Sardinia, in June-July 1953. This was much on the same lines but on a larger scale than the first one. Eighteen laboratories from European countries in addition to one from Australia took part in it. 25 balloons were launched and over 1000 emulsions were exposed 7 hours at altitudes between 25 and 30 km. This corresponds to 9.27 litres of nuclear emulsions weighting 37 kg. They were successfully recovered at sea on account of the employment of a seaplane and a corvette (Pomona) generously placed at disposal of the expedition by the Italian Airforce and the Italian Navy.

While in the 1952 expedition only glass-backed emulsions had been used, in 1953 expedition "stripped emulsions" were introduced. Thus, the tracks were followed almost always from one emulsion to the adjacent one, a point of great importance when high-energy events are studied. The development required special care and in particular the construction of special developing systems at the Universities of Bristol, Padua and Rome, where the emulsions of the whole expedition were processed.

In October 1953, a meeting was held at the Department of Physics of the University of Bern (Switzerland) for distributing the packages of exposed and developed emulsion among the participating universities and, in April 1954, an international conference was organized in Padua to discuss the first results and plan jointly the most efficient methods for the investigation.

On the 25 November 1959 J.Adams announced to the CERN Staff that on the evening before the PS has been put into operation.

Francis Perrin in Geneva, 12-15 February 1952.

Further results appeared in the normal scientific literature and in lectures given by various participants at the second Course of the Varenna International Summer School that took place during the summer 1954. At the same course, Fermi gave a series of "Lectures on pions and nucleons" which were his last contribution to the teaching of physics before his death in Chicago on November 29 of the same year.

Among the main results of the expeditions I can recall a number of determinations of the values of the masses of heavy mesons giving rise to different decays which brought suspicion that all these processes could be alternative decays of the same particle. Also the identification of the various hyperons and the determination of their masses and decay modes were considerably improved.

In July of the same year, 1953, a very important conference was held at Bagnéres de Bigorre in the French Pyrénées where the new heavy particles were discussed at length. At this meeting, for example, the Dalitz plot for the τ-meson was discussed by Dalitz, Michel presented the selection rules for decay processes and the Rome group introduced the logarithm of tgθ (later called rapidity) as a very appropriate variable for the description of high-energy events in which many secondary particles are emitted.

The speaker at the conference dinner was Cecil Powell who, in brilliant terms, pressed for the urgency of constructing big European accelerators for avoiding that in the old continent subnuclear particle research could be drawn by the mounting level of the sea of machine results already noticeable in the United States.

A third expedition was organized by the Universities of Bristol, Milan and Padua in October 1954 from Northern Italy (Novi Ligure). It consisted in the laughing of a single stack of 15 litres of nuclear emulsions (corresponding to a weight of 63 kg) which is indicated in the literature as "G-stack".

In the landing, which took place in the Apennines, the stack was partly damaged. The delicate operation of recovery of the damaged emulsions was made by the Milan group, while its processing, involving an exercise in small scale chemical engineering, was made at the Universities of Bristol and Padua.

The great advantage in employing a very large stack is evident for studying the different modes of decay of the heavy mesons produced in the collision of a high energy particle with a nucleus of the emulsions. A substantial fraction of the secondaries come to rest inside the stack so that their modes of decay can be observed and the energy determined with high accuracy from their range.

A great part of the results was presented at the International Conference on Elementary Particles held in Pisa in June 1955 to celebrate the Centenary of the Nuovo Cimento.

One of most important results of the G-stack study was the final recognition that the values of the masses of heavy mesons giving rise to different decay processes were identical within rather small experimental errors. Thus the interpretation of all these processes as alternative decays of the same particle was strengthened, contributing in an essential way to the general recognition of the so-called θ–τ puzzle. This found its explanation only in 1956 with the discovery by Lee and Yang of the nonconservation of parity by weak interactions.

A striking fact that emerged in Pisa was that the time for important contributions to subnuclear particle physics from the study of cosmic rays was very close to an end. A few papers presented by physicists from the U.S.A. showed clearly the advantage for the study of these particles presented by the Cosmotron of the Brookhaven National Laboratory (3 GeV) but even more by the Bevatron of the Lawrence Radiation Laboratory in Berkeley (6.3 GeV).

5. THE INTERIM STAGE

After this detour on the cosmic rays researches carried out, mainly under the influence of the Bristol group, in a wide international frame and with a strong support from Italian Universities, I return to the development of CERN.

The four groups of the provisional organization at the beginning had started to work at the institutions of the corresponding directors. In October 1953 the PS staff was assembled in Geneva, partly at the Institute of Physics of the University and partly in temporary huts built in its vicinity. At that time the transition took place from mainly theoretical work to experimentation and technical designing.

John and Hildred Blewett of the Brookhaven National Laboratory arrived in Geneva in September 1953 and joined the PS group, which benefited of their experience in design and construction of accelerators for about one year.

In October an International Conference on proton synchrotrons was organized at the Institute of Physics by the PS staff. In March 1954 Goward died after a short and tragic illness. He was succeeded by John Adams (1920-1984) who became Director of the PS Division in 1955 when Dahl went back to Norway to direct the design and construction of the first Norwegian nuclear research reactor.

During October 1953 an administrative nucleus began to function in a temporary Geneva office and, from January 1954 on, in the Villa Cointrin at the Geneva airport.

First instrumentation workshops, then a library and a few laboratories were gradually set up also at the airport in the summer 1954. On 12 August 1954 France completed its ratification procedure of CERN Convention and the situation appeared sufficiently promising for daring to start, on 13 August, 1954, mayor excavation work on the Meyrin site for the foundations of the first permanent buildings, in particular that to house the Synchro-Cyclotron. I had taken this decision with the agreement of the Council, under the pressure of the Geneva weather months before all ratification formalities were in place. Sir Ben Lockspeiser was clearly pleased and commented: "Now we have another task - keeping Amaldi out of jail".

To arrive at the appointment of the Director General of the permanent Organization the Council of the provisional Organization set up (in June 1953) a Nomination Committee composed of Niels Bohr, Sir John Cockcroft, G.Colonnetti, W.Heisenberg, F.Perrin and P.Scherrer with Perrin as convener. The attention of the Nomination Committee was pretty soon focused on the name of Felix Bloch (1905-1983), proposed by Niels Bohr and supported with enthusiasm by all the other members.

On occasion of a short visit to Europe paid by Felix Bloch in March 1954, on invitation of the Nomination Committee, Bloch and I examined together the steps to be taken for creating in Geneva, as soon as possible, the right atmosphere of a research institution and agreed that a few of them had to be started before the beginning of Bloch's tenure of office.

The first point regarded the gradual displacement from Copenhagen to Geneva of the Theoretical Group (or Division as it was called by now). As first step B. d'Espagnat was engaged from 1 September 1954 and J.Prentki from 3 January 1955. Both of them were on leave of absence from the "Centre National de la Recherche Scientifique" in Paris. They worked together on meson production and scattering.

Picture taken on December 1975 when the CERN directorship passed from Willy Jentschke to John Adams and Leon Van Hove. From left to right: L. Van Hove, W.Jentschke, Paul Leveau, President of the Council, and J.Adams.

Niels Bohr presses the botton to put the PS into operation on occasion of the inauguration of this machine on Febraury 5, 1960.

The other step agreed with Bloch was the creation of a two-pronged cosmic-ray experimental programme at CERN. One project consisted in setting up, in collaboration with the Scientific and Technical Services Division directed by L.Kowarski, a nucleus for an experimental physics group. The first scientists involved in this group, starting in summer 1954, were Y.Goldschmidt- Clermont from Belgium and G. von Dardel from Sweden, with C.Peyrou, then at the University of Bern, as consultant. Until the accelerators in CERN could be taken into operation, this group could produce scientific results and gain experience for the future by performing experiments.

The second project consisted of taking over, in August 1955, the U.K. cloud chamber group working at Jungfraujoch Station, led by J.A. Newth, and which had followed Blackett, in his transfer from Manchester University to Imperial College in London, in 1953.

The Geneva group collaborated with this team in a new experiment for measuring the lifetime of K-mesons, using a Cherenkov detector for triggering the cloud chamber on K- meson decays.

Felix Bloch arrived in Geneva on 11 September 1954. He was still only a candidate highly recommended by the Council of the provisional CERN. On September 29, 1954 when a prescribed point in the ratification procedure was reached, the provisional Organization came suddenly to an end while the permanent Organization was not yet in existence. Therefore all the assets of the provisional CERN became suddenly masterless. For eight days I had the honour - as Secretary General - of sole responsability of all properties and liabilities on behalf of a new born organization. Then the first meeting of the permanent Council assembled in Geneva on the 7 October 1954, the Secretary General presented his Final Report , Bloch was nominated Director General and CERN entered its final permanent form.

The Synchro-cyclotron was operated for the first time in 1958 and a proton beam at full energy circulated in the PS on 24 November 1959, i.e. almost one year before time schedule.

In concluding my speech, I will not try to summarize the many important results and the few discoveries made at CERN during the thirty years or so that passed since its creation. During this period CERN is grown enormously and, in my opinion, in a very satisfactory way, with scientific and technological contributions from all Member States, in particular from United Kingdom. Such a development has been possible only because an adequate funding was provided over the years by the Governments of the Member States. If in recent years Europe has regained a good position on the line where fundamental research is rapidly advancing, this is certainly in great part due to CERN. Criticisms also have been raised and they have received a hearing by a few authorities, in particular by those of United Kingdom, where even the interest and importance of this branch of science seems to be doubted. But if we look back to the history of the investigation of the last constituents of matter, we are struck by the number of first class people orginating from this country, or at least, grown up inside the cultural structures of this country, and have contributed over the centuries to it. Even limiting this historical outline to our century, everybody recognizes that the names of J.J. Thomson, Lord Rutherford, C.T.R. Wilson, P.A.M. Dirac, J. Chadwick, G.P. Thomson, Lord Blackett, C.F. Powell and J.D.Cockcroft, are all connected with fundamental steps forward in this branch of science, which is still flourishing and evolving not less than any other science.

It appears unbelievable that a long and glorious tradition like the one summarized by the names listed above may be voluntarily stopped. The younger generations should continue to have the challenge of participating in this kind of research at a continental level. This means that the senior scientists have the moral duty of trying to do anything possible for avoiding that such an opportunity may go lost.

With this message and wish, for the younger as well for the more senior generations of high energy physicists, from Great Britain and from all other Western European countries, I like to express to all of you my thanks for your patience and attention.

CERN 86 -04
30 May 1986

ORGANISATION EUROPÉENNE POUR LA RECHERCHE NUCLÉAIRE
CERN EUROPEAN ORGANIZATION FOR NUCLEAR RESEARCH

John Adams Memorial Lecture

JOHN ADAMS AND HIS TIMES

Lecture delivered at CERN on 2 December 1985

Edoardo Amaldi
Dipartimento di Fisica della Università 'La Sapienza',
Rome, Italy

GENEVA
1986

CERN—Service d'Information scientifique - RD/706 - 3500 - mai 1986

ABSTRACT

In this first John Adams' Memorial Lecture, an outline is given of his work, especially from the beginning of CERN in 1952 until his death in 1984. The historical survey covers John Adams' technical and managerial contributions to the development of CERN and its accelerators, as well as to fusion research in Britain and Europe. Exemplified by his role as member and president of the International Committee for Future Accelerators (ICFA), Adams' interest in international co-operation is also stressed. In the spirit of this great European, arguments are given for CERN to continue to be the first-rate high-energy physics laboratory which it has been in the past.

CONTENTS

1. MY FIRST ENCOUNTER WITH JOHN ADAMS

I met John Adams for the first time on 11 December 1952 in London at the Savile Club, where both of us were invited to lunch by Sir John Cockcroft. At that time I was Secretary-General of the provisional CERN. I had flown from Rome to London the day before, following a suggestion made on the telephone by Sir John, who was director of the British Atomic Energy Research Establishment at Harwell. I had known him since the summer of 1934, when I had spent July and a great part of August at the Cavendish Laboratory. On the morning of 11 December 1952 I had met for the first time Sir Ben Lockspeiser, Secretary of the Department of Scientific and Industrial Research. Both Sir Ben and Sir John were in favour of the participation of the United Kingdom in the CERN adventure, and for some reasons—still not clear to me—they had thought that I could help in convincing Lord Cherwell, scientific adviser to the then Prime Minister, Winston Churchill, to abandon, or at least to mitigate, his negative attitude towards this new continental commitment.

On the telephone I had expressed to Sir John the desire to exploit my trip to London also to meet some young British physicists or engineers who could be interested in participating in the construction of the European Laboratory.

When in London Sir John had asked me to join him for lunch, he mentioned that his arrangements also aimed at giving me the opportunity to meet a young engineer who had worked on the construction of the Harwell synchro-cyclotron. In a few words, typical of his style, Cockcroft succeeded in communicating to me his confidence in the capacities of this young man and that no objection would be raised from the British side to a possible offer by CERN, since the Harwell programme did not foresee, in the near future, developments requiring his specific abilities.

Thus, at lunch, I met John Adams. I was immediately impressed by his competence in accelerators, his open mind on a variety of scientific and technical subjects, and his interest in the problem of creating a new European Laboratory.

In the early afternoon I was received, together with Sir Ben and Sir John, by Lord Cherwell, and thus in some way I accomplished what I will call, emphatically, my political mission. According to the arrangements made by Cockcroft, shortly after 4 p.m. I left London by car with John Adams for Harwell, where at dinner, and after dinner, we had a very interesting conversation with Donald Fry, Frank Goward, John Lawson, and William Walkinshaw.

The conversation during the three hours drive to Harwell confirmed my first impression of John Adams: he was remarkable by any standard, and he was ready, incredibly ready, to come to work for CERN.

I was also very impressed by the other young people that I met at Harwell, not only because of their scientific abilities but also because of the detailed information they had about CERN. Contrary to the impression that I had got from Lord Cherwell early in the afternoon, they were not at all insularly minded.

Clearly Goward was, in great part, responsible for the rather complete information that all of them had about CERN. When, in May 1951, Pierre Auger, Director of the Department of Natural Sciences of UNESCO, convened in Paris the first meeting of experts to prepare a sufficiently well thought out proposal to be submitted to the representatives of the governments of the European States interested in the possible creation of a European Laboratory, Sir John Cockcroft had sent Goward, who participated in that as well as in the three successive meetings of the Board of Consultants.

When the agreement establishing the provisional organization came into force, the Provisional Council of CERN held in Paris its first meeting on 2 May 1952. In this phase the United Kingdom was not a Member of the provisional organization because Her Majesty's Government had preferred to keep the position of observer. In the meeting in Paris in May 1952 four groups were established, the first of which was the Proton Synchrotron (PS) Group with Odd Dahl from Bergen (Norway) as director and Frank Goward as deputy.

Initially this group consisted of full- and part-time staff members and of consultants, all working in their home institutions and meeting at regular intervals, usually in coincidence with Council meetings.

In June 1952 in Copenhagen, the task of the PS Group was defined as building a scaled-up version of the Cosmotron of Brookhaven with an energy of about 10 GeV.

In August 1952 Dahl, Goward, and Wideröe (a well-known accelerator expert, working at the time for the PS group as a consultant) visited Brookhaven to study in detail the Cosmotron, which was already close to completion. During their visit Courant, Livingston, and Snyder came out with the principle of strong focusing [1], which provided a considerable reduction of the magnet dimensions and thereby of the cost of the accelerator.

At its third meeting, held in Amsterdam in October 1952, the Provisional Council approved the proposal that the PS Group, instead of designing a 10 GeV weak-focusing machine, would concentrate its work on a strong-focusing proton synchrotron of highest possible energy—between 20 and 30 GeV—compatible with the same total cost.

My visit to London and Harwell, which I mentioned above, took place about two months after this fundamental Council decision, and all the young people that I met in Harwell were very much interested in the strong-focusing principle and in its implementation, which clearly was at the centre of their thoughts.

This great interest brought its fruit during the successive months, fruit that grew mostly within the ground of the PS Group that Odd Dahl had started to build up.

2. JOHN ADAMS AND MERVYN HINE

Before continuing I should like to remind you that John Bertram Adams was born in Kingston, Surrey, on 24 May 1920 and that, after leaving school, he went into apprenticeship with Siemens in the United Kingdom. Then, during the war, he worked at the British Telecommunications Research Establishment of the Ministry of Aircraft Production on the development of centimetric radar. He played an important role in the development of radar high-frequency components, particularly waveguides in the 10 and 3 cm bands. Towards the end of the war, he contributed to studies of telecommunication systems using centimetric waves.

According to Hine the responsibility that John Adams had had to take upon himself, at the age of about 23, in the radar work during the war was so heavy and complex, not only in its technical aspects but also in the many varied human contacts, as to determine the early maturation of all his qualities as a leader.

During the early days of atomic energy in Britain, John Adams joined the Ministry of Supply, which at the time was responsible for atomic research developments in the United Kingdom. In 1946 he went to the Atomic Energy Research Establishment, Harwell, to work initially on the design and later on the construction of the 110 inch synchro-cyclotron. This 175 MeV machine was the first high-energy proton accelerator built after the war; it was operated without interruption for many years after 1949.

Also in Harwell John developed, in collaboration with Mervyn Hine, from 1950 to 1952, high-frequency klystrons with 20 MW pulsed output, intended to power radar as well as linear accelerators.

I should recall that I had also met Hine during a visit to the Cavendish Laboratory in the late 1940's. When Professor Samuel Devons was showing me around and we arrived at a Cockcroft-Walton voltage multiplier for about 1 MeV, he called Hine—the best of his senior students—to come out of the 'bun' on top of the accelerating tube. He and Hugh Hereward had worked for about three years to rebuild the high-voltage set in preparation of theses on the angular distribution in (p,γ) and (p,α) reactions with light nuclei, parts of which were published in the international scientific press [2].

3. THE DESIGN AND CONSTRUCTION OF THE PS

When Odd Dahl was appointed Director of the PS Group, on 2 May 1952, he started to work on the design of the CERN PS at his Institute in Bergen, the Christian Michelsen Institute. He was helped by his assistant, Kjell Johnsen, and by F. Goward, who worked mainly in Harwell but went to Bergen from time to time.

Kjell Johnsen was then studying deeply the problems about synchrotron oscillations, and he worked out in detail the behaviour of the particle bunches at the transition energy, which was then a great source of anxiety.

The first meeting of the PS Group in which Kjell Johnsen also took part was held in June 1952 during the Copenhagen Scientific Conference, followed by the second meeting of the Provisional Council of CERN.

As I said before, in December 1952 the enthusiasm for the strong-focusing principle was great but there were also some points which were still not clear. In particular, at the beginning of winter 1952–53, John Lawson was the first to give the alarm—in a rather pessimistic form—that the effect of magnet imperfections would lead to steadily growing oscillations and the loss of the beam in a few revolutions. This problem was central for the construction of the PS or any other strong-focusing accelerator. It was studied theoretically and by means of computational or experimental models by a number of authors. In particular, Adams and Hine (then known as the 'Harwell twins') wrote a well-known series of theoretical reports which culminated in a paper published in *Nature,* where Adams, Hine, and Lawson [3] showed that the resonances can be managed provided the accuracy of the field and its index in each magnet, and the accuracy of alignment of the various magnet units around the machine are kept within very stringent limits. Similar conclusions were reached also by other authors at about the same time [4].

In May or June 1953 John and Hildred Blewett, of the Brookhaven National Laboratory, went to Bergen to help the PS group. During September and October 1953 the PS staff—including the Blewetts—was assembled in Geneva, partly at the Institute of Physics of the University and partly in temporary huts built in its vicinity. At that time the transition took place from mainly theoretical work to experimentation and technical design.

John and Hildred Blewett remained in Geneva until the end of spring 1954, so that the PS group could benefit from their experience in design and construction of accelerators for about one year.

In October 1953 an International Conference was organized at the Institute of Physics by the PS staff, who presented a set of parameters of the 25 GeV strong-focusing CERN PS.

The Council endorsed these parameters. At that time new staff members, amongst them John Adams and Mervyn Hine, were appointed and the group consisting of 12 persons, including the Blewetts, took up their work at the Institute of Physics.

In March 1954 Frank Goward died after a short and tragic illness. He was succeeded by John Adams, who became Director of the PS Division later in the same year when Odd Dahl went back to Norway to direct the design and construction of the first Halden reactor, located south of Oslo.

On 13 August 1954 major excavation work started on the Meyrin site for the foundations of the first buildings, in particular the one to house the Synchro-cyclotron (SC).

Starting from the time of Odd Dahl's direction, all the important decisions about the design and construction of the PS were taken at meetings of the 'Parameter Committee' composed of the senior staff of the PS Group. The importance of this Committee was even greater under Adams, who chaired it for hours every Monday morning, and exerted an extraordinary influence on all the participants by allowing the discussion to go around in a calm and rationally ordered atmosphere until a general consensus was reached on each specific point.

I participated in a few of these meetings and I was very much impressed—as was everybody else—by John Adams' style and efficiency, and also by the role played systematically by Mervyn Hine, whose function was that of putting together all the values agreed for the various parameters and the corresponding accuracies, in order to be sure that all of them fitted in a coherent scheme. All the people that participated in those meetings still recall with great admiration John Adams as director and leader of these vital discussions.

Already in 1957 John Adams started to give thought to the exploitation of the PS and its future, long before it was completed. There was, of course, the Scientific and Technical Services with L. Kowarski as Director, which was taking care of the instrumentation in general. There was also the SC Division, with W. Gentner as Director and G. Bernardini as Deputy Director for Research, which was carrying out experiments on the beams provided by the 600 MeV Synchro-cyclotron already in operation since the middle of 1956.

John Adams, however, did not feel completely sure of an adequate exploitation of the PS if he had not something to say about the preparation of this new phase of the work. Thus he arranged with Gentner and Bernardini to have a 'liaison' with the SC experimentalists in the person of Guy von Dardel, who was nominated member of the PS Parameter Committee, and in this new role took care of the preparation of the beams, including the design and construction of the first large Cherenkov counter, and of convincing a few of the experimentalists to start to prepare the first simple counter experiments.

Von Dardel also got in contact with Charles Peyrou concerning hydrogen bubble chambers. The corresponding group was transferred from the STS Division to the PS Division.

The Peyrou group was in the position of using the 30 cm hydrogen bubble chamber rather soon and they immediately started the design of the 2 m bubble chamber which was only ready much later. John Adams and Charles Peyrou, however, reached an agreement with Bernard P. Gregory of the École Polytechnique for the use of the 80 cm hydrogen bubble chamber of Saclay, which was exploited for a rather long period between the small (30 cm) and large (2 m) hydrogen bubble chambers of CERN. This gap was in part filled by the British 1.5 m chamber, which came to CERN in a similar way.

On 27 July 1959 the 100 units forming the magnet of the PS were energized for the first time.

On 16 September, while the International Conference on Accelerators was being held at CERN, the announcement was made that a proton beam injected into the machine had made a complete turn for the first time.

On 13 October the radiofrequency system, which had been designed by Chris Schmelzer, Pierre Germain, and Wolfgang Schnell, of the RF Group, was switched on for beam-trapping studies. The beam was successfully trapped on 15 October.

At the same time an accelerated beam was observed for several milliseconds. However, the first accelerator tests did not take place until 22 October. That evening a good-intensity proton beam was accelerated for 25 ms. The beam energy reached 400 MeV, namely eight times the injection energy.

Testing of the PS was interrupted during the first week of November in order to carry out a number of minor modifications which were mainly to improve the conditions under which the protons were injected into the ring of the PS.

On the evening of 24 November the proton beam at once reached 4.7 GeV, near the 'transition energy', at which level the pessimists had predicted beam loss.

At 7.22 p.m. the phase changer necessary for accelerating beyond the transition energy was switched on. Without visible loss of intensity the proton beam went up to about 6 GeV, the maximum energy possible with the magnet programme so far used. The magnet programme was changed to reach a top field of 12500 G in the hope of obtaining higher energies. At 7.40 p.m. on 24 November the beam was accelerated to approximately 24 GeV, i.e. the maximum energy under normal operating conditions [5].

This great achievement of John Adams and the PS Group was celebrated on 5 February 1960, with speeches and a press conference attended by one hundred and ten press correspondents, and with the participation of François de Rose, President of the Council, John Adams, Niels Bohr, John Cockcroft, E.M. McMillan, F. Perrin, J.R. Oppenheimer, C.J. Bakker, Director-General of CERN, and myself as President of the Scientific Policy Committee (SPC) [6] (Fig. 1).

The Alternating Gradient Synchrotron (AGS) of Brookhaven, which had been approved one or two weeks after the Council decision to make the CERN PS a strong-focusing machine (October 1952), was designed from the beginning for a maximum energy of 30 GeV, i.e. an energy appreciably higher than the 25 GeV—later brought up to 28 GeV—of the CERN PS.

The AGS entered into operation about six months after the PS. I should like to stress, and I am sure that John Adams would do the same if he could be present today, that the design and construction of these two machines went on in parallel in a climate of close and harmonious collaboration between the corresponding staff.

I should also like to recall that other machines built in the same years were of the weak-focusing type, mainly because not all the machine builders of the time had the courage to design and construct large and expensive pieces of equipment based on a principle, understood on paper but not yet tested experimentally on systems of comparable complexity and dimensions, particularly if high intensity was essential. For example Nimrod (7 GeV) at the Rutherford Laboratory and the Zero Gradient Synchrotron (ZGS) (12 GeV) at Argonne were both weak-focusing machines.

The PS entered into operation so smoothly and, in a certain sense, so suddenly at the end of November 1959 that, in spite of the efforts by all the people involved, the magnets for the construction of the secondary beams were not yet ready and could only be placed on the floor a few

Fig. 1 On the morning of 25 November 1959 John Adams announced to the CERN staff in the Main Auditorium the news that, on the evening before, the Proton Synchrotron had accelerated protons to 24 GeV. He had in his hands the famous vodka bottle given to him by Nikitin from Moscow a few months before, with the recommendation to drink it when the energy of the particles accelerated by the PS would surpass 10 GeV, the maximum energy of the Dubna machine. The same day John Adams sent back to Dubna the bottle with, inside it, the polaroid photograph of the oscillogram showing the successful operation of the PS. [*CERN Courier* **9**, 336 (1969)].

months later. In the meantime, however, simple experiments could be carried out either without magnets or using one or two spare magnets borrowed from the SC Division. Experiments of this type were the search for high-energy gamma-rays emitted by a beryllium target bombarded by protons of 24 GeV and the first determination of the cross-sections of $p + p$, $\pi + p$, $K + p$ and $\bar{p} + p$ collisions.

Finally I should like to recall that in the period 1958–59, in parallel with the work for the PS, Adams set up an Accelerator Research Group, with Arnold Schoch, Kjell Johnsen and Kees Zilverschoon as leaders, each for one year, in rotation.

This Group studied various advanced schemes, such as the Budker plasma accelerators, the Midwestern Universities Research Association (MURA) work, and colliding beams with storage rings. The study of the last approach paved the way to the design and construction of the Intersecting Storage Rings (ISR), carried out under Kjell's direction some years later (from 1965 to 1971).

4. ADAMS' FIRST INTERESTS IN PLASMA PHYSICS

Already in 1956 experimental work on plasma accelerators was carried out in the so-called Adams Hall (today building 70) and concerned the Budker proposal. The deep interest of John Adams in plasma physics and fusion was stimulated in 1958 by the first substantial publication of theoretical and experimental investigations on the possibility of reaching the required high temperature by using magnetic fields to confine ionized gases or plasmas.

While the PS was approaching completion, John Adams found the time to head a study group set up by the CERN Council to review the state of fusion research and to make proposals on how such research might develop as a European-wide programme [7]. This activity started informally in March 1958 with a presentation mainly of results from the Harwell experiment ZETA. The study group was set up formally in June of the same year, with the objectives 'to exchange information, to discuss the programmes of the various laboratories and to consider ways of facilitating fusion research work in Europe'. He mobilized about 50 scientists from 10 European countries, from the USA, from Euratom, from the OEEC, and from CERN. They had the formidable task of publishing the events of 1958, beginning with a review of the ZETA results in January 1958 and ending with the Second Geneva Conference on the Peaceful Uses of Atomic Energy held in September of the same year.

The situation was rather confused and people had difficulties in forming judgements about the next steps to take against the rapidly emerging picture produced by the declassification of all the main results obtained in years of secret research in the USA, the USSR, the United Kingdom, and France. Adams' first study on fusion recognized three main lines of research, namely toroidal pinches, stellarators, and mirror machines. It recognized also that the big machines had, on the one hand, demonstrated some of the basic principles, but, on the other hand, disappointed some of the chief sponsors, essentially because physics understanding had *not* advanced sufficiently. Therefore, one major task identified was the understanding of plasma physics.

The results of his analysis of the world-wide situation was summarized in a report of 1959, where he said that the European programme should aim to encourage the diverse activities at all levels and by whatever means were appropriate.

The study group under his chairmanship considered that 'unless it can be demonstrated that one (a European Laboratory) is needed in order to build larger facilities than can be built by national groups, the many other advantages of such a centre may prove insufficient to overcome the difficulties in its creation and maintenance'. The final recommendations included a regular review of European programmes, regular European scientific discussion meetings, exchange of staff, training, and close collaboration with the USA and USSR. John Adams concluded by sketching the constitution and statutes of a 'European Society of Controlled Thermonuclear Research' to give form to his ideas on European co-operation in this field.

At about the same time, Euratom was planning to set up a programme of joint research on plasma physics and fusion, and John Adams prepared, after consultation with and approval by the Director-General, C.J. Bakker, and by me (I was President of the SPC at the time), the proposal of carrying on this research, on Euratom money, in a building of the Meyrin Laboratory or in its vicinity. The proposal, approved by the SPC and by the major Member States of CERN, aroused very strong opposition from a few delegations in the Council, in particular from the Swedish Delegation, and was finally abandoned.

On 23 April 1960, the Director-General, Cornelius J. Bakker, from Amsterdam, died in an airplane accident, while on a mission in the USA. This tragic event required immediate action. The Council of CERN held an emergency meeting on 3 May 1960. Following this session, François de Rose, President of the Council, who had played an important role in the prehistory of CERN, announced the appointment of John Adams, Director of the PS Division, to the post of Acting Director-General [8].

This appointment was confirmed by the Council at its 16th regular session, held on 14 June 1960. The British authorities, however, had asked John Adams to go back to the United Kingdom to take up the direction of a new laboratory devoted to the national fusion programme.

The Council of CERN met again at a special session, on 11 July 1960, and appointed John Adams as Director-General of CERN until 1 August 1961. As from 1 October 1960, however, Adams was allowed to serve in this capacity on a part-time basis since, by agreement with the British authorities, he would devote part of his activity to the new Plasma Physics Laboratory at Culham, of which he had been nominated director-designate [8].

Shortly afterwards the Council decided to appoint Victor Frederick Weisskopf as Director-General of CERN, starting from immediately after John Adams' departure, i.e. from 1 August 1961. From 1 January 1961 Weisskopf, however, had been nominated a member of the directorate for research, in parallel with G. Bernardini who before that date had been Director of the SC Division. Weisskopf arrived even earlier (12 September 1960) to get direct experience of the structures, habits, and atmosphere, of the Meyrin Laboratory [8].

5. JOHN ADAMS' RETURN TO THE UK

When, at the beginning of August 1961, John Adams went back to the UK and started full-time direction of the Culham Laboratory, he had to fulfil a threefold task:

> 'Firstly, to bring together the UK teams from Aldermaston and from Harwell into a single unit;
> Secondly, to conduct research on plasma confinement physics on a broad front, investigating many different lines of magnetic confinement and plasma physics and diagnostic development while, at the same time, continuing research on the big machine ZETA;
> Thirdly, to conduct the research with the maximum of international collaboration using an open site.' [7]

His first job was to produce an efficient organization of the laboratory, which, in some respects, reflected the CERN requirements and practice rather than the previous Harwell practice. There were eight divisions, of which three were responsible for experimental research and one for theoretical research, supported by others for technological development, administration, site services, and engineering design. Adams persuaded the Authority to allow him to pursue the following pattern. He invented the term 'Experimental Assemblies' to describe what was needed for the experimental research, namely an assembly of components, such as power supplies which might be used for more than one experiment, the hardware peculiar to the experiment, and the diagnostic equipment needed to make measurements, movable from one experiment to the other.

As a result of the simple organization and of the far-sighted financial delegation to Adams by the UK Atomic Energy Authority—chiefly Sir William Penney—the development of Culham as a major force in fusion research was rapid and effective. By the start of 1967, when Adams left to take up the post of Member of the Authority for Research, some 17 Experimental Assemblies had been constructed and operated, 3 further Assemblies were under construction, whilst at the same time the one big machine, ZETA, continued to operate at Harwell and the studies of basic plasma physics included a natural-plasma research programme using rocket-borne UV spectroscopy. Finally, of course, he had built an entire new laboratory and completed it to time and cost.

The results were more encouraging to the specialists than to the outside world. The very success of the broad programme at Culham meant that no one line had been pushed very hard, the work being spread over a rather wide spectrum of basic research. This unfortunately coincided with the cultural revolution for government science introduced by the 1964 Wilson government, who had as advisors Blackett, Maddock, and Adams himself—appointed as part-time comptroller at the Ministry. Industrial strength through well-directed custom-oriented applied research became the required image and order of the day.

This led to a narrowing down of the fusion programme at Culham and the concentration of the efforts on what had emerged as the essential line of research: toroidal confinement.

6. BACK AGAIN AT CERN

Already in June 1960, less than eight months after the first successful operation of the 28 GeV PS, the idea of starting to study possible programmes for the far future of the Organization was considered and debated in the Meyrin Laboratories and also in the SPC and the Council.

Among other things, it was said in the Council that: 'CERN should plan to build a machine to replace the Proton Synchrotron in about 1970, when it would no longer be an up-to-date machine. It would probably take about seven years to build such a machine, therefore plans should be studied now (1960), to enable the Council to consider a plan in 1962 or 1963.'

As a result of discussions held by the SPC, a special study group was set up with the authorization of the Council, at the end of 1962, under the direction of Kjell Johnsen, to look into the feasibility of a proton synchrotron in the range of 150–300 GeV and a proton collider of about 2 × 25 GeV, and to arrive at realistic parameters and reliable estimates of cost, manpower, and time schedules.

In January 1963, the Director-General of CERN, V.F. Weisskopf, with the agreement and help of the SPC, invited physicists engaged in high-energy research in the laboratories and universities of the Member States to meet at CERN, to form the European Committee for Future Accelerators (ECFA).

For various reasons I became chairman of ECFA and E.H.S. Burhop, from London, became its secretary. A report on the European high-energy accelerator programme was prepared by a Working Party set up by ECFA. This '1963 Report' [9] and its conclusions, endorsed by ECFA and the SPC, were submitted to the Council at its session in June 1963.

As far as European joint action was concerned, two main conclusions were reached: the programme should include the construction of the Intersecting Storage Rings (ISR), in association with the CERN PS, and the construction of a new proton accelerator for an energy of about 300 GeV.

On the basis of the ECFA proposal the Council authorized a supplementary programme and budget for the preparation of the detailed design for these two projects, called at the time the 'summit projects'.

The special study group in the Accelerator Research Division of CERN, which had helped the Working Party of ECFA in the preparation of the '1963 Report', was temporarily enlarged and augmented with the addition of accelerator experts from laboratories in the Member States and from the USA and USSR, and produced two reports. The first, on the ISR project, was presented to the Council in June 1964. The second, regarding the 300 GeV Proton Synchrotron, was submitted to the Council in December of the same year.

On these occasions the SPC recommended to the Council a single coherent programme. This included the construction of the ISR to be fed with protons from the PS, and the creation of a new laboratory containing the 300 GeV accelerator.

In December 1965, the Council approved the first part of the European programme recommended by the SPC. The second part of the programme, the construction of the 300 GeV Laboratory, which was considered by many European high-energy physicists of even greater importance for the long-term future of this type of research, remained under discussion from December 1965 until February 1971.

In the meantime the construction of the ISR, directed by Kjell Johnsen, was finished, so that during spring 1971 the first experiments were started using this unique machine.

Since then, the ISR were intensively used for experimentation while their performance was steadily improved, thus allowing the collection of a number of extremely important results.

While the European programme for the 300 GeV machine was debated, similar ideas were discussed in the United States, and a study group was created at the Radiation Laboratory in Berkeley, which in 1965 published the design report for a 200 GeV proton synchrotron.

In 1966 ECFA was convened again to consider the European particle physics situation as it had developed since 1963. In this new phase of the work E.H.S. Burhop was replaced by A. Citron, from West Germany, as secretary.

About one year later, in May 1967, ECFA issued a full report on its studies during 1966–67 [10], which was submitted to the Council at its session in June 1967.

In this report ECFA recommended the building of a large proton accelerator in Europe with the least possible delay. It was said: 'This is essential if European scientists are to continue to contribute to the advances of high-energy physics into the 1980's.

The design of the accelerator, its cost, and laboratory manpower requirements were revised, and some changes and additions proposed to the original specifications of 1963–64, but substantially the concepts of the original propositions of CERN and ECFA were maintained.

The '1967 Report' insisted on the importance of reaching the highest possible energy and of a design allowing, at least at some later time, an intensity of 10^{13} protons per second to be obtained.

An extensive study of possible experiments to be made with the protons produced by the 300 GeV accelerator, and with the beams of secondary particles, was made during 1966–67 by working groups set up by ECFA.

The results of these studies were collected in two big volumes, which were presented to the Council as background information to the 1967 ECFA Report.

In the meantime a few very important decisions were reached in the United States; in particular the decision to construct, at Batavia (Chicago) the National Accelerator Laboratory (NAL, later to be known as Fermilab), centred on a proton synchrotron of augmentable energy (200–400 GeV); R.R. Wilson was nominated Director of the new project.

The machine was completed at the beginning of summer 1971 and since then had produced important results which, among other things, had shown the amplitude and richness of the explored high-energy domain.

Coming back from this short 'overseas excursion', I will recall that two fundamental problems, one of legal, the other of geographical nature, were faced by the Council while the technological and scientific studies were pursued inside and outside the Meyrin Laboratories.

The 300 GeV Programme clearly required a revision of the Convention of CERN. The preparation of a new Convention was started in March 1966; it reached an agreed form in December 1967, and was submitted to Member States for ratification in January 1968. The procedure of ratification by the Parliaments of Member States was completed in December 1970, so that the new Convention entered into force on 17 January 1971. Altogether about five years of legal work was involved. The same kind of work carried out for the original Convention creating CERN had required less than two years.

The other problem, that I have said was of geographical nature, was that of the site.

In June 1964 the Council had invited the Member States to make proposals for possible sites for the new machine and Laboratory. Among more than twenty possible sites proposed by Member States, five were selected in December 1967 by a 'Site Evaluation Panel' on the basis of a number of relevant factors [11].

The decision of which, among these five sites, should be the site where the construction of the SPS had to be started as soon as possible, immediately became a major difficulty, since the opinions of Member States were strongly divided.

This was the first and the only serious impasse met by the Council in this enterprise. The original enthusiasm among the people that had worked for about five years on the design of the machine and of its Laboratory, and on their exploitation for scientific work, could have started to diminish. Everybody, of course, was aware of the importance of the fact that, in June 1966, the President was authorized by the Council to place the 300 GeV Programme formally before the governments of Member States for discussion and decision, but something had to be done to gain time, during the search for a solution.

In an attempt to face the situation, in February 1968 a Steering Group was nominated on a joint proposal by the President of ECFA and the Director-General, Bernard Gregory, and with the approval of the SPC, presided over by Cecil Powell. The Group was composed of people who had directed successfully the construction of medium or large accelerators in Europe. The membership of the group was chosen to include possible future Directors of the Project and many of their main collaborators. The task of the group was that of proceeding in the study of those aspects of the 300 GeV project that, without binding the final design, would certainly represent important elements for the final decisions. Examples of this kind of problem are: a comparative study of different magnet designs, the use of semiautomatic corrections of the proton orbits, and so on.

This Steering Group led the work until the end of 1969 when the Council appointed, as Director of the 300 GeV Programme, John Adams, who was a member of the Steering Group. He started immediately to reconsider the main features of the machine as presented in the 1963–64 design study in the light of all later studies and experience [12].

At the end of 1969 the situation was as follows: six Member States had announced their intention to join the 300 GeV Programme and five of these Member States had offered sites for the new Laboratory. The Council had before it a Report on the project and all the documentation for the programme had been prepared.

Everybody hoped at that time to reach the decision to start the 300 GeV Programme at the Council Session of December 1969, but no agreement was reached on the choice of the site for the new laboratory and, consequently, on that occasion, no decision was taken on the start of the programme.

Apparently, the selection of a site for the erection of an important laboratory still involved problems which the European States had no recognized way of resolving.

John Adams, who had started to collect a small initial staff, went on with the design work and in June 1970 presented to the Council a new proposal for the 300 GeV Programme, which not only avoided the difficulties of site choice but also offered several advantages to the Organization. The

Fig. 2 The picture, taken in October 1971, shows John Adams chairing a meeting at the SPS [*CERN Courier* **12**, 347 (1971)].

300 GeV accelerator could be constructed on an extension of the Meyrin Laboratory, where the 28 GeV PS could be used as injector.

The proposal not only presented a number of technically and financially appealing features, but also constituted a very clever political move.

In the session which started in December 1970 the Council unanimously agreed that this alternative way of realizing the 300 GeV Programme should be studied further and adjourned its work for two months. When the Council met again in Geneva on 19 February 1971, the new proposal was approved by ten out of the twelve Member States. One of the two Member States that did not adhere in February joined the new project a few months later.

The design and construction of the 300 GeV PS of CERN was directed by John Adams, according to his usual and unique style, on the site put at the disposal of CERN by the French and Swiss governments (Fig. 2).

At the end of Willy Jentschke's mandate as Director-General, the Council appointed, starting from 1 January 1976, two Directors-General with specific responsibilities: Leon Van Hove as Director-General for Research and John Adams as Executive Director General—according to a suggestion that originated from John Adams himself (Fig. 3).

On 17 June of the same year, when addressing the summer session of the CERN Council, John Adams described the operation of the accelerators and concluded with the announcement that at midday the SPS had reached an energy of 300 GeV—the design energy specified in the programme approved by the Member States. He then consulted the Council about going to 400 GeV, in accordance with Council discussions three years earlier, in June 1973, and the higher energy was approved. At 15.35 h on that day the SPS accelerated protons to 400 GeV (Fig. 4) [13].

An important feature of the SPS that should be remembered is that it represented the first large accelerator based on the use of a distributed computer control system, where all the equipment was connected locally to mini-computers which were joined in a network to the main control room. Originally, the machine at Fermilab, for example, had also some microcomputers for individual systems which, however, were not integrated in a common network. Later the integrated system was also adopted there.

Another innovation of the SPS control system was the use of high-level languages for application programmes which could then be written by engineers and technicians as well as by professional programmers.

Fig. 3 Picture taken in December 1975 when the CERN directorship passed from Willy Jentschke to John Adams and Leon Van Hove. From left to right: Leon Van Hove, Willy Jentschke, Paul Leveau, President of Council, and John Adams.

Fig. 4 On the night of 17 June 1976 John Adams addressed the members of the SPS construction team who were joined by delegates of the Member States to celebrate the successful acceleration of protons, by the SPS, to an energy of 400 GeV, that had taken place in the early afternoon of that same day [*CERN Courier* **16**, 251 (1976)].

This last feature allowed, a few years later, a rapid change-over to the completely new mode of operation for the proton–antiproton collider.

A further important development in the SPS that I should recall is the use of a much higher accelerating frequency and a travelling wave structure, in place of the conventional ferrite-loaded cavities.

All his close collaborators always admired John Adams for his capacity of mastering the essential features of the project and of pushing on the work with a very strict systematic approach, as well as for his clear vision in spotting where the potential difficulties could arise and what steps could be taken to overcome them.

A few interesting examples can be mentioned. During the construction of the SPS a sudden and unexpected difficulty arose when a series of magnet coils broke down to earth under test, in spite of the good design and of the precautions taken. Was CERN also going to have the same difficulty as Fermilab? Several possible reasons were advanced, but John, with the help of his close collaborators, could determine quickly the reason in the use of an acid for cleaning the coil connections. The remedy required the revision of 300 bending magnets which, thanks to his immediate reorganization of the work, was done in record time, so that the overall schedule of the SPS construction was not affected.

A decision of paramount importance was taken jointly on 7 July 1978 by the two Directors-General of CERN; it concerned the construction of the proton–antiproton collider through a modification of the SPS, proposed and strongly pressed by Carlo Rubbia. The Director-General for Research, Leon Van Hove, supported this daring programme and the Executive Director-General, John Adams, was also convinced of its scientific importance, but could foresee better than anybody else the risks involved in the implementation of this new adventure. He finally accepted the great responsibility that such a decision was placing on his shoulders and arranged the start of the proton–antiproton collider construction as a part of the normal scientific and technical programme without asking for new credits from the Council.

As another example of his detailed understanding, I will recall that in 1978, there were two schools of thought about how these proton–antiproton collisions were to be realized. The first one was based on the direct injection of antiprotons from the accumulator in the SPS at 3.5 GeV. The second one insisted that the antiprotons had to be accelerated to 25 GeV before injection.

There was no way of reconciling these opposite views so that John decided to take it upon himself to re-elaborate in detail all the arguments. After a week of personal work he came to a final decision.

7. AT THE SERVICE OF THE EUROPEAN COMMUNITIES

The mandate of John Adams and Leon Van Hove as Directors General of CERN expired at the end of 1980. The Direction of CERN passed to Herwig Schopper.

The experience of the two past Directors-General, one in science, the other in technology, and of both of them as leaders of a big organization, was well known and too important not to be made use of. Thus, starting from 1981 John Adams was asked by the European Communities to contribute in various capacities to some of their activities.

First he became a member of the Beckurts Panel of 1981, which reviewed the European Fusion Programme and endorsed the Commission's main recommendations for strategy and their programme for 1982–86.

Later he was asked to chair the Governing Board of the Joint Research Centres of the European Communities (i.e. Ispra, Karlsruhe, Petten, and Geel) from 1982 to 1984, and in 1982–83 to chair the Ignitor Assessment Panel, also called the Adams Panel [including R. Bickerton from the Joint European Torus (JET), P. Reardon from Princeton (USA), P.H. Rebut from JET, and M.N. Rosenbluth from Austin (USA)], which re-examined the ignition experiments*), and the Ignitor in particular, bringing out the possibility of designing 10 MA machines, which, in addition to the $D + {}^3H$ reaction, may also ignite the reaction

*) The ignition of a plasma is defined as a regime in which the plasma temperature continues to rise because of alpha-particle heating when other forms of additional heating are switched off. The Ignitor has been studied and proposed by a group led by Bruno Coppi and composed of plasma experts from MIT (Boston) and Scuola Normale Superiore (Pisa).

$$D + {}^3He \rightarrow p + {}^4He,$$

the 14.7 MeV proton of which would remain confined.

This was his last scientific and technical work before illness struck him down*). Recognition of his great contributions had, however, come earlier, when he was knighted in the UK New Years' Honours list, in 1981.

8. FINAL CONSIDERATIONS AND REMARKS

During the 'Seminar on New Trends in Particle Accelerator Techniques', held in Capri at the beginning of June 1982, John Adams expressed to Giorgio Brianti, one of his close collaborators and friends for many years, that he considered it to have been very fortunate that his active life had begun at the emergence of three new technologies: radar, accelerators, and controlled fusion.

We agree, we should agree, in view of his remarkable achievements in the fields of controlled fusion and accelerators.

Paraphrasing his words, however, we should add that we, or better CERN, or even better Europe, were fortunate that when the first signs of the transition from small sciences to big sciences were appearing in various parts of the world, and in particular in Europe, John Adams was here, still very young, but already sufficiently mature to start immediately to make essential contributions to these developments, and to continue to do so for a period of more than 30 years.

John represents, and will continue to represent, one of the most brilliant examples of a new kind of man, typical of our century, during which science and technology, associated in various combinations, play different roles of increasing importance.

CERN, the European Space Agency (ESA), the European Southern Observatory (ESO), etc., are all examples of organizations where advanced technologies are at the service of the sciences. There are other places where certain advanced technologies are developed in view of military goals, and still others where they aim at civilian purposes.

All these combinations still constitute important parts of the activities of any one of the large groups in which mankind is divided.

It would, however, be inappropriate to talk of modern technologies with military goals on this occasion for anybody remembering how keen John Adams was on keeping friendly relations with any other scientist or technologist active in the fields of accelerators and controlled fusion, irrespective of the nation or group of nations to which he belonged. It may be enough to recall Adams' role in the International Committee for Future Accelerators (ICFA), which was presided over by him during the years 1978–1982 [14].

Thus, following his example, I will limit my remarks to the relative merit, or importance, of technologies at the service of sciences and technologies with civilian purposes. Of course, both these broad lines of human activity should be followed and developed with strength; both of them are essential for the cultural and economic development of a geographical region—or, maybe some time in the future, federation of nations or nation—such as Western Europe. One can quote a few cases in which CERN has already played an important role in helping other European efforts to come into existence or to consolidate their structures. Apart from the attempt—regrettably unsuccessful—to help Euratom at the beginning of its activity in the field of plasma physics, proposed by John Adams at the time when Cornelius Bakker was Director-General, I wish to recall the help given by CERN, at the time when Victor Weisskopf was Director-General, to the European molecular biologists, in particular to Sir John Kendrew, in the creation of EMBO and the setting up of its research laboratory in Heidelberg. The other case that I should also like to recall is the substantial technical help provided, at the time when Bernard Gregory was Director-General, in the design and construction of a 3.6 m reflector for ESO. At that time a formal agreement was signed by the Directors-General of CERN and ESO, according to which a high-level CERN technologist, in the person of Kees Zilverschoon, with appropriate subsidiary staff, devoted a few years of intensive activity to the Direction of the Organization, and the design and implementation of this project, including the preparation of the tenders and the supervision of the work carried out by the specialized industries.

*) John Adams died, in Geneva, on 3 March 1984.

Interventions of this kind by CERN are today even more important and urgent, in a Europe aiming at a 'Technological Renaissance'. The forms of intervention can be different and should, of course, be modelled according to the nature of each specific case.

In moving in this direction one should, however, be aware of various possible dangers, which can only be avoided if recognized in advance.

One danger is that some people, with a narrow technological view of the European development, may have in mind the unexpressed intention of gradually deviating CERN from its institutional goals, towards applied activities. This would be a mistake of undescribable dimensions. CERN, only now, after more than 30 years of full life, 30 years of thorough work of first-class scientists and technologists operating harmoniously together, has reached a first-line position in the sciences for which it was created.

It would be a crime towards the same idea of Europe, even to think of deviating CERN from its original function, as it is unbelievable that a substantial reduction of the normal funding of CERN could be applied when the completion of Europe's next large machine, LEP, is approaching.

The development of advanced technologies in Europe should be fostered, but by other means and mainly by other people. First of all the industries are the bodies that should be mobilized to play the principal role in this new venture.

The governments too should take part of the responsibility and burden of this new development on their shoulders, if they are really interested in it. One speaks in these days of 'market-led research', a phrase that sounds very good but that seems to indicate that the industries should proceed mainly—or even solely—by their own financial means. This is quite reasonable for developments which will carry financial benefits in a few years time and which, most probably, many European industries would have made in any case, even without a multigovernmental label.

Projects of this type, however, cannot compete with those proposed, for example, through the Strategic Defensive Initiative scheme, which carry considerable sums of public money.

Furthermore, 'a market-led research' approach can be reasonable only as a first step, but if the declaration of the beginning of a new period or 'era' should correspond to something concrete and valid in the sense of trying to change substantially the situation in a time of the order of 10 years, then one has to consider, *from now,* also technological projects *so advanced* that the most refined European technologists as some employed at CERN, at ESA, and in a few other places, can and should make substantial contributions.

Already, the industries which have worked with CERN have acknowledged the considerable technical and economical advantages they have gained in collaboration on advanced technological projects. This means that not only the mother organizations should get what is necessary for the continuation of their institutional goals, but also that their 'children-projects' should receive an adequate financial nourishment.

I am fully aware that these and many other points are matters for wide and thorough debates at various levels. These debates should start as soon as possible. I confess to being confident that the outcome will be in line and at the level of the most positive and successful European endeavours. But I regret—I can say we regret—that John Adams is not here to help us in the search for the most appropriate lines of approach to all these important and difficult problems.

REFERENCES

[1] E.D. Courant, M.S. Livingston and H.S. Snyder, The strong-focusing synchrotron—a new high-energy accelerator, *Phys. Rev.* **88,** 1190 (1952). The strong-focusing principle had been discovered by the Greek engineer N. Christofilos in 1950. In the same year he had applied for a US patent, but his application was not noticed by the experts in the field and therefore remained ignored.

[2] S. Devons and M.G.N. Hine, γ-radiation from ^{8}Be, *Phys. Rev.* **74,** 976 (1948). The angular distribution of γ-radiation from light nuclei—I. Experimental, *Proc. R. Soc. A* **199,** 56; II.Theoretical, *ibid.* **199,** 73 (1949).

[3] J.B. Adams, M.G.N. Hine and J.D. Lawson, Effect of magnet inhomogeneities in the strong-focusing synchrotron, *Nature* **171,** 926 (1953) (dated 9 March 1953).

[4] M. Sands and B. Touschek, Alignment errors in the strong focusing synchrotron, *Nuovo Cimento* **10,** 604 (1953) (received 13 March 1953).
E.R. Caianello, Non-linearities in the strong-focusing accelerator, *Nuovo Cimento* **13,** 581 (1953) (received 14 March 1953).
E.R. Caianello and A. Turin, Stability and periodicity in the strong focusing accelerator, *Nuovo Cimento* **10,** 594 (1953) (received 14 March 1953).

[5] J.B. Adams, The CERN Proton Synchrotron, *CERN Courier* **6–7,** 6 (1960). See also: *CERN Courier* **4,** 1 and 6–7 (1959). For a very vivid description of what happened the evening of 24 October 1959, see the article that Hildred Blewett wrote for the CERN Courier on the occasion of the tenth Anniversary of the first operation of the CERN Proton Synchrotron: 'Ten years ago... some personal reminiscences', *CERN Courier* **9,** 331 (1969).

[6] See *CERN Courier* **8,** 4 (1960).

[7] R.S. Pease, John Adams and the development of nuclear fusion research, Lecture delivered on 7 February 1985 at the dedication of the John Adams Lecture Theater, Culham Laboratory, UK, *Plasma Phys. and Controlled Fusion* **28,** 397 (1986).

[8] See *CERN Courier* **11,** 2, *ibid.* **12,** 10, *ibid.* **15,** 2 and 13 (1960), *ibid.* **1,** 10 (1962).

[9] Report of the Working Party on the European High Energy Accelerator Programme, FA/WP/23/Rev. 3, 12 June 1963.

[10] ECFA Report 1967, CERN 1700; CERN/ECFA 67/13/Rev. 2, Geneva, 15 May 1967.

[11] Report of the Site Evaluation Panel on the site offers by the Member States for the 300 GeV Accelerator Laboratory, CERN/761, Geneva, 4 December 1967; the Panel was presided over by J.H. Bannier, from the Netherlands.

[12] J.B. Adams and E.J.N. Wilson, Design studies for a large proton synchrotron and its laboratory, CERN 70-6 (1970).

[13] SPS reaches design energies: *CERN Courier* **16,** 247 (1976).

[14] J.B. Adams, Introductory Remarks to the Open Meeting of the Second Workshop of ICFA, Proc. Second ICFA Workshop on Possibilities and Limitations of Accelerators and Detectors, Les Diablerets (Switzerland), 1979, ed. U. Amaldi (CERN, Geneva, 1980), p. XI.

BIBLIOGRAPHY

[1] Enrico Fermi. *La ricerca scientifica* **25**(1), 1955, 3–13

[2] Commemorazione del socio Enrico Fermi. In *Enrico Fermi: Problemi attuali di scienza e di cultura (Quaderni)* **35**, ed. V. Arangio-Ruiz & E. Amaldi, 1955, pp. 8–22. Roma: Accademia Nazionale dei Lincei

[3] George Placzek. *La ricerca scientifica* **26**(7), 1956, 2037–2042

[4] The Fermi manuscripts at the Domus Galilaeana. *Physis* **1**(2), 1959, 69–72

[5] Galileo: 350 years later. *Ned. T. Natuurk* **30**, 1964, 217–227

[6] Realtà Naturale e Teorie Scientifiche. In *Saggi su Galileo*, 1967, pp. 1–66. Comitato Nazionale per le manifestazioni celebrative del IV Centenario della nascita di Galileo Galilei, Firenze, G. Barbera

[7] Nota biografica di Ettore Majorana. In *La vita e l'opera di Ettore Majorana*, 1966, pp. vii–xlix. Roma: Accademia Nazionale dei Lincei

[8] Influenza del pensiero di Albert Einstein sullo sviluppo della fisica moderna. *Cultura e scuola* **19**, July–September 1966, 5–12

[9] Ettore Majorana, man and scientist. In *Strong and Weak Interactions. Present problems*, ed. A. Zichichi, 1966, pp. 10–75. New York: Academic Press (English translation of [8])

[10] L'opera scientifica di Ettore Majorana. *Physis* **10**(3), 1968, 173–187

[11] Ricordo di Ettore Majorana. *Giornale di fisica* **9**(4), 1968, 300–318

[12] Venticinque anni dalla prima reazione a catena divergente controllata. *Cultura e scuola* **26**, April–June 1968, 5–13

[13] CERN past and future. In *Proceedings of the Conference on High Energy Collisions of Hadrons* (Geneva, 15–18 January 1968). CERN 68–7, 1968, Vol. I

[14] Personal recollections of early times in neutron physics. In *Proceedings of the International Conference on the Interactions of Neutrons with Nuclei* (Lowell (Mass.), 6–9 July 1976), ed. E. Sheldon. Technical Information Center, ERDA Conf. 760715–P2, 1976, Vol. 2, pp. 1492

[15] Personal notes on neutron work in Rome in the 30s and post–war European collaboration in high energy physics. In *History of Twentieth Century Physics*, LVII Course of the International School of Physics "Enrico Fermi" (Varenna, 31 July–12 August 1972), ed. C. Weiner, 1977, pp. 293–351. New York: Academic Press

[16] Attualità di Volta nella Fisica moderna. *Scientia* **112**, 1977, 171–179

[17] Radioactivity, a pragmatic pillar of probabilistic conceptions. In *Problems in the Foundations of Physics,* LXXII Course of the International School of Physics "Enrico Fermi" (Varenna, 25 July–6 August 1977), ed. G. Toraldo di Francia, 1979, pp. 1–28. New York: Academic Press

[18] Ricordo di Enrico Persico (with Franco Rasetti). *Accademia Nazionale dei Lincei,* 1979 (Celebrazioni Lincei, n. 115); *Giornale di fisica* **20**(4), 1979, 235–260

[19] Gli anni della ricostruzione. *Giornale di Fisica* **20**(3), 1979, 186–232; *Scientia* **114**, 1979, 29–50 & 421–437

[20] The years of reconstruction. In *Perspectives of Fundamental Physics* (Rome, 7–10 September 1978), ed. C. Schaerf, 1978, Harwood Academic Publishers; *Scientia* **114**, 1979, 51–68 & 439–451 (English translation of [19])

[21] *The Bruno Touschek Legacy*. CERN 81–19, 1981

[22] L'eredità di Bruno Touschek. *Quaderni del Giornale di Fisica* **7**, 1982, Vol. V. Società Italiana di Fisica (Italian version of [21])

[23] Gli Archivi per la Storia della Fisica Quantica. *Rendiconti della Accademia Nazionale delle Scienze detta dei XL* **100**, 1982, 221–225

[24] Beta Decay opens the way to weak interactions. *Journal de physique*, Colloque C8, supplément au n.12, Tome 43, 1982, 261–300

[25] Neutron work in Rome in 1934–36 and the discovery of uranium fission. *Rivista di storia della scienza* **1**(1), 1984, 1–24

[26] From the discovery of artificial radioactivity to the discovery of fission. In *La radioactivité artificielle a 50 ans*, 1984, pp. 1–24. Paris: CNRS

[27] From the discovery of the neutron to the discovery of nuclear fission. *Physics Reports* **111**(1–4), 1984, 1–331

[28] The Solvay conferences in physics. In *Aspects of Chemical Evolution, Proceedings of the XVII Solvay Conference in Chemistry,* ed. G. Nicolis, 1984, pp. 7–35. New York: John Wiley and Sons

[29] A cinquant'anni dalla radioattività artificiale provocata da neutroni. *Rendiconti della Accademia nazionale delle scienze detta dei XL*, s. V, Vol. VIII, p. II, 1984, 151–164

[30] Gian Carlo Wick during the Thirties. In *Old and New Problems in Fundamental Physics, Proceedings of the Conference in Honor of G.C. Wick*, 1984, pp. 5–17. Pisa

[31] A few flashes on Hans Bethe's contributions during the Thirties. In *From Nuclei to Stars*, XCI Course of the International School of Physics "Enrico Fermi" (Varenna, 18–23 June 1984), ed. A. Molinari & R.A. Ricci, 1986, pp. 1–16. Amsterdam: North Holland

[32] Introduction (on Niels Bohr and Italian physics). In *Rivista di storia della scienza: Proceedings of the International Symposium on Niels Bohr* **2**(3), 1985, 341–355

[33] Niels Bohr and the early history of CERN. In *Rivista di storia della scienza: Proceedings of the International Symposium on Niels Bohr* **2**(3), 1985, 511–526

[34] The history of CERN during the early 1950s. In *Proceedings of the 1985 International Symposium for Particle Physics in the 1950s* (Fermilab, 1985), pp. 508–518

[35] Nuclear physics at the University of Rome during the Thirties. In *Fifth Course of the International School of Intermediate Energy Nuclear Physics* (Verona, 1985), 1986

[36] *John Adams and His Times*. CERN 86–04, 1986

[37] The beginning of particle physics: from cosmic rays to CERN accelerators. In *40 years of Particle Physics,* ed. B. Foster & P.H. Fowler, 1987, pp. 109–119

[38] Physics in Rome in the 40's and 50's. In *Present Trends, Concepts and Instruments of Particle Physics. Symposium in Honor of Marcello Conversi's 70th Birthday* (Rome, 3–4 November 1987), pp. 139–169

[39] Ettore Majorana, a cinquant'anni dalla sua scomparsa. *Il nuovo Saggiatore. Bollettino della Società italiana di fisica* **4**(1), 1988, 13–26

[40] Ricordo di Ettore Pancini. *Il nuovo Saggiatore. Bollettino della Società italiana di fisica* **5**, 1989, 34–46

[41] Vicende dell'Accademia dei Lincei. In *Le conseguenze culturali delle leggi razziali* (Atti dei convegni lincei **81**), 1991, pp. 43–50. Roma: Accademia Nazionale dei Lincei

[42] Il caso della fisica. In *Le conseguenze culturali delle leggi razziali* (Atti dei convegni lincei **81**), 1991, pp. 107–133. Roma: Accademia Nazionale dei Lincei

[43] Ricordi di un fisico italiano. *Giano. Ricerche per la pace* **1**, 1989, 87–96

[44] Mario Ageno, ricercatore e trattatista. In *Fisica e Biofisica Oggi, Proceedings of the Conference in Honor of M. Ageno* (Rome, 1–2 October 1985), ed. E. Amaldi & L. Maiani, 1989, pp. 199–217. Bologna, Società Italiana di Fisica

[45] *The Adventurous Life of Friedrich Georg Houtermans, Physicist.* Unpublished

GIOVANNI BATTIMELLI (Voghera, 1948–) is a researcher at the Department of Physics of Rome University "La Sapienza". His main research interests in the history of science have focused on topics in late nineteenth and twentieth century theoretical physics, and more particularly on the Italian scenario. He is a member of the editorial board of *Rivista di Storia della Scienza*, and has recently edited, with M. De Maria, the memoirs of Edoardo Amaldi *Da via Panisperna all'America*, Rome 1997.

GIOVANNI PAOLONI (Rome, 1956–) is a researcher at the "Scuola Speciale per Archivisti e Bibliotecari" of Rome University "La Sapienza". He is interested in the history and archival heritage of R&D institutions in nineteenth and twentieth century Italy. He has been the editor of: *Vito Volterra e il suo tempo (1860–1940)*, Rome 1990; *Energia, ambiente, innovazione. Dal CNRN all'ENEA*, Rome 1992; and *Guglielmo Marconi e l'Italia*, Rome 1996, with R. Simili.